Gauge Theory of Weak Interactions

Fourth Edition

Greiner
Quantum Mechanics
An Introduction 4th Edition

Greiner
Quantum Mechanics
Special Chapters

Greiner · Müller
Quantum Mechanics
Symmetries 2nd Edition

Greiner
Relativistic Quantum Mechanics
Wave Equations 3rd Edition

Greiner · Reinhardt
Field Quantization

Greiner · Reinhardt
Quantum Electrodynamics
4th Edition

Greiner · Schramm · Stein
Quantum Chromodynamics
3rd Edition

Greiner · Maruhn
Nuclear Models

Greiner · Müller
Gauge Theory of Weak Interactions
4th Edition

Greiner
Classical Mechanics
Systems of Particles
and Hamiltonian Dynamics
2nd Edition

Greiner
Classical Mechanics
Point Particles and Relativity

Greiner
Classical Electrodynamics

Greiner · Neise · Stocker
Thermodynamics
and Statistical Mechanics

Walter Greiner · Berndt Müller

Gauge Theory of Weak Interactions

With a Foreword by
D.A. Bromley

Fourth Edition
With 121 Figures,
and 75 Worked Examples and Exercises

Springer

Prof. Dr. Dr. h. c. mult. Walter Greiner
Frankfurt Institute
for Advanced Studies (FIAS)
Johann Wolfgang Goethe-Universität
Ruth-Moufang-Str. 1
60438 Frankfurt am Main
Germany
greiner@fias.uni-frankfurt.de

Dr. Berndt Müller
Department of Physics
Duke University
Durham, NC 27708-0305
USA
mueller@phy.duke.edu

Title of the original German edition: *Theoretische Physik*, Ein Lehr- und Übungsbuch, Band 8: Eichtheorie der schwachen Wechselwirkung © Verlag Harri Deutsch, Thun, 1986

ISBN 978-3-540-87842-1
e-ISBN 978-3-540-87843-8
DOI 10.1007/978-3-540-87843-8
Springer Heidelberg Dordrecht London New York

Library of Congress Control Number: 2009936117

Cover design: eStudio Calamar S.L., Spain

Printed on acid-free paper

Springer is part of Springer Science+Business Media (www.springer.com)

Foreword to Earlier Series Editions

More than a generation of German-speaking students around the world have worked their way to an understanding and appreciation of the power and beauty of modern theoretical physics – with mathematics, the most fundamental of sciences – using Walter Greiner's textbooks as their guide.

The idea of developing a coherent, complete presentation of an entire field of science in a series of closely related textbooks is not a new one. Many older physicists remember with real pleasure their sense of adventure and discovery as they worked their ways through the classic series by Sommerfeld, by Planck and by Landau and Lifshitz. From the students' viewpoint, there are a great many obvious advantages to be gained through use of consistent notation, logical ordering of topics and coherence of presentation; beyond this, the complete coverage of the science provides a unique opportunity for the author to convey his personal enthusiasm and love for his subject.

The present five-volume set, *Theoretical Physics*, is in fact only that part of the complete set of textbooks developed by Greiner and his students that presents the quantum theory. I have long urged him to make the remaining volumes on classical mechanics and dynamics, on electromagnetism, on nuclear and particle physics, and on special topics available to an English-speaking audience as well, and we can hope for these companion volumes covering all of theoretical physics some time in the future.

What makes Greiner's volumes of particular value to the student and professor alike is their completeness. Greiner avoids the all too common "it follows that ..." which conceals several pages of mathematical manipulation and confounds the student. He does not hesitate to include experimental data to illuminate or illustrate a theoretical point and these data, like the theoretical content, have been kept up to date and topical through frequent revision and expansion of the lecture notes upon which these volumes are based.

Moreover, Greiner greatly increases the value of his presentation by including something like one hundred completely worked examples in each volume. Nothing is of greater importance to the student than seeing, in detail, how the theoretical concepts and tools under study are applied to actual problems of interest to a working physicist. And, finally, Greiner adds brief biographical sketches to each chapter covering the people responsible for the development of the theoretical ideas and/or the experimental data presented. It was Auguste Comte (1798–1857) in his *Positive Philosophy* who noted, "To understand a science it is necessary to know its history". This is all too often forgotten in modern physics teaching and the bridges that Greiner builds to the pioneering figures of our science upon whose work we build are welcome ones.

Greiner's lectures, which underlie these volumes, are internationally noted for their clarity, their completeness and for the effort that he has devoted to making physics an integral whole; his enthusiasm for his science is contagious and shines through almost every page.

These volumes represent only a part of a unique and Herculean effort to make all of theoretical physics accessible to the interested student. Beyond that, they are of enormous value to the professional physicist and to all others working with quantum phenomena. Again and again the reader will find that, after dipping into a particular volume to review a specific topic, he will end up browsing, caught up by often fascinating new insights and developments with which he had not previously been familiar.

Having used a number of Greiner's volumes in their original German in my teaching and research at Yale, I welcome these new and revised English translations and would recommend them enthusiastically to anyone searching for a coherent overview of physics.

Yale University
New Haven, CT, USA
1989

D. Allan Bromley
Henry Ford II Professor of Physics

Preface to the Fourth Edition

It is a pleasure to see the positive resonance of our book, which now necessitates a fourth edition. We have used this opportunity to correct misprints and errors, and to extend and improve the discussion of many of the exercises and examples.

The examples on neutrino mass (7.4), double β decay (7.8 and 7.9) and solar neutrinos (7.10) have been updated in the light of the experimental results obtained in the last years. Neutrino physics has become a tremendously active and exciting area of research, and we can expect to see conclusive experimental results on neutrino masses and oscillation parameters in the near future. We hope that our text will help students to appreciate these developments.

Again, we thank all colleagues and readers for their comments and information about misprints in the book, and we appreciate the cooperation with the publishing team at Springer-Verlag, and with Dr. Stefan Scherer, in the preparation of this fourth edition.

Frankfurt am Main
August 2009

Walter Greiner

Preface to the Third Edition

Again, we take this opportunity to correct misprints and errors and add new examples and exercises.

We thank several colleagues and students for helpful comments, particularly Dr. Joachim Reinhardt who helped me to improve some exercises and examples and Dipl.-Phys. Constantin Loizides who helped in the preparation of this third edition. Finally, we acknowledge the agreeable collaboration with Dr. H. J. Kölsch and his team at Springer-Verlag, Heidelberg.

Frankfurt am Main *Walter Greiner*
July 2000

Preface to the Second Edition

We are pleased to note that our text *Gauge Theory of Weak Interactions* has found many friends among physics students and researchers so that the need for a second edition has arisen. We have taken this opportunity to make several amendments and improvements to the text. A number of misprints and minor errors have been corrected and explanatory remarks have been added at various places. In addition to many other smaller changes the Sects. 6.4 on Cabibbo's theory of flavour mixing, 7.4 on the properties of allowed beta decay, 9.3 on the SU(5) Gauge Theory, and 9.5 on the scale of the SU(5) symmetry breaking have been expanded. Several new examples and exercises in Chaps. 6, 7, and 9 have been added, e.g., on parity violation in inelastic lepton–nucleon scattering or on the running coupling constant in quantum field theory.

We thank several colleagues and students for helpful comments. We also thank Dr. E. Stein and Dr. Steffen A. Bass who have supervised the preparation of the second edition of the book. Finally we acknowledge the agreeable collaboration with Dr. H. J. Kölsch and his team at Springer-Verlag, Heidelberg.

Frankfurt am Main and Durham, NC, USA — *Walter Greiner*
December 1995 — *Berndt Müller*

Preface to the First Edition

Modern theoretical physics has, over the past twenty years, made enormous progress, which may well be compared to the dramatic developments that occurred during the first few decades of this century. Whereas the discoveries of the early twentieth century (quantum mechanics, special and general relativity) concerned the foundations of modern physics, remaking the very concepts on which our view of the laws of nature are based, the recent breakthroughs have provided an almost complete understanding of the basic principles of the fundamental interactions among elementary particles. These principles are laid down in the so-called "Standard Model of Particle Physics" which successfully describes all established experimental data in physics.

At present, we know four fundamental interactions among elementary particles: the *strong nuclear* interaction (mediated by the exchange of mesons or – at a deeper level – of gluons), the *electromagnetic* interaction (mediated by photon exchange), the *weak nuclear* interaction (mediated by the exchange of the recently discovered W and Z bosons and, like the strong interaction, of short range), and *gravity*. Experimental searches have so far failed to uncover forces other than those four, although we cannot exclude the existence of other, very weak or short-ranged interactions.

The search for a common origin of all interactions is an ultimate (maybe *the* ultimate) goal of physics. Ever since Einstein's failed search for a unified field theory, it has been the dream of theoretical physicists to condense all laws of physics into a single fundamental equation, which contains all known interactions as special cases. This development had had its first dramatic success with Maxwell's theory of electromagnetism, which had combined the laws of electricity and magnetic interactions into a single set of equations which, in modern notation, take the beautifully simple form: $\partial_\nu F^{\mu\nu} = j^\mu$, $\partial_\nu \tilde{F}^{\mu\nu} = 0$. The disparate phenomena of electricity and magnetism suddenly had become recognized as inseparable parts of a more general interaction. Maxwell's equations had predicted the existence of electromagnetic waves. These were discovered shortly afterwards and today form the basis of the global communication network, without which modern life and modern science could not be imagined.

A comparable breakthrough occurred twenty years ago when Glashow, Salam, Weinberg, and others recognized a deep relation between the electromagnetic and the weak nuclear interaction and showed that they could be derived from a unified theory. These lecture notes deal with the ideas and insight that led to this modern unification, and introduce the student to the phenomena that played a central role in this development. We begin with a detailed exposition of Fermi's theory of beta decay and discuss the successes and shortcomings of this remarkable theory. The importance of the consideration of fundamental symmetries is illustrated by the violation of parity invariance, leading up to the (V–A) theory of weak interactions. Numerous solved problems and examples demonstrate various aspects of the weak interaction and provide an opportunity to apply the newly learned material.

The central part of the lectures introduces us to the concept of gauge theories, based on the generalization of the symmetry principle to local symmetries. The present volume may be regarded as continuation of volume 2 of this series: "Quantum Mechanics – Symmetries", extending the concepts of continuous symmetry groups to gauge transformations. The application of the gauge principle to weak isospin and hypercharge results in the unified electroweak gauge theory. The concepts of spontaneous symmetry breaking, charged and neutral currents, and mixing angles, are introduced and discussed in broad detail. Many aspects are illustrated with examples selected from current research fields, such as the problem of neutrino mixing with its application to the solar neutrino flux. Additional chapters are concerned with the applications of the electroweak gauge theory to hadronic decays and to the nuclear beta decay, where the presentation is systematically based on the quark model first introduced in volume 2. A separate chapter deals with the phenomenon of CP violation.

Only a few years after the formulation of the electroweak gauge theory, it was discovered that the strong interactions are also based on a set of equations that closely resembles those of the unified electroweak theory. This immediately fostered speculations that electroweak and strong interactions could be the low-energy manifestations of a "grand unified" gauge theory. The last section of our book contains an extended introduction on the principles underlying the search for such unified theories. We discuss the SU(5) model of Georgi and Glashow, the simplest unified gauge theory, and show how model building is constrained by experimental data. The presentation is broad and self-contained as usual in this series, introducing the student to the new concepts and formal techniques without unnecessary ballast. A detailed derivation of proton decay is presented, and the question of anomaly freedom is discussed. The book concludes with an outlook on supersymmetric unification in the light of recent precision measurements of the electroweak and strong gauge coupling constants.

These lectures make an attempt to familiarize the student with the developments of modern particle physics by providing a conceptually simple, yet rigorous introduction combined with hands-on experience through exercises and examples. They grew out of advanced graduate courses presented at the Johann Wolfgang Goethe-Universität in Frankfurt am Main and the Vanderbilt University in Nashville, Tennessee during the years 1982–85. The volume is designed as a self-contained introduction to gauge theories. Of course, much of the material is based on the framework of relativistic quantum field theory; it is desirable that the student has at least a working familiarity with the theory of quantum electrodynamics (volume 4 of this series). Some important and often used equations and relations are collected in appendices.

Our special gratitude goes to Dr. Matthias Grabiak and Professor Dr. Andreas Schäfer for their help with the examples and exercises. Several students have helped to convert the material from the stage of informal lecture notes to a textbook. For this first English edition we have enjoyed the help of Dipl.-Phys. Jürgen Augustin, Dipl.-Phys. Maria Berenguer, Dr. Oliver Graf, Dipl.-Phys. Christian Hofmann, cand. Phys. Markus Hofmann, Dipl.-Phys. André Jahns, Dipl.-Phys. Kyong-Ho Kang, Dipl.-Phys. Ullrich Katcher, Dipl.-Phys. Jürgen Klenner, cand. Phys. Yaris Pürsün, cand. Phys. Matthias Rosenstock, Dipl.-Phys. Jürgen Schaffner, Dipl.-Phys. Alexander Scherdin, cand. Phys. Christian Spieles and Dipl.-Phys. Mario Vidović. Miss Astrid Steidl drew the graphs and pictures. To all of them we express our sincere thanks.

We would especially like to thank Dipl. Phys. Raffaele Mattiello and Dr. Béla Waldhauser for their overall assistance in the preparation of the manuscript. Their organizational talent and advice in technical matters have contributed decisively to the successful completion of this work.

Frankfurt am Main, July 1993 *Walter Greiner*
Durham, USA, July 1993 *Berndt Müller*

Contents

Contents of Examples and Exercises

The Discovery of the Weak Interaction 1

All the known interactions that occur in nature can be reduced to four interactions between material particles. Listed in order of decreasing strength, these are: the strong (nuclear) interaction, electromagnetism, the weak (nuclear) interaction, and gravity (see Table 1.1). The interaction strength varies with distance between the force centers. This is illustrated in the figure accompanying Table 1.1.

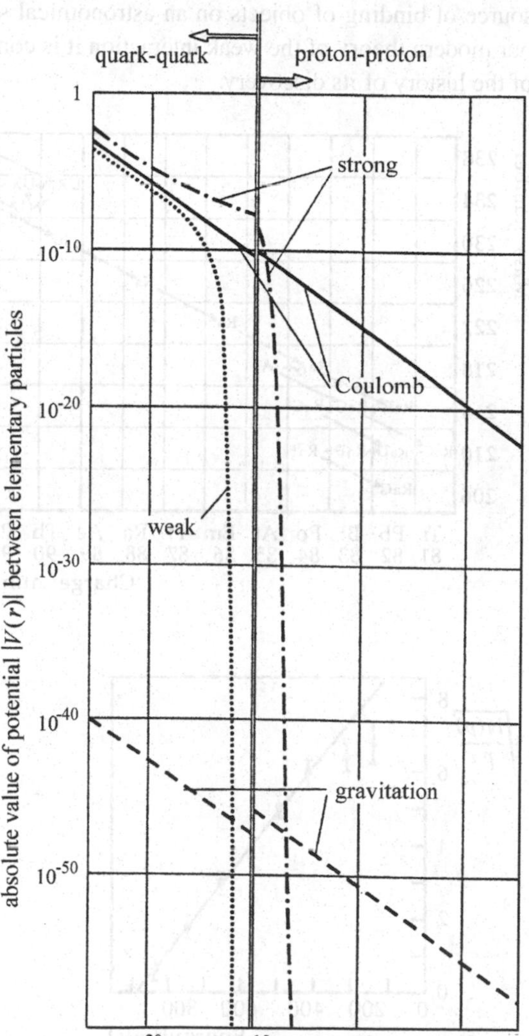

Interaction-potentials between force centers

W. Greiner, B. Müller, *Gauge Theory of Weak Interactions*,
DOI 10.1007/978-3-540-87843-8_1,

Table 1.1. Comparison of known interactions

Interaction	Relative strength	Range
Gravitation	10^{-41}	∞
Weak nuclear	10^{-15}	$\ll$1 fm
Electromagnetic	10^{-2}	∞
Strong nuclear	1	$\approx$1 fm

In contrast to the strong interaction, the weak interaction operates between all particles. It causes reactions which make these particles ultimately decay into the stable leptons and hadrons, that is, electrons, neutrinos, and protons. Characteristic in these reactions is the change in charge of the particles which undergo the reactions. The small strength of the weak interaction goes along with its small range. Unlike all other interactions, the weak interaction does not produce bound states, as do the strong interaction, which is responsible for the formation of atomic nuclei, the electromagnetic interaction, which binds together atoms and molecules, and gravity, which is the source of binding of objects on an astronomical scale. In order to better understand our modern theory of the weak interaction it is convenient to start with a short review of the history of its discovery.

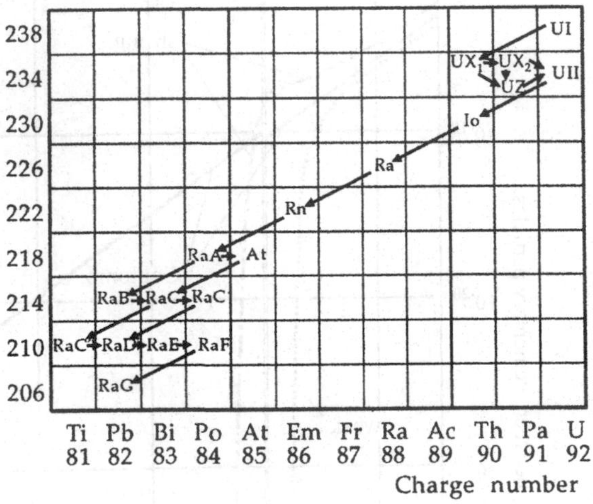

ig. 1.1. The uranium series. ʼithin the long decay chain it ʼas presumably the β decay f $^{234}_{91}$Pa to $^{234}_{92}$U (from UX_2 › UII) which was responsible for the blackening of Becuerel's photographic plate

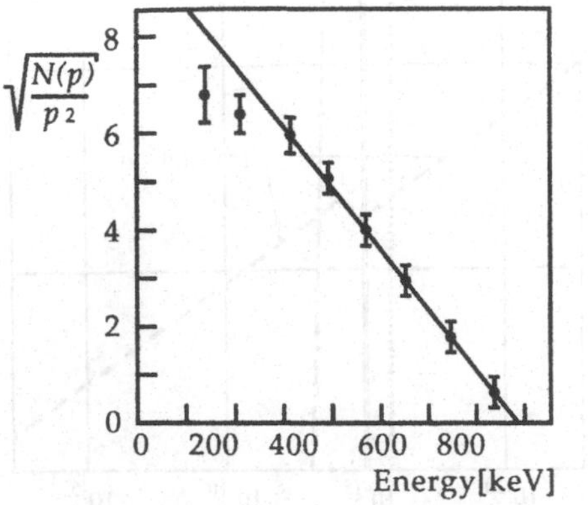

ig. 1.2. Form of the continu-ʼus β spectrum

1.1 The Universal Fermi Interaction

In the year 1896 **Henri Becquerel** discovered that uranium crystals are able to taint a photographic film if they are brought into contact with it. In subsequent years Becquerel, **Kaufmann**, and **Rutherford** succeeded in showing that uranium ore – like some other materials – emits fast, electrically charged rays (so-called beta rays), which are electrons moving with approximately the speed of light. As we know today, these beta rays originate from the decay of protactinium $^{234}_{91}$Pa into uranium $^{234}_{92}$U and other nuclear decays. After the establishment of the modern concept of atomic structure (nucleus and electron cloud) around 1910, it was first assumed that the electrons emitted in the beta decay with energies up to 2.5 MeV were present in the nucleus before the decay. This was difficult to understand, because according to Bohr's model of the atom the electrons ought to move predominantly in orbits well outside the nucleus. With the discovery of the neutron (**Chadwick** 1932) it became evident that the electron is created at the instant that the neutron transforms into a proton, which made this problem obsolete.

Another difficulty in the comprehension of β decay consisted in the fact that the electrons are emitted with a *continuous energy spectrum.* Since the initial and final nuclei have well-defined energies, this would mean a violation of energy conservation (see Exercise 1.1).

EXERCISE

1.1 Kinematics of Two-Body Decays

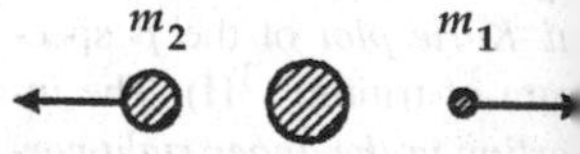

Fig. 1.3. Notation in two-particle decay

Problem. Using the relativistic energy–momentum relation, show that in the decay of a particle of mass M at rest into two-particles of masses m_1 and m_2, the two particles must have a definite energy.

Solution. For the solution of the problem we consider the rest system of the initial particle. Its energy–momentum vector is in this case

$$p^\mu = (M, \boldsymbol{O}) \quad . \tag{1}$$

After the decay the two decay products fulfill the relations

$$p_1^\mu = (E_1, \boldsymbol{p}_1) \quad \text{with} \quad E_1^2 = m_1^2 + \boldsymbol{p}_1^2 \quad , \tag{2a}$$

$$p_2^\mu = (E_2, \boldsymbol{p}_2) \quad \text{with} \quad E_2^2 = m_2^2 + \boldsymbol{p}_2^2 \quad . \tag{2b}$$

Energy–momentum conservation then requires that

$$p^\mu = p_1^\mu + p_2^\mu . \tag{3}$$

The spatial components yield

$$\boldsymbol{O} = \boldsymbol{p}_1 + \boldsymbol{p}_2 \quad , \quad \text{i.e.} \quad \boldsymbol{p}_2 = -\boldsymbol{p}_1 \quad . \tag{4}$$

Inserting this into the time component gives

$$M = E_1 + E_2 = (m_1^2 + \boldsymbol{p}_1^2)^{1/2} + (m_2^2 + \boldsymbol{p}_1^2)^{1/2} \quad . \tag{5}$$

xercise 1.1

Some transformations lead to

$$p_1^2 = \frac{M^2}{4} - \frac{m_1^2 + m_2^2}{2} + \frac{(m_1^2 - m_2^2)^2}{4M^2} \quad , \tag{6}$$

that is, p_1^2 is unequivocally determined. Then E_1 and E_2, according to (2a), (2b), are given by

$$E_1 = \frac{M}{2} + \frac{m_1^2 - m_2^2}{2M} \quad , \quad E_2 = \frac{M}{2} + \frac{m_2^2 - m_1^2}{2M} \quad . \tag{7}$$

Pauli therefore proposed in 1930 that an additional particle should be emitted besides the electron in the decay of the neutron. This particle, nowadays called the antineutrino[1] $\overline{\nu}$, should carry no electric charge and have a low mass:

$$\mathrm{n} \quad \rightarrow \quad \mathrm{p} + \mathrm{e}^- + \overline{\nu} \quad .$$

Since the neutron, proton, and electron each carry spin $\hbar/2$, conservation of angular momentum could be maintained if the neutrino was also assigned spin $\hbar/2$. We thus deal with a three-particle decay. The energy liberated in the decay of the free neutron, $(m_n - m_p)c^2 = 1.2934$ MeV, is statistically distributed over the neutrino and electron. According to today's experiments, however, the rest mass of the neutrino is less than 20 eV, but not equal to zero (see Fig. 1.4). For the latest considerations we refer to Chap. 7, Fig. 7.6, but for the time being, let us assume the neutrino to be massless.

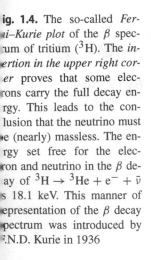

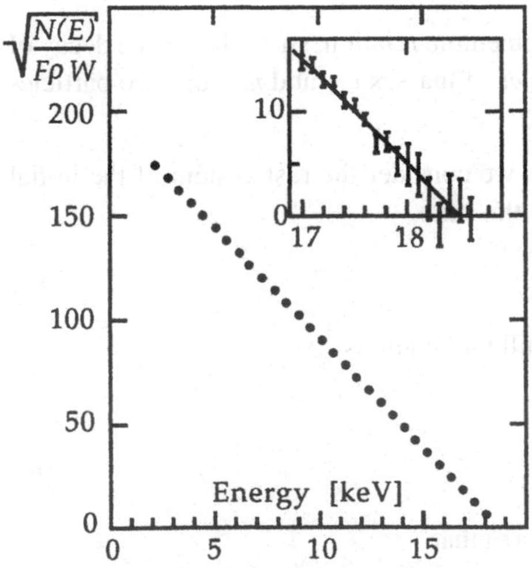

ig. 1.4. The so-called *Fer-i–Kurie plot* of the β spectrum of tritium (^{3}H). The *insertion in the upper right corner* proves that some electrons carry the full decay energy. This leads to the conclusion that the neutrino must be (nearly) massless. The energy set free for the electron and neutrino in the β decay of $^3\mathrm{H} \rightarrow {}^3\mathrm{He} + \mathrm{e}^- + \overline{\nu}$ is 18.1 keV. This manner of representation of the β decay spectrum was introduced by F.N.D. Kurie in 1936

The decay of the neutron is shown graphically in Fig. 1.5. In 1934 Fermi extended Pauli's idea to a quantitative theory of β decay. He did this by postulating that the

[1] It was discovered in the 1950s and 1960s that neutrinos form a family of particles with similar properties. Each neutrino is left-handed (see below) and accompanied by its antiparticle, the antineutrino, which is right-handed. In addition, there is a different type of neutrino associated with each species of charged lepton (electron, muon, tau-lepton). The particle originally postulated by Pauli to account for the properties of nuclear β decay, is the electron-antineutrino $\overline{\nu}_e$.

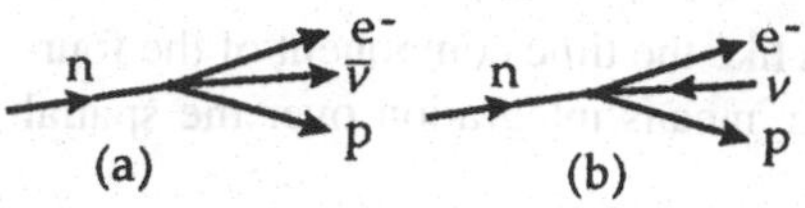

Fig. 1.5. Decay of the neutron into an electron e$^-$, an antineutrino $\overline{\nu}$, and a proton p: (**a**) in terms of physical particles, (**b**) in formal (field theoretical) terms

decay process can be described by adding to the Hamiltonian an interaction term containing the wave functions of the four free particles:

$$H_{\mathrm{F}} = H_{\mathrm{n}}^0 + H_{\mathrm{p}}^0 + H_{\mathrm{e}}^0 + H_{\nu}^0 + \underbrace{\sum_i C_i \int d^3x (\overline{u}_{\mathrm{p}} \hat{O}_i u_{\mathrm{n}})(\overline{u}_{\mathrm{e}} \hat{O}_i u_{\nu})}_{\text{interaction term}} \quad . \tag{1.1}$$

Here u_{p}, u_{n}, u_{e}, and u_ν denote the wave functions of the four particles, the bars on $\overline{u}_{\mathrm{p}}$ and $\overline{u}_{\mathrm{e}}$ indicate the (Dirac) adjoint, and the quantities $\hat{O}_i$ are appropriate operators which characterize the decay and which are weighted by the constants C_i. Since neutrinos are massless and the electron mass is low compared to the kinetic energies in β decay, the theory must be formulated relativistically, which means that the wave functions must be taken as solutions of the free *Dirac* equation

$$\left(\mathrm{i}\gamma^\mu \frac{\partial}{\partial x^\mu} - m_k\right) u_k(x) = 0 \quad (k = \mathrm{p}, \mathrm{n}, \mathrm{e}, \nu) \quad , \tag{1.2}$$

that is, as four-component spinors. The γ^μ are the *Dirac* matrices.[2]

The interaction term in (1.1) follows the current–current coupling, which is well known from electrodynamics. Here the term "current" has to be interpreted very generally, since $(\overline{u}_{\mathrm{p}} \hat{O}_i u_{\mathrm{n}})$ can be a vector current, for instance if $\hat{O}_i = \gamma^\mu$, but it can also be a scalar, for instance if $\hat{O}_i = 1$ (we refer to Exercise 1.2). In any case, however, the operators $\hat{O}_i$ must be 4×4 spin matrices. The question arises whether the elements of these matrices should be numbers or differential operators. It can be shown that in the interaction term the 4×4 differential operators can be reduced to (constant) 4×4 matrices. Indeed, since we are dealing with plane waves, (1.2) allows to express the differentials $\mathrm{i}\gamma^\mu \frac{\partial}{\partial x^\mu} u_k(x)$ by $m_k u_k(x)$.

Table 1.2. Elementary fermion transition operators

$\hat{O}_i$	Transformation property of $\overline{\Psi}\hat{O}_i\Psi$	Number of matrices
1	Scalar (S)	1
γ^μ	Vector (V)	4
$\sigma^{\mu\nu} = \frac{\mathrm{i}}{2}[\gamma^\mu, \gamma^\nu]_-$	Tensor (T)	6
$\gamma^\mu\gamma_5$	Axial vector (A)	4
$\gamma_5 = -\mathrm{i}\gamma_0\gamma_1\gamma_2\gamma_3 = \mathrm{i}\gamma^0\gamma^1\gamma^2\gamma^3$	Pseudoscalar (P)	1

[2] All conventions referring to the Dirac equation correspond to those W. Greiner: *Relativistic Quantum Mechanics – Wave Equations*, 3rd ed. (Springer, Berlin, Heidelberg, 2000). They are also used for instance, in J.D. Bjorken, S.D. Drell in *Relativistic Quantum Mechanics* (McGraw-Hill, New York, 1964); see also Appendices A.1 and A.2.

As is well known, the Hamiltonian transforms like the time component of the four-momentum vector. In the interaction term, $\int d^3x$ means integration over the spatial coordinates x_1, x_2, x_3. Then

$$(\overline{u}_p \hat{O}_i u_n)(\overline{u}_e \hat{O}_i u_\nu)$$

must be a Lorentz scalar. Of course the quantities $\overline{\Psi}\hat{O}_i\Psi$ must be well behaved under Lorentz transformations in order to make the above expression a scalar. In Exercise 1.2 we will show that under these general conditions the sole possibilities for the $\hat{O}_i$ are the 16 matrices $1, \gamma^\mu, \sigma^{\mu\nu}, \gamma^\mu\gamma_5$, and γ_5.

EXERCISE

1.2 Lorentz Transformation of Dirac Operators

Problem. Let $(\overline{u}_p \hat{O}_i u_n)(\overline{u}_e \hat{O}_i u_\nu) = \chi$ be a Lorentz scalar. Show that for the operators $\hat{O}_i$ there are the five possibilities given above (1, γ^μ, $\sigma^{\mu\nu}$, $\gamma^\mu\gamma_5$, γ_5) and no others.

Solution. Obviously it is necessary that $\overline{\Psi}\hat{O}_i\Psi$ has a well-defined transformation behavior. We therefore examine the behavior of these bilinear forms under Lorentz transformations $\hat{a}$:

$$x^{\nu'} = a^\nu_\mu x^\mu \quad . \tag{1}$$

As demonstrated in relativistic quantum mechanics,[3] a Dirac wave function transforms according to

$$\Psi'(x') = S(a)\Psi(x) \tag{2}$$

with the transformation matrix

$$S(a) = \exp[-\mathrm{i}/4\,\sigma_{\mu\nu}\,(a^{\mu\nu} - g^{\mu\nu})] \quad , \tag{3}$$

which has the following properties:

$$S^{-1}(a)\gamma^\nu S(a) = a^\nu_\mu \gamma^\mu \quad , \tag{4}$$

$$S^{-1}(a) = \gamma_0 S^\dagger \gamma_0 \quad . \tag{5}$$

Let us consider first $\overline{\Psi}\hat{1}\Psi = \overline{\Psi}\Psi$. With (1), (5) and $\overline{\Psi} = \Psi^\dagger\gamma_0$ we obtain

$$\begin{aligned}
\overline{\Psi'}(x')\Psi'(x') &= \Psi'^\dagger(x')\gamma_0\Psi'(x') \\
&= \Psi^\dagger(x)S^\dagger\gamma_0 S\Psi(x) \\
&= \Psi^\dagger(x)\gamma_0\gamma_0 S^\dagger\gamma_0 S\Psi(x) \\
&= \Psi^\dagger(x)\gamma_0 S^{-1} S\Psi(x) \\
&= \Psi^\dagger(x)\gamma_0\Psi(x) = \overline{\Psi}(x)\Psi(x) \quad .
\end{aligned} \tag{6}$$

[3] See W. Greiner: *Relativistic Quantum Mechanics – Wave Equations*, 3rd ed. (Springer, Berlin, Heidelberg, 2000).

This proves the first line of Table 1.2. Similarly we conclude for the other lines, for example,

$$\begin{aligned}
\overline{\Psi'}(x')\gamma^{\mu}\Psi'(x') &= \Psi'^{\dagger}(x')\gamma_0\gamma^{\mu}\Psi'(x') \\
&= \Psi^{\dagger}(x)S^{\dagger}\gamma_0\gamma^{\mu}S\Psi(x) \\
&= \Psi^{\dagger}(x)\gamma_0\gamma_0 S^{\dagger}\gamma_0\gamma^{\mu}S\Psi(x) \\
&= \Psi^{\dagger}(x)\gamma_0 S^{-1}\gamma^{\mu}S\Psi(x) \\
&= a^{\mu}_{\nu}\Psi^{\dagger}(x)\gamma_0\gamma^{\nu}\Psi(x) \\
&= a^{\mu}_{\nu}\overline{\Psi}(x)\gamma^{\nu}\Psi(x) \quad . \qquad (7)
\end{aligned}$$

Thus $\overline{\Psi}\gamma^{\mu}\Psi$ transforms as a vector. Only in the case of axial vectors and pseudoscalars does something new occur. We consider therefore

$$\begin{aligned}
\overline{\Psi'}(x')\gamma_5\Psi'(x') &= \Psi'^{\dagger}(x')\gamma_0\gamma_5\Psi'(x') \\
&= \Psi^{\dagger}(x)S^{\dagger}\gamma_0\gamma_5 S\Psi(x) \\
&= \Psi^{\dagger}(x)\gamma_0 S^{-1}\gamma_5 S\Psi(x) \quad . \qquad (8)
\end{aligned}$$

Here we must distinguish two different cases. We denote the proper Lorentz transformations by S_{ℓ} and the space inversion (parity transformation) by S_{p}.

1. $S = S_{\ell}$.
   Because $\gamma^{\mu}\gamma_5 + \gamma_5\gamma^{\mu} = 0$, using (3) one can easily convince oneself that, because $\sigma^{\mu\nu} = \mathrm{i}/2[\gamma^{\mu}, \gamma^{\nu}]$,

   $$[S_{\ell}, \gamma_5] = 0 \quad .$$

   Then we have immediately that

   $$\overline{\Psi}'(x')\gamma_5\overline{\Psi}'(x') = \overline{\Psi}(x)\gamma_5\overline{\Psi}(x) \quad . \qquad (9)$$

2. $S = S_{\mathrm{p}} = \gamma^0 = \begin{pmatrix} 1 & & & \\ & -1 & & \\ & & -1 & \\ & & & -1 \end{pmatrix}$.

   Here the commutation gives $S_{\mathrm{p}}\gamma_5 = \gamma^0\gamma_5 = -\gamma_5\gamma^0 = -\gamma_5 S_{\mathrm{p}}$. Thus the sign changes:

   $$\overline{\Psi}'(x')\gamma_5\Psi'(x') = -\overline{\Psi}(x)\gamma_5\Psi(x) \quad . \qquad (10)$$

We can combine these two results as follows:

$$\overline{\Psi'}(x')\gamma_5\Psi'(x') = \det(a)\overline{\Psi}(x)\gamma_5\Psi(x) \quad , \qquad (11)$$

because for proper Lorentz transformations $\det(a) = +1$. For the inversion, on the other hand, $\det(a) = -1$. These are exactly the *transformation properties of a pseudoscalar.* Analogously one can show that

$$\overline{\Psi'}(x')\gamma_5\gamma^\nu\Psi'(x') = \det(a)a^\nu_\mu\overline{\Psi}(x)\gamma_5\gamma^\mu\Psi(x) \quad , \tag{12}$$

which is thus an *axial vector.*

The 16 matrices 1, γ^μ, $\sigma^{\mu\nu}$, $\gamma^\mu\gamma_5$, and γ_5 are linearly independent and form a basis in the vector space of the 4×4 matrices. Any other 4×4 matrix can be expressed as a linear combination of those 16 matrices. Combinations like $\hat{O}_i = \gamma^\mu + \sigma^{\mu\nu}$, however, lead to mixed terms which are not scalars:

$$\begin{aligned}
&(\overline{u}_p(\gamma_\mu + \sigma_{\mu\alpha})u_n)(\overline{u}_e(\gamma^\mu + \sigma^{\mu\alpha})u_\nu) \\
&= (\overline{u}_p\gamma_\mu u_n)(\overline{u}_e\gamma^\mu u_\nu) + (\overline{u}_p\sigma_{\mu\alpha}u_n)(\overline{u}_e\sigma^{\mu\alpha}u_\nu) \\
&\quad + (\overline{u}_p\gamma_\mu u_n)(\overline{u}_e\sigma^{\mu\alpha}u_\nu) + (\overline{u}_p\sigma_{\mu\alpha}u_n)(\overline{u}_e\gamma^\mu u_\nu) \quad .
\end{aligned} \tag{13}$$

The expressions in the second line are the desired scalars, whereas those in the last line are vectors. Quantities for $\hat{O}_i$ different from 1, γ^μ, $\sigma^{\mu\nu}$, $\gamma^\mu\gamma_5$, and γ_5 are thus impossible, which is the desired result.

Since in nuclear β decay protons and neutrons move non-relativistically, the matrix elements can be simplified in the nucleonic part of the Hamiltonian. As is well known, a Dirac spinor can be decomposed into two two-component quantities ϕ and χ,

$$\Psi = \begin{pmatrix} \phi \\ \chi \end{pmatrix} \quad ,$$

where the components of the two-spinor ϕ in the non-relativistic limit are much larger than the components of χ, and in this limit

$$S, V \rightarrow \phi_p^\dagger\phi_n \quad , \quad T, A \rightarrow \phi_p^\dagger\boldsymbol{\sigma}\phi_n \quad , \quad P \rightarrow 0 \quad . \tag{1.3}$$

We convince ourselves of these facts in Exercise 1.3.

EXERCISE

1.3 The Non-relativistic Limit of the Transition Operators

Problem. Prove the limiting cases given above.

Solution. The γ^μ are defined as

$$\gamma^0 = \beta = \begin{pmatrix} 1 & 0 \\ 0 & -1 \end{pmatrix} \quad ,$$

$$\gamma_5 = \begin{pmatrix} 0 & 1 \\ 1 & 0 \end{pmatrix} \quad , \quad \gamma_i = \beta\alpha_i \quad , \tag{1}$$

$$\alpha_i = \begin{pmatrix} 0 & \sigma_i \\ \sigma_i & 0 \end{pmatrix} \quad , \quad i = 1, 2, 3 \quad .$$

Exercise 1.

We begin with the terms S and V, and systematically neglect all terms containing the small components χ_p and χ_n:

$$\begin{aligned} S &= \bar{u}_p u_n = (\phi_p^\dagger, \chi_p^\dagger)\begin{pmatrix}\phi_n\\ -\chi_n\end{pmatrix} \\ &= \phi_p^\dagger\phi_n - \chi_p^\dagger\chi_n \to \phi_p^\dagger\phi_n \quad , \end{aligned} \tag{2}$$

$$\begin{aligned} V &= \{\bar{u}_p\gamma^0 u_n\,,\, \bar{u}_p\boldsymbol{\gamma} u_n\} \\ &= \left\{(\phi_p^\dagger, \chi_p^\dagger)\begin{pmatrix}\phi_n\\ \chi_n\end{pmatrix}, (\phi_p^\dagger, \chi_p^\dagger)\boldsymbol{\alpha}\begin{pmatrix}\phi_n\\ \chi_n\end{pmatrix}\right\} \\ &= \{\phi_p^\dagger\phi_n + \chi_p^\dagger\chi_n\,,\, \phi_p^\dagger\boldsymbol{\sigma}\chi_n + \chi_p^\dagger\boldsymbol{\sigma}\phi_n\} \to \{\phi_p^\dagger\phi_n, \mathbf{0}\} \quad . \end{aligned} \tag{3}$$

This proves $S, V \to \phi_p^\dagger\phi_n$.

Let us next turn to the limit for the tensor coupling T. Using the definition $\sigma^{\mu\nu} = \mathrm{i}/2[\gamma^\mu, \gamma^\nu]_-$ and the relation $\sigma_i\sigma_j - \sigma_j\sigma_i = 2\mathrm{i}\epsilon_{ijk}\,\sigma_k$, one can easily check that $\sigma^{ij} = \begin{pmatrix}\sigma_k & 0\\ 0 & \sigma_k\end{pmatrix}$, where $i, j, k = 1, 2, 3$ or a cyclical permutation thereof. With this we have

$$\begin{aligned} (\bar{u}_p\sigma^{ij}u_n) &= (\phi_p^\dagger, \chi_p^\dagger)\begin{pmatrix}1 & 0\\ 0 & -1\end{pmatrix}\begin{pmatrix}\sigma_k & 0\\ 0 & \sigma_k\end{pmatrix}\begin{pmatrix}\phi_n\\ \chi_n\end{pmatrix} \\ &= \phi_p^\dagger\sigma_k\phi_n - \chi_p^\dagger\sigma_k\chi_n \quad . \end{aligned} \tag{4}$$

Furthermore, $\sigma^{0i} = \mathrm{i}\alpha_i = \mathrm{i}\begin{pmatrix}0 & \sigma_i\\ \sigma_i & 0\end{pmatrix}$ is valid, as well as

$$\bar{u}_p\sigma^{0i}u_n = \mathrm{i}\{\phi_p^\dagger\sigma_i\chi_n + \chi_p^\dagger\sigma_i\chi_n\} \to 0 \quad . \tag{5}$$

Thus, as claimed above, the non-relativistic limit yields

$$T = (\bar{u}_p\sigma^{\mu\nu}u_n) \to (\phi_p^\dagger\sigma_k\phi_n) \quad . \tag{6}$$

The axial vector coupling A is treated just as easily. In view of

$$\gamma_5\begin{pmatrix}\phi_n\\ \chi_n\end{pmatrix} = \begin{pmatrix}\chi_n\\ \phi_n\end{pmatrix} \quad , \tag{7}$$

the 0 component vanishes in the limit $\chi \to 0$:

$$\begin{aligned} \bar{u}_p\gamma_0\gamma_5 u_n &= (\phi_p^\dagger, \chi_p^\dagger)\begin{pmatrix}\chi_n\\ \phi_n\end{pmatrix} \\ &= \phi_p^\dagger\chi_n + \chi_p^\dagger\phi_n \to 0 \quad , \end{aligned} \tag{8}$$

and for the space-like components we obtain

$$\begin{aligned} \bar{u}_p\gamma^i\gamma_5 u_n &= (\phi_p^\dagger, \chi_p^\dagger)\alpha_i\begin{pmatrix}\chi_n\\ \phi_n\end{pmatrix} \\ &= \phi_p^\dagger\sigma_i\phi_n + \chi_p^\dagger\sigma_i\chi_n \to \phi_p^\dagger\sigma_i\phi_n \quad . \end{aligned} \tag{9}$$

Thus

$$A = \overline{u}_{\mathrm{p}}\gamma^{\mu}\gamma_5 u_{\mathrm{n}} = \{\overline{u}_{\mathrm{p}}\gamma_0\gamma_5 u_{\mathrm{n}}, \overline{u}_{\mathrm{p}}\gamma^{i}\gamma_5 u_{\mathrm{n}}\} \to \phi_{\mathrm{p}}^{\dagger}\sigma_i\phi_{\mathrm{n}} \quad . \tag{10}$$

Finally $P \to 0$ remains to be proved, which is trivial:

$$\begin{aligned} P &= \overline{u}_{\mathrm{p}}\gamma_5 u_{\mathrm{n}} = (\phi_{\mathrm{p}}^{\dagger}, \chi_{\mathrm{p}}^{\dagger})\gamma^0 \begin{pmatrix}\chi_{\mathrm{n}} \\ \phi_{\mathrm{n}}\end{pmatrix} \\ &= \phi_{\mathrm{p}}^{\dagger}\gamma^0\chi_{\mathrm{n}} + \chi_{\mathrm{p}}^{\dagger}\gamma^0\phi_{\mathrm{n}} \to 0 \quad , \end{aligned} \tag{11}$$

since all terms contain a small component of the Dirac wave function.

The relevant non-vanishing cases are called **Fermi** transitions:

$$S, V \to \phi_{\mathrm{p}}^{\dagger}\phi_{\mathrm{n}} \quad , \tag{1.4}$$

and **Gamow–Teller** transitions:

$$T, A \to \phi_{\mathrm{p}}^{\dagger}\boldsymbol{\sigma}\phi_{\mathrm{n}} \quad . \tag{1.5}$$

In the latter case obviously the spin of the decaying nucleus may change, whereas the nuclear spin remains unchanged in the case of a Fermi transition. Both cases are actually observed in nature, that is, the Fermi Hamiltonian (1.1) must contain some combination of S–V and T–A couplings. It is possible to show that oscillations would occur in the electron spectrum, if S and V couplings were simultaneously present, and the same would be true for T and A couplings at the same time. Since such effects are not observed, it follows that only the couplings S and T, or S and A, or V and T, or V and A are realized. Measurements of the lifetimes of several nuclei lead to the conclusion that the strength constants of Fermi and of Gamow–Teller transitions are about equal in magnitude, being nearly equal to

$$G \approx 10^{-4}\,\mathrm{MeV\,fm^3} \quad ,$$

or in natural units ($\hbar = c = 1$) – see Appendix A.1 –

$$G \approx 10^{-11}(\mathrm{MeV})^{-2} \approx 10^{-5} m_{\mathrm{p}}^{-2} \quad .$$

From 1938 on, more particles were discovered that decay by the weak interaction:

$$\begin{aligned} &\mu^{\pm} \to \mathrm{e}^{\pm} + \nu + \overline{\nu} \\ &\pi^{\pm} \to \mu^{\pm} + \nu/\overline{\nu} \\ &\mathrm{K}^{\pm} \to \pi^0 + \mu^{\pm} + \nu/\overline{\nu} \\ &\Lambda^0 \to \mathrm{p} + \mathrm{e}^{-} + \overline{\nu} \quad \text{etc.} \quad . \end{aligned} \tag{1.6}$$

In all cases almost the same constant G appears. This is why one speaks of the *universal Fermi interaction*, responsible for the β decay of many unstable elementary particles.

1.2 The Non-conservation of Parity

With the K mesons, however, a serious puzzle was soon encountered. It was found that the K^+ meson, apart from the final state $\pi^0\mu^+\nu$, can also decay into two pions, $\pi^+\pi^0$, and into three pions, $\pi^+\pi^+\pi^-$. On the other hand, it was well known that the pion has negative internal parity. Since all pions are emitted with angular momentum $\ell = 0$ (this follows from their angular distribution), the spatial part of the wave function of pions has positive parity. Thus the total parity of the final state is determined by the number of pions:

$$\begin{aligned} &\pi^{\pm}\pi^0 \text{ has positive parity } (-1)^2 = +1 \quad , \\ &\pi^+\pi^+\pi^- \text{ has negative parity } (-1)^3 = -1 \quad . \end{aligned} \tag{1.7}$$

At first it was supposed that there were two different particles, with the mass and charge of the K^+ meson, one with positive and the other with negative internal parity. They were called τ and θ and one talked about the τ–θ puzzle.

T.D. Lee and **C.N. Yang**, however, pointed out[4] that there was another, revolutionary way out of the dilemma: the violation of parity conservation in K^+ decay. Indeed, a serious examination revealed that there was no evidence for parity being conserved in β decay. A short time after that, **C.S. Wu, E. Ambler, R.W. Hayward, D.D. Hoppes**, and **R.P. Hudson** proved, in a now famous experiment,[5] that parity conservation is indeed violated in the β decay of atomic nuclei. Wu and her collaborators examined the decay of $^{60}_{27}$Co into $^{60}_{28}$Ni under emission of an electron and an antineutrino. To understand the experiment, one must know that a $^{60}_{27}$Co nucleus has spin $5\hbar$ and positive parity ($J^P = 5^+$) in the ground state, whereas $^{60}_{28}$Ni has spin $4\hbar$ and also positive parity ($J^P = 4^+$). During β decay the nuclear spin thus changes by one unit and therefore after what we learned at the end of Exercise 1.3, it must be a Gamow–Teller transition. In addition one must know that the transition from ^{60}Co to ^{60}Ni is a so-called *allowed decay*, that is, a decay which occurs with the fullest possible strength. As we will learn in Chap. 7, this means that the wave function of the emitted particles (e and $\bar{\nu}$) must be large in the region of the nucleus. On the other hand, this is possible only if their total angular momentum with respect to the nucleus is $1/2\,\hbar$ (only $s_{1/2}$ and $p_{1/2}$ waves have a non-vanishing probability to be found at the place of the nucleus). In order to probe parity invariance, one must first prepare an initial situation which under reflection does not pass over into itself. To ensure this, the cobalt nuclei were arranged in a strong magnetic field at a temperature of 0.01 K (this was the most difficult part of the whole experiment; it was carried out by Ambler and Hayward of the National Bureau of Standards at Washington, DC). To conserve angular momentum, the spins of the emitted electron and neutrino must then point in the same direction. When the angular distribution of the electrons was measured, it was found that the electrons are predominantly emitted opposite to the nuclear spin, that is, they showed an anisotropy in their emission probability relative to the directed magnetic field. This result is in clear contradiction to parity invariance, whereas, during a reflection, the momentum vector of the electron reverses its direction, whereas the (axial) angular momentum

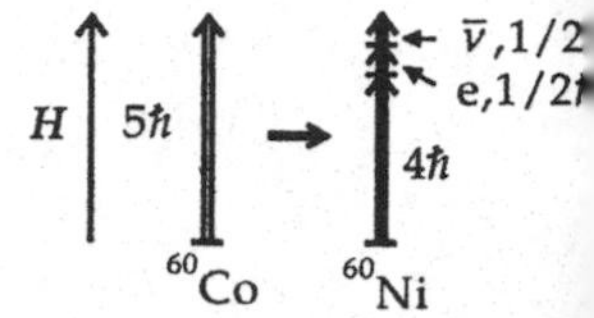

Fig. 1.6. Angular-momentum balance in the β decay of ^{60}C

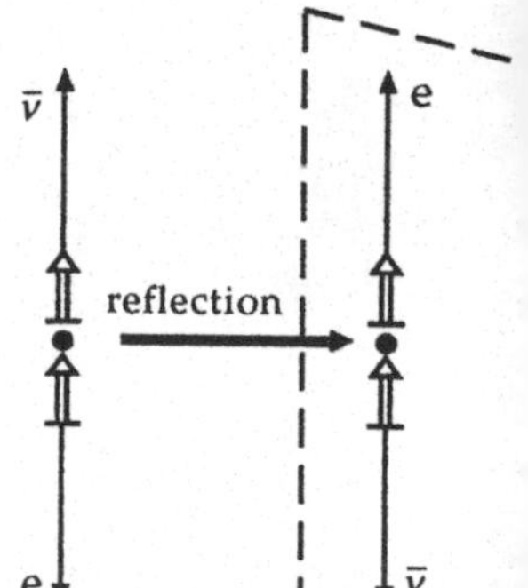

Fig. 1.7. During a reflection the directions of the angular momentum vectors and the magnetic field remain the same whereas the directions of emission change

---

[4] T.D. Lee, C.N. Yang: Phys. Rev. **104**, 254 (1956); *Nobel Lectures on Physics (1942–62)* (Elsevier, Amsterdam, 1964), p. 387.

[5] C.S. Wu, E. Ambler, R.W. Hayward, D.D. Hoppes R.P. Hudson: Phys. Rev. **105**, 1413 (1957).

vector of the nuclear spin, as well as the magnetic field, keeps its direction. The experiment thus showed that the β decay of atomic nuclei proceeds in a manner which is not reflection invariant. Owing to conservation of angular momentum, the spin of the electron emitted in the experiment must always point in the direction of nuclear polarization (see Fig. 1.6). The fact that the direction of emission of the electron is predominantly opposite to the nuclear spin can also be expressed as follows: electrons are predominantly polarized opposite to the direction of their motion, that is, they have *negative helicity*.

EXERCISE

1.4 Properties of the Helicity Operator

Problem. Show that the helicity operator $\hat{\Lambda} = \hat{\boldsymbol{\Sigma}} \cdot \boldsymbol{p}/p$ (with $p = |\boldsymbol{p}|$ and $\hat{\boldsymbol{\Sigma}} = \left(\begin{smallmatrix} \sigma & 0 \\ 0 & \sigma \end{smallmatrix}\right)$) commutes with the Dirac Hamiltonian $\hat{H} = \boldsymbol{\alpha} \cdot \boldsymbol{p} + \beta m$ and has the eigenvalues ± 1.

Solution. The 4×4 Pauli matrices $\hat{\Sigma}_i$ obey the same commutation relations as the normal 2×2 matrices σ_i:

$$\hat{\Sigma}_i \hat{\Sigma}_k = \mathrm{i}\varepsilon_{ikl} \hat{\Sigma}_l + \delta_{ik} \quad , \tag{1a}$$

$$\hat{\Sigma}_i \hat{\Sigma}_k + \hat{\Sigma}_k \hat{\Sigma}_i = 2\delta_{ik} \quad . \tag{1b}$$

With this we obtain

$$\begin{aligned} \hat{\Lambda}^2 p^2 = (\hat{\boldsymbol{\Sigma}} \cdot \boldsymbol{p})^2 &= \sum_{i,k=1}^{3} (\hat{\Sigma}_i \hat{\Sigma}_k p_i p_k) \\ &= \frac{1}{2} \sum_{i,k=1}^{3} (\hat{\Sigma}_i \hat{\Sigma}_k + \hat{\Sigma}_k \hat{\Sigma}_i) p_i p_k = p^2 \quad , \end{aligned} \tag{2}$$

where we have split the expression into two identical parts and exchanged the labeling of the indices i, k in the second one. $\hat{\Lambda}^2 = 1$ now implies that the eigenvalues can only be ± 1.

The first part of the problem is approached as follows. First we note that

$$\boldsymbol{\alpha} = \gamma_5 \hat{\boldsymbol{\Sigma}} = \hat{\boldsymbol{\Sigma}} \gamma_5 \quad , \tag{3}$$

which leads to

$$\begin{aligned} [\alpha_i, \hat{\Sigma}_k] &= \alpha_i \hat{\Sigma}_k - \hat{\Sigma}_k \alpha_i \\ &= \gamma_5 \hat{\Sigma}_i \hat{\Sigma}_k - \hat{\Sigma}_k \hat{\Sigma}_i \gamma_5 \\ &= \mathrm{i} \sum_{l=1}^{3} (\varepsilon_{ikl} \gamma_5 \hat{\Sigma}_l - \varepsilon_{kil} \hat{\Sigma}_l \gamma_5) \\ &= 2\mathrm{i} \sum_{l=1}^{3} \varepsilon_{ikl} \alpha_l \quad . \end{aligned} \tag{4}$$

We now obtain

Exercise 1.

$$[\boldsymbol{\alpha}\cdot\boldsymbol{p}, \hat{\boldsymbol{\Sigma}}\cdot\boldsymbol{p}] = \sum_{i,k=1}^{3}[\alpha_i, \hat{\Sigma}_k]p_i p_k = 2\mathrm{i}\sum_{ikl=1}^{3}\varepsilon_{ikl}\alpha_l p_i p_k = 0 \tag{5}$$

because of the antisymmetry of ε_{ikl}. Furthermore, since $\hat{\Sigma}_i$ commutes with β, it follows that $[\hat{\Lambda}, \hat{H}] = 0$.

Now the question arises whether the *preference of the negative value of helicity* has anything to do with the initial polarization of the ^{60}Co nuclei, or whether it is a *general property* of β decay. Systematic measurements of the polarization of emitted electrons reveal that they *always* have negative helicity. In β^+ *decay*, where positrons are emitted (for example $^{22}\mathrm{Na} \rightarrow {}^{22}\mathrm{Ne} + \mathrm{e}^+ + \nu_\mathrm{e}$), the opposite is true: the positrons have *positive* helicity.

Under the assumption that *only* electrons of negative helicity are emitted, the angular distribution in the experiment of Wu et al. can be explained. To do so, it is sufficient to restrict the argument to two-spinors. Let the ^{60}Co nuclei be polarized in the direction of the negative z axis (see Fig. 1.8). If the electrons are emitted at an angle θ relative to the z axis, then (provided the x-axis direction is chosen conveniently) their momentum is

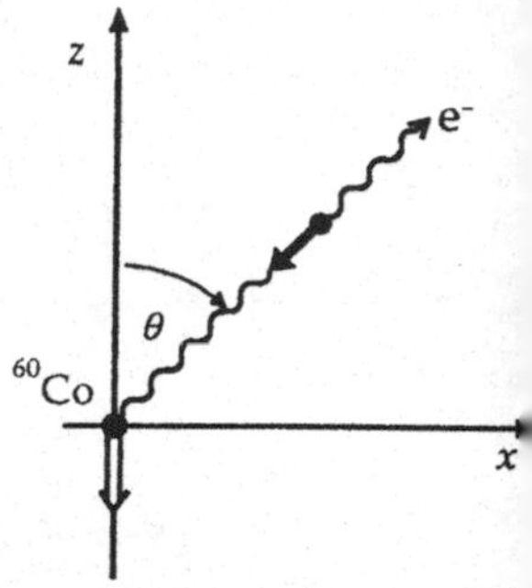

Fig. 1.8. The angular distrib ution of the electrons in the ^{60}Co decay

$$\boldsymbol{p} = p(\boldsymbol{e}_z\cos\theta + \boldsymbol{e}_x\sin\theta) \quad . \tag{1.8}$$

The helicity operator is then

$$\begin{aligned}\hat{\Lambda}_\theta &= \frac{\hat{\boldsymbol{\Sigma}}\cdot\boldsymbol{p}}{p} = \hat{\Sigma}_z\cos\theta + \hat{\Sigma}_x\sin\theta \\ &= \begin{pmatrix}\sigma_z\cos\theta + \sigma_x\sin\theta & 0 \\ 0 & \sigma_z\cos\theta + \sigma_x\sin\theta\end{pmatrix} \\ &= \begin{pmatrix}\cos\theta & \sin\theta & 0 & 0 \\ \sin\theta & -\cos\theta & 0 & 0 \\ 0 & 0 & \cos\theta & \sin\theta \\ 0 & 0 & \sin\theta & -\cos\theta\end{pmatrix} \quad .\end{aligned} \tag{1.9}$$

The condition

$$\hat{\Lambda}_\theta\chi_\theta^{(\pm)} = \pm\chi_\theta^{(\pm)} \tag{1.10}$$

leads to the eigenfunctions $\chi_\theta^{(\pm)}$ (we only need to consider the upper two components of the 4-spinor; the two lower ones are exactly equal). They are

$$\chi_\theta^{(+)} = \begin{pmatrix}\cos\frac{1}{2}\theta \\ \sin\frac{1}{2}\theta\end{pmatrix}, \qquad \chi_\theta^{(-)} = \begin{pmatrix}-\sin\frac{1}{2}\theta \\ \cos\frac{1}{2}\theta\end{pmatrix} \quad . \tag{1.11}$$

Because of spin conservation, as we have already noted, the only part that can contribute is that polarized in the same direction as the ^{60}Co nuclei, that is, in the direction of the negative z axis. The amplitude of the emission of an electron of negative helicity, and thus with the spinor $\chi_\theta^{(-)}$, is therefore given by the overlap with the spinor $\chi_{\theta=0}^{(-)} = \binom{0}{1}$. The emission probability is consequently proportional to

$$W(\theta) = \left|\langle \chi_{\theta=0}^{(-)} | \chi_\theta^{(-)} \rangle\right|^2 = \cos^2 \frac{1}{2}\theta = \frac{1}{2}(1+\cos\theta) \quad . \tag{1.12}$$

$W(\theta)$ is largest in the direction of the positive z axis; however, in the opposite direction it vanishes. This is just what Wu et al. observed in their experiment.

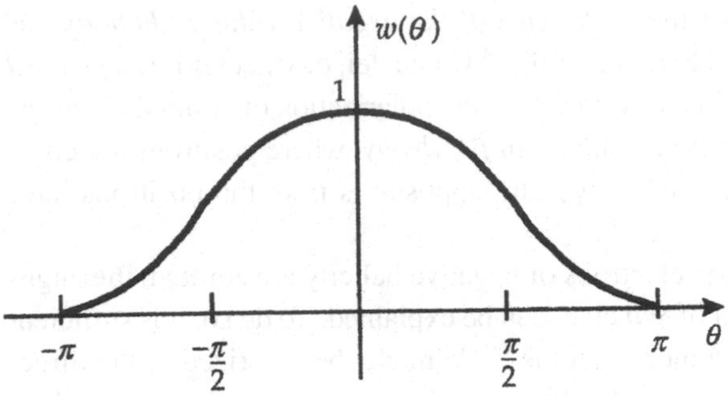

g. 1.9. Angular distribution f electrons in the decay of Co relative to the plane of olarization

EXERCISE

1.5 Rotation of Helicity Eigenfunctions

Problem. Show that the eigenvectors $\chi_\theta^{(\pm)}$ of $\hat{\Lambda}_\theta$ are obtained from the eigenvectors of σ_z by a rotation about the y axis through the angle θ.

Solution. The quantum-mechanical operator, which rotates a spin-$\frac{1}{2}$ spinor through an angle θ about an axis given by the unit vector $\boldsymbol{n}$, is given by

$$\begin{aligned}\hat{R}_\xi(\theta, \boldsymbol{n}) &= \exp\left(-\frac{\mathrm{i}}{2}\theta \boldsymbol{n}\cdot\boldsymbol{\sigma}\right)\\ &= \boldsymbol{n}\cos\frac{\theta}{2} - \mathrm{i}(\boldsymbol{n}\cdot\boldsymbol{\sigma})\sin\frac{\theta}{2} \quad .\end{aligned} \tag{1}$$

The second form follows from the Taylor-series expansion of the exponential function together with $(\boldsymbol{n}\cdot\boldsymbol{\sigma})^2 = 1$. For a rotation around the y axis we thus obtain

$$\hat{R}_y(\theta) = \cos\frac{\theta}{2} - \mathrm{i}\sigma_y \sin\frac{\theta}{2} = \begin{pmatrix} \cos\frac{\theta}{2} & -\sin\frac{\theta}{2} \\ \sin\frac{\theta}{2} & \cos\frac{\theta}{2} \end{pmatrix} \quad . \tag{2}$$

Application of this operator to the unit spinors

$$\chi^{(+)} = \begin{pmatrix}1\\0\end{pmatrix}, \quad \chi^{(-)} = \begin{pmatrix}0\\1\end{pmatrix} \tag{3}$$

yields, as asserted,

Exercise 1.

$$\hat{R}_y(\theta)\chi^{(+)} = \begin{pmatrix} \cos\frac{\theta}{2} \\ \sin\frac{\theta}{2} \end{pmatrix} = \chi_\theta^{(+)} \quad , \tag{4a}$$

$$\hat{R}_y(\theta)\chi^{(-)} = \begin{pmatrix} -\sin\frac{\theta}{2} \\ \cos\frac{\theta}{2} \end{pmatrix} = \chi_\theta^{(-)} \quad . \tag{4b}$$

The helicity of the emitted neutrino or antineutrino cannot be measured directly, because neutrinos are not influenced by magnetic fields. One must therefore rely on inferring the neutrino spin and momentum from the spin and recoil momentum of the daughter nucleus. **M. Goldhaber**, **L. Grodzins**, and **A. Sunyar** succeeded in doing this in a beautiful experiment.[6] They observed the decay of ^{152}Eu into ^{152}Sm, in which the nucleus captures an electron from the K shell and emits a neutrino:

$$^{152}_{63}\mathrm{Eu}(0^-) + \mathrm{e}^- \rightarrow ^{152}_{62}\mathrm{Sm}^*(1^-) + \nu \quad . \tag{1.13}$$

The excited state of ^{152}Sm, characterized by * at the symbol Sm for samarium, subsequently emits a photon, thus passing over to the ground state which has spin and parity 0^+:

$$^{152}\mathrm{Sm}^*(1^-) \rightarrow ^{152}\mathrm{Sm}(0^+) + \gamma \quad . \tag{1.14}$$

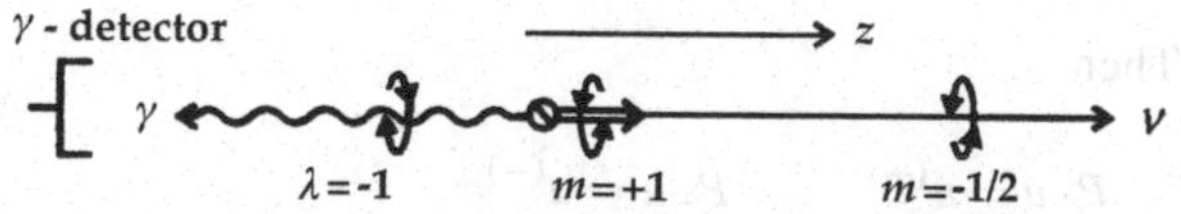

Fig. 1.10. Schematic representation of the experiment of Goldhaber, Grodzins, and Sunyar

Let us assume that the neutrino is emitted to the right, in the direction of the positive z axis (see Fig. 1.10); the nucleus then recoils to the left. Goldhaber et al. then looked at γ rays, which had the full Doppler shift corresponding to the recoiling nucleus, and thus had been emitted along the negative z axis (in doing so, they spared themselves the effort of detecting the recoil nucleus!). They observed that these γ quanta are *always right circularly polarized*,[7] that is, they have helicity $\lambda = -1$. In the case of a right circularly polarized wave, if one looks towards the wave, the electric field vector runs clockwise, whereas in the case of a left circularly polarized wave it runs counter-clockwise. Note that helicity $\lambda = -1$ for the photon means that the projection of its spin on the z axis is $+1$! The projection of nuclear spin on the right-hand side of reaction (1.14) must therefore be $m = +1$, because of conservation of the z component of angular momentum. The captured electron had angular momentum $+\frac{1}{2}$; only in this way can the captured electron e^- together with the emitted neutrino ν give the nuclear

[6] M. Goldhaber, L. Grodzins, A.W. Sunyar: Phys. Rev. **109**, 1015 (1958).

[7] For a definition of circular polarization see, for example, J.D. Jackson: *Classical Electrodynamics* (Wiley, New York, 1975) or J.M. Eisenberg and Walter Greiner: *Nuclear Theory, Vol. II: Excitation Mechanisms* (North-Holland, Amsterdam, 1985); see also Walter Greiner: *Classical Electrodynamics* (Springer, Berlin, Heidelberg, New York, 1998).

spin $m = +1$. The projection of the neutrino spin on the z axis must equal $-\frac{1}{2}$; because of the inverse direction of motion relative to the photon its helicity is thus also $\lambda = -1$.

Antineutrinos, correspondingly, have positive helicity, as other experiments revealed. If parity were conserved in β decay, particles of both helicities would be emitted with equal probability. In the case of a slight violation of parity invariance, one of the helicities would be slightly preferred. However, as we have discussed, the experiments revealed that only a single helicity appears: electrons and neutrinos are always created left-handed ($\lambda = -1$), positrons and antineutrinos are always created right-handed ($\lambda = +1$). This corresponds to the strongest possible violation of parity: one says that *parity violation is maximal*.

Now we return to the question of the precise form of the interaction term in the Hamiltonian (1.1). The experiments showed that in the part containing the electron and neutrino spinors,

$$\bar{u}_e \hat{O}_i u_\nu \quad , \tag{1.15}$$

only those parts of the wave function that have negative helicity can appear. This can be formally achieved with the help of the projection operators

$$\hat{P}_\pm = \frac{1 \pm \hat{\Lambda}}{2} \quad , \tag{1.16}$$

which project the components of positive and negative helicity, respectively, out of an arbitrary spinor. Let us decompose a four-spinor into these two components:

$$u = u^{(+)} + u^{(-)} \quad , \quad \hat{\Lambda} u^{(\pm)} = \pm u^{(\pm)} \quad . \tag{1.17}$$

Then

$$\hat{P}_+ u = u^{(+)} \quad , \quad \hat{P}_- u = u^{(-)} \tag{1.18}$$

holds. Now the difficulty is that $\hat{\Lambda} = \boldsymbol{\Sigma} \cdot \boldsymbol{p}/p$, and hence also $\hat{P}_\pm$, is *not Lorentz invariant*. To obtain the correct expression, we bear in mind that the electrons emitted in β decay are highly relativistic, that is $E \approx p \gg m_0$ (for the massless neutrinos this will be true exactly). Using the Dirac equation

$$(\boldsymbol{\alpha} \cdot \boldsymbol{p} + \beta m_0) u = E u \tag{1.19}$$

and Exercise 1.4, (3),

$$\boldsymbol{\alpha} = \gamma_5 \hat{\boldsymbol{\Sigma}} = \hat{\boldsymbol{\Sigma}} \gamma_5 \quad , \tag{1.20}$$

we obtain

$$\begin{aligned}
\hat{P}_\pm u &= \frac{1}{2}(1 \pm \hat{\Lambda}) u = \frac{1}{2}(1 \pm \hat{\boldsymbol{\Sigma}} \cdot \boldsymbol{p}/p) u \\
&= \frac{1}{2}\left(1 \pm \gamma_5 \frac{\boldsymbol{\alpha} \cdot \boldsymbol{p}}{p}\right) u = \frac{1}{2}\left(1 \pm \gamma_5 \frac{E - \beta m_0}{p}\right) u \\
&\approx \frac{1}{2}(1 \pm \gamma_5) u \quad .
\end{aligned} \tag{1.21}$$

We can thus save Lorentz invariance if, instead of the projection operators $\hat{P}_{\pm}$ (1.16), we use the operators

$$\hat{P}'_{\pm} = \frac{1}{2}(1 \pm \gamma_5) \quad , \tag{1.22}$$

which are called projection operators on states of *positive and negative chirality*. For the neutrinos (if they really have vanishing rest mass) relation (1.22) is exact, that is, the replacement of $\hat{P}_{\pm}$ by $\hat{P}'_{\pm}$ makes no difference. For electrons, $\hat{P}_{\pm}$ and $\hat{P}'_{\pm}$ differ considerably if the electron momentum is small. The statement that electrons always have negative helicity can thus be only approximately correct. However, in the case of nuclear β decay the fraction with positive helicity is usually so small that it is hard to detect. Later, when we discuss the β decay of the pion (Sect. 4.2), we will realize that it is just this small contribution of the component with *positive helicity* but *negative chirality* which allows the pion decay at all. Once again we emphasize that *these conclusions are based solely on the requirement of Lorentz invariance of the β-decay Hamiltonian.* Lorentz invariance is considered the superior symmetry guiding us to the correct form of the β-decay Hamiltonian.

According to these considerations we must replace the spinors in (1.17) by their components with negative chirality,

$$\bar{u}_{\mathrm{e}} \hat{O}_i u_\nu \rightarrow \overline{(\hat{P}'_{-} u_{\mathrm{e}})} \hat{O}_i (\hat{P}'_{-} u_\nu) \quad . \tag{1.23}$$

We find that

$$\begin{aligned} \overline{(\hat{P}'_{-} u)} &= (\hat{P}'_{-} u)^\dagger \gamma^0 = u^\dagger \hat{P}'^\dagger_{-} \gamma^0 \\ &= u^\dagger \left(\frac{1-\gamma_5}{2}\right)^\dagger \gamma^0 = u^\dagger \frac{1-\gamma_5}{2} \gamma^0 \\ &= u^\dagger \gamma^0 \frac{1+\gamma_5}{2} = \bar{u} \hat{P}'_{+} \quad , \end{aligned} \tag{1.24}$$

because $\gamma_5^\dagger = \gamma_5$ and $\gamma_5 \gamma^0 = -\gamma^0 \gamma_5$. The modified form of the electron–neutrino term thus reads

$$\bar{u}_{\mathrm{e}} \hat{O}'_i u_\nu \tag{1.25}$$

with

$$\hat{O}'_i = \hat{P}'_{+} \hat{O}_i \hat{P}'_{-} \quad . \tag{1.26}$$

Now it is our task to calculate the helicity-projected parts $\hat{O}'_i$ of the five operators S, V, T, A, and P (compare also Exercise 1.6). The results are summarized in Table 1.3. Thus we conclude that only V and A coupling can be of relevance in the description of the weak interaction.

The restriction to the experimentally observed helicity components excludes all couplings except V and A, which yield the same result up to a sign (see Table 1.3). If we neglect the factor $1/2$, the only possible coupling is thus

$$\gamma^\mu (1 - \gamma_5) = \gamma^\mu - \gamma^\mu \gamma_5 \quad . \tag{1.27}$$

Table 1.3. Properties of helicity-projected fermion transition operators

	$\hat{O}_i$	$\hat{O}'_i$
S	1	0
V	γ^μ	$\gamma^\mu \hat{P}'_- = \frac{1}{2}\gamma^\mu(1-\gamma_5)$
T	$\sigma^{\mu\nu}$	0
A	$\gamma^\mu\gamma_5$	$-\gamma^\mu \hat{P}'_- = -\frac{1}{2}\gamma^\mu(1-\gamma_5)$
P	γ_5	0

It is obviously of V–A type (pronounced "V minus A"); one thus speaks of V–A coupling.[8]

Lorentz invariance of the total Hamiltonian (1.1) requires the nucleonic part

$$\bar{u}_\mathrm{p}\hat{O}_i u_\mathrm{n}$$

to be a combination of V and A couplings as well. To determine the exact form, one must carefully compare the life-times of Fermi transitions (these measure the V component, see Exercise 1.3) and Gamow–Teller transitions (A component), and, in addition, one must measure the angular distribution of the emitted electrons. Extensive experimental analysis has led to the conclusion that the experiments are correctly described if the coupling assumes the form[9]

$$\bar{u}_\mathrm{p}\gamma^\mu(C_V + C_A\gamma_5)u_\mathrm{n} \quad , \tag{1.28}$$

with

$$C_A/C_V = -1.255 \pm 0.006 \quad . \tag{1.29}$$

The complete expression for the interaction is therefore given by

$$H_\mathrm{int}(\mathrm{n},\mathrm{p},\mathrm{e},\nu) = \frac{G}{\sqrt{2}}\int \mathrm{d}^3x \big[\bar{u}_\mathrm{p}\gamma^\mu(C_V + C_A\gamma_5)u_\mathrm{n}\big] \times \big[\bar{u}_\mathrm{e}\gamma_\mu(1-\gamma_5)u_\nu\big] \quad . \tag{1.30}$$

The appearance of the factor $\sqrt{2}$ has historical reasons; it could readily be absorbed in the (coupling) constant G. The constant of nuclear vector coupling, C_V, is often combined with G into a single coupling constant

$$G_\beta = GC_V \quad , \tag{1.31}$$

which is called the *Fermi constant of nuclear β decay*.

That the violation of parity is not only a property of nuclear β decay, but generally occurs in decays mediated by the weak interaction, had been shown shortly after Wu's

---

[8] R.P. Feynman, M. Gell-Mann: Phys. Rev. **109**, 193 (1958); R.E. Marshak, E.C.G. Sudarshan: Phys. Rev. **109**, 1860 (1958); J.J. Sakurai: Nuovo Cimento **7**, 649 (1958).

[9] J.C. Hardy, I.S. Towner: Nucl. Phys. **A 254**, 221 (1975).

experiment in two brilliant experiments.[10] Both groups started with the decay chain

$$\begin{aligned} \pi^+ &\to \mu^+ + \nu_\mu \\ \mu^+ &\to e^+ + \nu_e + \bar{\nu}_\mu \quad , \end{aligned} \tag{1.32}$$

where they made use of the twofold action of parity violation. Here we introduced the distinction between electron neutrinos, ν_e, and muon neutrinos, ν_μ, which will be further explained at the end of this section. Lepton quantum numbers l_e (for the electronic family) and l_μ (for the muonic family) are listed in Table 1.4. Note that each of these lepton quantum numbers is conserved in the reactions (1.32). In the π^+ decay of (1.32) the neutrino has purely negative helicity. Owing to angular-momentum conservation, the muon must have negative helicity, too, in the pion rest frame (see Fig. 1.11). As we have learned, this is possible because the muon, as a heavy particle, moves non-relativistically ($v/c \approx 0.27$). If one selects muons according to their direction of motion, they are completely polarized. The subsequent decay of the polarized muons leads to an anisotropic angular distribution of the positrons, in analogy to Wu's experiment with polarized ^{60}Co nuclei. This can be easily measured. In β decay only left-handed neutrinos are produced. Since the neutrino interacts neither strongly nor electromagnetically, there is consequently no practical source of right-handed neutrinos and no way of proving or disproving their existence. (Of course, right-handed neutrinos would interact by gravitation, if they existed, and could by this means be produced, for example in the cosmic big bang or by black holes. However, their detection would be virtually impossible.)

Fig. 1.11. Helicities in the de cay of the pion. Since it is two-body decay, the neutrin (ν_μ) and positive muon (μ^+ move in opposite directions

Table 1.4. Lepton quantum numbers

	e^-	e^+	ν_e	$\bar{\nu}_e$	μ^-	μ^+	ν_μ	$\bar{\nu}_\mu$
ℓ_e	+1	−1	+1	−1	0	0	0	0
ℓ_μ	0	0	0	0	+1	−1	+1	−1

Finally, as an experimental result[11] it was found in 1962 that there are two kinds of neutrinos, ν_e and ν_μ, which differ in their electron and muon number, ℓ_e and ℓ_μ respectively. In (1.32) we have already introduced this distinction. When muon neutrinos interact with matter, they always produce muons but never electrons:

$$\nu_\mu + n \to \mu^- + p \quad , \qquad \nu_\mu + n \not\to e^- + p \quad .$$

The existence of separate muon and electron quantum numbers had been predicted earlier by **Schwinger** and **Nishijima**,[12] because muons decay weakly ($\mu \to e\bar{\nu}_e\nu_\mu$) but not electromagnetically ($\mu \to e + \gamma$). The origin of this quantum number is still not understood. However, it was found that there is at least one more separate "generation" of leptons: τ (1784 MeV) and ν_τ (see Sect. 2.6 for an extended discussion).

[10] R.L. Garwin, L.M. Ledermann, M. Weinrich: Phys. Rev. **105**, 1415 (1957); J.I. Friedmann, V.L. Telegdi: Phys. Rev. **105**, 1681 (1957).

[11] G. Danby, J.M. Gaillard, E. Goulianos, L.M. Ledermann, M. Mistry, M. Schwartz, J. Steinberger: Phys. Rev. **9**, 36 (1962).

[12] K. Nishijima: Nuovo Cimento **5**, 732 (1957).

EXERCISE

1.6 Left-Handed Dirac Operators

Problem. Calculate the operators $\hat{O}'_i$.

Solution. First we need the two relations

$$\begin{aligned} \gamma_\mu\gamma_5 &= -\gamma_5\gamma_\mu \quad , \\ (\gamma_5)^2 &= 1 \quad . \end{aligned} \tag{1}$$

With their aid the following equations, which quite generally hold for projection operators, can be proved:

$$\begin{aligned} \hat{P}'_+\hat{P}'_- &= \frac{1}{2}(1+\gamma_5)\frac{1}{2}(1-\gamma_5) = 0 \quad , \\ \hat{P}'_-\hat{P}'_+ &= 0 \quad , \\ \left(\hat{P}'_+\right)^2 &= \frac{1}{4}(1+\gamma_5)^2 = \frac{1}{2}(1+\gamma_5) = \hat{P}'_+ \quad , \\ \left(\hat{P}'_-\right)^2 &= \hat{P}'_- \quad . \end{aligned} \tag{2}$$

We further see that

$$\begin{aligned} \hat{P}'_+\gamma^\mu &= \frac{1}{2}(1+\gamma_5)\gamma^\mu = \gamma^\mu\frac{1}{2}(1-\gamma_5) \\ &= \gamma^\mu\hat{P}'_- \quad , \\ \gamma_5\hat{P}'_- &= \gamma_5\frac{1}{2}(1-\gamma_5) = \frac{1}{2}(\gamma_5-1) \\ &= -\hat{P}'_- \quad . \end{aligned} \tag{3}$$

Now we can start calculating:

$$\hat{P}'_+\mathbb{1}\hat{P}'_- = 0 \quad , \tag{4a}$$

$$\hat{P}'_+\gamma^\mu\hat{P}'_- = \gamma^\mu(\hat{P}'_-)^2 = \gamma^\mu\hat{P}'_- \quad , \tag{4b}$$

$$\begin{aligned} \hat{P}'_+\sigma^{\mu\nu}\hat{P}'_- &= \frac{\mathrm{i}}{2}\hat{P}'_+\left(\gamma^\mu\gamma^\nu - \gamma^\nu\gamma^\mu\right)\hat{P}'_- \\ &= \frac{\mathrm{i}}{2}\left(\gamma^\mu\hat{P}'_-\hat{P}'_+\gamma^\nu - \gamma^\nu\hat{P}'_-\hat{P}'_+\gamma^\mu\right) \\ &= 0 \quad , \end{aligned} \tag{4c}$$

$$\begin{aligned} \hat{P}'_+\gamma^\mu\gamma_5\hat{P}'_- &= \gamma^\mu\hat{P}'_-\gamma_5\hat{P}'_- = -\gamma^\mu\left(\hat{P}'_-\right)^2 \\ &= -\gamma^\mu\hat{P}'_- \quad , \end{aligned} \tag{4d}$$

$$\hat{P}'_+\gamma_5\hat{P}'_- = -\hat{P}'_+\hat{P}'_- = 0 \quad . \tag{4e}$$

EXERCISE

1.7 The Weyl Equation

Problem. Show that the four-spinor $\Psi' = \hat{P}'_{-}\Psi$ has only two linearly independent components. Use this to deduce the *Weyl equation*,

$$\mathrm{i}\dot{\psi} = \boldsymbol{\sigma} \cdot \nabla \psi \quad ,$$

for the two-spinor ψ, which composes the four-spinor $\Psi' = \binom{+\psi}{-\psi}$.

Solution. We look at the eigenvalues and eigenvectors of the matrix $\hat{P}'_{-} = (1-\gamma_5)/2$:

$$\begin{aligned}
0 &= \det\left(\hat{P}'_{-} - \lambda \mathbb{1}\right) = \det\left(\frac{1-\gamma_5}{2} - \lambda \mathbb{1}\right) \\
&= \det\begin{pmatrix} \frac{1}{2}-\lambda & 0 & -\frac{1}{2} & 0 \\ 0 & \frac{1}{2}-\lambda & 0 & -\frac{1}{2} \\ -\frac{1}{2} & 0 & \frac{1}{2}-\lambda & 0 \\ 0 & -\frac{1}{2} & 0 & \frac{1}{2}-\lambda \end{pmatrix} \\
&= \left(\frac{1}{2}-\lambda\right)^4 - \frac{1}{4}\left(\frac{1}{2}-\lambda\right)^2 + \left(\frac{1}{4}\right)^2 - \frac{1}{4}\left(\frac{1}{2}-\lambda\right)^2 \\
&= \left[\left(\frac{1}{2}-\lambda\right)^2 - \frac{1}{4}\right]^2 = \lambda^2(\lambda-1)^2 \quad . \qquad (1)
\end{aligned}$$

The eigenvalues are thus $\lambda = 0$ and $\lambda = 1$, each occurring twice, that is, the matrix is of rank two. The eigenvectors of

$$\frac{1}{2}(1-\gamma_5) = \frac{1}{2}\begin{pmatrix} 1 & 0 & -1 & 0 \\ 0 & 1 & 0 & -1 \\ -1 & 0 & 1 & 0 \\ 0 & -1 & 0 & 1 \end{pmatrix} \quad , \qquad (2)$$

for $\lambda = 1$ are

$$\phi_1 = \begin{pmatrix} 1 \\ 0 \\ -1 \\ 0 \end{pmatrix} \quad , \quad \phi_2 = \begin{pmatrix} 0 \\ 1 \\ 0 \\ -1 \end{pmatrix} \quad , \qquad (3a)$$

and for $\lambda = 0$

$$\phi_3 = \begin{pmatrix} 1 \\ 0 \\ 1 \\ 0 \end{pmatrix} \quad , \quad \phi_4 = \begin{pmatrix} 0 \\ 1 \\ 0 \\ 1 \end{pmatrix} \quad . \qquad (3b)$$

If we expand Ψ into ϕ_1 to ϕ_4, we obtain

$$\Psi' = \frac{1}{2}(1-\gamma_5)\Psi(x) = \frac{1}{2}(1-\gamma_5)\sum_{i=1}^{4}\psi_i(x)\phi_i$$

$$= \sum_{i=1}^{4}\psi_i(x)\lambda_i\phi_i = \psi_1(x)\phi_1 + \psi_2(x)\phi_2$$

$$= \begin{pmatrix} \psi_1(x) \\ \psi_2(x) \\ -\psi_1(x) \\ -\psi_2(x) \end{pmatrix} \quad . \tag{4}$$

Thus, in $\psi_1(x)$ and $\psi_2(x)$, Ψ' has two independent components. If we decompose the Dirac spinor Ψ' into two two-spinors ϕ and χ,

$$\Psi' = \begin{pmatrix} \phi \\ \chi \end{pmatrix} \quad , \tag{5}$$

then obviously, according to (4)

$$\phi = -\chi = \begin{pmatrix} \psi_1 \\ \psi_2 \end{pmatrix} \quad . \tag{6}$$

The Dirac equation for massless particles (which are the only kind of particle that can have the left-handed spinor $\hat{P}'_-\Psi$ as an eigenfunction, see (1.21)), is

$$\mathrm{i}\dot{\Psi}' = -\boldsymbol{\alpha}\cdot\nabla\Psi' \quad , \tag{7}$$

or, written as two-spinors,

$$\mathrm{i}\dot{\phi} = -\boldsymbol{\sigma}\cdot\nabla\chi \quad , \tag{8a}$$

$$\mathrm{i}\dot{\chi} = -\boldsymbol{\sigma}\cdot\nabla\phi \quad . \tag{8b}$$

It can thus, because of (6), be rewritten as a *single* equation,

$$\mathrm{i}\dot{\phi} = \boldsymbol{\sigma}\cdot\nabla\phi \quad , \tag{9}$$

which is called the **Weyl** *equation.* Instead of calculating with the helicity-projected Dirac spinor Ψ' for the neutrinos, we can also state: *the neutrinos obey the Weyl-equation.*

1.3 Biographical Notes

BECQUEREL, Henri Antoine, physicist, *15.12.1852 in Paris, †25.8.1908 in Le Croisie (France), professor at the Ecole Polytechnique. He discovered natural radioactivity in uranium and its salts; also discovered the Faraday effect in gases. He shared the 1903 Nobel Prize in physics with Pierre and Marie Curie.

BOHR, Niels Henrik, theoretical physicist, *7.10 1885, †18.11.1962, in Copenhagen. Bohr spent most of his scientific career in Copenhagen, where he received his doctorate in 1911 and became university professor in 1916. His Institute for Theoretical Physics (supported by the Carlsberg brewery) became the focus of the development of quantum mechanics in the 1920s and 1930s. He developed the first quantum model of the atom, the Copenhagen interpretation of quantum mechanics and the liquid drop model of the atomic nucleus. He received the Nobel Prize for physics in 1922. During the second World War Bohr saved many Jewish scientists from persecution by the Nazi regime.

CHADWICK, James, *20.10.1891 in Manchester, †24.7.1974 in Cambridge, professor at the University of Liverpool, was a student of Ernest Rutherford and Hans Geiger. In 1932 he discovered the neutron by bombarding beryllium with alpha particles in an ionization chamber. For this he was awarded the Nobel prize in 1935.

FERMI, Enrico, *29.9.1901 in Rome, †28.11.1954 in Chicago, from 1927 he was professor of theoretical physics at the University of Rome. He was awarded the Nobel prize in 1938 for the discovery of artificial radioactive elements and the nuclear reactions induced by slow neutrons. In 1935 he had to emigrate to the USA, where from 1941 on he was the scientific leader of the nuclear reactor/atomic bomb projects in New York, Chicago, and Los Alamos. Fermi holds an eminent place in the physics of the twentieth century, because he made equally outstanding contributions in theoretical and experimental physics.

FEYNMAN, Richard, theoretical physicist, *11.5.1918 in New York, †15.2.1988 in Pasadena, received his doctorate at Princeton in 1942. After working on the nuclear bomb in Los Alamos, he became professor at Cornell and since 1950 at the California Institute of Technology. Feynman made seminal contributions to many areas of theoretical physics: he developed quantum electrodynamics, the V–A theory of weak interactions and the quark-parton model. He invented the path-integral formulation of quantum mechanics, and contributed to the theory of liquid helium. The Nobel Prize in physics was awarded to him in 1965 together with Schwinger and Tomonaga.

GAMOW, George, *4.3.1904 in Odessa, †19.8.68 in Boulder (Colorado), professor of physics at George Washington University (St. Louis) and the University of Colorado. Besides his work on beta decay he made pioneering contributions to the explanation of alpha decay of nuclei (the tunneling effect) and applications of nuclear physics to astrophysical problems. In connection with the problem discussed here, see Phys. Rev. **51**, 288 (1937).

GOLDHABER, Maurice, physicist, *18.4.1911 in Lemberg (then in Austria), studied in Berlin, after 1933 in Cambridge, emigrated to the United States in 1938. From 1945 he was professor at the University of Illinois, after 1950 senior scientist at Brookhaven National Laboratory, since 1961 its director. He made many important discoveries in nuclear and particle physics, including the moderation of neutrons by certain materials.

KAUFMANN, Walter, *5.6.1871 Elberfeld (Germany), †1.1.1947 in Freiburg im Breisgau (Germany), professor at the Physics Institute at the University of Königsberg (Prussia), determined the charge-to-mass ratio of β particles; in 1901 he showed that the electron mass rises with increasing velocity.

LEE, Tsung-Dao, theoretical physicist, *25.11.1926 in Shanghai, came to the United States in 1946, received his Ph.D. from the University of Chicago in 1950. Since 1953 professor at Columbia University. Together with Yang he refuted the law of parity conservation for which he shared in the 1957 Nobel Prize in physics. He also made many contributions to particle physics and statistical physics.

NISHIJIMA, Kazuhiko, physicist, *4.9.1926 in Tsuchiura (Japan), †15.2.2009 in Tokyo (Japan). After his education in Tokyo and Osaka, he worked in Göttingen, Princeton and Illinois. In 1966 N. became professor at the University of Tokyo; later director of the Yukawa Hall at Kyoto University. N. has made numerous contributions to theoretical particle physics, most notably the suggestion of the strangeness quantum number, the two-neutrino hypothesis and the field theory of bound states.

PAULI, Wolfgang, *25.4.1900 in Vienna, †15.12.1958 Zürich, professor at the "Eidgenössische Technische Hochschule" (Federal Technical University) in Zürich since 1928, was a student of Arnold Sommerfeld and Max Born. In 1945 he was awarded the Nobel prize for the discovery of the exclusion principle which carries his name. He also developed the first theory of the electron spin, which led to the Pauli equation.

RUTHERFORD, Sir Ernest, physicist, *30.8.1871 in Spring Grove (New Zealand), †19.10.1937 in Cambridge. Professor at McGill University in Montreal and Victoria University in Manchester, after 1919 director of the Cavendish Laboratory in Cambridge, which became the foremost nuclear physics laboratory of its time. His scattering experiments with alpha particles established the existence of the atomic nucleus, and he first observed the artificial transmutation of nuclei by ion bombardment. He was awarded the 1908 Nobel Prize in chemistry.

SCHWINGER, Julian, theoretical physicist, *12.2.1918 in New York, †16.7.1994 in Los Angeles. He received his Ph.D. at Columbia University in 1939, and as early as 1947 became a full professor at Harvard University. Since 1975 professor at the University of California, Los Angeles. Schwinger's work laid the foundations for modern quantum field theory and in 1947 predicted the anomalous magnetic moment of the electron. He shared in the 1965 Nobel Prize in physics with Feynman and Tomonaga for the development of quantum electrodynamics.

TELLER, Edward, *15.1.1908 in Budapest, †9.9.2003 in Palo Alto, California, studied in Germany obtaining his Ph.D. in 1930 as a student of Arnold Sommerfeld. From 1935, he was professor of physics at George Washington University (St. Louis), and subsequently at the University of California, in particular, at Livermore Laboratory. He is considered the "father of the hydrogen bomb", the construction of which he decisively influenced.

WEYL, Hermann, *9.11.1885 in Elmshorn (Germany), †5.12.1955 in Zürich, received his Ph.D. in mathematics in Göttingen in 1908 as a student of D. Hilbert. He was professor at the Universities of Zürich, Göttingen and Princeton, He was one of the most prominent mathematicians of his time, in 1925 he was awarded the Lobachevsky prize.

WU, Chien-Shiung, *31.5.1912 in Shanghai, †16.2.1997 in New York, has been professor of physics at Columbia University since 1952, discovered parity violation in nuclear beta decay, received the Wolf Price in physics 1974.

YANG, Chen-Ning, theoretical physicist,*22.9.1922 in Hofei (China), came to the United States in 1945 and received his Ph.D. in 1948 at the University of Chicago under Fermi. Became professor at the Institute for Advanced Study at Princeton in 1955, now professor at the State University of New York at Stony Brook. Together with Lee he predicted the possibility of parity nonconservation. Yang also made important contributions to quantum field theory (Yang–Mills equations) and to statistical physics. He shared the Nobel Prize in physics with Lee in 1957.

Leptonic Interactions

2

2.1 The Current–Current Interaction (Charged Currents)

Let us first consider only the weak interactions between leptons. Today three leptonic hierarchies (e, μ, τ) are known; the experimental data are listed in Table 2.1. To recall, parity violation in nuclear β decay suggested an interaction of the form (see (1.30))

$$H_{\text{int}} = \frac{G}{\sqrt{2}} \int d^3x \left[\bar{u}_p(x)\gamma_\alpha(C_V + C_A\gamma_5)u_n(x)\right]\left[\bar{u}_e(x)\gamma^\alpha(1-\gamma_5)u_{\nu_e}(x)\right] \tag{2.1}$$

where the leptonic contribution

$$\bar{u}_e(x)\gamma^\alpha(1-\gamma_5)u_{\nu_e}(x) \tag{2.2}$$

contains terms that resemble the electromagnetic current

$$j^\alpha(x) = e\bar{\Psi}(x)\gamma^\alpha\Psi(x) \quad . \tag{2.3}$$

Table 2.1. Experimental data for leptons

Lepton	e	ν_e	μ	ν_μ	τ	ν_τ
mass (MeV)	0.511	$<17\times10^{-6}$	105.66	<0.27	1784 ± 4	<35
lifetime (s)	∞	∞	2.2×10^{-6}	∞?	3×10^{-13}	?

By analogy with the electromagnetic current, we therefore introduce the total *weak leptonic current* by adding the currents of the three leptonic families:

$$\begin{aligned} J_\alpha^{(\mathrm{L})}(x) &= \bar{u}_e(x)\gamma_\alpha(1-\gamma_5)u_{\nu_e}(x) + \bar{u}_\mu(x)\gamma_\alpha(1-\gamma_5)u_{\nu_\mu}(x) \\ &\quad + \bar{u}_\tau(x)\gamma_\alpha(1-\gamma_5)u_{\nu_\tau}(x) \\ &= J_\alpha^{(e)}(x) + J_\alpha^{(\mu)}(x) + J_\alpha^{(\tau)}(x) \quad . \end{aligned} \tag{2.4}$$

To describe the mutual weak interaction of leptons we generalize (2.1) by postulating that

$$H_{\text{int}}^{(\mathrm{L})} = \frac{G}{\sqrt{2}} \int d^3x \, J_\alpha^{(\mathrm{L})\dagger}(x) J_{(\mathrm{L})}^\alpha(x) \quad . \tag{2.5}$$

The consequences of this step are non-trivial. Since $H_{\text{int}}^{(\mathrm{L})}$ is quadratic in $J_\alpha^{(\mathrm{L})}$, each leptonic hierarchy interacts with itself as well as with each of the other two. The following diagrams are some examples for such possible processes (see also Exercise 2.1).

W. Greiner, B. Müller, *Gauge Theory of Weak Interactions*,
DOI 10.1007/978-3-540-87843-8_2, © Springer-Verlag Berlin Heidelberg 2009

Neutrino–electron scattering:

$$J^{(e)\dagger}_{\alpha} J^{\alpha}_{(e)} = \left[\bar{u}_{\nu_e}\gamma_\alpha(1-\gamma_5)u_e\right]\left[\bar{u}_e\gamma^\alpha(1-\gamma_5)u_{\nu_e}\right] \quad .$$

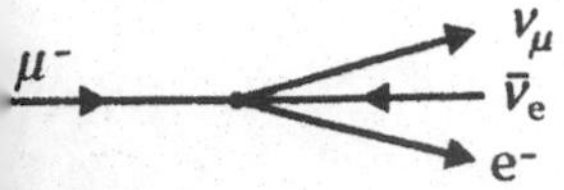

Muon decay:

$$J^{(\mu)\dagger}_{\alpha} J^{\alpha}_{(e)} = \left[\bar{u}_{\nu_\mu}\gamma_\alpha(1-\gamma_5)u_\mu\right]\left[\bar{u}_e\gamma^\alpha(1-\gamma_5)u_{\nu_e}\right] \quad .$$

Muon production in muon-neutrino–electron scattering:

$$J^{(e)\dagger}_{\alpha} J^{\alpha}_{(\mu)} = \left[\bar{u}_{\nu_e}\gamma_\alpha(1-\gamma_5)u_e\right]\left[\bar{u}_\mu\gamma^\alpha(1-\gamma_5)u_{\nu_\mu}\right] \quad .$$

On the other hand, a process like

is *not* allowed. This means that ν_μ and e can interact *only* via the creation of a muon, which is an immediate consequence of the specific form of the currents $J^{(i)}_\mu$, allowing for a neutrino converting into a charged lepton (or vice versa!), but prohibiting an interaction without a conversion of particles. This property of the interaction is usually expressed by calling the currents (2.4) *charged currents* (more accurate by *charged transition currents*) since the charge of the particle of a particular leptonic hierarchy changes by one unit. In the electromagnetic current (2.3) the charge of the particle does not change, it is therefore called a *neutral current*. We shall later see that neutral currents also appear in the context of the gauge theory of weak interaction.

EXERCISE

2.1 Neutrino–Electron Exchange Current

Problem. Prove that $J^{\mu\,\dagger}_{(e)} = \bar{u}_{\nu_e}\gamma_\mu(1-\gamma_5)u_e$.

Solution. With $\gamma_5{}^\dagger = \gamma_5$ we find

$$\begin{aligned} J^{\mu\,\dagger}_{(e)} &= \left[\bar{u}_e\gamma^\mu(1-\gamma_5)u_{\nu_e}\right]^\dagger \\ &= u_{\nu_e}{}^\dagger(1-\gamma_5)\gamma^{\mu\dagger}\bar{u}_e^\dagger \\ &= \bar{u}_{\nu_e}\gamma^0(1-\gamma_5)\gamma^{\mu\dagger}\gamma^{0\dagger}u_e \quad . \end{aligned} \tag{1}$$

Using the identity

$$\gamma^{\mu\dagger} = \gamma^0\gamma^\mu\gamma^0 \quad , \tag{2}$$

that is, $\gamma^{i\dagger} = -\gamma^i$, $\gamma^{0\dagger} = \gamma^0$, yields the desired result:

$$J^{\mu\,\dagger}_{(e)} = \bar{u}_{\nu_e}\gamma^0(1-\gamma_5)\gamma^0\gamma^\mu u_e$$

$$= \bar{u}_{\nu_e}(1+\gamma_5)\gamma^\mu u_e$$
$$= \bar{u}_{\nu_e}\gamma^\mu(1-\gamma_5)u_e \quad , \tag{3}$$

where we have used the fact that γ_5 anticommutes with all other γ matrices.

2.2 The Decay of the Muon

Of all pure leptonic processes, muon decay was the first to be investigated with high accuracy. Muon decay occurs because of the general hypothesis (2.5) for the weak interaction of leptons. Its observation, therefore, is a very important check of the generalization (2.5) of the original Fermi theory of weak interactions. It is therefore appropriate to begin our study with this particular process. Since the decay implies a change in the state of the muon, and because the interaction that causes it is weak, it can be described in the framework of time-dependent perturbation theory.

The quantum mechanical wavefunction obeys a Schrödinger equation,

$$\mathrm{i}\frac{\partial \Psi(\boldsymbol{x},t)}{\partial t} = \hat{H}(\boldsymbol{x},t)\Psi(\boldsymbol{x},t)$$

which – after eliminating the space coordinates $\boldsymbol{x}$ – simply reads

$$\mathrm{i}\frac{\partial \Psi(t)}{\partial t} = \hat{H}(t)\Psi(t) \quad . \tag{2.6}$$

We now study the time development appropriate for our case (2.5) of weak interaction. Starting at t_0 with the initial wavefunction $\Psi_\mathrm{i} = \Psi(t_0)$, we obtain after a time step Δt_0

$$\begin{aligned}\Psi(t_1) &= \Psi(t_0+\Delta t_0) = \Psi(t_0) - \mathrm{i}\Delta t_0 \hat{H}(t_0)\Psi(t_0)\\ &= (1-\mathrm{i}\hat{H}(t_0)\Delta t_0)\Psi(t_0) \quad .\end{aligned}$$

After a next time step Δt_1 we get

$$\begin{aligned}\Psi(t_2) &= \Psi(t_0+\Delta t_0+\Delta t_1) = \Psi(t_1) - \mathrm{i}\Delta t_1 \hat{H}(t_1)\Psi(t_1)\\ &= \Psi(t_0) - \mathrm{i}\Delta t_0 \hat{H}(t_0)\Psi(t_0)\\ &\quad - \mathrm{i}\Delta t_1 \hat{H}(t_1)\Psi(t_1) \quad ,\end{aligned}$$

and after N steps

$$\begin{aligned}\Psi(t) = \Psi(t_N) &= \Psi(t_0+\Delta t_0+\Delta t_1+\cdots+\Delta t_{N-1})\\ &= \Psi(t_0) - \mathrm{i}\Delta t_0 \hat{H}(t_0)\Psi(t_0)\\ &\quad - \mathrm{i}\Delta t_1 \hat{H}(t_1)\Psi(t_1)\\ &\quad \vdots\\ &\quad - \mathrm{i}\Delta t_{N-1} \hat{H}(t_{N-1})\Psi(t_{N-1})\end{aligned}$$

$$= \Psi(t_0) - \mathrm{i}\sum_{i=0}^{N-1} \Delta t_i \hat{H}(t_i)\Psi(t_i)$$

$$\approx \Psi(t_0) - \mathrm{i}\int_{t_0}^{t} \hat{H}(t')\Psi(t')\mathrm{d}t' \quad , \tag{2.7}$$

where higher-order terms in $\hat{H}$ were neglected.

Now, the S-matrix element S_{fi} for a transition from an initial state $|\Psi_{\mathrm{i}}\rangle$ to a final state $|\Psi_{\mathrm{f}}\rangle \neq |\Psi_{\mathrm{i}}\rangle$ is defined as[1]

$$\begin{aligned} S_{\mathrm{fi}} &= \lim_{t\to\infty} \langle \Psi_{\mathrm{f}}|\Psi_{\mathrm{i}}(t)\rangle \\ &= \lim_{\substack{t\to\infty \\ t_0\to-\infty}} \left\langle \Psi_{\mathrm{f}} \middle| \Psi_{\mathrm{i}}(t_0) - \mathrm{i}\int_{t_0}^{t} \hat{H}(t')\Psi_{\mathrm{i}}(t')\mathrm{d}t' \right\rangle \\ &= \delta_{\mathrm{fi}} - \mathrm{i}\int_{-\infty}^{\infty} \langle \Psi_{\mathrm{f}}|\hat{H}(t')|\Psi_{\mathrm{i}}(t')\rangle \mathrm{d}t' \quad . \end{aligned} \tag{2.8}$$

Only $\hat{H}_{\mathrm{int}}$ of $\hat{H} = \hat{H}_0 + \hat{H}_{\mathrm{int}}$ contributes to the integral, because of the supposed orthogonality of initial and final state, $\langle \Psi_{\mathrm{f}}|\Psi_{\mathrm{i}}\rangle = 0$.

Specializing to the case of the muon decay, the lowest-order transition amplitude is

$$S_{\mathrm{fi}} = -\mathrm{i}\int_{-\infty}^{+\infty} \mathrm{d}t\, H_{\mathrm{int}}^{(\mathrm{L})}(\mu^- \to \mathrm{e}^-\bar{\nu}_{\mathrm{e}}\nu_\mu) \quad . \tag{2.9}$$

As discussed in Sect. 2.1, the relevant part of $H_{\mathrm{int}}^{(\mathrm{L})}$ contributing to this process is

$$-\mathrm{i}\frac{G}{\sqrt{2}}\int \mathrm{d}^3x \big[\bar{u}_{\nu_\mu}(x)\gamma_\mu(1-\gamma_5)u_\mu(x)\big]\big[\bar{u}_{\mathrm{e}}(x)\gamma^\mu(1-\gamma_5)u_{\nu_{\mathrm{e}}}(x)\big] \quad . \tag{2.10}$$

For this first-order approximation we may choose free wave functions to describe the four particles with four-momenta p, p', k, k' and spins s, s', t, t', respectively. According to the Feynman rules the (outgoing) antineutrino is represented by an (incoming) wave function with negative energy (see Fig. 2.1). Employing the form of the plane

---

[1] See W. Greiner: *Quantum Mechanics – An Introduction*, 4th ed. (Springer, Berlin, Heidelberg, 2001), and W. Greiner and J. Reinhardt: *Field Quantization*, 1st ed. (Springer, Berlin, Heidelberg, 1996).

waves of Appendix A.2 we have[2]

$$
\begin{aligned}
u_\mu(x) &= (2E_\mu V)^{-1/2} u_\mu(\boldsymbol{p}', s') \exp(-\mathrm{i}p'_\mu x^\mu) \quad , \\
u_\mathrm{e}(x) &= (2E_\mathrm{e} V)^{-1/2} u_\mathrm{e}(\boldsymbol{p}, s) \exp(-\mathrm{i}p_\mu x^\mu) \quad , \\
u_{\bar{\nu}_\mathrm{e}}(x) &= (2E_{\nu_\mathrm{e}} V)^{-1/2} v_{\nu_\mathrm{e}}(\boldsymbol{k}, t) \exp(+\mathrm{i}k_\mu x^\mu) \quad , \\
u_{\nu_\mu}(x) &= (2E_{\nu_\mu} V)^{-1/2} u_{\nu_\mu}(\boldsymbol{k}', t') \exp(-\mathrm{i}k'_\mu x^\mu) \quad ,
\end{aligned}
\tag{2.11}
$$

where

$$
E_\mu = p'^0, \quad E_\mathrm{e} = p^0, \quad E_{\nu_\mathrm{e}} = k^0, \quad E_{\nu_\mu} = k'^0 \tag{2.12}
$$

and $u(\boldsymbol{p}, s)$, $v(\boldsymbol{p}, s)$ denote the spinor parts (E positive!)

$$
\begin{aligned}
u(\boldsymbol{p}, s) &= (E+m)^{\frac{1}{2}} \begin{pmatrix} \chi_s \\ \frac{\boldsymbol{\sigma}\cdot\boldsymbol{p}}{E+m}\chi_s \end{pmatrix} \quad , \\
v(\boldsymbol{p}, s) &= (E+m)^{\frac{1}{2}} \begin{pmatrix} \frac{\boldsymbol{\sigma}\cdot\boldsymbol{p}}{E+m}\chi_s \\ \chi_s \end{pmatrix}
\end{aligned}
\tag{2.13}
$$

with the two-component unit spinors χ_s. Substituting this expression into the matrix element (2.9) yields

$$
\begin{aligned}
S(\mu^- \to \mathrm{e}^- \bar{\nu}_\mathrm{e} \nu_\mu) &= -\frac{\mathrm{i}G}{\sqrt{2}} \int \mathrm{d}^4x \frac{\exp[\mathrm{i}(k'_\mu - p'_\mu + p_\mu + k_\mu)x^\mu]}{[16(k'^0V)(p'^0V)(p^0V)(k^0V)]^{\frac{1}{2}}} \\
&\quad \times \left[\bar{u}_{\nu_\mu}(\boldsymbol{k}', t')\gamma_\mu(1-\gamma_5)u_\mu(\boldsymbol{p}', s')\right] \\
&\quad \times \left[\bar{u}_\mathrm{e}(\boldsymbol{p}, s)\gamma^\mu(1-\gamma_5)v_{\nu_\mathrm{e}}(\boldsymbol{k}, t)\right] \\
&= -\mathrm{i}(2\pi)^4 \frac{G}{\sqrt{2}} \frac{\delta^4(p+k+k'-p')}{[16V^4k'^0k^0p'^0p^0]^{\frac{1}{2}}} \\
&\quad \times \left[\bar{u}_{\nu_\mu}(\boldsymbol{k}', t')\gamma_\mu(1-\gamma_5)u_\mu(\boldsymbol{p}', s')\right] \\
&\quad \times \left[\bar{u}_\mathrm{e}(\boldsymbol{p}', s)\gamma^\mu(1-\gamma_5)v_{\nu_\mathrm{e}}(\boldsymbol{k}, t)\right] \quad .
\end{aligned}
\tag{2.14}
$$

To obtain the transition probability, (2.14) must be multiplied with its Hermitian conjugate. This gives a factor

$$
\left[\delta^4(p+k+k'-p')\right]^2 = \delta^4(p+k+k'-p')\delta^4(0) \quad , \tag{2.15}
$$

which is replaced by

$$
\frac{VT}{(2\pi)^4}\delta^4(p+k+k'-p') \tag{2.16}
$$

according to the usual prescription, which can be derived heuristically (although mathematically oversimplified) as follows:

$$
\delta^4(0) = \lim_{q\to 0}\delta^4(q) = \lim_{q\to 0}\int \frac{\mathrm{d}^4y}{(2\pi)^4}\mathrm{e}^{\mathrm{i}y_\mu q^\mu} = \int \frac{\mathrm{d}^4y}{(2\pi)^4} = \frac{VT}{(2\pi)^4} \quad . \tag{2.17}
$$

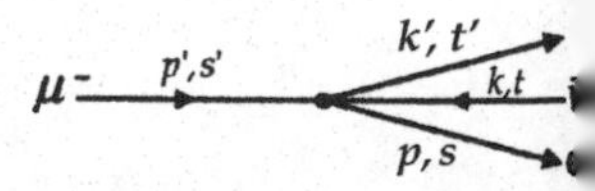

Fig. 2.1. Momenta and spi[n]s for the muon decay. The ant[i]neutrino $\bar{\nu}_\mathrm{e}$ is represented b[y] an incoming wave with nega[-]tive energy and negative mo[-]mentum, i.e. negative four momentum

---

[2] Note that we are using the index "μ" for two different purposes: it denotes the muon wave function u_μ and energy E_μ, and it occurs as a four-vector index, such as in p_μ, x_μ, γ_μ. Although this is somewhat unfortunate, we must get used to this double meaning.

V and T are understood to be macroscopic quantities so that the physical process takes place entirely within the finite space-time volume VT. In practice, the two neutrinos cannot be observed, that is, we need to sum or integrate over all possible final states. Furthermore, to obtain the transition probability within a small interval of momentum, we multiply by the density of the electron final states within an interval $V\,\mathrm{d}^3p/(2\pi)^3$. Finally we divide by T to get the *decay rate*, that is, the transition probability per unit time interval. Following these steps we find that

$$\begin{aligned}
\mathrm{d}W &= \frac{1}{T}\frac{V\mathrm{d}^3p}{(2\pi)^3} V\int \frac{\mathrm{d}^3k}{(2\pi)^3} V\int \frac{\mathrm{d}^3k'}{(2\pi)^3}\sum_{t,t'}|S(\mu^-\to \mathrm{e}^-\bar{\nu}_\mathrm{e}\nu_\mu)|^2 \\
&= \frac{G^2}{2}\frac{1}{(2\pi)^5}\frac{\mathrm{d}^3p}{2p'^0 2p^0}\int\frac{\mathrm{d}^3k}{2k^0}\int\frac{\mathrm{d}^3k'}{2k'^0}\delta^4(p+k+k'-p')\sum_{t,t'}|M|^2 \quad , \qquad (2.18)
\end{aligned}$$

where

$$M = \left[\bar{u}_{\nu_\mu}\gamma^\mu(1-\gamma_5)u_\mu\right]\left[\bar{u}_\mathrm{e}\gamma_\mu(1-\gamma_5)v_{\nu_\mathrm{e}}\right] \quad . \qquad (2.19a)$$

The expression $|M|^2$ consists of two similar factors for the muonic and electronic transition currents. If we write $M = M^\mu E_\mu$ with $M^\mu = (\bar{u}_{\nu\mu}\gamma^\mu(1-\gamma_5)u_\mu)$ and $E_\mu = (\bar{u}_\mathrm{e}\gamma_\mu(1-\gamma_5)u_{\nu_\mathrm{e}})$, (2.19a) becomes

$$\sum_{t,t'}|M|^2 = \sum_{t,t'}(M^\mu E_\mu)(M^\nu E_\nu)^\dagger = \sum_{t,t'}(M^\mu M^{\nu\dagger})(E_\mu E_\nu^\dagger) \quad . \qquad (2.19b)$$

Let us first focus on the muonic factor, making use of Exercise 2.1:

$$\begin{aligned}
X^{\mu\nu}(\mu) &= M^\mu M^{\nu\dagger} \\
&= \sum_{t'}\left[\bar{u}_{\nu_\mu}(\boldsymbol{k}',t')\gamma^\mu(1-\gamma_5)u_\mu(\boldsymbol{p}',s')\right] \\
&\quad\times\left[\bar{u}_{\nu_\mu}(\boldsymbol{k}',t')\gamma^\nu(1-\gamma_5)u_\mu(\boldsymbol{p}',s')\right]^\dagger \\
&= \sum_{t'}\bar{u}_{\nu_\mu}(\boldsymbol{k}',t')\gamma^\mu(1-\gamma_5)u_\mu(\boldsymbol{p}',s')\bar{u}_\mu(\boldsymbol{p}',s') \\
&\quad\times\gamma^\nu(1-\gamma_5)u_{\nu_\mu}(\boldsymbol{k}',t') \quad . \qquad (2.20)
\end{aligned}$$

In order to evaluate this expression we make use of some helpful formulas for Dirac spinors and γ matrices (see Appendix A.2),[3]

$$\sum_{t'}u_{\nu_\mu}(k',t')_\alpha\bar{u}_{\nu_\mu}(k',t')_\beta = (\not{k}'+m_\nu)_{\alpha\beta} = \not{k}'_{\alpha\beta} \quad , \qquad (2.21)$$

where α, β denote the spinor indices and $m_\nu = 0$. Since the summation is not over the initial muon states, we have (see Appendix A.2)

$$u_\mu(p',s')_\alpha\bar{u}_\mu(p',s')_\beta = \left[(\not{p}'+m_\mu)\left(\frac{1+\gamma_5\not{s}'}{2}\right)\right]_{\alpha\beta} \quad , \qquad (2.22)$$

---

[3] See W. Greiner and J. Reinhardt: *Quantum Electrodynamics*, 4th ed. (Springer, Berlin, Heidelberg, 2009).

with the spin four-vector

$$s'_\mu = \left(\frac{\boldsymbol{p}'\cdot\boldsymbol{s}'}{m}, \boldsymbol{s}' + \frac{(\boldsymbol{p}'\cdot\boldsymbol{s}')\boldsymbol{p}'}{m(E'+m)}\right) , \tag{2.23}$$

where $\boldsymbol{s}'$ is the spin vector with respect to the rest frame. Here $\boldsymbol{s}'$ is a unit vector so that $s'^\mu s'_\mu = -1$. Inserting the relations (2.21) and (2.22) into (2.20) we obtain the following expression for the muonic contribution to the transition currents $X^{\mu\nu}(\mu)$:

$$\begin{aligned} X^{\mu\nu}(\mu) &= \sum_{t'} \bar{u}_{\nu_\mu}(k',t')_\pi \gamma^\mu_{\pi\varrho}(1-\gamma_5)_{\varrho\alpha}\left[(\not{p}'+m_\mu)\left(\frac{1+\gamma_5\not{s}'}{2}\right)\right]_{\alpha\beta} \\ &\quad \times \gamma^\nu_{\beta\sigma}(1-\gamma_5)_{\sigma\tau}u_{\nu_\mu}(k',t')_\tau \\ &= \gamma^\mu_{\pi\varrho}(1-\gamma_5)_{\varrho\alpha}\left[(\not{p}'+m_\mu)\left(\frac{1+\gamma_5\not{s}'}{2}\right)\right]_{\alpha\beta}\gamma^\nu_{\beta\sigma}(1-\gamma_5)_{\sigma\tau}\not{k}'_{\tau\pi} \quad . \end{aligned} \tag{2.24}$$

Summing over the first and last index π means that we have to evaluate the trace of the (4×4) matrix:

$$X^{\mu\nu}(\mu) = \mathrm{Tr}\left\{\gamma^\mu(1-\gamma_5)(\not{p}'+m_\mu)\left(\frac{1+\gamma_5\not{s}'}{2}\right)\gamma^\nu(1-\gamma_5)\not{k}'\right\} \quad . \tag{2.25}$$

Since $\gamma^\alpha\gamma_5 = -\gamma_5\gamma^\alpha$ and $\mathrm{Tr}\{AB\} = \mathrm{Tr}\{BA\}$ this yields

$$X^{\mu\nu}(\mu) = \frac{1}{2}\mathrm{Tr}\{(\not{p}'+m_\mu)(1+\gamma_5\not{s}')\gamma^\nu\not{k}'(1+\gamma_5)\gamma^\mu(1-\gamma_5)\} \quad . \tag{2.26}$$

Now we make use of the property that any trace of a product of an odd number of γ matrices vanishes (see Appendix A.2). Since $\gamma_5 = \mathrm{i}\gamma^0\gamma^1\gamma^2\gamma^3$ it consists of an even number of γ matrices. Furthermore it holds that $(1-\gamma_5)^2 = 2(1-\gamma_5)$, so that (2.26) becomes

$$\begin{aligned} X^{\mu\nu}(\mu) &= \mathrm{Tr}\{(\not{p}'+m_\mu)(1+\gamma_5\not{s}')\gamma^\nu\not{k}'\gamma^\mu(1-\gamma_5)\} \\ &= \mathrm{Tr}\{\not{p}'\gamma^\nu\not{k}'\gamma^\mu(1-\gamma_5) + \not{p}'\gamma_5\not{s}'\gamma^\nu\not{k}'\gamma^\mu(1-\gamma_5) \\ &\quad + m_\mu\gamma^\nu\not{k}'\gamma^\mu(1-\gamma_5) + m_\mu\gamma_5\not{s}'\gamma^\nu\not{k}'\gamma^\mu(1-\gamma_5)\} \quad . \end{aligned} \tag{2.27}$$

Obviously the second and the third terms are "odd"; therefore they do not contribute. The remaining first and last terms are "even". Taking into account that $\gamma_5(1-\gamma_5) = -(1-\gamma_5)$, we find that

$$\begin{aligned} X^{\mu\nu}(\mu) &= \mathrm{Tr}\{\not{p}'\gamma^\nu\not{k}'\gamma^\mu(1-\gamma_5) - m_\mu\not{s}'\gamma^\nu\not{k}'\gamma^\mu(1-\gamma_5)\} \\ &= \mathrm{Tr}\{(\not{p}'-m_\mu\not{s}')\gamma^\nu\not{k}'\gamma^\mu(1-\gamma_5)\} \quad . \end{aligned} \tag{2.28}$$

In Appendix A.2 it is shown that successive application of $\gamma^\mu\gamma^\nu + \gamma^\nu\gamma^\mu = 2g^{\mu\nu}$ yields the general relations

$$\begin{aligned} \mathrm{Tr}\{\gamma^\alpha\gamma^\beta\gamma^\sigma\gamma^\tau\} &= 4(g^{\alpha\beta}g^{\sigma\tau} - g^{\alpha\sigma}g^{\beta\tau} + g^{\alpha\tau}g^{\beta\sigma}) \\ \mathrm{Tr}\{\gamma^\alpha\gamma^\beta\gamma^\sigma\gamma^\tau\gamma_5\} &= -4\mathrm{i}\varepsilon^{\alpha\beta\sigma\tau} \quad . \end{aligned} \tag{2.29}$$

Using this for the trace, (2.28) gives the final result for the muonic part of the transition currents:

$$X^{\mu\nu}(\mu) = 4\big[(p' - m_\mu s')^\nu k'^\mu - (p' - m_\mu s')^\alpha k'_\alpha g^{\mu\nu} + (p' - m_\mu s')^\mu k'^\nu$$
$$+ \mathrm{i}\varepsilon^{\alpha\nu\beta\mu}(p' - m_\mu s')_\alpha k'_\beta\big] \quad . \tag{2.30}$$

The electronic contribution (2.19)–(2.20) is evaluated in a similar manner, which gives

$$\begin{aligned} X_{\mu\nu}(\mathrm{e}) &= E_\mu E_\nu^\dagger \\ &= \sum_t \big[\bar{u}_\mathrm{e}(p,s)\gamma_\mu(1-\gamma_5)v_{\nu_\mathrm{e}}(k,t)\big]\big[\bar{u}_\mathrm{e}(p,s)\gamma_\nu(1-\gamma_5)v_{\nu_\mathrm{e}}(k,t)\big]^\dagger \\ &= \mathrm{Tr}\big\{(\not{p} - m_\mathrm{e}\not{s})\gamma_\mu \not{k}\gamma_\nu(1-\gamma_5)\big\} \\ &= 4\big[(p - m_\mathrm{e}s)_\mu k_\nu - (p - m_\mathrm{e}s)^\alpha k_\alpha g_{\mu\nu} + (p - m_\mathrm{e}s)_\nu k_\mu \\ &\quad - \mathrm{i}\varepsilon_{\alpha\mu\beta\nu}(p - m_\mathrm{e}s)^\alpha k^\beta\big] \quad . \end{aligned} \tag{2.31}$$

The final result for the squared invariant matrix element (2.19b) is the product of the two expressions (2.30) and (2.31) which, after some work, is formed to be (see Exercise 2.3)

$$\sum_{t,t'} |M|^2 = X^{\mu\nu}(\mu)X_{\mu\nu}(\mathrm{e}) = 64(p' - m_\mu s')^\alpha k_\alpha (p - m_\mathrm{e}s)^\beta k'_\beta \quad . \tag{2.32}$$

EXERCISE

2.2 Proof of (2.31)

Problem. Prove the first part of (2.31)

$$X_{\mu\nu}(\mathrm{e}) = \mathrm{Tr}\big\{(\not{p} - m_\mathrm{e}\not{s})\gamma_\mu\not{k}\gamma_\nu(1-\gamma_5)\big\} \quad .$$

Solution. Starting from the expression (2.31) and performing the t summation, we arrive at

$$\begin{aligned} X_{\mu\nu}(\mathrm{e}) &= \sum_t E_\mu E_\nu^\dagger \\ &= \sum_t \big[\bar{u}_\mathrm{e}(\boldsymbol{p},s)\gamma_\mu(1-\gamma_5)v_{\nu_\mathrm{e}}(\boldsymbol{k},t)\big]\big[\bar{u}_\mathrm{e}(\boldsymbol{p},s)\gamma_\nu(1-\gamma_5)v_{\nu_\mathrm{e}}(\boldsymbol{k},t)\big]^\dagger \\ &= \sum_t \big[\bar{u}_\mathrm{e}(\boldsymbol{p},s)\gamma_\mu(1-\gamma_5)v_{\nu_\mathrm{e}}(\boldsymbol{k},t)\bar{v}_{\nu_\mathrm{e}}(\boldsymbol{k},t)\gamma_\nu(1-\gamma_5)u_\mathrm{e}(\boldsymbol{p},s)\big] \\ &= \bar{u}_\mathrm{e}(\boldsymbol{p},s)\gamma_\mu(1-\gamma_5)\underbrace{\left[\sum_t v_{\nu_\mathrm{e}}(\boldsymbol{k},t)\bar{v}_{\nu_\mathrm{e}}(\boldsymbol{k},t)\right]}_{=\not{k}-m_{\nu_\mathrm{e}}=\not{k}}\gamma_\nu(1-\gamma_5)u_\mathrm{e}(\boldsymbol{p},s) \\ &= \bar{u}_\mathrm{e}(\boldsymbol{p},s)_\pi(\gamma_\mu)_{\pi\varrho}(1-\gamma_5)_{\varrho\alpha}\not{k}_{\alpha\beta}(\gamma_\nu)_{\beta\sigma}(1-\gamma_5)_{\sigma\tau}u_\mathrm{e}(\boldsymbol{p},s)_\tau \quad . \end{aligned} \tag{1}$$

With the identity (see (2.22))

$$\bar{u}_\mathrm{e}(\boldsymbol{p},s)_\pi u_\mathrm{e}(\boldsymbol{p},s)_\tau = \left[(\not{p} + m_\mathrm{e})\frac{(1+\gamma_5\not{s})}{2}\right]_{\tau\pi} \tag{2}$$

we obtain

$$\begin{aligned} X_{\mu\nu}(\mathrm{e}) &= (\gamma_\mu)_{\pi\varrho}(1-\gamma_5)_{\varrho\alpha}\not{k}_{\alpha\beta}(\gamma_\nu)_{\beta\sigma}(1-\gamma_5)_{\sigma\tau}\left[(\not{p}+m_\mathrm{e})\left(\frac{1+\gamma_5\not{s}}{2}\right)\right]_{\tau\pi} \\ &= \mathrm{Tr}\left\{\gamma_\mu(1-\gamma_5)\not{k}\gamma_\nu(1-\gamma_5)\left[(\not{p}+m_\mathrm{e})\left(\frac{1+\gamma_5\not{s}}{2}\right)\right]\right\} \\ &= \mathrm{Tr}\{(\not{p}+m_\mathrm{e})(1+\gamma_5\not{s})\gamma_\mu\not{k}\gamma_\nu(1-\gamma_5)\} \end{aligned} \tag{3}$$

where we have used the relation

$$\begin{aligned} \gamma_\mu(1-\gamma_5)\not{k}\gamma_\nu(1-\gamma_5) &= \gamma_\mu\not{k}(1+\gamma_5)\gamma_\nu(1-\gamma_5) \\ &= \gamma_\mu\not{k}\gamma_\nu(1-\gamma_5)^2 \\ &= \gamma_\mu\not{k}\gamma_\nu 2(1-\gamma_5) \end{aligned}$$

and the trace identity $\mathrm{Tr}\{AB\} = \mathrm{Tr}\{BA\}$. This expression (3) transforms to (2.27) if we replace

$$\begin{aligned} \not{p} &\to \not{p}' \ , \\ m_\mathrm{e} &\to m_\mu \ , \\ \not{s} &\to \not{s}' \ , \\ \gamma_\mu &\to \gamma^\nu \ , \\ \not{k} &\to \not{k}' \ , \\ \gamma_\nu &\to \gamma^\mu \ . \end{aligned} \tag{4}$$

Therefore we may simply rewrite (2.28) by substituting for the muonic quantities the corresponding electron quantities:

$$X_{\mu\nu}(\mathrm{e}) = \mathrm{Tr}\{(\not{p} - m_\mathrm{e}\not{s})\gamma_\mu\not{k}\gamma_\nu(1-\gamma_5)\} \ . \tag{5}$$

EXERCISE

2.3 Calculation of the Averaged Decay Matrix Element

Problem. Evaluate $\sum|M|^2$ in (2.32) by using the following relation for the antisymmetric Levi-Civita tensor,

$$\varepsilon^{\alpha\beta\mu\nu}\varepsilon_{\bar{\alpha}\bar{\beta}\mu\nu} = 2(\delta^\alpha_{\bar{\beta}}\delta^\beta_{\bar{\alpha}} - \delta^\alpha_{\bar{\alpha}}\delta^\beta_{\bar{\beta}}) \ ,$$

and the property that any product of $\varepsilon^{\alpha\beta\mu\nu}$ with a tensor that is symmetric in the indices μ, ν vanishes (see also Exercise 2.4).

Solution. Introducing the following abbreviations

$$(p' - m_\mu s')_\nu \equiv q'_\nu \ , \tag{1}$$

$$(p - m_\mathrm{e} s)_\mu \equiv q_\mu \ , \tag{2}$$

Exercise 2.3

$$q_\alpha k^\alpha \equiv (q \cdot k) \quad , \tag{3}$$

and using the relations (2.30)–(2.32) we rewrite $\sum_{t,t'} |M|^2$ as follows:

$$\begin{aligned}
\sum_{t,t'} |M|^2 &= 16\big[q'^\nu k'^\mu - (q' \cdot k') g^{\mu\nu} + q'^\mu k'^\nu - \mathrm{i}\varepsilon^{\alpha\nu\beta\mu} q'_\alpha k'_\beta\big] \\
&\quad \times \big[q_\mu k_\nu - (q \cdot k) g_{\mu\nu} + q_\nu k_\mu + \mathrm{i}\varepsilon_{\alpha\mu\beta\nu} q^\alpha k^\beta\big] \\
&= 16\big[(q' \cdot k)(q \cdot k') - (q \cdot k)(q' \cdot k') \\
&\quad + (q' \cdot q)(k' \cdot k) - \mathrm{i}\varepsilon_{\alpha\mu\beta\nu} q'^\nu k'^\mu q^\alpha k^\beta \\
&\quad - (q' \cdot k')\{(q \cdot k) - (q \cdot k) \cdot 4 + (q \cdot k)\} \\
&\quad + (q' \cdot q)(k' \cdot k) - (q \cdot k)(q' \cdot k') + (q' \cdot k)(k' \cdot q) \\
&\quad - \mathrm{i}\varepsilon_{\alpha\mu\beta\nu} q'^\mu k'^\nu q^\alpha k^\beta + \mathrm{i}\varepsilon^{\alpha\nu\beta\mu} q'_\alpha k'_\beta q_\mu k_\nu \\
&\quad + \mathrm{i}\varepsilon^{\alpha\nu\beta\mu} q'_\alpha k'_\beta q_\nu k_\mu - \varepsilon^{\alpha\beta\nu\mu} q'_\alpha k'_\beta \varepsilon_{\bar\alpha\bar\beta\mu\nu} q^{\bar\alpha} k^{\bar\beta}\big] \\
&= 16\big[2(q' \cdot k)(q \cdot k') + 2(q' \cdot q)(k' \cdot k) \\
&\quad \underbrace{-\mathrm{i}\varepsilon_{\alpha\mu\beta\nu}(q'^\mu k'^\nu + q'^\nu k'^\mu) q^\alpha k^\beta}_{=0} \underbrace{+\mathrm{i}\varepsilon^{\alpha\nu\beta\mu}(q_\mu k_\nu + q_\nu k_\mu) q'_\alpha k'_\beta}_{=0} \\
&\quad + \varepsilon^{\alpha\beta\mu\nu} \varepsilon_{\bar\alpha\bar\beta\mu\nu} q'_\alpha k'_\beta q^{\bar\alpha} k^{\bar\beta}\big] \\
&= 32\big[(q' \cdot k)(q \cdot k') + (q' \cdot q)(k' \cdot k) + (\delta^\alpha_{\bar\beta}\delta^\beta_{\bar\alpha} - \delta^\alpha_{\bar\alpha}\delta^\beta_{\bar\beta}) q'_\alpha k'_\beta q^{\bar\alpha} k^{\bar\beta}\big] \\
&= 32\big[(q' \cdot k)(q \cdot k') + (q' \cdot q)(k' \cdot k) \\
&\quad + (q' \cdot k)(k' \cdot q) - (q' \cdot q)(k' \cdot k)\big] \\
&= 64 (q' \cdot k)(k' \cdot q) \quad .
\end{aligned} \tag{4}$$

Returning to the original notation (3) this result is equivalent to (2.32).

EXERCISE

2.4 A Useful Relation for the Levi-Civita Tensor

Problem. Prove the formula

$$\varepsilon^{\alpha\beta\mu\nu} \varepsilon_{\bar\alpha\bar\beta\mu\nu} = 2(\delta^\alpha_{\bar\beta}\delta^\beta_{\bar\alpha} - \delta^\alpha_{\bar\alpha}\delta^\beta_{\bar\beta})$$

that was applied in the Exercise 2.3.

Solution. The totally antisymmetric Levi-Civita tensor $\varepsilon^{\alpha\beta\mu\nu}$ is defined as

$$\varepsilon^{\alpha\beta\mu\nu} = \begin{cases} \mathrm{sgn}(\hat P) & \text{if } (\alpha\beta\mu\nu) = \hat P(0123) \\ 0 & \text{otherwise} \end{cases} \quad , \tag{1}$$

where $\hat P$ denotes a permutation of the indices (0123). $\varepsilon^{\alpha\beta\mu\nu}$ vanishes if two of its indices are equal.

Exercise 2.

Since $\varepsilon_{\bar{\alpha}\bar{\beta}\bar{\mu}\bar{\nu}} = g_{\bar{\alpha}\alpha} g_{\bar{\beta}\beta} g_{\bar{\mu}\mu} g_{\bar{\nu}\nu} \varepsilon^{\alpha\beta\mu\nu}$ is non-zero if and only if $(\bar{\alpha}\bar{\beta}\bar{\mu}\bar{\nu})$ is a permutation of (0123), we then have for example for $(\bar{\alpha}\bar{\beta}\bar{\mu}\bar{\nu}) = (0123)$,

$$\begin{aligned}\varepsilon_{0123} &= g_{0\alpha} g_{1\beta} g_{2\mu} g_{3\nu} \varepsilon^{\alpha\beta\mu\nu} \\ &= 1 \cdot (-1)^3 \varepsilon^{0123} = 1 \times (-1)^3 = -1 \quad ,\end{aligned} \tag{2}$$

and a similar relation for all other non-vanishing components of the covariant Levi-Civita tensor, that is,

$$\varepsilon^{\alpha\beta\mu\nu} = -\varepsilon_{\alpha\beta\mu\nu} \quad . \tag{3}$$

Now consider the desired contraction with respect to the indices μ, ν,

$$\varepsilon^{\alpha\beta\mu\nu} \varepsilon_{\bar{\alpha}\bar{\beta}\mu\nu} \quad . \tag{4}$$

For fixed values of $\alpha, \bar{\alpha}, \beta, \bar{\beta}$, only those terms contribute that contain tensor components with third and fourth indices different from $\alpha, \bar{\alpha}, \beta, \bar{\beta}$. Furthermore, since the third and fourth indices are the same for both the covariant and the contravariant tensor, an additional condition is that either

$$\alpha = \bar{\alpha} \quad , \quad \beta = \bar{\beta} \tag{5a}$$

or

$$\alpha = \bar{\beta} \quad , \quad \beta = \bar{\alpha} \quad . \tag{5b}$$

In each of the two cases (5a) and (5b) only two possible combinations for the values of the indices μ, ν remain, namely those of the two numbers (0123) that differ from α and β. We then have the following relations:

Case A:

$$\begin{aligned}\alpha = \bar{\alpha} \quad , \quad \beta = \bar{\beta} \quad : \quad \varepsilon^{\alpha\beta\mu\nu} \varepsilon_{\bar{\alpha}\bar{\beta}\mu\nu} &= \sum_{\mu,\nu} \varepsilon^{\alpha\beta\mu\nu} \varepsilon_{\bar{\alpha}\bar{\beta}\mu\nu} \\ &= \sum_{\mu,\nu} \varepsilon^{\alpha\beta\mu\nu} \varepsilon_{\alpha\beta\mu\nu} \\ &= 2 \cdot 1 \cdot (-1) \\ &= -2 \quad .\end{aligned} \tag{6a}$$

Case B:

$$\begin{aligned}\alpha = \bar{\beta} \quad , \quad \beta = \alpha \quad : \quad \varepsilon^{\alpha\beta\mu\nu} \varepsilon_{\bar{\alpha}\bar{\beta}\mu\nu} &= \sum_{\mu,\nu} \varepsilon^{\alpha\beta\mu\nu} \varepsilon_{\beta\alpha\mu\nu} \\ &= -\sum_{\mu\nu} \varepsilon^{\alpha\beta\mu\nu} \varepsilon_{\alpha\beta\mu\nu} \\ &= -(-2) \\ &= +2 \quad .\end{aligned} \tag{6b}$$

Therefore the final result is

$$\varepsilon^{\alpha\beta\mu\nu}\varepsilon_{\bar{\alpha}\bar{\beta}\mu\nu} = \begin{cases} -2 \quad , & \alpha=\bar{\alpha}, \beta=\bar{\beta} \\ +2 \quad , & \alpha=\bar{\beta}, \beta=\bar{\alpha} \\ 0 \quad , & \text{otherwise} \end{cases}$$

$$= 2\{\delta^{\alpha}_{\bar{\beta}}\delta^{\beta}_{\bar{\alpha}} - \delta^{\alpha}_{\bar{\alpha}}\delta^{\beta}_{\bar{\beta}}\} \quad , \tag{7}$$

which was to be shown.

For calculating the decay rate $\mathrm{d}W$ we proceed by inserting the result of (2.32) into the expression for $\mathrm{d}W$ (2.18). In order to perform the required integration, we need to evaluate the integral

$$I_{\alpha\beta} = \int \frac{\mathrm{d}^3k}{2k^0} \int \frac{\mathrm{d}^3k'}{2k'^0} k_\alpha k'_\beta \delta^4(p+k+k'-p') \quad . \tag{2.33}$$

$I_{\alpha\beta}$ is manifestly Lorentz covariant. This is obvious because $\delta^4(p+k+k'-p')$ and $\mathrm{d}^3p/2p_0 = \int_{-\infty} \mathrm{d}^4p\,\delta(p^2-m_0^2)\theta(p_0)$ are Lorentz invariant. The latter has been shown in *Quantum Electrodynamics*.[4] Since the variables k and k' are integrated over only the two second-rank tensors $g_{\alpha\beta}$ and $(p'-p)_\alpha(p'-p)_\beta = q_\alpha q_\beta$ can occur in the result. Note that the vector $q=(p'-p)$ is different from that defined in Exercise 2.3! We keep this in mind and proceed with the ansatz

$$I_{\alpha\beta} = Aq^2 g_{\alpha\beta} + Bq_\alpha q_\beta \quad , \tag{2.34}$$

where $q^2 = q^\alpha q_\alpha$ was split off in order to have A and B dimensionless. From (2.34) we construct the following invariants:

$$g^{\alpha\beta} I_{\alpha\beta} = (4A+B)q^2 \quad , \tag{2.35a}$$

$$q^\alpha q^\beta I_{\alpha\beta} = (A+B)q^4 \quad . \tag{2.35b}$$

To proceed, we now distinguish two cases:

(i) The vector $q = p'-p$ is time-like, that is $q^2 > 0$. With this condition we can always perform a proper Lorentz transformation, such that

$$\tilde{q}^\nu := a^\nu{}_\mu q^\mu = (\tilde{q}^0, 0) \tag{2.36}$$

defines the reference system. With respect to this reference frame we have

$$\begin{aligned} g^{\alpha\beta} I_{\alpha\beta} &= \int \frac{\mathrm{d}^3k}{2k^0} \int \frac{\mathrm{d}^3k'}{2k'^0} k_\alpha k'^\alpha \delta^3(\boldsymbol{k}+\boldsymbol{k}')\delta(k^0+k'^0-\tilde{q}^0) \\ &= \int \frac{\mathrm{d}^3k}{2k^0} \int \frac{\mathrm{d}^3k'}{2k'^0} [(k^0)^2 - (\boldsymbol{k}'\cdot\boldsymbol{k})]\delta^3(\boldsymbol{k}+\boldsymbol{k}')\delta(k^0+k'^0-\tilde{q}^0) \end{aligned}$$

[4] See W. Greiner and J. Reinhardt: *Quantum Electrodynamics*, 4th ed. (Springer, Berlin, Heidelberg 2009), equation (3.72).

$$= \int \frac{d^3k}{4(k^0)^2} 2(k^0)^2 \delta(2k^0 - \tilde{q}^0) \quad , \tag{2.37}$$

since $\boldsymbol{k}' = -\boldsymbol{k}$ and consequently $k'^0 = k^0 = |\boldsymbol{k}| = |\boldsymbol{k}'|$. The integral can further be simplified by substituting $x = 2k^0$,

$$g^{\alpha\beta} I_{\alpha\beta} = 2\pi \int_0^\infty (k^0)^2 \, dk_0 \delta(2k^0 - \tilde{q}^0) = \frac{\pi}{4} \int_0^\infty x^2 \, dx \delta(x - \tilde{q}^0) \quad . \tag{2.38}$$

For positive $\tilde{q}^0$ the argument of the δ function has its zero value within the integration interval. By means of the Θ function

$$\Theta(x) = \begin{cases} 1 & \text{for} \quad x > 0 \\ 0 & \text{for} \quad x < 0 \end{cases} \quad , \tag{2.39}$$

the above result can be expressed as

$$g^{\alpha\beta} I_{\alpha\beta} = \frac{\pi}{4} (\tilde{q}^0)^2 \Theta(\tilde{q}^0) \quad . \tag{2.40}$$

In order to rewrite this in a Lorentz invariant form we remark that for time-like four-vectors q^μ the sign of the zeroth component q^0 remains unchanged under proper Lorentz transformations, that is, $\Theta(\tilde{q}^0) = \Theta(q^0)$. Furthermore, with respect to our chosen reference frame we have $\tilde{q}^2 = (\tilde{q}^0)^2 = q^2$. Hence the result (2.40) can be stated in the Lorentz invariant form

$$g^{\alpha\beta} I_{\alpha\beta} = \frac{\pi}{4} q^2 \Theta(q^0) \quad \text{for} \quad q^2 > 0 \quad . \tag{2.41}$$

Similarly we obtain

$$\begin{aligned} q^\alpha q^\beta I_{\alpha\beta} &= (\tilde{q}^0)^2 I_{00} \\ &= (\tilde{q}^0)^2 \int \frac{d^3k}{2} \int \frac{d^3k'}{2} \delta^3(\boldsymbol{k} + \boldsymbol{k}') \delta(k^0 + k'^0 - \tilde{q}^0) \\ &= \frac{1}{4} (\tilde{q}^0)^2 \int d^3k \delta(2k^0 - \tilde{q}^0) \\ &= \pi (\tilde{q}^0)^2 \int (k^0)^2 \, dk^0 \delta(2k^0 - \tilde{q}^0) = \frac{\pi}{8} (\tilde{q}^0)^4 \Theta(\tilde{q}^0) \\ &= \frac{\pi}{8} q^4 \Theta(q^0) \quad \text{for} \quad q^2 > 0 \quad . \end{aligned} \tag{2.42}$$

(ii) The vector q_μ is space-like, that is $q^2 < 0$. In this case the argument of the δ function, $(k + k' - q)$, is non-zero everywhere. This property can be understood by recalling that, owing to the vanishing mass of the neutrinos, it holds that

$$\begin{aligned} k^2 &= k'^2 = 0 \quad , \\ \boldsymbol{k} \cdot \boldsymbol{k}' &= k^0 k'^0 \cos\theta \quad , \end{aligned} \tag{2.43}$$

where θ is the angle between $\boldsymbol{k}$ and $\boldsymbol{k}'$. Consequently we have

$$(k + k')^2 = 2(k^0 k'^0 - \boldsymbol{k} \cdot \boldsymbol{k}') = 2k^0 k'^0 (1 - \cos\theta) \geq 0 \quad , \tag{2.44}$$

which implies that $q^\mu = k^\mu + k'^\mu$ cannot be satisfied. Therefore

$$I_{\alpha\beta} = 0 \quad \text{for} \quad q^2 < 0 \quad . \tag{2.45}$$

The results of (2.41), (2.42) and (2.45) may be summarized as follows:

$$g^{\alpha\beta} I_{\alpha\beta} = \frac{\pi}{4} q^2 \Theta(q^0)\Theta(q^2) \tag{2.46a}$$

$$q^\alpha q^\beta I_{\alpha\beta} = \frac{\pi}{8} q^4 \Theta(q^0)\Theta(q^2) \quad . \tag{2.46b}$$

Equating these expressions with (2.35) gives

$$4A + B = \frac{\pi}{4}\Theta(q^0)\Theta(q^2) \quad , \tag{2.47a}$$

$$A + B = \frac{\pi}{8}\Theta(q^0)\Theta(q^2) \quad , \tag{2.47b}$$

yielding the solution

$$A = \frac{\pi}{24}\Theta(q^0)\Theta(q^2) \quad , \tag{2.48a}$$

$$B = \frac{\pi}{12}\Theta(q^0)\Theta(q^2) \quad . \tag{2.48b}$$

Substituting in (2.34) we finally obtain

$$I_{\alpha\beta} = \frac{\pi}{24}(q^2 g_{\alpha\beta} + 2q_\alpha q_\beta)\Theta(q^0)\Theta(q^2) \quad . \tag{2.49}$$

The decay rate of a muon with polarization s' into an electron with polarization s is given in terms of (2.18), (2.32) and (2.49); thus we find that

$$\begin{aligned}
\mathrm{d}W &= \frac{G^2}{2}\frac{1}{(2\pi)^5}\frac{\mathrm{d}^3 p}{2p'^0 2p^0}\int \frac{\mathrm{d}^3 k}{2k^0}\int \frac{\mathrm{d}^3 k'}{2k'^0}\sum_{t,t'} \delta^4(p+k+k'-p')|M|^2 \\
&= \frac{G^2}{2}\frac{1}{(2\pi)^5}\frac{64\mathrm{d}^3 p}{2p'^0 2p^0} I_{\alpha\beta}(p' - m_\mu s')^\alpha (p - m_\mathrm{e} s)^\beta \\
&= \frac{G^2}{3}\frac{\pi \mathrm{d}^3 p}{(2\pi)^5 p'^0 p^0}\big[(p'-p)^2 (p' - m_\mu s')^\alpha (p - m_\mathrm{e} s)_\alpha \\
&\quad + 2(p'-p)_\alpha (p' - m_\mu s')^\alpha (p'-p)_\beta (p - m_\mathrm{e} s)^\beta\big] \\
&\quad \times \Theta(p'^0 - p^0)\Theta((p'-p)^2) \quad .
\end{aligned} \tag{2.50}$$

Note that the effect of time dilatation, which accompanies the observation of the muon lifetime, becomes obvious from (2.50). For a moving muon we have $p'^0 = \gamma m_\mu$ with the Lorentz factor $\gamma = (1 - v^2/c^2)^{-1/2}$. As can be seen from the expression for $\mathrm{d}W$ (2.50), $\mathrm{d}W \propto 1/\gamma$, implying that the decay rate decreases considerably for fast-moving muons, that is, the life-time $\tau_\mu \propto \gamma$ is prolonged. To proceed we switch to the *rest frame of the muon*, which is characterized by $p^{\alpha\prime} - p^\alpha = (m_\mu - p^0, -\boldsymbol{p})$. Since

$$\begin{aligned}
(p'-p)^2 &= (m_\mu - p^0)^2 - \boldsymbol{p}^2 = (m_\mu - p^0)^2 - (p^{02} - m_\mathrm{e}^2) \\
&= -2p^0 m_\mu + m_\mu^2 + m_\mathrm{e}^2 \quad ,
\end{aligned} \tag{2.51}$$

the condition $(p'-p)^2>0$ for a non-vanishing $\mathrm{d}W$ yields the restriction

$$p^0 < p^0_{\max} = (m_\mu{}^2 + m_e{}^2)/2m_\mu \quad , \tag{2.52}$$

which consequently requires $p'^0 - p^0 > 0$, since

$$p'^0 - p^0 = m_\mu - p^0 > m_\mu - p^0_{\max} = (m_\mu{}^2 - m_e{}^2)/2m_\mu > 0 \quad . \tag{2.53}$$

The condition $p^0 < p^0_{\max}$ in (2.52) and (2.53) assures that the first $\Theta(p'_0 - p_0)$ function in (2.50) is automatically fulfilled. Therefore we may replace the product of the two Θ functions in (2.50) by $\Theta(p^0_{\max} - p_0)$. Furthermore, with respect to the rest frame of the muon, it holds that $s^{\mu\prime} = (0, \boldsymbol{s}')$, so that the final result is

$$\begin{aligned}\mathrm{d}W(s') = {} & \frac{G^2}{3}\frac{\pi\,\mathrm{d}^3 p}{(2\pi)^5 p^0}\{[(m_\mu - p^0)^2 - \boldsymbol{p}^2][(p_0 - m_e s^0) + \boldsymbol{s}'\cdot(\boldsymbol{p} - m_e\tilde{\boldsymbol{s}})] \\ & + 2[m_\mu - p^0 - \boldsymbol{s}'\cdot\boldsymbol{p}][(m_\mu - p^0)(p^0 - m_e s^0) + \boldsymbol{p}\cdot(\boldsymbol{p} - m_e\tilde{\boldsymbol{s}})]\} \\ & \times \Theta(p^0_{\max} - p^0) \quad .\end{aligned} \tag{2.54}$$

Here $\tilde{\boldsymbol{s}} = \boldsymbol{s} + \frac{(\boldsymbol{p}\boldsymbol{s})\boldsymbol{p}}{m_e(p^0+m_e)}$ is the space component of the electron spin vector (2.23).

EXERCISE

2.5 The Endpoint of the Electron Energy Spectrum in Muon Decay

Problem. Show that the highest electron energy is given in terms of (2.52) by energy and momentum conservation.

Solution. The highest energy of the electron corresponds to the largest value of its momentum. The latter is obtained if both neutrinos are emitted in one direction while the electron is scattered in the other direction (Fig. 2.2), that is,

$$\boldsymbol{p} = -(\boldsymbol{k} + \boldsymbol{k}') \quad . \tag{1}$$

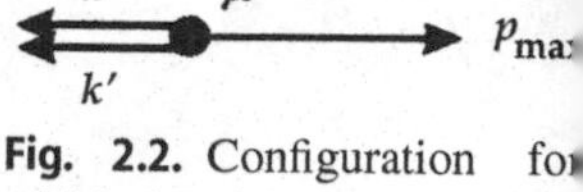

Fig. 2.2. Configuration for which the electron reaches its maximum value of momentum

Because $k^0 = |\boldsymbol{k}|$ and $k'^0 = |\boldsymbol{k}'|$ it holds that

$$\begin{aligned} m_\mu &= p^0_{\max} + k^0 + k'^0 = p^0_{\max} + |\boldsymbol{k}| + |\boldsymbol{k}'| \\ &= p^0_{\max} + |\boldsymbol{p}_{\max}| \\ &= p^0_{\max} + [(p^0_{\max})^2 - m_e^2]^{\frac{1}{2}} \quad . \end{aligned} \tag{2}$$

Inverting this relation gives

$$\begin{aligned} p^0_{\max} &= \frac{m_\mu^2 + m_e^2}{2m_\mu} = 52.83\ \mathrm{MeV} \quad , \\ |\boldsymbol{p}_{\max}| &= \frac{m_\mu^2 - m_e^2}{2m_\mu} \quad , \end{aligned} \tag{3}$$

which agrees exactly with the conditions (2.52) and (2.53).

2.3 The Lifetime of the Muon

To determine the muon lifetime τ_μ we sum over the electron spin orientations s, average over the spin orientation of the muon s', and integrate over the electron momentum $\boldsymbol{p}$:

$$\begin{aligned}\frac{1}{\tau_\mu} &= W_\mu = \frac{1}{2}\sum_{s,s'}\int \mathrm{d}W \\ &= 2\frac{G^2}{3}\frac{\pi}{(2\pi)^5}\int \frac{\mathrm{d}^3 p}{p^0}\{[(m_\mu - p^0)^2 - \boldsymbol{p}^2]p^0 + 2(m_\mu - p^0) \\ &\times [(m_\mu - p^0)p^0 + \boldsymbol{p}^2]\}\Theta(p^0_{\mathrm{max}} - p^0) \\ &= 2\frac{G^2}{3}\int \frac{\pi\,\mathrm{d}^3 p}{(2\pi)^5 p^0}[-4m_\mu(p^0)^2 + 3p^0(m_\mu^2 + m_\mathrm{e}^2) - 2m_\mu m_\mathrm{e}^2] \\ &\times \Theta(p^0_{\mathrm{max}} - p^0) \quad .\end{aligned} \tag{2.55}$$

In deriving (2.55) we used the fact that the averaging over $\boldsymbol{s}$ gives $\langle \boldsymbol{s}\rangle = 0$ so that also $\langle s^0\rangle = \frac{1}{m}\langle \boldsymbol{p}\cdot\boldsymbol{s}\rangle = 0$ (cf. (2.23)). If we employ the following identity:

$$\int \mathrm{d}^3\boldsymbol{p}[\ldots]\Theta(p^0_{\mathrm{max}} - p^0) = 4\pi \int_0^{|\boldsymbol{p}_{\mathrm{max}}|} |\boldsymbol{p}|^2\,\mathrm{d}|\boldsymbol{p}|[\ldots] \quad , \tag{2.56}$$

and take into account that $\boldsymbol{p}^2 = (p_0^2 - m_\mathrm{e}^2)$ and therefore that

$$\mathrm{d}|\boldsymbol{p}|/\mathrm{d}p^0 = p^0/|\boldsymbol{p}| \quad ,$$

we can rewrite W_μ in the form

$$\begin{aligned}W_\mu &= \frac{2G^2}{3(2\pi)^3}\int_{m_\mathrm{e}}^{p^0_{\mathrm{max}}} \mathrm{d}p^0\sqrt{(p^0)^2 - m_\mathrm{e}^2}[-4m_\mu(p^0)^2 + 3p^0(m_\mu^2 + m_\mathrm{e}^2) - 2m_\mu m_\mathrm{e}^2] \\ &= \frac{G^2 m_\mu{}^5}{192\pi^3}[1 - 8y + 8y^3 - y^4 - 12y^2\ln y] \quad ,\end{aligned} \tag{2.57}$$

with the abbreviation $y \equiv m_\mathrm{e}^2/m_\mu^2$. The contributions involving y lead only to small corrections, namely

$$W_\mu = \frac{G^2 m_\mu^5}{192\pi^3}(1 - 1.87\times 10^{-4}) \quad . \tag{2.58}$$

From (2.57) it is obvious that the decay rate would vanish if $y = 1$. This reflects the fact that in this (academic) case the muon would be stable since $|\boldsymbol{p}_{\mathrm{max}}| = 0$, so that there would be no phase space available for the final-state electron.

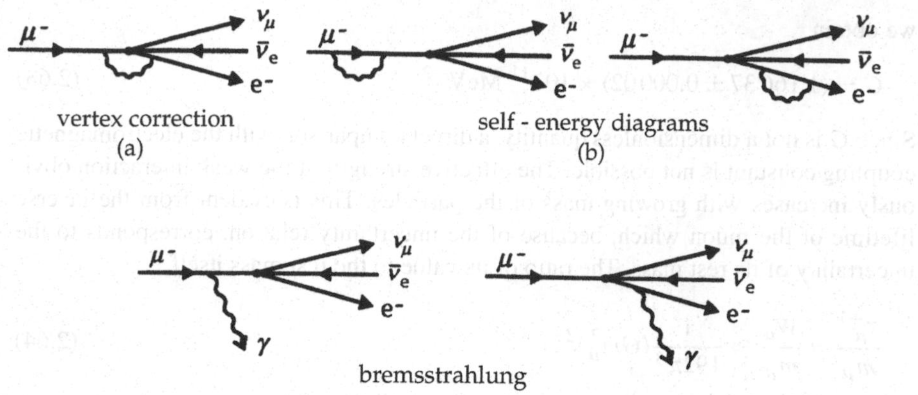

Fig. 2.3. Vertex correction (a), self-energy (b) and bremsstrahlung (c) contributions

The result (2.58) does not include the so-called radiative corrections, which also need to be considered. These effects are represented by diagrams in which one of the charged particles interacts with the electromagnetic field (see Fig. 2.3). The bremsstrahlung diagram has to be included since, owing to the vanishing photon mass, photons with arbitrary small energies may be emitted. On the other hand, because of the limited experimental resolution, it is impossible to distinguish the muon decay accompanied by emission of an extremely "soft" photon from a decay without radiation. This contribution exactly cancels the divergent terms in the self-energy diagrams for very soft photons (infrared divergence).[5] The calculation of these contributions leads to a modification of the decay rate W by a factor[6]

$$1-\frac{\alpha}{2\pi}\left(\pi^2-\frac{25}{4}\right)=0.9958\ldots \quad . \tag{2.59}$$

Hence, the radiative corrections are of greater importance than the influence of the finite mass of the electron. The final result for the muon decay rate is now given by

$$W_\mu=\frac{1}{\tau_\mu}=\frac{G^2 m_\mu^5}{192\pi^3}\left(1-\frac{\alpha}{2\pi}\left(\pi^2-\frac{25}{4}\right)-8\frac{m_e^2}{m_\mu^2}\quad\ldots\right) \quad . \tag{2.60}$$

Using this formula we may calculate the value of the Fermi coupling constant G by taking into account the experimental value for the average life time of the muon

$$\tau_\mu=(2.19703\pm 0.00004)\times 10^{-6}\ \mathrm{s} \quad ,$$

i.e.

$$W_\mu=\tau_\mu^{-1}=2.996\times 10^{-16}\ \mathrm{MeV} \quad . \tag{2.61}$$

With the most accurate value for the muon mass

$$m_\mu=(105.658387\pm 0.000034)\ \mathrm{MeV} \tag{2.62}$$

[5] See W. Greiner and J. Reinhardt: *Quantum Electrodynamics*, 4th ed. (Springer, Berlin, Heidelberg, 2009).

[6] S.M. Berman: Phys. Rev. **112**, 267 (1958); M. Roos and A. Sirling: Nucl. Phys. **B 29**, 296 (1971); L.D. Landau, E.M. Lifschitz: *Theoretical Physics* (Pergamon, Oxford, 1974), Vol. IVb, p. 147.

we obtain

$$G = (1.16637 \pm 0.00002) \times 10^{-11}\ \mathrm{MeV}^{-2} \quad . \tag{2.63}$$

Since G is not a dimensionless quantity, a direct comparison with the electromagnetic coupling constant is not possible. The effective strength of the weak interaction obviously increases with growing mass of the particles. This is evident from the inverse lifetime of the muon which, because of the uncertainty relation, corresponds to the uncertainty of its rest mass. The ratio of its value to the rest mass itself,

$$\frac{\tau_\mu^{-1}}{m_\mu} = \frac{W_\mu}{m_\mu} \approx \frac{1}{192\pi^3}(G m_\mu^2)^2 \tag{2.64}$$

manifests the role of Gm_μ^2 as an effective coupling strength. For curiosity's sake we may now evaluate the mass M for which the effective coupling constant equals the fine-structure constant α:

$$GM^2 = \alpha \quad \rightarrow \quad M = \sqrt{\alpha/G} = 25\ \mathrm{GeV} \quad . \tag{2.65}$$

The experimental investigation of this energy region has become possible with the large particle accelerators of DESY (Hamburg), SLAC (Stanford), CERN (Geneva), and Fermilab (Chicago). As we will soon see, these investigations have revealed new information concerning the nature of the weak interaction.

EXERCISE

2.6 Myon Decay for Finite Neutrino Masses

Problem. Generalize the relation (2.49),

$$\begin{aligned} I_{\alpha\beta} &= \int \frac{\mathrm{d}^3k}{2k^0} \int \frac{\mathrm{d}^3k'}{2k'^0} k_\alpha k'_\beta \delta^4(k + k' - q) \\ &= \frac{\pi}{24}(q^2 g_{\alpha\beta} + 2q_\alpha q_\beta) \quad , \end{aligned}$$

which is valid for $q^2, q^0 > 0$, to the case of non-vanishing rest masses m, m', of the two decay products with the four-momenta k_α, k'_β.

Solution. As in (2.34) we make the ansatz

$$I_{\alpha\beta} = A q^2 g_{\alpha\beta} + B q_\alpha q_\beta \quad , \tag{1}$$

which implies the relations (2.35),

$$g^{\alpha\beta} I_{\alpha\beta} = (4A + B)q^2 \quad , \tag{2a}$$

$$q^\alpha q^\beta I_{\alpha\beta} = (A + B)q^4 \quad . \tag{2b}$$

For the calculation of these two Lorentz invariants, we take the frame of reference in which q^α consists of a time-like component only,

$$\tilde{q}^\alpha = \left(\tilde{q}^0 = \sqrt{q^2}, \boldsymbol{0}\right) \quad .$$

Thus, with

$$k^0 = [\boldsymbol{k}^2 + m^2]^{1/2} \quad , \quad k'^0 = [\boldsymbol{k}'^2 + m'^2]^{1/2} \quad , \tag{3}$$

and after performing the k' integration, we get

$$\begin{aligned} g^{\alpha\beta} I_{\alpha\beta} &= \int \frac{\mathrm{d}^3k}{2k^0} \int \frac{\mathrm{d}^3k'}{2k'^0} k_\alpha k'^\alpha \delta^3(\boldsymbol{k}+\boldsymbol{k}')\delta(k^0 + k'^0 - \tilde{q}^0) \\ &= \frac{1}{4} \int \mathrm{d}^3k \Big[\sqrt{(\boldsymbol{k}^2+m^2)}\sqrt{(\boldsymbol{k}^2+m'^2)}\Big]^{-1} \\ &\quad \times \Big[\sqrt{(\boldsymbol{k}^2+m^2)}\sqrt{(\boldsymbol{k}^2+m'^2)} + \boldsymbol{k}^2\Big] \\ &\quad \times \delta\Big(\sqrt{(\boldsymbol{k}^2+m^2)} + \sqrt{(\boldsymbol{k}^2+m'^2)} - \tilde{q}^0\Big) \quad . \end{aligned} \tag{4}$$

We substitute the sum of the two square roots in the argument of the δ function by x,

$$x = [\boldsymbol{k}^2 + m^2]^{1/2} + [\boldsymbol{k}^2 + m'^2]^{1/2} \quad , \tag{5}$$

and transform to polar coordinatesmentum space. The volume element transforms into

$$\frac{|\boldsymbol{k}|\,\mathrm{d}|\boldsymbol{k}|}{\sqrt{(\boldsymbol{k}^2+m^2)}\sqrt{(\boldsymbol{k}^2+m'^2)}} = \frac{\mathrm{d}x}{x} \quad , \tag{6}$$

and by squaring (5), we have

$$\begin{aligned} m^2 - m'^2 - x^2 &= -2x[\boldsymbol{k}^2 + m'^2]^{1/2} \quad , \\ \boldsymbol{k}^2 &= \frac{(x^2 - m^2 + m'^2)^2}{4x^2} - m'^2 \\ &= \frac{(x^2 - m^2 - m'^2)^2}{4x^2} - \frac{m^2 m'^2}{x^2} \quad . \end{aligned} \tag{7}$$

Squaring (5) also yields the relation

$$\boldsymbol{k}^2 + [\boldsymbol{k}^2 + m^2]^{1/2}[\boldsymbol{k}^2 + m'^2]^{1/2} = \frac{(x^2 - m^2 - m'^2)}{2} \quad . \tag{8}$$

Equations (5)–(8) now give

$$\begin{aligned} g^{\alpha\beta} I_{\alpha\beta} &= \pi \int_0^\infty \frac{\mathrm{d}x}{x} |\boldsymbol{k}| \frac{1}{2}(x^2 - m^2 - m'^2)\delta(x - \tilde{q}^0) \\ &= \frac{\pi}{4(\tilde{q}^0)^2}((\tilde{q}^0)^2 - m^2 - m'^2)\big[((\tilde{q}^0)^2 - m^2 - m'^2)^2 - 4m^2m'^2\big]^{1/2} \\ &= \frac{\pi}{4q^2}(q^2 - m^2 - m'^2)\big[(q^2 - m^2 - m'^2)^2 - 4m^2m'^2\big]^{1/2} \quad , \end{aligned} \tag{9}$$

where the expression in its last form again is written in a manifestly Lorentz invariant form. In the same way we get for the second invariant

$$q^\alpha q^\beta I_{\alpha\beta} = \frac{(\tilde{q}^0)^2}{4} \int \mathrm{d}^3k \int \mathrm{d}^3k' \delta(\boldsymbol{k}+\boldsymbol{k}')\delta(k^0 + k'^0 - \tilde{q}^0)$$

ercise 2.6

$$= \frac{(\tilde{q}^0)^2}{4} \int d^3k \delta\left(\sqrt{(\boldsymbol{k}^2+m^2)} + \sqrt{(\boldsymbol{k}^2+m'^2)} - \tilde{q}^0\right)$$

$$= \pi (\tilde{q}^0)^2 \int_0^\infty \frac{dx}{x} |\boldsymbol{k}| \left[\boldsymbol{k}^2+m^2\right]^{1/2} \left[\boldsymbol{k}^2+m'^2\right]^{1/2} \delta(x-\tilde{q}^0) \quad . \tag{10}$$

Combining (7) and (8) gives

$$\left[\boldsymbol{k}^2+m^2\right]^{1/2}\left[\boldsymbol{k}^2+m'^2\right]^{1/2}$$

$$= \frac{x^2-m^2-m'^2}{2} - \frac{(x^2-m^2-m'^2)^2}{4x^2} + \frac{m^2m'^2}{x^2}$$

$$= (4x^2)^{-1}\left[x^4-(m^2+m'^2)^2+4m^2m'^2\right]$$

$$= (4x^2)^{-1}\left[x^4-(m^2-m'^2)^2\right] \quad , \tag{11}$$

which facilitates the final calculation,

$$q^\alpha q^\beta I_{\alpha\beta} = \frac{\pi}{8(\tilde{q}^0)^2}\left[(\tilde{q}^0)^4-(m^2-m'^2)^2\right]$$

$$\times \left[\left((\tilde{q}^0)^2-m^2-m'^2\right)^2-4m^2m'^2\right]^{1/2}$$

$$= \frac{\pi}{8q^2}\left[q^4-(m^2-m'^2)^2\right]$$

$$\times \left[(q^2-m^2-m'^2)^2-4m^2m'^2\right]^{1/2} \quad . \tag{12}$$

In addition, the δ function of (4) tells us that the results (9) and (12) are valid only for $\tilde{q}^0 = \sqrt{q^2} > m+m'$. This is expressed by the fact, that the expression under the square root in (9) and (12) may be written as follows:

$$(q^2-m^2-m'^2)^2-4m^2m'^2$$

$$= \left[q^2-(m+m')^2\right]\left[q^2-(m-m')^2\right] \quad , \tag{13}$$

which is easily checked. The radicand in (12) becomes negative for $q^2 < (m+m')^2$.

With the aid of definition (2) the quantities A and B can be determined:

$$A = (3q^2)^{-1}\left(g^{\alpha\beta} I_{\alpha\beta} - \frac{q^\alpha q^\beta}{q^2} I_{\alpha\beta}\right)$$

$$= \frac{\pi}{24q^6}\left[q^2-(m+m')^2\right]^{1/2}\left[q^2-(m-m')^2\right]^{1/2}$$

$$\times \left[2q^2(q^2-m^2-m'^2)-q^4+(m^2-m'^2)^2\right]$$

Exercise 2.

$$= \frac{\pi}{24q^6}\left[q^2-(m+m')^2\right]^{3/2}\left[q^2-(m-m')^2\right]^{3/2}$$

$$= \frac{\pi}{24}\left[1-\frac{(m+m')^2}{q^2}\right]^{3/2}\left[1-\frac{(m-m')^2}{q^2}\right]^{3/2} , \tag{14}$$

$$B = (3q^2)^{-1}\left(4\frac{q^\alpha q^\beta}{q^2}I_{\alpha\beta} - g^{\alpha\beta}I_{\alpha\beta}\right)$$

$$= \frac{\pi}{12q^6}\left[q^2-(m+m')^2\right]^{1/2}\left[q^2-(m-m')^2\right]^{1/2}$$

$$\times\left[2q^4-2(m^2-m'^2)^2-q^2(q^2-m^2-m'^2)\right]$$

$$= \frac{\pi}{12}\left[1-\frac{(m+m')^2}{q^2}\right]^{1/2}\left[1-\frac{(m-m')^2}{q^2}\right]^{1/2}$$

$$\times\left[1+\frac{m^2+m'^2}{q^2}-2\frac{(m^2-m'^2)^2}{q^4}\right] . \tag{15}$$

The final result is thus

$$I_{\alpha\beta} = \frac{\pi}{24}\left[1-\frac{(m+m')^2}{q^2}\right]^{1/2}\left[1-\frac{(m-m')^2}{q^2}\right]^{1/2}$$

$$\times\left[g_{\alpha\beta}q^2\left(1-\frac{(m+m')^2}{q^2}\right)\left(1-\frac{(m-m')^2}{q^2}\right)\right.$$

$$\left.+2q_\alpha q_\beta\left(1+\frac{m^2+m'^2}{q^2}-2\frac{(m^2-m'^2)^2}{q^4}\right)\right]$$

$$\times\,\Theta\left(q^2-(m+m')^2\right) . \tag{16}$$

In the limit $m=m'=0$ one again gets (2.49) as is to be expected. For later use we note the special case $m'=0$ (that is, one of the two particles is a neutrino),

$$I_{\alpha\beta}(m) = \frac{\pi}{24}\left(1-\frac{m^2}{q^2}\right)^2\left[q^2\left(1-\frac{m^2}{q^2}\right)g_{\alpha\beta}\right.$$

$$\left.+2\left(1+\frac{2m^2}{q^2}\right)q_\alpha q_\beta\right]\Theta(q^2-m^2) . \tag{17}$$

2.4 Parity Violation in the Muon Decay

We now want to discuss two experiments which prove the violation of reflection invariance in muon decay. The first experiment observes the *decay of unpolarized muons* and measures the average helicity of the emitted electrons. In the second experiment one starts with *polarized muons*, which are produced by pion decay (see (1.32) and the subsequent equations), and measures only the angular distribution of the electron momenta with respect to the spin direction of the decaying muons.

Let us start with the first experiment. For unpolarized muons the expression (2.54) has to be averaged over $\boldsymbol{s}'$. Therefore all terms containing $\boldsymbol{s}'$ vanish:

$$\begin{aligned} \mathrm{d}\tilde{W} &= \frac{1}{2}\sum_{s'} \mathrm{d}W(s') \\ &= \frac{G^2}{3}\frac{\pi\,\mathrm{d}^3 p}{(2\pi)^5 p^0}\Theta(p^0_{\max} - p^0)\{[3(m_\mu - p^0)^2 - \boldsymbol{p}^2] \\ &\quad\times (p^0 - m_\mathrm{e}s^0) + 2(m_\mu - p^0)(\boldsymbol{p}^2 - m_\mathrm{e}\boldsymbol{p}\cdot\tilde{\boldsymbol{s}})\} \quad , \end{aligned} \tag{2.66}$$

where $\tilde{\boldsymbol{s}}$ is the space-like component of the spin four-vector s^α. The four-vector of the electron spin also satisfies (2.23), and we get

$$\begin{aligned} m_\mathrm{e}s^0 &= \boldsymbol{p}\cdot\boldsymbol{s} \quad , \\ \boldsymbol{p}\cdot\tilde{\boldsymbol{s}} &= (\boldsymbol{p}\cdot\boldsymbol{s})\left[1 + \frac{|\boldsymbol{p}|^2}{m_\mathrm{e}(p^0 + m_\mathrm{e})}\right] = \frac{p^0}{m_\mathrm{e}}(\boldsymbol{p}\cdot\boldsymbol{s}) \quad , \end{aligned} \tag{2.67}$$

where $\boldsymbol{s}$ is the spin vector defined in the muon rest frame. The two possible eigenstates of the helicity operator $\hat{\Lambda} = \boldsymbol{\sigma}\cdot\boldsymbol{p}/|\boldsymbol{p}|$ correspond to the values

$$\boldsymbol{p}\cdot\boldsymbol{s} = \pm|\boldsymbol{p}| \quad . \tag{2.68}$$

Remember that the spin vector $\boldsymbol{s}$ within relativistic quantum mechanics (2.23) is normalized to 1, i.e. $\boldsymbol{s}\cdot\boldsymbol{s} = 1$, so that $s_\mu s^\mu = -1$.[7] In the first case, electron spin and direction of motion are parallel and in the second case antiparallel; the corresponding helicities are $(+1)$ and (-1), respectively. Because of (2.66) we obtain the following average value of the helicity operator:

$$\begin{aligned} \langle\Lambda\rangle &= \frac{\mathrm{d}\tilde{W}(\boldsymbol{p}\cdot\boldsymbol{s} = |\boldsymbol{p}|) - \mathrm{d}\tilde{W}(\boldsymbol{p}\cdot\boldsymbol{s} = -|\boldsymbol{p}|)}{\mathrm{d}\tilde{W}(\boldsymbol{p}\cdot\boldsymbol{s} = |\boldsymbol{p}|) + \mathrm{d}\tilde{W}(\boldsymbol{p}\cdot\boldsymbol{s} = -|\boldsymbol{p}|)} \\ &= \frac{-2|\boldsymbol{p}|[3(m_\mu - p^0)^2 - \boldsymbol{p}^2 + 2(m_\mu - p^0)p^0]}{2[3(m_\mu - p^0)^2 p^0 - \boldsymbol{p}^2 p^0 + 2(m_\mu - p^0)\boldsymbol{p}^2]} \quad . \end{aligned} \tag{2.69}$$

We demonstrate this simply for the nominator only:

$$\begin{aligned} &\left[3(m_\mu - p^0)^2 - |\boldsymbol{p}|^2\right](p^0 - |\boldsymbol{p}|) + 2(m_\mu - p^0)(\boldsymbol{p}^2 - p^0|\boldsymbol{p}|) \\ &\quad - \left[\left[3(m_\mu - p^0)^2 - |\boldsymbol{p}|^2\right](p^0 + |\boldsymbol{p}|) + 2(m_\mu - p^0)(\boldsymbol{p}^2 + p^0|\boldsymbol{p}|)\right] \\ &= \left[3(m_\mu - p^0)^2 - |\boldsymbol{p}|^2\right](-2|\boldsymbol{p}|) + 2(m_\mu - p^0)(-2p^0|\boldsymbol{p}|) \\ &= -2|\boldsymbol{p}|\left[3(m_\mu - p^0)^2 - |\boldsymbol{p}|^2 + 2(m_\mu - p^0)p^0\right] \quad . \end{aligned}$$

---

[7] See W. Greiner: *Relativistic Quantum Mechanics – Wave Equations*, 3rd ed. (Springer, Berlin, Heidelberg, 2000).

Applying the value (2.52) for the maximum electron energy, the expression (2.69) can be written in the form

$$\frac{-|\boldsymbol{p}|m_\mu(3p^0_{\max}-2p^0-m_e^2/m_\mu)}{m_\mu p^0(3p^0_{\max}-2p^0-m_e^2/p^0)} = -1+O\left(\frac{m_e^2}{(p^0)^2}\right) \quad . \tag{2.70}$$

Let us quickly verify this result by inserting

$$\boldsymbol{p}^2=(p^0)^2-m_e^2$$

and

$$p^0_{\max}=\frac{m_\mu^2+m_e^2}{2m_\mu} \quad \text{or} \quad m_\mu=2p^0_{\max}-\frac{m_e^2}{m_\mu},$$

which yields for nominator $\mathcal{N}$:

$$\begin{aligned}
\mathcal{N} &= -2|\boldsymbol{p}|(3m_\mu^2-6m_\mu p^0+3(p^0)^2-(p^0)^2+m_e^2+2m_\mu p^0-2(p^0)^2)\\
&= -2|\boldsymbol{p}|\left(3m_\mu^2-4m_\mu p^0+\frac{m_e^2}{m_\mu}m_\mu\right)\\
&= -2|\boldsymbol{p}|m_\mu\left(3\left(2p^0_{\max}-\frac{m_e^2}{m_\mu}\right)-4p^0+\frac{m_e^2}{m_\mu}\right)\\
&= -4|\boldsymbol{p}|m_\mu\left(3p^0_{\max}-2p^0-\frac{m_e^2}{m_\mu}\right),
\end{aligned}$$

and for the denominator $\mathcal{D}$:

$$\begin{aligned}
\mathcal{D} &= 2\Big(3m_\mu^2p^0-6m_\mu(p^0)^2+3(p^0)^3-(p^0)^3+p^0m_e^2\\
&\qquad +2m_\mu(p^0)^2-2m_\mu m_e^2-2(p^0)^3+2p^0m_e^2\Big)\\
&= 2m_\mu p^0\left(3m_\mu-4p^0+3\frac{m_e^2}{m_\mu}-2\frac{m_e^2}{p^0}\right)\\
&= 2m_\mu p^0\left(3\cdot 2p^0_{\max}-4p^0-2\frac{m_e^2}{p^0}\right)\\
&= 4m_\mu p^0\left(3p^0_{\max}-2p^0-\frac{m_e^2}{p^0}\right)
\end{aligned}$$

and, therefore,

$$\langle\Lambda\rangle=\frac{-|\boldsymbol{p}|}{p^0}\,\frac{3p^0_{\max}-2p^0-\frac{m_e^2}{m_\mu}}{3p^0_{\max}-2p^0-\frac{m_e^2}{p^0}}.$$

The result (2.70) is most interesting. We notice that for energies $p^0 \gg m_e$ the electron is predicted to be in an almost completely left-handed state. For the average electron helicity in the kinematically allowed energy interval $[m_e, p^0_{max}]$, with $p^0_{max} \approx 100 m_e$, the experimentally observed value is[8]

$$\overline{\langle \Lambda \rangle} = -1.00 \pm 0.13 \quad . \tag{2.71}$$

The fact that the electron is limited to a left-handed state follows directly from the interaction (2.10), because the electronic transition current can be written as $\bar{U}_e \gamma^\mu (1 - \gamma_5) U_{\nu_e} = \overline{\frac{1}{2} U_2 (1 - \gamma_5)} \gamma^\mu U_{\nu_e}$. Thus, the electron, like the electron neutrino, has negative helicity. High-energy electrons are thus negatively polarized.

Next we consider the experiment in which the angular distribution of the electrons emitted in the decay of polarized muons is measured. Since the electron helicity is not observed, we must sum over the electron spin in (2.54). The value of the muon spin is assumed to be fixed. Let us begin with the expression (2.50), which we denote once more

$$\begin{aligned} \mathrm{d}W = \frac{G^2}{3} \frac{\pi \mathrm{d}^3 p}{(2\pi)^5 p'^0 p^0} \big[&(p' - p)^2 (p' - m_\mu s')^\alpha (p - m_e s)_\alpha \\ &+ 2(p' - p)_\alpha (p' - m_\mu s')^\alpha (p' - p)_\beta (p - m_e s)^\beta \big] \Theta(p^0_{max} - p^0) \quad . \end{aligned}$$

We remember that due to the discussion following (2.51)–(2.53) the two step functions in (2.50) can be abbreviated by $\Theta(p^0_{max} - p^0)$. Now the summation of the expression in the bracket [...] over the electron spins yields

$$\sum_{\pm s} [\ldots] = 2(p' - p)^2 (p' - m_\mu s')^\alpha p_\alpha + 4(p' - p)_\alpha (p' - m_\mu s')^\alpha (p' - p)_\beta p^\beta \quad .$$

It is easier to continue the calculation in the rest frame of the muon, for which $p'^\nu = (m_\mu, 0)$, $s'^\nu = (0, \boldsymbol{s}')$ holds. Then

$$\begin{aligned} \sum_{\pm s} [\ldots] = {} & 2\big[(m_\mu - p^0)^2 - \boldsymbol{p}^2\big]\big[m_\mu p^0 + m_\mu \boldsymbol{s}' \cdot \boldsymbol{p}\big] \\ & + 4\big[(m_\mu - p^0) m_\mu - m_\mu \boldsymbol{s}' \cdot \boldsymbol{p}\big]\big[(m_\mu - p^0) p^0 + \boldsymbol{p}^2\big] \quad . \end{aligned}$$

Inserting $\boldsymbol{p}^2 = p_0^2 - m_e^2$, $p^0_{max} = (m_\mu^2 + m_e^2)/2m_\mu$, $\boldsymbol{s}' \cdot \boldsymbol{p} = 1 \cdot |\boldsymbol{p}| \cos\theta$ and separating terms proportional to $\cos\theta$ yields

$$\begin{aligned} \sum_{\pm s} [\ldots] &= 4 m_\mu^2 \left[\frac{m_\mu^2 + m_e^2}{2 m_\mu} - p^0 \right] [p^0 + \boldsymbol{s}' \cdot \boldsymbol{p}] \\ &\quad + 4 m_\mu^2 \big[(m_\mu - p^0) - \boldsymbol{s}' \cdot \boldsymbol{p}\big] \left[p^0 - \frac{m_e^2}{m_\mu} \right] \\ &= 4 m_\mu^2 \left[p^0 \left(p^0_{max} - p^0 + m_\mu - p^0 - \frac{m_e^2}{p^0} + \frac{m_e^2}{m_\mu} \right) \right. \\ &\quad \left. + |\boldsymbol{p}| \cos\theta \left(p^0_{max} - p^0 - p^0 + \frac{m_e^2}{m_\mu} \right) \right] \end{aligned}$$

---

[8] Review of Particle Properties in: Review of Modern Physics (April 1988); J. Duclos, J. Heintze, A. de Rujula, V. Soergel: Phys. Lett. **9**, 62 (1964).

$$= 4m_\mu^2 p^0 \left[\left(3p^0_{\max} - 2p^0 - \frac{m_e^2}{p^0} \right) + \frac{|\boldsymbol{p}|}{p^0} \cos\theta \left(p^0_{\max} - 2p^0 + \frac{m_e^2}{m_\mu} \right) \right] .$$

Therefore, the decay rate summed over the electron spin is given by

$$\mathrm{d}\bar{W} = \sum_s \mathrm{d}W(s)$$

$$= \frac{2G^2}{3(2\pi)^3} m_\mu |\boldsymbol{p}| p^0 \mathrm{d}p^0 \sin\theta \mathrm{d}\theta \left[3p^0_{\max} - 2p^0 - \frac{m_e^2}{p^0} \right.$$

$$\left. + \frac{|\boldsymbol{p}|}{p^0} \cos\theta \left(p^0_{\max} - 2p^0 + \frac{m_e^2}{m_\mu} \right) \right] \Theta(p^0_{\max} - p^0) \quad . \tag{2.72}$$

Here θ denotes the angle between the muon spin $\boldsymbol{s}'$ and the electron momentum $\boldsymbol{p}$. The volume element of the electron momentum space has been used according to (2.56) in the form

$$\mathrm{d}^3 p = 2\pi |\boldsymbol{p}|^2 \mathrm{d}|\boldsymbol{p}| \sin\theta \mathrm{d}\theta = 2\pi |\boldsymbol{p}| p^0 \mathrm{d}p^0 \sin\theta \mathrm{d}\theta \quad . \tag{2.73}$$

Equation (2.72) does not yet contain the electromagnetic corrections. If one considers the corrections of the order $\alpha = e^2/\hbar c \simeq 1/137$, some terms are added to $\mathrm{d}\bar{W}$. But the parity-violating structure, which is expressed in the factor $\cos\theta$ in (2.72), is not changed. The agreement between the predicted angular distribution $\mathrm{d}\bar{W}$ and the experimentally measured one is better than 0.5%.

EXERCISE

2.7 Average Helicity and Parity Violation

Problem. Calculate the helicity expectation value averaged over the whole energy region and show that the result $\overline{\langle \Lambda \rangle} = -1$ is evidence for the violation of parity invariance.

Solution. (a) We set $|\boldsymbol{p}| = p$ and

$$\mathrm{d}\bar{W}(\boldsymbol{p} \cdot \boldsymbol{s} = \pm|\boldsymbol{p}|) = \mathrm{d}\tilde{W}^{\pm}(p) \quad . \tag{1}$$

The probability of an electron being emitted with momentum $\boldsymbol{p}$ is

$$\mathrm{d}\tilde{W}^{+}(p) + \mathrm{d}\tilde{W}^{-}(p) \quad . \tag{2}$$

The average of the expectation values is therefore

$$\overline{\langle \Lambda \rangle} = \frac{\int [\overline{\langle \Lambda \rangle}(\mathrm{d}\tilde{W}^{+}(p)) + \overline{\langle \Lambda \rangle}(\mathrm{d}\tilde{W}^{-}(p))]}{\int [\mathrm{d}\tilde{W}^{+}(p) + \mathrm{d}\tilde{W}^{-}(p)]}$$

$$= \frac{\int [\mathrm{d}\tilde{W}^{+}(p) - \mathrm{d}\tilde{W}^{-}(p)]}{W_\mu} \quad . \tag{3}$$

Exercise 2.7

W_μ is already known from (2.57); we only need to calculate the numerator. With the help of (2.66)–(2.68) we get

$$
\begin{aligned}
&\int \left[\mathrm{d}\tilde{W}^+(p) - \mathrm{d}\tilde{W}^-(p)\right] \\
&= -\frac{2}{3}G^2\pi(2\pi)^{-5}\int \frac{\mathrm{d}^3 p}{p^0}\Theta(p^0_{\max} - p^0)p\left[3(m_\mu - p^0)^2 - p^2 + 2(m_\mu - p^0)p^0\right] \\
&= -\frac{2}{3}G^2\pi(2\pi)^{-5}4\pi \int\limits_{m_\mathrm{e}}^{p^0_{\max}} p^2 \mathrm{d}p^0\left[3(m_\mu - p^0)^2 - p^2 + 2(m_\mu - p^0)p^0\right] \\
&= -\frac{2}{3}G^2(2\pi)^{-3}m_\mu \int\limits_{m_\mathrm{e}}^{p^0_{\max}} \mathrm{d}p^0\left[(p^0)^2 - m_\mathrm{e}^2\right]\left[3m_\mu - 4p^0 + \frac{m_\mathrm{e}^2}{m_\mu}\right] \ . \qquad (4)
\end{aligned}
$$

Performing the integral yields

$$
\begin{aligned}
&\int \left[\mathrm{d}\tilde{W}^+(p) - \mathrm{d}\tilde{W}^-(p)\right] \\
&= -\frac{2}{3}G^2(2\pi)^{-3}m_\mu\left\{\left(m_\mu + \frac{m_\mathrm{e}^2}{3m_\mu}\right)\left[(p^0_{\max})^3 - 3m_\mathrm{e}^2 p^0_{\max} + 2m_\mathrm{e}^3\right]\right. \\
&\quad \left. - (p^0_{\max})^4 + 2m_\mathrm{e}^2(p^0_{\max})^2 - m_\mathrm{e}^4\right\} \\
&= -\frac{G^2 m_\mu^5}{24(2\pi)^3}\left(1 - \frac{40}{3}y + 2\sqrt{y^3} - 30y^2 + \frac{32}{3}y\sqrt{y^5} - \frac{1}{3}y^4\right) \ , \qquad (5)
\end{aligned}
$$

where again $y = (m_\mathrm{e}/m_\mu)^2$. Applying (2.57) we obtain in lowest order in y

$$
\begin{aligned}
\langle \Lambda \rangle &\simeq -\frac{1 - \frac{40}{3}y}{1 - 8y} \\
&\simeq -1 + \frac{16}{3} \times \frac{m_\mathrm{e}^2}{m_\mu^2} + \cdots \\
&\simeq -0.999\,88 \ . \qquad (6)
\end{aligned}
$$

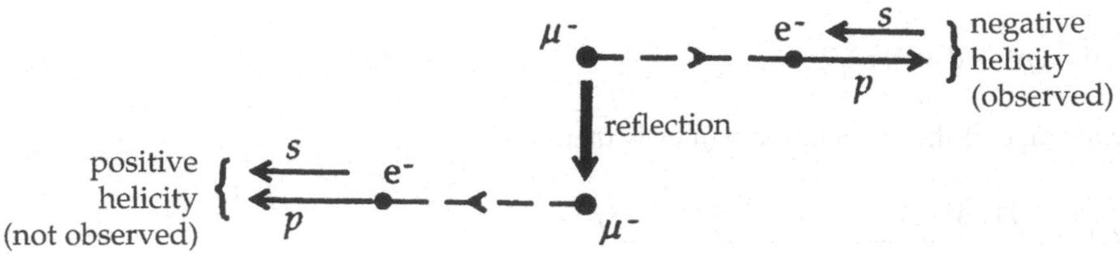

Fig. 2.4. Parity violation in muon decay

(b) Obviously (almost) all the electrons emitted in muon decay have negative helicity ($\lambda = -1$). A space reflection (see Fig. 2.4) would give the electrons positive helicity ($\lambda = +1$). In the case of parity invariance of the process, one would therefore measure equal numbers of electrons with positive and negative helicity. This is not the case; thus parity invariance must be broken.

EXERCISE

2.8 Angular Distribution and Parity Violation

Problem. Show that the violation of parity is due to the appearance of $\cos\theta$ in (2.72), describing the angular distribution.

Solution. We can write the angular distribution as

$$\frac{\mathrm{d}\bar{W}}{\mathrm{d}p^0\mathrm{d}\theta} = \sin\theta\left[A(p^0) + B(p^0)\cos\theta\right] \quad , \tag{1}$$

where $A(p^0)$ and $B(p^0)$ are given by comparison with (2.72). The geometry is displayed in Fig. 2.5. If we perform a space reflection, θ changes to $\theta_s = \pi - \theta$, and

$$\begin{aligned} \sin\theta &\to \sin\theta_s = \sin\theta \quad , \\ \cos\theta &\to \cos\theta_s = -\cos\theta \end{aligned} \tag{2}$$

thus the angular distribution becomes

$$\frac{\mathrm{d}\bar{W}}{\mathrm{d}p^0\mathrm{d}\theta_s} = \sin\theta\{A(p^0) - B(p^0)\cos\theta\} \quad . \tag{3}$$

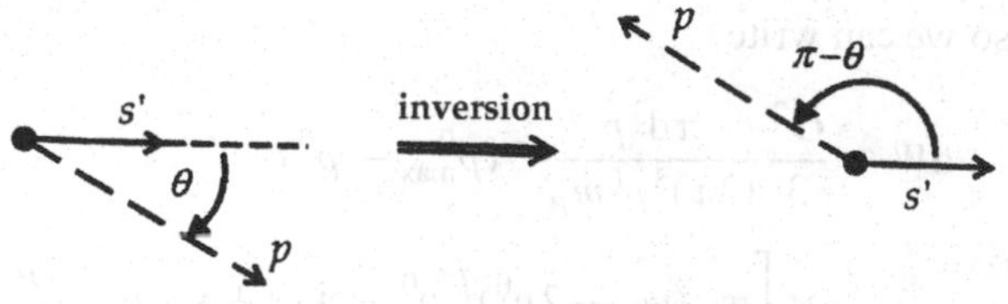

Fig. 2.5. Parity violation in an angular distribution

$\mathrm{d}\bar{W}/\mathrm{d}p^0\mathrm{d}\theta_s$ and $\mathrm{d}\bar{W}/\mathrm{d}p^0\mathrm{d}\theta$ differ from each other in the sign of the term proportional to $\cos\theta$: the angular distribution is not parity invariant. This argument is supported by geometrical considerations. The figure shows the intensity of the emitted electrons for $B(p^0) = 0$: no electrons are emitted in the direction of $\boldsymbol{s}'$. If $B(p^0) > 0$ and $B(p^0) <$

$A(p^0)$, the distribution is deformed and not reflection invariant, because $\sin\theta\cos\theta = \frac{1}{2}\sin(2\theta)$.

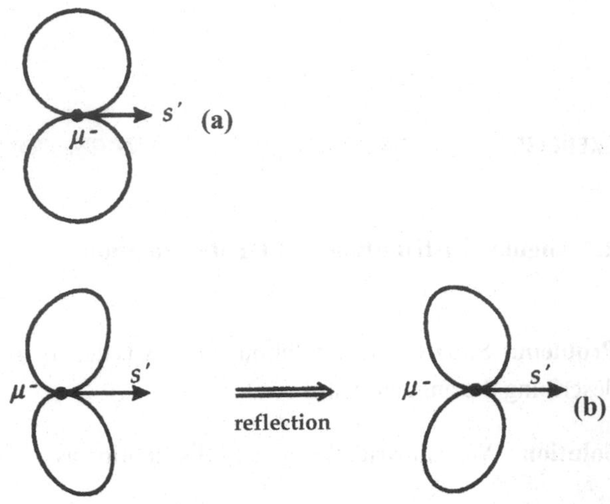

g. 2.6. Angular distribution f electrons with respect to the uon spin: (**a**) without violaon of parity; (**b**) with violaon of parity

EXERCISE

2.9 Electron Helicity in Muon Decay

Problem. Show, for the limit $p^0 \gg m_e$, that for the decay of a muon with spin s' in its rest system, the emission of an electron with spin s is given by $\mathrm{d}W \sim \sin^2(\theta/2)$, where $\mathrm{d}W$ is calculated in the limit $p^0 \gg m_e$ and θ denotes the angle between the electron spin s and momentum p of the electron.

Solution. We start from (2.23) inserted in (2.54) ($m = m_e$) and neglect systematically all terms with m_e. It is important to recognize that terms like $m_e s^0$ or $m_e \boldsymbol{p}\cdot\tilde{\boldsymbol{s}}$ do not contain the effective electron mass. Remember, $\tilde{\boldsymbol{s}}$ is the space component of the electron spin vector (2.23)! In this spirit we have

$$|\boldsymbol{p}| = \sqrt{(p^0)^2 + m_e^2} \approx p^0 \quad , \tag{1}$$

so we can write

$$\begin{aligned}
\mathrm{d}W &\approx \frac{G^2}{3}\frac{\pi\mathrm{d}^3p}{(2\pi)^5 p^0 m_\mu}\Theta(p^0_{\max} - p^0) \\
&\quad \times\left[m_\mu(m_\mu - 2p^0)\left(p^0 - \boldsymbol{p}\cdot\boldsymbol{s} + \boldsymbol{s}'\cdot\boldsymbol{p} - \frac{(\boldsymbol{p}\cdot\boldsymbol{s})(\boldsymbol{p}\cdot\boldsymbol{s}')}{p^0}\right)\right. \\
&\quad \left. + 2(m_\mu - p^0 - \boldsymbol{p}\cdot\boldsymbol{s}')[(m_\mu - p^0)(p^0 - \boldsymbol{p}\cdot\boldsymbol{s}) + p^0(p^0 - \boldsymbol{p}\cdot\boldsymbol{s})]\right] \\
&= \frac{G^2}{3}\frac{\pi\mathrm{d}^3p}{(2\pi)^5 m_\mu}\Theta(p^0_{\max} - p^0)\left(1 - \frac{\boldsymbol{p}\cdot\boldsymbol{s}}{p^0}\right)m_\mu
\end{aligned}$$

$$\times \left[(m_\mu - 2p^0)\left(1 - \frac{\boldsymbol{p}\cdot\boldsymbol{s}'}{p^0}\right) + 2(m_\mu - p^0 - \boldsymbol{p}\cdot\boldsymbol{s}') \right]$$

$$= \frac{G^2}{3}\frac{\mathrm{d}^3 p}{(2\pi)^5}\Theta(p^0{}_{\max} - p^0)\left(1 - \frac{\boldsymbol{p}\cdot\boldsymbol{s}}{p^0}\right)$$

$$\times \left[3p^0_{\max} - 2p^0 - \frac{\boldsymbol{p}\cdot\boldsymbol{s}'}{p^0} p^0_{\max} \right] \quad . \tag{2}$$

In the last step we have applied (2.52) for the matrix elements of the electron

$$p^0_{\max} = \frac{m_\mu^2 + m_e^2}{2m_\mu} \simeq \frac{1}{2} m_\mu \quad . \tag{3}$$

The coefficient in (2) which contains the spin $\boldsymbol{s}$ of the electron gives the desired angular dependence:

$$1 - \frac{\boldsymbol{p}\cdot\boldsymbol{s}}{p^0} = 1 - \frac{|\boldsymbol{p}|}{p^0}\cos\theta \simeq 2\sin^2\frac{\theta}{2} \quad . \tag{4}$$

The maximum of the distribution is at $\theta = \pi$, that is the electrons are preferentially polarized against their momentum (negative helicity). The result (4) is in accordance with the angular distribution in (1.12), which we calculated from the β decay of cobalt, if we take $\theta = \pi - \theta$ (here the z axis points downwards!). This is another confirmation of the heuristic consideration in relation to the experiment of Wu et al. (see Sect. 1.2).

In the limit $p^0 \to p^0_{\max}$ the last factor takes the form

$$3p^0_{\max} - 2p^0 - p^0_{\max}\frac{\boldsymbol{p}\cdot\boldsymbol{s}'}{p^0} \to p^0_{\max}\left(1 - \frac{\boldsymbol{p}\cdot\boldsymbol{s}'}{p^0}\right) , \tag{5}$$

that is, the preferential emission of the electron is opposite to the polarization of the muons. This is easy to see for the case $p^0 = p^0_{\max}$ in Fig. 2.2, where the two neutrinos are emitted in the same direction while the electron goes in the opposite direction. Because $\bar{\nu}_e$ and ν_μ have opposite helicities, the sum of their angular momenta is equal to zero. The result is that the electron must acquire the spin of the decaying muon. Because of its negative helicity the electron is preferentially emitted opposite to the muon spin.

EXERCISE

2.10 CP Invariance in Muon Decay

Problem. The term $J^{\alpha\dagger}_{(e)} J^{(\mu)}_\alpha$ in the current–current coupling is responsible for the decay of the positive muon, μ^+. Show that this leads to a change of the sign of the spin-dependent terms in the squared transition amplitude (2.32). On the basis of these results discuss the connection between violation of the invariance under spatial reflection and the invariance under charge conjugation.

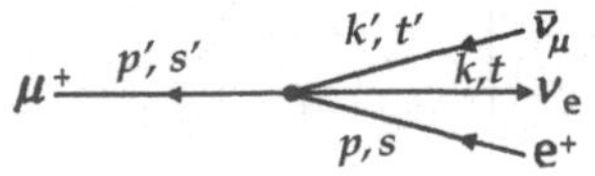

g. 2.7. Decay of a μ^+ parti- e[7]

Solution. (a) The S-matrix element for the μ^+ decay is given by[9]

$$\begin{aligned} &S(\mu^+ \to \mathrm{e}^+ \nu_\mathrm{e} \bar{\nu}_\mu) \\ &= -\mathrm{i} \int \mathrm{d}^4 x \frac{G}{\sqrt{2}} \left[\bar{u}_\mu(x) \gamma^\mu (1-\gamma_5) u_{\nu_\mu}(x)\right]\left[\bar{u}_{\nu_\mathrm{e}}(x) \gamma_\mu (1-\gamma_5) u_\mathrm{e}(x)\right] \quad . \end{aligned} \tag{1}$$

As now all particles are antiparticles, except the ν_e, the spinors are given in analogy to (2.11) by

$$\begin{aligned} u_\mu(x) &= v_\mu(\boldsymbol{p}', s') \exp(\mathrm{i} p'_\mu x^\mu)(2p'^0 V)^{-1/2} \quad , \\ u_\mathrm{e}(x) &= v_\mathrm{e}(\boldsymbol{p}, s) \exp(\mathrm{i} p_\mu x^\mu)(2p^0 V)^{-1/2} \quad , \\ u_{\nu_\mathrm{e}}(x) &= u_{\nu_\mathrm{e}}(\boldsymbol{k}, t) \exp(-\mathrm{i} k_\mu x^\mu)(2k^0 V)^{-1/2} \quad , \\ u_{\nu_\mu}(x) &= v_{\nu_\mu}(\boldsymbol{k}', t') \exp(\mathrm{i} k'_\mu x^\mu)(2k'^0 V)^{-1/2} \quad . \end{aligned} \tag{2}$$

The calculation proceeds exactly as before up to (2.18), (2.19), because the δ function does not change when the sign of its argument is inverted. Thus we obtain

$$\begin{aligned} \mathrm{d}W(\mu^+) = &\frac{G^2}{2} \frac{1}{(2\pi)^5} \frac{\mathrm{d}^3 p}{2p'^0 2p^0} \\ &\times \int \frac{\mathrm{d}^3 k}{2k^0} \int \frac{\mathrm{d}^3 k'}{2k'^0} \delta^4(p + k + k' - p') \sum_{t,t'} |M^2| \quad , \end{aligned} \tag{3}$$

with

$$M = \left[\bar{v}_\mu \gamma^\mu (1-\gamma_5) v_{\nu_\mu}\right]\left[\bar{u}_{\nu_\mathrm{e}} \gamma_\mu (1-\gamma_5) v_\mathrm{e}\right] \quad . \tag{4}$$

The only difference compared to μ^- decay is in the spinors which enter into the transition amplitude M, where all particles are replaced by antiparticle spinors and vice versa. $\sum_{t,t'} |M|^2$ separates again into two similar contributions for the muonic and electronic particles. First we repeat the calculation from (2.20) to (2.30) for the muonic part. Here we need the analogous relation to (2.21) (see Appendix A.2),

$$\sum_{t'} v_{\nu_\mu}(k', t')_\alpha \bar{v}_{\nu_\mu}(k', t')_\beta = \not{k}'_{\alpha\beta} \quad , \tag{5}$$

and to (2.22),

$$\bar{v}_\mu(p', s')_\pi v_\mu(p', s')_\tau = \left[(\not{p}' - m_\mu)\frac{1 + \gamma_5 \not{s}'}{2}\right]_{\tau\pi} \quad . \tag{6}$$

With these expressions we find that

$$\begin{aligned} X^{\mu\nu}(\mu) &= \sum_{t'} \bar{v}_\mu(p', s')_\pi \gamma^\mu_{\pi\rho}(1-\gamma_5)_{\rho\alpha} \\ &\quad \times v_{\nu_\mu}(k', t')_\alpha \bar{v}_{\nu_\mu}(k', t')_\beta \gamma^\nu_{\beta\sigma}(1-\gamma_5)_{\sigma\tau} v_\mu(p', s')_\tau \\ &= \mathrm{Tr}\left\{\gamma^\mu(1-\gamma_5)\not{k}'\gamma^\nu(1-\gamma_5)(\not{p}' - m_\mu)\left(\frac{1+\gamma_5\not{s}'}{2}\right)\right\} \quad . \end{aligned} \tag{7}$$

---

[9] M.L. Perl: Ann. Rev. Nucl. Part. Science **30**, 299 (1980).

With a cyclical permutation we take the last two factors to the front; we also permute $\not{k}'$ with $(1-\gamma_5)$, changing the sign in $(1-\gamma_5)$. We thus obtain

$$X^{\mu\nu}(\mu) = \frac{1}{2}\,\mathrm{Tr}\{(\not{p}' - m_\mu)(1+\gamma_5\not{s}')\gamma^\mu \not{k}'(1+\gamma_5)\gamma^\nu(1-\gamma_5)\} \quad . \tag{8}$$

This result is distinct from (2.26) by the sign of m_μ and also by permutation of the Lorentz indices μ and ν. Hence we can skip all subsequent calculations and write down directly the analogue of (2.30):

$$\begin{aligned} X^{\mu\nu}(\mu) = 4\big[&(p'+m_\mu s')^\mu k'^\nu - (p'+m_\mu s')^\alpha k'_\alpha g^{\nu\mu} \\ &+(p'+m_\mu s')^\nu k'^\mu + \mathrm{i}\epsilon^{\alpha\mu\beta\nu}(p'+m_\mu s')_\alpha k'_\beta\big] \quad . \end{aligned} \tag{9}$$

For the electronic part the same relation holds; compared to (2.31) it changes the sign of m_e, and μ and ν have to be permuted:

$$\begin{aligned} X_{\mu\nu}(\mathrm{e}) &= \sum_t \bar{u}_{\nu_e}(k,t)\gamma_\mu(1-\gamma_5)v_e(p,s)\bar{v}_e(p,s)\gamma_\nu(1-\gamma_5)u_{\nu_e}(k,t) \\ &= 4\big[(p+m_e s)_\nu k_\mu - (p+m_e s)^\alpha k_\alpha g_{\nu\mu} \\ &\quad + (p+m_e s)_\mu k_\nu + \mathrm{i}\epsilon_{\alpha\nu\beta\mu}(p+m_e s)^\alpha k^\beta\big] \quad . \end{aligned} \tag{10}$$

The permutation can be reversed, because by constructing $|M|^2$ we sum over the indices μ and ν. What remains is just the change of the sign of m_μ and m_e in (2.32):

$$\begin{aligned} \sum_{t,t'} |M|^2 &= X^{\mu\nu}(\mu)X_{\mu\nu}(\mathrm{e}) \\ &= 64(p'+m_\mu s')^\alpha k_\alpha (p+m_e s)^\beta k'_\beta \quad . \end{aligned} \tag{11}$$

As the spin vectors s^μ and s'^μ enter only in the combination $m_\mu s'$ or $m_e s$, we can also easily get the result (11) from (2.32) by inversion of the spin vectors: $s, s' \to -s, -s'$.

(b) Equation (11) follows from (2.32) if we invert the sign of the charge of the decaying muon, that is, it follows from the operation of charge conjugation. We thus see that *the β decay of the muon is not invariant against charge conjugation.*

An interesting point is that (11) could also be obtained by space reflection. On being reflected, the momentum vector $\boldsymbol{p}'$ changes its sign, whereas the axial spin vector (in the rest system) $\boldsymbol{s}'$ does not change:

$$p'^\alpha = (p'^0, \boldsymbol{p}') \to (p'^0, -\boldsymbol{p}') \quad , \tag{12}$$

$$\begin{aligned} s'^\alpha &= \left(\frac{\boldsymbol{p}'\cdot\boldsymbol{s}'}{m_\mu}, \boldsymbol{s}' + \frac{(\boldsymbol{p}'\cdot\boldsymbol{s}')}{m_\mu(p'^0+m_\mu)}\boldsymbol{p}'\right) \\ &\to \left(-\frac{\boldsymbol{p}'\cdot\boldsymbol{s}'}{m_\mu}, \boldsymbol{s}' + \frac{(\boldsymbol{p}'\cdot\boldsymbol{s}')}{m_\mu(p'^0+m_\mu)}\boldsymbol{p}'\right) \quad , \end{aligned} \tag{13}$$

and

$$k^\alpha = (k^0, \boldsymbol{k}) \to (k^0, -\boldsymbol{k})$$

or

$$k_\alpha = (k^0, -\boldsymbol{k}) \to (k^0, \boldsymbol{k}) \quad . \tag{14}$$

With these we obtain the following change to (2.32):

$$\begin{aligned}
&(p' - m_\mu s')^\alpha k_\alpha \\
&= \left(p'^0 - \frac{p'\cdot s'}{m_\mu}\right)k^0 - \left(\boldsymbol{p}' - m_\mu \boldsymbol{s}' - \frac{\boldsymbol{p}'\cdot\boldsymbol{s}'}{p^0+m_\mu}\boldsymbol{p}'\right)\cdot\boldsymbol{k} \\
&\to \left(p'^0 + \frac{\boldsymbol{p}'\cdot\boldsymbol{s}'}{m_\mu}\right)k^0 + \left(-\boldsymbol{p}' - m_\mu \boldsymbol{s}' - \frac{(\boldsymbol{p}'\cdot\boldsymbol{s}')}{p'^0+m_\mu}\boldsymbol{p}'\right)\cdot\boldsymbol{k} \\
&= (p' + m_\mu s')^\alpha k_\alpha
\end{aligned} \tag{15}$$

and analogously

$$(p - m_e s)^\beta k'_\beta \to (p + m_e s)^\beta k'_\beta \quad . \tag{16}$$

We can thus conclude that the weak interaction (of the leptons) behaves under charge conjugation $\hat{C}$ in the same way as under space reflection $\hat{P}$. Since according to this the simultaneous application of $\hat{C}$ and $\hat{P}$ yields the identity, that is, everything remains invariant, it means that the weak interaction is invariant under the product $\hat{C}\hat{P}$. (We shall see later, in Chap. 8, how the weak interaction among quarks can lead to a slight violation of CP invariance.)

2.5 The Michel Parameters

We now ask how far the muon decay confirms the V–A theory. For this purpose we write down the most general form of the coupling matrix element,

$$\begin{aligned}
\tilde{H}_{\mu^-\to e^-\bar{\nu}_e\nu_\mu} = \frac{G}{\sqrt{2}}\int d^3x \sum_i &\left[\bar{u}_{\nu_\mu}(x)\hat{O}_i u_\mu(x)\right] \\
&\times \left[\bar{u}_e(x)\hat{O}^i(A_i + A'_i\gamma_5)u_{\nu_e}(x)\right] \quad ,
\end{aligned} \tag{2.74}$$

and allow this time every type of coupling $\hat{O}_i = (\mathrm{S, V, T, A, P})$. It is customary to use other constants C_i, C'_i instead of A_i, A'_i. The two sets of constants are related to each other through the transformation (the so-called Fierz transformation, see Supplement 2.12):

$$\begin{aligned}
&C_i = \sum_j \Lambda_{ij}A_j \quad , \quad C'_i = \sum_j \Lambda_{ij}A'_j \quad , \\
&(\Lambda_{ij}) = \frac{1}{4}\begin{pmatrix} 1 & 4 & 6 & 4 & 1 \\ 1 & -2 & 0 & 2 & -1 \\ 1 & 0 & -2 & 0 & 1 \\ 1 & 2 & 0 & -2 & -1 \\ 1 & -4 & 6 & -4 & 1 \end{pmatrix} \quad .
\end{aligned} \tag{2.75}$$

With C_i and C'_i, we can write the coupling in the form

$$\tilde{H} = \frac{G}{\sqrt{2}}\int d^3x \sum_i \left[\bar{u}_e(x)\hat{O}_i u_\mu(x)\right]\left[\bar{u}_{\nu_\mu}(x)\hat{O}^i(C_i + C'_i\gamma_5)u_{\nu_e}(x)\right] \quad . \tag{2.76}$$

One easily checks that pure vector coupling minus axial vector coupling in (2.74) results also in a vector minus axial vector coupling in (2.76) up to the minus sign, while the other couplings in (2.74) result in complicated superpositions in (2.76). This circumstance may seem to endow the V–A law with particular significance. However, four other combinations have comparable properties, as can be found by diagonalizing the matrix Λ_{ij}. The possibility of using the invariance under transpositions among the fields as a basis for singling out the correct coupling was explored extensively without decisive results.[10]

The advantage of the notation (2.76) is that the wave function of the observable particles – the electron and the muon – are connected in one matrix element, whereas the wave functions of the two neutrinos are separated in the second spinor matrix element. In the interaction (2.74) or (2.76) only the conservation of electron and muon number, and Lorentz invariance, is assumed. Let us introduce the abbreviation

$$a_i = |C_i|^2 + |C_i'|^2 \quad . \tag{2.77}$$

Since the factor G stands in front of the expression (2.76), the proper coupling constants are given by GC_i or GC_i', respectively. It is obvious that a variation of the value of G can be compensated by a multiplication of all constants C_i, C_i' with a common factor. If we determine G by experiment, the C_i, C_i' are no longer independent, that is, they must satisfy a normalization condition. We choose this condition to be

$$a_{\mathrm{S}} + 4a_{\mathrm{V}} + 6a_{\mathrm{T}} + 4a_{\mathrm{A}} + a_{\mathrm{P}} = 16 \quad . \tag{2.78}$$

It is necessary to calculate the muon decay once more, but now with all types of coupling allowed. We assume that the μ^- is polarized before the decay, but we do not observe the polarizations of the three decay products (see (2.54) and also Exercise 2.9, but remember that those results were valid for V–A coupling only). With the abbreviation

$$x = \frac{p^0}{p^0_{\max}} = \frac{2m_\mu p^0}{m_\mu^2 + m_{\mathrm{e}}^2} \tag{2.79}$$

and the emission angle θ of the electron with respect to the muon spin

$$\cos\theta = \frac{\boldsymbol{p}\cdot\boldsymbol{s}}{|\boldsymbol{p}|} \quad , \tag{2.80}$$

we get after a lengthy calculation the following electron spectrum:

$$\begin{aligned} \frac{\mathrm{d}W}{\mathrm{d}\Omega\,\mathrm{d}p^0} = \frac{G^2 m_\mu}{12\pi^4}|\boldsymbol{p}|p^0 \Bigg\{ & 3(p^0_{\max} - p^0) + \frac{2}{3}\rho\left(4p^0 - 3p^0_{\max} - \frac{1}{3}\frac{m_{\mathrm{e}}^2}{m_\mu}\right) \\ & + 3\frac{m_{\mathrm{e}}}{p^0}\eta(p^0_{\max} - p^0) - \xi\frac{|\boldsymbol{p}|}{p^0}\cos\theta\bigg[(p^0_{\max} - p^0) \\ & + \frac{2}{3}\delta(4p^0 - 3p^0_{\max} - m_{\mathrm{e}}^2/m_\mu)\bigg]\Bigg\}\theta(p^0_{\max} - p^0) \quad . \end{aligned} \tag{2.81a}$$

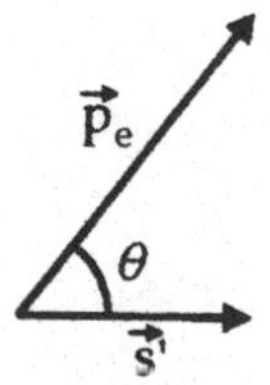

Fig. 2.8. The angle of electron emission relative to the spin s' of the muon

[10] See many papers beginning with C. Gitchfield: Phys. Rev. **63**, 417 (1943) through to E. Caianello: Nuovo Cimento **8**, 749 (1952), in which references to earlier work can be found. See also E.J. Konopinski: *The Theory of Beta Radioactivity* (Oxford University Press, London, 1966).

The details of this calculation are layed down in Exercise 2.11. If we neglect the mass of the electron and make use of definition (2.79) this becomes

$$\frac{\mathrm{d}W}{\mathrm{d}\Omega\,\mathrm{d}x} = \frac{G^2 m_\mu{}^5}{192\pi^4} x^2 \left\{ \frac{1}{1+4\eta\frac{m_e}{m_\mu}} \left[4(x-1) + \frac{2}{3}\rho(4x-3) + 6\frac{m_e}{m_\mu}\frac{1-x}{x}\eta \right] \right.$$
$$\left. - \xi\cos\theta \left[(1-x) + \frac{2}{3}\delta(4x-3) \right] \right\} . \tag{2.81b}$$

In this formula ρ, η, ξ, and δ are the so-called **Michel** parameters

$$\begin{aligned} \rho &= \frac{1}{16}(3a_\mathrm{V} + 6a_\mathrm{T} + 3a_\mathrm{A}) \quad , \quad \xi = \frac{-1}{16}(4b' + 3a' - 14c') \quad , \\ \eta &= \frac{1}{16}(a_\mathrm{S} - 2a_\mathrm{V} + 2a_\mathrm{A} - a_\mathrm{P}) \quad , \quad \delta = \frac{-1}{16\xi}(3b' - 6c') \quad , \end{aligned} \tag{2.82}$$

where

$$\begin{aligned} a' &= 2\,\mathrm{Re}\{C_\mathrm{S} C_\mathrm{P}'^{\,*} + C_\mathrm{S}' C_\mathrm{P}^*\} \quad , \\ b' &= 2\,\mathrm{Re}\{C_\mathrm{V} C_\mathrm{A}'^{\,*} + C_\mathrm{V}' C_\mathrm{A}^*\} \quad , \\ c' &= 2\,\mathrm{Re}\{C_\mathrm{T} C_\mathrm{T}'^{\,*}\} \quad . \end{aligned} \tag{2.83}$$

The parameters are chosen in such a way that if one integrates over x from 0 to 1 then ρ and δ disappear. Therefore the lifetime of the muon is independent of ρ and δ. For a pure V–A coupling, which was assumed during the discussion in Sects. 2.2, 2.3, and 2.4, we get

$$\begin{aligned} C_\mathrm{S} &= C_\mathrm{S}' = C_\mathrm{T} = C_\mathrm{T}' = C_\mathrm{P} = C_\mathrm{P}' = 0 \quad , \\ C_\mathrm{V} &= C_\mathrm{V}' = -C_\mathrm{A} = -C_\mathrm{A}' = 1 \quad . \end{aligned} \tag{2.84}$$

Considering Supplement 2.12, (23), one gets

$$\begin{aligned} M &= [\bar{u}_e\gamma_\mu u_\mu][\bar{u}_{\nu_\mu}\gamma^\mu(1-\gamma_5)u_{\nu_e}] + [u_e\gamma_5\gamma_\mu u_\mu][u_{\nu_\mu}\gamma_5\gamma^\mu(1-\gamma_5)u_{\nu_e}] \\ &= \left[\bar{u}_e\gamma_\mu(1-\gamma_5)u_\mu\right]\left[\bar{u}_{\nu_\mu}\gamma^\mu(1-\gamma_5)u_{\nu_e}\right] \quad . \end{aligned} \tag{2.85}$$

By inserting this value into (2.82) we obtain the prediction of the V–A theory for the Michel parameters:

$$\rho = \frac{3}{4} \quad , \quad \xi = 1 \quad , \quad \eta = 0 \quad , \quad \delta = \frac{3}{4} \quad . \tag{2.86a}$$

The experimental values are derived from a careful measurement of the electron spectrum (or the positron spectrum in the case of the μ^+ decay) and of the angular distribution. Equation (2.81) tells us that ρ must be fitted to the whole spectrum, whereas η is mainly sensitive to low energies ($x \to 0$). It is not surprising, therefore, that η is the most uncertain of the parameters. ξ can be obtained by integrating the angular distribution over the energy, whereas δ can be determined by measuring the energy dependence of this distribution. The best experimental values are

$$\begin{aligned} \rho &= 0.7517 \pm 0.0026 \quad , \\ \eta &= -0.12 \pm 0.21 \quad , \end{aligned}$$

$$\xi = 0.972 \pm 0.013 \quad , \tag{2.86b}$$

$$\delta = 0.7551 \pm 0.0085 \quad .$$

These values are in very good agreement with the predictions (2.85) of the V–A theory; see also Fig. 2.9.

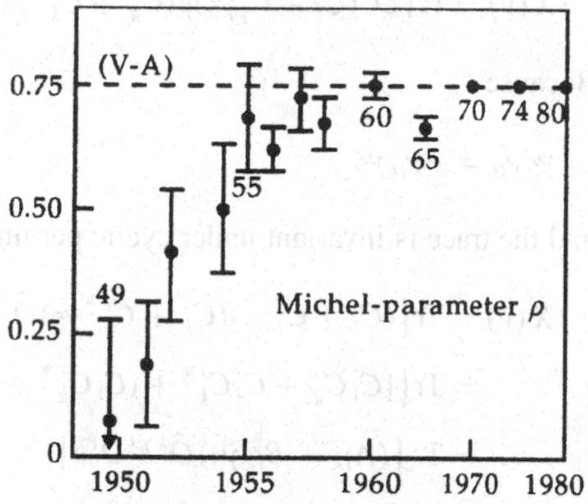

Fig. 2.9. Experimental determination of the Michel parameter ϱ since 1950. The curve shows the improvement of the experiments, but perhaps also the prejudice of the experimentalists

EXERCISE

2.11 Muon Decay and the Michel Parameters

Problem. Calculate the muon decay with the general interaction (2.76) in the same manner as in Sect. 2.2 and derive (2.81) by summing over the polarizations of the outgoing particles.

Solution. To derive (2.81) we repeat the steps which lead us from (2.10) to (2.18). The normalization and the phase-space factor are obtained in the same manner. The only difference occurs in the matrix element M. With (2.76) this is given by

$$M = \sum_i [\bar{u}_e \hat{O}_i u_\mu][\bar{u}_{\nu_\mu} \hat{O}^i (C_i - C_i' \gamma_5) u_{\nu_e}] \quad . \tag{1}$$

First we calculate the part of $|M|^2$ which stems from the neutrinos. We sum over the unobservable neutrino spins and get

$$\begin{aligned} X(\nu) &= \sum_{t,t'} [\bar{u}_{\nu_\mu}(k',t') \hat{O}^i (C_i - C_i' \gamma_5) u_{\nu_e}(k,t)] \\ &\quad \times [\bar{u}_{\nu_\mu}(k',t') \hat{O}^k (C_k - C_k' \gamma_5) u_{\nu_e}(k,t)] \\ &= \sum_{t,t'} [\bar{u}_{\nu_\mu}(k',t') \hat{O}^i (C_i - C_i' \gamma_5) u_{\nu_e}(k,t)] \\ &\quad \times [\bar{u}_{\nu_e}(k,t) \gamma_0 (C_k^* - C_k'^* \gamma_5^\dagger)(\hat{O}^k)^\dagger \gamma_0 u_{\nu_\mu}(k',t')] \\ &= \sum_{t,t'} [\bar{u}_{\nu_\mu}(k',t') \hat{O}^i (C_i - C_i' \gamma_5) u_{\nu_e}(k,t)] \\ &\quad \times [\bar{u}_{\nu_e}(k,t) (C_k^* + C_k'^* \gamma_5)(\hat{O}^k) u_{\nu_\mu}(k',t')] \end{aligned} \tag{2}$$

where we have inserted $\gamma_0^2 = 1$ in front of $(\hat{O}^k)^\dagger$ and used Supplement 2.12, (3), which yields

$$\gamma_0 \gamma_5^\dagger \gamma_0 = -\gamma_5 \quad . \tag{3}$$

With (2.21) we get

$$X(\nu) = \mathrm{Tr}\{\hat{O}^i (C_i - C_i' \gamma_5) k\!\!\!/ (C_k^* + C_k'^* \gamma_5) \hat{O}^k k\!\!\!/'\} \quad . \tag{4}$$

Because

$$\gamma_5 \gamma_\mu = -\gamma_\mu \gamma_5 \tag{5}$$

and the trace is invariant under cyclic permutations, this form may be transcribed to

$$\begin{aligned} X(\nu) &= \mathrm{Tr}\{(C_i + C_i'\gamma_5)(C_k^* + C_k'^*\gamma_5)\hat{O}^k k\!\!\!/' \hat{O}^i k\!\!\!/\} \\ &= \mathrm{Tr}\{\{C_i C_k^* + C_i' C_k'^* + (C_i C_k'^* + C_i' C_k^*)\gamma_5\}\hat{O}^k k\!\!\!/' \hat{O}^i k\!\!\!/\} \\ &= \mathrm{Tr}\{(A_{ik} + B_{ik}\gamma_5)\hat{O}^k k\!\!\!/' \hat{O}^i k\!\!\!/\} \\ &= \mathrm{Tr}\{(A_{ik} \pm B_{ik}\gamma_5)\hat{O}^i k\!\!\!/ \hat{O}^k k\!\!\!/'\} \quad , \end{aligned} \tag{6}$$

where we have $+$ for $\hat{O} = \hat{S}, \hat{P}, \hat{T}$ and $-$ for $\hat{O} = V, A$, and the following abbreviations have been introduced:

$$\begin{aligned} A_{ik} &= C_i C_k^* + C_i' C_k'^* \quad , \\ B_{ik} &= C_i C_k'^* + C_i' C_k^* \quad . \end{aligned} \tag{7}$$

Notice that $X(\nu)$ is non-zero only if both $\hat{O}^i$ and $\hat{O}^k$ contain either an even or an odd number of γ matrices. Otherwise the trace in (6) vanishes. This property will be useful for the evaluation of $X(\mu, \mathrm{e})$, since we can then restrict our consideration to the corresponding combinations of $\hat{O}^i$ and $\hat{O}^k$. In determining $X(\mu, \mathrm{e})$ we assume that the electron spin is not observed, and we therefore sum over the spin orientations. Furthermore we make the approximation of neglecting the electron mass.

We then find

$$\begin{aligned} X(\mu, \mathrm{e}) &= \sum_s [\bar{u}_\mathrm{e}(\boldsymbol{p}, s)\hat{O}_i u_\mu(\boldsymbol{p}', s')][\bar{u}_\mathrm{e}(\boldsymbol{p}, s)\hat{O}_k u_\mu(\boldsymbol{p}', s')]^\dagger \\ &= \sum_s [\bar{u}_\mathrm{e}(\boldsymbol{p}, s)\hat{O}_i u_\mu(\boldsymbol{p}', s')][\bar{u}_\mu(\boldsymbol{p}', s')\hat{O}_k u_\mathrm{e}(\boldsymbol{p}, s)] \\ &= \mathrm{Tr}\left[\hat{O}_i (p\!\!\!/' + m_\mu)\frac{1 + \gamma_5 s\!\!\!/'}{2}\hat{O}_k p\!\!\!/\right] \quad . \end{aligned} \tag{8}$$

If both $\hat{O}_i$ and $\hat{O}_k$ contain an even or an odd number of γ matrices, $X(\mu, \mathrm{e})$ reduces to

$$X(\mu, \mathrm{e}) = \frac{1}{2}\mathrm{Tr}\{\hat{O}_i p\!\!\!/' \hat{O}_k p\!\!\!/\} + \frac{1}{2} m_\mu \mathrm{Tr}\{\hat{O}_i \gamma_5 s\!\!\!/' \hat{O}_k p\!\!\!/\} \quad . \tag{9}$$

All other terms in (8) do not contribute, since a trace consisting of an uneven number of γ matrices vanishes.

Let us now consider the particular combinations of $\hat{O}_i$ and $\hat{O}_k$ in detail. For this purpose we again employ the formulas listed in Appendix A.2, especially

$$(\gamma_5)^2 = 1 \quad \text{and} \quad \gamma_5\gamma_\mu = -\gamma_\mu\gamma_5 \quad .$$

- $\hat{O}^i = \hat{O}^k = 1$:

$$\begin{aligned} X(\mu, \mathrm{e}) &= \frac{1}{2}\,\mathrm{Tr}\{\not{p}\not{p}'\} = 2(p\cdot p') \quad , \\ X(\nu) &= \mathrm{Tr}\{(A_{\mathrm{SS}} - B_{\mathrm{SS}}\gamma_5)\not{k}\not{k}'\} = 4A_{\mathrm{SS}}(k\cdot k') \quad . \end{aligned} \tag{10}$$

- $\hat{O}^i = \mathrm{i}\gamma_5,\ \hat{O}^k = 1$:

$$\begin{aligned} X(\mu, \mathrm{e}) &= \frac{1}{2}m_\mu\,\mathrm{Tr}\{\mathrm{i}\gamma_5\gamma_5\not{s}'\not{p}\} = 2\mathrm{i}m_\mu(p\cdot s') \quad , \\ X(\nu) &= \mathrm{Tr}\{(A_{\mathrm{PS}} - B_{\mathrm{PS}}\gamma_5)\mathrm{i}\gamma_5\not{k}\not{k}'\} \\ &= -4\mathrm{i}B_{\mathrm{PS}}(k\cdot k') \quad . \end{aligned} \tag{11}$$

- $\hat{O}^i = 1,\ \hat{O}^k = \mathrm{i}\gamma_5$:

$$\begin{aligned} X(\mu, \mathrm{e}) &= \frac{1}{2}m_\mu\,\mathrm{Tr}\{\gamma_5\not{s}'\cdot\mathrm{i}\gamma_5\not{p}\} = -2\mathrm{i}m_\mu(p\cdot s') \quad , \\ X(\nu) &= \mathrm{Tr}\{(A_{\mathrm{SP}} - B_{\mathrm{SP}}\gamma_5)\not{k}\mathrm{i}\gamma_5\not{k}'\} \\ &= +4\mathrm{i}B_{\mathrm{SP}}(k\cdot k') \quad . \end{aligned} \tag{12}$$

- $\hat{O}^i = \mathrm{i}\gamma_5,\ \hat{O}^k = \mathrm{i}\gamma_5$:

$$\begin{aligned} X(\mu, \mathrm{e}) &= \frac{1}{2}\,\mathrm{Tr}\{\mathrm{i}\gamma_5\not{p}'\cdot\mathrm{i}\gamma_5\not{p}\} = 2(p\cdot p') \quad , \\ X(\nu) &= \mathrm{Tr}\{(A_{\mathrm{PP}} - B_{\mathrm{PP}}\gamma_5)\mathrm{i}\gamma_5\not{k}\mathrm{i}\gamma_5\not{k}'\} \\ &= 4A_{\mathrm{PP}}(k\cdot k') \quad . \end{aligned} \tag{13}$$

Collecting together (10)–(13), we obtain

$$\begin{aligned} \sum_{i,k=\mathrm{S,P}} X(\mu, \mathrm{e})X(\nu) = 8(k\cdot k')\big[&(A_{\mathrm{SS}} + A_{\mathrm{PP}})(p\cdot p') \\ &+ (B_{\mathrm{PS}} + B_{\mathrm{SP}})m_\mu(p\cdot s')\big] \quad , \end{aligned} \tag{14}$$

or, adopting the abbreviations (2.77) and (2.83),

$$\sum_{i,k=\mathrm{S,P}} X(\mu, \mathrm{e})X(\nu) = 8(k\cdot k')\big[(a_{\mathrm{S}} + a_{\mathrm{P}})(p\cdot p') + a'm_\mu(p\cdot s')\big] \quad . \tag{15}$$

(Note that according to the convention (23) of Supplement 2.12 it holds that $B_{\mathrm{PS}} + B_{\mathrm{SP}} = -a'$.)

- $\hat{O}^i = \gamma^\mu,\ \hat{O}^k = \gamma^\nu$:

$$\begin{aligned} X(\mu, \mathrm{e}) &= \frac{1}{2}\,\mathrm{Tr}\{\gamma_\mu\not{p}'\gamma_\nu\not{p}\} + \frac{1}{2}m_\mu\,\mathrm{Tr}\{\gamma_\mu\gamma_5\not{s}'\gamma_\nu\not{p}\} \\ &= 2(p'_\mu p_\nu + p_\mu p'_\nu - g_{\mu\nu}\,p\cdot p') + 2\mathrm{i}m_\mu\varepsilon_{\sigma\nu\tau\mu}s'^\sigma p^\tau , \end{aligned} \tag{16}$$

Exercise 2.11

$$
\begin{aligned}
X(\nu) &= \mathrm{Tr}\{(A_{VV} + B_{VV}\gamma_5)\gamma^{\mu}\not{k}\gamma^{\nu}\not{k}'\} \\
&= 4A_{VV}\left[k^{\mu}k'^{\nu} + k^{\nu}k'^{\mu} - g^{\mu\nu}(k\cdot k')\right] + 4\mathrm{i}B_{VV}\varepsilon^{\mu\alpha\nu\beta}k_{\alpha}k'_{\beta} \quad .
\end{aligned} \tag{17}
$$

Evaluating the products leads to

$$
\begin{aligned}
X(\mu,\mathrm{e})X(\nu) &= 8A_{VV}\big[(p'\cdot k)(p\cdot k') + (p'\cdot k')(p\cdot k) \\
&\quad - (p\cdot p')(k\cdot k') + (p\cdot k)(p'\cdot k') \\
&\quad + (p\cdot k')(p'\cdot k) - (p\cdot p')(k\cdot k') \\
&\quad - (p\cdot p')(k\cdot k') - (p\cdot p')(k\cdot k') \\
&\quad + 4(p\cdot p')(k\cdot k')\big] \\
&\quad - 8m_{\mu}B_{VV}\varepsilon_{\sigma\nu\tau\mu}\varepsilon^{\mu\alpha\nu\beta}s'^{\sigma}p^{\tau}k_{\alpha}k'_{\beta} \\
&= 16A_{VV}\big[(p'\cdot k)(p\cdot k') + (p'\cdot k')(p\cdot k)\big] \\
&\quad - 8m_{\mu}B_{VV}\varepsilon_{\sigma\nu\tau\mu}\varepsilon^{\mu\alpha\nu\beta}s'^{\sigma}p^{\tau}k_{\alpha}k'_{\beta} \quad .
\end{aligned} \tag{18}
$$

The last term does not contribute, since, in the course of the further evaluation, $k_{\alpha}k'_{\beta}$ yields the symmetric tensor $I_{\alpha\beta}$ which is contracted with $\varepsilon^{\mu\alpha\nu\beta}$; thus there is no need to evaluate this term further. The next three cases may be treated in just the same way.

- $\hat{O}^i = \gamma_5\gamma^{\mu}$, $\hat{O}^k = \gamma^{\nu}$:

$$
\begin{aligned}
X(\mu,\mathrm{e}) &= \frac{1}{2}\mathrm{Tr}\{\gamma_5\gamma_{\mu}\not{p}'\gamma_{\nu}\not{p}\} + \frac{1}{2}m_{\mu}\mathrm{Tr}\{\gamma_5\gamma_{\mu}\gamma_5\not{s}'\gamma_{\nu}\not{p}\} \\
&= -2\mathrm{i}\varepsilon_{\mu\sigma\nu\tau}p'^{\sigma}p^{\tau} - 2m_{\mu}\left[s'_{\mu}p_{\nu} + s'_{\nu}p_{\mu} - g_{\mu\nu}(s'\cdot p)\right] \quad , \\
X(\nu) &= \mathrm{Tr}\{(A_{AV} + B_{AV}\gamma_5)\gamma_5\gamma^{\mu}\not{k}\gamma^{\nu}\not{k}'\} \\
&= -4\mathrm{i}A_{AV}\varepsilon^{\mu\alpha\nu\beta}k_{\alpha}k'_{\beta} + 4B_{AV}\left[k^{\mu}k'^{\nu} + k^{\nu}k'^{\mu} - g^{\mu\nu}(k\cdot k')\right] \quad , \\
X(\mu,\mathrm{e})X(\nu) &= -16m_{\mu}B_{AV}\left[(k\cdot p)(k'\cdot s') + (k\cdot s')(k'\cdot p)\right] \\
&\quad - 8A_{AV}\varepsilon_{\mu\sigma\nu\tau}\varepsilon^{\mu\alpha\nu\beta}p'^{\sigma}p^{\tau}k_{\alpha}k'_{\beta} \quad .
\end{aligned} \tag{19}
$$

- $\hat{O}^i = \gamma^{\mu}$, $\hat{O}^k = \gamma_5\gamma^{\nu}$:

$$
\begin{aligned}
X(\mu,\mathrm{e}) &= \frac{1}{2}\mathrm{Tr}\{\gamma_{\mu}\not{p}'\gamma_5\gamma_{\nu}\not{p}\} + \frac{1}{2}m_{\mu}\mathrm{Tr}\{\gamma_{\mu}\gamma_5\not{s}'\gamma_5\gamma_{\nu}\not{p}\} \\
&= 2\mathrm{i}\varepsilon_{\nu\sigma\mu\tau}p'^{\sigma}p^{\tau} - 2m_{\mu}\left[s'_{\mu}p_{\nu} + s'_{\nu}p_{\mu} - g_{\mu\nu}(s'\cdot p)\right] \quad , \\
X(\nu) &= \mathrm{Tr}\{(A_{VA} + B_{VA}\gamma_5)\gamma^{\mu}\not{k}\gamma_5\gamma^{\nu}\not{k}'\} \\
&= -4\mathrm{i}A_{VA}\varepsilon^{\nu\beta\mu\alpha}k_{\alpha}k'_{\beta} + 4B_{VA}\left[k^{\mu}k'^{\nu} + k^{\nu}k'^{\mu} - g^{\mu\nu}(k\cdot k')\right] \quad , \\
X(\mu,\mathrm{e})X(\nu) &= -16m_{\mu}B_{VA}\left[(k\cdot p)(k'\cdot s') + (k\cdot s')(k'\cdot p)\right] \\
&\quad + 8A_{VA}\varepsilon_{\nu\sigma\mu\tau}\varepsilon^{\nu\beta\mu\alpha}p'^{\sigma}p^{\tau}k_{\alpha}k'_{\beta} \quad .
\end{aligned} \tag{20}
$$

- $\hat{O}^i = \gamma_5\gamma^{\mu}$, $\hat{O}^k = \gamma_5\gamma^{\nu}$:

$$
\begin{aligned}
X(\mu,\mathrm{e}) &= \frac{1}{2}\mathrm{Tr}\{\gamma_5\gamma_{\mu}\not{p}'\gamma_5\gamma_{\nu}\not{p}\} + \frac{1}{2}m_{\mu}\mathrm{Tr}\{\gamma_5\gamma_{\mu}\gamma_5\not{s}'\gamma_5\gamma_{\nu}\not{p}\} \\
&= 2\left[p'_{\mu}p_{\nu} + p'_{\nu}p_{\mu} - g_{\mu\nu}(p\cdot p')\right] + 2\mathrm{i}m_{\mu}\varepsilon_{\mu\sigma\nu\tau}s'^{\sigma}p^{\tau} \quad ,
\end{aligned}
$$

Exercise 2.1

$$X(\nu) = \mathrm{Tr}\{(A_{\mathrm{AA}} + B_{\mathrm{AA}}\gamma_5)\gamma_5\gamma^\mu \not{k}\gamma_5\gamma^\nu \not{k}'\}$$
$$= 4A_{\mathrm{AA}}[k^\mu k'^\nu + k^\nu k'^\mu - g^{\mu\nu}(k\cdot k')] - 4\mathrm{i}B_{\mathrm{AA}}\varepsilon^{\nu\beta\mu\alpha}k_\alpha k'_\beta \quad , \tag{21}$$

$$X(\mu,\mathrm{e})X(\nu) = 16A_{\mathrm{AA}}[(k\cdot p)(k'\cdot p') + (k'\cdot p)(k\cdot p')]$$
$$- 8m_\mu B_{\mathrm{AA}}\varepsilon_{\mu\sigma\nu\tau}\varepsilon^{\nu\beta\mu\alpha}s'^\sigma p^\tau k_\alpha k'_\beta \quad .$$

Combining the last four results, we find that

$$\sum_{i,k=\mathrm{V,A}} X(\mu,e)X(\nu)$$
$$= 16(A_{\mathrm{VV}} + A_{\mathrm{AA}})[(k\cdot p)(k'\cdot p') + (k\cdot p')(k'\cdot p)]$$
$$- 16m_\mu(B_{\mathrm{AV}} + B_{\mathrm{VA}})[(k\cdot p)(k'\cdot s') + (k\cdot s')(k'\cdot p)] + X^{\alpha\beta}k_\alpha k'_\beta$$
$$= 16(a_{\mathrm{V}} + a_{\mathrm{A}})[(k\cdot p)(k'\cdot p') + (k\cdot p')(k'\cdot p)]$$
$$+ 16m_\mu b'[(k\cdot p)(k'\cdot s') + (k\cdot s')(k'\cdot p)] + X^{\alpha\beta}k_\alpha k'_\beta \,. \tag{22}$$

Here $X^{\alpha\beta}$ contains all terms which are antisymmetric in the indices α and β. In the course of further evaluation $k_\alpha k'_\beta$ yields the symmetric tensor $I_{\alpha\beta}$ and therefore the term containing $X^{\alpha\beta}$ will vanish. Again, with respect to the convention (23) of Supplement 2.12, we have $b' = -(B_{\mathrm{VA}} + B_{\mathrm{AV}})$.

• $\hat{O}^i = \sigma^{\mu\nu}$, $\hat{O}^k = \sigma^{\bar{\mu}\bar{\nu}}$:
This case requires the evaluation of

$$\mathrm{Tr}\{\sigma^{\mu\nu}\gamma^\alpha\sigma^{\bar{\mu}\bar{\nu}}\gamma^\beta\} \quad . \tag{23}$$

For this purpose we first consider

$$\mathrm{Tr}\{\mathrm{i}\gamma^\mu\gamma^\nu\gamma^\alpha\mathrm{i}\gamma^{\bar{\mu}}\gamma^{\bar{\nu}}\gamma^\beta\} \quad . \tag{24}$$

We use

$$\sigma^{\mu\nu} = \frac{\mathrm{i}}{2}(\gamma^\mu\gamma^\nu - \gamma^\nu\gamma^\mu) \tag{25}$$

and antisymmetrize (24) with respect to the indices μ and ν (that is, exchange μ and ν, subtract the result from the original term, and finally divide by 2) and then with respect to the indices $\bar{\mu}$ and $\bar{\nu}$. Finally, by repeated application of (A.33), we obtain

$$-\mathrm{Tr}\{\gamma^\mu\gamma^\nu\gamma^\alpha\gamma^{\bar{\mu}}\gamma^{\bar{\nu}}\gamma^\beta\}$$
$$= -(g^{\mu\nu}\,\mathrm{Tr}\{\gamma^\alpha\gamma^{\bar{\mu}}\gamma^{\bar{\nu}}\gamma^\beta\} - g^{\mu\alpha}\,\mathrm{Tr}\{\gamma^\nu\gamma^{\bar{\mu}}\gamma^{\bar{\nu}}\gamma^\beta\}$$
$$+ g^{\mu\bar{\mu}}\,\mathrm{Tr}\{\gamma^\nu\gamma^\alpha\gamma^{\bar{\nu}}\gamma^\beta\} - g^{\mu\bar{\nu}}\,\mathrm{Tr}\{\gamma^\nu\gamma^\alpha\gamma^{\bar{\mu}}\gamma^\beta\} + g^{\mu\beta}\,\mathrm{Tr}\{\gamma^\nu\gamma^\alpha\gamma^{\bar{\mu}}\gamma^{\bar{\nu}}\})$$
$$= -4[g^{\mu\nu}(g^{\alpha\bar{\mu}}g^{\bar{\nu}\beta} - g^{\alpha\nu}g^{\mu\bar{\beta}} + g^{\alpha\beta}g^{\bar{\mu}\bar{\nu}})$$
$$- g^{\mu\alpha}(g^{\nu\bar{\mu}}g^{\bar{\nu}\beta} - g^{\nu\bar{\nu}}g^{\bar{\mu}\beta} + g^{\nu\beta}g^{\bar{\mu}\bar{\nu}})$$
$$+ g^{\mu\bar{\mu}}(g^{\nu\alpha}g^{\bar{\nu}\beta} - g^{\nu\bar{\nu}}g^{\alpha\beta} + g^{\nu\beta}g^{\alpha\bar{\nu}})$$
$$- g^{\mu\bar{\nu}}(g^{\nu\alpha}g^{\bar{\mu}\beta} - g^{\nu\bar{\mu}}g^{\alpha\beta} + g^{\nu\beta}g^{\alpha\bar{\mu}})$$
$$+ g^{\mu\beta}(g^{\nu\alpha}g^{\bar{\mu}\beta} - g^{\nu\bar{\mu}}g^{\alpha\bar{\nu}} + g^{\nu\bar{\nu}}g^{\alpha\bar{\mu}})] \quad . \tag{26}$$

Exercise 2.11

Owing to the procedure of antisymmetrization with respect to μ and ν, as well as to $\bar{\mu}$ and $\bar{\nu}$, all terms proportional to $g^{\mu\nu}$ and $g^{\bar{\mu}\bar{\nu}}$ vanish, so that we are left with

$$\begin{aligned}
&-\mathrm{Tr}\{\sigma^{\mu\nu}\gamma^{\alpha}\sigma^{\bar{\mu}\bar{\nu}}\gamma^{\beta}\} \\
&=4\big[(g^{\mu\bar{\mu}}g^{\nu\bar{\nu}}-g^{\mu\bar{\nu}}g^{\nu\bar{\mu}})g^{\alpha\beta}-g^{\mu\bar{\mu}}(g^{\nu\alpha}g^{\bar{\nu}\beta}+g^{\bar{\nu}\alpha}g^{\mu\beta}) \\
&\quad+g^{\mu\bar{\nu}}(g^{\nu\alpha}g^{\bar{\mu}\beta}+g^{\bar{\mu}\alpha}g^{\nu\beta})+g^{\bar{\mu}\nu}(g^{\mu\alpha}g^{\bar{\nu}\beta}+g^{\bar{\nu}\alpha}g^{\mu\beta}) \\
&\quad-g^{\bar{\nu}\nu}(g^{\mu\alpha}g^{\bar{\mu}\beta}+g^{\bar{\mu}\alpha}g^{\mu\beta})\big] \quad . \qquad (27)
\end{aligned}$$

For later purposes it is worth mentioning that this term is simply the antisymmetrized form of

$$8g^{\mu\bar{\mu}}\{g^{\nu\bar{\nu}}g^{\alpha\beta}-2g^{\nu\alpha}g^{\bar{\nu}\beta}-2g^{\bar{\nu}\alpha}g^{\nu\beta}\} \quad . \qquad (28)$$

This is easily checked by multiplying (27) or (28) by a term which itself is antisymmetric with respect to μ, ν and $\bar{\mu}$, $\bar{\nu}$; the two corresponding results are identical.

In order to evaluate the quantity $X(\mu,\mathrm{e})X(\nu)$ we need to consider

$$\begin{aligned}
&\mathrm{Tr}\{\sigma^{\mu\nu}\gamma^{\alpha}\sigma^{\bar{\mu}\bar{\nu}}\gamma^{\beta}\}\,\mathrm{Tr}\{\sigma_{\mu\nu}\gamma_{\varrho}\sigma_{\bar{\mu}\bar{\nu}}\gamma_{\sigma}\} \\
&\quad=8g^{\bar{\mu}\bar{\mu}}(g^{\nu\bar{\nu}}g^{\alpha\beta}-2g^{\nu\alpha}g^{\bar{\nu}\beta}-2g^{\bar{\nu}\alpha}g^{\nu\beta}) \\
&\qquad\times 4\big[(g_{\mu\bar{\mu}}g_{\nu\bar{\nu}}-g_{\mu\bar{\nu}}g_{\nu\bar{\mu}})g_{\varrho\sigma} \\
&\qquad-g_{\mu\bar{\mu}}(g_{\nu\varrho}g_{\bar{\nu}\sigma}+g_{\bar{\nu}\varrho}g_{\nu\sigma})+g_{\mu\bar{\nu}}(g_{\nu\varrho}g_{\bar{\mu}\sigma}+g_{\bar{\mu}\varrho}g_{\nu\sigma}) \\
&\qquad+g_{\bar{\mu}\nu}(g_{\mu\varrho}g_{\bar{\nu}\sigma}+g_{\bar{\nu}\varrho}g_{\mu\sigma})-g_{\bar{\nu}\nu}(g_{\mu\varrho}g_{\bar{\mu}\sigma}+g_{\bar{\mu}\varrho}g_{\mu\sigma})\big] \\
&\quad=32(g^{\nu\bar{\nu}}g^{\alpha\beta}-2g^{\nu\alpha}g^{\bar{\nu}\beta}-2g^{\bar{\nu}\alpha}g^{\nu\beta}) \\
&\qquad\times\big[3g_{\nu\bar{\nu}}g_{\varrho\sigma}-4(g_{\nu\varrho}g_{\bar{\nu}\sigma}+g_{\bar{\nu}\varrho}g_{\nu\sigma}) \\
&\qquad+g_{\nu\varrho}g_{\bar{\nu}\sigma}+g_{\bar{\nu}\varrho}g_{\nu\sigma}+g_{\nu\varrho}g_{\bar{\nu}\sigma}+g_{\bar{\nu}\varrho}g_{\nu\sigma}-2g_{\nu\bar{\nu}}g_{\varrho\sigma}\big] \\
&\quad=32(g^{\nu\bar{\nu}}g^{\alpha\beta}-2g^{\nu\alpha}g^{\bar{\nu}\beta}-2g^{\bar{\nu}\alpha}g^{\nu\beta})(g_{\nu\bar{\nu}}g_{\varrho\sigma}-2g_{\nu\varrho}g_{\bar{\nu}\sigma}-2g_{\bar{\nu}\varrho}g_{\nu\sigma}) \\
&\quad=32(4g^{\alpha\beta}g_{\varrho\sigma}-2g^{\alpha\beta}g_{\varrho\sigma}-2g^{\alpha\beta}g_{\varrho\sigma}-2g^{\alpha\beta}g_{\varrho\sigma} \\
&\qquad+4\delta^{\alpha}_{\varrho}\delta^{\beta}_{\sigma}+4\delta^{\alpha}_{\sigma}\delta^{\beta}_{\varrho}-2g^{\alpha\beta}g_{\varrho\sigma}+4\delta^{\alpha}_{\sigma}\delta^{\beta}_{\varrho}+4\delta^{\alpha}_{\varrho}\delta^{\beta}_{\sigma}) \\
&\quad=128(-g^{\alpha\beta}g_{\varrho\sigma}+2\delta^{\alpha}_{\varrho}\delta^{\beta}_{\sigma}+2\delta^{\alpha}_{\varrho}\delta^{\beta}_{\sigma}) \quad . \qquad (29)
\end{aligned}$$

The evaluation of (6), or of (9), furthermore contains terms of the form

$$\mathrm{Tr}\{\gamma_5\sigma^{\mu\nu}\gamma^{\alpha}\sigma^{\bar{\mu}\bar{\nu}}\gamma^{\beta}\} \quad . \qquad (30)$$

However, these can be reduced to (27) by employing the relation

$$\gamma_5\sigma^{\mu\nu}=\frac{\mathrm{i}}{2}\varepsilon^{\mu\nu\varrho\tau}\sigma_{\varrho\tau} \quad , \qquad (31)$$

so that we obtain

$$\begin{aligned}
&\mathrm{Tr}\{\gamma_5\sigma^{\mu\nu}\gamma^{\alpha}\sigma^{\bar{\mu}\bar{\nu}}\gamma^{\beta}\}=\frac{\mathrm{i}}{2}\varepsilon^{\mu\nu\varrho\tau}\,\mathrm{Tr}\{\sigma_{\varrho\tau}\gamma^{\alpha}\sigma^{\bar{\mu}\bar{\nu}}\gamma^{\beta}\} \\
&\quad=\frac{\mathrm{i}}{2}\varepsilon^{\mu\nu\varrho\tau}\times 4\big[(\delta^{\bar{\mu}}_{\varrho}\delta^{\bar{\nu}}_{\tau}-\delta^{\bar{\mu}}_{\tau}\delta^{\bar{\nu}}_{\varrho})g^{\alpha\beta}-\delta^{\bar{\mu}}_{\varrho}(\delta^{\alpha}_{\tau}g^{\bar{\nu}\beta}+\delta^{\beta}_{\tau}g^{\bar{\nu}\alpha})
\end{aligned}$$

$$+\delta^{\bar\nu}_{\varrho}(\delta^{\alpha}_{\tau}g^{\bar\mu\beta}+\delta^{\beta}_{\tau}g^{\bar\mu\alpha})+\delta^{\bar\mu}_{\tau}(\delta^{\alpha}_{\varrho}g^{\bar\nu\beta}+\delta^{\beta}_{\varrho}g^{\bar\nu\alpha})$$
$$-\delta^{\bar\nu}_{\tau}(\delta^{\alpha}_{\sigma}g^{\bar\mu\beta}+\delta^{\beta}_{\sigma}g^{\bar\mu\alpha})]$$
$$=4\mathrm{i}(\varepsilon^{\mu\nu\bar\mu\bar\nu}g^{\alpha\beta}-\varepsilon^{\mu\nu\bar\mu\alpha}g^{\bar\nu\beta}-\varepsilon^{\mu\nu\bar\mu\beta}g^{\bar\nu\alpha}+\varepsilon^{\mu\nu\bar\nu\alpha}g^{\bar\mu\beta}+\varepsilon^{\mu\nu\bar\nu\beta}g^{\bar\mu\alpha}) \quad . \tag{32}$$

Another typical term that occurs in $X(\mu,\mathrm{e})X(\nu)$ is

$$\mathrm{Tr}\{\gamma_5\sigma^{\mu\nu}\gamma^{\alpha}\sigma^{\bar\mu\bar\nu}\gamma^{\beta}\}\,\mathrm{Tr}\{\sigma_{\mu\nu}\gamma_{\varrho}\sigma_{\bar\mu\bar\nu}\gamma_{\sigma}\} \quad . \tag{33}$$

However, it is easily verified that this contribution vanishes. The first factor is again antisymmetric with respect to μ and ν and also to $\bar\mu$ and $\bar\nu$. Thus, for the second trace we may substitute the expression (28), which leads to

$$\mathrm{Tr}\{\gamma_5\sigma^{\mu\nu}\gamma^{\alpha}\sigma^{\bar\mu\bar\nu}\gamma^{\beta}\}\cdot\mathrm{Tr}\{\sigma_{\mu\nu}\gamma_{\varrho}\sigma_{\bar\mu\bar\nu}\gamma_{\sigma}\}$$
$$=4\mathrm{i}(\varepsilon^{\mu\nu\bar\mu\bar\nu}g^{\alpha\beta}-\varepsilon^{\mu\nu\bar\mu\alpha}g^{\bar\nu\beta}-\varepsilon^{\mu\nu\bar\mu\beta}g^{\bar\nu\alpha}+\varepsilon^{\mu\nu\bar\nu\alpha}g^{\bar\mu\beta}+\varepsilon^{\mu\nu\bar\nu\beta}g^{\bar\mu\alpha})$$
$$\times 8g_{\mu\bar\mu}(g_{\nu\bar\nu}g_{\varrho\sigma}-2g_{\nu\varrho}g_{\bar\nu\sigma}-2g_{\nu\sigma}g_{\bar\nu\varrho})$$
$$=-32\mathrm{i}(\varepsilon^{\beta\nu\bar\nu\alpha}+\varepsilon^{\alpha\nu\bar\nu\beta})(g_{\nu\bar\nu}g_{\varrho\sigma}-2g_{\nu\varrho}g_{\bar\nu\sigma}-2g_{\nu\sigma}g_{\bar\nu\varrho})=0 \quad , \tag{34}$$

Using (28) and the relation

$$\varepsilon_{\mu\nu\gamma\omega}\varepsilon^{\mu\nu\lambda\tau}=2(\delta^{\lambda}_{\omega}\delta^{\tau}_{\gamma}-\delta^{\lambda}_{\gamma}\delta^{\tau}_{\omega}) \quad , \tag{35}$$

we finally evaluate the following expression

$$\mathrm{Tr}\{\gamma_5\sigma_{\mu\nu}\gamma_{\varrho}\sigma_{\bar\mu\bar\nu}\gamma_{\sigma}\}\,\mathrm{Tr}\{\gamma_5\sigma^{\mu\nu}\gamma^{\alpha}\sigma^{\bar\mu\bar\nu}\gamma^{\beta}\}$$
$$=-\frac{1}{4}\varepsilon_{\mu\nu\gamma\omega}\varepsilon^{\mu\nu\lambda\tau}\,\mathrm{Tr}\{\sigma^{\gamma\omega}\gamma_{\varrho}\sigma_{\bar\mu\bar\nu}\gamma_{\varrho}\}\,\mathrm{Tr}\{\sigma_{\lambda\tau}\gamma^{\alpha}\sigma^{\bar\mu\bar\nu}\gamma^{\beta}\}$$
$$=\frac{1}{2}(\delta^{\lambda}_{\gamma}\delta^{\tau}_{\omega}-\delta^{\lambda}_{\omega}\delta^{\tau}_{\gamma})\,\mathrm{Tr}\{\sigma^{\gamma\omega}\gamma_{\varrho}\sigma_{\bar\mu\bar\nu}\gamma_{\sigma}\}\,\mathrm{Tr}\{\sigma_{\lambda\tau}\gamma^{\alpha}\sigma^{\bar\mu\bar\nu}\gamma^{\beta}\}$$
$$=\mathrm{Tr}\{\sigma^{\gamma\omega}\gamma_{\varrho}\sigma_{\bar\mu\bar\nu}\gamma_{\sigma}\}\,\mathrm{Tr}\{\sigma_{\gamma\omega}\gamma^{\alpha}\sigma^{\bar\mu\bar\nu}\gamma^{\beta}\}$$
$$=\mathrm{Tr}\{\sigma_{\mu\nu}\gamma_{\varrho}\sigma_{\bar\mu\bar\nu}\gamma_{\sigma}\}\,\mathrm{Tr}\{\sigma^{\mu\nu}\gamma^{\alpha}\sigma^{\bar\mu\bar\nu}\gamma^{\beta}\} \quad . \tag{36}$$

This result exactly coincides with the one we previously obtained in (29). Now we have all the ingredients necessary to consider the contribution of tensor coupling $\hat O^i=\sigma^{\mu\nu}$, $\hat O^k=\sigma^{\bar\mu\bar\nu}$.
With respect to (6) we obtain

$$X(\nu)=A_{\mathrm{TT}}\,\mathrm{Tr}\{\sigma^{\mu\nu}\gamma^{\alpha}\sigma^{\bar\mu\bar\nu}\gamma^{\beta}\}k_{\alpha}k'_{\beta}-B_{\mathrm{TT}}\,\mathrm{Tr}\{\gamma_5\sigma^{\mu\nu}\gamma^{\alpha}\sigma^{\bar\mu\bar\nu}\gamma^{\beta}\}k_{\alpha}k'_{\beta}$$
$$=a_{\mathrm{T}}\,\mathrm{Tr}\{\sigma^{\mu\nu}\gamma^{\alpha}\sigma^{\bar\mu\bar\nu}\gamma^{\beta}\}k_{\alpha}k'_{\beta}-c'\,\mathrm{Tr}\{\gamma_5\sigma^{\mu\nu}\gamma^{\alpha}\sigma^{\bar\mu\bar\nu}\gamma^{\beta}\}k_{\alpha}k'_{\beta} \quad , \tag{37}$$

where we have again adopted the abbreviations (2.77) and (2.83). The contribution of the massive leptons is given by

$$X(\mu,\mathrm{e})=\frac{1}{2}\,\mathrm{Tr}\{\sigma_{\mu\nu}\gamma_{\varrho}\sigma_{\bar\mu\bar\nu}\gamma_{\sigma}\}p'^{\varrho}p^{\sigma}+\frac{1}{2}m_{\mu}\,\mathrm{Tr}\{\gamma_5\sigma_{\mu\nu}\gamma_{\varrho}\sigma_{\bar\mu\bar\nu}\gamma_{\sigma}\}s'^{\varrho}p^{\sigma} \quad . \tag{38}$$

Exercise 2.11

All other terms vanish, since they contain an uneven number of γ matrices. We recall that the expression (33) does not contribute, so that we obtain

$$\begin{aligned}\sum_{\mathrm{T}} X(\nu)X(\mu,\mathrm{e}) = \frac{1}{4}\Bigg[&\frac{1}{2}a_{\mathrm{T}}k_\alpha k'_\beta p'^\varrho p^\sigma\, \mathrm{Tr}\{\sigma^{\mu\nu}\gamma^\alpha\sigma^{\bar{\mu}\bar{\nu}}\gamma^\beta\}\,\mathrm{Tr}\{\sigma_{\mu\nu}\gamma_\varrho\sigma_{\bar{\mu}\bar{\nu}}\gamma_\sigma\}\\ &-\frac{1}{2}m_\mu c' k_\alpha k'_\beta s'^\varrho p^\sigma\, \mathrm{Tr}\{\gamma_5\sigma^{\mu\nu}\gamma^\alpha\sigma^{\bar{\mu}\bar{\nu}}\gamma^\beta\}\\ &\times \mathrm{Tr}\{\gamma_5\sigma_{\mu\nu}\gamma_\varrho\sigma_{\bar{\mu}\bar{\nu}}\gamma_\sigma\}\Bigg] \quad . \end{aligned} \tag{39}$$

Here we have introduced a factor $\frac{1}{4}$ in order to avoid double counting of $\sigma^{\mu\nu}$, or $\sigma^{\bar{\mu}\bar{\nu}}$, since the sum includes $\sigma^{\mu\nu}$ as well as $\sigma^{\nu\mu} = -\sigma^{\mu\nu}$!

Equation (39) may be further reduced by using (29) and (36):

$$\begin{aligned}&\sum_{\mathrm{T}} X(\nu)X(\mu,\mathrm{e})\\ &\quad = 16a_{\mathrm{T}}k_\alpha k'_\beta p'^\varrho p^\sigma(-g^{\alpha\beta}g_{\varrho\sigma} + 2\delta^\alpha_\varrho\delta^\beta_\sigma + 2\delta^\alpha_\sigma\delta^\beta_\varrho)\\ &\quad\quad - 16m_\mu c' k_\alpha k'_\beta s'^\varrho p^\sigma(-g^{\alpha\beta}g_{\varrho\sigma} + 2\delta^\alpha_\varrho\delta^\beta_\sigma + 2\delta^\alpha_\sigma\delta^\beta_\varrho)\\ &\quad = 16a_{\mathrm{T}}\big[-(k\cdot k')(p\cdot p') + 2(k\cdot p)(k'\cdot p') + 2(k\cdot p')(k'\cdot p)\big]\\ &\quad\quad - 16m_\mu c'\big[-(k\cdot k')(s'\cdot p') + 2(k\cdot s')(k'\cdot p) + 2(k\cdot p)(k'\cdot s')\big] \quad . \end{aligned} \tag{40}$$

With the following argument we can conclude that all other combinations of $\hat{O}^i$ and $\hat{O}^k$ do not contribute: if for example we identify $\hat{O}^i$ with V or A, then $\hat{O}^k$ can neither be S nor P nor T, since otherwise $X(\nu)$ in (6) would contain an uneven number of γ matrices. On the other hand, all remaining combinations lead to an $X(\nu)$ which is antisymmetric with respect to the exchange of k and k', for example, for the combination "ST",

$$\begin{aligned}&\mathrm{Tr}\big\{(A_{\mathrm{ST}} - B_{\mathrm{ST}}\gamma_5)\cdot 1\cdot \not{k}\sigma^{\mu\nu}\not{k}'\big\}\\ &\quad = 4\mathrm{i}A_{\mathrm{ST}}(k^\mu k'^\nu - k^\nu k'^\mu) + 4B_{\mathrm{ST}}\varepsilon^{\alpha\mu\nu\beta}k_\alpha k'_\beta \quad . \end{aligned} \tag{41}$$

As we have already mentioned in connection with (18), such terms do not contribute to the decay rate.

Combining the previous results (15), (22) and (40) as well as the terms of (41), we find that

$$\begin{aligned}&\sum_{i,k} X(\nu)X(\mu,\mathrm{e})\\ &\quad = \big\{8g^{\alpha\beta}\big[(a_{\mathrm{S}} + a_{\mathrm{P}})(p\cdot p') - a' m_\mu(p\cdot s')\big]\\ &\quad\quad + 16(a_{\mathrm{V}} + a_{\mathrm{A}})\big[p^\alpha p'^\beta + p'^\alpha p^\beta\big] + 16m_\mu b'\big[p^\alpha s'^\beta + s'^\alpha p^\beta\big]\\ &\quad\quad + 16a_{\mathrm{T}}\big[-g^{\alpha\beta}(p\cdot p') + 2p^\alpha p'^\beta + 2p'^\alpha p^\beta\big]\\ &\quad\quad - 16m_\mu c'\big[-g^{\alpha\beta}(s'\cdot p) + 2s'^\alpha p^\beta + 2p^\alpha s'^\beta\big]\\ &\quad\quad + Y^{\alpha\beta}\big\}k_\alpha k'_\beta \quad , \end{aligned} \tag{42}$$

where $Y^{\alpha\beta}$ is an antisymmetric tensor which contains terms like $X^{\alpha\beta}$ of (22) as well as the contribution corresponding to (41).

In (2.18) we now replace

$$\sum_{t,t'} |M|^2$$

by $X(\mu, e)X(\nu)$ of (42) thereby abbreviating the last term by $Z^{\alpha\beta}k_\alpha k_{\beta'}$. This results in

$$\mathrm{d}W = \frac{G^2}{2}\frac{1}{(2\pi)^5}\frac{d^3p}{2p'^0 2p^0}\int\frac{\mathrm{d}^3k}{2k^0}\int\frac{\mathrm{d}^3k'}{2k'^0}\delta^4(p+k+k'-p')Z^{\alpha\beta}k_\alpha k'_\beta \ . \tag{43}$$

Now, utilizing (2.51)–(2.53), we employ (2.49), according to which it holds that

$$\begin{aligned} I_{\alpha\beta} &\equiv \int\frac{\mathrm{d}^3k}{2k^0}\int\frac{\mathrm{d}^3k'}{2k'^0}k_\alpha k'_\beta\delta^4(k+k'-q) \\ &= \frac{\pi}{24}(q^2 g_{\alpha\beta} + 2q_\alpha q_\beta)\Theta(p^0_{\max} - p^0) \ . \end{aligned} \tag{44}$$

The quantity

$$\mathrm{d}W = \frac{G^2}{2(2\pi)^5}\frac{d^3p}{(2p'^0 2p^0)}I_{\alpha\beta}Z^{\alpha\beta} \tag{45}$$

is now easily evaluated. Neglecting the electron rest mass, the rest frame of the muon is again characterized by

$$\begin{aligned} &p'^0 = m_\mu \ , \quad \boldsymbol{p}' = 0 \ , \\ &(p\cdot p') = m_\mu p^0 \ , \quad (p'\cdot s') = 0 \ , \\ &(p\cdot s') = -\boldsymbol{p}\cdot\boldsymbol{s}' = -|\boldsymbol{p}|\cos\theta = -p^0\cos\theta \ . \end{aligned} \tag{46}$$

Introducing $q = p' - p$, from these relations we obtain

$$\begin{aligned} &\frac{\pi}{24}(q^2 g_{\alpha\beta} + 2q_\alpha q_\beta)g^{\alpha\beta} = \frac{\pi}{24}6m_\mu(m_\mu + 2p^0) \ , \\ &\frac{\pi}{24}(q^2 g_{\alpha\beta} + 2q_\alpha q_\beta)p^\alpha p'^\beta = \frac{\pi}{24}(3m_\mu^3 p^0 - 4m_\mu^2(p^0)^2) \ , \\ &\frac{\pi}{24}(q^2 g_{\alpha\beta} + 2q_\alpha q_\beta)p^\alpha s'^\beta = -\frac{\pi}{24}m_\mu(m_\mu - 4p^0)(p^0)\cos\theta \end{aligned} \tag{47}$$

and finally

$$\begin{aligned} \mathrm{d}W = {} & \frac{G^2}{2}\frac{1}{(2\pi)^5}\frac{d^3p}{2p'^0 2p^0}\frac{\pi}{24}\Theta(p^0 - p^0_{\max}) \\ & \times\{[48(a_\mathrm{S} + a_\mathrm{P}) + 96a_\mathrm{T} + 96(a_\mathrm{V} + a_\mathrm{A})]m_\mu^3 p^0 \\ & - [96(a_\mathrm{S} + a_\mathrm{P}) + 64a_\mathrm{T} + 128(a_\mathrm{V} + a_\mathrm{A})]m_\mu^2(p^0)^2 \\ & + (48a' - 32b' - 32c')m_\mu^3 p^0\cos\theta \\ & - (96a' - 128b' + 64c')m_\mu^2(p^0)^2\cos\theta\} \ . \end{aligned} \tag{48}$$

Neglecting the electron mass implies that

$$p^0_{\text{max}} = \frac{m_\mu}{2} \quad , \tag{49}$$

so that

$$p^0 = x p^0_{\text{max}} = x \frac{m_\mu}{2} \tag{50}$$

and

$$\mathrm{d}^3 p = |\boldsymbol{p}|^2 \mathrm{d}|\boldsymbol{p}| \mathrm{d}\Omega = (p^0)^2 \,\mathrm{d}p^0 \mathrm{d}\Omega = \frac{m_\mu^3}{8} x^2 \mathrm{d}x \,\mathrm{d}\Omega \quad . \tag{51}$$

We collect all these expressions and substitute them into (48). The final result is

$$\begin{aligned} \mathrm{d}W = {} & \frac{G^2 m_\mu^5}{192\pi^4} x^2 \mathrm{d}x \,\mathrm{d}\Omega \frac{1}{16} \Big\{ \big[3(a_\mathrm{S} + a_\mathrm{P}) + 6a_\mathrm{T} + 6(a_\mathrm{V} + a_\mathrm{A})\big] \\ & - \big[3(a_\mathrm{S} + a_\mathrm{P}) + 2a_\mathrm{T} + 4(a_\mathrm{V} + a_\mathrm{A})\big] \cdot x + [3a' - 2b' - 2c'] \cos\theta \\ & - [3a' - 4b' + 2c'] x \cos\theta \Big\} \Theta(1 - x) \quad . \end{aligned} \tag{52}$$

This agrees with (2.81b), as is easily verified by inserting the Michel parameters and using (2.78).

MATHEMATICAL SUPPLEMENT

2.12 The Fierz Transformation

Within the framework of the Fermi theory there are two different but equivalent ways of describing a reaction $\psi_1 + \psi_2 \to \psi_3 + \psi_4$, namely

$$(\bar{\psi}_3 \hat{\Gamma} \psi_1)(\bar{\psi}_4 \hat{\Gamma} \psi_2) \quad \text{and} \quad (\bar{\psi}_4 \hat{\Gamma} \psi_1)(\bar{\psi}_3 \hat{\Gamma} \psi_2) \quad . \tag{1}$$

The properties of the Clifford algebra[11] allow us to form 16 matrices

$$\{1, \gamma_\mu, \sigma_{\mu\nu}, \gamma_5\gamma_\mu, \mathrm{i}\gamma_5\} =: \{\hat{O}_1, \ldots, \hat{O}_{16}\} \quad . \tag{2a}$$

$$\{1, \gamma^\mu, \sigma^{\mu\nu}, \gamma_5\gamma^\mu, \mathrm{i}\gamma_5\} =: \{\hat{O}^1, \ldots, \hat{O}^{16}\} \quad . \tag{2b}$$

which form a basis for any 4×4 matrix. Furthermore it holds that

$$\gamma_0 \hat{O}_i^\dagger \gamma_0 = \hat{O}_i \quad . \tag{3}$$

Hence we may expand $\hat{\Gamma}$ in terms of the $\hat{O}^i$.

$$\sum_{i=1}^{16} C_i \bar{\psi}_3 \hat{O}_i \psi_1 \bar{\psi}_4 \hat{O}^i \psi_2$$

[11] See W. Greiner: *Relativistic Quantum Mechanics – Wave Equations*, 3rd ed. (Springer, Berlin, Heidelberg, 2000).

Mathematical Supplement 2.1

or

$$\sum_{i=1}^{16} C'_i \bar{\psi}_4 \hat{O}_i \psi_1 \bar{\psi}_3 \hat{O}^i \psi_2 \quad . \tag{4a}$$

The requirement for Lorentz invariance demands that

$$C_2 = \cdots = C_5 \quad , \quad C_6 = C_7 = \cdots = C_{11} \quad , \quad C_{12} = \cdots = C_{15} \quad . \tag{4b}$$

Since the two representations (4a) are equivalent, these expressions must be identical for arbitrary values of ψ_1, ψ_2 ,ψ_3, and ψ_4. In terms of the particular components this implies that

$$\sum_{i=1}^{16} C_i (\hat{O}_i)_{\gamma\alpha} (\hat{O}^i)_{\delta\beta} = \sum_{i=1}^{16} C'_i (\hat{O}_i)_{\delta\alpha} (\hat{O}^i)_{\gamma\beta} \quad . \tag{5}$$

In the following steps we will solve this equation for C_i, which requires the determination of the transformation matrix Λ_{ij} connecting the two representations, that is,

$$C_i = \sum_j \Lambda_{ij} C'_j \quad . \tag{6}$$

The transformation from the C'_j to the C_i (or vice versa) is called the **Fierz** transformation.

Multiplying (5) by $(O^l)^{\alpha\gamma}(O_l)^{\beta\delta}$ and summing over α, β, γ, and δ yields

$$\sum_{i=1}^{16} C_i \operatorname{Tr}\{\hat{O}_i \hat{O}^l\} \operatorname{Tr}\{\hat{O}^i \hat{O}_l\} = \sum_{j=1}^{16} \operatorname{Tr}\{\hat{O}_j \hat{O}^l \hat{O}^j \hat{O}_l\} C'_j \quad . \tag{7}$$

We employ the following formulas (see Appendix A.2):

$$\begin{aligned}
\operatorname{Tr}\{1\} &= 4 \quad , \\
\operatorname{Tr}\{\sigma_{\mu\nu}\} &= \operatorname{Tr}\{\gamma_\mu\} = \operatorname{Tr}\{\mathrm{i}\gamma_5\} \\
&= \operatorname{Tr}\{\gamma_5\gamma_\nu\} = 0 \quad , \\
\operatorname{Tr}\{\gamma_\mu \gamma^\nu\} &= 4 g_\mu{}^\nu \quad , \\
\operatorname{Tr}\{\gamma^\mu \sigma_{\mu\nu}\} &= \operatorname{Tr}\{\gamma^\mu \mathrm{i}\gamma_5\} = \operatorname{Tr}\{\gamma^\mu \gamma_5 \gamma_\nu\} = 0 \quad , \\
\operatorname{Tr}\{\sigma_{\mu\nu}\sigma^{\lambda\varrho}\} &= -\frac{1}{4} \operatorname{Tr}\{[\gamma_\mu, \gamma_\nu]\gamma^\lambda\gamma^\varrho - [\gamma_\mu, \gamma_\nu]\gamma^\varrho\gamma^\lambda\} \\
&= -\{g_{\mu\nu} g^{\lambda\varrho} + g_\mu{}^\varrho g_\nu{}^\lambda - g_\mu{}^\lambda g_\nu{}^\varrho \\
&\quad - g_{\mu\nu} g^{\lambda\varrho} - g_\nu{}^\varrho g_\mu{}^\lambda + g_\nu{}^\lambda g_\mu{}^\varrho \\
&\quad - g_{\mu\nu} g^{\varrho\lambda} - g_\mu{}^\lambda g_\nu{}^\varrho + g_\mu{}^\varrho g_\nu{}^\lambda \\
&\quad + g_{\mu\nu} g^{\varrho\lambda} + g_\nu{}^\lambda g_\mu{}^\varrho - g_\nu{}^\varrho g_\mu{}^\lambda\} \\
&= 4\{g_\mu{}^\lambda g_\nu{}^\varrho - g_\nu{}^\lambda g_\mu{}^\varrho\} \quad , \\
\operatorname{Tr}\{\sigma_{\mu\nu}\gamma_5\gamma^\lambda\} &= \operatorname{Tr}\{\sigma_{\mu\nu}\mathrm{i}\gamma_5\} = 0 \quad ,
\end{aligned}$$

$$\mathrm{Tr}\{\mathrm{i}\gamma_5\gamma_\mu\mathrm{i}\gamma_5\gamma^\nu\} = \mathrm{Tr}\{\gamma_\mu\gamma^\nu\} = 4g_\mu{}^\nu = 4\delta_{\mu\nu} \quad ,$$

$$\mathrm{Tr}\{\mathrm{i}\gamma_5\gamma_\mu\gamma_5\} = 0 \quad ,$$

$$\mathrm{Tr}\{\gamma_5\gamma_5\} = 4 \quad .$$

All these relations may be combined to give

$$\mathrm{Tr}\{\hat{O}^i\,\hat{O}_l\} = 4\delta^i{}_l\varepsilon_l \quad , \quad \varepsilon_l = \begin{cases} +1 & \text{for} \quad l = 1, \ldots, 11 \\ -1 & \text{for} \quad l = 12, \ldots, 16 \end{cases} \quad . \tag{8}$$

Inserting (8) into (7), we then have

$$C_l = \frac{1}{16}\sum_{j=1}^{16} C'_j\,\mathrm{Tr}\{\hat{O}_j\hat{O}^l\hat{O}^j\hat{O}_l\} \quad . \tag{9}$$

There remains the evaluation of

$$\tilde{\Lambda}_{lj} = \frac{1}{16}\,\mathrm{Tr}\{\hat{O}_j\hat{O}^l\hat{O}^j\hat{O}_l\} = \tilde{\Lambda}_{jl} \quad . \tag{10}$$

In order to solve for $\tilde{\Lambda}_{jl}$ we consider the particular cases separately.

- $j = 1$:

$$\tilde{\Lambda}_{1l} = 4\varepsilon_l\frac{1}{16} = \frac{1}{4}\varepsilon_l \quad , \tag{11}$$

according to (8).

- $j = 2, \ldots, 5;\, l = 2, \ldots, 5$:

$$\tilde{\Lambda}_{jl} = \frac{1}{16}\,\mathrm{Tr}\{\gamma_{j-2}\gamma^{l-2}\gamma^{j-2}\gamma_{l-2}\} = \frac{1}{4}\{2\delta_{jl} - 1\} \quad ,$$

$$(\tilde{\Lambda}_{jl}) = \frac{1}{4}\begin{pmatrix} +1 & -1 & -1 & -1 \\ -1 & +1 & -1 & -1 \\ -1 & -1 & +1 & -1 \\ -1 & -1 & -1 & +1 \end{pmatrix} \quad , \tag{12}$$

where the index j labels the rows and l the columns.

- $l = 6, \ldots, 11$:

$$\hat{O}_l = \mathrm{i}\gamma_\nu\gamma_\mu \quad (\nu \neq \mu) \quad ,$$

$$\hat{O}_6 = \mathrm{i}\gamma_0\gamma_1 \quad , \quad \hat{O}_7 = \mathrm{i}\gamma_0\gamma_2 \quad ,$$

$$\hat{O}_8 = \mathrm{i}\gamma_0\gamma_3 \quad , \quad \hat{O}_9 = \mathrm{i}\gamma_1\gamma_2 \quad ,$$

$$\hat{O}_{10} = \mathrm{i}\gamma_1\gamma_3 \quad , \quad \hat{O}_{11} = \mathrm{i}\gamma_2\gamma_3 \quad ,$$

$$\begin{aligned} \tilde{\Lambda}_{jl} &= -\frac{1}{16}\,\mathrm{Tr}\{\gamma_{j-2}\gamma^\nu\gamma^\mu\gamma^{j-2}\gamma_\nu\gamma_\mu\} \\ &= -\frac{1}{8}\delta^\nu_{j-2}\,\mathrm{Tr}(\gamma^\mu\gamma^{j-2}\gamma_\nu\gamma_\mu) + \frac{1}{16}\,\mathrm{Tr}(\gamma^\nu\gamma_{j-2}\gamma^\mu\gamma^{j-2}\gamma_\nu\gamma_\mu) \\ &= -\frac{1}{8}\delta^\nu_{j-2}\,\mathrm{Tr}(\gamma^{j-2}\gamma_\nu\gamma_\mu\gamma^\mu) + \frac{1}{8}\delta^\mu_{j-2}\,\mathrm{Tr}(\gamma^\nu\gamma^{j-2}\gamma_\nu\gamma_\mu) \end{aligned} \tag{13}$$

Mathematical Supplement 2.1

$$-\frac{1}{16}\,\mathrm{Tr}(\gamma^{\nu}\gamma^{\mu}\gamma_{j-2}\gamma^{j-2}\gamma_{\nu}\gamma_{\mu})$$

$$= -\frac{1}{8}\delta^{\nu}_{j-2}\,\mathrm{Tr}(\gamma^{j-2}\gamma_{\nu}) + \frac{1}{8}\delta^{\mu}_{j-2}\,\mathrm{Tr}(\gamma^{\nu}\gamma^{j-2}\gamma_{\nu}\gamma_{\mu}) - \frac{1}{16}\,\mathrm{Tr}(\gamma^{\nu}\gamma^{\mu}\gamma_{\nu}\gamma_{\mu}) \quad .$$

For the first term we get

$$-\frac{1}{8}\delta^{\nu}_{j-2}\,\mathrm{Tr}(\gamma^{j-2}\gamma_{\nu}) = -\frac{1}{8}\delta^{\nu}_{j-2}\cdot 4\delta^{j-2}_{\nu} = -\frac{1}{2}\delta^{\nu}_{j-2} \quad .$$

To evaluate the last two terms we take into consideration that $\mu \neq \nu$ and therefore $\gamma_{\mu}\gamma_{\nu} = -\gamma_{\nu}\gamma_{\mu}$, yielding

$$\begin{aligned}
\frac{1}{8}\delta^{\mu}_{j-2}\,\mathrm{Tr}(\gamma^{\nu}\gamma^{j-2}\gamma_{\nu}\gamma_{\mu}) &= -\frac{1}{8}\delta^{\mu}_{j-2}\,\mathrm{Tr}(\gamma^{\nu}\gamma^{j-2}\gamma_{\mu}\gamma_{\nu}) \\
&= -\frac{1}{8}\delta^{\mu}_{j-2}\,\mathrm{Tr}(\gamma_{\nu}\gamma^{\nu}\gamma^{j-2}\gamma_{\mu}) \\
&= -\frac{1}{8}\delta^{\mu}_{j-2}\,\mathrm{Tr}(\gamma^{j-2}\gamma_{\mu}) \\
&= -\frac{1}{8}\delta^{\mu}_{j-2}\cdot 4\delta^{j-2}_{\mu} \\
&= -\frac{1}{2}\delta^{\mu}_{j-2} \quad ,
\end{aligned}$$

in which we have used the fact that the trace is constant under cyclic permutation. For the third term we then obtain

$$\begin{aligned}
-\frac{1}{16}\,\mathrm{Tr}(\gamma^{\nu}\gamma^{\mu}\gamma_{\nu}\gamma_{\mu}) &= \frac{1}{16}\,\mathrm{Tr}(\gamma^{\nu}\gamma_{\nu}\gamma^{\mu}\gamma_{\mu}) \\
&= \frac{1}{16}\,\mathrm{Tr}(1) = \frac{1}{4} \quad ,
\end{aligned}$$

and, in summary, we finally have

$$\tilde{A}_{jl} = \frac{1}{4}(1 - 2\delta_{j-2,\nu} - 2\delta_{j-2,\mu}) \quad ,$$

$$(\tilde{A}_{jl}) = \frac{1}{4}\begin{pmatrix} -1 & -1 & -1 & +1 & +1 & +1 \\ -1 & +1 & +1 & -1 & -1 & +1 \\ +1 & -1 & +1 & -1 & +1 & -1 \\ +1 & +1 & -1 & +1 & -1 & -1 \end{pmatrix} \quad . \tag{14}$$

- $l = 12, \ldots, 15$:

$$\tilde{A}_{jl} = \frac{1}{16}\,\mathrm{Tr}\{\gamma_{j-2}\gamma_5\gamma^{l-12}\gamma^{j-2}\gamma_5\gamma_{l-12}\} \quad ,$$

$$(\tilde{A}_{jl}) = \frac{1}{4}\begin{pmatrix} +1 & -1 & -1 & -1 \\ -1 & +1 & -1 & -1 \\ -1 & -1 & +1 & -1 \\ -1 & -1 & -1 & +1 \end{pmatrix} \quad . \tag{15}$$

- $l = 16$:

$$\tilde{A}_{jl} = -\frac{1}{16}\,\mathrm{Tr}\{\gamma_{j-2}\gamma_5\gamma^{j-2}\gamma_5\} = \frac{1}{4} \quad . \tag{16}$$

athematical Supplement 2.12 • $j=6,\ldots,11;\, l=6,\ldots,11$:

$$\hat{O}_j = \mathrm{i}\gamma_\nu\gamma_\mu \quad \text{with} \quad \mu>\nu \quad ,$$
$$O_l = \mathrm{i}\gamma_\varrho\gamma_\lambda \quad \text{with} \quad \lambda>\varrho \quad ,$$

$$\begin{aligned}
\tilde{\Lambda}_{jl} &= \frac{1}{16}\,\mathrm{Tr}\{\gamma_\nu\gamma_\mu\gamma^\varrho\gamma^\lambda\gamma^\nu\gamma^\mu\gamma_\varrho\gamma_\lambda\}\\
&= \frac{1}{16}\big[-2\delta^\nu{}_\mu\,\mathrm{Tr}\{\gamma_\mu\gamma^\lambda\gamma^\nu\gamma^\mu\gamma_\varrho\gamma_\lambda\} + 2\delta^\lambda{}_\nu\,\mathrm{Tr}\{\gamma_\mu\gamma^\varrho\gamma^\nu\gamma^\mu\gamma_\varrho\gamma_\lambda\}\\
&\quad - 2\delta_\mu{}^\varrho\,\mathrm{Tr}\{\gamma^\lambda\gamma^\mu\gamma_\varrho\gamma_\lambda\} + 2\delta_\mu{}^\lambda\,\mathrm{Tr}\{\gamma^\varrho\gamma^\mu\gamma_\varrho\gamma_\lambda\} - \mathrm{Tr}\{\gamma^\varrho\gamma^\lambda\gamma_\varrho\gamma_\lambda\}\big]\\
&= \frac{1}{16}\big[+2\delta_{\mu\varrho}\cdot 4(2\delta_{\nu\lambda}-1) + 2\delta_{\nu\lambda}\cdot 4(2\delta_{\mu\varrho}-1) - 8\delta_{\mu\varrho} - 8\delta_{\mu\lambda} + 4\big]\\
&= \frac{1}{4}\big[1 - 2(\delta_{\nu\varrho}+\delta_{\nu\lambda}+\delta_{\mu\varrho}+\delta_{\mu\lambda}) + 4\delta_{\nu\varrho}\delta_{\mu\lambda} + 4\delta_{\nu\lambda}\delta_{\mu\varrho}\big] \quad ,
\end{aligned} \tag{17}$$

$$\tilde{\Lambda}_{jl} = \frac{1}{4}\begin{pmatrix} +1 & -1 & -1 & -1 & -1 & +1\\ -1 & +1 & -1 & -1 & +1 & -1\\ -1 & -1 & +1 & +1 & -1 & -1\\ -1 & -1 & +1 & +1 & -1 & -1\\ -1 & +1 & -1 & -1 & +1 & -1\\ +1 & -1 & -1 & -1 & -1 & +1 \end{pmatrix} \quad .$$

• $l=12,\ldots,15$:

$$\tilde{\Lambda}_{jl} = -\frac{1}{16}\,\mathrm{Tr}\{\gamma_\nu\gamma_\mu\gamma_5\gamma^{l-12}\gamma^\nu\gamma^\mu\gamma_5\gamma_{l-12}\} \quad .$$

Together with (14) it follows that

$$\tilde{\Lambda}_{jl} = \frac{1}{4}\begin{pmatrix} +1 & +1 & -1 & -1\\ +1 & -1 & +1 & -1\\ +1 & -1 & -1 & +1\\ -1 & +1 & +1 & -1\\ -1 & +1 & -1 & +1\\ -1 & -1 & +1 & +1 \end{pmatrix} \quad . \tag{18}$$

• $l=16$:
From (8) it follows that

$$\tilde{\Lambda}_{j,16} = -\frac{1}{16}\,\mathrm{Tr}(\hat{O}_j\hat{O}^j) = -\frac{1}{4} \quad . \tag{19}$$

• $j=12,\ldots,15;\, l=12,\ldots,15$:

$$\begin{aligned}
\tilde{\Lambda}_{jl} &= \frac{1}{16}\,\mathrm{Tr}\{\gamma_{j-12}\gamma^{l-12}\gamma^{j-12}\gamma_{l-12}\}\\
&= \frac{1}{4}\{2\delta_{jl}-1\} \quad ,
\end{aligned} \tag{20}$$

$$\tilde{\Lambda}_{jl} = \frac{1}{4}\begin{pmatrix} +1 & -1 & -1 & -1\\ -1 & +1 & -1 & -1\\ -1 & -1 & +1 & -1\\ -1 & -1 & -1 & +1 \end{pmatrix} \quad .$$

Mathematical Supplement 2.1

• $l = 16$:

$$\tilde{\Lambda}_{j,16} = -\frac{1}{16}\,\mathrm{Tr}\{\gamma_5\gamma_{j-12}\gamma_5\gamma_5\gamma^{j-12}\gamma_5\} = -\frac{1}{4} \quad . \tag{21}$$

• $j = 16;\ l = 16$:

$$\tilde{\Lambda}_{16,16} = \frac{1}{16}4 = \frac{1}{4} \quad . \tag{22}$$

According to (4a) we may combine the C_i , C_j as follows:

$$\begin{aligned} C_\mathrm{S} &= C_1 \quad , \\ C_\mathrm{V} &= C_2 = C_3 = C_4 = C_5 \quad , \\ C_\mathrm{T} &= C_6 = \cdots = C_{11} \quad , \\ C_\mathrm{A} &= -C_{12} = \cdots = -C_{15} \quad , \\ C_\mathrm{P} &= -C_{16} \quad , \end{aligned} \tag{23}$$

where the negative signs correspond to the convention. Similarly (6) now reads as follows:

$$\begin{aligned} C_i &= \sum_j C'_j \tilde{\Lambda}_{ij} \\ &= \tilde{\Lambda}_{i1}C'_1 + \sum_{j=2}^{5}\tilde{\Lambda}_{ij}C'_j + \sum_{j=6}^{11}\tilde{\Lambda}_{ij}C'_j + \sum_{j=12}^{15}\tilde{\Lambda}_{ij}C'_j + \tilde{\Lambda}_{i,16}C'_{16} \\ &= \tilde{\Lambda}_{i1}C_\mathrm{S} + \left(\sum_{j=2}^{5}\tilde{\Lambda}_{ij}\right)C_\mathrm{V} + \left(\sum_{j=6}^{11}\tilde{\Lambda}_{ij}\right)C_\mathrm{T} - \left(\sum_{j=12}^{15}\tilde{\Lambda}_{ij}\right)C_\mathrm{A} - \tilde{\Lambda}_{i,16}C_\mathrm{P} \quad . \end{aligned}$$

Thus it follows that

$$\Lambda_{IJ} = \sum_{j \text{ in } J} \varepsilon^I \varepsilon^J \tilde{\Lambda}_{ij} \quad \text{with } i \text{ in } I \quad , \tag{24}$$

where

$$\varepsilon^I = \begin{cases} +1 & \text{for} \quad \mathrm{S, V, T} \\ -1 & \text{for} \quad \mathrm{A, P} \end{cases} \quad . \tag{25}$$

From (11)–(22) it follows that

$$\Lambda_{IJ} = \frac{1}{4}\begin{pmatrix} 1 & 4 & 6 & 4 & 1 \\ 1 & -2 & 0 & 2 & -1 \\ 1 & 0 & -2 & 0 & 1 \\ 1 & 2 & 0 & -2 & -1 \\ 1 & -4 & 6 & -4 & 1 \end{pmatrix} \quad . \tag{26}$$

This is the standard representation of the Fierz transformation and the one most commonly used. It is easily checked that $\Lambda^2 = 1$, and therefore $\Lambda = \Lambda^{-1}$, i.e. Λ is its own inverse.

athematical Supplement 2.12

An alternative form is obtained from (22) and (23) by introducing

$$
\begin{aligned}
S(3,1;4,2) &:= (\bar{\psi}_3 1 \psi_1)(\bar{\psi}_4 1 \psi_2) \quad , \\
V(3,1;4,2) &:= (\bar{\psi}_3 \gamma_\mu \psi_1)(\bar{\psi}_4 \gamma^\mu \psi_2) \quad , \\
T(3,1;4,2) &:= \frac{1}{2}(\bar{\psi}_3 \sigma_{\mu\nu} \psi_1)(\bar{\psi}_4 \sigma^{\mu\nu} \psi_2) \quad , \\
A(3,1;4,2) &:= (\bar{\psi}_3 \gamma_5 \gamma_\mu \psi_1)(\bar{\psi}_4 \gamma^\mu \gamma_5 \psi_2) \quad , \\
P(3,1;4,2) &:= (\bar{\psi}_3 \gamma_5 \psi_1)(\bar{\psi}_4 \gamma_5 \psi_2) \quad ,
\end{aligned}
\tag{27}
$$

and replacing (5) by

$$
\sum_{I=S,V,T,A,P} C_I I(3,1;4,2) = \sum_{J=S,V,T,A,P} C'_J \, J(4,1;3,2) \quad . \tag{28}
$$

The transformation of the matrices I and J is then given by

$$
I(3,1;4,2) = \sum_J \Lambda_{JI} J(4,1;3,2) \quad ,
$$

and because Λ is self-inversive it also follows that

$$
J(4,1;32) = \sum \Lambda_{IJ} I(3,1;42) \quad .
$$

2.6 The Tau Lepton

In the year 1975 a further lepton was discovered at Stanford (SLAC) by **Perl**, which has been named the τ lepton.[12] With a mass of 1784 $\pm$ 3 MeV it is almost 20 times heavier than the muon. Its lifetime is

$$
T_\tau = (3.4 \pm 0.5) \times 10^{-13}\ \text{s} \quad . \tag{2.87}
$$

The scheme of τ lepton decay is completely analogous to muon decay, which we have discussed in detail. Since both the electron and the muon have smaller masses than the τ lepton, both decay processes are possible:

$$
\tau^- \to \mu^- + \bar{\nu}_\mu + \nu_\tau \quad , \tag{2.88a}
$$

$$
\tau^- \to e^- + \bar{\nu}_e + \nu_\tau \quad . \tag{2.88b}
$$

In addition, the τ lepton may also decay into strongly interacting particles, especially into three or more pions together with a τ neutrino. These hadronic processes contribute about 65% to the total decay probability of the τ lepton (see Table 2.2); however, we will not consider them here but will rather focus on the leptonic processes.

---

12 M.L. Perl et al.: Phys. Rev. Lett. **35**, 148 (1975); M.L. Perl: Ann. Rev. Nucl. Part. Science **30**, 299 (1980); G.S. Abrams, M.L. Perl et al.: Phys. Rev. Lett **43**, 1555 (1979).

Table 2.2. Decay probabilities of the τ lepton

Decay	%
$\tau^- \to e^- \bar{\nu}_e \nu_\tau$	16.4 ± 1.8
$\mu^- \bar{\nu}_\mu \nu_\tau$	16.0 ± 1.7
$\pi^- \nu_\tau$	10.3 ± 1.2
$\rho^- \nu_\tau$	22.1 ± 2.4
$K^- \nu_\tau$	1.3 ± 0.5
$\pi^- \rho^0 \nu_\tau$	5.4 ± 1.7
further hadronic decays	26.0 ± 1.3

Besides the properties that result from its rather large mass, the τ lepton behaves just like an electron or muon. This fact is sometimes termed e–μ–τ *universality*. For example, the τ lepton is observed with large accuracy to be point-like. Its internal extension amounts to less than 0.004 fm. Furthermore the electron and muon spectra observed in the decay processes (2.88) may be analyzed in terms of Michel parameters, in analogy to the case of muon decay, which we have already treated. The result is[13]

$$\rho_\tau = 0.742 \pm 0.035 \pm 0.020 \quad , \tag{2.89}$$

which is a strong argument for V–A coupling ($\rho = 0.75$) and unambiguously excludes V+A coupling ($\rho = 0$), as well as pure V or A coupling ($\rho = 0.375$). This behavior also becomes obvious from Fig. 2.10, which compares the observed electron spectrum with the predictions of the V–A and V+A theory.

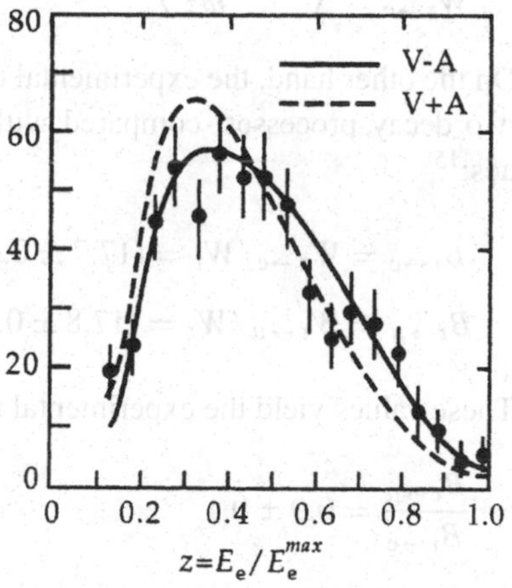

Fig. 2.10. Electron spectrum of the τ decay

From a detailed analysis of the shape of the high-energy end of the muon spectrum, an upper limit for the mass of the τ neutrino can be inferred (see Fig. 2.11). The most accurate value today is

$$m_{\nu_\tau} \leq 70\,\text{MeV} \tag{2.90}$$

[13] H. Albrecht et al. [ARGUS Collaboration]: Phys. Lett. **B246** (1990) 278–284.

However, it is not unlikely that its rest mass vanishes. If we assume that the weak current of the τ particles is of the familiar form

$$J_\mu^{(\tau)}(x) = \bar{u}_\tau(x)\gamma_\mu(1-\gamma_5)u_{\nu_\tau}(x) \quad , \tag{2.91}$$

we immediately obtain the decay rates into the leptonic channels (2.88) by simply adopting the formula (2.57) for the muonic decay. Again, we set $m_e = 0$, but do not neglect the muon mass:

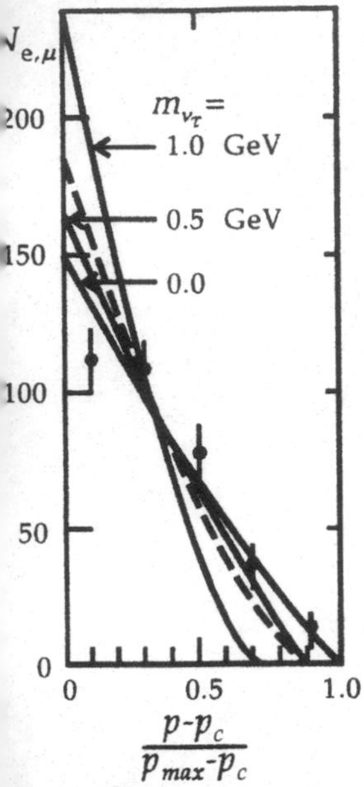

ig. 2.11. The number of the bserved electrons and muons depicted as a function of ne momentum.[14] Here $p_c =$ .65 GeV is the lower limit f the momentum observed in ne experiment. Each curve tands for one value of the nass of the τ neutrino. The ashed curve is for V+A couling and $m_{\nu_\tau} = 0$

$$W_{\tau^-\to e^-\nu_e\nu_\tau} = \frac{G^2 m_\tau^5}{192\pi^3} = W_{\mu^-\to e^-\nu_e\nu_\mu}\left(\frac{m_\tau}{m_\mu}\right)^5 \quad , \tag{2.92a}$$

$$\begin{aligned} W_{\tau^-\to\mu^-\nu_\mu\nu_\tau} &= \frac{G^2 m_\tau^5}{192\pi^3}\left(1-8\frac{m_\mu^2}{m_\tau^2}\right) \\ &= W_{\mu^-\to e^-\nu_e\nu_\mu}\left(\frac{m_\tau}{m_\mu}\right)^5\left(1-8\frac{m_\mu^2}{m_\tau^2}\right) \quad . \end{aligned} \tag{2.92b}$$

Inserting the value $m_\tau/m_\mu = 16.86$, we obtain

$$W_{\tau\to e} = 0.620\times 10^{+12}\,\mathrm{s}^{-1} \quad , \tag{2.93a}$$

$$W_{\tau\to\mu} = 0.603\times 10^{+12}\,\mathrm{s}^{-1} \quad . \tag{2.93b}$$

The ratio of these quantities is

$$\frac{W_{\tau\to\mu}}{W_{\tau\to e}} = \left(1-8\frac{m_\mu^2}{m_\tau^2}\right) = 0.972 \quad . \tag{2.94}$$

On the other hand, the experimental determination of the relative probability for these two decay processes, compared with the total decay rate, yields the following values:[15]

$$\begin{aligned} B_{\tau\to e} &= W_{\tau\to e}/W_\tau = (17.7\pm 0.4) \quad , \\ B_{\tau\to\mu} &= W_{\tau\to\mu}/W_\tau = (17.8\pm 0.4) \quad . \end{aligned} \tag{2.95}$$

These values yield the experimental ratio

$$\frac{B_{\tau\to\mu}}{B_{\tau\to e}} = 0.9\pm 0.1 \quad , \tag{2.96}$$

which agrees with the theoretical prediction (2.94) within the accuracy of the experiment. By inserting (2.95) in (2.92a) we can give a theoretical prediction for the lifetime of the τ lepton:

$$T_\tau = \frac{B_{\tau\to e}}{W_{\tau\to e}} = (2.6\pm 0.2)\times 10^{-13}\mathrm{s} \quad , \tag{2.97}$$

---

[14] M.L. Perl: Ann. Rev. Nucl. Part. Science **30**, 299 (1980).

[15] Review of particle properties in M. Aguilar-Benitez et al.: Phys. Rev. D **45**, Part II (June 1992).

which at least does not contradict the experimental value (2.87). From this it follows that the coupling constant G occurring in (2.92a) cannot differ significantly from the coupling constant G of muon decay.

To summarize, we conclude that according to the actual data the τ lepton fits perfectly into the family of leptons (e, μ, τ). The only differences between these leptons are their masses and a quantum number that guarantees the separate conservation of the electronic, muonic, and τ-leptonic particle numbers. In particular, the leptons exhibit a completely universal behavior in electromagnetic and weak interactions.

EXAMPLE

2.13 The Discovery of the Tau Lepton

The τ lepton was discovered at the SPEAR storage ring by the magnetic detector of the SLAC–LBL collaboration (Stanford Linear Accelerator Center – Lawrence Berkeley Laboratory).[16] The principle of a storage ring is that particle and antiparticle beams circulate within the ring in opposite directions and are forced to overlap in the region of the detector (see Fig. 2.12). The detector was constructed in such a way that electrons, muons, and photons, as well as hadrons, could be detected and identified within a large solid angle. In addition, the trajectories of the charged particles in the magnetic field allowed for a determination of their momentum.

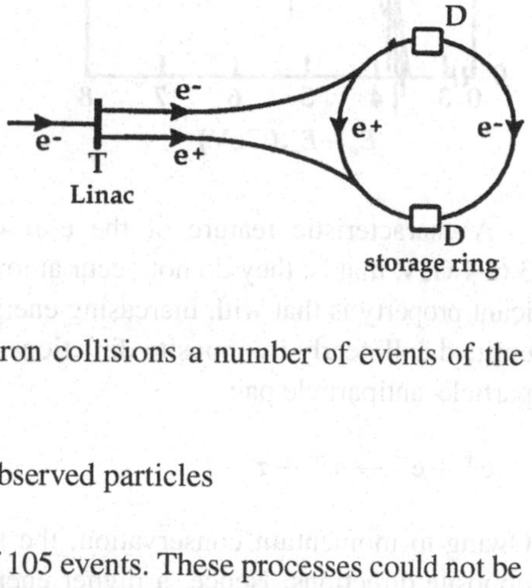

Fig. 2.12. Schematic picture of the storage ring facility SPEAR at SLAC

Through investigations of electron–positron collisions a number of events of the form

$$e^+ + e^- \to \begin{Bmatrix} e^+ + \mu^- \\ e^- + \mu^+ \end{Bmatrix} + \text{at least 2 unobserved particles}$$

were observed – until 1975 a total number of 105 events. These processes could not be understood in terms of a conventional interpretation, especially since the possible uncertainty in particle identification by the detector had already been taken into account, that is to say, the most unfavorable assumption was made, namely that all processes with three observed charged particles implied the production of hadrons only. Thus every "electron" or "muon" was claimed to be a misinterpretation of the detector. This allowed an estimate to be made of how reliable particle identification was. It was therefore possible to evaluate, from the number of observed events in which a lepton and a hadron, or two hadrons, occurred, the number of misinterpreted e–μ events. It followed that of the 139 events originally observed, 34 were spurious and had to

[16] G.J. Feldman and M.L. Perl: Phys. Rep. **19**, 233 (1975).

be subtracted. The immediate conclusion was that the uncertainty in particle identification could not explain all these events. One might argue that at least one of the observed particles was a charged particle, a photon, or a neutral pion decaying into two photons, but one not reaching the effective region of the detector. On the other hand, such processes would imply the occurrence of corresponding events in which the particle is actually detected. However, this was not the case.

The sole remaining explanation was the interpretation of the e–μ events in terms of the production of hardly detectable particles such as neutrons, K_L^0 (see Chap. 8), or neutrinos. However, the K_L^0 is ruled out by the reasonable assumption that the production rates for K_L^0 and K_S^0 are equal. The latter could easily be identified by its decay products π^+ and π^-. However, up until 1976 only a single event had been observed.

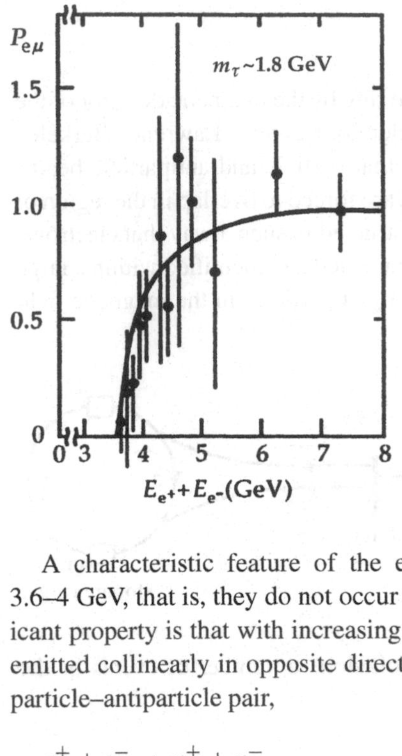

ig. 2.13. Cross section for ectron–muon events in storge ring experiments

A characteristic feature of the e–μ events is their threshold energy of about 3.6–4 GeV, that is, they do not occur at lower energies (see Fig 2.13). Another significant property is that with increasing energy the electron and muon are preferentially emitted collinearly in opposite directions. This strongly suggests the production of a particle–antiparticle pair,

$$e^+ + e^- \rightarrow \tau^+ + \tau^- \quad .$$

Owing to momentum conservation, the two particles should be emitted in exactly opposite directions. Hence, a higher energy implies that the particles have a larger momentum. Subsequently, the two particles decay into an electron (positron) or a muon which is emitted isotropically with respect to the rest frame of the corresponding τ particle. However, the larger the velocity of the τ particle, the less the direction of emission with respect to the τ particle's rest frame contributes to the emission actually observed within the lab system, whereby the latter is then essentially determined by the direction of emission of the τ particle.

The observed threshold energy leads to the conclusion that the mass of the τ particle lies in the range 1.6–2 GeV. In order to characterize the nature of the τ particle, there were in practice two options: either it is a lepton that decays according to

$$\tau^- \rightarrow \nu_\tau + e^- + \bar{\nu}_e \quad , \qquad \tau^- \rightarrow \nu_\tau + \mu^- + \bar{\nu}_\mu \quad ,$$

Example 2.1

and similarly for the antiparticle τ^+, or it is a boson with the following decay channels:

$$\tau^- \to e^- + \bar{\nu}_e \quad , \quad \tau^- \to \mu^- + \bar{\nu}_\mu \quad ,$$

as are observed for example, in the decay of negatively charged pions. The latter interpretation, however, could certainly be ruled out by the analysis of the momentum distribution of produced electrons (positrons) and muons (the specific form of this distribution also contradicts the interpretation of the observed particle in terms of a neutron).

Therefore the sole explanation that remained was the classification of the τ particle as a new, heavy lepton. Figure 2.14 illustrates how the total process results in the observed e–μ events. Since 1975 the properties of the τ lepton have been extensively studied, its mass has been accurately determined to be 1784 ± 3 MeV, its Michel parameters were obtained[17] as $\rho = 0.731 \pm 0.031$, $\xi = 1.03 \pm 0.11$, $\xi\delta = 0.63 \pm 0.09$, and thus the V–A coupling of its decay has been verified in detail.

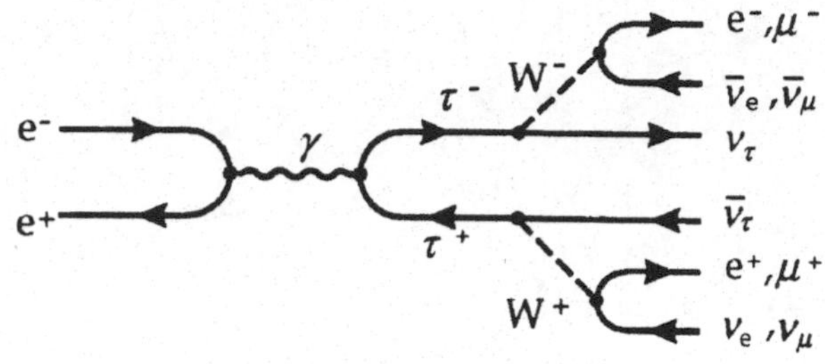

Fig. 2.14. Production and decay of the τ lepton

2.7 Biographical Notes

FIERZ, Markus, *20.6.1912 in Basel (Switzerland), †20.6.2006 in Küsnacht (Switzerland), professor at the University of Basel 1944–1960, since 1960 successor of W. Pauli at the ETH Zürich, in 1969 appointed director of the Theoretical Division at CERN, Geneva.

LEVI-CIVITA, Tullio, mathematician, *29.3.1873 in Padua (Italy), †29.12.1941 in Rome. In 1898 he became professor of mechanics in Padua, since 1918 at the University of Rome. He developed differential and tensor calculus, which laid the basis for Einstein's general theory of relativity. He introduced the idea of parallel transport and developed the theory of curved spaces.

MICHEL, Louis, *4.5.1923 in Roanne (France), †30.12.1999 in Bures-sur-Yvette (France), professor at the Ecole Polytechnique in Paris, since 1962.

PERL, Martin, L. *1927 in New York. Attended New York city schools. After military services in World War II, he received a Bachelor in Chemical Engineering degree from the Polytechnic Institute of Brooklyn in 1948. After several years working for the General Electric Co. as a chemical engineer, he went to graduate school in physics at Columbia University, studied under

[17] H. Albrecht et al. (51 authors): The ARGUS Collaboration, DESY-preprint 97-194.

I. I. Rabi, and received his Ph.D. in 1955. From 1955 to 1963 he did research and taught at the University of Michigan. Since 1963 he has been at the Stanford Linear Accelerator Center at Stanford University where he is a Professor of Physics and Group Leader. In 1990–1992 he was a Distinguished Visiting Professor at the University of Michigan.
His major research interest is experimental elementary particle physics. Other research interests are optical and electronic devices, and the application of small drop technology. He is also interested in applying these technologies to industry, biology, and medicine.
He received the 1995 Nobel Prize in Physics in recognition of his discovery of the tau lepton, the heaviest known member of the electron–muon–tau sequence of charged leptons. Finding the tau lepton subsequently led to the discovery of the three generations of elementary particles, an essential ingredient in what has now become the Standard Model of fundamental particles and interactions. He has published 250 papers in physics and science editions and edited or authored five books including Reflectors on Experimental Science which he wrote in 1996.
In addition to the 1995 Nobel Prize in Physics, he received the 1982 Wolf Prize in Physics. He is a fellow of the American Physical Society and a member of the U.S. National Academy of Science and American Academy of Arts and Science. He holds honorary degrees from the University of Chicago in 1990 and Polytechnic University in 1996.

Limitations of Fermi Theory 3

The Fermi theory of weak interactions is patterned according to the well-known current–current coupling of quantum electrodynamics. All observations could – up to now – be classified and understood within this scheme. Nevertheless, as we shall see, the Fermi theory contains severe difficulties and is therefore unsatisfactory. In order to reveal these difficulties we shall discuss first another phenomenon of weak interactions, i.e. the neutral currents.

3.1 Neutral Currents

We have noted that there are no scattering processes of the form $\nu_\mu e^- \to \nu_\mu e^-$ or $\bar{\nu}_\mu c^- \to \bar{\nu}_\mu e^-$ in the framework of Fermi's theory with V–A coupling. One therefore has to carefully investigate experimentally whether such scattering occurs in nature. These experiments are extremely difficult, because the expected cross sections (if any) lie in the range 10^{-41}–10^{-44} cm^2 (10^{-17}–10^{-20} barn). Only with the high neutrino currents in modern accelerators (Fermilab near Chicago, CERN-SPS) and high neutrino energies (several hundred GeV) did such experiments become practical at all.

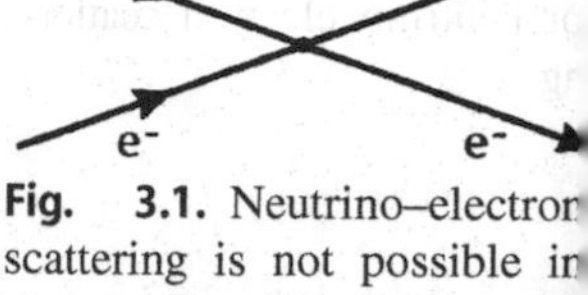

Fig. 3.1. Neutrino–electron scattering is not possible in the context of V–A theory, as developed so far

In fact many such processes were observed; the best experimental values for the cross sections are:

$$\frac{1}{E_{\nu_\mu}}\sigma(\nu_\mu e^- \to \nu_\mu e^-) = (1.45 \pm 0.26) \times 10^{-42}\ \mathrm{cm}^2/\mathrm{GeV} \quad , \tag{3.1a}$$

$$\frac{1}{E_{\bar{\nu}_\mu}}\sigma(\bar{\nu}_\mu e^- \to \bar{\nu}_\mu e^-) = (1.3 \pm 1.0) \times 10^{-42}\ \mathrm{cm}^2/\mathrm{GeV} \quad . \tag{3.1b}$$

The existence of such so-called "neutral" currents can therefore be regarded as being firmly established. Here the name "neutral current" has the following origin. If one starts from the conservation of electron and muon numbers separately, the only possible interpretation of the scattering process $\nu_\mu e^- \to \nu_\mu e^-$ is that at the interaction point the incoming electron turns into the outgoing electron and the incoming μ neutrino turns into the outgoing μ neutrino. The obvious method to implement this process in our theory is therefore to supplement the leptonic current $J_\mu^{(L)}$ by expressions of the form

$$\bar{u}_{\nu_\mu}\gamma_\alpha(1-\gamma_5)u_{\nu_\mu} \quad , \tag{3.2a}$$

$$\bar{u}_e\gamma_\alpha(g_V - g_A\gamma_5)u_e \quad . \tag{3.2b}$$

Here we have made use of the fact that in any event neutrinos must have negative helicity. The current (3.2a) does not contain a charged particle at all, that is it is really

W. Greiner, B. Müller, *Gauge Theory of Weak Interactions*,
DOI 10.1007/978-3-540-87843-8_3, © Springer-Verlag Berlin Heidelberg 2009

"neutral", while in (3.2b) the charge of the particle is conserved, which one also somewhat sloppily refers to as "neutral" (in this sense the electromagnetic current $\bar{u}\gamma^\alpha u$ is "neutral" for all particles!). Important is the fact that the incoming particle changes its charge in the charged transition currents as occurs in (2.1) and (2.4). This is not the case with neutral currents.

3.2 Scattering of a Muon Neutrino by an Electron[1]

We now calculate the cross section for $\nu_\mu e^- \to \nu_\mu e^-$, starting from the currents (3.2). The relevant interaction term is:

$$H_{\text{int}}(\nu_\mu e^- \to \nu_\mu e^-) = \frac{G}{\sqrt{2}} \int d^3x \left[\bar{u}_{\nu_\mu} \gamma^\alpha (1-\gamma_5) u_{\nu_\mu}\right]\left[\bar{u}_e \gamma_\alpha (g_V - g_A\gamma_5) u_e\right] \quad . \tag{3.3}$$

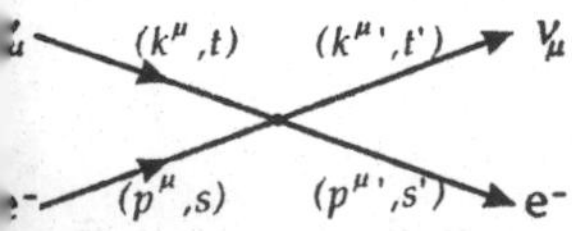

ig. 3.2. Momenta and spins or neutrino–electron scattering

Because the cross section is incredibly small, it is impossible to observe the helicity of the electrons before or after the scattering process. Hence it is sufficient to calculate the averaged cross section, where one averages or sums over all helicities. In addition, it is convenient to integrate over all momenta of the outgoing particles, since these are also not measurable in practice (the momentum p' of the electron could in principle be measured, but the statistics of such a differential experiment would be completely insufficient).

We then obtain the following result (the details of the calculation are the subject of Exercise 3.1):

$$\bar{\sigma} = \frac{G^2}{8\pi^2}\frac{1}{16(k\cdot p)} \int \frac{d^3k'}{k_0'} \int \frac{d^3p'}{p_0'} \delta^4(p'+k'-p-k)\frac{1}{2}\sum_{\substack{s,s'\\t,t'}} |M|^2 \quad , \tag{3.4}$$

where

$$M = \left[\bar{u}_{\nu_\mu}(\boldsymbol{k}',t')\gamma^\alpha(1-\gamma_5)u_{\nu_\mu}(\boldsymbol{k},t)\right]\left[\bar{u}_e(\boldsymbol{p}',s')\gamma_\alpha(g_V - g_A\gamma_5)u_e(\boldsymbol{p},s)\right] \quad . \tag{3.5}$$

The individual parts of the matrix elements are now evaluated exactly as in the case of muon decay (cf. Chap. 2). We start with that of the neutrino:

$$\begin{aligned}
&\sum_{t,t'} \bar{u}_{\nu_\mu}(\boldsymbol{k}',t')\gamma^\alpha(1-\gamma_5)u_{\nu_\mu}(\boldsymbol{k},t)\bar{u}_{\nu_\mu}(\boldsymbol{k},t)\gamma^\beta(1-\gamma_5)u_{\nu_\mu}(\boldsymbol{k}',t') \\
&= \mathrm{Tr}\{\gamma^\alpha(1-\gamma_5)\not{k}\gamma^\beta(1-\gamma_5)\not{k}'\} = 2\,\mathrm{Tr}\{\gamma^\alpha\not{k}\gamma^\beta(1-\gamma_5)\not{k}'\} \\
&= 2k_\mu k_\nu'\,\mathrm{Tr}\{\gamma^\alpha\gamma^\mu\gamma^\beta\gamma^\nu(1+\gamma_5)\} \\
&= 8\left[k^\alpha k'^\beta - g^{\alpha\beta}(k\cdot k') + k^\beta k'^\alpha - \mathrm{i}\varepsilon^{\alpha\mu\beta\nu}k_\mu k_\nu'\right] \quad .
\end{aligned} \tag{3.6}$$

[1] F.J. Hasert, H. Faissner, et al.: Phys. Lett. **46B**, 121 (1973).

Analogously we find that for the electronic part

$$\sum_{s,s'} \bar{u}_e(p',s')\gamma_\alpha(g_V - g_A\gamma_5)u_e(p,s)\bar{u}_e(p,s)\gamma_\beta(g_V - g_A\gamma_5)u_e(p',s')$$
$$= \mathrm{Tr}\{\gamma_\alpha(g_V - g_A\gamma_5)(\not{p} + m_e)\gamma_\beta(g_V - g_A\gamma_5)(\not{p}' + m_e)\}$$
$$= \mathrm{Tr}\{\gamma_\alpha\not{p}\gamma_\beta\not{p}'(g_V + g_A\gamma_5)^2 + \gamma_\alpha\gamma_\beta m_e^2(g_V + g_A\gamma_5)(g_V - g_A\gamma_5)\}$$
$$= \mathrm{Tr}\{\gamma_\alpha\not{p}\gamma_\beta\not{p}'(g_V^2 + g_A^2 + 2g_Ag_V\gamma_5)\} + m_e^2\,\mathrm{Tr}\{\gamma_\alpha\gamma_\beta\}(g_V^2 - g_A^2)$$
$$= 4\big[(g_V^2 + g_A^2)(p_\alpha p'_\beta + p_\beta p'_\alpha - g_{\alpha\beta}p_\mu p'^\mu) + 2\mathrm{i}g_Ag_V\varepsilon_{\alpha\mu\beta\nu}p^\mu p'^\nu$$
$$+ m_e^2(g_V^2 - g_A^2)g_{\alpha\beta}\big] \quad , \tag{3.7}$$

where we have made use of the fact that the trace of an odd number of γ matrices vanishes.

The product of both results yields (we again write $(p \cdot k)$ instead of $p_\alpha k^\alpha$ etc.)

$$\frac{1}{2}\sum_{s,s',t,t'} |M| = 32(g_V^2 + g_A^2)\big[(p\cdot k)(p'\cdot k') + (p'\cdot k)(p\cdot k')\big]$$
$$+ 64 g_A g_V\big[(p\cdot k)(p'\cdot k') - (p'\cdot k)(p\cdot k')\big]$$
$$- 32(g_V^2 - g_A^2)m_e^2(k\cdot k')$$
$$= 32(g_V + g_A)^2(p\cdot k)(p'\cdot k') + 32(g_V - g_A)^2(p'\cdot k)(p\cdot k')$$
$$- 32(g_V^2 - g_A^2)m_e^2(k\cdot k') \quad . \tag{3.8}$$

Now we again need the integral (2.33), which is found in Exercise 2.6, (17):

$$I_{\alpha\beta} = \int \frac{\mathrm{d}^3k'}{2k'_0}\frac{\mathrm{d}^3p'}{2p'_0} k'_\alpha p'_\beta \delta^4(k' + p' - k - p)$$
$$= \frac{\pi}{24}\left[1 - \frac{m_e^2}{(k+p)^2}\right]^2$$
$$\times \left\{g_{\alpha\beta}\big((k+p)^2 - m_e^2\big) + 2(k+p)_\alpha(k+p)_\beta\left[1 + \frac{2m_e^2}{(k+p)^2}\right]^2\right\}$$
$$\times\,\Theta\big((k+p)^2 - m_e^2\big) \quad . \tag{3.9}$$

If we set $\alpha = \beta$ and sum over α, we get

$$\int \frac{\mathrm{d}^3k'}{2k'_0}\frac{\mathrm{d}^3p'}{2p'_0}(k'\cdot p')\delta^4(k' + p' - k - p) = \frac{\pi}{4}(k+p)^2\left[1 - \frac{m_e^2}{(k+p)^2}\right]^2$$
$$= \frac{\pi}{4}s\left(1 - \frac{m_e^2}{s}\right)^2 \quad , \tag{3.10}$$

where $s = (k+p)^2$. Furthermore, we need the integral

$$I = \int \frac{\mathrm{d}^3k'}{2k'_0}\frac{\mathrm{d}^3p'}{2p'_0}\delta^4(p' + k' - p - k)(k\cdot k') \quad . \tag{3.11}$$

In the centre-of-mass frame of the incoming particles $\boldsymbol{k}+\boldsymbol{p}=0$ and $k_0+p_0=E$. Also for neutrinos $k'^{\mu}=(k'^0,\boldsymbol{k}')$ with $|\boldsymbol{k}'|=k'_0$, similarly $|\boldsymbol{k}|=k_0$. For $\boldsymbol{k}'$ we introduce polar coordinates with respect to $\boldsymbol{k}$, with

$$\boldsymbol{k}\cdot\boldsymbol{k}'=k_0k'_0\cos\theta \quad . \tag{3.12}$$

Then it follows that

$$\begin{aligned} I &= \int \frac{\mathrm{d}^3k'}{2k'_0}\frac{\delta\left(\sqrt{\boldsymbol{k}'^2+m_\mathrm{e}^2}+k'_0-E\right)}{2\sqrt{\boldsymbol{k}'^2+m_\mathrm{e}^2}}k_0k'_0(1-\cos\theta) \\ &= k_0\int \mathrm{d}\Omega(1-\cos\theta)\int_0^\infty \mathrm{d}k'_0\,k'^2_0\frac{\delta\left(\sqrt{k'^2_0+m_\mathrm{e}^2}+k'_0-E\right)}{4\sqrt{k'^2_0+m_\mathrm{e}^2}} \quad . \end{aligned} \tag{3.13}$$

To evaluate the integral over k'_0 we need the root of the argument of the δ function:

$$g(k'_0)=\sqrt{k'^2_0+m_\mathrm{e}^2}+k'_0-E=0 \quad\Longrightarrow\quad k'_0=\frac{E^2-m_\mathrm{e}^2}{2E} \quad . \tag{3.14}$$

Using the chain rule for the δ function (assuming that $g(x)$ has only a single root),

$$\int \mathrm{d}x f(x)\delta(g(x))=\left.\frac{f(x)}{|g'(x)|}\right|_{g(x)=0} \quad , \tag{3.15}$$

we obtain the following result:

$$\begin{aligned} I &= \frac{1}{4}k_0\int \mathrm{d}\Omega(1-\cos\theta)\frac{k'^2_0}{\sqrt{k'^2_0+m_\mathrm{e}^2}}\left.\frac{\sqrt{k'^2_0+m_\mathrm{e}^2}}{k'_0+\sqrt{k'^2_0+m_\mathrm{e}^2}}\right|_{k'_0=\frac{E^2-m_\mathrm{e}^2}{2E}} \\ &= \frac{1}{4}k_0\int \mathrm{d}\Omega(1-\cos\theta)\frac{(E^2-m_\mathrm{e}^2)^2}{4E^3} \\ &= \frac{4\pi}{4}\frac{k_0E}{E^2}\frac{(E^2-m_\mathrm{e}^2)^2}{4E^2}=\pi\frac{k_0E}{E^2}\frac{(E^2-m_\mathrm{e}^2)^2}{4E^2} \quad . \end{aligned} \tag{3.16}$$

This result, which is up to now only valid in the centre-of-mass frame, can be written in a Lorentz invariant fashion if we consider that (since $k^2=0$)

$$k_\alpha p^\alpha=k_\alpha(k+p)^\alpha \quad\xrightarrow{\text{c.m.frame}}\quad k_0(k^0+p^0)=k_0E \tag{3.17}$$

and

$$s\equiv(k+p)^2 \quad\xrightarrow{\text{c.m.frame}}\quad (k_0+p_0)^2=E^2 \quad . \tag{3.18}$$

Thus the variable s is the square of the centre-of-mass energy, it fulfills the relation

$$s=(k+p)^2=k^2+p^2+2k_\alpha p^\alpha=2k_\alpha p^\alpha+m_\mathrm{e}^2 \quad . \tag{3.19}$$

Hence in the rest frame of the electron before the collision, it holds that

$$s=2k_0p^0+m_\mathrm{e}^2=2E_{\nu_\mu}m_\mathrm{e}+m_\mathrm{e}^2=m_\mathrm{e}(2E_{\nu_\mu}+m_\mathrm{e}) \quad . \tag{3.20}$$

With the relations (3.17)–(3.18), (3.16) takes the form

$$I = \pi(k \cdot p)\frac{(s - m_e^2)^2}{4s^2} \quad , \tag{3.21}$$

which is valid in all frames of reference.

The step function in (3.9) is obviously irrelevant, since we trivially have

$$(k + p)^2 = s > m_e^2 > 0 \quad . \tag{3.22}$$

We therefore drop it. Inserting the averaged transition probability (3.8) into (3.4) and using (3.9), (3.10) (3.21), (3.22), we get

$$\begin{aligned}
\bar{\sigma} &= \frac{G^2}{4\pi^2}\left\{(g_V + g_A)^2 \pi s\left(1 - \frac{m_e^2}{s}\right)^2 + (g_V - g_A)^2 \frac{\pi}{6}\left(1 - \frac{m_e^2}{s}\right)^2\right. \\
&\quad \times \left[(s - m_e^2) + \frac{2[(k+p)\cdot p][(k+p)\cdot k](1 + \frac{2m_e^2}{s})}{(k \cdot p)}\right] \\
&\quad \left. - (g_V^2 - g_A^2)\frac{\pi(s - m_e^2)^2}{s^2} m_e^2\right\} \\
&= \frac{G^2}{4\pi}s\left(1 - \frac{m_e^2}{s}\right)^2\left[(g_V + g_A)^2 + \frac{1}{3}(g_V - g_A)^2\left(1 + \frac{m_e^2}{s} + \frac{m_e^4}{s^2}\right)\right. \\
&\quad \left. - (g_V^2 - g_A^2)\frac{m_e^2}{s}\right] \quad .
\end{aligned} \tag{3.23}$$

In the last step we have applied relation (3.18) several times, as well as the relations $k^2 = 0$ and $p^2 = m_e^2$. Ordering terms yields the result

$$\begin{aligned}
\bar{\sigma} &= \frac{G^2}{3\pi}s\left(1 - \frac{m_e^2}{s}\right)^2\left\{(g_V^2 + g_A g_V + g_A^2) - (g_V^2 + 4g_A g_V - 5g_A^2)\frac{m_e^2}{8s}\right. \\
&\quad \left. + \frac{1}{2}(g_V - g_A)^2\frac{m_e^4}{s^2}\right\} \quad .
\end{aligned} \tag{3.24}$$

For large scattering energies, m_e can be neglected compared to the centre-of-mass energy $E = \sqrt{s}$:

$$\bar{\sigma}(\nu_\mu + e^- \to \nu_\mu + e^-) \simeq \frac{G^2 s}{3\pi}(g_V^2 + g_V g_A + g_A^2) \quad . \tag{3.25}$$

3.3 $\bar{\nu}_\mu$–e$^-$ Scattering

For $\bar{\nu}_\mu$–e$^-$ scattering one simply has to exchange k_μ and k'_μ (the incoming *anti*neutrino corresponds to an outgoing neutrino of negative energy). Then $\bar{\sigma}(\bar{\nu}_\mu + e^- \to \bar{\nu}_\mu + e^-)$ follows in analogy to (3.8):

$$\begin{aligned}
\frac{1}{2}\sum_{s,s',t,t'} |M|^2 &= 32(g_V + g_A)^2(p \cdot k')(p' \cdot k) \\
&\quad + 32(g_V - g_A)^2(p' \cdot k')(p \cdot k) - 32(g_V^2 - g_A^2)m_e^2(k \cdot k') \quad .
\end{aligned} \tag{3.26}$$

We would have obtained the same expression if we had replaced g_A by $-g_A$ in the final results of the previous section. This procedure yields, for example,

$$\bar{\sigma}(\bar{\nu}_\mu + \mathrm{e}^- \to \bar{\nu}_\mu + \mathrm{e}^-) = \frac{G^2 s}{3\pi}(g_V^2 - g_A g_V + g_A^2) \qquad (3.27)$$

for large scattering energies.

Now, according to (3.20), $s \approx 2m_\mathrm{e}E_{\nu_\mu}$ or $2m_\mathrm{e}E_{\bar{\nu}_\mu}$, if E_{ν_μ} or $E_{\bar{\nu}_\mu}$, respectively, denotes the scattering energy of the (anti)neutrino in the laboratory frame. The experiment therefore has to be compared with the theoretical prediction

$$\frac{\bar{\sigma}}{E_{\nu/\bar{\nu}}} = \frac{2G^2 m_\mathrm{e}}{3\pi}(g_V^2 \pm g_A g_V + g_A^2) \quad . \qquad (3.28)$$

With the value of G from muon decay we have $G^2 m_\mathrm{e} = 2.707 \times 10^{-41}\ \mathrm{cm}^2/\mathrm{GeV}$, and hence comparison with the experiment (3.1) yields

$$\begin{aligned} g_V^2 + g_V g_A + g_A^2 &= 0.252 \pm 0.045 \quad , \\ g_V^2 - g_V g_A + g_A^2 &= 0.226 \pm 0.174 \quad . \end{aligned} \qquad (3.29)$$

Both expressions are of the same order in the limits of experimental errors. This tells us that $g_V g_A$ has to be small compared to $(g_A^2 + g_V^2)$, that is either g_A or g_V is small compared to one. We therefore obtain two possible solutions:

$$g_V \approx 0 \quad , \quad g_A \approx \pm 0.5 \quad , \qquad (3.30a)$$

$$g_V \approx \pm 0.5 \quad , \quad g_A \approx 0 \quad . \qquad (3.30b)$$

Which of the two coupling types, (mainly) pure A or pure V coupling, is realized in nature we cannot decide on the basis of experiments concerning electron–neutrino scattering alone. For this we need further experiments which are sensitive to neutral weak currents (see examples for further explanation). At present the best data values are[2]

$$g_V = 0.043 \pm 0.063 \quad , \quad g_A = -0.545 \pm 0.056 \quad . \qquad (3.31)$$

Hence the experiments give evidence of a nearly pure axial vector coupling for the neutral weak current of the electron. The values for g_V, g_A in (3.30), (3.31) will be naturally understood within the Weinberg–Salam theory – see the discussion after (4.108).

EXERCISE

3.1 Muon Neutrino–Electron Scattering Cross Section

Problem. Derive (3.4) for the neutrino–electron scattering cross section.

Solution. The coordinate-dependent spinor wave functions in the incoming and outgoing channel read:

[2] T.E. Kim, P. Langacker, M. Levine, H.H. Williams: Rev. Mod. Phys. **53**, 211 (1981).

Exercise 3.

$$\text{incoming} = \begin{cases} u_e(x) = (2p^0V)^{-1/2} u_e(\boldsymbol{p}, s) \exp(-\mathrm{i}p_\mu x^\mu) \\ u_{\nu_\mu}(x) = (2k^0V)^{-1/2} u_{\nu_\mu}(\boldsymbol{k}, t) \exp(-\mathrm{i}k_\mu x^\mu) \end{cases} , \tag{1a}$$

$$\text{outgoing} = \begin{cases} u_e(x) = (2p'^0V)^{-1/2} u_e(\boldsymbol{p}', s') \exp(-\mathrm{i}p'_\mu x^\mu) \\ u_{\nu_\mu}(x) = (2k'^0V)^{-1/2} u_{\nu_\mu}(\boldsymbol{k}', t') \exp(-\mathrm{i}k'_\mu x^\mu) \end{cases} . \tag{1b}$$

With these wave functions we obtain the following expression for the scattering matrix element

$$\begin{aligned} S(\nu_\mu e^- \to \nu_\mu e^-) &= -\mathrm{i} \int \mathrm{d}t\, H_{\text{int}}(\nu_\mu e^- \to \nu_\mu e^-) \\ &= -\mathrm{i}\frac{G}{\sqrt{2}} \int \mathrm{d}^4x \left[\bar{u}_{\nu_\mu}(x)\gamma_\alpha(1-\gamma_5)u_{\nu_\mu}(x)\right] \\ &\quad \times \left[\bar{u}_e(x)\gamma^\alpha(g_V - g_A\gamma_5)u_e(x)\right] \\ &= -\mathrm{i}\frac{G}{\sqrt{2}}(2\pi)^4 \frac{\delta^4(p'+k'-p-k)}{\sqrt{16V^4k^0p^0k'^0p'^0}} \\ &\quad \times \left(\bar{u}_{\nu\mu}(\boldsymbol{k}', t')\gamma_\alpha(1-\gamma_5)u_{\nu_\mu}(\boldsymbol{k}, t)\right) \\ &\quad \times \left(\bar{u}_e(\boldsymbol{p}', s')\gamma^\alpha(g_V - g_A\gamma_5)u_e(\boldsymbol{p}', s')\right) . \end{aligned} \tag{2}$$

Here we have used (3.3) for the interaction Hamiltonian and performed the space-time integration over the plane waves, yielding the δ function.

According to the standard rules for Feynman graphs (see Appendix A.3) we get the scattering cross section by first calculating the transition rate per unit volume,

$$W(\nu_\mu e^- \to \nu_\mu e^-) = \frac{1}{VT}|S(\nu_\mu e^- \to \nu_\mu e^-)|^2 , \tag{3}$$

and then normalizing on unit flux of incoming neutrinos, as well as dividing by the density of target electrons. The neutrino flux in the laboratory system is given by (Appendix A.3)

$$J_\nu = \frac{\sqrt{(k_\nu p^\nu)^2 - m_\nu^2 m_e^2}}{k^0p^0V} = \frac{k_\mu p^\mu}{k^0p^0V} . \tag{4}$$

The normalization of the spinors (1) is chosen such that one particle is present in volume V:

$$\varrho_e = \frac{1}{V} . \tag{5}$$

To obtain a physical cross section we still have to sum over all observed final states. Since one has to sum over all scattered particles the formula for the total cross section is given by

$$\sigma_{st}(\nu_\mu e^- \to \nu_\mu e^-) = V \int \frac{\mathrm{d}^3p'}{(2\pi)^3} V \int \frac{\mathrm{d}^3k'}{(2\pi)^3} \sum_{s',t'} \frac{W(\nu_\mu e^- \to \nu_\mu e^-)}{J_\nu V^{-1}} . \tag{6}$$

Here the indices s, t denote that this is the cross section for certain polarizations of electron and neutrino before scattering. When averaging over initial spin states we have to pay attention to the fact that electrons can assume two spin states, while neutrinos appear in only one state of negative helicity:

$$\bar{\sigma} = \frac{1}{2} \sum_{s,t} \sigma_{st}(\nu_\mu e^- \to \nu_\mu e^-) \quad . \tag{7}$$

If we combine (2)–(7) and furthermore make use of the relation (cf. (2.15), (2.16))

$$\left((2\pi)^4 \delta^4(p' + k' - p - k)\right)^2 \to VT(2\pi)^4 \delta^4(p' + k' - p - k) \quad , \tag{8}$$

we obtain

$$\bar{\sigma} = \frac{G^2}{2\pi^2} \frac{1}{16(k \cdot p)} \int \frac{d^3k'}{2k'_0} \int \frac{d^3p'}{2p'_0} \delta^4(p' + k' - p - k) \frac{1}{2} \sum_{ss',tt'} |M|^2 \quad , \tag{9}$$

where M is given by (3.5).

3.4 High-Energy Behavior of Neutrino–Electron Scattering

The first problem of Fermi's theory of beta decay is the existence of weak neutral currents, which it did not predict. Nevertheless, we have observed that these can be easily introduced into the theory. The generalization, however, appears to be quite crude and not very elegant. The second problem of Fermi's theory lies in the fact that the cross section for neutrino-lepton scattering in general increases with the square of the centre-of-mass energy, that is with s (see (3.17), (3.25), (3.27)). This holds also for the "normal" processes with charged currents, like the two displayed in Fig. 3.3.

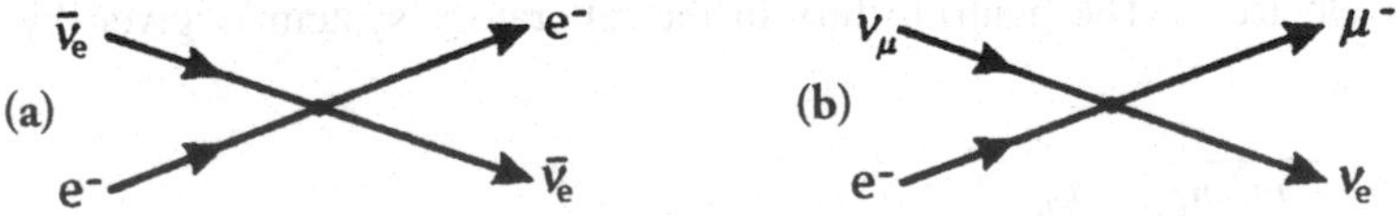

Fig. 3.3. Neutrino–electron scattering in Fermi theory. Each scattered particle changes its charge

In these two cases one finds the following expressions for the averaged cross sections (cf. Exercises 3.2 and 3.3):

$$\bar{\sigma}(\bar{\nu}_e e^- \to e^- \bar{\nu}_e) = \frac{G^2}{3\pi} s \left(1 - \frac{m_e^2}{s}\right)\left(1 + \frac{m_e^6}{s^3}\right) \quad , \tag{3.32}$$

$$\bar{\sigma}(\nu_i e^- \to l_i^- \nu_e) = \frac{G^2}{\pi} s \left(1 - \frac{m_i^2}{s}\right)^2 \quad , \quad i = e, \mu, \tau \quad . \tag{3.33}$$

When calculating the cross section for $\nu_i + e^- \to l_i^- + \nu$ one furthermore observes that it is completely isotropic (this is *not* valid for $\bar{\nu}_e + e^- \to e^- + \bar{\nu}_e$). Hence only the partial wave with angular momentum zero (s wave) contributes to the scattering! This can be intuitively understood: the current–current coupling of the Fermi interaction allows scattering only if both particles are located at the same point. Thus neutrino and electron have to come very close together during the scattering process; the collision must be central. This demands a vanishing relative angular momentum.

The general partial wave analysis is explicitly treated in a later section. Here we only use the result, that is (3.80) and (3.81). The differential cross section in the centre-of-mass system is

$$\left(\frac{\mathrm{d}\sigma}{\mathrm{d}\Omega_{cm}}\right)_{\lambda_1',\lambda_2'\lambda_1,\lambda_2} = \frac{1}{p^2}\left|\sum_J (2J+1) d^J_{\lambda\lambda'}(\theta) T^J_{\lambda_1',\lambda_2',\lambda_1,\lambda_2}\right|^2 , \tag{3.34}$$

if the two particles have helicities $\lambda_1\lambda_2$ before and $\lambda_1'\lambda_2'$ after the collision. Here p is the absolute value of the momentum of the particles in the centre-of-mass frame (cf. Exercise 3.4)

$$p^2 = \frac{1}{4s}\left((s-m_1^2-m_2^2)^2 - 4m_1^2m_2^2\right) \overset{s\to\infty}{\longrightarrow} \frac{s}{4} , \tag{3.35}$$

and $\lambda = \lambda_1 - \lambda_2$, $\lambda' = \lambda_1' - \lambda_2'$. T denotes the matrix elements of the transition operator $\hat{T}$, which is defined in (3.67).

As we already know, an important consequence of the V–A coupling is that massless particles experience an interaction only if they have negative helicity. For particles with non-vanishing mass, the operator $(1-\gamma_5)$ projects approximately on negative helicity if the particles move relativistically, that is $p \gg m$, as was shown in (1.21). It therefore holds that

$$T^J_{\lambda_1'\lambda_2',\lambda_1'\lambda_1'} \overset{s\to\infty}{\longrightarrow} T^J_{-1-1,-1-1}\delta_{\lambda_1'-1}\delta_{\lambda_2'-1}\delta_{\lambda_1-1}\delta_{\lambda_2-1} . \tag{3.36}$$

In the following we therefore simply write T without indices.

Hence, after averaging over initial helicities, we see that the scattering cross section in the limit of high energies is given by

$$\begin{aligned}\frac{\mathrm{d}\sigma}{\mathrm{d}\Omega_{CM}} &\overset{s\to\infty}{\longrightarrow} \frac{1}{4p^2}\left|\sum_J (2J+1) d^J_{00}(\theta) T^J\right|^2 \\ &= \frac{1}{s}\left|\sum_J (2J+1) P_J(\cos\theta) T^J\right|^2 = \frac{1}{s}\left|\sum_J \sqrt{\frac{4\pi}{2J+1}} Y_{J0}(\cos\theta) T^J\right|^2 ,\end{aligned} \tag{3.37}$$

where (3.35) and $Y_{J0} = \sqrt{(2J+1)/4\pi}\, P_J$ was used. The unitarity of the scattering operator, $\hat{S}^\dagger\hat{S} = 1$, means that the scattering probability in each single partial wave reaches at most the value 1. The contribution of the s wave ($J=0$) to the total cross section is therefore limited:

$$\bar{\sigma} = 4\pi \frac{\mathrm{d}\bar{\sigma}}{\mathrm{d}\Omega}(J=0) = \frac{4\pi}{s}|T^{J=0}|^2 \le \frac{4\pi}{s} . \tag{3.38}$$

This value is called the *unitarity bound*.

Since the calculated averaged cross section of the reaction $\nu_i e^- \to l_i^- \nu_e$ is exactly isotropic, the sole contribution to the scattering comes from the s wave according to (3.37). We therefore have the physical constraint ($s \gg m_i^2$)

$$\sigma(\nu_i e^- \to l_i \nu_e) = \frac{G^2 s}{\pi} \le \frac{4\pi}{s} ,$$

yielding the condition

$$s \le \frac{2\pi}{G} = (734\ \mathrm{GeV})^2 . \tag{3.39}$$

The critical value is reached when every particle in the centre-of-mass frame obtains an energy of about 367 GeV. We are, however, still far from reaching this value experimentally; because of (3.20), $s \approx 2m_e E_\nu^{\text{lab}}$ in the electron rest frame, the critical neutrino energy in the laboratory system is

$$E_\nu^{\text{crit}} = 5.27 \times 10^8 \text{ GeV} \quad . \tag{3.40}$$

The fact that the cross section $\sigma \propto G^2 s/\pi$ can formally exceed the unitarity bound does not imply that unitarity is actually violated, since σ was calculated from perturbation theory. Hence it only implies that one has to consider higher-order terms of the perturbation series (multiple scattering) or, in other words, that one cannot simply use plane waves for the scattering particles. Typical processes are depicted in Fig. 3.4. To study the high-energy behavior of the theory, one therefore has to calculate higher-order processes. This will lead us to another difficulty of the Fermi theory, as we shall see in a moment.

ig. 3.4. Contributions of
e higher order to neutrino–
lectron scattering

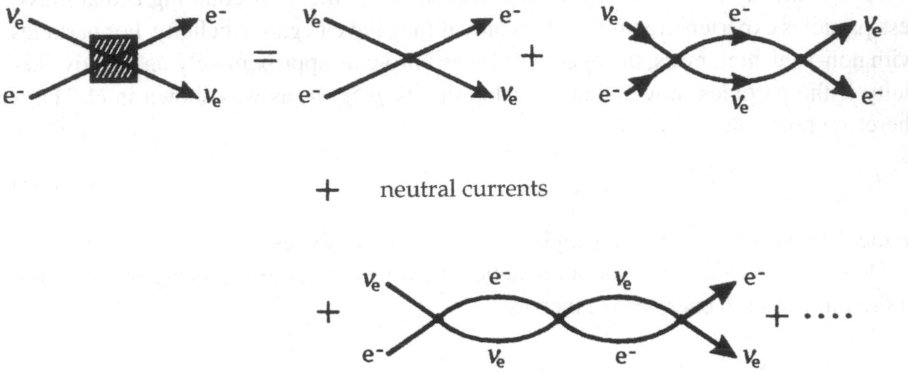

EXERCISE

3.2 The Spin-Averaged Cross Section for Antineutrino–Electron Scattering

Problem. Calculate the spin-averaged cross section for antineutrino–electron scattering according to Fig. 3.3a and the angular distribution in the centre-of-mass frame.

Solution. We use the same notation as in the first section of this chapter. The spinors for the incoming and outgoing particles are then as follows

$$\text{incoming} = \begin{cases} u_e(x) = (2p^0 V)^{-1/2} u_e(\boldsymbol{p}, s) \exp(-\mathrm{i} p_\mu x^\mu) \\ u_{\bar{\nu}_e}(x) = (2k^0 V)^{-1/2} v_{\nu_e}(\boldsymbol{k}, t) \exp(+\mathrm{i} k_\mu x^\mu) \end{cases} , \tag{1a}$$

$$\text{outgoing} = \begin{cases} u_e(x) = (2p'^0 V)^{-1/2} u_e(\boldsymbol{p}', s') \exp(-\mathrm{i} p'_\mu x^\mu) \\ u_{\bar{\nu}_e}(x) = (2k'^0 V)^{-1/2} v_{\nu_e}(\boldsymbol{k}', t') \exp(+\mathrm{i} k'_\mu x^\mu) \end{cases} . \tag{1b}$$

Writing down the interaction Hamiltonian, one has to pay attention to the fact that in the spirit of the Feynman rules the antineutrino with $(k^{\mu\prime}, t')$ is an "incoming" particle of negative energy, while the incoming antineutrino with (k^μ, t) is an "outgoing" particle. Here the only possibility is

$$H_{\text{int}}(\bar{\nu}_e e^- \to \bar{\nu}_e e^-) = \frac{G}{\sqrt{2}} \int d^3x \, J_\alpha^{(e)\dagger}(x) J_{(e)}^\alpha(x) \tag{2}$$

with the scattering matrix element

$$S(\bar{\nu}_e e^- \to \bar{\nu}_e e^-) = -i\frac{G}{\sqrt{2}}(2\pi)^4 \frac{\delta^4(p'+k'-p-k)}{\sqrt{16V^4 k_0 p_0 k'_0 p'_0}} M \,, \tag{3}$$

where

$$M = \left[\bar{u}_{\nu_e}(\boldsymbol{k},t)\gamma_\alpha(1-\gamma_5)u_e(\boldsymbol{p},s)\right]\left[\bar{u}_e(\boldsymbol{p}',s')\gamma^\alpha(1-\gamma_5)u_{\nu_e}(\boldsymbol{k}',t')\right] \,. \tag{4}$$

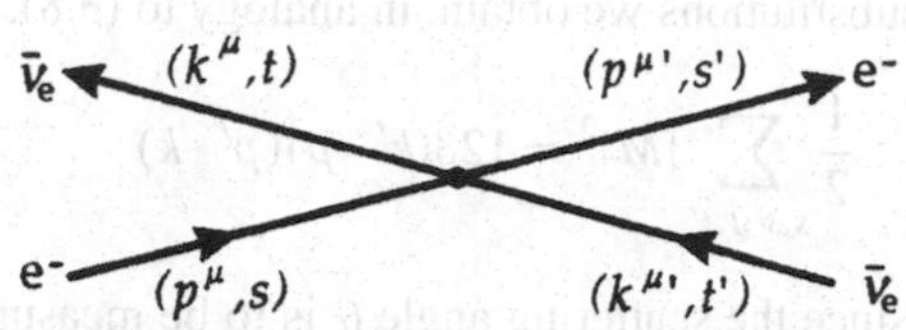

Fig. 3.5. Notation for the calculation of antineutrino–electron scattering

Here we have already performed the space-time integration. We can literally copy from Exercise 3.1 all steps which lead from (3) to the averaged cross section. If we want to get the angular distribution of the scattering, we must not, however, integrate over the scattering angle of the antineutrino and therefore have to separate the integral over d^3k' into an integral over $d|\boldsymbol{k}'|$ and the angular part $d\Omega$ (here we work in the centre-of-mass frame):

$$\begin{aligned}
&\frac{d\bar{\sigma}}{d\Omega}(\bar{\nu}_e e^- \to \bar{\nu}_e e^-) \\
&= V \int \frac{d^3p'}{(2\pi)^3} V \int \frac{d|\boldsymbol{k}'|}{(2\pi)^3} |\boldsymbol{k}'|^2 \frac{1}{2} \sum_{s,s',t,t'} \frac{|S|^2 k^0 p^0 V}{(k \cdot p) V^{-1}(VT)} \\
&= \frac{G^2}{2\pi^2} \frac{1}{16(k\cdot p)} \int \frac{d^3p'}{2p'_0} \int \frac{|\boldsymbol{k}'|^2 d|\boldsymbol{k}'|}{2k'_0} \\
&\quad \times \delta^4(p'+k'-p-k) \frac{1}{2} \sum_{s,s',t,t'} |M|^2
\end{aligned} \tag{5}$$

with the matrix element M from (4). The result (5) corresponds exactly to Exercise 3.1, (3). The matrix element splits into parts, containing physically incoming and outgoing particles. For the first we find, analogously to (3.6)

$$\begin{aligned}
&\sum_{s,t} \bar{v}_{\nu_e}(\boldsymbol{k},t)\gamma_\alpha(1-\gamma_5)u_e(\boldsymbol{p},s)\bar{u}_e(\boldsymbol{p},s)\gamma_\beta(1-\gamma_5)v_{\nu_e}(\boldsymbol{k},t) \\
&= \text{Tr}\{\gamma_\alpha(1-\gamma_5)(\not{p}+m_e)\gamma_\beta(1-\gamma_5)\not{k}\} \\
&= 2\,\text{Tr}\{\gamma_\alpha(\not{p}+m_e)\gamma_\beta(1-\gamma_5)\not{k}\} \\
&= 2p^\mu k_\nu \,\text{Tr}\{\gamma_\alpha\gamma_\mu\gamma_\beta\gamma_\nu(1+\gamma_5)\} \\
&= 8\left[p_\alpha k_\beta - g_{\alpha\beta}(p\cdot k) + p_\beta k_\alpha + i\varepsilon_{\alpha\mu\beta\nu}p^\mu k^\nu\right] \,,
\end{aligned} \tag{6}$$

Exercise 3.2

where we have dropped the term proportional to m_e since it contains an odd number of γ matrices. The second term of M yields

$$\begin{aligned}
&\sum_{s',t'} \bar{u}_e(\boldsymbol{p}',s')\gamma^\alpha(1-\gamma_5)v_{\nu_e}(\boldsymbol{k}',t')\bar{v}_{\nu_e}(\boldsymbol{k}',t')\gamma^\beta(1-\gamma_5)u_e(\boldsymbol{p}',s') \\
&\quad = \mathrm{Tr}\{\gamma^\alpha(1-\gamma_5)\not{k}'\gamma^\beta(1-\gamma_5)(\not{p}'+m_e)\} \\
&\quad = 2\,\mathrm{Tr}\{\gamma^\alpha\not{k}'\gamma^\beta(1-\gamma_5)(\not{p}'+m_e)\} \\
&\quad = 2k'_\mu p'_\nu\,\mathrm{Tr}\{\gamma^\alpha\gamma^\mu\gamma^\beta\gamma^\nu(1+\gamma_5)\} \\
&\quad = 8\left[k'^\alpha p'^\beta - g^{\alpha\beta}(k'\cdot p') + k'^\beta p'^\alpha - \mathrm{i}\varepsilon^{\alpha\mu\beta\nu}k'_\mu p'_\nu\right] \quad . \qquad (7)
\end{aligned}$$

A comparison shows that (6) emerges from (3.6) by substituting $k \to p$, $k' \to k$. Analogously, (7) follows from (3.7) by substituting $p \to k'$, with $g_V = g_A = 1$. With these substitutions we obtain, in analogy to (3.8),

$$\frac{1}{2}\sum_{s,s',t,t'} |M|^2 = 128(k'\cdot p)(p'\cdot k) \quad . \qquad (8)$$

Since the scattering angle θ is to be measured in the centre-of-mass system, we continue the calculation in this reference system:

$$\boldsymbol{p}+\boldsymbol{k} = \boldsymbol{p}'+\boldsymbol{k}' = 0 \quad , \qquad (9)$$

$$s = (p+k)^2 = (p_0+k_0)^2 = (p'_0+k'_0)^2 \quad . \qquad (10)$$

The identity of both expressions before and after scattering is guaranteed by the δ function in (5). For the same reason, and because $k^2 = k'^2 = 0$, it follows that

$$(k'\cdot p) = \left[k'\cdot(p'+k'-k)\right] = (k'\cdot p') - (k'\cdot k) \quad , \qquad (11a)$$

$$(p'\cdot k) = \left[(p+k-k')\cdot k\right] = (p\cdot k) - (k'\cdot k) \quad . \qquad (11b)$$

By taking the squares, we find that the definition of s (10) yields

$$(p\cdot k) = \frac{1}{2}(s - m_e^2) = (k'\cdot p') \quad . \qquad (12)$$

Finally, because $k_0 = |\boldsymbol{k}|$, $k'_0 = |\boldsymbol{k}'|$, we have

$$(k\cdot k') = k_0k'_0 - \boldsymbol{k}\cdot\boldsymbol{k}' = k_0k'_0(1-\cos\theta) \quad . \qquad (13)$$

With the help of (11)–(13) we obtain:

$$(k'\cdot p)(p'\cdot k) = \left[\frac{1}{2}(s-m_e^2) - k_0k'_0(1-\cos\theta)\right]^2 \quad . \qquad (14)$$

To express k_0 and k'_0 in terms of s we combine (9) and (10),

$$\begin{aligned}
k_0 &= \sqrt{s} - \sqrt{\boldsymbol{p}^2 + m_e^2} \\
&= \sqrt{s} - \sqrt{\boldsymbol{k}^2 + m_e^2} \\
&= \sqrt{s} - \sqrt{k_0^2 + m_e^2} \qquad (15)
\end{aligned}$$

and solve for k_0, giving

$$k_0 = (2\sqrt{s})^{-1}(s - m_e^2) = k'_0 \quad . \tag{16}$$

The last equality holds, since (15) is also valid for k'_0. Then we have

$$\begin{aligned}(k' \cdot p)(p' \cdot k) &= \frac{1}{4}(s - m_e^2)^2\left[1 - \frac{1}{2}\left(1 - \frac{m_e^2}{s}\right)(1 - \cos\theta)\right]^2 \\ &= \frac{1}{16}(s - m_e^2)^2\left[1 + \frac{m_e^2}{s} + \left(1 - \frac{m_e^2}{s}\right)\cos\theta\right]^2 \quad . \end{aligned}\tag{17}$$

The differential cross section is therefore given, according to (5) and (8), by

$$\begin{aligned}\frac{d\bar{\sigma}}{d\Omega} &= \frac{G^2}{8\pi^2}(s - m_e^2)\left[1 + \frac{m_e^2}{s} + \left(1 - \frac{m_e^2}{s}\right)\cos\theta\right]^2 \\ &\times \int \frac{d^3 p'}{p'_0} \int_0^\infty k'_0\, dk'_0\, \delta^4(p' + k' - p - k) \quad . \end{aligned}\tag{18}$$

Here we have again made use of (12) and the relation $k'_0 = |\boldsymbol{k}'|$. The remaining integrals are easily performed by splitting the space-time δ function into time-like and the three-dimensional spatial parts:

$$\delta(p'_0 + k'_0 - p_0 - k_0) = \delta(p'_0 + k'_0 - \sqrt{s}) \quad , \tag{19a}$$

$$\delta^3(\boldsymbol{p}' + \boldsymbol{k}' - \boldsymbol{p} - \boldsymbol{k}) = \delta^3(\boldsymbol{p}' + \boldsymbol{k}') \quad , \tag{19b}$$

where we have exploited (9) and (10). Now, because $|\boldsymbol{k}'| = k'_0$, the momentum integral yields

$$\begin{aligned}&\int_0^\infty k'_0\, dk'_0 \int d^3 p' (\boldsymbol{p}'^2 + m_e^2)^{-1/2} \delta\left(k'_0 + \sqrt{\boldsymbol{p}^2 + m_e^2} - \sqrt{s}\right)\delta^3(\boldsymbol{p}' + \boldsymbol{k}') \\ &= \int_0^\infty k'_0\, dk'_0 (k_0'^2 + m_e^2)^{-\frac{1}{2}} \delta\left(k'_0 + \sqrt{k_0'^2 + m_e^2} - \sqrt{s}\right) \\ &= \int_{m_e}^\infty \frac{(x^2 - m_e^2)}{2x^2}\delta(x - \sqrt{s})dx = \frac{(s - m_e^2)}{2s} \quad , \end{aligned}\tag{20}$$

where $x = k'_0 + \sqrt{k'_0 + m_e^2}$ has been substituted. The differential cross section is then

$$\frac{d\bar{\sigma}}{d\Omega}(\bar{\nu}_e e^- \to \bar{\nu}_e e^-) = \frac{G^2}{16\pi^2}s\left(1 - \frac{m_e^2}{s}\right)^2\left[1 + \frac{m_e^2}{s} + \left(1 - \frac{m_e^2}{s}\right)\cos\theta\right]^2 . \tag{21}$$

In the limit $s \gg m_e^2$ we have

$$\left.\frac{d\bar{\sigma}}{d\Omega}\right|_{\theta=0} \approx \frac{G^2 s}{4\pi^2} \quad , \quad \left.\frac{d\bar{\sigma}}{d\Omega}\right|_{\theta=\pi} \approx 0 \quad . \tag{22}$$

This is easily understood if we consider that the antineutrino has positive helicity, and in the limit of high energies only the negative helicity state of the electron participates

xercise 3.2

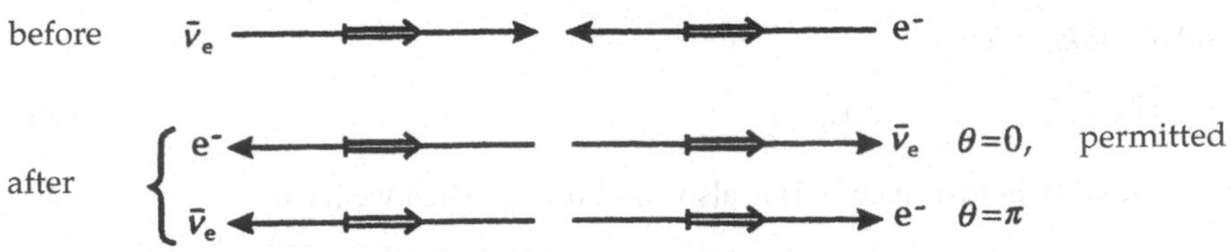

Fig. 3.6. Allowed and forbidden antineutrino–electron scattering. Note that spins cannot flip, because the projection of the angular momentum on the scattering axis is conserved. In the case of $\nu_\mu - e^-$ scattering this argument is not applicable, since the total helicity in the initial and final state vanishes

in the interaction. The situation is depicted in the Fig. 3.6 as seen for the centre-of-mass frame. The total cross section is obtained via the formula

$$\begin{aligned}\int \mathrm{d}\Omega (a + b\cos\theta)^2 &= 2\pi \int_{-1}^{+1} \mathrm{d}(\cos\theta)(a + b\cos\theta)^2 \\ &= 2\pi \left(a^2 \cos\theta + ab\cos^2\theta + \frac{b^2/3}{\cos^3\theta} \right) \Bigg|_{-1}^{+1} \\ &= 4\pi \left(a^2 + \frac{b^2}{3} \right) \quad ,\end{aligned} \tag{23}$$

as

$$\begin{aligned}\bar{\sigma}(\nu_e e^- \to \nu_e e^-) &= \frac{G^2}{3\pi} s \left(1 - \frac{m_e^2}{s} \right)^2 \left(1 + \frac{m_e^2}{s} + \frac{m_e^4}{s^2} \right) \\ &= \frac{G^2}{3\pi} s \left(1 - \frac{m_e^2}{s} \right) \left(1 + \frac{m_e^6}{s^3} \right) \quad .\end{aligned} \tag{24}$$

Thus (3.32) has been derived.

EXERCISE

3.3 The Spin-Averaged Cross Section of Muon Neutrino–Electron Scattering

Problem. Calculate the spin-averaged cross section for muon neutrino–electron scattering according to Fig. 3.3b. Show that the cross section is isotropic in the centre-of-mass frame.

Solution. In this case we deal exclusively with particles; the spinors are therefore given by

$$\begin{aligned}u_e(x) &= (2p_0 V)^{-1/2} u_e(\boldsymbol{p}, s) \exp(-\mathrm{i} p_\mu x^\mu) \quad , \\ u_{\nu_\mu}(x) &= (2k_0 V)^{-1/2} u_{\nu_\mu}(\boldsymbol{k}, t) \exp(-\mathrm{i} k_\mu x^\mu) \quad , \\ u_\mu(x) &= (2p'_0 V)^{-1/2} u_\mu(\boldsymbol{p}', s') \exp(-\mathrm{i} p'_\mu x^\mu) \quad , \\ u_{\nu_e}(x) &= (2k'_0 V)^{-1/2} u_{\nu_e}(\boldsymbol{k}', t') \exp(-\mathrm{i} k'_\mu x^\mu) \quad .\end{aligned} \tag{1}$$

Exercise 3.

Scattering according to the Hamiltonian

$$H_{\text{int}}(\nu_\mu e^- \to \mu^- \nu_e) = \frac{G}{\sqrt{2}} \int \mathrm{d}^3x \, J_\alpha^{(e)+}(x) J_{(\mu)}^\alpha(x) \tag{2}$$

yields the scattering matrix element

$$S(\nu_\mu e^- \to \mu^- \nu_e) = -\mathrm{i}\frac{G}{\sqrt{2}}(2\pi)^4 \frac{\delta^4(p' + k' - p - k)}{\sqrt{16V^4 k_0 p_0 k_0' p_0'}} M \quad , \tag{3}$$

where

$$M = \left[\bar{u}_{\nu_e}(\boldsymbol{k}', t')\gamma_\alpha(1-\gamma_5)u_e(\boldsymbol{p}, s)\right]\left[\bar{u}_\mu(\boldsymbol{p}', s')\gamma^\alpha(1-\gamma_5)u_{\nu_\mu}(\boldsymbol{k}, t)\right] \quad . \tag{4}$$

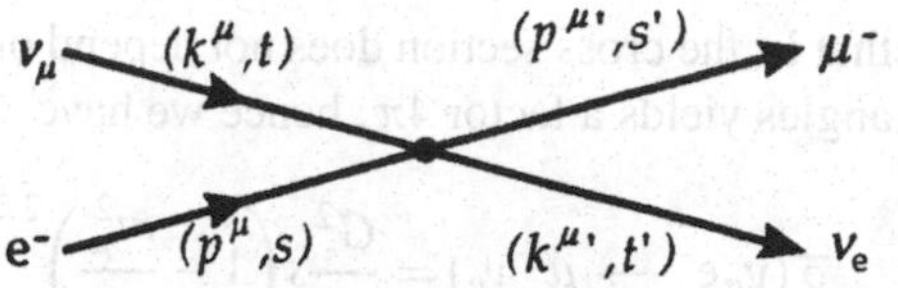

Fig. 3.7. Notations for muon neutrino–electron scattering

The computation proceeds completely analogously to that in Exercise 3.2; hence we obtain ($|\boldsymbol{k}'| = k_0'$)

$$\begin{aligned}\frac{\mathrm{d}\bar{\sigma}}{\mathrm{d}\Omega}&(\nu_\mu e^- \to \mu^- \nu_e) \\ &= \frac{G^2}{8\pi^2} \frac{1}{16(k \cdot p)} \int \frac{\mathrm{d}^3p'}{p_0'} \int k_0' \, \mathrm{d}k_0' \, \delta^4(p' + k' - p - k) \frac{1}{2} \sum_{s,s',t,t'} |M|^2 \quad . \end{aligned}\tag{5}$$

Since the neutrino is massless, the calculation for the electronic part of $\sum |M|^2$ is similar to (6) of Exercise 3.2. We only have to replace $\boldsymbol{k}$ by $\boldsymbol{k}'$:

$$\begin{aligned}&\sum_{s,t'} \bar{u}_{\nu_e}(\boldsymbol{k}', t')\gamma_\alpha(1-\gamma_5)u_e(\boldsymbol{p}, s)\bar{u}_e(\boldsymbol{p}', t')\gamma_\beta(1-\gamma_5)u_{\nu_e}(\boldsymbol{k}', t') \\ &= 8\left[p_\alpha k_\beta' - g_{\alpha\beta}(p \cdot k') + p_\beta k_\alpha' + \mathrm{i}\varepsilon_{\alpha\mu\beta\nu} p^\mu k'^\nu\right] \quad . \end{aligned}\tag{6}$$

For the muonic part we are able to use Exercise 3.2, if we substitute $\boldsymbol{k}$ for $\boldsymbol{k}'$ and m_μ by m_e. This yields

$$\begin{aligned}&\sum_{s',t} \bar{u}_\mu(\boldsymbol{p}', s')\gamma^\alpha(1-\gamma_5)u_{\nu_\mu}(\boldsymbol{k}, t)\bar{u}_{\nu_\mu}(\boldsymbol{k}, t)\gamma^\beta(1-\gamma_5)u_\mu(\boldsymbol{p}', s') \\ &= 8\left[k^\alpha p'^\beta - g^{\alpha\beta}(k \cdot p') + k^\beta p'^\alpha - \mathrm{i}\varepsilon^{\alpha\mu\beta\nu} k_\mu p_\nu'\right] \quad . \end{aligned}\tag{7}$$

Combining (6) and (7), it follows as in Exercise 3.2 that

$$\frac{1}{2} \sum_{s,s',t,t'} |M|^2 = 128 \times (k \cdot p)(p' \cdot k') \quad . \tag{8}$$

As in Exercise 3.2, (12), we have $(p' \cdot k') = \frac{1}{2}(s - m_\mu^2)$; thus we obtain, this time without lengthy intermediate calculation,

$$\frac{\mathrm{d}\bar{\sigma}}{\mathrm{d}\Omega}(\nu_\mu e^- \to \mu^- \nu_e) = \frac{G^2}{\pi^2} \frac{1}{2}(s - m_\mu^2) \int \frac{\mathrm{d}^3p'}{p_0} \int \mathrm{d}k_0' \, \delta^4(p' + k' - p - k) \quad . \tag{9}$$

Exercise 3.3

To compute the momentum integral we refer to Exercise 3.2, and consider the fact that the muon mass m_μ in the final state enters instead of m_e:

$$\int_0^\infty dk_0' \, k_0' \int \frac{d^3 p'}{\sqrt{p'^2 + m_\mu^2}} \delta\left(k_0' + \sqrt{p'^2 + m_\mu^2} - \sqrt{s}\right) \delta^3(\boldsymbol{p}' + \boldsymbol{k}')$$

$$= \frac{s - m_\mu^2}{2s} \quad . \tag{10}$$

As a final result we therefore obtain in the centre-of-mass frame

$$\frac{d\overline{\sigma}}{d\Omega}(\nu_\mu e^- \to \mu^- \nu_e) = \frac{G^2 s}{4\pi^2}\left(1 - \frac{m_\mu^2}{s}\right)^2 \quad , \tag{11}$$

that is, the cross section does not depend on the scattering angle θ. Integrating over all angles yields a factor 4π; hence we have

$$\overline{\sigma}(\nu_\mu e^- \to \mu^- \nu_e) = \frac{G^2}{\pi} s \left(1 - \frac{m_\mu^2}{s}\right)^2 \tag{12}$$

in agreement with (3.33) for $i = \mu$.

For neutrinos of other leptonic families (e, τ) the calculation proceeds identically, but the Hamiltonian (2) is in general given by

$$H_{\mathrm{int}}(\nu_i e^- \to l_i \bar{\nu}_e) = \frac{G}{\sqrt{2}} \int d^3 x \, J_\alpha^{(e)\dagger} J_{(i)}^\alpha \quad , \quad i = e, \mu, \tau \quad . \tag{13}$$

For the final result one simply has to replace the muon mass everywhere by the mass m_i of the charged lepton l_i.

One more remark: In the case $i = e$, that is for $\nu_e e \to \nu_e e$, the above equation describes only the scattering by the charged weak current. In principle one has to add the scattering by the neutral weak current, which cannot be experimentally distinguished from the former. A similar scattering process was studied in Exercise 3.1. Since the final states cannot be distinguished, both scattering matrix elements have to be added coherently!

EXERCISE

3.4 High-Energy Scattering

Problem. Derive (3.35).

Solution. The variable s is defined as the square of the total energy in the centre-of-mass frame (cf. (3.18)); hence it holds that

$$s = (E_1 + E_2)^2 = \left(\sqrt{m_1^2 + p^2} + \sqrt{m_2^2 + p^2}\right)^2 \quad , \tag{1}$$

since $|\boldsymbol{p}_1| = |\boldsymbol{p}_2| = p$ in the centre-of-mass frame. This relation must now be solved for p. By performing the multiplication on the right-hand side, we first of all obtain

$$s = m_1^2 + m_2^2 + 2p^2 + 2\sqrt{m_1^2 + p^2}\sqrt{m_2^2 + p^2} \quad , \tag{2}$$

and from this

$$\begin{aligned}(s-m_1^2-m_2^2-2p^2)^2 &= 4(m_1^2+p^2)(m_2^2+p^2) \\ &= (s-m_1^2-m_2^2)^2-4p^2(s-m_1^2-m_2^2)^2+4p^4 \\ &= 4m_1^2m_2^2+4(m_1^2+m_2^2)p^2+4p^4 \end{aligned} \tag{3}$$

Here some terms cancel, leaving

$$(s-m_1^2-m_2^2-2p^2)-4p^2s=4m_1^2m_2^2 \tag{4}$$

and hence

$$p^2=\frac{1}{4s}[(s-m_1^2-m_2^2)^2-4m_1^2m_2^2] \quad . \tag{5}$$

In the limit $s\to\infty$ one can neglect the masses of the particles, and (3.35) follows from (5):

$$p^2\to\frac{s}{4} \quad (s\to\infty) \quad . \tag{6}$$

3.5 Supplement: Scattering Formalism for Spin-$\frac{1}{2}$ Particles

The description of the scattering of spin-$\frac{1}{2}$ particles is more complicated than that of spinless particles owing to the possibility of spin flips. In a relativistic description we have the further complication that a Lorentz transformation in general changes the direction of the spin vector. When considering the polarization of the electron in the muon decay we took this into account by defining the spin direction in the rest frame and then properly transforming to the laboratory frame (2.23)

$$s^\mu_{(0)}=(0,\boldsymbol{s}) \quad \to \quad s^\mu=\left(\frac{\boldsymbol{p}\cdot\boldsymbol{s}}{m},\boldsymbol{s}+\frac{(\boldsymbol{p}\cdot\boldsymbol{s})\boldsymbol{p}}{m(E+m)}\right) \quad . \tag{3.41}$$

Defining the spin vector in the rest frame of the particle has the disadvantage that in a scattering process between two particles four different spins appear (two before and two after the collision), which are defined in four different reference frames and first have to be transformed to a common reference frame, for example the centre-of-mass frame. To define the spin vector in the respective rest frame has the additional disadvantage of failing for neutrinos, since one cannot define a rest frame for massless particles.

For these reasons it is convenient in relativistic particle physics to classify states of a particle by means of the helicity instead of its spin vector. We have already encountered the helicity operator

$$\hat{\Lambda}=\frac{\boldsymbol{\Sigma}\cdot\boldsymbol{p}}{|\boldsymbol{p}|} \quad \text{with} \quad 2\boldsymbol{J}=\boldsymbol{\Sigma} \tag{3.42}$$

in Chap. 1 when we discussed the neutrino states. The helicity specifies the component of the spin-$\frac{1}{2}$ $\boldsymbol{\Sigma}$ in the direction of the particle momentum $\boldsymbol{p}$. Note that $\boldsymbol{J}=\frac{1}{2}\boldsymbol{\Sigma}$, which implies that the eigenvalues of the helicity operator are ± 1. It is true that the helicity is not a Lorentz invariant quantity, but being a scalar product of two 3-vectors it is at least rotationally invariant, and it also has the advantage of being properly defined for massless particles. The trick is simply to use the direction of motion of the particle in the observer's frame as a reference axis to fix the direction of the spin vector.

For massive particles one can use both types of description. To this end we start from the rest frame of the particle. We assume that the particle has rest mass m and spin s. Then the particle states are described by the projection μ of the spin onto the z axis; we denote it in the form

$$|\boldsymbol{p}=0,\mu\rangle \quad , \quad -s \le \mu \le s \quad . \tag{3.43}$$

For Dirac particles, in the coordinate representation, these states are given by

$$|\boldsymbol{p}=0,\mu\rangle = (2mV)^{-\frac{1}{2}}\begin{pmatrix}\chi_\mu\\ 0\end{pmatrix} \quad , \quad \chi_{+\frac{1}{2}} = \begin{pmatrix}1\\ 0\end{pmatrix} \quad , \quad \chi_{-\frac{1}{2}} = \begin{pmatrix}0\\ 1\end{pmatrix} \quad . \tag{3.44}$$

In the standard treatment one obtains the corresponding state, where the particle moves with non-vanishing momentum $\boldsymbol{p}$, by a Lorentz transformation $L(\boldsymbol{p})$. According to Appendix A.4, for Dirac particles this transformation is accomplished by a unitary transformation

$$\hat{U}(L(\boldsymbol{p})) = \exp(\mathrm{i}\xi \boldsymbol{n}\cdot\hat{\boldsymbol{K}}) = \exp\left[\xi\boldsymbol{n}\cdot\left(t\nabla + \boldsymbol{x}\frac{\partial}{\partial t} - \frac{1}{2}\boldsymbol{\alpha}\right)\right] \quad , \tag{3.45}$$

where $\xi = \operatorname{artanh}(p/p_0)$ is the rapidity and $\boldsymbol{n} = \boldsymbol{p}/p$ is the unit vector in the direction of motion ($p \equiv |\boldsymbol{p}|$). Therefore we have

$$|\boldsymbol{p},\mu\rangle = \hat{U}(L(\boldsymbol{p}))|\boldsymbol{p}=0,\mu\rangle \quad . \tag{3.46}$$

We now look for a transformation that transforms eigenstates of the z component of the spin into helicity eigenstates. We proceed in several steps. At first we consider the special case where the particle moves in the z direction as seen from the observer's frame. The helicity eigenstates in this system are just the eigenstates of the spin projection onto the z axis. One obtains them from the corresponding states in the rest frame by applying the transformation $\hat{U}(L(p\boldsymbol{e}_z))$, since this transformation commutes with the operator $\hat{J}_z$. We therefore have

$$|p\boldsymbol{e}_z,\mu\rangle = \hat{U}(L(p\boldsymbol{e}_z))|\boldsymbol{p}=0,\mu=\lambda\rangle \quad . \tag{3.47}$$

From this state one can generate the helicity eigenstate of a particle moving in an arbitrary direction by applying a convenient rotation operator. Let

$$\boldsymbol{n} = \frac{\boldsymbol{p}}{p} = (\sin\theta\cos\phi, \sin\theta\sin\phi, \cos\theta) \tag{3.48}$$

be the corresponding direction vector. Then one can define the desired helicity state

$$|\boldsymbol{p},\lambda\rangle = \hat{R}(\phi,\theta,-\phi)\hat{U}(L(p\boldsymbol{e}_z))|\boldsymbol{p}=0,\mu=\lambda\rangle \quad , \tag{3.49}$$

where according to Appendix C the rotation is described by

$$\hat{R}(\phi,\theta,-\phi) = \exp\left(-\frac{\mathrm{i}}{2}\phi\hat{s}_z\right)\exp\left(-\frac{\mathrm{i}}{2}\theta\hat{s}_y\right)\exp\left(+\frac{\mathrm{i}}{2}\phi\hat{s}_z\right) \quad . \tag{3.50}$$

The Euler angles of the so-defined rotation are chosen in such a way that $\boldsymbol{e}_z$ transforms into the unit vector $\boldsymbol{n}$. The third angle which describes the first rotation around the z axis can be arbitrarily chosen and is fixed here by demanding that $\hat{R}(\phi, 0, -\phi) = 1$ for all ϕ. Now we still have to construct the spin eigenstate corresponding to the rest frame, which has to be inserted into (3.49). This state is found by inverting (3.46). Combining everything one gets

$$|\boldsymbol{p}, \lambda\rangle = \hat{R}(\phi, \theta, -\phi)\hat{U}(L(p\boldsymbol{e}_z))\hat{U}(L^{-1}(\boldsymbol{p}))|\boldsymbol{p}, \mu = \lambda\rangle \quad , \tag{3.51}$$

where, of course, $L^{-1}(\boldsymbol{p}) = L(-\boldsymbol{p})$. Since $\hat{R}(\phi, \theta, -\phi)$ exactly transforms the vector $p\boldsymbol{e}_z$ into the vector $\boldsymbol{p}$, we have

$$\hat{R}(\phi, \theta, -\phi)\hat{U}(L(p\boldsymbol{e}_z))\hat{R}^{-1}(\phi, \theta, -\phi) = \hat{U}(L(\boldsymbol{p})) \quad . \tag{3.52}$$

Hence we can also write (3.51) in the form

$$|\boldsymbol{p}, \lambda\rangle = \hat{U}(L(\boldsymbol{p}))\hat{R}(\phi, \theta, -\phi)\hat{U}(L^{-1}(\boldsymbol{p}))|\boldsymbol{p}, \mu = \lambda\rangle \quad . \tag{3.53}$$

The interpretation of this formula is obvious: one obtains the helicity eigenstate from the corresponding spin state (in the observer's frame) by first transforming into the rest frame, then rotating in the direction of motion, and finally returning to the original reference frame.

The next step consists in constructing states of good helicity for two particles. Here we are interested in the centre-of-mass representation, where $\boldsymbol{p}_1 + \boldsymbol{p}_2 = 0$. Since both momenta point in opposite directions and have equal absolute values, this state is uniquely determined by the quantities

$$\begin{aligned} p = |\boldsymbol{p}_1| &= |\boldsymbol{p}_2| \quad , \\ (\theta, \phi) &= (\theta_1, \phi_1) = (\pi - \theta_2, \phi_2 \pm \pi) \quad , \\ \lambda_1 \quad &, \quad \lambda_2 \quad . \end{aligned} \tag{3.54}$$

For this state the following relation holds:

$$|p, \theta, \phi, \lambda_1, \lambda_2\rangle = |\boldsymbol{p}_1, \lambda_1\rangle|\boldsymbol{p}_2, \lambda_2\rangle = |\boldsymbol{p}_1, \lambda_1\rangle| - \boldsymbol{p}_1, \lambda_2\rangle \quad . \tag{3.55}$$

As the total helicity of the two-particle state one defines

$$\lambda = \lambda_1 - \lambda_2 \quad . \tag{3.56}$$

Note that λ changes sign when the two particles are exchanged. Our Dirac wave functions (see Appendix A.2) are normalized in such a way that for one-particle states the relation

$$\langle \boldsymbol{p}', \lambda' | \boldsymbol{p}, \lambda\rangle = \frac{(2\pi)^3}{V}\delta^3(\boldsymbol{p} - \boldsymbol{p}')\delta_{\lambda\lambda'} \tag{3.57}$$

is fulfilled. Accordingly the two-particle states satisfy

$$\begin{aligned} &\langle \boldsymbol{p}_1', \lambda_1' | \boldsymbol{p}_1, \lambda_1\rangle\langle \boldsymbol{p}_2', \lambda_2' | \boldsymbol{p}_2, \lambda_2\rangle \\ &= \frac{(2\pi)^6}{V^2}\delta^3(\boldsymbol{p}_1 - \boldsymbol{p}_1')\delta^3(\boldsymbol{p}_2 - \boldsymbol{p}_2')\delta_{\lambda_1\lambda_1'}\delta_{\lambda_2\lambda_2'} \quad . \end{aligned} \tag{3.58}$$

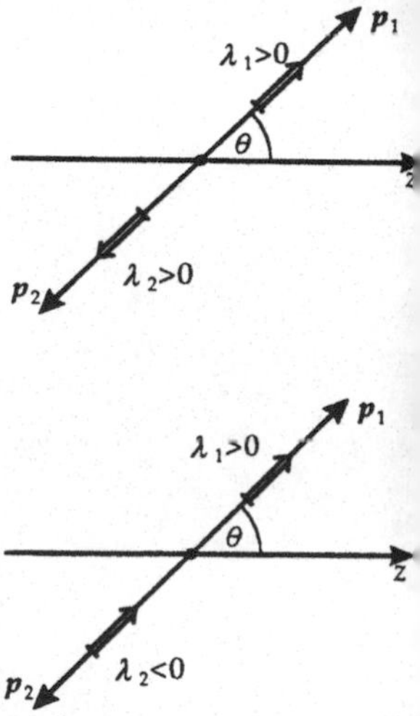

Fig. 3.8. Two-body helicity states for spin-$\frac{1}{2}$ particles. $\lambda = 0$ if the spin points along the direction of motion. The second case shows $\lambda = 1$

Transforming to new coordinates $P^\alpha = (p_1^\alpha + p_2^\alpha)$ and $k^\alpha = \frac{1}{2}(p_1^\alpha - p_2^\alpha)$, we have

$$\delta^3(\boldsymbol{p}_1 - \boldsymbol{p}_1')\delta^3(\boldsymbol{p}_2 - \boldsymbol{p}_2') = \delta^3(\boldsymbol{P} - \boldsymbol{P}')\delta^3(\boldsymbol{k} - \boldsymbol{k}') \quad , \tag{3.59}$$

since θ, ϕ and θ', ϕ' are just the polar angles of the relative momenta of the two particles. This transformation holds, since for a coordinate transformation from the set of N variables x_i to the set y_α, from the normalization condition

$$\int \delta(y_\alpha)\mathrm{d}^N y = \int \delta(x_i)\mathrm{d}^N x$$

one can derive the relation

$$\delta(y_\alpha(x_i)) = \left|\det\left(\frac{\partial y_\alpha}{\partial x_i}\right)\right|^{-1} \delta(x_i) \quad .$$

Hence the centre-of-mass systems of the incoming and outgoing particles coincide. In the common centre-of-mass system it then holds that $|\boldsymbol{k}| = p$ (cf. (3.54)), $|\boldsymbol{k}'| = p'$, and therefore

$$\delta(|\boldsymbol{k}| - |\boldsymbol{k}'|) = \delta(P^0 - P^{0'})\frac{\mathrm{d}P^0}{\mathrm{d}|\boldsymbol{k}|} \tag{3.60}$$

with

$$\frac{\mathrm{d}P^0}{\mathrm{d}|\boldsymbol{k}|} = \frac{\mathrm{d}}{\mathrm{d}p}\left(\sqrt{M_1^2 + p^2} + \sqrt{M_2^2 + p^2}\right) = \frac{pP^0}{p_1^0 p_2^0} = \frac{p\sqrt{s}}{p_1^0 p_2^0} \quad , \tag{3.61}$$

where $s = P^2$ as usual.

Inserting this into (3.59), we obtain a four-dimensional δ function for the centre-of-mass momentum. It is therefore convenient to separate the plane-wave part of the centre-of-mass motion:

$$|p, \theta, \phi, \lambda_1, \lambda_2\rangle = \frac{(2\pi)^3}{V}\left(\frac{\sqrt{s}}{pp_1^0 p_2^0}\right)^{\frac{1}{2}} |\theta, \phi, \lambda_1, \lambda_2\rangle |P\rangle \quad , \tag{3.62}$$

with the orthogonality relations

$$\langle\theta', \phi', \lambda_1', \lambda_2'|\theta, \phi, \lambda_1, \lambda_2\rangle = \delta(\cos\theta - \cos\theta')\delta(\phi - \phi')\delta_{\lambda_1\lambda_1'}\delta_{\lambda_2\lambda_2'} \quad , \tag{3.63a}$$

$$\langle P'|P\rangle = \delta^4(P - P') \quad . \tag{3.63b}$$

Thus the norm of the state (3.62) agrees with the right-hand side of (3.58), as one may readily prove by calculation.

According to our general rules, the differential cross section is given by the square of the transition amplitude integrated over all final states and divided by the space-time volume VT, as well as by the flux of incoming particles $\boldsymbol{J}_p$ and the density of target particles. In the centre-of-mass frame ($\boldsymbol{p}_2 = -\boldsymbol{p}_1 \equiv -\boldsymbol{p}$)

$$\begin{aligned} \boldsymbol{J}_p &= \frac{\boldsymbol{v}_1 - \boldsymbol{v}_2}{V}V^{-1} = \frac{1}{V^2}\left(\frac{\boldsymbol{p}_1}{p_1^0} - \frac{\boldsymbol{p}_2}{p_2^0}\right) \\ &= \frac{\boldsymbol{p}}{V^2}\left(\frac{1}{p_1^0} + \frac{1}{p_2^0}\right) = \frac{\boldsymbol{p}\sqrt{s}}{V^2 p_1^0 p_2^0} \quad . \end{aligned} \tag{3.64}$$

Hence in the centre-of-mass frame we obtain

$$\frac{\mathrm{d}\sigma}{\mathrm{d}\Omega} = \frac{V^2 p_1^0 p_2^0}{p\sqrt{s}VT} V \int \frac{\mathrm{d}^3 p_1'}{(2\pi)^3} V \int_0^\infty \frac{\mathrm{d}|\boldsymbol{p}_2'||\boldsymbol{p}_2'|^2}{(2\pi)^3}$$
$$\times |\langle p', \theta, \phi, \lambda_1', \lambda_2'|\hat{S} - 1|0, 0, \lambda_1, \lambda_2\rangle|^2$$
$$= \frac{V^4 p_1^0 p_2^0}{p\sqrt{s}(2\pi)^6 VT} \int \mathrm{d}^3 P' \int_0^\infty \mathrm{d}p' p'^2 \frac{(2\pi)^{12}}{V^4} \left(\frac{\sqrt{s}}{p p_1^0 p_2^0}\right)^2$$
$$\times |\langle P'|\langle \theta, \phi, \lambda_1', \lambda_2'|\hat{S} - 1|0, 0, \lambda_1, \lambda_2\rangle|P\rangle|^2 \quad . \tag{3.65}$$

Here we have assumed that the particles enter along the z axis $\theta = \phi = 0$, and are scattered in the direction (θ, ϕ). $\hat{S}$ is the scattering operator and is given by

$$\hat{S} = \lim_{t\to\infty} \mathrm{T}\left[\exp\left(-\mathrm{i}\int_{-t}^{t} \mathrm{d}t'\hat{H}\right)\right] \quad , \tag{3.66}$$

where T denotes the time-ordered product of operators. Since the centre-of-mass energy is conserved during the scattering process if external fields are absent, the scattering matrix element differs from zero only if $P' = P$. Therefore it is convenient to introduce the so-called *transition matrix* (T matrix) by the following definition

$$\langle P', \alpha_f|\hat{S} - 1|P, \alpha_i\rangle = 2\mathrm{i}\delta^4(P - P')\langle \alpha_f|\hat{T}|\alpha_i\rangle \quad , \tag{3.67}$$

where we have abbreviated further quantum numbers of the initial and final state by a_i and a_f respectively. If we insert this definition into (3.65) and again make use of the symbolic relation $(2\pi)^4\delta^4(0) = VT$, we obtain

$$\frac{\mathrm{d}\sigma}{\mathrm{d}\Omega} = \frac{s^{\frac{1}{2}}(2\pi)^2 \cdot 4}{p^3 p_1^0 p_2^0} \int \mathrm{d}^3 P' \int_0^\infty \mathrm{d}p' p'^2 \delta^4(P - P')$$
$$\times |\langle \theta, \phi, \lambda_1', \lambda_2'|\hat{T}|0, 0, \lambda_1, \lambda_2\rangle|^2 \quad . \tag{3.68}$$

Because of (3.60), (3.61) we have

$$\delta^4(P - P') = \delta^3(\boldsymbol{P} - \boldsymbol{P}')\delta(p - p')\frac{p_1^0 p_2^0}{p\sqrt{s}} \tag{3.69}$$

so that we get the differential cross section in the centre-of-mass system after performing the momentum-space integrals:

$$\frac{\mathrm{d}\sigma}{\mathrm{d}\Omega} = \frac{(4\pi)^2}{p^2}|\langle \theta, \phi, \lambda_1', \lambda_2'|\hat{T}|0, 0, \lambda_1, \lambda_2\rangle|^2 \quad . \tag{3.70}$$

We have therefore found the general expression for the relativistic two particle scattering cross section.

The dependence of the transition matrix element in (3.70) on the scattering angle becomes clearer, if one performs a partial-wave decomposition. This is done by transforming the functions $|\theta, \phi, \lambda_1, \lambda_2\rangle$, while specifying a certain direction of the relative momentum, into a superposition of functions with good angular momentum, or, in other words, by projecting on good angular momentum. The functions which accomplish this are the Wigner D functions,[3]

$$|J,M,\lambda_1,\lambda_2\rangle = \sqrt{\frac{2J+1}{4\pi}} \int_0^{2\pi} \mathrm{d}\phi \times \int_{-1}^{+1} \mathrm{d}\cos\theta\, D^{J*}_{M\lambda}(\phi,\theta,-\phi)\,|\theta,-\phi,\lambda_1,\lambda_2\rangle \quad , \tag{3.71}$$

where again $\lambda = \lambda_1 - \lambda_2$.

The inversion is given by

$$|\theta,\phi,\lambda_1,\lambda_2\rangle = \sqrt{\frac{2J+1}{4\pi}} \sum_{J,M} D^J_{M\lambda}(\phi,\theta,-\phi)\,|J,M,\lambda_1,\lambda_2\rangle \quad . \tag{3.72}$$

The new functions fulfill the eigenvalue equations

$$\hat{J}^2|J,M,\lambda_1,\lambda_2\rangle = J(J+1)|J,M,\lambda_1,\lambda_2\rangle \quad , \tag{3.73a}$$

$$\hat{J}_z|J,M,\lambda_1,\lambda_2\rangle = M|J,M,\lambda_1,\lambda_2\rangle \quad , \tag{3.73b}$$

and their orthogonality relation is given by

$$\langle J',M',\lambda_1',\lambda_2'|J,M,\lambda_1,\lambda_2\rangle = \delta_{JJ'}\delta_{MM'}\delta_{\lambda_1\lambda_1'}\delta_{\lambda_2\lambda_2'} \quad . \tag{3.74}$$

Since angular momentum is conserved in a scattering process of two particles, only terms with $J = J'$ and $M = M'$ contribute to the T matrix. Furthermore the value of the matrix element has to be independent of M, otherwise a certain spatial distribution would be preferred. Hence we can write

$$\begin{aligned}\langle\theta,\phi,\lambda_1',\lambda_2'|\hat{T}|0,0,\lambda_1,\lambda_2\rangle &= \sum_{J',M'}\sum_{J,M}\sqrt{\frac{(2J'+1)}{4\pi}\frac{(2J+1)}{4\pi}} \\ &\quad\times D^{J'*}_{M'\lambda'}(\phi,\theta,-\phi)D^J_{M\lambda}(0,0,0)\langle J',M',\lambda_1',\lambda_2'|\hat{T}|J,M,\lambda_1,\lambda_2\rangle \\ &= \sum_{J,M}\frac{(2J+1)}{4\pi}D^{J*}_{M\lambda'}(\phi,\theta,-\phi)D^J_{M\lambda}(0,0,0) \\ &\quad\times\langle J',M',\lambda_1',\lambda_2'|\hat{T}|J,M,\lambda_1,\lambda_2\rangle \quad .\end{aligned} \tag{3.75}$$

With

$$D^J_{m'm}(\alpha,\beta,\gamma) = \mathrm{e}^{-\mathrm{i}\alpha m'}d^J_{m'm}(\beta)\mathrm{e}^{-\mathrm{i}\gamma m} \tag{3.76}$$

and

$$d^J_{m'm}(0) = \delta_{m'm} \tag{3.77}$$

this can be further simplified to yield

$$\sum_J \frac{(2J+1)}{4\pi}\mathrm{e}^{-\mathrm{i}(\lambda-\lambda')\phi}d^J_{\lambda\lambda'}(\theta)\langle J,\lambda,\lambda_1',\lambda_2'|\hat{T}|J,\lambda,\lambda_1,\lambda_2\rangle \quad . \tag{3.78}$$

[3] See E.M. Rose: *Elementary Theory of Angular Momentum* (Wiley, New York, 1957).

As already mentioned, the matrix element cannot depend on the value of M; hence we can write it symbolically in the form

$$\langle J, \lambda_1', \lambda_2' || \hat{T} || J, \lambda_1, \lambda_2 \rangle \tag{3.79}$$

and choose the most convenient value of M, for example $M = \lambda$ as in (3.78), to evaluate it. In the centre-of-mass frame the differential cross section therefore assumes the simple form

$$\frac{d\sigma}{d\Omega} = |f(\theta, \phi)|^2 \quad \text{(centre-of-mass frame)} \tag{3.80}$$

with

$$f(\theta, \phi) = \frac{1}{p} e^{-i(\lambda-\lambda')\phi} \sum_J (2J+1) d^J_{\lambda\lambda'}(\theta) \langle J, \lambda_1', \lambda_2' || \hat{T} || J, \lambda_1, \lambda_2 \rangle \quad . \tag{3.81}$$

The unitarity of the scattering operator (3.66), $\hat{S}^\dagger \hat{S} = 1$, leads to the following formal equation for the operator of the $\hat{T}$ matrix, which is defined by (3.67):

$$\hat{S} = \hat{1} + 2i\hat{T} \quad \rightarrow \quad \hat{T} - \hat{T}^\dagger = 2i\hat{T}^\dagger \hat{T} \quad . \tag{3.82}$$

The matrix elements of the scattering operator satisfy the unitary constraint

$$|T_{fi}| \leq 1 \quad . \tag{3.83}$$

This can be readily shown by calculating the matrix elements of the unitary relation $\hat{S}^\dagger \hat{S} = 1$, given explicitly by $\sum_n S^*_{ni} S_{nf} = \delta_{if}$, and estimating the following

(a) $i = f$:

$$\delta_{ij} = 1 = \sum_n |S_{ni}|^2 \geq |S_{ii}|^2 = |1 + 2iT_{ii}|^2 \quad . \tag{3.84}$$

Let $T_{ii} = x + iy$. Then it follows that $|1 + 2iT_{ii}|^2 = |1 + 2ix - 2y|^2 = (1-2y)^2 + 4x^2 \leq 1$ or $-2y + 4y^2 + 4x^2 \leq 0$. For fixed y this is fulfilled if $|x| \leq \sqrt{\frac{1}{2}y(1-2y)}$. Hence one concludes that $y \leq \frac{1}{2}$ and, therefore, $|T_{ii}|^2 = x^2 + y^2 \leq \frac{1}{2}y \leq \frac{1}{4}$ or $|T_{ii}| < \frac{1}{2} < 1$.

(b) $i \neq f$:

$$1 = \sum_n |S_{ni}|^2 \geq |S_{fi}|^2 = 4|T_{fi}|^2 \rightarrow |T_{fi}| \leq \frac{1}{2} < 1 \tag{3.85}$$

Of course, one has to assume orthonormality and the closure relation for the functions which form the basis of the representation of $\hat{S}$.

3.6 Divergences in Higher-Order Processes

We concluded in one of the previous sections of this chapter that the high-energy behavior of Fermi's theory is determined by higher-order processes. However, if one tries to calculate such processes a new difficulty arises. Let us for instance consider the contribution of the diagram in Fig. 3.9 to neutrino–electron scattering. The interaction Hamiltonian at both vertices is

$$H_{\mathrm{int}}(\nu_{\mathrm{e}}\mathrm{e}^- \to \mathrm{e}^-\nu_{\mathrm{e}}) = \frac{G}{\sqrt{2}} \int \mathrm{d}^3x\, J_\alpha^{(\mathrm{e})\dagger} J_{(\mathrm{e})}^\alpha(x) \quad . \tag{3.86}$$

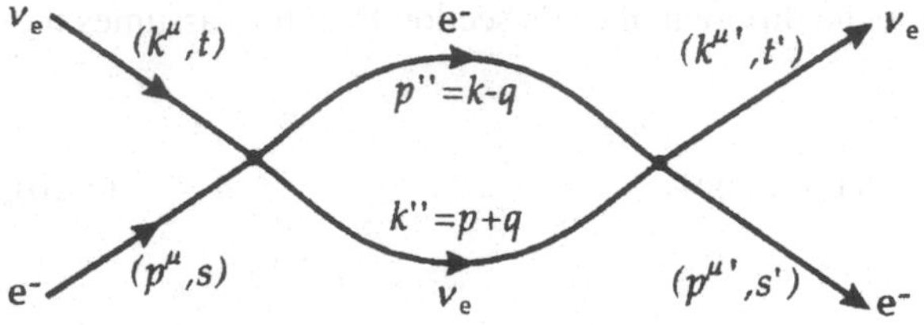

g. 3.9. Neutrino–electron :attering in second order

According to standard Feynman rules (see Appendix A.3), inner fermion lines are represented by Dirac propagators; the contribution to the scattering matrix element is therefore

$$\begin{aligned} &S^{(2)}(\nu_{\mathrm{e}}\mathrm{e}^- \to \mathrm{e}^-\nu_{\mathrm{e}}) \\ &= -\mathrm{i}\frac{G^2}{2}(2\pi)^4 \frac{\delta^4(p'+k'-p-k)}{\sqrt{16V^4 k_0 p_0 k_0' p_0'}} \\ &\quad \times \int \frac{\mathrm{d}^4q}{(2\pi)^4}\left[\overline{u}_{\mathrm{e}}(\boldsymbol{p}',s')\gamma_\alpha(1-\gamma_5)\frac{1}{\not p + \not q}\gamma_\beta(1-\gamma_5)u_{\mathrm{e}}(\boldsymbol{p},s)\right] \\ &\quad \times \left[\overline{u}_{\nu_{\mathrm{e}}}(\boldsymbol{k}',t')\gamma^\alpha(1-\gamma_5)\frac{1}{\not k - \not q - m_{\mathrm{e}}}\gamma^\beta(1-\gamma_5)u_{\nu_{\mathrm{e}}}(\boldsymbol{k},t)\right] . \end{aligned} \tag{3.87}$$

As required, we have integrated over the 4-momentum q^μ exchanged in the first scattering. One observes that this integral over q diverges, since only two powers of q appear in the denominator

$$S^{(2)} \simeq G^2 \int \frac{\mathrm{d}^4q}{q^2} \simeq G^2 \int_0^\infty \frac{q^3\mathrm{d}q}{q^2} \simeq G^2 \int_0^\infty q\,\mathrm{d}q \quad . \tag{3.88}$$

Such divergences in Feynman graphs of higher order in the Fermi coupling constant G are in principle nothing new; they also appear in quantum electrodynamics.[4] A typical divergent graph is the self-energy of the photon in lowest order ("vacuum polarization"), where a quadratically divergent integral over the intermediate momentum q also arises (Fig. 3.10). In this case, however, the quadratic divergence can be eliminated by demanding gauge invariance for the photon propagator. What remains is a logarithmic divergence of the form $\int \mathrm{d}q/q$ which can be absorbed by renormalizing the electric charge ("charge renormalization"). This procedure is not applicable to the divergence (3.88), since there is no gauge principle in Fermi's theory which could reduce the order of divergence. In higher-order processes even worse divergences appear. One says that *Fermi's theory is not renormalizable*.[5] The reason for the divergence of the integral in (3.87) lies in the nature of the interaction vertex between the

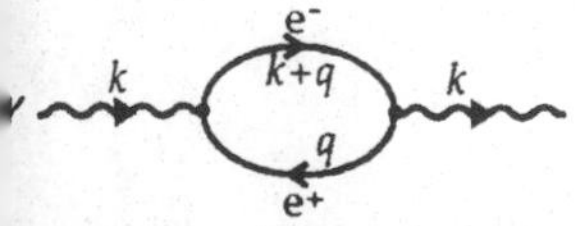

ig. 3.10. Vacuum polarization n QED

[4] See W. Greiner and J. Reinhardt: *Quantum Electrodynamics*, 2nd ed. (Springer, Berlin, Heidelberg, 1994).

[5] This statement should be handled with care. Strictly speaking it means only that Fermi's theory is not renormalizable in successive orders of perturbation theory. If one were able to sum all orders ex-

electron and neutrino, namely that both particles interact only if they are at the same place. The sole q dependence therefore arises from the propagators of the particles in the intermediate state, but the current–current coupling does not contribute any q dependence. However, if electron and neutrino were interacting by means of an "intermediate" boson W (as in the electron–electron scattering of QED), as illustrated in Fig. 3.11, one would have an effective Fermi coupling constant which would depend on the momentum transfer:

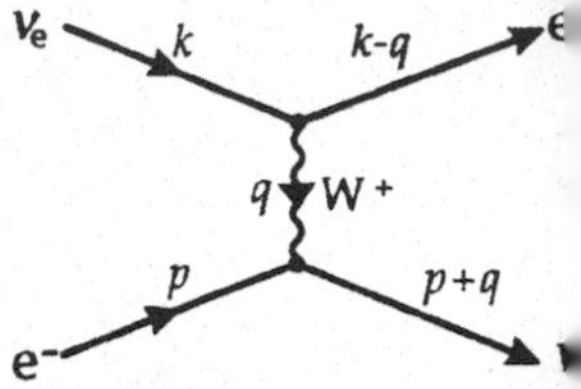

Fig. 3.11. Electron–neutrino scattering by means of a positively charged intermediate boson (W^+)

$$G_{\text{eff}}(q^2) = \frac{g^2}{q^2 - M_W^2} \to \frac{g^2}{q^2} \quad (q \to \infty) \quad . \tag{3.89}$$

Here g is a dimensionless coupling constant for the $\nu_e - e - W$ vertex. The second interaction point in a graph of type Fig. 3.9 contributes another q dependence through the momentum transfer $(p' - p - q)$:

$$G_{\text{eff}}((p' - p - q)^2) = \frac{g^2}{(p' - p - q)^2 - M_W^2} \to \frac{g^2}{q^2} \quad (q \to \infty) \tag{3.90}$$

Now we would have four additional powers of q in the denominator and the diagram would finally behave as:

$$S^{(2)} \simeq g^4 \int \frac{d^4q}{q^2} \left(\frac{1}{q^2}\right)^2 \simeq g^4 \int \frac{dq}{q^3} \quad , \tag{3.91}$$

which is no longer divergent at the upper boundary $(q \to \infty)$. Hence with the help of an intermediate boson one could construct a consistent theory of weak interactions.

Unfortunately we know from Fermi's theory that the coupling must have vector character (or axial-vector character). The intermediate boson W has to couple to the vector currents j_μ and hence must be described by a vector field, that is, it must be a spin-1 particle similar to the photon in quantum electrodynamics. As shown in Example 4.6, the propagator for a spin-1 particle has the form

$$D_{\mu\nu}^{(M)}(q^2) = -\frac{g_{\mu\nu} - \frac{q_\mu q_\nu}{M_W^2}}{q^2 - M_W^2} \to \frac{q_\mu q_\nu}{q^2 M_W^2} \to \frac{\text{const}}{M_W^2} \quad (q^2 \to \infty) \quad . \tag{3.92}$$

The appearance of the momentum q in the numerator just destroys the whole effect of introducing an intermediate propagator. Such a theory would therefore also be nonrenormalizable. The renormalizability of quantum electrodynamics derives from the fact that the photon has vanishing rest mass; in this case the propagator again has a different form (see Example 4.6), namely

$$D_{\mu\nu}^{(M=0)}(q^2) = -\frac{g_{\mu\nu} - \frac{q_\mu q_\nu}{q^2}}{q^2} \quad \to \quad \frac{\text{const}}{q^2} \quad (q^2 \to \infty) \quad . \tag{3.93}$$

---

plicitly, it might eventually happen that the complete theory contains no divergences. (As an example of such behavior we refer to the integral $\int_0^\infty dx e^{-x} = 1$. If we expand the function e^{-x} in terms of the power series $e^{-x} = 1 - x + x^2/2! - x^3/3! \ldots$, which is convergent for every x, the integral over each single term of the series diverges!) In a somewhat different context it was recently conjectured that point-like four-fermion interactions may, indeed, be renormalizable under certain conditions.

Now we are in a dilemma: the theory would only be renormalizable with massless vector bosons, but we require the mass $M_W \neq 0$ for the intermediate bosons to regain Fermi's theory in the limit of small momentum transfer. For $q^2 \to 0$ it should, of course, hold that

$$G_{\text{eff}}(q^2) \equiv g^2 D_{\mu\nu}(q^2) \to G \quad . \tag{3.94}$$

This condition is not fulfilled for the propagator with $M_W = 0$, whereas for a massive boson, with conveniently chosen values of g and M_W, we have

$$G_{\text{eff}}(q^2) \to \frac{g^2}{M_W^2} = G \quad (q^2 \to 0) \quad . \tag{3.95}$$

A way out is provided by an ingenious trick often called the *Higgs mechanism*, which should actually be referred to as the Brout–Englert–Higgs mechanism.[6] One starts with a massless boson W but couples it to a new spin-0 field ϕ, the Higgs field, by means of an interaction term $g^2\phi^2 W_\mu$. If now the scalar field ϕ assumes a constant value $\phi_0 = M_W/g$ everywhere in space, one obtains in this way a term $g^2\phi_0^2 W_\mu(x)$ in the wave equation for the W boson which plays the same role as a mass term $M_W^2 W_\mu(x)$.

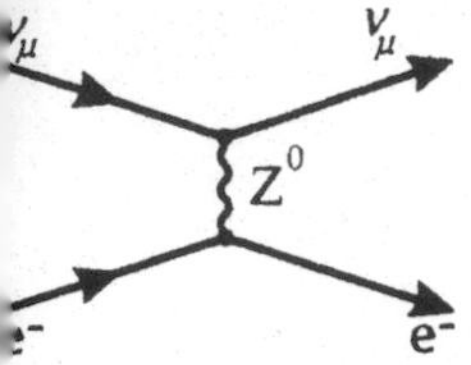

ig. 3.12. Neutral current oupling by means of a neu- :al intermediate boson Z^0

We also have to remember the existence of the neutral currents. To describe the process $\nu_\mu e^- \to \nu_\mu e^-$ one obviously requires a neutral boson Z^0 (see Fig. 3.12). In all we therefore need the charged bosons W^+, W^- and the neutral Z^0. S. Weinberg and A. Salam realized that, since the Z^0 boson is originally massless and neutral, it has nearly the same properties as the photon! One can therefore expect the theory really to make sense only if we start from the beginning with the three intermediate bosons of the weak interaction W^+, W^-, Z^0 together with the photon γ and develop a unified theory of *"electroweak"* interactions from there.

3.7 Biographical Notes

SALAM, Abdus, theoretical physicist, *29.1.1926 in Jhang (Pakistan), †21.11.1996 in Oxford (England). He received his Ph.D. in 1952 from the University of Cambridge and became professor of Imperial College, London in 1957; since 1964 he is director of the International Center of Theoretical Physics at Trieste (Italy). His main work is on the two-component theory of the neutrino and on symmetry properties of elementary particles, for which he shared in the 1979 Nobel Prize in physics with Glashow and Weinberg.

WEINBERG, Steven, theoretical physicist, *3.5.1933 in New York. He received his Ph.D. from Princeton in 1957, became professor at the University of California in Berkeley in 1960, later at the Massachusetts Institute of Technology, Harvard University, and since 1982 professor at the University of Texas in Austin. He made many contributions to the theory of elementary particles, including the unified electroweak theory, the theory of strong interactions, and to particle cosmology. He shared in the 1979 Nobel Prize in physics with Glashow and Salam.

---

[6] This technique was independently "invented" by several theorists: (a) F. Englert and R. Brout: Phys. Rev. Lett. **13**, 321 (1964) first; (b) P.W. Higgs (Phys. Rev. Lett. **13**, 508 (1964) and Phys. Rev. **145**, 1156 (1966)). The paper by Englert and Brout is more general and detailed. A paper by Kibble is pedagogically useful: T.W. Kibble: Phys. Rev. **155**, 1554 (1967).

The Salam–Weinberg Theory 4

4.1 The Higgs Mechanism

In this chapter we shall reformulate Fermi theory as a quantum field theory involving the exchange of massive vector bosons that mediate the weak interactions, and incorporating charged and neutral currents in a unified way. First we have to learn how to convert a theory involving massless vector bosons into a theory of massive particles, without disturbing the favorable high-energy behavior of a massless propagator. We shall proceed by discussing the methodical approach first, before performing a detailed calculation.

The propagator of a massless spin-1 particle,

$$\mathrm{i}D_{\mu\nu}(q^2, M^2 = 0) = -\mathrm{i}\frac{g_{\mu\nu} - q_\mu q_\nu / q^2}{q^2 + \mathrm{i}\varepsilon} \equiv -\mathrm{i}\frac{P_{\mu\nu}(q)}{q^2 + \mathrm{i}\varepsilon} \quad , \tag{4.1}$$

differs from the Feynman propagator of a massive spin-1 particle,

$$\mathrm{i}D_{\mu\nu}(q^2, M) = -\mathrm{i}\frac{g_{\mu\nu} - q_\mu q_\nu / M^2}{q^2 - M^2 + \mathrm{i}\varepsilon} \quad , \tag{4.2}$$

in that the former contains the *operator projecting onto transverse states*

$$P_{\mu\nu}(q) = g_{\mu\nu} - \frac{q_\mu q_\nu}{q^2} \quad , \tag{4.3}$$

since

$$q^\mu P_{\mu\nu}(q) = P_{\mu\nu}(q) q^\nu = 0 \quad , \tag{4.4}$$

whereas in the case of *massive* spin-1 particles we get the relation

$$q^\mu D_{\mu\nu}(q^2, M) = +\mathrm{i}\frac{q_\nu}{M^2} \quad . \tag{4.5}$$

That is, the propagator contains longitudinal parts which increase, even for $M \to 0$.

The longitudinal parts, which increase together with q, yield the disturbing property of a theory of massive spin-1 particles, namely that Feynman graphs including closed loops diverge faster than logarithmically. As a result, the divergences cannot be absorbed into renormalization constants, that is, the massive spin-1 quantum field theory is not renormalizable.[1] Therefore we have to think of a 'trick' to incorporate

[1] A more detailed study reveals that the theory remains renormalizable if the massive vector bosons couple only to a conserved fermion current. However, as we have learned in Chap. 1, the weak interaction couples to a mixture of vector and axial-vector currents, which is not conserved.

W. Greiner, B. Müller, *Gauge Theory of Weak Interactions*,
DOI 10.1007/978-3-540-87843-8_4, © Springer-Verlag Berlin Heidelberg 2009

boson masses without destroying transversality. To this end we add an interaction of the massless spin-1 particle with another field (which we will specify later on). Therefore we get the following graphical representation of the exact propagator $G_{\mu\nu}(q^2)$:

where $\Pi_{\mu\nu}(q^2)$ denotes the tensor of vacuum polarization. In terms of Feynman graphs $\Pi_{\mu\nu}$ contains all one-particle irreducible graphs, that is, graphs which cannot be split up into two separated parts by cutting just a single line. For instance, the following graphs are irreducible:

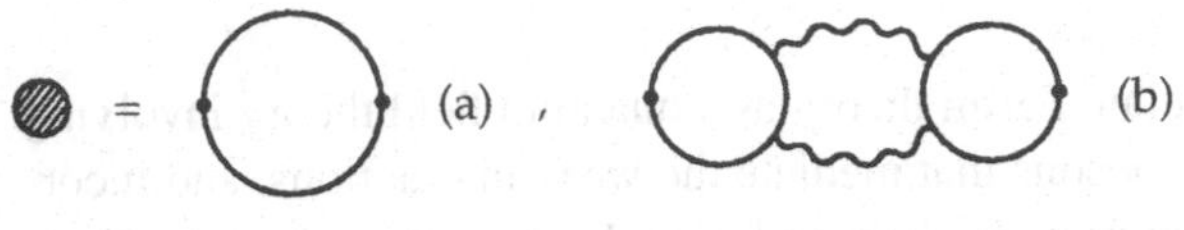

whereas the polarization graph

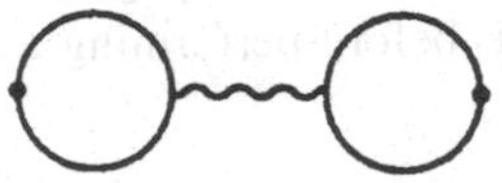
(c)

is reducible. This can easily be seen by cutting the photon line in the centre of the diagram. The contribution from this graph (c) is included in the third term $\mathrm{i}D(\mathrm{i}\Pi)\mathrm{i}D(\mathrm{i}\Pi)\mathrm{i}D$ of the expanded exact propagator $G_{\mu\nu}(q^2)$. Taking (4.1) we have explicitly

$$\begin{aligned}\mathrm{i}G_{\mu\nu}(q^2) &= \mathrm{i}D_{\mu\nu}(q^2) + \mathrm{i}D_{\mu\alpha}(q^2)\mathrm{i}\Pi^{\alpha\beta}(q^2)\mathrm{i}D_{\beta\nu}(q^2) + \cdots \\ &= -\mathrm{i}\frac{P_{\mu\nu}(q)}{q^2} + (-\mathrm{i})\frac{P_{\mu\alpha}(q)}{q^2}\mathrm{i}\Pi^{\alpha\beta}(q^2)(-\mathrm{i})\frac{P_{\beta\mu}(q)}{q^2} + \cdots \quad . \end{aligned} \tag{4.6}$$

In QED it can be shown[2] that gauge invariance, that is charge conservation, requires that the vacuum polarization tensor $\Pi_{\mu\nu}$ is purely transverse; the common notation is therefore

$$\Pi_{\mu\nu}(q^2) = (g_{\mu\nu}q^2 - q_\mu q_\nu)\Pi(q^2) = q^2 P_{\mu\nu}(q^2)\Pi(q^2) \quad . \tag{4.7}$$

When inserting this result into (4.6) we can make use of $P_{\mu\nu}$'s being a projection operator, which can therefore be applied several times without changing the result:

$$\begin{aligned}P^\alpha_\mu(q)P_{\alpha\nu}(q) &= \left(g^\alpha_\mu - \frac{q_\mu q^\alpha}{q^2}\right)\left(g_{\alpha\nu} - \frac{q_\alpha q_\nu}{q^2}\right) \\ &= g_{\mu\nu} - \frac{q_\mu q_\nu}{q^2} - \frac{q_\mu q_\nu}{q^2} + \frac{q_\mu q^\alpha}{q^2}\frac{q_\alpha q_\nu}{q^2} \\ &= g_{\mu\nu} - \frac{q_\mu q_\nu}{q^2} = P_{\mu\nu}(q) \quad . \end{aligned} \tag{4.8}$$

[2] See W. Greiner and J. Reinhard: *Quantum Electrodynamics*, 4th ed. (Springer, Berlin, Heidelberg, 2009).

We then sum the resulting geometrical series:

$$\begin{aligned} \mathrm{i}G_{\mu\nu}(q^2) &= -\mathrm{i}\frac{P_{\mu\nu}(q)}{q^2} - \mathrm{i}\frac{P_{\mu\nu}(q)}{q^2}\Pi(q^2) - \mathrm{i}\frac{P_{\mu\nu}(q)}{q^2}\Pi^2(q^2)\ldots \\ &= -\mathrm{i}\frac{P_{\mu\nu}}{q^2}\left[1-\Pi(q^2)\right]^{-1} = \mathrm{i}D_{\mu\nu}(q^2)\left[1-\Pi(q^2)\right]^{-1} \quad . \end{aligned} \tag{4.9}$$

This result is important: *even the exact propagator remains transverse!* This behavior was caused by the transversality of the polarization tensor $\Pi_{\mu\nu}$. In fact, this property results from the gauge invariance of the theory (in this case QED), which directly yields current conservation and consequently the relation[3]

$$q^\mu \Pi_{\mu\nu}(q^2) = q^\nu \Pi_{\mu\nu}(q^2) = 0 \quad . \tag{4.10}$$

Thus we draw the conclusion that we have to formulate the theory of the intermediate boson in a gauge-invariant way. If it were possible to construct the theory in a such way that for $q^2 \to 0$ we had

$$\Pi(q^2) \to M^2/q^2 \quad , \tag{4.11}$$

we would obtain the low-momentum limit $q^2 \to 0$ of the propagator,

$$\mathrm{i}G_{\mu\nu}(q^2) = -\mathrm{i}\frac{P_{\mu\nu}(q)}{q^2(1-M^2/q^2)} = -\mathrm{i}\frac{P_{\mu\nu}(q)}{q^2-M^2} \quad . \tag{4.12}$$

In effect, for small q^2 we would have a transverse propagator with mass! In the case of $q^2 \gg M^2$, $\Pi(q^2)$ would then have a totally different behavior, but this would not be in contradiction to the experimental data obtained at low energies. All we must demand is that in the $q^2 \to 0$ limit the propagator behaves like a propagator of a massive particle.

The effect just mentioned, of a totally different behavior for small momenta compared to large ones, is well known from the classical electrodynamics of continuous media. In a conducting liquid (electrolyte) charges are screened; the characteristic length (Debye length) corresponds to the inverse of the mass M: $\lambda_D = \hbar c/M$. This means that in an electrolyte the longitudinal photons, being massless at the beginning, gain mass in the static case ($\omega \to 0$). For large frequencies this is no longer the case, because the particles of the electrolyte cannot follow the rapidly changing fields.

EXERCISE

4.1 The Debye Effect

Problem. Show that in an electrolyte every charge is screened according to

$$\phi(r) \propto \frac{\mathrm{e}^{-\mu r}}{r}$$

[3] More rigorously, in QED one demands that the vacuum polarization tensor obey (4.10), in order to eliminate quadratic divergences.

ercise 4.1

with

$$\mu^2 = 8\pi e^2 n_0/(k_B T)$$

given by the average density n_0 of the charged particles in the electrolyte and by the temperature T of the system.

Solution. We consider a point charge e in the electrolyte (which could even be one of the ions of the electrolyte!) with

$$\phi(r) \quad , \quad r = |\boldsymbol{x} - \boldsymbol{x}'| \quad , \tag{1}$$

being the electrostatic potential in the vicinity of the charge, which has to be determined. The density of like-charged particles, $n_+(\boldsymbol{r})$, around the charge is given by the Boltzmann distribution

$$n_+(r) = n_0 \exp\left(-\frac{e\phi(r)}{k_B T}\right) \quad , \tag{2a}$$

whereas the density of the oppositely charged particles is given by

$$n_-(r) = n_0 \exp\left(+\frac{e\phi(r)}{k_B T}\right) \quad , \tag{2b}$$

where n_0 denotes the equilibrium density of the charged particles in the electrolyte. The term $\pm e\phi(r)$ is just the local potential energy of the charged particles.

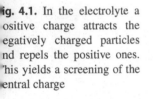

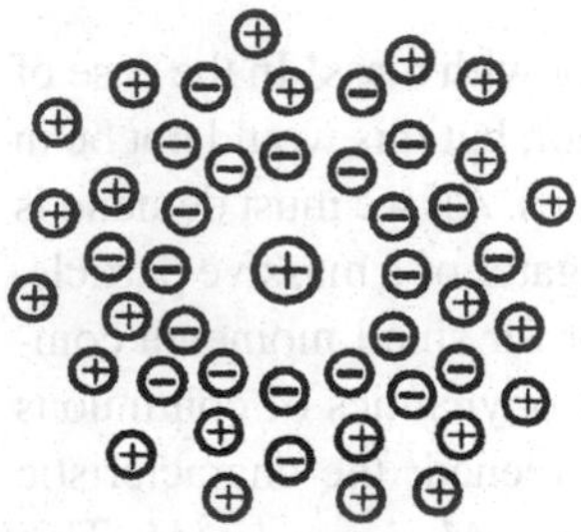

ig. 4.1. In the electrolyte a ositive charge attracts the egatively charged particles nd repels the positive ones. his yields a screening of the entral charge

A second relation between ϕ and n originates from the Poisson equation,

$$\nabla^2\phi(r) = -4\pi e\big(n_+(r) - n_-(r)\big) - 4\pi e\delta^3(\boldsymbol{r}) \quad . \tag{3}$$

Inserting (2) into (3) and expanding the exponential for small ϕ yields

$$\begin{aligned}\nabla^2\phi(r) &= +4\pi e n_0 2\sinh\left(\frac{e\phi(r)}{k_B T}\right) - 4\pi e\delta^3(\boldsymbol{r}) \\ &\approx \frac{8\pi e^2 n_0}{k_B T}\phi(r) - 4\pi e\delta^3(\boldsymbol{r}) \quad .\end{aligned} \tag{4}$$

With

$$\mu^2 = \frac{8\pi e^2 n_0}{k_B T} \tag{5}$$

we obtain the modified Poisson equation

$$(\nabla^2 - \mu^2)\phi(r) = -4\pi e\delta^3(\boldsymbol{r}) \quad , \tag{6}$$

with the solution

$$\phi(r) = \frac{e}{r}\exp(-\mu r) \quad . \tag{7}$$

Thus the medium screens the charge which generates the field. Formally, (6) corresponds to the equation of a potential due to a massive field, for instance occurring in Yukawa theory, where μ denotes the pion mass. The (longitudinal) photons obtain a mass through the interaction with the charged particles of the electrolyte. (This behavior is only valid in the static limit. In the case of rapidly changing fields the situation changes.)

The simplest model which is capable of generating massive intermediate bosons is the Higgs–Kibble model.[4] Before turning to a discussion of this model, we must learn how to identify masses of physical particles in a non-linear boson theory.

EXAMPLE

4.2 Creation of Mass in Interacting Fields

We start with a set of N interacting scalar fields ϕ_i, for which the Lagrangian reads

$$L(x) = \sum_{j=1}^{N} \frac{1}{2}(\partial_\mu \phi_j(x))(\partial^\mu \phi_j^*) - U(\phi_i(x)) \quad . \tag{1}$$

The function U is some polynomial of the fields ϕ_i. Now we shall determine the particle spectrum of the theory in the simplest limit of weak excitations. A simple mechanical example may serve as an intuitive guide to the solution of this problem. We consider a non-relativistic particle moving in a potential $V(x)$. In order to determine the lowest quantum-mechanical states, we approximate the potential $V(x)$ in the vicinity of its minimum by a harmonic oscillator (see figure) and determine the excited states of the oscillator. The potential $V(x)$ of this example corresponds to

$$\int d^3r \left[\sum_j \frac{1}{2}|\nabla\phi_j|^2 + U(\phi_i)\right] \quad . \tag{2}$$

The analogue of the coordinate x of the particle is given by the fields $\phi_i(\boldsymbol{r})$. The term (2) has a minimum for fields $\phi_i(x) = \phi_i^{(0)}$, being constant in space, chosen in such a way that $U(\phi_i^{(0)})$ attains a minimum. For varying fields, $U(\phi_i(x))$ is generally larger than $U(\phi_i^{(0)})$. Furthermore, the first part of (2), denoting the kinetic energy of ϕ_i, yields an additional positive contribution.

[4] P.W. Higgs: Phys. Lett. **12**, 132 (1964), Phys. Rev. Lett. **13**, 508 (1964), and Phys. Rev. **145**, 1156 (1966); T.W. Kibble: Phys. Rev. **155**, 1554 (1967); F. Englert and R. Brout: Phys. Rev. Lett. **13**, 321 (1964).

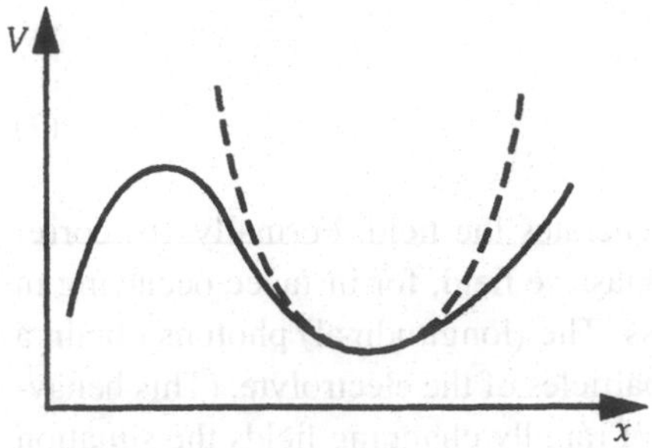

g. 4.2. Approximation by
ı oscillator potential

We now expand $U(\phi_i)$ around $\phi_i^{(0)}$ in terms of a Taylor series:

$$U(\phi_i) = U(\phi_i^{(0)}) + \frac{1}{2}\sum_{j,k=1}^{N} \left.\frac{\partial^2 U}{\partial\phi_j\partial\phi_k}\right|_{\phi_i=\phi_i^{(0)}} (\phi_j - \phi_j^{(0)})(\phi_k - \phi_k^{(0)})$$
$$+\frac{1}{6}\sum_{j,k,l=1}^{N} \left.\frac{\partial^3 U}{\partial\phi_j\partial\phi_k\partial\phi_l}\right|_{\phi_i=\phi_i^{(0)}} (\phi_j - \phi_j^{(0)})(\phi_k - \phi_k^{(0)})(\phi_l - \phi_l^{(0)})$$
$$+\cdots \qquad (3)$$

The term linear in $\phi_i - \phi_i^{(0)}$ vanishes, since $\phi_i^{(0)}$ determine a minimum of U, and hence

$$\left.\frac{\partial U}{\partial\phi_j}\right|_{\phi_i=\phi_i^{(0)}} = 0 \quad . \qquad (4)$$

Now we transform the symmetric matrix

$$(M^2)_{jk} \equiv \left.\frac{\partial^2 U}{\partial\phi_j\partial\phi_k}\right|_{\phi_i=\phi_i^{(0)}} \qquad (5)$$

onto its principal axes by introducing the fields

$$\tilde{\phi}_\alpha = \sum_{j=1}^{N} R_{\alpha j}(\phi_j - \phi_j^{(0)}) \quad . \qquad (6)$$

The rotational matrix $R_{\alpha j}$ obeys the relation[5]

$$\sum_\alpha R_{\alpha j}R_{\alpha k} = \delta_{jk} \quad ,$$
$$\sum_j R_{\alpha j}R_{\beta j} = \delta_{\alpha\beta} \quad . \qquad (7)$$

Therefore we can directly derive

$$\sum_{i=1}^{N}[\dot{\phi}_i^2 - |\nabla\phi_i|^2] = \sum_{\alpha=1}^{N}[\dot{\tilde{\phi}}_\alpha^2 - |\nabla\tilde{\phi}_\alpha|^2] \quad . \qquad (8)$$

---
5 See W. Greiner and B. Müller: *Quantum Mechanics – Symmetries*, 2nd ed. 3rd printing (Springer, Berlin, Heidelberg, 2001).

Furthermore, from (7) we can deduce the inverse of (6):

$$\phi_j - \phi_j^{(0)} = \sum_{\alpha=1}^{N} R_{\alpha j}\tilde{\phi}_\alpha \quad . \tag{9}$$

Thus we get

$$\begin{aligned}\sum_{j,k=1}^{N}&(M^2)_{jk}(\phi_j - \phi_j^{(0)})(\phi_k - \phi_k^{(0)}) \\ &= \sum_{\alpha,\beta=1}^{N}\left(\sum_{j,k}(M^2)_{jk}R_{\alpha j}R_{\beta k}\right)\tilde{\phi}_\alpha\tilde{\phi}_\beta \quad .\end{aligned} \tag{10}$$

By supposition, the transformation $R_{\alpha j}$ diagonalizes the matrix M^2, so we write

$$\sum_{j,k=1}^{N}(M^2)_{jk}R_{\alpha j}R_{\beta k} = M_\alpha^2\delta_{\alpha\beta} \quad . \tag{11}$$

M_α^2 are the eigenvalues of M^2. Using (3), (5), (10), and (11) we can denote the change of the function U for an infinitesimal variation of the fields $\tilde{\phi}_\alpha$:

$$\delta U = \frac{1}{2}\sum_{\alpha=1}^{N} M_\alpha^2(\delta\tilde{\phi}_\alpha)^2 \quad . \tag{12}$$

For a negative eigenvalue M_γ^2 of the matrix M^2, the special choice

$$\delta\tilde{\phi}_\alpha \begin{cases} \neq 0 & \text{for } \alpha = \gamma \\ = 0 & \text{otherwise} \end{cases} \tag{13}$$

would yield a negative δU. This contradicts the requirements of U attaining a minimum at $\phi_i - \phi_i^{(0)}$. Thus the eigenvalues M_α^2 have to be greater than or equal to zero. The Lagrangian can be written in terms of $\tilde{\phi}$:

$$L = \frac{1}{2}\sum_{\alpha=1}^{N}[\dot{\tilde{\phi}}_\alpha^2 - |\nabla\tilde{\phi}_\alpha|^2 - M_\alpha^2\tilde{\phi}_\alpha^2] + O(\tilde{\phi}^3) + U(\phi_i^{(0)}) \quad . \tag{14}$$

The last term is just a negligible constant. By neglecting all terms of third or higher order in $\tilde{\phi}$, the variation of the action with respect to $\tilde{\phi}_\alpha$ yields the equations of motion:

$$\ddot{\tilde{\phi}}_\alpha - \Delta\tilde{\phi}_\alpha + M_\alpha^2\tilde{\phi}_\alpha \equiv \Box\tilde{\phi}_\alpha + M_\alpha^2\tilde{\phi}_\alpha = 0 \quad . \tag{15}$$

Taking no account of higher-order terms, (14) describes a set of N Klein–Gordon fields $\tilde{\phi}_\alpha$, yielding N particles with masses M_α by quantization. The third-order and higher-order terms can then be treated as perturbations generating interactions among the N particles.

In the Higgs–Kibble model the initially massless vector field is coupled to a complex (charged) scalar field ϕ called the Higgs field. We choose minimal coupling in order to proceed analogously to gauge-invariant electrodynamics. Introducing $F_{\mu\nu} = \partial_\mu A_\nu - \partial_\nu A_\mu$, we have for the Lagrangian

$$L = -\frac{1}{4} F_{\mu\nu} F^{\mu\nu} + |(\partial_\mu - \mathrm{i} g A_\mu)\phi|^2 - U(|\phi|^2) \quad , \tag{4.13}$$

$U(|\phi|^2)$ being some convenient polynomial of $|\phi|^2$ given below. In order to formulate the *quantum theory of the fields* A_μ, ϕ, we first neglect the electromagnetic interaction and consider just the part of the Lagrangian containing the scalar field:

$$\begin{aligned} L_\phi = |\partial_\mu \phi|^2 - U(|\phi|^2) &= |\dot{\phi}|^2 - |\nabla\phi|^2 - U(|\phi|^2) \\ &= \dot{\phi}\dot{\phi}^* - \nabla\phi \cdot \nabla\phi^* - U(\phi\phi^*) \quad . \end{aligned} \tag{4.14}$$

Now we can determine the particle spectrum according to Example 4.2. To this end we calculate the classical vacuum by determining the minimum value of the function

$$U(|\phi|) = -\mu^2 |\phi|^2 + h|\phi|^4 \quad , \tag{4.15}$$

shown in Fig. 4.3. It is given by

$$|\phi_0| = \sqrt{\mu^2/2h} \equiv \lambda/\sqrt{2} \quad . \tag{4.16}$$

Therefore the general solution is

$$\phi_0 = \frac{1}{\sqrt{2}} \lambda \mathrm{e}^{\mathrm{i}\alpha} \quad . \tag{4.17}$$

We choose $\alpha = 0$, that is, we consider the solution $\phi_0 = \lambda/\sqrt{2}$. This does not imply some limitation, since the general solution (4.17) can be generated by applying a phase transformation

$$\phi \to \mathrm{e}^{\mathrm{i}\alpha}\phi \quad . \tag{4.18}$$

This does not change the physics, because only real quantities such as $|\phi|^2$ are measurable.

Nevertheless, by fixing $\alpha = 0$ we have chosen a distinct phase. We therefore cannot perform a phase transformation without simultaneously changing the vacuum expectation value of the Higgs field. Though the Lagrangian density is invariant with respect to a phase transformation, *the vacuum state is not*, which breaks the symmetry. This effect is commonly called *spontaneous symmetry breaking*. (It has nothing to do with the violation of parity invariance by the weak interaction. The latter is explicitly incorporated in the Lagrangian.)

Now we expand ϕ around the value $\lambda/\sqrt{2}$, setting

$$\phi(x) = \frac{1}{\sqrt{2}} (\lambda + \chi(x) + \mathrm{i}\theta(x)) \quad , \tag{4.19}$$

or

$$\mathrm{Re}\{\phi(x)\} = \frac{1}{\sqrt{2}} (\lambda + \chi(x)) \quad ,$$

$$\mathrm{Im}\{\phi(x)\} = \frac{1}{\sqrt{2}}\theta(x) \quad , \tag{4.20}$$

with two real fields χ and θ. Hence we get

$$|\dot{\phi}|^2 = \frac{1}{2}|\dot{\chi} + \mathrm{i}\dot{\theta}|^2 = \frac{1}{2}\dot{\chi}^2 + \frac{1}{2}\dot{\theta}^2 \quad , \tag{4.21}$$

and correspondingly

$$|\nabla\phi|^2 = \frac{1}{2}\left(|\nabla\chi|^2 + |\nabla\theta|^2\right) \quad . \tag{4.22}$$

Furthermore,

$$\begin{aligned}
U(\phi) &= h|\phi|^4 - \mu^2|\phi|^2 \\
&= \frac{1}{4}h|\lambda + \chi + \mathrm{i}\theta|^4 - \frac{1}{2}\mu^2|\lambda + \chi + \mathrm{i}\theta|^2 \\
&= \frac{1}{4}h[(\lambda + \chi)^2 + \theta^2]^2 - \frac{1}{2}\mu^2[(\lambda + \chi)^2 + \theta^2] \\
&= \frac{1}{4}h[\lambda^2 + 2\lambda\chi + \chi^2 + \theta^2]^2 - \frac{1}{2}\mu^2[\lambda^2 + 2\lambda\chi + \chi^2 + \theta^2] \\
&= \left(\frac{1}{4}h\lambda^4 - \frac{1}{2}\mu^2\lambda^2\right) + \left(\frac{1}{4}h2\lambda^2 - \frac{1}{2}\mu^2\right)(2\lambda\chi + \chi^2 + \theta^2) \\
&\quad + \frac{1}{4}h4\lambda^2\chi^2 + \frac{1}{4}h4\lambda\chi(\chi^2 + \theta^2) + \frac{1}{4}h(\chi^2 + \theta^2)^2 \quad .
\end{aligned} \tag{4.23}$$

Owing to (4.16), the second term of the sum vanishes. The first term equals $-\frac{\mu^4}{4h}$; it is a constant. Therefore, omitting the constant, the Lagrangian density can be written in terms of χ and θ in the form

$$\begin{aligned}
L_\phi &= \frac{1}{2}\dot{\chi}^2 - \frac{1}{2}|\nabla\chi|^2 - \frac{1}{2}\left(\sqrt{2h}\lambda\right)^2\chi^2 + \frac{1}{2}\dot{\theta}^2 - \frac{1}{2}|\nabla\theta|^2 \\
&\quad - h\lambda\chi(\chi^2 + \theta^2) - \frac{h}{4}(\chi^2 + \theta^2)^2 \quad .
\end{aligned} \tag{4.24}$$

From (4.23), for $\chi = 0$ and $\theta = 0$ we get

$$\frac{\partial^2 U}{\partial\chi^2} = 2h\lambda^2 \quad , \quad \frac{\partial^2 U}{\partial\chi\partial\theta} = 0 \quad , \quad \frac{\partial^2 U}{\partial\theta^2} = 0 \; . \tag{4.25}$$

By setting

$$\chi = \phi_1 \quad , \quad \theta = \phi_2 \quad , \tag{4.26}$$

we recognize the matrix

$$\frac{\partial^2 U}{\partial\phi_i\partial\phi_k} \tag{4.27}$$

to be already diagonal, because of the second condition (4.25). Thus the fields χ and θ describe the physical particles whose masses are given by

$$m_\chi^2 = 2h\lambda^2 \quad , \quad m_\theta^2 = 0 \tag{4.28}$$

according to Example 4.2.

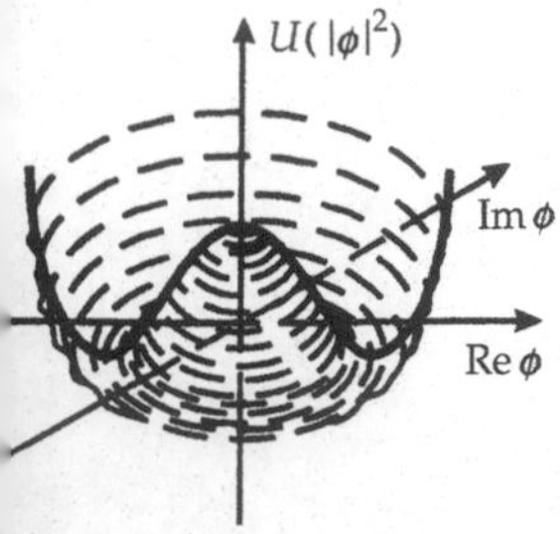

ig. 4.3. Behavior of the unction $U(|\phi|^2)$ (4.23) in the omplex ϕ plane

The intuitive meaning is that for $\phi = \lambda/\sqrt{2}$ a change of the real part of ϕ effects a change of the field χ according to (4.20), describing an outward motion in Fig. 4.3. However, in this direction the potential changes, with $U(|\phi|)$ being approximately parabolic in the vicinity of the minimum. Thus a field excitation of this kind costs potential energy. However, moving perpendicularly to the radial direction by changing the field θ in (4.20), the system stays in the 'bottom of the potential valley'. Consequently $U(\phi)$ does not change, resulting in the second equation of (4.28). This behavior is related to the fact that the Lagrangian density, and therefore also the potential $U(\phi)$, is invariant under phase transformations, which correspond to rotations around the point $\phi = 0$ in Fig. 4.3. Although we chose a special phase as characterizing the ground state of the field ϕ, all other phases yield equivalent vacuum states, which are therefore not connected with a change of potential energy. So we recognize how massless particles are automatically produced by spontaneous symmetry breaking, as can be seen in (4.28). This kind of particle is called a *Goldstone boson.* As we will see in the following there are *no* Goldstone bosons in the complete theory (4.13).

Indeed, the Lagrangian density (4.13) is invariant with respect to gauge transformations given by

$$\begin{aligned} A_\mu(x) &\to A'_\mu(x) = A_\mu(x) + \partial_\mu \Lambda(x) \quad , \\ \phi(x) &\to \phi'(x) = \phi(x)\exp(\mathrm{i}g\Lambda(x)) \quad . \end{aligned} \tag{4.29}$$

We again replace the complex field $\phi(x)$ by two *real* fields $\chi(x)$ and $\theta(x)$,

$$\begin{aligned} \phi(x) &= \frac{1}{\sqrt{2}}(\lambda + \chi(x))\exp\left(\mathrm{i}\frac{\theta(x)}{\lambda}\right) \\ &\approx \frac{\lambda}{\sqrt{2}} + \frac{1}{\sqrt{2}}(\chi(x) + \mathrm{i}\theta(x)) \quad \text{for} \quad \chi, \theta \ll \lambda \quad , \end{aligned} \tag{4.30}$$

similar to (4.19), where λ is defined in (4.16). Inserting this expression into the Lagrangian density (4.13) and taking all derivatives, we obtain

$$\begin{aligned} L = &-\frac{1}{4}F_{\mu\nu}F^{\mu\nu} + \frac{1}{2}(\partial_\mu\chi)^2 + \frac{1}{2}\left(1+\frac{\chi}{\lambda}\right)^2(\partial_\mu\theta)^2 - \lambda\left(1+\frac{\chi}{\lambda}\right)^2 gA_\mu(\partial_\mu\theta) \\ &+ \frac{1}{2}\lambda^2\left(1+\frac{\chi}{\lambda}\right)^2 g^2 A_\mu A^\mu + \frac{1}{2}\mu^2\lambda^2\left(1+\frac{\chi}{\lambda}\right)^2 - \frac{h}{4}\lambda^4\left(1+\frac{\chi}{\lambda}\right)^4 \quad . \end{aligned} \tag{4.31}$$

Let us just, as an example, confirm the second and third terms of this expression by calculating

$$\begin{aligned} |\partial_\mu\phi|^2 &= (\partial_\mu\phi^*)(\partial^\mu\phi) \\ &= \frac{1}{\sqrt{2}}\left(\partial_\mu\chi - \mathrm{i}(\lambda+\chi)\frac{\partial_\mu\theta}{\lambda}\right)\cdot\frac{1}{\sqrt{2}}\left(\partial^\mu\chi + \mathrm{i}(\lambda+\chi)\frac{\partial^\mu\theta}{\lambda}\right) \\ &= \frac{1}{2}(\partial_\mu\chi)^2 + \frac{1}{2}\frac{(\lambda+\chi)^2}{\lambda^2}(\partial_\mu\theta)^2 \\ &= \frac{1}{2}(\partial_\mu\chi)^2 + \frac{1}{2}\left(1+\frac{\chi}{\lambda}\right)^2(\partial_\mu\theta)^2 \quad . \end{aligned}$$

However, the field $\theta(x)$ which appears in L is spurious, that is to say, if we perform a gauge transformation with $\Lambda(x) = -\theta(x)/g\lambda$ before inserting (4.30), we get

$$\phi = \frac{1}{\sqrt{2}}(\lambda+\chi)\mathrm{e}^{\mathrm{i}\theta/\lambda} \rightarrow \phi\mathrm{e}^{\mathrm{i}g\Lambda} = \phi\mathrm{e}^{-\mathrm{i}\theta/\lambda} = \frac{1}{\sqrt{2}}(\lambda+\chi) \quad , \tag{4.32}$$

totally eliminating θ. Hence the field θ *only contains gauge degrees of freedom* and therefore does not enter into the physics. The special gauge (4.32) is called the *U gauge* (unitary gauge). The corresponding Lagrangian is given by

$$\begin{aligned} L = &-\frac{1}{4}F_{\mu\nu}F^{\mu\nu} + \frac{1}{2}(\partial_\mu\chi)^2 + \frac{1}{2}\lambda^2\left(1+\frac{\chi}{\lambda}\right)^2(\mu^2 + g^2A_\mu A^\mu) \\ &-\frac{h\lambda^4}{4}\left(1+\frac{\chi}{\lambda}\right)^4 , \end{aligned} \tag{4.33}$$

or, ordered with respect to powers of the fields,

$$\begin{aligned} L = &\left(\frac{\lambda^2\mu^2}{2} - \frac{h\lambda^4}{4}\right) + (\lambda\mu^2\chi - h\lambda^3\chi) \\ &+\left(\frac{1}{2}(\partial_\mu\chi)^2 + \frac{\mu^2\chi^2}{2} - \frac{3}{2}h\lambda^2\chi^2 - \frac{1}{4}F_{\mu\nu}F^{\mu\nu} + \frac{1}{2}g^2\lambda^2A_\mu A^\mu\right) \\ &+\left(g^2\lambda\chi A_\mu A^\mu - h\lambda\chi^3 + \frac{1}{2}g^2\chi^2A_\mu A^\mu - \frac{h}{4}\chi^4\right) . \end{aligned} \tag{4.34}$$

The term linear in χ vanishes, owing to the relation $h\lambda^2 = \mu^2$ (the system is in the minimum of the potential U). The constant term $\frac{1}{2}\lambda^2\mu^2 - \frac{1}{4}h\lambda^4 = \frac{1}{4}h\lambda^4$ is of no importance and can therefore be neglected. The term in front of χ^2 reads $\frac{\mu^2}{2} - \frac{3}{2}h\lambda^2 = \frac{h\lambda^2}{2} - \frac{3}{2}h\lambda^2 = -h\lambda^2$. Thus we get[6]

$$\begin{aligned} L' = &-\frac{1}{4}F_{\mu\nu}F^{\mu\nu} + \frac{1}{2}g^2\lambda^2A_\mu A^\mu + \frac{1}{2}(\partial_\mu\chi)^2 - h\lambda^2\chi^2 \\ &+\left[\left(\lambda\chi + \frac{1}{2}\chi^2\right)g^2A_\mu A^\mu - h\chi^2\left(\lambda\chi + \frac{1}{4}\chi^2\right)\right] . \end{aligned} \tag{4.35}$$

Referring to the mass terms of the various fields, that is, terms quadratic in the fields not containing derivatives, which occur in the above Lagrangian, we can state that

(a) the vector field has the mass $(g\lambda)$,
(b) the scalar field χ has the mass term $\frac{1}{2}m_\chi\chi^2$ and therefore the mass $\lambda\sqrt{2h}$,
(c) the auxiliary field θ without a mass term in (4.31) is completely eliminated.

This is the desired result, especially point (a). Point (c) is also important. If we started with a non-gauge-invariant theory, there would have been no possibility of eliminating the massless field θ. This would have been a severe shortcoming of our model, since

[6] We emphasize the sign of the vector field. The quantized part (the physical spin-1 particles) of $A^\mu = (A^0, \boldsymbol{A})$ is given by the transverse 3-vector $\boldsymbol{A}_\mathrm{T}$. By splitting up A_μ into A^0, $\boldsymbol{A}_\mathrm{L}$ and $\boldsymbol{A}_\mathrm{T}$ the contribution of $\boldsymbol{A}_\mathrm{T}$ has the opposite sign, resulting from the Minkowskian metric, $-\frac{1}{4}F_{\mu\nu}F^{\mu\nu} + \frac{1}{2}g^2\lambda^2A_\mu\frac{1}{2}A^\mu = \dot{\boldsymbol{A}}^2 - \frac{1}{2}(\nabla\times\boldsymbol{A})^2 - \frac{1}{2}g^2\lambda^2\boldsymbol{A}^2 +$ (terms containing A^0).

all presumably massless fields occurring in nature have a spin different from zero (neutrinos: spin $\frac{1}{2}$; photons: spin 1; gravitons: spin 2)! A massless scalar field has not yet been observed, suggesting no field of that kind exists.

It is important to reflect on the creation of the mass terms again. We have started with a theory describing massless vector particles (4.13) and extended the theory by introducing a background field (Higgs field) ϕ. This Higgs field tends to a minimum value ϕ_0 owing to self-interaction given by $|\phi|^4$ terms in the 'potential' $U(|\phi|)$. Therefore everywhere ϕ should be near to the value $\phi = \phi_0 = \sqrt{\mu^2/2h}$. The interaction of the vector field A_μ with ϕ_0 staying in its ground state induces the vector particles to act as if they had mass. Of course, this description is only valid if the energy of the system (momentum transfer) is small and the ϕ field does not deviate too much from its minimum value. At high energy the ϕ field need no longer attain its minimum value. Then the ϕ fields vary above the minimum of $U(|\phi|^2)$ in Fig. 4.1 with a large amplitude, again yielding massless vector particles. This behavior is just what we expected of the behavior of the vector particles as a function of momentum transfer q^2 (compare the discussion related to (4.6) and (4.11)). The mass-producing mechanism by spontaneous symmetry breaking is very similar to the case discussed in Exercise 4.1, where we saw that in an electrolyte photons obtain an effective mass owing to the influence of the Debye cloud.

EXERCISE

4.3 Gauge Invariance of the Lagrangian Corresponding to the Kinetic Energy of the Meson Fields

Problem. Show that the Lagrangian (4.13) is invariant with respect to gauge transformations (4.29).

Solution. We prove this by explicit calculation of each term in (4.13).

(a)

$$\begin{aligned}
&\partial_\mu(A_\nu + \partial_\nu \Lambda) - \partial_\nu(A_\mu + \partial_\mu \Lambda) \\
&\quad = \partial_\mu A_\nu - \partial_\nu A_\mu + \partial_\mu \partial_\nu \Lambda - \partial_\nu \partial_\mu \Lambda \\
&\quad = \partial_\mu A_\nu - \partial_\nu A_\mu = F_{\mu\nu} \quad , \qquad (1)
\end{aligned}$$

because the second derivatives of continuously differentiable fields can be commuted.

(b)

$$\begin{aligned}
&\left[\partial_\mu - \mathrm{i}g(A_\mu + \partial_\mu \Lambda)\right]\phi \mathrm{e}^{\mathrm{i}g\Lambda} \\
&\quad = (\partial_\mu \phi)\mathrm{e}^{\mathrm{i}g\Lambda} + \mathrm{i}g(\partial_\mu \Lambda)\phi \mathrm{e}^{\mathrm{i}g\Lambda} - \mathrm{i}g A_\mu \phi \mathrm{e}^{\mathrm{i}g\Lambda} - \mathrm{i}g(\partial_\mu \Lambda)\phi \mathrm{e}^{\mathrm{i}g\Lambda} \\
&\quad = \mathrm{e}^{\mathrm{i}g\Lambda}(\partial_\mu - \mathrm{i}g A_\mu)\phi \quad . \qquad (2)
\end{aligned}$$

Analogously,

(c)

$$\left[\partial_\mu + \mathrm{i}g(A_\mu + \partial_\mu \Lambda)\right]\phi^* \mathrm{e}^{-\mathrm{i}g\Lambda} = \mathrm{e}^{-\mathrm{i}g\Lambda}(\partial_\mu + \mathrm{i}g A_\mu)\phi^* \quad . \qquad (3)$$

That is, $|\phi|^2$ and $|(\partial_\mu - \mathrm{i}g A_\mu)\phi|^2$ remain unchanged.

4.2 The Yang–Mills Field

The vector field A_μ considered up to now does not describe the field of an intermediate boson W^+ or W^-, since A_μ is not charged. If the vector field itself is charged (even in the sense of the "charged" weak interaction) it should also couple to itself, that is, the Lagrangian has to contain at least a third-order term of the field. How do we construct a theory of that kind?

We can start by considering the analogous problem of the triplet of pions π^-, π^0, π^+ which can be regarded as the three substates of an isospin-1 multiplet. It was introduced in this spirit by Yukawa in order to describe the strong interaction within the nucleon isospin doublet (p, n).

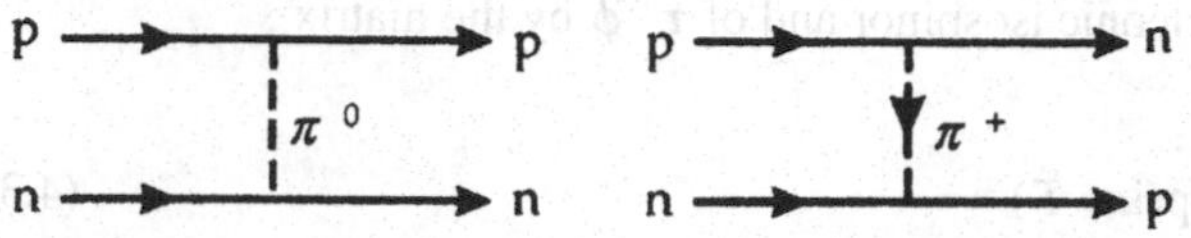

The analogy to the diagrams of the weak interaction is obvious.

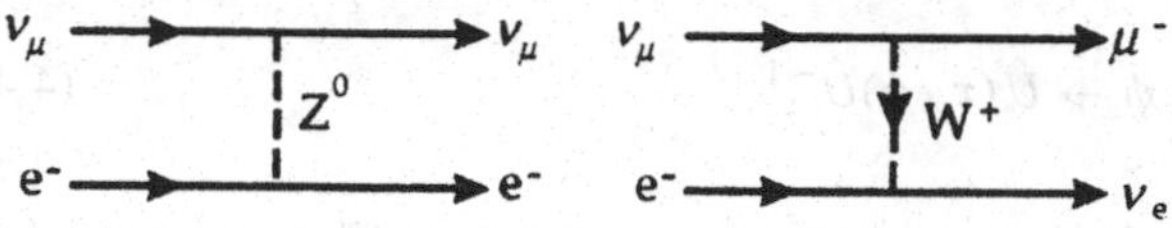

In Yukawa theory the nucleon field is given by an *isospinor* $\boldsymbol{\Psi}$, while the pion field is described in terms of an isovector $\boldsymbol{\Phi}$:

$$\boldsymbol{\Psi} = \begin{pmatrix} \psi_p \\ \psi_n \end{pmatrix} \quad , \quad \boldsymbol{\Phi} = \begin{pmatrix} \phi_{\pi^+} \\ \phi_{\pi^0} \\ \phi_{\pi^-} \end{pmatrix} \quad . \tag{4.36a}$$

Here ψ_p, ψ_n are four-spinors and the components $\phi_{\pi^\pm}$, ϕ_{π^0} are pseudoscalar functions. They are the components of the isovector in spherical representation, that is,

$$\begin{aligned} \phi_{\pi^+} &= \phi_{(-)} = \frac{1}{\sqrt{2}}(\phi_x - i\phi_y) \quad , \\ \phi_{\pi^0} &= \phi_0 = \phi_z \quad , \\ \phi_{\pi^-} &= \phi_{(+)} = \frac{1}{\sqrt{2}}(\phi_x + i\phi_y) \quad . \end{aligned} \tag{4.36b}$$

The coupling between nucleons and pions is given by the interaction

$$\begin{aligned} L_{\text{int}} &= g_{\pi N}\bar{\boldsymbol{\Psi}}\gamma_5(\boldsymbol{\tau}\cdot\boldsymbol{\phi})\boldsymbol{\Psi} = g_{\pi N}\bar{\boldsymbol{\Psi}}\gamma_5(\tau_+\phi_{(-)} + \tau_-\phi_{(+)} + \tau_0\phi_{\pi^0})\boldsymbol{\Psi} \\ &= g_{\pi N}\bar{\boldsymbol{\Psi}}\gamma_5(\tau_+\phi_{\pi^+} + \tau_-\phi_{\pi^-} + \tau_0\phi_{\pi^0})\boldsymbol{\Psi} \\ &= g_{\pi N}(\bar{\psi}_p, \bar{\psi}_n)\gamma_5 \begin{pmatrix} \phi_{\pi^0} & \sqrt{2}\phi_{\pi^+} \\ \sqrt{2}\phi_{\pi^-} & -\phi_{\pi^0} \end{pmatrix} \begin{pmatrix} \psi_p \\ \psi_n \end{pmatrix} \\ &= g_{\pi N}\big[\phi_{\pi^0}\bar{\psi}_p\gamma_5\psi_p + \sqrt{2}\phi_{\pi^+}\bar{\psi}_p\gamma_5\psi_n \\ &\quad + \sqrt{2}\phi_{\pi^-}\bar{\psi}_n\gamma_5\psi_p - \phi_{\pi^0}\bar{\psi}_n\gamma_5\psi_n\big] \quad . \end{aligned} \tag{4.37}$$

In spherical representation the Pauli vector $\boldsymbol{\tau}$ is given by

$$\tau_+ = \begin{pmatrix} 0 & \sqrt{2} \\ 0 & 0 \end{pmatrix} \quad , \quad \tau_- = \begin{pmatrix} 0 & 0 \\ \sqrt{2} & 0 \end{pmatrix} \quad , \quad \tau_0 = \begin{pmatrix} 1 & 0 \\ 0 & -1 \end{pmatrix} \quad . \tag{4.38}$$

These are connected to the Cartesian components through the well-known relation

$$\tau_\pm = \frac{1}{\sqrt{2}}(\tau_x \pm \mathrm{i}\tau_y) \quad .$$

The interaction (4.37) is invariant with respect to a rotation in isospin space, generating a transformation of the nucleonic isospinor and of $\boldsymbol{\tau} \cdot \boldsymbol{\phi}$ by the matrix

$$\hat{U} = \exp\left(\frac{\mathrm{i}}{2}\boldsymbol{a} \cdot \boldsymbol{\tau}\right) = \exp(\mathrm{i}\boldsymbol{a} \cdot \hat{\boldsymbol{T}}) \tag{4.39}$$

according to

$$\Psi \to \Psi' = \hat{U}\Psi \quad , \quad \boldsymbol{\tau} \cdot \boldsymbol{\phi} \to \hat{U}(\boldsymbol{\tau} \cdot \boldsymbol{\phi})\hat{U}^{-1} \quad . \tag{4.40}$$

This is immediately obvious. Let's now turn to weak interactions. The same procedure, as outlined above for Yukawa theory can be applied to the W bosons and fermion doublets

$$\begin{pmatrix} \psi_{\nu_e} \\ \psi_{e^-} \end{pmatrix} \quad , \quad \begin{pmatrix} \psi_{\nu_\mu} \\ \psi_{\mu^-} \end{pmatrix} \quad , \quad \dots \quad .$$

To this end we consider a *triplet of vector mesons*, whose fields we denote by A^i_μ $(i = 1, 2, 3)$. The superscript i labels the field, that is the three isospin components, analogously to ϕ_i in (4.36). *The index μ expresses the fact that every isospin component itself is a space-time four-vector* $(\mu = 0, 1, 2, 3)$.

We denote the isospherical combinations $(A^1_\mu \pm \mathrm{i}A^2_\mu)/\sqrt{2}$ by $A^\pm_\mu$, and A^3_μ by A^0_μ. Furthermore we consider a fermion doublet $\Psi = \binom{\psi_1}{\psi_2}$, where both components are Dirac spinors. By close analogy with Yukawa theory, we formulate the interaction

$$L_{\mathrm{int}} = g\sum_{i=1}^{3} \bar{\Psi}\gamma^\mu \frac{\tau^i}{2} A^i_\mu \Psi \equiv g\bar{\Psi}\gamma^\mu \boldsymbol{A}_\mu \cdot \frac{\boldsymbol{\tau}}{2}\Psi \quad . \tag{4.41}$$

The term $\boldsymbol{\tau} \cdot \boldsymbol{A}$ denotes a scalar product in isospace, whereas $\gamma^\mu A_\mu$ is a scalar product of four-vectors in space-time. Often we write $\hat{T}^i$ instead of $\tau^i/2$, the $\hat{T}^i$ (and of course also the $\hat{\tau}^i$) are the generators of the isospin group SU(2). They obey the relation

$$\sum_{i=1}^{3}(\hat{T}^i)^2 = \frac{3}{4}\mathbb{1} = \frac{1}{2}\left(\frac{1}{2}+1\right)\mathbb{1} \quad , \tag{4.42}$$

where $\mathbb{1}$ denotes the unit matrix. Furthermore the $\hat{T}^i$ obey the commutation relations of the SU(2) *Lie algebra*[7]

$$[\hat{T}^i, \hat{T}^j] = \mathrm{i}\sum_{i=1}^{3} \varepsilon_{ijk}\hat{T}^k \equiv \mathrm{i}\varepsilon_{ijk}\hat{T}^k \quad . \tag{4.43}$$

The spherical components of $\hat{T}$ are

$$\hat{T}_+ = \hat{T}_1 + \mathrm{i}\hat{T}_2 = \frac{1}{2}(\tau_1 + \mathrm{i}\tau_2) = \frac{1}{\sqrt{2}}\tau_+ = \begin{pmatrix} 0 & 1 \\ 0 & 0 \end{pmatrix} \quad ,$$

$$\hat{T}_0 = \hat{T}_z = \frac{1}{2}\begin{pmatrix} 1 & 0 \\ 0 & -1 \end{pmatrix} \quad ,$$

$$\hat{T}_- = \hat{T}_1 - \mathrm{i}\hat{T}_2 = \begin{pmatrix} 0 & 0 \\ 1 & 0 \end{pmatrix} .$$

It would be convenient to denote the free Lagrangian of the vector fields A^i_μ by

$$L_{\mathrm{free}} = -\frac{1}{4}F^i_{\mu\nu}F^{\mu\nu}_i \quad , \tag{4.44a}$$

with

$$F^i_{\mu\nu} = \partial_\mu A^i_\nu - \partial_\nu A^i_\mu \quad , \tag{4.44b}$$

but this expression violates (local) gauge invariance. In order to see this, we must first derive the properties of the vector potentials A^i_μ under a local gauge transformation (4.39), that is, with $\boldsymbol{a}(x)$ being a function of space and time. In other words, if we again transform $\Psi \to \Psi' = \hat{U}\Psi$ with $\hat{U} = \exp(\mathrm{i}\boldsymbol{a}\cdot\boldsymbol{T})$, the derivatives of the fermionic fields are

$$\begin{aligned}\partial_\mu\Psi \to \hat{U}\partial_\mu\Psi &= \hat{U}\partial_\mu(\hat{U}^{-1}\hat{U}\Psi) = \partial_\mu(\hat{U}\Psi) + \hat{U}(\partial_\mu\hat{U}^{-1})\hat{U}\Psi \\ &= \partial_\mu\Psi' + \hat{U}(\partial_\mu\hat{U}^{-1})\Psi' = \left[\partial_\mu + \hat{U}(\partial_\mu\hat{U}^{-1})\right]\Psi' \quad .\end{aligned} \tag{4.45}$$

As we shall see soon, the additional term $\hat{U}(\partial_\mu\hat{U}^{-1})$, occurring for $\partial_\mu\boldsymbol{a}(x) \neq 0$, can be absorbed by gauging the fields A^i_μ simultaneously. By adding the coupling (4.41) to the free Dirac Lagrange function L_F the expression to be cast into gauge-invariant form reads

$$L = L_\mathrm{F} + L_{\mathrm{int}} = \mathrm{i}\bar{\Psi}\gamma^\mu\partial_\mu\Psi + g\bar{\Psi}\gamma^\mu\boldsymbol{A}_\mu\cdot\hat{\boldsymbol{T}}\Psi \quad . \tag{4.46}$$

Here we intentionally avoided introducing an explicit electron mass term. The gauge symmetry implies that, ab initio, the upper and lower components of the leptonic doublet, that is, neutrino and electron, are indistinguishable. This property of the theory would be destroyed by attaching different masses to the electron and neutrino. As can be seen later the physical difference between the electron and neutrino regarding different charges and masses can be related to the spontaneous symmetry breaking through the Higgs field.

[7] See W. Greiner and B. Müller: *Quantum Mechanics – Symmetries*, 2nd ed. (Springer, Berlin, Heidelberg, 1994).

We again stress the following point: *it is demanded as a principle that we construct the Lagrangian* (4.46) *in a gauge-invariant way with respect to local* ($\boldsymbol{a} = \boldsymbol{a}(x)$) *gauge transformations* (4.40). Now (4.46) does not incorporate the full Lagrangian density of the fields involved. The kinetic-energy term of the A_μ fields is missing, for example, of the form (4.44a). As can be guessed by analogy with electrodynamics this term can be separately constructed without difficulty. In electrodynamics $F_{\mu\nu}$ itself is gauge-invariant.

Let us begin with the Lagrangian (4.46). We know that the Ψ fields can be transformed by (cf. (4.40))

$$\Psi \to \Psi' = \hat{U}\Psi(x) = \exp(\mathrm{i}\boldsymbol{a}(x)\cdot\hat{\boldsymbol{T}})\Psi(x) \quad . \tag{4.47}$$

However, at the moment we do not know how to transform the $\boldsymbol{A}_\mu$ fields simultaneously. As we shall show in the following, this can be derived by demanding L to be gauge invariant. This implies that the original Lagrangian density

$$L = \mathrm{i}\bar{\Psi}\gamma^\mu\partial_\mu\Psi + g\bar{\Psi}\gamma^\mu\boldsymbol{A}_\mu\cdot\hat{\boldsymbol{T}}\Psi \tag{4.48a}$$

and the gauged Lagrangian density

$$L' = \mathrm{i}\bar{\Psi}'\gamma^\mu\partial_\mu\Psi' + g\bar{\Psi}'\gamma^\mu\boldsymbol{A}'_\mu\cdot\hat{\boldsymbol{T}}\Psi' \tag{4.48b}$$

should be identical in form and value.

In (4.48b) we know the field Ψ' (from (4.40)), but the dependence of A'_μ on A_μ and on the gauge operator $\hat{U}$ remains unknown. However, the gauge dependence of Ψ can be exploited by writing

$$\begin{aligned}
L &= \mathrm{i}\bar{\Psi}\gamma^\mu\partial_\mu\Psi + g\bar{\Psi}\gamma^\mu\boldsymbol{A}_\mu\cdot\hat{\boldsymbol{T}}\Psi \\
&= \mathrm{i}\bar{\Psi}\hat{U}^{-1}\hat{U}\gamma^\mu\partial_\mu\hat{U}^{-1}\hat{U}\Psi + g\bar{\Psi}\hat{U}^{-1}\hat{U}\gamma^\mu\boldsymbol{A}_\mu\cdot\hat{\boldsymbol{T}}\hat{U}^{-1}\hat{U}\Psi \\
&= \mathrm{i}\bar{\Psi}'\hat{U}\gamma^\mu\partial_\mu\hat{U}^{-1}\Psi' + g\bar{\Psi}'\hat{U}\gamma^\mu\boldsymbol{A}_\mu\cdot\hat{\boldsymbol{T}}\hat{U}^{-1}\Psi' \\
&= \mathrm{i}\bar{\Psi}'\gamma^\mu\hat{U}\partial_\mu\hat{U}^{-1}\Psi' + g\bar{\Psi}'\gamma^\mu\hat{U}\boldsymbol{A}_\mu\cdot\hat{\boldsymbol{T}}\hat{U}^{-1}\Psi' \\
&= \mathrm{i}\bar{\Psi}'\gamma^\mu\partial_\mu\Psi' + \mathrm{i}\bar{\Psi}'\gamma^\mu\big[\hat{U}(\partial_\mu\hat{U}^{-1})\big]\Psi' + g\bar{\Psi}'\gamma^\mu\hat{U}\boldsymbol{A}_\mu\cdot\hat{\boldsymbol{T}}\hat{U}^{-1}\Psi' \\
&= \mathrm{i}\bar{\Psi}'\gamma^\mu\partial_\mu\Psi' + g\bar{\Psi}'\gamma^\mu\left[\hat{U}\boldsymbol{A}_\mu\cdot\hat{\boldsymbol{T}}\hat{U}^{-1} + \frac{\mathrm{i}}{g}\hat{U}(\partial_\mu\hat{U}^{-1})\right]\Psi' \quad .
\end{aligned} \tag{4.48c}$$

Here we made use of the unitarity of $\hat{U}$, i.e. $\hat{U}^+ = \hat{U}^{-1}$, and the fact that $\hat{U}$ commutes with γ^μ, of course. This expression (4.48c) should be identical to L' in (4.48b). Thus it follows that

$$\boldsymbol{A}'_\mu\cdot\hat{\boldsymbol{T}} = \hat{U}\boldsymbol{A}_\mu\cdot\hat{\boldsymbol{T}}\hat{U}^{-1} + \frac{\mathrm{i}}{g}\hat{U}(\partial_\mu\hat{U}^{-1}) \quad . \tag{4.49}$$

Consequently we are forced to incorporate the term $\hat{U}(\partial_\mu\hat{U}^{-1})$, which is generated by gauging the kinetic energy of the Ψ field into the gauge transformation of the $\boldsymbol{A}_\mu$ fields. It is useful to discuss the significance of this term in electrodynamics. In that case $\hat{U}$ is just $\hat{U} = \exp\{\mathrm{i}a(x)\}$ with the single function $a(x)$. Thus (4.49) reads

$$A'_\mu(x) = A_\mu(x) + \frac{1}{g}\partial_\mu a(x) \quad .$$

Obviously the prescription (4.49) is reduced to the well-known gauge transformation of electromagnetic fields. In non-Abelian gauge theories the more complicated transformation (4.49) describes the gauge properties of the fields involved. By referring to (4.48) we can write L concisely in the form

$$\begin{aligned} L &= \mathrm{i}\bar{\Psi}\gamma^\mu(\partial_\mu - \mathrm{i}g\boldsymbol{A}_\mu \cdot \hat{\boldsymbol{T}})\Psi \\ &\equiv \mathrm{i}\bar{\Psi}\gamma^\mu \hat{D}_\mu \Psi \quad . \end{aligned} \tag{4.50a}$$

Here we have introduced the *gauge-covariant derivative*

$$\hat{D}_\mu = \partial_\mu - \mathrm{i}g\boldsymbol{A}_\mu \cdot \hat{\boldsymbol{T}} \quad . \tag{4.50b}$$

With its help the gauge-transformation properties of the $\boldsymbol{A}_\mu$ fields (4.49) can be summarized as

$$\hat{D}_\mu \to \hat{D}'_\mu = \hat{U}\hat{D}_\mu\hat{U}^{-1} \quad . \tag{4.50c}$$

Indeed,

$$\begin{aligned} \hat{U}\hat{D}_\mu\hat{U}^{-1} &= \hat{U}(\partial_\mu - \mathrm{i}g\boldsymbol{A}_\mu \cdot \hat{\boldsymbol{T}})\hat{U}^{-1} \\ &= \partial_\mu + \hat{U}(\partial_\mu\hat{U}^{-1}) - \mathrm{i}g\hat{U}\boldsymbol{A}_\mu \cdot \hat{\boldsymbol{T}}\hat{U}^{-1} \\ &= \partial_\mu - \mathrm{i}g\left(\hat{U}\boldsymbol{A}_\mu \cdot \hat{\boldsymbol{T}}\hat{U}^{-1} + \frac{\mathrm{i}}{g}\hat{U}(\partial_\mu\hat{U}^{-1})\right) \\ &= \partial_\mu - \mathrm{i}g\boldsymbol{A}'_\mu\hat{\boldsymbol{T}} \\ &= \hat{D}'_\mu \quad . \end{aligned} \tag{4.50d}$$

Now we are ready to turn to the gauge-invariant construction of the kinetic energy of the $\boldsymbol{A}_\mu$ fields. By analogy with electrodynamics, we expect a term similar to (4.44a). We will see that this is indeed the case. First we define the gauge-invariant field-strength tensor matrix $\hat{F}_{\mu\nu} = \boldsymbol{F}_{\mu\nu} \cdot \hat{\boldsymbol{T}}$. We call $\hat{F}_{\mu\nu}$ *the operator field strength,* in contrast to $F^i_{\mu\nu}$, which is just the usual isovector of field strengths. Note the operator symbol "^" on top of $\hat{F}_{\mu\nu}$ on the left-hand but not on the right-hand side. We have

$$\begin{aligned} \hat{F}_{\mu\nu} = \boldsymbol{F}_{\mu\nu} \cdot \hat{\boldsymbol{T}} = \sum_{i=1}^{3} F^i_{\mu\nu}\hat{T}^i &= \hat{D}_\mu(\boldsymbol{A}_\nu \cdot \hat{\boldsymbol{T}}) - \hat{D}_\nu(\boldsymbol{A}_\mu \cdot \hat{\boldsymbol{T}}) \\ &= \partial_\mu(\boldsymbol{A}_\nu \cdot \hat{\boldsymbol{T}}) - \partial_\nu(\boldsymbol{A}_\mu \cdot \hat{\boldsymbol{T}}) - \mathrm{i}g[\boldsymbol{A}_\mu \cdot \hat{\boldsymbol{T}}, \boldsymbol{A}_\nu \cdot \hat{\boldsymbol{T}}] \\ &= (\partial_\mu \boldsymbol{A}_\nu) \cdot \hat{\boldsymbol{T}} - (\partial_\nu \boldsymbol{A}_\mu) \cdot \hat{\boldsymbol{T}} + g\sum_{ijk} A^i_\mu A^j_\nu \varepsilon_{ijk}\hat{T}^k \\ &= (\partial_\mu \boldsymbol{A}_\nu - \partial_\nu \boldsymbol{A}_\mu + g\boldsymbol{A}_\mu \times \boldsymbol{A}_\nu) \cdot \hat{\boldsymbol{T}} \quad . \end{aligned} \tag{4.51a}$$

Be reminded again that the (bold) vector notation refers to the isospin degrees of freedom. The isovector of field strengths is thus

$$\boldsymbol{F}_{\mu\nu} = \partial_\mu \boldsymbol{A}_\nu - \partial_\nu \boldsymbol{A}_\mu + g\boldsymbol{A}_\mu \times \boldsymbol{A}_\nu \quad . \tag{4.51b}$$

We obtain a gauge-invariant Lagrangian by performing the trace over the isospin indices:

$$\begin{aligned} L_A &= -\frac{1}{2}\mathrm{Tr}\{\hat{F}_{\mu\nu}\hat{F}^{\mu\nu}\} \\ &= -\frac{1}{2}\mathrm{Tr}\{(\boldsymbol{F}_{\mu\nu}\cdot\hat{\boldsymbol{T}})(\boldsymbol{F}^{\mu\nu}\cdot\hat{\boldsymbol{T}})\} \\ &= -\frac{1}{2}F^i_{\mu\nu}F^{\mu\nu\,j}\mathrm{Tr}\{\hat{T}_i\hat{T}_j\} \\ &= -\frac{1}{4}F^i_{\mu\nu}F^{\mu\nu\,j}\delta_{ij} = -\frac{1}{4}\boldsymbol{F}_{\mu\nu}\cdot\boldsymbol{F}^{\mu\nu} \quad , \end{aligned} \tag{4.52}$$

which makes use of the relation $\mathrm{Tr}\{\hat{T}^i\hat{T}^k\} = \frac{1}{2}\delta_{ik}$. It is gauge invariant since here we make use of the commutation of matrices under the trace, $\mathrm{Tr}\{AB\} = \mathrm{Tr}\{BA\}$:

$$\begin{aligned} L'_A &= -\frac{1}{2}\mathrm{Tr}\{(\boldsymbol{F}'_{\mu\nu}\cdot\hat{\boldsymbol{T}})(\boldsymbol{F}'^{\mu\nu}\cdot\hat{\boldsymbol{T}})\} \\ &= -\frac{1}{2}\mathrm{Tr}\{\hat{U}(\boldsymbol{F}_{\mu\nu}\cdot\hat{\boldsymbol{T}})\hat{U}^{-1}\hat{U}(\boldsymbol{F}^{\mu\nu}\cdot\hat{\boldsymbol{T}})\hat{U}^{-1}\} \\ &= -\frac{1}{2}\mathrm{Tr}\{\hat{U}^{-1}\hat{U}(\boldsymbol{F}_{\mu\nu}\cdot\hat{\boldsymbol{T}})(\boldsymbol{F}^{\mu\nu}\cdot\hat{\boldsymbol{T}})\} = L_A \quad . \end{aligned} \tag{4.53}$$

A more detailed analysis of this procedure, especially of the property

$$\hat{F}'_{\mu\nu} = \boldsymbol{F}'_{\mu\nu}\cdot\hat{\boldsymbol{T}} = \hat{U}\boldsymbol{F}_{\mu\nu}\cdot\hat{\boldsymbol{T}}\hat{U}^{-1} = \hat{U}\hat{F}_{\mu\nu}\hat{U}^{-1} \quad ,$$

can be found in Exercise 4.5. The complete Lagrangian reads

$$\begin{aligned} L &= L_F + L_{\mathrm{int}} + L_A \\ &= \bar{\Psi}\gamma^\mu(\mathrm{i}\partial_\mu + g\boldsymbol{A}_\mu\cdot\hat{\boldsymbol{T}})\Psi - \bar{\Psi}\hat{M}\Psi - \frac{1}{4}\boldsymbol{F}_{\mu\nu}\cdot\boldsymbol{F}^{\mu\nu} \quad . \end{aligned} \tag{4.54}$$

Here we have introduced an additional constant-mass matrix $\hat{M}$, which can be generated by a coupling to Higgs fields according to Example 4.2. In Sect. 4.5 we shall discuss in more detail gauge-invariant mass generation in the frame of the Higgs mechanism. By using the expressions for $\boldsymbol{F}_{\mu\nu}$ from (4.51a) we have explicitly

$$\begin{aligned} \boldsymbol{F}_{\mu\nu}\cdot\boldsymbol{F}^{\mu\nu} = &(\partial_\mu\boldsymbol{A}_\nu - \partial_\nu\boldsymbol{A}_\mu)\cdot(\partial^\mu\boldsymbol{A}^\nu - \partial^\nu\boldsymbol{A}^\mu) \\ &+ 2g(\partial_\mu\boldsymbol{A}_\nu - \partial_\nu\boldsymbol{A}_\mu)\cdot(\boldsymbol{A}^\mu\times\boldsymbol{A}^\nu) \\ &+ g^2(\boldsymbol{A}_\mu\times\boldsymbol{A}_\nu)\cdot(\boldsymbol{A}^\mu\times\boldsymbol{A}^\nu) \quad . \end{aligned} \tag{4.55}$$

The last term can also be written in the form

$$g^2(\boldsymbol{A}_\mu\cdot\boldsymbol{A}^\mu)^2 - g^2(\boldsymbol{A}_\mu\cdot\boldsymbol{A}^\nu)(\boldsymbol{A}^\mu\cdot\boldsymbol{A}_\nu) \quad . \tag{4.56}$$

EXERCISE

4.4 Isospin Rotations

Problem. Show that the transformation (4.40)

$$\boldsymbol{\tau}\cdot\boldsymbol{\phi}\to\boldsymbol{\tau}\cdot\boldsymbol{\phi}'=\exp\left(\frac{\mathrm{i}}{2}\boldsymbol{a}\cdot\boldsymbol{\tau}\right)(\boldsymbol{\tau}\cdot\boldsymbol{\phi})\exp\left(-\frac{\mathrm{i}}{2}\boldsymbol{a}\cdot\boldsymbol{\tau}\right)$$

effects a rotation of the ϕ field in isospin space around the axis $\boldsymbol{n}=\boldsymbol{a}/|\boldsymbol{a}|$ by an angle $|\boldsymbol{a}|$.

Solution. First we calculate

$$\begin{aligned}(\boldsymbol{a}\cdot\boldsymbol{\tau})^2&=\sum_{i,k=1}^{3}a_i a_k\tau_i\tau_k\\&=\frac{1}{2}\sum_{i,k=1}^{3}a_i a_k(\tau_i\tau_k+\tau_k\tau_i)\\&=\sum_{i,k=1}^{3}a_i a_k\delta_{ik}=|\boldsymbol{a}|^2\quad,\end{aligned}\tag{1}$$

since the Pauli matrices obey the relation

$$\tau_i\tau_k+\tau_k\tau_i=2\delta_{ik}\quad.$$

Therefore a Taylor series expansion of the exponential yields

$$\begin{aligned}\exp\left(\pm\frac{\mathrm{i}}{2}\boldsymbol{a}\cdot\boldsymbol{\tau}\right)&=\sum_{n=0}^{\infty}\frac{1}{n!}\left(\pm\frac{\mathrm{i}}{2}\boldsymbol{a}\cdot\boldsymbol{\tau}\right)^n\\&=\sum_{n=0}^{\infty}\frac{(-1)^n}{(2n)!2^{2n}}(\boldsymbol{a}\cdot\boldsymbol{\tau})^{2n}\pm\mathrm{i}\sum_{n=0}^{\infty}\frac{(-1)^n}{(2n+1)!2^{2n+1}}(\boldsymbol{a}\cdot\boldsymbol{\tau})^{2n+1}\\&=\sum_{n=0}^{\infty}\frac{(-1)^n}{(2n)!2^{2n}}|\boldsymbol{a}|^{2n}\pm\mathrm{i}\left(\frac{\boldsymbol{a}\cdot\boldsymbol{\tau}}{|\boldsymbol{a}|}\right)\sum_{n=0}^{\infty}\frac{(-1)^n}{(2n+1)!2^{2n+1}}(\boldsymbol{a}\cdot\boldsymbol{\tau})^{2n+1}\\&=\cos\frac{|\boldsymbol{a}|}{2}\pm\mathrm{i}\frac{(\boldsymbol{a}\cdot\boldsymbol{\tau})}{|\boldsymbol{a}|}\sin\frac{|\boldsymbol{a}|}{2}\\&=\cos\frac{|\boldsymbol{a}|}{2}\pm\mathrm{i}(\boldsymbol{n}\cdot\boldsymbol{\tau})\sin\frac{|\boldsymbol{a}|}{2}\quad.\end{aligned}\tag{2}$$

By use of the relation

$$(\boldsymbol{a}\cdot\boldsymbol{\tau})(\boldsymbol{b}\cdot\boldsymbol{\tau})=\boldsymbol{a}\cdot\boldsymbol{b}+\mathrm{i}\boldsymbol{\tau}\cdot(\boldsymbol{a}\times\boldsymbol{b})\quad,\tag{3}$$

which is easily proved from the commutation relation (4.43)

$$\tau_i\tau_k-\tau_k\tau_i=2\mathrm{i}\sum_{l=1}^{3}\varepsilon_{ikl}\tau_l\quad,\tag{4}$$

xercise 4.4

we derive ($\theta \equiv |\boldsymbol{a}|/2$)

$$\begin{aligned}
&\left[\cos\theta + \mathrm{i}(\boldsymbol{n}\cdot\boldsymbol{\tau})\sin\theta\right](\boldsymbol{\tau}\cdot\boldsymbol{\phi})\left[\cos\theta - \mathrm{i}(\boldsymbol{n}\cdot\boldsymbol{\tau})\sin\theta\right] \\
&\quad = (\boldsymbol{\tau}\cdot\boldsymbol{\phi})\cos^2\theta + \mathrm{i}\sin\theta\cos\theta\left[(\boldsymbol{n}\cdot\boldsymbol{\tau})(\boldsymbol{\tau}\cdot\boldsymbol{\phi}) - (\boldsymbol{\tau}\cdot\boldsymbol{\phi})(\boldsymbol{n}\cdot\boldsymbol{\tau})\right] \\
&\qquad + \sin^2\theta(\boldsymbol{n}\cdot\boldsymbol{\tau})(\boldsymbol{\tau}\cdot\boldsymbol{\phi})(\boldsymbol{n}\cdot\boldsymbol{\tau}) \\
&\quad = (\boldsymbol{\tau}\cdot\boldsymbol{\phi})\cos^2\theta - 2\sin\theta\cos\theta\,\boldsymbol{\tau}\cdot(\boldsymbol{n}\times\boldsymbol{\phi}) \\
&\qquad + \sin^2\theta\left[\boldsymbol{n}\cdot\boldsymbol{\phi} + \mathrm{i}\boldsymbol{\tau}\cdot(\boldsymbol{n}\times\boldsymbol{\phi})\right](\boldsymbol{n}\cdot\boldsymbol{\tau}) \\
&\quad = (\boldsymbol{\tau}\cdot\boldsymbol{\phi})\cos^2\theta - \sin 2\theta\left[\boldsymbol{\tau}\cdot(\boldsymbol{n}\times\boldsymbol{\phi})\right] \\
&\qquad + \sin^2\theta\left\{(\boldsymbol{n}\cdot\boldsymbol{\phi})(\boldsymbol{n}\cdot\boldsymbol{\tau}) - \boldsymbol{\tau}\cdot\left[(\boldsymbol{n}\times\boldsymbol{\phi})\times\boldsymbol{n}\right]\right\} \\
&\quad = \boldsymbol{\tau}\cdot\left[\boldsymbol{\phi}\cos^2\theta - (\boldsymbol{n}\times\boldsymbol{\phi})\sin 2\theta + \boldsymbol{n}(\boldsymbol{n}\cdot\boldsymbol{\phi})\sin^2\theta - (\boldsymbol{n}\times\boldsymbol{\phi})\times\boldsymbol{n}\sin^2\theta\right] \\
&\quad = \boldsymbol{\tau}\cdot\left[\boldsymbol{n}(\boldsymbol{n}\cdot\boldsymbol{\phi}) - (\boldsymbol{n}\times\boldsymbol{\phi})\sin 2\theta + (\boldsymbol{n}\times\boldsymbol{\phi})\times\boldsymbol{n}\cos 2\theta\right] \equiv \boldsymbol{\tau}\cdot\boldsymbol{\phi}' \quad . \qquad (5)
\end{aligned}$$

Here we have applied the relation

$$\boldsymbol{n}\times(\boldsymbol{n}\times\boldsymbol{\phi}) = \boldsymbol{n}(\boldsymbol{n}\cdot\boldsymbol{\phi}) - \boldsymbol{\phi} \quad . \qquad (6)$$

In the last line of (5) the expression $\boldsymbol{\phi}'$ in brackets is just the vector $\boldsymbol{\phi}$ rotated by an angle $|\boldsymbol{a}| = 2\theta$ around the $\boldsymbol{n}$ axis. For instance for $\boldsymbol{n} = \boldsymbol{e}_z$ we get

$$\begin{aligned}
\phi'_x &= \phi_x\cos 2\theta + \phi_y\sin 2\theta \quad , \\
\phi'_y &= -\phi_x\sin 2\theta + \phi_y\cos 2\theta \quad , \qquad (7) \\
\phi'_z &= \phi_z \quad .
\end{aligned}$$

EXERCISE

4.5 Gauge Covariance of Minimal Coupling and of the Field-Strength Tensor

Problem. (a) Show that the expression

$$\bar{\Psi}\gamma^\mu(\mathrm{i}\partial_\mu + g\boldsymbol{A}_\mu\cdot\hat{\boldsymbol{T}})\Psi$$

is gauge invariant.

(b) Prove the gauge covariance of the field-strength tensor $\boldsymbol{F}_{\mu\nu}$, that is, show that

$$\hat{\boldsymbol{T}}\cdot\boldsymbol{F}'_{\mu\nu} = \hat{U}\boldsymbol{F}_{\mu\nu}\cdot\hat{\boldsymbol{T}}\hat{U}^{-1} \quad .$$

Solution. We abbreviate

$$\gamma^\mu\partial_\mu \equiv \hat{\partial} \quad , \quad \gamma^\mu\boldsymbol{A}_\mu\cdot\hat{\boldsymbol{T}} \equiv \hat{A} \quad . \qquad (1)$$

$\hat{\partial}$ and $\hat{A}$ are matrices in the space of Dirac spinors and of isospinors. More accurately, $\hat{\partial}$ should be denoted $\hat{\partial}\mathbb{1}$, where $\mathbb{1}$ is the 2×2 unit matrix in isospace, but here we adopt the short-hand notation. Then we get the identity

$$\begin{aligned}
\bar{\Psi}(\mathrm{i}\hat{\partial} + g\hat{A})\Psi &= \bar{\Psi}\hat{U}^{-1}\hat{U}(\mathrm{i}\hat{\partial} + g\hat{A})\hat{U}^{-1}\hat{U}\Psi \\
&= \bar{\Psi}'\hat{U}(\mathrm{i}\hat{\partial} + g\hat{A})\hat{U}^{-1}\Psi'
\end{aligned}$$

$$= \bar{\Psi}'\left[\mathrm{i}\hat{\partial} + \mathrm{i}\hat{U}(\hat{\partial}\hat{U}^{-1}) + g\hat{U}\hat{A}\hat{U}^{-1}\right]\Psi'$$
$$= \bar{\Psi}'\left[\mathrm{i}\hat{\partial} + g\hat{U}\hat{A}\hat{U}^{-1} + \mathrm{i}\hat{U}(\hat{\partial}\hat{U}^{-1})\right]\Psi'$$
$$= \bar{\Psi}'(\mathrm{i}\hat{\partial} + g\hat{A}')\Psi' \quad , \tag{2}$$

with

$$\hat{A}' = \hat{U}\hat{A}\hat{U}^{-1} + \frac{\mathrm{i}}{g}\hat{U}(\hat{\partial}\hat{U}^{-1}) \quad . \tag{3}$$

The prescription (3) concerning the gauged field $\hat{A}'$ also expresses the rule for gauge transforming the A field.

Since the γ matrices are linearly independent and $\hat{U}$ does not act in spinor space, because of (1) this result (3) corresponds to

$$\boldsymbol{A}'_\mu \cdot \hat{\boldsymbol{T}} = \hat{U}(\boldsymbol{A}_\mu \cdot \hat{\boldsymbol{T}})\hat{U}^{-1} + \frac{\mathrm{i}}{g}\hat{U}(\partial_\mu \hat{U}^{-1}) \quad , \tag{4}$$

yielding the gauge transformation (4.49) of the potential field. Thus (2) demonstrates the gauge invariance of the kinetic part of the Lagrangian.

(b) In Exercise 4.4 we showed that the isovector $\boldsymbol{F}'_{\mu\nu}$, defined by

$$\hat{\boldsymbol{T}} \cdot \boldsymbol{F}'_{\mu\nu} = \hat{U}(\hat{\boldsymbol{T}} \cdot \boldsymbol{F}_{\mu\nu})\hat{U}^{-1} \quad , \tag{5}$$

results from a rotation of $\boldsymbol{F}_{\mu\nu}$ in isospace. Therefore, if we can prove (5) we will have shown that in fact $\boldsymbol{F}_{\mu\nu}$ acts like a vector in isospace. This is just the meaning of "covariance" of a gauge transformation (rotation in isospin space). Thus covariance means form invariance (in our case with respect to gauge transformations).

Furthermore we mention that in contrast to electrodynamics, where $F_{\mu\nu} = F'_{\mu\nu}$, here we have $\boldsymbol{F}_{\mu\nu} \neq \boldsymbol{F}'_{\mu\nu}$; nevertheless the kinetic energy $L'_A = L_A$ (see (4.53)) is invariant with respect to the gauge transformation $\hat{U}$ owing to the trace operation.

For the calculation we need a relation between $(\partial_\mu \hat{U})$ and $(\partial_\mu \hat{U}^{-1})$. This can be derived by differentiating the relation $\hat{U}\hat{U}^{-1} = \mathbb{1}$:

$$0 = \partial_\mu(\hat{U}\hat{U}^{-1}) = (\partial_\mu \hat{U})\hat{U}^{-1} + \hat{U}(\partial_\mu \hat{U}^{-1}) \quad , \tag{6}$$

and therefore

$$\partial_\mu \hat{U} = -\hat{U}(\partial_\mu \hat{U}^{-1})\hat{U} \quad , \tag{7a}$$
$$\partial_\mu \hat{U}^{-1} = -\hat{U}^{-1}(\partial_\mu \hat{U})\hat{U}^{-1} \quad . \tag{7b}$$

The relation between $\boldsymbol{A}'_\mu$ and $\boldsymbol{A}_\mu$ in (4) can then be written in the form

$$\boldsymbol{A}'_\mu \cdot \hat{\boldsymbol{T}} = \hat{U}(\boldsymbol{A}_\mu \cdot \hat{\boldsymbol{T}})\hat{U}^{-1} - \frac{\mathrm{i}}{g}(\partial_\mu \hat{U})\hat{U}^{-1} \quad . \tag{4'}$$

We can simplify the explicit calculation if we write the expression $(\boldsymbol{A}_\mu \cdot \hat{\boldsymbol{T}})$ in the short-hand form

Exercise 4.5

$$\hat{A}_\mu = \boldsymbol{A}_\mu \cdot \hat{\boldsymbol{T}} = \frac{1}{\sqrt{2}}(A_\mu^{(+)}\hat{T}_- + A_\mu^{(-)}\hat{T}_+) + A_\mu^{(0)}\hat{T}_0$$
$$= \frac{1}{2}\begin{pmatrix} A_\mu^3 & \dfrac{A_\mu^1 - \mathrm{i}A_\mu^2}{\sqrt{2}} \\ \dfrac{A_\mu^1 + \mathrm{i}A_\mu^2}{\sqrt{2}} & -A_\mu^3 \end{pmatrix} \quad , \tag{8a}$$

and analogously

$$\hat{F}_{\mu\nu} = \boldsymbol{F}_{\mu\nu} \cdot \hat{\boldsymbol{T}} = \frac{1}{2}\begin{pmatrix} F_{\mu\nu}^3 & \dfrac{F_{\mu\nu}^1 - \mathrm{i}F_{\mu\nu}^2}{\sqrt{2}} \\ \dfrac{F_{\mu\nu}^1 + \mathrm{i}F_{\mu\nu}^2}{\sqrt{2}} & -F_{\mu\nu}^3 \end{pmatrix} \quad , \tag{8b}$$

that is, we represent the isovector fields $\boldsymbol{A}_\mu$, $\boldsymbol{F}_{\mu\nu}$ by 2×2 matrix fields $\hat{A}_\mu$, $\hat{F}_{\mu\nu}$ in isospace. This implies solely a convenient abbreviation. Following (4.51a) and adopting this notation, we get

$$\hat{F}_{\mu\nu} = \partial_\mu \hat{A}_\nu - \partial_\nu \hat{A}_\mu - \mathrm{i}g[\hat{A}_\mu, \hat{A}_\nu]_- \quad . \tag{9}$$

Now we check the validity of (5):

$$\begin{aligned}\hat{F}'_{\mu\nu} &= \partial_\mu \hat{A}'_\nu - \partial_\nu \hat{A}'_\mu - \mathrm{i}g[\hat{A}'_\mu, \hat{A}'_\nu]_- \\ &= \partial_\mu\left(\hat{U}\hat{A}_\nu\hat{U}^{-1} - \frac{\mathrm{i}}{g}(\partial_\nu\hat{U})\hat{U}^{-1}\right) - \partial_\nu\left(\hat{U}\hat{A}_\mu\hat{U}^{-1} - \frac{\mathrm{i}}{g}(\partial_\mu\hat{U})\hat{U}^{-1}\right) \\ &\quad - \mathrm{i}g\left[\hat{U}\hat{A}_\mu\hat{U}^{-1} - \frac{\mathrm{i}}{g}(\partial_\mu\hat{U})\hat{U}^{-1}, \hat{U}\hat{A}_\nu\hat{U}^{-1} - \frac{\mathrm{i}}{g}(\partial_\nu\hat{U})\hat{U}^{-1}\right]_- \quad . \end{aligned} \tag{10}$$

Using (7b) several times, we find that these three terms yield

$$\begin{aligned}&\partial_\mu\left(\hat{U}\hat{A}_\nu\hat{U}^{-1} - \frac{\mathrm{i}}{g}(\partial_\nu\hat{U})\hat{U}^{-1}\right) \\ &= \hat{U}(\partial_\mu\hat{A}_\nu)\hat{U}^{-1} + (\partial_\mu\hat{U})\hat{A}_\nu\hat{U}^{-1} - \hat{U}\hat{A}_\nu\hat{U}^{-1}(\partial_\mu\hat{U})\hat{U}^{-1} \\ &\quad - \frac{\mathrm{i}}{g}(\partial_\mu\partial_\nu\hat{U})\hat{U}^{-1} + \frac{\mathrm{i}}{g}(\partial_\nu\hat{U})\hat{U}^{-1}(\partial_\mu\hat{U})\hat{U}^{-1} \quad , \end{aligned} \tag{11a}$$

$$\begin{aligned}&\partial_\nu\left(\hat{U}\hat{A}_\mu\hat{U}^{-1} - \frac{\mathrm{i}}{g}(\partial_\mu\hat{U})\hat{U}^{-1}\right) \\ &= \hat{U}(\partial_\nu\hat{A}_\mu)\hat{U}^{-1} + (\partial_\nu\hat{U})\hat{A}_\mu\hat{U}^{-1} - \hat{U}\hat{A}_\mu\hat{U}^{-1}(\partial_\nu\hat{U})\hat{U}^{-1} \\ &\quad - \frac{\mathrm{i}}{g}(\partial_\nu\partial_\mu\hat{U})\hat{U}^{-1} + \frac{\mathrm{i}}{g}(\partial_\mu\hat{U})\hat{U}^{-1}(\partial_\nu\hat{U})\hat{U}^{-1} \quad , \end{aligned} \tag{11b}$$

$$\begin{aligned}&\mathrm{i}g\left[\hat{U}\hat{A}_\mu\hat{U}^{-1} - \frac{\mathrm{i}}{g}(\partial_\mu\hat{U})\hat{U}^{-1}, \hat{U}\hat{A}_\nu\hat{U}^{-1} - \frac{\mathrm{i}}{g}(\partial_\nu\hat{U})\hat{U}^{-1}\right]_- \\ &= \mathrm{i}g\hat{U}\hat{A}_\mu\hat{A}_\nu\hat{U}^{-1} + (\partial_\mu\hat{U})\hat{A}_\nu\hat{U}^{-1} + \hat{U}\hat{A}_\mu\hat{U}^{-1}(\partial_\nu\hat{U})\hat{U}^{-1} \\ &\quad - \frac{\mathrm{i}}{g}(\partial_\mu\hat{U})\hat{U}^{-1}(\partial_\nu\hat{U})\hat{U}^{-1} - \mathrm{i}g\hat{U}\hat{A}_\nu\hat{A}_\mu\hat{U}^{-1} - (\partial_\nu\hat{U})\hat{A}_\mu\hat{U}^{-1} \\ &\quad - \hat{U}\hat{A}_\nu\hat{U}^{-1}(\partial_\mu\hat{U})\hat{U}^{-1} + \frac{\mathrm{i}}{g}(\partial_\nu\hat{U})\hat{U}^{-1}(\partial_\mu\hat{U})\hat{U}^{-1} \quad . \end{aligned} \tag{11c}$$

On inserting (11a)–(11c) into (10), we cancel most of the terms. If we also use the fact that partial derivatives commute, all that remains is

$$\begin{aligned}\hat{F}'_{\mu\nu} &= \hat{U}(\partial_\mu \hat{A}_\nu)\hat{U}^{-1} - \hat{U}(\partial_\nu \hat{A}_\mu)\hat{U}^{-1} \\ &\quad - \mathrm{i}g\hat{U}\hat{A}_\mu\hat{A}_\nu\hat{U}^{-1} + \mathrm{i}g\hat{U}\hat{A}_\nu\hat{A}_\mu\hat{U}^{-1} \\ &= \hat{U}\hat{F}_{\mu\nu}\hat{U}^{-1} \quad , \end{aligned} \tag{12}$$

as was to be shown. According to Exercise 4.4, (5), this means that $\hat{U} = \exp(\mathrm{i}\boldsymbol{a}\cdot\hat{\boldsymbol{T}})$ generates a rotation of the isovector components of $\boldsymbol{F}_{\mu\nu}$ in isospace:

$$\begin{aligned}\boldsymbol{F}'_{\mu\nu} &= \frac{\boldsymbol{a}}{|\boldsymbol{a}|}\left(\frac{\boldsymbol{a}}{|\boldsymbol{a}|}\cdot\boldsymbol{F}_{\mu\nu}\right) - \left(\frac{\boldsymbol{a}}{|\boldsymbol{a}|}\times\boldsymbol{F}_{\mu\nu}\right)\sin|\boldsymbol{a}| \\ &\quad + \left(\frac{\boldsymbol{a}}{|\boldsymbol{a}|}\times\boldsymbol{F}_{\mu\nu}\right)\times\frac{\boldsymbol{a}}{|\boldsymbol{a}|}\cdot\cos|\boldsymbol{a}| \quad . \end{aligned} \tag{13}$$

Since the isovector $\boldsymbol{F}_{\mu\nu}$ is rotated in isospace by a gauge transformation, we again recognize in a most simple way that the kinetic energy of the $\boldsymbol{A}_\mu$ fields, that is, $-\frac{1}{4}\boldsymbol{F}_{\mu\nu}\cdot\boldsymbol{F}^{\mu\nu}$ as a scalar product in isospin space, is invariant with respect to gauge transformations.

There also exists a more elegant method to prove relation (5). According to (4.50c) the operator of the covariant derivative transforms like

$$\hat{D}_\mu \to \hat{D}'_\mu = \hat{U}\hat{D}_\mu\hat{U}^{-1} \quad . \tag{14}$$

With this we also have

$$[\hat{D}'_\mu, \hat{D}'_\nu]_- = \hat{U}[\hat{D}_\mu, \hat{D}_\nu]_-\hat{U}^{-1} \tag{15}$$

An explicit calculation of the commutator yields

$$\begin{aligned}[\hat{D}_\mu, \hat{D}_\nu]_- &= [\partial_\mu - \mathrm{i}g\boldsymbol{A}_\mu\cdot\hat{\boldsymbol{T}}, \partial_\nu - \mathrm{i}g\boldsymbol{A}_\nu\cdot\hat{\boldsymbol{T}}]_- \\ &= [\partial_\mu, \partial_\nu]_- + \partial_\mu(-\mathrm{i}g\boldsymbol{A}_\nu\cdot\hat{\boldsymbol{T}}) - (-\mathrm{i}g\boldsymbol{A}_\nu\cdot\hat{\boldsymbol{T}})\partial_\mu \\ &\quad + (-\mathrm{i}g\boldsymbol{A}_\mu\cdot\hat{\boldsymbol{T}})\partial_\nu - \partial_\nu(-\mathrm{i}g\boldsymbol{A}_\mu\cdot\hat{\boldsymbol{T}}) + (-\mathrm{i}g)^2[\boldsymbol{A}_\mu\cdot\hat{\boldsymbol{T}}, \boldsymbol{A}_\nu\cdot\hat{\boldsymbol{T}}]_- \\ &= -\mathrm{i}g\{((\partial_\mu\boldsymbol{A}_\nu)\cdot\hat{\boldsymbol{T}}) - ((\partial_\nu\boldsymbol{A}_\mu)\cdot\hat{\boldsymbol{T}}) - \mathrm{i}g[\boldsymbol{A}_\mu\cdot\hat{\boldsymbol{T}}, \boldsymbol{A}_\nu\cdot\hat{\boldsymbol{T}}]_-\} \quad . \end{aligned} \tag{16}$$

Notice that in the last expression the derivatives act only on the gauge fields. The term in the curly brackets is just $\hat{F}_{\mu\nu}$ according to (4.51a), and therefore (15) expresses the same as (5) or (12). In other words, the relation

$$\hat{F}_{\mu\nu} = \frac{\mathrm{i}}{g}[\hat{D}_\mu, \hat{D}_\nu]_- \tag{17}$$

holds and (15) then reads

$$\hat{F}'_{\mu\nu} = \hat{U}\hat{F}_{\mu\nu}\hat{U}^{-1} \quad , \tag{18}$$

which is just the same as (12) or (5).

4.3 The Feynman Rules for Yang–Mills Theory

The Lagrangian (4.54) of the coupled Dirac and Yang–Mills fields Ψ and $\boldsymbol{A}_\mu$ contains, apart from the second-order contributions, terms of third and higher order in the fields. They arise from the field energy $\boldsymbol{F}_{\mu\nu} \cdot \boldsymbol{F}^{\mu\nu}$, as is obvious from (4.55). On a closer look we see that all terms of higher than second order contain the coupling constant g (see (4.55)). Since we want to construct a theory describing the 'weak' interaction, we can assume g to be small. Therefore it is convenient to split the full Lagrangian for electrons, neutrinos, and the gauge field into a part L_0 containing the bilinear terms (with no g occurring) and a part L_I containing the remainder, which is proportional to g:

$$L_0 = \mathrm{i}\bar{\Psi}\gamma^\mu \partial_\mu \Psi - \frac{1}{4}(\partial_\mu \boldsymbol{A}_\nu - \partial_\nu \boldsymbol{A}_\mu) \cdot (\partial^\mu \boldsymbol{A}^\nu - \partial^\nu \boldsymbol{A}^\mu) - \bar{\Psi}\hat{M}\Psi \quad , \tag{4.57}$$

$$L_\mathrm{I} = g\bar{\Psi}\gamma^\mu \boldsymbol{A}_\mu \cdot \hat{\boldsymbol{T}}\Psi - \frac{g}{2}(\partial_\mu \boldsymbol{A}_\nu - \partial_\nu \boldsymbol{A}_\mu) \cdot (\boldsymbol{A}^\mu \times \boldsymbol{A}^\nu)$$
$$- \frac{g^2}{4}(\boldsymbol{A}_\mu \times \boldsymbol{A}_\nu) \cdot (\boldsymbol{A}^\mu \times \boldsymbol{A}^\nu) \quad , \tag{4.58}$$

with the mass matrix

$$\hat{M} = \begin{pmatrix} 0 & 0 \\ 0 & m_\mathrm{e} \end{pmatrix} = \frac{1}{2}(1 - \tau_3)m_\mathrm{e}$$

giving a mass to the electron. Although the mass matrix violates gauge invariance, we anticipate the results of Sect. 4.5, where we shall show how to attribute a mass to the electron via the Higgs mechanism without destroying gauge symmetry. Let us assume for the moment that the result of this procedure can be effectively described in terms of the matrix $\hat{M}$. (For other fermions participating in the weak interaction, such as heavier leptons and quarks, the mass term has to be correspondingly modified and this will be discussed later.)

L_0 is called the Lagrangian of the "free fields". In a strict sense this nomenclature is physically incorrect, because L_0 itself is not gauge invariant and therefore defines no physically meaningful theory. The terms of L_0 which violate gauge invariance are just compensated by corresponding terms of L_I. The coupling constant g may be arbitrarily small but it must not be zero, since then gauge invariance is lost. This can also be clearly seen by considering the transformation law of the $\boldsymbol{A}_\mu$ field (4.49), which contains a factor $1/g$. From Hamilton's principle,

$$\delta \int \mathrm{d}^4x L_0(\bar{\Psi}, \Psi, \boldsymbol{A}_\mu) = 0 \quad , \tag{4.59}$$

the linear field equations can be derived:

$$\mathrm{i}\gamma^\mu \partial_\mu \Psi(x) - \hat{M}\Psi(x) = 0 \quad , \tag{4.60}$$

$$\partial_\nu(\partial^\mu \boldsymbol{A}^\nu(x) - \partial^\nu \boldsymbol{A}^\mu(x)) = 0 \quad . \tag{4.61}$$

Formally these are equivalent to the free-field equations of quantum electrodynamics (QED), with the difference that the Dirac field has two and the vector field three isospin indices and $\hat{M}$ is a 2×2 mass matrix. We assume $\hat{M}$ to be diagonal, that is, the two isospin components to be eigenstates of $\hat{M}$. Otherwise we could diagonalize $\hat{M}$

by a constant unitary transformation $\Psi \to \hat{V}\Psi$. This can be done because, despite the mass term L_0 is gauge invariant with respect to spatially constant isospin rotations $\hat{V}$.

By analogy with QED, the propagation of perturbations of the free fields generated by L_{I} are described by Feynman propagators which are defined for a point-like perturbation:

$$(\mathrm{i}\gamma^\mu \partial_\mu - \hat{M}) S_{\mathrm{F}}(x - x') = \delta^4(x - x') \quad , \tag{4.62}$$

$$\partial^\alpha \partial_\mu \overset{\leftrightarrow}{D}_{\mathrm{F}\alpha\nu}(x - x') - \partial_\alpha \partial^\alpha \overset{\leftrightarrow}{D}_{\mathrm{F}\mu\nu} = -g_{\mu\nu} \mathbb{1}\, \delta^4(x - x') \quad . \tag{4.63}$$

In contrast to QED, here S_{F} is a 2×2 matrix in isospin space as well as a 4×4 Dirac matrix. Furthermore, $\overset{\leftrightarrow}{D}_{\mathrm{F}\mu\nu}$ is not just a second-rank tensor in space-time with indices μ, ν, but also a 3×3 matrix in isospin space, which is denoted by the double arrow ($\leftrightarrow$). However, the operators which act on S_{F} and $D_{\mu\nu}$ are diagonal in isospin coordinates (which we explicitly demanded for $\hat{M}$). Therefore S_{F} and $D_{\mu\nu}$ are represented by unit matrices in isospin space. Exploiting translational invariance, we go over to momentum space by Fourier transformation, obtaining the equations

$$(\gamma^\mu p_\mu - \hat{M}) S_{\mathrm{F}}(p) = 1 \quad , \tag{4.62$'$}$$

$$(p^2 g^\alpha_\mu - p^\alpha p_\mu) \overset{\leftrightarrow}{D}_{\alpha\nu}(p) = -\overset{\leftrightarrow}{\mathbb{1}}\, g_{\mu\nu} \quad . \tag{4.63$'$}$$

The first of the two equations can be solved by (see Example 4.6 for a detailed discussion of propagators):

$$S_{\mathrm{F}}(p) = (\gamma^\mu p_\mu - \hat{M} + \mathrm{i}\varepsilon)^{-1} = \frac{\gamma^\mu p_\mu + \hat{M}}{p^2 - \hat{M}^2 + \mathrm{i}\varepsilon} \quad . \tag{4.64}$$

As usual, the iε in the denominator determines the Feynman propagator, propagating particles forward in time and antiparticles backward in time.[8] Equation (4.63′) cannot be solved for $D_{\mu\nu}$ without further discussion, because the matrix

$$p^2 g^\alpha_\mu - p^\alpha p_\mu = p^2 P^\alpha_\mu(p) \quad , \tag{4.65}$$

like the projection operator (4.3), has eigenvalue zero (because $P_{\mu\nu} q_\nu = 0$) and therefore cannot be inverted. This means that we must not assume a perturbation with tensorial structure of $g_{\mu\nu}$ on the right-hand side, but only one which has the same properties as $P_{\mu\nu}(p)$, expressed in (4.4). Since $P_{\mu\nu}(p)$ projects onto transverse vector mesons (see (4.4)), the source of the vector boson propagator must also be transverse. This can be most simply achieved by taking the perturbation itself to be proportional to $P_{\mu\nu}(p)$. Then (4.63′) takes the form

$$(p^2 g^\alpha_\mu - p^\alpha p_\mu) \overset{\leftrightarrow}{D}_{\alpha\nu} = -\overset{\leftrightarrow}{\mathbb{1}}\, P_{\mu\nu}(p) \quad , \tag{4.63$''$}$$

or

$$p^2 P^\alpha_\mu(p) \overset{\leftrightarrow}{D}_{\alpha\nu} = -\overset{\leftrightarrow}{\mathbb{1}}\, P_{\mu\nu}(p) \quad .$$

---

[8] See W. Greiner and J. Reinhardt: *Quantum Electrodynamics*, 4th ed. (Springer, Berlin, Heidelberg, 2009).

Since $D_{\mu\nu}$ should be proportional to $P_{\mu\nu}$, (4.1) is one solution of the equation for the propagator:

$$\overset{\leftrightarrow}{D}_{\mathrm{F}\mu\nu} = -\overset{\leftrightarrow}{\mathbb{1}} \frac{P_{\mu\nu}(p)}{p^2 + \mathrm{i}\varepsilon} \quad . \tag{4.66}$$

Alternative solutions are discussed in Example 4.6. The special choice (4.66) is called the *Landau gauge*.

In the propagator language, the interaction between free particles is described by so-called *vertex functions*. For instance, in QED the interaction term of the Lagrangian is given by

$$L_{\mathrm{int}}^{\mathrm{EM}} = -e\bar{\Psi}\gamma^\mu \Psi A_\mu \quad . \tag{4.67}$$

It contains field variables which are either described by the wave functions of incoming and outgoing particles or by the corresponding propagators in the case of virtual particles. The positive elementary charge is denoted by e. Thus, the vertex function is just

$$\Gamma_{\mathrm{EM}}^{\mu}(\Psi, A) = -e\gamma^\mu \quad . \tag{4.68}$$

In the same manner we can read off from the interaction Lagrange density L_{I} (4.58), that the vertex function of the coupling between fermions and the Yang–Mills field takes the form

$$\Gamma_i^{\mu}(\Psi, \boldsymbol{A}) = g\gamma^\mu \hat{T}_i \quad , \tag{4.69}$$

the index i denoting the isospin component.

In order to obtain the vertex functions related to the self-couplings of the Yang–Mills field, we have to transform the corresponding interaction terms into momentum space. The third-order term in $\boldsymbol{A}_\mu$ (see (4.58)) is given by

$$\begin{aligned} &-\frac{g}{2}\varepsilon_{lmn}\left[-\mathrm{i}p_\mu g_{\nu\sigma} A_l^\sigma(p) + \mathrm{i}p_\nu g_{\mu\sigma} A_l^\sigma(p)\right] A_m^\mu(q) A_n^\nu(k) \\ &= -\frac{\mathrm{i}g}{2}\varepsilon_{lmn}[p_\nu g_{\mu\sigma} - p_\mu g_{\nu\sigma}] A_l^\sigma(p) A_m^\mu(q) A_n^\nu(k) \quad , \end{aligned} \tag{4.70}$$

because we have to introduce distinct momenta for every field. Here we have explicitly written the vector product in isospin space with the summation indices l, m, n. From (4.70) we read off the form of the triple boson vertex (Fig. 4.4),

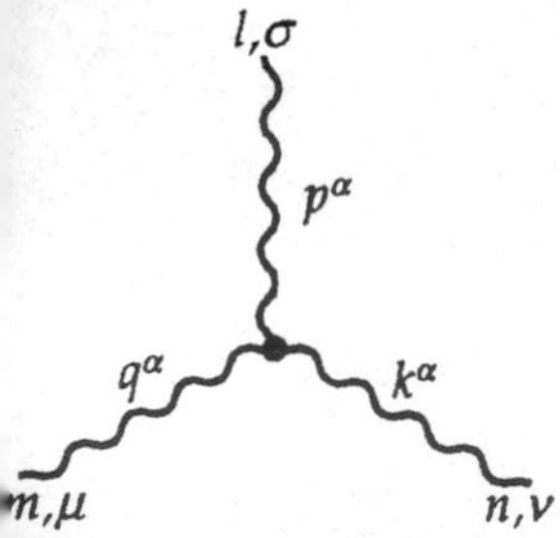

Fig. 4.4. The triple vertex in Yang–Mills theory, with our notation of four-momenta and indices of vector and isospin. The first index (m, n, or l) denotes isospin. The second (μ, ν, or σ) characterizes the space-time component of the corresponding vector meson

$$\bar{\Gamma}_{lmn}^{\sigma\mu\nu}(A^3) = -\frac{\mathrm{i}g}{2}\varepsilon_{lmn}(p_\nu g_{\mu\sigma} - p_\mu g_{\nu\sigma}) \quad . \tag{4.71}$$

This is not yet the final result because the three Yang–Mills fields occurring in (4.70) have the same rank. In total there are six possible ways of distributing them to the three 'legs' of the vertex. By adding to (4.71) all possible expressions derived by exchanging the indices and momenta, we see that every term occurs twice. Then the final result is (see Exercise 4.7)

$$\Gamma_{lmn}^{\sigma\mu\nu}(A^3) = -\mathrm{i}g\varepsilon_{lmn}\left[(p_\nu - q_\nu)g_{\mu\sigma} + (k_\mu - p_\mu)g_{\nu\sigma} + (q_\sigma - k_\sigma)g_{\mu\nu}\right] \quad . \tag{4.72}$$

The quartic vertex (Fig. 4.5) can be determined in the same procedure by transforming the last term of L_{I} (4.58) into momentum spaces. Explicit notation of the isospin indices yields

$$-\frac{1}{4}g^2\varepsilon_{ijm}\varepsilon_{klm}g_{\mu\sigma}g_{\nu\tau}A_k^\sigma(q)A_l^\tau(r)A_i^\mu(k)A_j^\nu(p) \quad . \tag{4.73}$$

We easily read off that

$$\tilde{\Gamma}_{ijkl}^{\mu\nu\sigma\tau}(A^4) = -\frac{g^2}{4}\varepsilon_{ijm}\varepsilon_{klm}g_{\mu\sigma}g_{\nu\tau} \quad . \tag{4.74}$$

However, now there are 24 combinations of the four Yang–Mills fields, and every different term occurs four times, i.e the vector meson characterized by $(i\mu)$ can carry the momenta k, p, q or r (4 possibilities), for the vector meson $(j\nu)$ there are then only 3 possibilities left, etc. The final result is

$$\begin{aligned}\Gamma_{ijkl}^{\mu\nu\sigma\tau}(A^4) = -g^2[&\varepsilon_{ijm}\varepsilon_{klm}(g_{\mu\sigma}g_{\nu\tau} - g_{\mu\tau}g_{\nu\sigma})\\ &+\varepsilon_{ikm}\varepsilon_{jlm}(g_{\mu\nu}g_{\sigma\tau} - g_{\mu\tau}g_{\nu\sigma})\\ &+\varepsilon_{ilm}\varepsilon_{jkm}(g_{\mu\sigma}g_{\nu\tau} - g_{\mu\nu}g_{\tau\sigma})] \quad .\end{aligned} \tag{4.75}$$

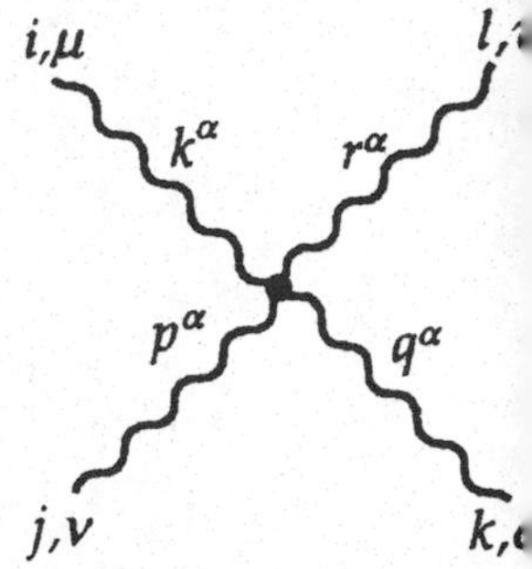

Fig. 4.5. The quartic vertex i Yang–Mills theory. The firs index at each leg (i, j, k, l) de notes the isospin of the vec tor boson concerned; the sec ond index (μ, ν, σ, τ) charac terizes the space-time compo nents

We would now have a complete list of the Feynman rules if we could neglect the technical problems of Yang–Mills theory connected with gauge invariance. Owing o the singularity of the wave operator $(\Box g_{\mu\nu} - \partial_\mu\partial_\nu)$, we have to choose a special gauge in order to derive a unique expression for the propagator $D_{\mu\nu}$ of the Yang–Mills field, as is done in (4.66) (for a detailed discussion see Example 4.6). In order to derive manifestly covariant Feynman rules, it is convenient to choose the *Lorentz gauge*,

$$\partial_\mu A^\mu = 0 \quad . \tag{4.76}$$

This gauge yields the propagator (4.66). More generally we could choose the gauge condition

$$\hat{L}_\mu A^\mu = 0 \quad , \tag{4.77}$$

with an arbitrary linear operator $\hat{L}_\mu$. The Lorentz gauge corresponds to the choice

$$\hat{L}_\mu = \partial_\mu \quad . \tag{4.78}$$

To obtain the *Coulomb gauge*, that is,

$$\nabla \cdot \boldsymbol{A}_i = 0 \tag{4.79}$$

(here the vector notation in $\boldsymbol{A}$ refers to the space components, whereas i denotes the components in isospin space!), we would choose

$$L_\mu = (0, \nabla) \quad . \tag{4.80}$$

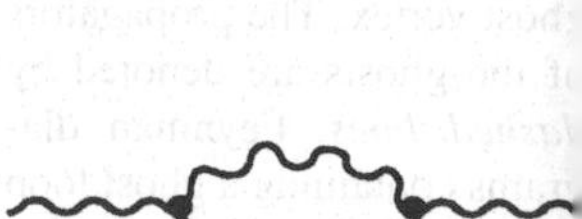

Fig. 4.6. Loop diagram of the vector field

However, a rigorous analysis (which we will not discuss here) shows that Feynman graphs including at least one loop diagram of the vector field (like the example in Fig. 4.6) are no longer gauge invariant. Different values are obtained even for matrix elements related to physical processes, depending on the choice of $\hat{L}_\mu$. This is, of course, not acceptable. As shown by Fadeev and Popov,[9] the basic reason for this failure is the fact that the Feynman rules describing Yang–Mills theory as obtained so far are incomplete. We must take into account the additional contributions of so-called *ghost fields*. These fields are described by the additional Lagrangian

$$L_{\text{ghost}} = -\bar{\chi}_i \hat{L}_\mu (\partial^\mu \delta_{ij} + g\varepsilon_{ijk} A_k^\mu)\chi_j \quad , \tag{4.81}$$

the ghost fields are denoted by χ_i, where i is an isospin (vector) index. The corresponding anti-ghost fields are denoted by $\bar{\chi}_i$. Taking the Lorentz gauge (4.78) we get

$$\begin{aligned} L_{\text{ghost}} &= -\bar{\chi}_i \partial_\mu (\partial^\mu \delta_{ij} + g\varepsilon_{ijk} A_k^\mu)\chi_j \\ &= -\bar{\chi}_i \Box \chi_i - g\varepsilon_{ijk} A_k^\mu \bar{\chi}_i \partial_\mu \chi_j \quad , \end{aligned} \tag{4.82}$$

where we have exploited the fact that the four-divergence of $\boldsymbol{A}^\mu$ vanishes owing to (4.76). Despite the ghost fields carrying only an isospin index, and being therefore spin-zero fields, they must be attributed Fermi–Dirac statistics in contrast to the usual spin-statistic theorem, so that every ghost loop in a Feynman diagram carries an additional factor -1. Moreover, ghost propagators may not occur as external lines of a Feynman diagram, since they correspond to unphysical modes; therefore only closed ghost loops contribute.

We can split the Lagrangian (4.82) into a free part and one containing the interaction between ghosts and vector bosons. The latter is given by

$$L_{\text{FP,A}} = -g\varepsilon_{ijk} \bar{\chi}_i \partial_\mu \chi_j A_k^\mu \quad , \tag{4.83}$$

where the index "FP" reminds us of Fadeev and Popov, who derived the extension of the Feynman rules in order to cancel unphysical contributions. In the same way as above, we derive the vertex function by writing $L_{\text{FP,A}}$ in momentum space

$$+\mathrm{i}g\varepsilon_{ikj} q_\mu \bar{\chi}_i(q)\chi_j(p) A_k^\mu(k) \quad , \tag{4.84}$$

yielding the vertex

$$\Gamma_\mu^{ijk}(\text{FP}, A) = -\mathrm{i}g\varepsilon_{ijk} q_\mu \quad , \tag{4.85}$$

shown in Fig. 4.7a. The vertex shows that ghosts can only be created and annihilated in pairs. Without detailed discussion, we mention that the unphysical contributions of the graph in Fig. 4.7c is compensated by the graph in Fig. 4.7b, if the ghost loop contains a factor -1. If we take into account both Feynman diagrams the total result is gauge invariant.

Finally, the "free" ghost propagator is given by

$$L_{\text{FP}} = -\bar{\chi}_i \Box \chi_i \quad , \tag{4.86}$$

according to (4.82). Apart from a pure divergence, it can also be written in the form

$$L_{\text{FP}} = (\partial_\mu \bar{\chi}_i)(\partial^\mu \chi_i) \quad . \tag{4.87}$$

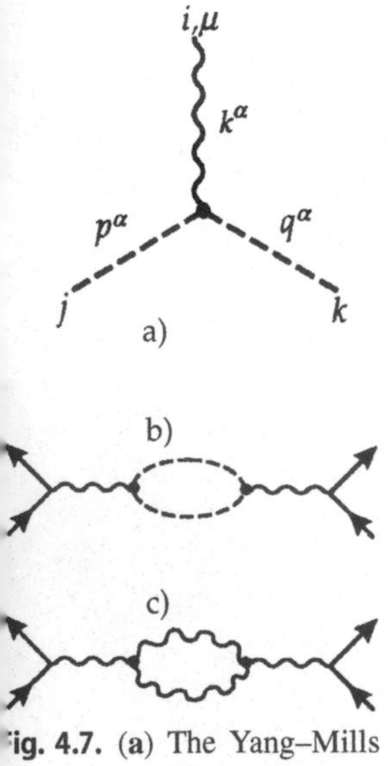

Fig. 4.7. (**a**) The Yang–Mills ghost vertex. The propagators of the ghosts are denoted by *dashed lines*. Feynman diagrams containing a ghost loop (**b**) and a Yang–Mills particle loop (**c**), whose unphysical contributions cancel each other

[9] L.D. Fadeev, V.N. Popov: Phys. Lett. **25B**, 29 (1967). The procedure is discussed extensively in W. Greiner, S. Schramm, and E. Stein: *Quantum Chromodynamics*, 3rd ed. (Springer, Berlin, Heidelberg, 2007).

Thus the ghost field propagates like a massless spin-zero field; its propagator is given by

$$\Delta(p) = \frac{1}{p^2 + \mathrm{i}\varepsilon} \quad . \tag{4.88}$$

This completes our set of rules. Including ghost fields, the full Lagrange function reads

$$\begin{aligned} L &= L_{\mathrm{F}} + L_{\mathrm{int}} + L_{\mathrm{A}} + L_{\mathrm{FP}} + L_{\mathrm{FP,A}} \\ &= \bar{\Psi}\gamma^\mu(\mathrm{i}\partial_\mu + g\boldsymbol{A}_\mu \cdot \hat{\boldsymbol{T}})\Psi - \bar{\Psi}\hat{M}\Psi - \frac{1}{4}\boldsymbol{F}_{\mu\nu}\cdot\boldsymbol{F}^{\mu\nu} \\ &\quad + \partial_\mu\bar{\chi}_i\partial^\mu\chi_i - g\varepsilon_{ikj}\bar{\chi}_i\partial_\mu\chi_j A_k^\mu \quad , \end{aligned} \tag{4.89}$$

the part L_{A} is written out explicitly in (4.55). The Feynman rules are listed in Table 4.1. There the propagators and vertices are multiplied by the imaginary unit i, as required by the Feynman rules.

EXAMPLE

4.6 Propagators and Gauge Invariance

Non-relativistic Perturbation Theory. Let us assume that the solutions ϕ_n of the free stationary Schrödinger equation

$$\hat{H}_0\phi_n = E_n\phi_n \quad ,$$

with

$$\int\limits_{L^3} \phi_m^*\phi_n d^3x = \delta_{mn} \quad , \tag{1}$$

are known. The Hamiltonian H_0 should be time independent, and the ϕ_n are normalized in a box with volume L^3. We look for solutions of the Schrödinger equation

$$\frac{\mathrm{i}\partial\psi(x,t)}{\partial t} = (\hat{H}_0 + V(\boldsymbol{x},t))\psi(\boldsymbol{x},t) \tag{2}$$

for a particle in a time-dependent potential $V(\boldsymbol{x},t)$. Every solution of (2) can be expanded in terms of the stationary states (1):

$$\psi(x,t) = \sum_n a_n(t)\phi_n(\boldsymbol{x})\exp(-\mathrm{i}E_n t) \quad . \tag{3}$$

Inserting (3) into (2) yields

$$\mathrm{i}\sum_n \frac{\mathrm{d}a_n(t)}{\mathrm{d}t}\phi_n(\boldsymbol{x})\exp(-\mathrm{i}E_n t) = \sum_n V(\boldsymbol{x},t)a_n(t)\phi_n(\boldsymbol{x})\exp(-\mathrm{i}E_n t) \quad . \tag{4}$$

After multiplying by $\phi_f^*(\boldsymbol{x},t)$ from the left and subsequently integrating, we get

$$\frac{\mathrm{d}a_f}{\mathrm{d}t} = -\mathrm{i}\sum_n a_n(t)\int \phi_f^* V\phi_n \mathrm{d}^3x \exp(\mathrm{i}(E_f - E_n)t) \quad . \tag{5}$$

xample 4.6

Table 4.1. Feynman rules of Yang–Mills theory in the covariant Landau gauge. Every closed ghost loop gets an additional factor (-1). The indices i, j, k, l denote isospin space, and the indices μ, ν, σ, τ, denote space-time

Fermion:	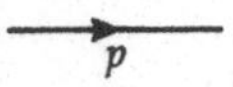	$iS_F(p) = \dfrac{i}{\gamma^\mu p_\mu - M + \mathrm{i}\varepsilon} = i\dfrac{\gamma^\mu p_\mu + M}{p^2 - M^2 + \mathrm{i}\varepsilon}$
Boson:		$i(D_{F_{\mu\nu}}(p))_{ik} = \dfrac{-\mathrm{i}P_{\mu\nu}(p)\delta_{ik}}{p^2 + \mathrm{i}\varepsilon}$
Fermion vertex:		$(\Gamma^\mu)^i = \mathrm{i}g\gamma^\mu T^i$
triple vertex:	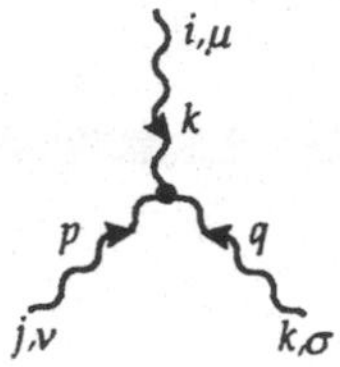	$(\Gamma^{\mu\nu\sigma})_{ijk} = \mathrm{i}g\varepsilon_{ijk}[g_{\mu\nu}(k-p)_\sigma + g_{\nu\sigma}(p-q)_\mu + g_{\sigma\mu}(q-k)_\nu]$
quadruple vertex:	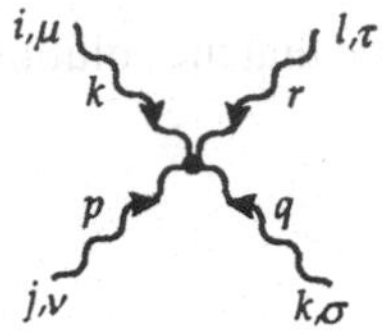	$(\Gamma^{\mu\sigma\nu\tau})_{ijkl} = -\mathrm{i}g^2[\varepsilon_{ijm}\varepsilon_{klm}(g_{\mu\sigma}g_{\nu\tau} - g_{\mu\tau}g_{\nu\sigma}) + \varepsilon_{ikm}\varepsilon_{jlm}(g_{\mu\nu}g_{\sigma\tau} - g_{\mu\tau}g_{\nu\sigma}) + \varepsilon_{ilm}\varepsilon_{jkm}(g_{\mu\sigma}g_{\nu\tau} - g_{\mu\nu}g_{\sigma\tau})]$
Ghost:	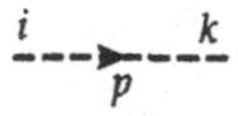	$i(\Delta(p))_{ik} = \frac{-\mathrm{i}\delta_{ik}}{p^2+\mathrm{i}\varepsilon}$
Ghost-vertex:	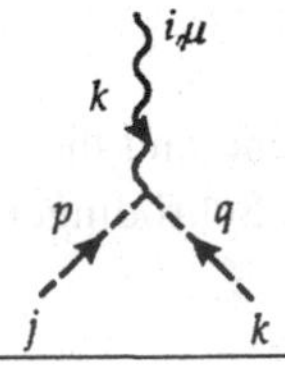	$(\Gamma^\mu(q))_{ijk} = g\varepsilon_{ijk}q_\mu$

This system of coupled equations is now solved with the following initial conditions: the particle should be in the eigenstate ϕ_i before the time $t = -T/2$; the potential $V(\boldsymbol{x}, t)$ should act only after $t = -T/2$. Mathematically this implies that at time $t = -T/2$

$$a_n\left(-\frac{1}{2}T\right) = \delta_{ni} \quad . \tag{6}$$

For weak potentials we have approximately (first-order successive approximation) that

$$\frac{\mathrm{d}a_f}{\mathrm{d}t} = -\mathrm{i}\int \mathrm{d}^3x\,\phi_f^* V(\boldsymbol{x}, t)\phi_i \exp\big(\mathrm{i}(E_f - E_i)t\big) \quad . \tag{7}$$

Example 4.

Equation (7) is the result of first-order perturbation theory for weak, transient potentials. It can be integrated yielding

$$a_f(t) = -\mathrm{i} \int\limits_{-T/2}^{t} \mathrm{d}t' \int\limits_{L^3} \mathrm{d}^3x \phi_f^*(x) V(\boldsymbol{x}, t') \phi_i(x) \exp\big(\mathrm{i}(E_f - E_i)t'\big) \quad . \tag{8}$$

In particular, at time $t = T/2$, when the interaction vanishes again, we have

$$\begin{aligned} S_{fi} &\equiv a_f\left(\frac{1}{2}T\right) \\ &= -\mathrm{i} \int\limits_{-\frac{1}{2}T}^{\frac{1}{2}T} \mathrm{d}t \int\limits_{L^3} \mathrm{d}^3x \big[\phi_f(\boldsymbol{x}) \exp(-\mathrm{i}E_f t)\big]^* V(\boldsymbol{x}, t) \big[\phi_i(\boldsymbol{x}) \exp(-\mathrm{i}E_i t)\big] \quad . \end{aligned} \tag{9}$$

Clearly, the transition amplitude T_{fi} can be written in covariant form:

$$S_{fi} = -\mathrm{i} \int \mathrm{d}^4x \phi_f^*(x) V(x) \phi_i(x) \quad . \tag{10}$$

First-order perturbation theory only makes sense when $|a_f(t)| \ll 1$, which we assumed in (5)–(7). It is tempting to interpret $|S_{fi}|^2$ as the probability for a particle in state ϕ_i to be scattered into state ϕ_f. This is true but also implies that the states ϕ_i and ϕ_f are separated by an infinite time interval. We can understand this remark by supposing the potential to be merely space dependent, that is, $V = V(\boldsymbol{x})$. Then (10) reads

$$\begin{aligned} S_{fi} &= -\mathrm{i}V_{fi} \int\limits_{-\infty}^{\infty} \mathrm{d}t \exp(\mathrm{i}(E_f - E_i)t) \\ &= -2\pi \mathrm{i} V_{fi} \delta(E_f - E_i) \quad , \end{aligned} \tag{11}$$

with

$$V_{fi} = \int\limits_{L^3} \mathrm{d}^3x \phi_f^*(\boldsymbol{x}) V(\boldsymbol{x}) \phi_i(\boldsymbol{x}) \quad .$$

In (11), the resulting δ function expresses energy conservation for the transition $i \to f$. The uncertainty relation requires an infinite time interval between ϕ_i and ϕ_f. For stationary potentials it is therefore more convenient to consider the transition probability per unit time,

$$W = \lim_{T\to\infty} \left(\frac{1}{T} |S_{fi}|^2\right) \quad . \tag{12}$$

With (11) we get

$$\begin{aligned} W &= \lim_{T\to\infty} \frac{2\pi}{T} |V_{fi}|^2 \delta(E_f - E_i) \int\limits_{-T/2}^{T/2} \mathrm{d}t \exp(\mathrm{i}(E_f - E_i)t) \\ &= \lim_{T\to\infty} \frac{2\pi}{T} |V_{fi}|^2 \delta(E_f - E_i) \int\limits_{-T/2}^{T/2} \mathrm{d}t \\ &= 2\pi |V_{fi}|^2 \delta(E_f - E_i) \quad . \end{aligned} \tag{13}$$

Here, the now well-known trick for the square of the delta function has been used, i.e.

$$(2\pi\delta(E_f - E_i))^2 = 2\pi\delta(E_f - E_i) \lim_{T\to\infty} \int_{-T/2}^{T/2} \exp(\mathrm{i}(E_f - E_i)t)\mathrm{d}t \quad .$$

Equation (13) is only physically applicable if we sum (or integrate) over a number of final states. In fact, often in physics, especially in particle physics, the system starts in a discrete state but is then scattered into a whole continuum of final states. With $\varrho(E_f)dE_f$ denoting the density of final states, the transition probability per unit time into these states is given by

$$W = 2\pi \int \mathrm{d}E_f \varrho(E_f)|V_{fi}|^2\delta(E_f - E_i) = 2\pi|V_{fi}|^2\varrho(E_i) \quad . \tag{14}$$

This relation is well known as *Fermi's golden rule.*

In second order we can improve our calculation by inserting $a_n(t)$, as given by (8) and (6), into (5). Indeed, after the first step of successive approximation $a_n^{(1)}(t)$ reads

$$a_n^{(1)}(t) = \delta_{ni} - \mathrm{i} \int_{-T/2}^{t} \mathrm{d}t' V_{ni} \exp(i(E_n - E_i)t') \quad .$$

Here V_{ni}, as denoted at $-T/2$ above, is taken to be time independent. This yields

$$\begin{aligned}\frac{\mathrm{d}a_f}{\mathrm{d}t} &= -\mathrm{i} \int \mathrm{d}^3x \phi_f^* V \phi_i \exp(\mathrm{i}(E_f - E_i)t) \\ &\quad + (-\mathrm{i})^2 \sum_{n\neq i} \left[V_{ni} \int_{-T/2}^{t} \mathrm{d}t' \exp(\mathrm{i}(E_n - E_i)t') \right] \\ &\quad \times V_{fn} \exp(\mathrm{i}(E_f - E_n)t) \quad ,\end{aligned} \tag{15}$$

where we have added the first- and second-order amplitudes in the usual manner of perturbation expansion. Now, according to (9) the transition amplitudes S_{fi} are given by

$$\begin{aligned}S_{fi} &= \lim_{T\to\infty} \left[-\mathrm{i} \int_{-T/2}^{T/2} \mathrm{d}t \exp(\mathrm{i}(E_f - E_i)t) V_{fi} \right. \\ &\quad \left. + (-\mathrm{i})^2 \sum_{n\neq i} V_{fn} V_{ni} \int_{-T/2}^{T/2} \mathrm{d}t \exp(\mathrm{i}(E_f - E_n)t) \int_{-T/2}^{t} \mathrm{d}t' \exp(\mathrm{i}(E_n - E_i)t') \right] \\ &= -\mathrm{i} \left[V_{fi} \int_{-\infty}^{\infty} \exp(\mathrm{i}(E_f - E_i)t)\mathrm{d}t \right. \\ &\quad - \mathrm{i} \sum_{n\neq i} V_{fn} V_{ni} \int_{-\infty}^{\infty} \mathrm{d}t \exp(\mathrm{i}(E_f - E_n)t) \\ &\quad \left. \times \int_{-\infty}^{t} \exp\big(\mathrm{i}(E_n - E_i - \mathrm{i}\varepsilon)t'\big)\mathrm{d}t' \right] \quad .\end{aligned} \tag{16}$$

In the last integral we have inserted a *convergence factor*

$$\exp(\varepsilon t) = \exp\big(\mathrm{i}(-\mathrm{i}\varepsilon)t\big)$$

Example 4.

in order to "switch on the integrand adiabatically" in the limit $t \to -\infty$. Later we let ε tend to zero. This is another way of saying that the propagation of the states involved is directed into the future.[10] We get

$$\int_{-\infty}^{t} \mathrm{d}t' \exp(\mathrm{i}(E_n - E_i - \mathrm{i}\varepsilon)t') = \mathrm{i}\frac{\exp(\mathrm{i}(E_n - E_i - \mathrm{i}\varepsilon)t)}{E_i - E_n + \mathrm{i}\varepsilon} \quad . \tag{17}$$

Hence (17) results in the form

$$S_{fi} = -\mathrm{i}\left[V_{fi} + \sum_{n \neq i} \frac{V_{fn}V_{ni}}{E_i - E_n + \mathrm{i}\varepsilon}\right] 2\pi\delta(E_f - E_i) \quad . \tag{18}$$

This equation has the same structure as (11), and we can now guess at the rule for determining higher-order transition amplitudes in the perturbation series. The transition amplitude always has the structure

$$S_{fi} = -\mathrm{i}[\tilde{V}_{fi}] 2\pi\delta(E_f - E_i) \tag{19}$$

with

$$\tilde{V}_{fi} = V_{fi} + \sum_{n \neq i} \frac{V_{fn}V_{ni}}{E_i - E_n + \mathrm{i}\varepsilon} + \cdots \quad . \tag{20}$$

The form of Fermi's golden rule remains unchanged, only V_{fi} is replaced by $\tilde{V}_{fi}$.

The Electron Propagator. The transition amplitude (18) can also be written in the form

$$S_{fi} = 2\pi\delta(E_f - E_i)\langle f|(-\mathrm{i}V) + (-\mathrm{i}V)\frac{\mathrm{i}}{E_i - \hat{H}_0 + \mathrm{i}\varepsilon}(-\mathrm{i}V) + \cdots|i\rangle \quad . \tag{21}$$

Here we have used the completeness relation

$$\sum_n |n\rangle\langle n| = 1 \tag{22}$$

of the eigenstates of $\hat{H}_0$. The prescription of adding a "$+\mathrm{i}\varepsilon$" term is extensively discussed elsewhere.[11] Obviously it is convenient to take $-iV$ as the scattering potential (perturbational parameter) instead of V. The factor $-i$ stems from the Schrödinger equation

$$\mathrm{i}\partial\psi/\partial t = (\hat{H}_0 + V)\psi$$

being written in the form

$$\partial\psi/\partial t = (-\mathrm{i}\hat{H}_0 - \mathrm{i}V)\psi \quad . \tag{23}$$

[10] W. Greiner and J. Reinhardt: *Quantum Electrodynamics*, 4th ed. (Springer, Berlin, Heidelberg, 2009).

[11] W. Greiner and J. Reinhardt: *Quantu Electrodynamics*, 4th ed. (Springer, Berlin, Heidelberg, 2009).

xample 4.6

In the *interaction representation*

$$\psi'(x,t) = e^{+i\hat{H}_0 t}\psi(x,t)$$

this is especially apparent. There the Schrödinger equation (23) reads

$$\partial\psi'/\partial t = (-iV)\psi' \quad , \tag{24}$$

yielding the time dependence

$$\psi' \propto \exp(-iVt) \quad . \tag{25}$$

The operator

$$\frac{i}{E_i - \hat{H}_0 + i\varepsilon} \tag{26}$$

in (21) is the non-relativistic electron propagator with the *vertex function* $(-iV)$. In fact, from the Schrödinger equation (2),

$$E_i\psi = (\hat{H}_0 + V)\psi \quad ,$$

we get

$$-i(E_i - \hat{H}_0)\psi = (-iV)\psi \tag{27}$$

and therefore by computing the inverse operator,

$$\begin{aligned}\psi &= [-i(E_i - \hat{H}_0)]^{-1}(-iV)\psi\\ &= i[E_i - \hat{H}_0]^{-1}(-iV)\psi \quad ,\end{aligned} \tag{28}$$

formally confirming our interpretation. For real values of E the inverse operator $(E - \hat{H}_0)^{-1}$ is mathematically not well defined, because the Hermitian operator $\hat{H}_0$ has real eigenvalues. Adding an infinitesimal imaginary part $i\varepsilon$ cures this problem. From a mathematical point of view there are many alternative ways of defining this operator uniquely. Thus physical reasons (adiabatically switching on the potential) are required for us to be able to decide here.

The Relativistic Propagator of a Spin-0 Particle. A spinless massive particle obeys the Klein–Gordon equation

$$(\Box + m^2)\phi = -V\phi \quad , \tag{29}$$

which, by analogy with (27), can be written as

$$i(\Box + m^2)\phi = (-iV)\phi \quad . \tag{30}$$

The propagator is the inverse of the differential operator on the left-hand side of the equation,

$$\left[i(\Box + m^2)\right]^{-1} \quad , \tag{31}$$

which in momentum space, that is, taking a basis of plane waves $\phi = N\exp(\mathrm{i}px)$, is given by

$$\left[\mathrm{i}(-p^2+m^2-\mathrm{i}\varepsilon)\right]^{-1} = \frac{\mathrm{i}}{p^2-m^2+\mathrm{i}\varepsilon} \quad . \tag{32}$$

The Relativistic Electron Propagator. The electron in an electromagnetic field A_μ satisfies the Dirac equation,

$$(\not{p}-m)\psi = +e\gamma^\mu A_\mu \psi \quad . \tag{33}$$

Again we multiply by $+\mathrm{i}$ and get

$$+\mathrm{i}(\not{p}-m)\psi = \mathrm{i}e\gamma^\mu A_\mu \psi \quad , \tag{34}$$

yielding the propagator

$$\begin{aligned} \frac{1}{[+\mathrm{i}(\not{p}-m)]} &= \frac{-\mathrm{i}}{(\not{p}-m)} = \frac{-\mathrm{i}(\not{p}+m)}{(p^2-m^2)} \\ &= \frac{-\mathrm{i}(\sum_s u_s\bar{u}_s)}{(p^2-m^2)} \end{aligned} \tag{35}$$

and the related vertex function

$$\mathrm{i}e\gamma^\mu \quad . \tag{36}$$

Note that we have defined e as the *positive* elementary charge, that is, the charge of the electron is given by

$$-e \quad . \tag{37}$$

The Photon Propagator. The wave equation of the photon is the Maxwell equation,

$$(g^{\lambda\nu}\Box - \partial^\nu\partial^\lambda)A_\lambda = j^\nu \quad . \tag{38}$$

The operator on the left-hand side has no inverse, even if we apply the iε prescription. This becomes clear in momentum space, where the left-hand side of (38) reads

$$-g^{\nu\lambda}q^2 + q^\nu q^\lambda = -q^2\left(g^{\nu\lambda} - \frac{q^\nu q^\lambda}{q^2}\right) \equiv -q^2\hat{P}^{\nu\lambda} \quad . \tag{39}$$

Let us assume that an inverse operator exists. It should have the structure

$$A(q^2)g_{\lambda\mu} + B(q^2)q_\lambda q_\mu \quad , \tag{40}$$

because there are only two second-rank tensors that can be formed: namely the metric tensor $g_{\lambda\mu}$ and the direct product $q_\lambda q_\mu$. The functions $A(q^2)$ and $B(q^2)$ can be determined from the relation

$$(-g^{\nu\lambda}q^2 + q^\nu q^\lambda)\left(A(q^2)g_{\lambda\mu} + B(q^2)q_\lambda q_\mu\right) = \delta^\nu_\mu \quad . \tag{41}$$

Expanding the equations yields

$$-A(q^2)q^2\delta^\nu_\mu + A(q^2)q^\nu q_\mu - B(q^2)q^2q^\nu q_\mu + B(q^2)q^2q^\nu q_\mu = \delta^\nu_\mu \quad . \tag{42}$$

xample 4.6

The terms proportional to $B(q^2)$ cancel, and the remaining equation for $A(q^2)$ is

$$A(q^2)\left[q^\nu q_\mu - q^2\delta^\nu_\mu\right] = -\delta^\nu_\mu \quad . \tag{43}$$

This has no solution, while $B(q^2)$ is completely undetermined. The basic reason for this is that the operator $\hat{P}^{\nu\lambda}$ in (39),

$$\hat{P}^{\nu\lambda} = g^{\nu\lambda} - q^\nu q^\lambda/q^2 \quad , \tag{44}$$

is a projection operator ($\hat{P}^\nu_\lambda \hat{P}^\lambda_\mu = \hat{P}^\nu_\mu$; see (4.8)) with eigenvalue zero and therefore cannot be inverted (compare the discussion of (4.3), (4.4)). That an eigenvalue of $\hat{P}^{\nu\mu}$ is zero follows from $P^{\mu\nu}q_\nu = 0$.

How then can a photon propagator be found which provides an effective inversion of (38)? We can solve this problem rigorously by coupling the massless vector field A^ν to a so-called Higgs field, a complex scalar field with a real-valued constant part,

$$\phi = \frac{1}{\sqrt{2}}\left(f + \chi_1(x) + \mathrm{i}\chi_2(x)\right) \quad . \tag{45}$$

The coupling is performed through the gauge-invariant Klein–Gordon current[12]

$$\Box A^\nu - \partial^\nu(\partial_\mu A^\mu) = \mathrm{i}e(\phi^*\partial^\nu\phi - (\partial^\nu\phi^*)\phi) - 2e^2A^\nu|\phi|^2 \tag{46}$$

yielding

$$\begin{aligned}(\Box + M^2)A^\nu &- \partial^\nu(\partial_\mu A^\mu) \\ &= -M\partial^\nu\chi_2 + e(\chi_2\partial^\nu\chi_1 - \chi_1\partial^\nu\chi_2) - e^2A^\nu(\chi_1^2 + 2f\chi_1 + \chi_2^2) \quad ,\end{aligned} \tag{47}$$

with $M = ef$. Here e is the charge of the ϕ field. Now the photon has acquired a mass M. Thus the gauge invariance of the theory seems to be broken, but actually it is just hidden, as we shall see.

The problem related to the photon propagator can now be solved by taking a special (but quite general) choice of gauge, that is,

$$\partial_\mu A^\mu = M\zeta\chi_2 \quad . \tag{48}$$

Here ζ is an *arbitrary* parameter. The gauge (48), which was first proposed by *'t Hooft*[13] in 1971 is manifestly covariant, because M, ζ and χ_2 are scalars, as is $\partial_\mu A^\mu$. Through the gauge the number of degrees of freedom of the A^μ field is reduced by one. Using this gauge in (47), we can express the term linear in χ_2 by

$$\begin{aligned}\chi_2 &= \frac{1}{M\zeta}\partial_\mu A^\mu \quad , \\ -M\partial^\nu\chi_2 &= -\frac{1}{\zeta}\partial^\nu\partial_\mu A^\mu \quad .\end{aligned} \tag{49}$$

---

[12] See W. Greiner and J. Reinhardt: *Quantum Electrodynamics*, 4th ed. (Springer, Berlin, Heidelberg, 2009), equation (8.11).

[13] G. 't Hooft: Nucl. Phys. **B35**, 167 (1971).

Therefore (47) results in the form

$$(\Box + M^2)A^\nu - \partial^\nu(\partial_\mu A^\mu)\left(1 - \frac{1}{\zeta}\right)$$
$$= e(\chi_2\partial^\nu\chi_1 - \chi_1\partial^\nu\chi_2) - e^2A^\nu(\chi_1^2 + 2f\chi_1 + \chi_2^2) \quad . \tag{50}$$

The terms on the right-hand side are interactions between vector and Higgs-field χ_1 as well as self-interactions. The operator on the left-hand side of (50) can be easily inverted with the same technique which we tried to apply to (40) without success. In momentum space the left-hand side of (50) reads

$$\left[g^{\nu\lambda}(-q^2 + M^2) + q^\nu q^\lambda\left(1 - \frac{1}{\zeta}\right)\right]A_\lambda = \cdots \quad . \tag{51}$$

Using the ansatz (40) for the propagator, that is, for the inverse operator of the bracket on the left-hand side of (51), we find that the equation determining the functions $A(q^2)$ and $B(q^2)$ is given by

$$\left[g^{\nu\lambda}(-q^2 + M^2) + q^\nu q^\lambda\left(1 - \frac{1}{\zeta}\right)\right]\left[A(q^2)g_{\lambda\mu} + B(q^2)q_\lambda q_\mu\right]$$
$$= A(q^2)\left[(-q^2 + M^2)\delta^\nu_\mu + q^\nu q_\mu\left(1 - \frac{1}{\zeta}\right)\right]$$
$$+ B(q^2)\left[q^\nu q_\mu(-q^2 + M^2) + q^2 q^\nu q_\mu\left(1 - \frac{1}{\zeta}\right)\right]$$
$$= \delta^\nu_\mu \quad . \tag{52}$$

Thus we get for the terms proportional to g^ν_μ

$$A(q^2)(-q^2 + M^2) = 1 \quad ,$$

and for those proportional to $q^\mu q^\nu$

$$A(q^2)\left[q^\nu q_\mu\left(1 - \frac{1}{\zeta}\right)\right] + B(q^2)\left[q^\nu q_\mu(-q^2 + M^2) + q^2 q^\nu q_\mu\left(1 - \frac{1}{\zeta}\right)\right] = 0 \quad .$$

Hence,

$$A(q^2) = -(q^2 - M^2)^{-1} \quad , \tag{53}$$

and

$$B(q^2) = (q^2 - M^2)^{-1}\left(1 - \frac{1}{\zeta}\right)\left[(-q^2 + M^2) + q^2\left(1 - \frac{1}{\zeta}\right)\right]^{-1} \quad , \tag{54}$$

and the propagator results in the form

$$A(q^2)g^{\lambda\mu} + B(q^2)q^\lambda q^\nu$$
$$= \frac{-g^{\lambda\mu}}{q^2 - M^2} - \frac{[q^\lambda q^\mu(1 - \zeta^{-1})](q^2 - M^2)^{-1}}{(q^2 - M^2) - q^2(1 - \zeta^{-1})}$$
$$= \frac{1}{q^2 - M^2}\left[-g^{\lambda\mu} + \frac{q^\lambda q^\mu(1 - \zeta)}{q^2 - \zeta M^2}\right] \quad . \tag{55}$$

xample 4.6

This propagator has interesting features. In the limit $\zeta \to \infty$ it takes the usual form of the propagator of massive vector bosons, namely

$$\mathrm{i}D^{\mu\nu} = \mathrm{i}\frac{-g^{\mu\nu} + q^\mu q^\nu/M^2}{q^2 - M^2} \quad . \tag{56}$$

As we learned in Sect. 3.6, this propagator produces all the high-energy problems which complicate the renormalization of the theory. However, even for this gauge a consistent set of Feynman rules can be formulated, so that all the cancellations of divergent terms occur which are required for renormalizability. We can understand this point without extensive calculations in the following manner. For a finite ζ the *'t Hooft propagator* (55),

$$\mathrm{i}D_T^{\mu\nu}(\zeta) = \mathrm{i}\frac{-g^{\mu\nu} + q^\mu q^\nu (1-\zeta)(q^2 - \zeta M^2)^{-1}}{q^2 - M^2} \quad , \tag{57}$$

tends to $1/q^2$ in the high-energy limit $q^2 \to \infty$. Therefore no problems occur, since then the propagator (57) behaves like the propagator of the renormalizable theory of QED in the Lorentz gauge. So, if no problems occur for any finite ζ, it seems plausible that this also holds in the limit $\zeta \to \infty$.

We must also discuss the second pole of the propagator (57) at $q^2 = \zeta M^2$. This pole has to be unphysical because it depends on the arbitrary parameter ζ. In fact, in his cited publication, *'t Hooft* showed that the contributions of this pole are exactly canceled by the contributions of a similar pole of the propagator of the Higgs field.

In the literature the gauge $\zeta \to \infty$ is described as the *unitary gauge* or *U gauge*, because only physical poles occur in this gauge. The propagators (57) given by the ζ gauge are also called *R gauge* propagators, since renormalizability can easily be shown in that gauge. As mentioned before, the latter contain unphysical Higgs fields such as χ_2.

In order to calculate Feynman diagrams without closed loops (so-called *tree diagrams*) it is most convenient to use the graphical rules of the U gauge. Technical problems within this gauge only occur in the calculations of *loop diagrams* as shown in Fig. 4.8, which are easier to calculate in the R gauge.

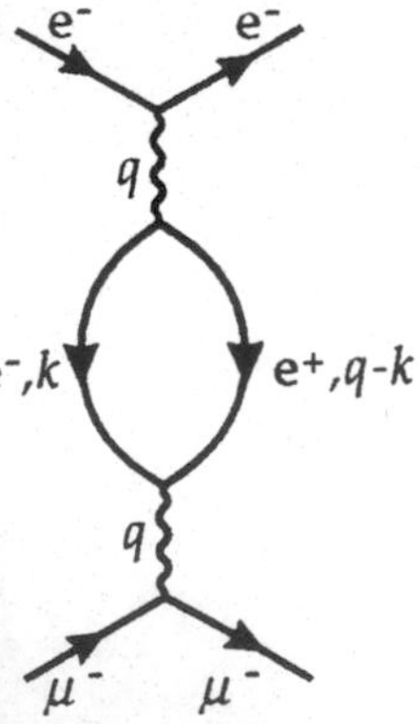

Fig. 4.8. Loop diagrams

We mention that our general formula (57) for the propagator also contains the propagator of a *massless vector boson*, that is, a photon, in the limit $M \to 0$ for an arbitrary gauge:

$$\mathrm{i}D^{\mu\nu}(\zeta) = \frac{\mathrm{i}}{q^2}\left[-g^{\mu\nu} + (1-\zeta)\frac{q^\mu q^\nu}{q^2}\right] \quad . \tag{58}$$

The well-known photon propagator in the *Feynman gauge* is obtained by choosing $\zeta = 1$:

$$\mathrm{i}D^{\mu\nu}_{\text{photon}} = \mathrm{i}\frac{-g^{\mu\nu}}{q^2} \quad . \tag{59}$$

EXERCISE

4.7 Self-Interaction of Gauge Fields

Problem. Derive the expression for the triple self-interaction vertex of the gauge field (4.72).

Exercise 4.

Solution. In Feynman's propagator formalism the interaction vertex can be constructed by variational derivation of the corresponding term in the action with respect to the fields, which is performed in momentum space. According to (4.58) the part of the Lagrangian density containing three gauge fields is given by

$$\begin{aligned} L^{(3)}_{\text{int}} &= -\frac{1}{2}g(\partial_\alpha A^a_\beta - \partial_\beta A^a_\alpha)\varepsilon_{abc}A^\alpha_b A^\beta_c \\ &= -g\varepsilon_{abc}\partial_\alpha A^a_\beta(x)A^\alpha_b(x)A^\beta_c(x) \quad . \end{aligned} \tag{1}$$

Introducing the Fourier representation

$$A^i_\mu(x) = \int \frac{\mathrm{d}^4 p}{(2\pi)^4} A^i_\mu(p)\mathrm{e}^{-\mathrm{i}p\cdot x} \tag{2}$$

and further exploiting the rule that the product of two functions in space corresponds to a convolution in momentum space we obtain

$$\begin{aligned} S^{(3)}_{\text{int}} &= \int \mathrm{d}^4 x L^{(3)}_{\text{int}}(x) \\ &= \int \mathrm{d}^4 x \frac{\mathrm{d}^4 p \mathrm{d}^4 k \mathrm{d}^4 q}{(2\pi)^{12}}(-g\varepsilon_{abc})(-\mathrm{i}p_\alpha)A^a_\beta(p)A^\alpha_b(q)A^\beta_c(k)\mathrm{e}^{-\mathrm{i}(p+q+k)\cdot x} \\ &= \mathrm{i}g\varepsilon_{abc}\int \frac{\mathrm{d}^4 p \mathrm{d}^4 k \mathrm{d}^4 q}{(2\pi)^8} p_\alpha A^a_\beta(p)A^\alpha_b(q)A^\beta_c(k)\delta(p+k+q) \quad . \end{aligned} \tag{3}$$

By renaming the three variables of integration p, q, k we can write the integrand in six different forms. Hence variation with respect to

$$\delta A^l_\sigma(p)\cdot\delta A^m_\mu(q)\cdot\delta A^n_\nu(k) \tag{4}$$

yields altogether six terms. To clarify this point we write the integrand in its six representations explicitly (without the momentum-conserving δ function):

$$\begin{aligned} &p_\alpha A^a_\beta(p)A^\alpha_b(q)A^\beta_c(k) \quad , \\ &p_\alpha A^a_\beta(p)A^\alpha_b(k)A^\beta_c(q) \quad , \\ &q_\alpha A^a_\beta(q)A^\alpha_b(k)A^\beta_c(p) \quad , \\ &q_\alpha A^a_\beta(q)A^\alpha_b(p)A^\beta_c(k) \quad , \\ &k_\alpha A^a_\beta(k)A^\alpha_b(p)A^\beta_c(q) \quad , \\ &k_\alpha A^a_\beta(k)A^\alpha_b(q)A^\beta_c(p) \quad . \end{aligned} \tag{5}$$

Variation yields

$$\frac{\delta A^a_\beta(p)}{\delta A^l_\sigma(p)} = g^\sigma_\beta\,\delta_{al} \quad , \tag{6}$$

and so on. We therefore get the following terms:

Exercise 4.7

$$
\begin{aligned}
& \mathrm{i} g\varepsilon_{abc}\big[p_\alpha g^\sigma_\beta \delta_{al} g^{\alpha\mu}\delta_{bm} g^{\beta\nu}\delta_{cn} + p_\alpha g^\sigma_\beta \delta_{al} g^{\beta\mu}\delta_{cm} g^{\alpha\nu}\delta_{bn} \\
& \quad + q_\alpha g^{\beta\sigma}\delta_{cl} g^\mu_\beta \delta_{am} g^{\alpha\nu}\delta_{bn} + q_\alpha g^{\alpha\sigma}\delta_{bl} g^\mu_\beta \delta_{am} g^{\beta\nu}\delta_{cn} \\
& \quad + k_\alpha g^{\alpha\sigma}\delta_{bl} g^{\beta\mu}\delta_{cm} g^\nu_\beta \delta_{an} + k_\alpha g^{\beta\sigma}\delta_{cl} g^{\alpha\mu}\delta_{bm} g^\nu_\beta \delta_{an}\big] \\
& = \mathrm{i} g(\varepsilon_{lmn} p^\mu g^{\sigma\nu} + \varepsilon_{lnm} p^\nu g^{\sigma\mu} + \varepsilon_{mnl} q^\nu g^{\sigma\mu} + \varepsilon_{mln} q^\sigma g^{\mu\nu} \\
& \quad + \varepsilon_{nlm} k^\sigma g^{\mu\nu} + \varepsilon_{nml} k^\mu g^{\sigma\nu}) \\
& = \mathrm{i} g\varepsilon_{lmn}(p^\mu g^{\sigma\nu} - p^\nu g^{\sigma\mu} + q^\nu g^{\sigma\mu} - q^\sigma g^{\mu\nu} + k^\sigma g^{\mu\nu} - k^\mu g^{\sigma\nu}) \quad , \qquad (7)
\end{aligned}
$$

where we have exploited the antisymmetry of ε_{lmn} in the last step. By collecting terms with identical indices of the metric tensor, we derive expression (4.2) of the three-gluon vertex,

$$
-\mathrm{i} g\varepsilon_{lmn}\big[(p^\nu - q^\nu)g^{\sigma\mu} + (k^\mu - p^\mu)g^{\sigma\nu} + (q^\sigma - k^\sigma)g^{\mu\nu}\big] \quad . \qquad (8)
$$

4.4 The Glashow–Salam–Weinberg Model of Leptons

We now have sufficient knowledge to formulate the Glashow–Salam–Weinberg theory of weak and electromagnetic interactions among leptons and to study its properties. Let us first state the starting point and the aim of our study:

1. There exist charged and neutral weak currents.
2. The charged currents contain only couplings between left-handed leptons.
3. The bosons W^+, W^-, and Z^0 mediating the weak interaction must be very massive.
4. Nevertheless we shall begin with massless bosons which receive masses through the Higgs mechanism. At that point we want to simultaneously include the photon field.

In order to fulfill these conditions we introduce two vector fields, one isospin triplet A^i_μ $(i = 1,2,3)$ and one singlet B_μ which should finally result as fields of the physical particles W^+, W^-, Z^0, and the photon through the symmetry breaking induced by the Higgs mechanism. The leptonic fields have to be distinguished according to their helicity. Every *fermion generation* (e, μ, τ) contains two related left-handed leptons. These form an *"isospin" doublet* of *left-handed leptons*, denoted by L_i $(i = \mathrm{e}, \mu, \tau)$:

$$
L_\mathrm{e} = \frac{1-\gamma_5}{2}\begin{pmatrix}\psi_{\nu_\mathrm{e}} \\ \psi_\mathrm{e}\end{pmatrix} \quad , \quad L_\mu = \frac{1-\gamma_5}{2}\begin{pmatrix}\psi_{\nu_\mu} \\ \psi_\mu\end{pmatrix} \quad , \quad L_\tau = \frac{1-\gamma_5}{2}\begin{pmatrix}\psi_{\nu_\tau} \\ \psi_\tau\end{pmatrix} \quad . \qquad (4.90)
$$

There are also right-handed components of the charged massive leptons. A right-handed neutrino does not exist (at least in the framework of weak and electromagnetic interactions), therefore right-handed leptons can be represented by singlets:

$$
R_\mathrm{e} = \left(\frac{1+\gamma_5}{2}\right)\psi_\mathrm{e} \quad , \quad R_\mu = \left(\frac{1+\gamma_5}{2}\right)\psi_\mu \quad , \quad R_\tau = \left(\frac{1+\gamma_5}{2}\right)\psi_\tau \quad . \qquad (4.91)
$$

Now we consider the various currents successively (charged, neutral, and electromagnetic currents) by adopting the new notation. Here we rewrite solely the terms including e and ν_e; the two remaining generations can be treated in the same manner. In our calculation we have used the property of γ_5 anticommuting with every γ^α, and we further utilize the property that $(\gamma_5)^2 = 1$. This yields, for example,

$$(1-\gamma_5)^2 = 2(1-\gamma_5)$$

and

$$\gamma^\alpha(1-\gamma_5) = \frac{1}{2}\gamma^\alpha(1-\gamma_5)^2 = \frac{1}{2}(1+\gamma_5)\gamma^\alpha(1-\gamma_5) = 2\frac{1+\gamma_5}{2}\gamma^\alpha\frac{1-\gamma_5}{2} \quad .$$

Finally we apply relation (1.24), which implies that

$$\bar{L}_e = (\bar{\psi}_{\nu_e}, \bar{\psi}_e)\frac{1+\gamma_5}{2} \quad , \tag{4.92}$$

and so on. The charged weak currents have the form (see Sect. 2.1)

$$\begin{aligned} J_-^{(e)\alpha} &= \bar{\psi}_e\gamma^\alpha(1-\gamma_5)\psi_{\nu_e} = 2\bar{\psi}_e\frac{1+\gamma_5}{2}\gamma^\alpha\frac{1-\gamma_5}{2}\psi_{\nu_e} \\ &= 2\bar{L}_e\gamma^\alpha\begin{pmatrix}0 & 0\\ 1 & 0\end{pmatrix}L_e = 2\bar{L}_e\gamma^\alpha\hat{T}_-L_e \quad , \end{aligned} \tag{4.93a}$$

$$J_+^{(e)\alpha} \equiv (J_-^{(e)\alpha})^\dagger = 2\bar{L}_e\gamma^\alpha\hat{T}_+L_e \quad , \tag{4.93b}$$

with $\hat{T}_\pm = \hat{T}_1 \pm \mathrm{i}\hat{T}_2$ and $(\hat{T}_-)^+ = \hat{T}_+$. The electromagnetic current exists only for the electron (as well as for the muon and τ lepton) since it is charged. It is given by

$$\begin{aligned} J_{\mathrm{EM}}^{(e)\alpha} &= \bar{\psi}_e\gamma^\alpha\psi_e = \frac{1}{2}\bar{\psi}_e\gamma^\alpha(1-\gamma_5)\psi_e + \frac{1}{2}\bar{\psi}_e\gamma^\alpha(1+\gamma_5)\psi_e \\ &= \bar{\psi}_e\frac{1+\gamma_5}{2}\gamma^\alpha\frac{1-\gamma_5}{2}\psi_e + \bar{\psi}_e\frac{1-\gamma_5}{2}\gamma^\alpha\frac{1+\gamma_5}{2}\psi_e \\ &= \bar{L}_e\gamma^\alpha\begin{pmatrix}0 & 0\\ 0 & 1\end{pmatrix}L_e + \bar{R}_e\gamma^\alpha R_e \\ &= \bar{L}_e\gamma^\alpha\left(\frac{1}{2} - \hat{T}_3\right)L_e + \bar{R}_e\gamma^\alpha R_e \quad . \end{aligned} \tag{4.94}$$

This electromagnetic current splits up into a part $-\bar{L}_e\gamma^\mu T_3 L_e$ belonging to an isotriplet like (4.93a), (4.93b) and a further part $\frac{1}{2}\bar{L}_e\gamma^\alpha L_e + \bar{R}_e\gamma^\alpha R_e$ representing an isosinglet current. The relative weighting of the left-handed parts of the isosinglet by a factor $\frac{1}{2}$ enters through the electromagnetic current (4.94). We may expect the neutral weak current $J_0^{(e)\alpha}$ to have a similar form, unknown to us at the moment. We rearrange the currents into an isotriplet by analogy with the isovector gauge field $\boldsymbol{A}_\mu$,

$$\bar{L}_e\gamma^\alpha \boldsymbol{T} L_e \quad , \tag{4.95a}$$

and an isosinglet to be associated with the gauge field B_μ,

$$\frac{1}{2}\bar{L}_e\gamma_\alpha L_e + \bar{R}_e\gamma_\alpha R_e \quad . \tag{4.95b}$$

These currents are minimally coupled to the corresponding gauge fields:

$$L_{\mathrm{int}}^{(\mathrm{e})} = g(\bar{L}_{\mathrm{e}}\gamma^{\alpha}\boldsymbol{T}L_{\mathrm{e}})\cdot\boldsymbol{A}_{\alpha} - g'\left[\frac{1}{2}(\bar{L}_{\mathrm{e}}\gamma^{\alpha}L_{\mathrm{e}}) + (\bar{R}_{\mathrm{e}}\gamma^{\alpha}R_{\mathrm{e}})\right]B_{\alpha} \quad . \tag{4.96}$$

This equation characterizes the structure of interaction as demanded by our general group-theoretical considerations concerning the two gauge fields $\boldsymbol{A}_{\alpha}$ and B_{α}. Here we have introduced two undetermined coupling constants g and g' whose significance will become clear in the following (see (4.101)).

The real (physical) photon does not couple to the singlet current (4.95b) but to J^{α}_{EM}, (4.94). Therefore it has to be represented by a mixture of B_{μ} and A^3_{μ} fields. We write the photon field in the following form:

$$A_{\mu} = \cos\theta\, B_{\mu} + \sin\theta\, A^3_{\mu} \quad . \tag{4.97a}$$

It is characterized solely by the index μ of a four-vector and can thereby be discerned from the isovector gauge field $\boldsymbol{A}_{\mu}$. The combination orthogonal to A_{μ},

$$Z_{\mu} = -\sin\theta\, B_{\mu} + \cos\theta\, A^3_{\mu} \tag{4.97b}$$

must describe the (physical) neutral intermediate boson of weak interactions. The inverses of (4.97a), (4.97b) are

$$B_{\mu} = \cos\theta\, A_{\mu} - \sin\theta\, Z_{\mu} \quad , \tag{4.98a}$$

$$A^3_{\mu} = \sin\theta\, A_{\mu} + \cos\theta\, Z_{\mu} \quad . \tag{4.98b}$$

The mixing angle θ is called the *Weinberg angle*. Finally we write (pay attention to the signs!)[14]

$$W^{(\pm)}_{\mu} = \frac{1}{\sqrt{2}}(A^1_{\mu} \mp \mathrm{i}A^2_{\mu}) \quad . \tag{4.99}$$

The field W_{μ} describes an *incoming negative* or an *outgoing positive* W boson, W^*_{μ} describes an *incoming positive* or an *outgoing negative* W boson. Sometimes in the literature the notation $W^{(+)} \equiv W^*$ is also used. We shall also sometimes adopt this in the following. By inserting (4.98), (4.99) into the Lagrange density $L^{(\mathrm{e})}_{\mathrm{int}}$ (4.96) we obtain

$$\begin{aligned}
\bar{L}^{(\mathrm{e})}_{\mathrm{int}} &= \frac{g}{\sqrt{2}}\bar{L}_{\mathrm{e}}\gamma^{\alpha}(\hat{T}_{-}W^{(-)}_{\alpha} + \hat{T}_{+}W^{(+)}_{\alpha})L_{\mathrm{e}} \\
&\quad + \left[g\cos\theta\,\bar{L}_{\mathrm{e}}\gamma^{\alpha}\hat{T}_3 L_{\mathrm{e}} + g'\sin\theta\left(\frac{1}{2}\bar{L}_{\mathrm{e}}\gamma^{\alpha}L_{\mathrm{e}} + \bar{R}_{\mathrm{e}}\gamma^{\alpha}R_{\mathrm{e}}\right)\right]Z_{\alpha} \\
&\quad + \left[-g'\cos\theta\left(\frac{1}{2}\bar{L}_{\mathrm{e}}\gamma^{\alpha}L_{\mathrm{e}} + \bar{R}_{\mathrm{e}}\gamma^{\alpha}R_{\mathrm{e}}\right) + g\sin\theta\,\bar{L}_{\mathrm{e}}\gamma^{\alpha}\hat{T}_3 L_{\mathrm{e}}\right]A_{\alpha} \\
&\equiv \frac{g}{2\sqrt{2}}\left(J^{(\mathrm{e})\alpha}_{-}W^{(-)}_{\alpha} + J^{(\mathrm{e})\alpha}_{+}W^{(+)}_{\alpha} + J^{(\mathrm{e})\alpha}_{0}Z_{\alpha}\right) - eJ^{(\mathrm{e})\alpha}_{\mathrm{EM}}A_{\alpha}
\end{aligned} \tag{4.100}$$

[14] This is completely analogous to the definition of charged pions (4.36b).

with

$$J_-^{(e)\alpha} = 2\bar{L}_e\gamma^\alpha\hat{T}_- L_e \quad , \quad J_+^{(e)\alpha} = 2\bar{L}_e\gamma^\alpha\hat{T}_+ L_e \quad ,$$

$$J_0^{(e)\alpha} = 2\sqrt{2}\left[\cos\theta\bar{L}_e\gamma^\alpha\hat{T}_3 L_e + \frac{g'}{g}\sin\theta\left(\frac{1}{2}\bar{L}_e\gamma^\alpha L_e + \bar{R}_e\gamma^\alpha R_e\right)\right] \quad ,$$

$$-e\,J_{EM}^{(e)\alpha} = -e\left[\bar{L}_e\gamma^\alpha\left(\frac{1}{2}-\hat{T}_3\right)L_e + \bar{R}_e\gamma^\alpha R_e\right]$$

$$= -\left[\bar{L}_e\gamma^\alpha\left(\frac{g'\cos\theta}{2} - g\sin\theta\hat{T}_3\right)L_e + g'\cos\theta\bar{R}_e\gamma^\alpha R_e\right] \quad .$$

Comparing the latter result, which is the correct expression of the electromagnetic current, we find that (4.94) yields directly the relations

$$e = g\sin\theta = g'\cos\theta \quad , \tag{4.101}$$

and

$$\tan\theta = \frac{\sin\theta}{\cos\theta} = \frac{g'}{g} \quad . \tag{4.102}$$

Separately, the angle θ can be expressed in terms of the original coupling constants g, g':

$$\sin\theta = \frac{g'}{\sqrt{g^2+g'^2}} \quad , \quad \cos\theta = \frac{g}{\sqrt{g^2+g'^2}} \quad . \tag{4.103}$$

Thus we can connect the elementary electric charge of the electron with the two coupling constants g, g':

$$e = \frac{gg'}{\sqrt{g^2+g'^2}} \quad , \tag{4.104}$$

or

$$\frac{1}{e^2} = \frac{1}{g^2} + \frac{1}{g'^2} \quad .$$

Equation (4.100) contains the explicit form of the neutral weak current in the framework of the Glashow–Salam–Weinberg theory. Using (4.101) we can write

$$J_0^{(e)\alpha} = 2\sqrt{2}\left[\cos\theta\bar{L}_e\gamma^\alpha\hat{T}_3 L_e + \frac{g'}{g}\sin\theta\left(\frac{1}{2}\bar{L}_e\gamma^\alpha L_e + \bar{R}_e\gamma^\alpha R_e\right)\right]$$

$$= \frac{2\sqrt{2}}{\cos\theta}\left[\bar{L}_e\gamma^\alpha\left(\hat{T}_3\cos^2\theta + \frac{1}{2}\sin^2\theta\right)L_e + \sin^2\theta\bar{R}_e\gamma^\alpha R_e\right]$$

$$= \frac{\sqrt{2}}{\cos\theta}\left[\bar{L}_e\gamma^\alpha\begin{pmatrix}1 & 0\\ 0 & -\cos 2\theta\end{pmatrix}L_e + 2\sin^2\theta\bar{R}_e\gamma^\alpha R_e\right] \quad . \tag{4.105}$$

Written in terms of neutrino and electron fields separately, the neutral weak current becomes

$$
\begin{aligned}
J_0^{(\mathrm{e})\alpha} &= \frac{\sqrt{2}}{\cos\theta}\left[\bar{\psi}_{\nu_\mathrm{e}}\frac{1+\gamma_5}{2}\gamma^\alpha\frac{1-\gamma_5}{2}\psi_{\nu_\mathrm{e}} - (1-2\sin^2\theta)\bar{\psi}_\mathrm{e}\frac{1+\gamma_5}{2}\gamma^\alpha\frac{1-\gamma_5}{2}\psi_\mathrm{e}\right.\\
&\quad \left. + 2\sin^2\theta\,\bar{\psi}_\mathrm{e}\frac{1-\gamma_5}{2}\gamma^\alpha\frac{1+\gamma_5}{2}\psi_\mathrm{e}\right]\\
&= (\sqrt{2}\cos\theta)^{-1}\left[\bar{\psi}_{\nu_\mathrm{e}}\gamma^\alpha(1-\gamma_5)\psi_{\nu_\mathrm{e}} - (1-2\sin^2\theta)\bar{\psi}_\mathrm{e}\gamma^\alpha(1-\gamma_5)\psi_\mathrm{e}\right.\\
&\quad \left. + 2\sin^2\theta\,\bar{\psi}_\mathrm{e}\gamma^\alpha(1+\gamma_5)\psi_\mathrm{e}\right]\\
&= (\sqrt{2}\cos\theta)^{-1}\left[\bar{\psi}_{\nu_\mathrm{e}}\gamma^\alpha(1-\gamma_5)\psi_{\nu_\mathrm{e}} - \bar{\psi}_\mathrm{e}\gamma^\alpha(g'_V - g'_A\gamma_5)\psi_\mathrm{e}\right] \ , \qquad (4.106)
\end{aligned}
$$

where

$$
g'_A = 1 \quad , \quad g'_V = 1 - 4\sin^2\theta \quad . \qquad (4.107)
$$

Comparing this result with the former ansatz of the neutral current, which we introduced in order to describe neutrino–electron scattering (3.2), we recognize that the two agree with each other. In fact, the neutrino part of the neutral current has the same form as in (3.2a), that is, pure V–A coupling. This is hardly surprising, since the theory contains only left-handed components of the neutrino field; hence the neutrino coupling must have this form. On inspecting the electronic part we see that in (4.106) the relations (4.107) hold, apart from a normalizing factor fixing the total coupling strength. The investigation of the experimental data in Sects. 3.1, 3.2, and 3.3 showed that either g_V or g_A must be very small (see (3.30)). Now relation (4.107) states that the Glashow–Salam–Weinberg theory excludes the possibility of a vanishing g_A. Therefore we must have

$$
\sin^2\theta \approx 0.25 \quad , \qquad (4.108)
$$

yielding a very small vector contribution to the electronic neutral weak current. The value of the Weinberg angle, given by (4.108), yields $\cos\theta \approx 1$. Therefore, in (4.105) there is roughly a prefactor $1/\sqrt{2}$ for the total neutral current compared to the charged currents. This becomes particularly evident from the explicit form of the various currents given later explicitly in (4.139).

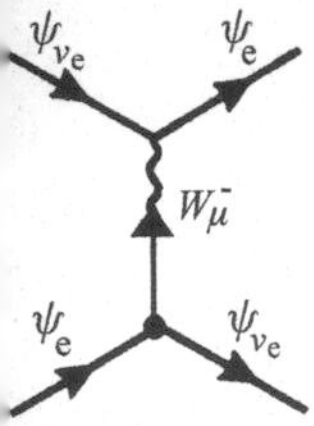

The effective Hamiltonian for neutrino–electron scattering (3.3) contains the neutral neutrino and the neutral electron current appearing in $J_0^{(\mathrm{e})}$; see (4.106) and also (4.139). The interaction between charged currents contributes with the two vertices

$$
\frac{g}{2\sqrt{2}} \cdot \frac{g}{2\sqrt{2}} = \frac{g^2}{8} ,
$$

the interaction between the neutral currents contributes with the two vertices

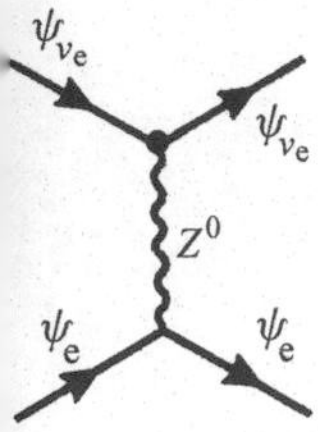

$$
\frac{g}{4\cos\theta} \cdot \frac{g}{4\cos\theta} \approx \frac{g^2}{16} .
$$

Therefore compared with the interaction strength between the charged currents, the total coupling strength between neutral currents has to be smaller by a factor $\approx (1/\sqrt{2})^2 = 1/2$. Thus $g'_A = 1$ in (4.107) yields the effective value $g_A \approx 1/2$ for the Fermi interaction with neutral currents given in (3.3). This is in excellent agreement with the experimental data; see (3.30a). Thus the Glashow–Salam–Weinberg theory passes a first important test.

EXAMPLE

4.8 The Gauge-Covariant Formulation of the GSW Theory: Weak Isospin and Weak Hypercharge

Let us summarize the main points of our discussion from a slightly different point of view. In the theory of weak interactions, the leptons are divided into left-handed *isodoublets* ($T = \frac{1}{2}, T_3 = \pm\frac{1}{2}$),

$$L_\ell = \begin{pmatrix} \psi_{\nu_\ell} \\ \psi_\ell \end{pmatrix}_{\mathrm{L}} = \frac{1-\gamma_5}{2} \begin{pmatrix} \psi_{\nu_\ell} \\ \psi_\ell \end{pmatrix} , \tag{1}$$

and right-handed *isosinglets* $T = 0$,

$$R_\ell = (\psi_\ell)_{\mathrm{R}} = \frac{1+\gamma_5}{2}\psi_\ell , \tag{2}$$

where ℓ runs over the "generations" e, μ, τ. Starting from (1) and (2) we demand the invariance of the theory with respect to gauge transformations, that is,

$$\begin{aligned} \begin{pmatrix} \psi_{\nu_\ell} \\ \psi_\ell \end{pmatrix}_{\mathrm{L}} \to \begin{pmatrix} \psi_{\nu_\ell} \\ \psi_\ell \end{pmatrix}_{\mathrm{L}}' &= \exp(\mathrm{i}\boldsymbol{a}(x)\cdot\hat{\boldsymbol{T}}) \begin{pmatrix} \psi_{\nu_\ell} \\ \psi_\ell \end{pmatrix}_{\mathrm{L}} \\ &= \exp\left(\mathrm{i}\boldsymbol{a}(\boldsymbol{x})\cdot\frac{\hat{\boldsymbol{\tau}}}{2}\right) \begin{pmatrix} \psi_{\nu_\ell} \\ \psi_\ell \end{pmatrix}_{\mathrm{L}} \equiv \hat{U}_2 L_\ell \\ (\psi_\ell)_{\mathrm{R}} \to (\psi_\ell)'_{\mathrm{R}} &= \exp(\mathrm{i}a(x)\hat{Y})(\psi_\ell)_{\mathrm{R}} \equiv \hat{U}_1 R_\ell \quad . \end{aligned} \tag{3}$$

Here $\hat{Y} = Y$ is the generator (a constant) of the U(1) group. By making the simple ansatz

$$\bar{L}_\ell \gamma^\mu \mathrm{i}\partial_\mu L_\ell + \bar{R}_\ell \gamma^\mu \mathrm{i}\partial_\mu R_\ell , \tag{4}$$

for the kinetic-energy term of the Lagrange density, we find that the latter is only gauge invariant with respect to (3) if we replace the derivative ∂_μ by the covariant derivative

$$\hat{D}_\mu = \partial_\mu - \mathrm{i}g\hat{\boldsymbol{T}}\cdot\boldsymbol{A}_\mu - \mathrm{i}\frac{g'}{2}\hat{Y}B_\mu \quad . \tag{5}$$

$\hat{\boldsymbol{T}} = (\hat{T}_1, \hat{T}_2, \hat{T}_3)$ are the three generators of the SU(2) group, $\hat{Y}$ is the generator of the U(1) group. The former fulfill the commutation relations

$$[\hat{T}_i, \hat{T}_j] = \mathrm{i}\varepsilon_{ijk}\hat{T}_k , \tag{6}$$

whereas $\hat{Y}$ is just a number. Obviously

$$[\hat{T}_i, \hat{Y}] = 0 \tag{7}$$

holds, because the generators belong to different groups.

In the following we shall see that it is convenient to denote the coupling constant to the B_μ field by $\frac{1}{2}g'$ (instead of g'), because the interaction will then have the form (4.96). In (5) we were forced to introduce the *gauge fields* $\boldsymbol{A}_\mu$ (isovector, four-vector in

xample 4.8

space-time) and B_μ (isoscalar, four-vector in space-time) with gauge transformations (see (4.49))

$$\begin{aligned} \boldsymbol{A}_\mu \cdot \hat{\boldsymbol{T}} \to \boldsymbol{A}'_\mu \cdot \hat{\boldsymbol{T}} &= \hat{U}_2 \boldsymbol{A}_\mu \cdot \hat{\boldsymbol{T}} \hat{U}_2^{-1} + \frac{\mathrm{i}}{g} \hat{U}_2 (\partial_\mu \hat{U}_2^{-1}) \quad , \\ B_\mu \to B'_\mu &= B_\mu + \frac{2\mathrm{i}}{g'} U_1 (\partial_\mu U_1^{-1}) \quad , \end{aligned} \tag{8}$$

in order to ensure gauge invariance. Here U_1 and $\hat{U}_2$ are U(1) and SU(2) transformations, respectively. The new degrees of freedom connected to the gauge fields $\boldsymbol{A}_\mu$ and B_μ now have to be supplemented by gauge-invariant "kinetic-energy" terms for these gauge fields in the Lagrange density. This can be done as in electrodynamics by using field-strength tensors (see (4.51a))

$$\begin{aligned} \hat{F}_{\mu\nu} \equiv \boldsymbol{F}_{\mu\nu} \cdot \hat{\boldsymbol{T}} &= \hat{D}_\mu (\boldsymbol{A}_\nu \cdot \hat{\boldsymbol{T}}) - \hat{D}_\nu (\boldsymbol{A}_\mu \cdot \hat{\boldsymbol{T}}) \\ &= (\partial_\mu \boldsymbol{A}_\nu - \partial_\nu \boldsymbol{A}_\mu + g \boldsymbol{A}_\mu \times \boldsymbol{A}_\nu) \cdot \hat{\boldsymbol{T}} , \\ \boldsymbol{F}_{\mu\nu} &= \partial_\mu \boldsymbol{A}_\nu - \partial_\nu \boldsymbol{A}_\mu + g \boldsymbol{A}_\mu \times \boldsymbol{A}_\nu \end{aligned} \tag{9}$$

and

$$B_{\mu\nu} = \partial_\mu B_\nu - \partial_\nu B_\mu \tag{10}$$

with gauge-transformation properties (see (18) of Exercise 4.5)

$$\hat{F}_{\mu\nu} \to \hat{F}'_{\mu\nu} = \hat{U}_2 \hat{F}_{\mu\nu} \hat{U}_2^{-1} = \hat{U}_2 (\boldsymbol{F}_{\mu\nu} \cdot \hat{\boldsymbol{T}}) \hat{U}_2^{-1}$$

and

$$B_{\mu\nu} \to B'_{\mu\nu} = B_{\mu\nu} \quad , \tag{11}$$

respectively. The gauge-invariant kinetic energies of the gauge fields $\boldsymbol{A}_\mu$ and B_μ are given by

$$\begin{aligned} L'_A &= -\frac{1}{2} \operatorname{Tr} \left\{ \hat{F}_{\mu\nu} \hat{F}^{\mu\nu} \right\} = -\frac{1}{2} \operatorname{Tr} \left\{ (\boldsymbol{F}_{\mu\nu} \cdot \hat{\boldsymbol{T}})(\boldsymbol{F}^{\mu\nu} \cdot \hat{\boldsymbol{T}}) \right\} \\ &= -\frac{1}{4} \boldsymbol{F}_{\mu\nu} \cdot \boldsymbol{F}^{\mu\nu} \end{aligned} \tag{12}$$

and

$$L'_B = -\frac{1}{4} B_{\mu\nu} B^{\mu\nu} \quad . \tag{13}$$

Summarizing our result we get

$$L'_{\text{int}} = \bar{L}_\ell \gamma^\mu \mathrm{i} \hat{D}_\mu L_\ell + \bar{R}_\ell \gamma^\mu \mathrm{i} \hat{D}_\mu R_\ell - \frac{1}{4} \boldsymbol{F}_{\mu\nu} \cdot \boldsymbol{F}^{\mu\nu} - \frac{1}{4} B_{\mu\nu} B^{\mu\nu} \quad , \tag{14}$$

or explicitly

$$\begin{aligned} L' &= \mathrm{i} \bar{L}_\ell \gamma^\mu \left[\partial_\mu - \mathrm{i} g \hat{\boldsymbol{T}} \cdot \boldsymbol{A}_\mu - \mathrm{i} \frac{g'}{2} \hat{Y} B_\mu \right] L_\ell \\ &\quad + \mathrm{i} \bar{R}_\ell \gamma^\mu \left[\partial_\mu - \mathrm{i} g \hat{\boldsymbol{T}} \cdot \boldsymbol{A}_\mu - \mathrm{i} \frac{g'}{2} \hat{Y} B_\mu \right] R_\ell - \frac{1}{4} \boldsymbol{F}_{\mu\nu} \cdot \boldsymbol{F}^{\mu\nu} - \frac{1}{4} B_{\mu\nu} B^{\mu\nu} . \end{aligned} \tag{15}$$

Example 4.

With the help of Table 4.2 we can easily verify that, by taking the appropriate quantum numbers T and Y for L_e and R_e, respectively, the interaction contained in (15) agrees with $L^{(e)}_{int}$ (4.96). In particular, $\hat{T}R_\ell = 0$, because R_ℓ is an isosinglet.

Now we also understand the choice taken in (4.96) when we wrote the term proportional to g' (that is, the interaction with the field B_μ) with a negative sign. In that way the terms correspond directly to the general scheme (15) using gauge-covariant derivatives.

At this stage the gauge fields are still massless. However, some of them acquire a mass, since we know the vector bosons must be massive. We shall achieve this in the next section by applying the Higgs mechanism. But let us first discuss the physical meaning of *"weak isospin"* $\hat{T}$ and *"weak hypercharge"* $\hat{Y}$. Glashow[15] proposed that the Gell-Mann–Nishijima relation for the electric charge Q should also be valid in the case of the weak interaction, that is,

$$eQ = e\left(T_3 + \frac{1}{2}Y\right) \quad . \tag{16}$$

Table 4.2. Quantum numbers of weak isospin and hypercharge of leptons and quarks

Fermion	T	T_3	Y	Q
ν_e, ν_μ, ν_τ	1/2	1/2	−1	0
e_L, μ_L, τ_L	1/2	−1/2	−1	−1
e_R, μ_R, τ_R	0	0	−2	−1
u_L, c_L	1/2	1/2	1/3	2/3
$(d_C)_L, (s_C)_L$	1/2	−1/2	1/3	−1/3
u_R, c_R	0	0	4/3	2/3
$(d_C)_R, (s_C)_R$	0	0	−2/3	−1/3

Since $\hat{T}_3$ and $\hat{Y}$ commute, both can be diagonal simultaneously. Therefore we replace $\hat{T}_3$ and $\hat{Y}$ by their eigenvalues in (16). From the known charge of the neutrino ($Q = 0$) and leptons ($Q = -1$) and from their classification with respect to isodoublets and isosinglets we can directly determine the T_3 and Y values of the various particles, as shown in Table 4.2. In addition to the quantum numbers of the leptons we also include the quarks, although their "weak properties" will be discussed later in Chap. 6. In particular some of the quarks carry the index "C" referring to Cabibbo mixing of the quarks (see Sect. 6.4).

4.5 Spontaneous Symmetry Breaking: The Higgs Sector

So far, the vector bosons W^+_μ, W^-_μ, and Z_μ have been treated as if they were massless. In order to give them mass we apply the Higgs mechanism discussed in Sect. 4.1. As the left-handed leptons form an isodoublet and the gauge fields $\boldsymbol{A}_\mu$ an isovector, we now need an isodoublet of Higgs fields,

$$\Phi = \begin{pmatrix} \phi^{(+)} \\ \phi^{(0)} \end{pmatrix} \quad , \quad |\Phi|^2 = |\phi^{(+)}|^2 + |\phi^{(0)}|^2 \quad , \tag{4.109}$$

15 S.L. Glashow: Nucl. Phys. **22**, 579 (1961).

consisting of a positively charged and a neutral spin-zero particle. With the help of the weak Gell-Mann–Nishijima relation discussed in Example 4.8, (16), we find that the isospin T and hypercharge Y of the Higgs field must be given by

$$T = \frac{1}{2} \quad , \quad Y = 1 \quad . \tag{4.110}$$

In order to obtain a non-vanishing vacuum expectation value of the Higgs field, we add a potential term with the "wrong" sign of the mass term to the Higgs Lagrangian,

$$U(\Phi) = -\mu^2 |\Phi|^2 + h|\Phi|^4 \quad . \tag{4.111}$$

In addition there is a gradient term for the kinetic energy of the Higgs field, where it is minimally coupled to the gauge fields $\boldsymbol{A}_\mu$ and B_μ in the same way as the leptonic doublet:

$$\left| \left(\mathrm{i}\partial_\mu + g\hat{\boldsymbol{T}} \cdot \boldsymbol{A}_\mu + \frac{g'}{2} B_\mu \hat{Y} \right) \Phi \right|^2 = \left| \left(\partial_\mu - \mathrm{i}g\hat{\boldsymbol{T}} \cdot \boldsymbol{A}_\mu - \mathrm{i}\frac{g'}{2} B_\mu \hat{Y} \right) \Phi \right|^2 \tag{4.112}$$

(Example 4.8, (5), (15)). We recognize that the relative sign of the coupling to the electrically neutral fields A_μ^3 and B_μ differs for the two components of the Higgs field. Therefore the electromagnetic field A_μ does *not* couple to the lower component of the Higgs field, as we shall see in the following. The hypercharge of the Higgs doublet, fixed by (4.110), is the origin of this change of sign. We can say that the special choice of the Higgs doublet (4.109), which seems to be unmotivated at first glance, just yields the effect that the $A_\mu^3 - B_\mu$ mixture representing the photon field A_μ does not couple to the neutral component ϕ^0. This must be so since the photon must remain massless. With the help of (4.98) we can easily prove that

$$\begin{aligned} gT^3 A_\mu^3 + \frac{g'}{2} B_\mu &= \frac{1}{2} \begin{pmatrix} +gA_\mu^3 + g'B_\mu & 0 \\ 0 & -gA_\mu^3 + g'B_\mu \end{pmatrix} \\ &= \frac{1}{2} \begin{pmatrix} A_\mu(g\sin\theta + g'\cos\theta) + Z_\mu(+g\cos\theta - g'\sin\theta) & 0 \\ 0 & A_\mu(-g\sin\theta + g'\cos\theta) - Z_\mu(g\cos\theta + g'\sin\theta) \end{pmatrix} \\ &= \begin{pmatrix} eA_\mu + g\dfrac{\cos 2\theta}{2\cos\theta} Z_\mu & 0 \\ 0 & -\dfrac{g}{2\cos\theta} Z_\mu \end{pmatrix} \quad . \end{aligned} \tag{4.113}$$

This result shows that the lower component of the Higgs field does not couple to the photon field, that is, it carries no electric charge. Formally, the upper component has the opposite charge of the electron so that it is positively charged corresponding to (4.109). We are always able to set the upper component $\phi^{(+)}$ to zero by an appropriate gauge (rotation in isospin space). Without restricting generality we write

$$\Phi = \frac{1}{\sqrt{2}} \big(\lambda + \chi(x)\big) \exp\left(\frac{\mathrm{i}}{\lambda} \boldsymbol{\Theta}(x) \cdot \boldsymbol{T} \right) \begin{pmatrix} 0 \\ 1 \end{pmatrix} \tag{4.114}$$

with the isovector $\boldsymbol{\Theta} = (\Theta_1, \Theta_2, \Theta_3)$. The four fields $\boldsymbol{\Theta}(x)$ and $\chi(x)$ are real-valued. $\lambda = \sqrt{\mu^2/h}$ describes the vacuum expectation value of the Higgs field, known from Sect. 4.1, and $\chi(x)$ denotes the local deviation of the Higgs field from that value.

With regard to SU(2) gauge invariance of the Yang–Mills theory, we are now able to eliminate the exponential, because Φ can be written in the form

$$\Phi = \frac{1}{\sqrt{2}}(\lambda + \chi(x))\hat{U}_2 \begin{pmatrix} 0 \\ 1 \end{pmatrix} = \hat{U}_2 \frac{1}{\sqrt{2}}(\lambda + \chi(x)) \begin{pmatrix} 0 \\ 1 \end{pmatrix} = \hat{U}_2 \Phi_0 \quad , \tag{4.115}$$

$\hat{U}_2$ being the SU(2) gauge transformation

$$\hat{U}_2 = \exp\left(+\frac{\mathrm{i}}{\lambda} \boldsymbol{\Theta}(x) \cdot \boldsymbol{T} \right) \quad . \tag{4.116}$$

Now we perform a gauge transformation $\Phi \to \Phi' = U_2^{-1}\Phi$. Of course also the gauge fields have to be transformed accordingly, $A_\mu \to A'_\mu$; see (8) in Example 4.8. As a result the Higgs field we obtain the simple form

$$\Phi' = \frac{1}{\sqrt{2}}(\lambda + \chi(x)) \begin{pmatrix} 0 \\ 1 \end{pmatrix} \quad . \tag{4.117}$$

In the following, we will adopt this gauge condition and for simplicity drop the primes. This choice is called the *unitary gauge* and was introduced earlier in the case of a one-dimensional Higgs field; see (4.32). Without this choice of gauge, the theory would be more difficult to interpret. The field Φ then apparently would have more degrees of freedom which, however, are spurious since they can be removed by a gauge transformation. Nevertheless, the unitary gauge is not always the most convenient choice and other conditions can be imposed, e.g. for the purpose of renormalization of the theory.

The vacuum expectation value of the Higgs field is fixed by the condition that $U(\Phi)$ attains a minimum. To this end we set $\chi(x) = 0$ and, using the gauge (4.117), we get the vacuum expectation value of the Higgs field operator $\hat{\Phi}$

$$\langle 0|\hat{\Phi}|0\rangle = \frac{\lambda}{\sqrt{2}} \begin{pmatrix} 0 \\ 1 \end{pmatrix} \quad , \quad |\langle 0|\hat{\Phi}|0\rangle|^2 = \frac{\lambda^2}{2} \quad . \tag{4.118}$$

The Higgs potential results in the form

$$U(\langle 0|\hat{\Phi}|0\rangle) = -\frac{\mu^2}{2}\lambda^2 + \frac{h}{4}\lambda^4 \equiv V(\lambda) \quad . \tag{4.119}$$

Demanding $\mathrm{d}V/\mathrm{d}\lambda = 0$ we get

$$\lambda^2 = \frac{\mu^2}{h} \quad , \tag{4.120}$$

yielding the vacuum expectation value of the Higgs field

$$\langle 0|\hat{\Phi}|0\rangle = \frac{\mu}{\sqrt{2h}} \begin{pmatrix} 0 \\ 1 \end{pmatrix} \quad . \tag{4.121}$$

Again we mention that the lower component of the Higgs field does not couple to the electromagnetic field according to (4.113). Therefore the non-vanishing vacuum expectation value does not influence the photon field, in other words, in spite of the symmetry breaking the photon remains massless.

Now we collect all parts of the Lagrangian density, that is, the contributions of the free fields (leptons, Yang–Mills field and neutral vector field B_μ), writing

$$B_{\mu\nu} = \partial_\mu B_\nu - \partial_\nu B_\mu \quad , \qquad \boldsymbol{F}_{\mu\nu} = \partial_\mu \boldsymbol{A}_\nu - \partial_\nu \boldsymbol{A}_\mu + g\boldsymbol{A}_\mu \times \boldsymbol{A}_\nu \quad , \tag{4.122}$$

and also the interaction terms (4.100) and the contributions of the Higgs field (4.111), (4.112). Finally we add a term ($\ell = \mathrm{e}, \mu, \tau$)

$$-\sqrt{2} f_\ell (\bar{R}_\ell \Phi^\dagger L_\ell + \bar{L}_\ell \Phi R_\ell) = -f_\ell(\lambda + \chi)\bar{\psi}_\ell \psi_\ell \tag{4.123}$$

for each lepton generation, in order to give the charged leptons the mass $m_\ell = f_\ell \lambda$. A detailed discussion of these terms can be found in Exercise 4.9. Here $\Phi^\dagger = (\Phi^{(+)^*}, \Phi^{(0)^*})$ denotes the Hermitian conjugate of the isodoublet vector. Obviously the combinations

$$\Phi^\dagger L_\ell \quad \text{and} \quad \bar{L}_\ell \Phi \tag{4.124}$$

are isoscalars and spinors or adjoint spinors, respectively, in space-time and can be combined with the isoscalar spinors R_ℓ and adjoint isoscalar spinors $\bar{R}_\ell$, respectively, forming the invariants

$$\bar{R}_\ell \Phi^\dagger L_\ell \quad \text{and} \quad \bar{L}_\ell \Phi R_\ell \quad , \tag{4.125}$$

which are related to each other by

$$(\bar{R}\Phi^\dagger L)^\dagger = (\bar{L}\Phi R) \quad . \tag{4.126}$$

Hence the form of the interaction chosen in (4.123) is Hermitian. Each of the terms in (4.123) is also invariant with respect to gauge transformations. This can be seen by using the notation of gauge transformations $\hat{U} = \hat{U}_1\hat{U}_2 = \hat{U}_2\hat{U}_1$ introduced in Example 4.8, (3). $\hat{U}_1$ is the U(1) gauge transformation, whereas $\hat{U}_2$ denotes the SU(2) transformation. Obviously both commute. We can then write

$$\begin{aligned} \bar{R}_\ell \Phi^\dagger L_\ell &= \bar{R}_\ell \hat{U}^{-1}\hat{U}\Phi^\dagger \hat{U}^{-1}\hat{U} L_\ell \\ &= \bar{R}_\ell \hat{U}_1^{-1}\hat{U}_1 \Phi^\dagger \hat{U}_1^{-1}\hat{U}_2^{-1}\hat{U}_1\hat{U}_2 L_\ell \\ &= \bar{R}'_\ell \Phi'^\dagger L'_\ell \quad , \end{aligned} \tag{4.127}$$

with

$$R'_\ell = \hat{U}_1 R_\ell \quad , \quad L'_\ell = \hat{U}_1\hat{U}_2 L_\ell \quad , \quad \Phi' = \hat{U}_1\hat{U}_2\Phi\hat{U}_1^{-1} = \hat{U}_2\Phi \quad . \tag{4.128}$$

We denote the neutrino components by ψ_{ν_ℓ} ($\ell = \mathrm{e}, \mu, \tau$) and the massive leptons by ψ_ℓ. Collecting all contributions (see Exercise 4.10), the Lagrangian of the weak interactions is given by

$$\begin{aligned} L = &-\frac{1}{4}\boldsymbol{F}_{\mu\nu}\cdot\boldsymbol{F}^{\mu\nu} - \frac{1}{4}B_{\mu\nu}B^{\mu\nu} - e\Big(\sum_\ell \bar{\psi}_\ell \gamma^\mu \psi_\ell\Big)A_\mu \\ &+ \sum_\ell \mathrm{i}\Big(\bar{\psi}_{\nu_\ell}\gamma^\mu \frac{1}{2}(1-\gamma_5)\partial_\mu \psi_{\nu_\ell} + \mathrm{i}\bar{\psi}_\ell \gamma^\mu \partial_\mu \psi_\ell - f_\ell \bar{\psi}_\ell \psi_\ell (\lambda + \chi)\Big) \end{aligned}$$

$$+\frac{g}{2\sqrt{2}}\sum_{\ell}[\bar{\psi}_{\ell}\gamma^{\mu}(1-\gamma_5)\psi_{\nu_\ell}W_{\mu}^{(-)}+\bar{\psi}_{\nu_\ell}\gamma^{\mu}(1-\gamma_5)\psi_{\ell}W_{\mu}^{(+)}]$$
$$+\frac{g}{4\cos\theta}\sum_{\ell}[\bar{\psi}_{\nu_\ell}\gamma^{\mu}(1-\gamma_5)\psi_{\nu_\ell}-\bar{\psi}_{\ell}\gamma^{\mu}(g_V'-\gamma_5)\psi_{\ell}]Z_{\mu}$$
$$+\frac{h\lambda^4}{4}+\frac{1}{2}(\partial_{\mu}\chi)^2-h\lambda^2\chi^2-h\chi^2\left(\lambda\chi+\frac{1}{4}\chi^2\right)$$
$$+\frac{g^2}{8}\left(2W_{\mu}^{(+)}W^{(-)\mu}+\frac{Z_{\mu}Z^{\mu}}{\cos^2\theta}\right)(\lambda+\chi)^2 \qquad (4.129)$$

Remember, $g_V'=1-\sin^2\theta$, see (4.107). Here we have used (4.120) in order to eliminate the term linear in χ. The fourth and fifth terms describe the kinetic energies of the (left-handed) neutrinos and massive leptons. The last term represents the coupling of the W and Z bosons to the Higgs fields, which generates the W and Z boson masses. We evaluate this term in Exercise 4.10, (7)–(9). The constants f_ℓ are given by the lepton masses $m_\ell=\lambda f_\ell$. Furthermore, we see that the masses of the W and Z boson result in the form

$$M_{\mathrm{W}}=\frac{g\lambda}{2}\ , \quad M_{\mathrm{Z}}=\frac{M_{\mathrm{W}}}{\cos\theta}\ . \qquad (4.130)$$

Finally, according to our discussion in Sect. 3.6 (see (3.86)) we can identify the Fermi coupling constant

$$\frac{G}{\sqrt{2}}=\frac{g^2}{8M_{\mathrm{W}}^2}=\frac{e^2}{8M_{\mathrm{W}}^2\sin^2\theta}\approx\frac{e^2}{2M_{\mathrm{W}}^2}\ . \qquad (4.131)$$

This equation originates from the vertex coupling strength $\bar{g}=g/2\sqrt{2}$ of the charged weak current in (4.129), compared to the vertex coupling $G/\sqrt{2}$ in (3.86). Furthermore, we have used the asymptotic (low-energy) behavior for the boson propagator

$$\frac{(-\mathrm{i}\bar{g})^2}{q^2-M_{\mathrm{W}}^2}\xrightarrow{q\to 0}\frac{g^2}{8M_{\mathrm{W}}^2}\ . \qquad (4.132)$$

This is expressed graphically in Fig. 4.9.

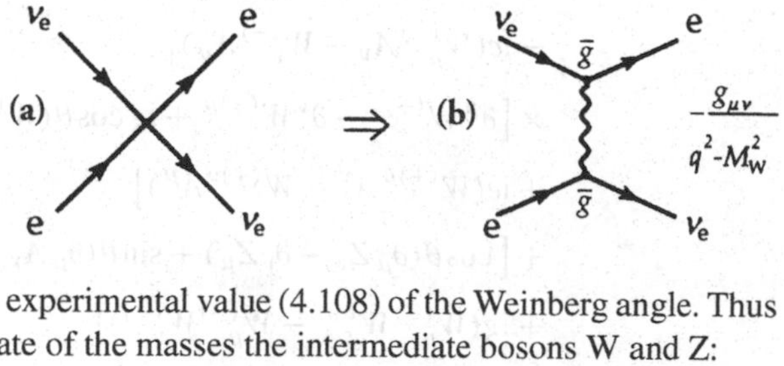

Fig. 4.9. Interaction of weak currents (**a**) in the Fermi theory, (**b**) in the standard model

In (4.131) we have used the experimental value (4.108) of the Weinberg angle. Thus we obtain the following estimate of the masses the intermediate bosons W and Z:

$$M_{\mathrm{W}}=\left(\frac{e^2}{4\sqrt{2}G\sin^2\theta}\right)^{1/2}\approx\left(\frac{e^2}{\sqrt{2}G}\right)\approx 75\ \mathrm{GeV}\ , \qquad (4.133a)$$
$$M_{\mathrm{Z}}=86\ \mathrm{GeV}\ . \qquad (4.133b)$$

Here we have used the value given by (2.63) for G. (Owing to higher-order radiative corrections the effective value of G is renormalized at high energies. The value for G in the 80–90 GeV mass range is somewhat lower than (2.63) and therefore the precise predictions of the theory for M_W and M_Z are about 6 percent higher; see also Exercises 5.2, 5.3.) At the time of invention of the Weinberg–Salam theory the masses of the intermediate bosons as predicted by the new theory (75 and 86 MeV, respectively), were very high, making their production difficult.[16] For example, the electron–positron storage ring PETRA at Hamburg had a maximum available energy of 45 GeV. There the influence of the Z^0 boson on the scattering could only be seen indirectly, since the conversion of an electron–positron pair into a muon pair (or τ pair) can occur with either a virtual photon or a virtual Z^0 boson in the intermediate state (see Fig. 4.10). Both processes are in principle indistinguishable, since they yield the same final state, and their amplitudes have to be added coherently. However, as the neutral weak current of the charged leptons has different parity properties (nearly pure axial vector current – cf. (4.106), (4.107)) to the electromagnetic one (pure vector current), the superposition of the two processes yields a characteristic shape of the angular distributions of the particles produced, which allows for a crude determination of the mass of the Z boson. (Further details can be found in Sect. 5.2.) The generation of electron–positron colliders available since the 1990s, with centre-of-mass energies of around 100 GeV, has allowed for copious production of Z bosons, making a detailed investigation of their properties possible (see Example 5.3).

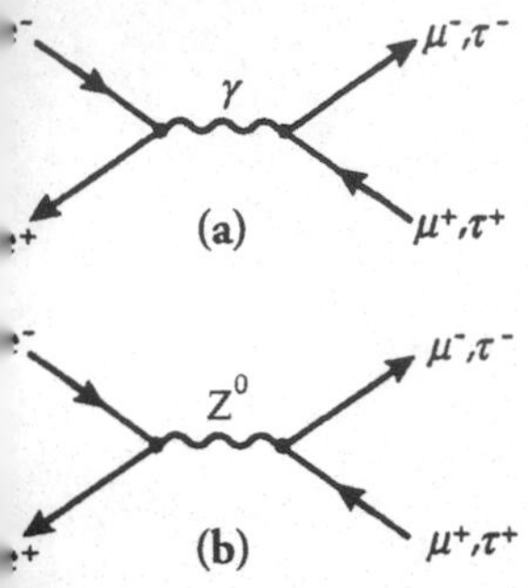

Fig. 4.10. Muon and τ lepton pair production, with a photon or the neutral boson Z^0 as intermediate state

Finally we have to rewrite the free-field parts $\boldsymbol{F}_{\mu\nu}\cdot\boldsymbol{F}^{\mu\nu}$ and $B_{\mu\nu}B^{\mu\nu}$ in terms of the physical fields A_μ, Z_μ, and $W_\mu^{\pm}$. We find that

$$
\begin{aligned}
\boldsymbol{F}_{\mu\nu}\cdot\boldsymbol{F}^{\mu\nu} &= \sum_i \left(\partial_\mu A^i_\mu - \partial_\nu A^i_\mu + g\varepsilon_{ikl}A^k_\mu A^l_\nu\right)\left(\partial^\mu A^{i\nu} - \partial^\nu A^{i\mu} + g\varepsilon_{ikl}A^{k\mu}A^{l\nu}\right) \\
&= 2\left[\partial_\mu W^{(-)}_\nu - \partial_\nu W^{(-)}_\mu - \mathrm{i}g(W^{(-)}_\mu A^3_\nu - W^{(-)}_\nu A^3_\mu)\right] \\
&\quad \times \left[\partial^\mu W^{(+)\nu} - \partial^\nu W^{(+)\mu} + \mathrm{i}g(W^{(+)\mu}A^{3\nu} - W^{(+)\nu}A^{3\mu})\right] \\
&\quad + \left[\partial_\mu A^3_\nu - \partial_\nu A^3_\mu + \mathrm{i}g(W^{(-)}_\mu W^{(+)}_\nu - W^{(-)}_\nu W^{(+)}_\mu)\right] \\
&\quad \times \left[\partial^\mu A^{3\nu} - \partial^\nu A^{3\mu} + \mathrm{i}g(W^{(-)\mu}W^{(+)\nu} - W^{(-)\nu}W^{(+)\mu})\right] \\
&= 2\left[\partial_\mu W^{(-)}_\nu - \partial_\nu W^{(-)}_\mu - \mathrm{i}g\cos\theta(W^{(-)}_\mu Z_\nu - W^{(-)}_\nu Z_\mu)\right. \\
&\quad \left. - \mathrm{i}e(W^{(-)}_\mu A_\nu - W^{(-)}_\nu A_\mu)\right] \\
&\quad \times \left[\partial^\mu W^{(+)\nu} - \partial^\nu W^{(+)\mu} + \mathrm{i}g\cos\theta(W^{(+)\mu}Z^\nu - W^{(+)\nu}Z^\mu)\right. \\
&\quad \left. + \mathrm{i}e(W^{(+)\mu}A^\nu - W^{(+)\nu}A^\mu)\right] \\
&\quad + \left[\cos\theta(\partial_\mu Z_\nu - \partial_\nu Z_\mu) + \sin\theta(\partial_\mu A_\nu - \partial_\nu A_\mu)\right. \\
&\quad \left. + \mathrm{i}g(W^{(-)}_\mu W^{(+)}_\nu - W^{(-)}_\nu W^{(+)}_\mu)\right] \\
&\quad \times \left[\cos\theta(\partial^\mu Z^\nu - \partial^\nu Z^\mu) + \sin\theta(\partial^\mu A^\nu - \partial^\nu A^\mu)\right. \\
&\quad \left. + \mathrm{i}g(W^{(-)\mu}W^{(+)\nu} - W^{(+)\mu}W^{(-)\nu})\right]
\end{aligned}
\tag{4.134}
$$

[16] The discovery of the intermediate bosons at the proton–antiproton collider of CERN in Geneva, with two colliding beams of 270 GeV each, will be discussed in Example 5.2.

where we have used (4.101), setting

$$e = g\sin\theta = g'\cos\theta \quad . \tag{4.135}$$

The free part of the isosinglet field is

$$\begin{aligned} B_{\mu\nu}B^{\mu\nu} &= (\partial_\mu B_\nu - \partial_\nu B_\mu)(\partial^\mu B^\nu - \partial^\nu B^\mu) \\ &= \cos^2\theta(\partial_\mu A_\nu - \partial_\nu A_\mu)(\partial^\mu A^\nu - \partial^\nu A^\mu) \\ &\quad + \sin^2\theta(\partial_\mu Z_\nu - \partial_\nu Z_\mu)(\partial^\mu Z^\nu - \partial^\nu Z^\mu) \\ &\quad - 2\sin\theta\cos\theta(\partial_\mu A_\nu - \partial_\nu A_\mu)(\partial^\mu Z^\nu - \partial^\nu Z^\mu) \quad . \end{aligned} \tag{4.136}$$

By adding both contributions (4.134), (4.136), we find that those terms which mix the photon A_μ and the *neutral* intermediate boson Z_μ disappear.

Now we have determined all terms of the Glashow–Salam–Weinberg Lagrangian describing the electromagnetic and weak ("electroweak") interaction of the leptons. We write the total expression in a closed form, omitting the constant term in the Higgs sector, that is, $\frac{1}{4}h\lambda^4$:

$$L_{SW} = L^{(2)}_{\mathrm{SW}} + \sum_\ell L^{(3L)}_{\mathrm{SW},\ell} + L^{(3B)}_{\mathrm{SW}} + L^{(4B)}_{\mathrm{SW}} + L^{(H)}_{\mathrm{SW}} \quad . \tag{4.137}$$

Here $L^{(2)}_{\mathrm{SW}}$ describes the part of the free boson and lepton fields, $L^{(3L)}_{\mathrm{SW},l}$ represents the coupling between the leptons of the generation $l = (\mathrm{e}, \mu, \tau)$ and the intermediate bosons, $L^{(3B)}_{\mathrm{SW}}$ and $L^{(4B)}_{\mathrm{SW}}$ are the third- and fourth-order terms of the bosonic fields describing their self-coupling, and lastly $L^{(\mathrm{H})}_{\mathrm{SW}}$ contains all terms of the Higgs field which are not contained in the mass terms. In detail, the complete Lagrangian consists of the following parts:

(a) free fields (massive vector bosons, photons, leptons):

$$\begin{aligned} L^{(2)}_{\mathrm{SW}} &= -\frac{1}{2}(\partial_\mu W^{(+)}_\nu - \partial_\nu W^{(+)}_\mu)(\partial^\mu W^{(-)\nu} - \partial^\nu W^{(-)\mu}) + M^2_{\mathrm{W}} W^{(+)}_\mu W^{(-)\mu} \\ &\quad - \frac{1}{4}(\partial_\mu Z_\nu - \partial_\nu Z_\mu)(\partial^\mu Z^\nu - \partial^\nu Z^\mu) + \frac{1}{2}M^2_{\mathrm{Z}} Z_\mu Z^\mu \\ &\quad - \frac{1}{4}(\partial_\mu A_\nu - \partial_\nu A_\mu)(\partial^\mu A^\nu - \partial^\nu A^\mu) \\ &\quad + \sum_{\ell=\mathrm{e},\mu,\tau} \left[\bar{\psi}_{\nu_\ell} \mathrm{i}\gamma^\mu \partial_\mu \frac{1-\gamma_5}{2}\psi_{\nu_\ell} + \bar{\psi}_\ell(\mathrm{i}\gamma^\mu\partial_\mu - m_\ell)\psi_\ell \right] \quad , \end{aligned} \tag{4.138}$$

(b) the lepton–boson interaction:

$$\begin{aligned} L^{(3L)}_{\mathrm{SW},l} &= \frac{g}{2\sqrt{2}}[\bar{\psi}_\ell\gamma^\mu(1-\gamma_5)\psi_{\nu_\ell} W^{(-)}_\mu + \bar{\psi}_{\nu_\ell}\gamma^\mu(1-\gamma_5)\psi_\ell W^{(+)}_\mu] \\ &\quad + \frac{g}{4\cos\theta}[\bar{\psi}_{\nu_\ell}\gamma^\mu(1-\gamma_5)\psi_{\nu_\ell} - \bar{\psi}_\ell\gamma^\mu(\underbrace{1-4\sin^2\theta}_{=g'_V} - \gamma_5)\psi_\ell]Z_\mu \\ &\quad - e\bar{\psi}_\ell\gamma^\mu\psi_\ell A_\mu \quad , \end{aligned} \tag{4.139}$$

(c) third-order interactions of vector bosons:

$$\begin{aligned}L_{\mathrm{SW}}^{(3B)} &= \mathrm{i}g\cos\theta\big[(\partial_\mu W_\nu^{(-)} - \partial_\nu W_\mu^{(-)})W^{(+)\mu}Z^\nu - (\partial_\mu W_\nu^{(+)} - \partial_\nu W_\mu^{(+)})W^{(-)\mu}Z^\nu\big]\\ &\quad - \mathrm{i}e(\partial_\mu W_\nu^{(-)} - \partial_\nu W_\mu^{(-)})W^{(+)\mu}A^\nu + \mathrm{i}e(\partial_\mu W_\nu^{(+)} - \partial_\nu W_\mu^{(+)})W^{(-)\mu}A^\nu\\ &\quad + \mathrm{i}g\cos\theta(\partial_\mu Z_\nu - \partial_\nu Z_\mu)W^{(+)\mu}W^{(-)\nu}\\ &\quad - \mathrm{i}e(\partial_\mu A_\nu - \partial_\nu A_\mu)W^{(+)\mu}W^{(-)\nu} \quad ,\end{aligned} \tag{4.140}$$

(d) fourth-order interactions of vector bosons:

$$\begin{aligned}L_{\mathrm{SW}}^{(4B)} &= -g^2\cos^2\theta(W_\mu^{(+)}W^{(-)\mu}Z_\nu Z^\nu - W_\mu^{(+)}W_\nu^{(-)}Z^\mu Z^\nu)\\ &\quad - e^2(W_\mu^{(+)}W^{(-)\mu}A_\nu A^\nu - W_\mu^{(+)}W_\nu^{(-)}A^\mu A^\nu)\\ &\quad + eg\cos\theta(2W_\mu^{(+)}W^{(-)\mu}Z_\nu A^\nu - W_\mu^{(+)}W_\nu^{(-)}Z^\mu A^\nu - W_\mu^{(+)}W_\nu^{(-)}Z^\nu A^\mu)\\ &\quad + g^2(W_\mu^{(+)}W^{(-)\mu}W_\nu^{(+)}W^{(-)\nu} - W_\mu^{(-)}W^{(-)\mu}W_\nu^{(+)}W^{(+)\nu}) \quad ,\end{aligned} \tag{4.141}$$

(e) the Higgs sector:

$$\begin{aligned}L_{\mathrm{SW}}^{(\mathrm{H})} &= \frac{1}{2}(\partial_\mu\chi)(\partial^\mu\chi) - h\lambda^2\chi^2\\ &\quad + \frac{1}{4}g^2\big[W_\mu^{(+)}W^{(-)\mu} + (2\cos\theta)^{-1}Z_\mu Z^\mu\big](2\lambda\chi + \chi^2) - h\chi^2\left(\lambda\chi + \frac{1}{4}\chi^2\right)\\ &\quad - \sum_\ell f_\ell\bar\psi_\ell\psi_\ell\chi \quad .\end{aligned} \tag{4.142}$$

Remember that sometimes in the literature $W^* \equiv W^{(+)}$ and $W \equiv W^{(-)}$, as we have earlier remarked before (4.100). The following relations are valid for the masses:

$$M_\mathrm{W} = \frac{g\lambda}{2} = \frac{e}{\sqrt{G\sqrt{2}}\cdot 2\sin\theta} \quad , \quad M_\mathrm{Z} = \frac{M_\mathrm{W}}{\cos\theta} \quad , \quad m_\ell = f_\ell\lambda \quad , \tag{4.143}$$

and for the coupling constants

$$e = g\sin\theta \quad , \quad \frac{G}{\sqrt{2}} = \frac{1}{2\lambda^2} = \frac{g^2}{8M_\mathrm{W}^2} \quad . \tag{4.144}$$

Thus the Lagrange density contains a fairly large number of terms; its conceptual simplicity, however, originates from the underlying gauge principle of the SU(2) × U(1) group.

EXERCISE

4.9 Lepton Masses

Problem. Evaluate relation (4.123) explicitly.

Solution. Omitting the index ℓ in Example 4.8, (1), and with (4.118), we obtain

$$\begin{aligned}\bar R\Phi^\dagger L + \bar L\Phi R &= \bar\psi_\mathrm{R}\left(0, \frac{\lambda}{\sqrt{2}}\right)\begin{pmatrix}\psi_\nu\\ \psi\end{pmatrix}_\mathrm{L} + (\bar\psi_\nu, \bar\psi)_\mathrm{L}\begin{pmatrix}0\\ \frac{\lambda}{\sqrt{2}}\end{pmatrix}\psi_\mathrm{R}\\ &= \frac{\lambda}{\sqrt{2}}(\bar\psi_\mathrm{R}\psi_\mathrm{L} + \bar\psi_\mathrm{L}\psi_\mathrm{R}) \quad .\end{aligned} \tag{1}$$

Since

$$\psi_{\mathrm{L}} = \frac{1-\gamma_5}{2}\psi \quad , \quad \bar{\psi}_{\mathrm{L}} = \bar{\psi}\frac{1+\gamma_5}{2} \quad , \tag{2}$$

and also

$$\begin{gathered}\psi_{\mathrm{R}} = \frac{1+\gamma_5}{2}\psi \quad , \quad \bar{\psi}_{\mathrm{R}} = \bar{\psi}\frac{1-\gamma_5}{2} \quad , \\ \left(\frac{1-\gamma_5}{2}\right)^2 = \frac{1-\gamma_5}{2} \quad , \quad \left(\frac{1+\gamma_5}{2}\right)^2 = \frac{1+\gamma_5}{2} \quad ,\end{gathered} \tag{3}$$

we find the result

$$\frac{\lambda}{\sqrt{2}}\left(\bar{\psi}\frac{1-\gamma_5}{2}\psi + \bar{\psi}\frac{1+\gamma_5}{2}\psi\right) = \frac{\lambda}{\sqrt{2}}\bar{\psi}\psi \tag{4}$$

for each generation, where we have used the identities (3). In the standard model, one combines the initially independent components ψ_{R} and ψ_{L} belonging to different SU(2) multiplets to a single fermion field ψ. Expressing the Lagrangian in terms of this combined field ψ, we obtain a mass term for ψ. That is, the ψ_{L} and ψ_{R} behave like the left-handed and right-handed components of a massive fermion field owing to their coupling to the Higgs field. The intermingling of terms belonging to different SU(2) multiplets (i.e. ψ_{L}, ψ_{R} where ψ_{L} belongs to the isodoublet and ψ_{R} is an isosinglet) is an expression of the spontaneous breaking of SU(2) symmetry by the Higgs field.

EXERCISE

4.10 The Glashow–Salam–Weinberg Lagrangian

Problem. Collect all the terms contributing to the Glashow–Salam–Weinberg Lagrangian.

Solution. The Lagrangian contains free parts describing the leptons $(l, \nu_{\mathrm{e}}) = (e, \nu_{\mathrm{e}}), (\mu, \nu_\mu), (\tau, \nu_\tau)$, the massive vector gauge bosons (W^+, W^-, Z_0), the photon, the scalar Higgs boson, and their mutual interaction terms.

(1) The Free bosonic part has the standard form

$$\begin{aligned}L_{\text{free}}^{\text{Boson}} &= -\frac{1}{2}(\partial_\mu W_\nu^{(+)} - \partial_\nu W_\mu^{(+)})(\partial^\mu W^{(-)\nu} - \partial^\nu W^{(-)\mu}) + M_{\mathrm{W}}^2 W_\mu^{(+)} W^{(-)\mu} \\ &\quad - \frac{1}{4}(\partial_\mu Z_\nu - \partial_\nu Z_\mu)(\partial^\mu Z^\nu - \partial^\nu Z^\mu) + \frac{1}{2}M_{\mathrm{Z}}^2 Z_\mu Z^\mu \\ &\quad - \frac{1}{4}F_{\mu\nu}F^{\mu\nu} + \frac{1}{2}\partial_\mu\chi\partial^\mu\chi - h\lambda^2\chi^2 \quad ,\end{aligned} \tag{1}$$

with

$$M_{\mathrm{W}} = \frac{g\lambda}{2} \quad , \quad M_{\mathrm{Z}} = \frac{M_{\mathrm{W}}}{\cos\theta} \quad , \quad M_{\mathrm{H}} = \sqrt{2h}\lambda \quad .$$

Exercise 4.10

(2) The free Leptonic part was given in (4) of Example 4.8:

$$\begin{aligned} L_{\text{free}}^{\text{Fermion}} &= \sum_l \mathrm{i}\bar{L}_l\gamma^\mu\partial_\mu L_l + \sum_l \mathrm{i}\bar{R}_l\gamma^\mu\partial_\mu R_l \\ &= \sum_l \mathrm{i}\left[\frac{1-\gamma_5}{2}\begin{pmatrix}\psi_{\nu_l}\\ \psi_l\end{pmatrix}\right]^+\gamma^0\gamma^\mu\partial_\mu\frac{1-\gamma_5}{2}\begin{pmatrix}\psi_{\nu_l}\\ \psi_l\end{pmatrix} \\ &\quad + \sum_l \mathrm{i}\left[\frac{1+\gamma_5}{2}\psi_l\right]^+\gamma^0\gamma^\mu\partial_\mu\frac{1+\gamma_5}{2}\psi_l \quad , \end{aligned} \tag{2}$$

where the left-handed isodoublet and right-handed isosinglet spinors, L_l and R_l, have been expressed in terms of the lepton and neutrino spinors ψ_l and ψ_{ν_l}. The summation runs over the families $l = e, \mu, \psi$. Anticommuting the γ_5 matrix, we change (2) into

$$\begin{aligned} L_{\text{free}}^{\text{Fermion}} &= \sum_l \mathrm{i}\begin{pmatrix}\psi_{\nu_l}\\ \psi_l\end{pmatrix}^+\gamma^0\gamma^\mu\left(\frac{1-\gamma_5}{2}\right)^2\partial_\mu\begin{pmatrix}\psi_{\nu_l}\\ \psi_l\end{pmatrix} \\ &\quad + \sum_l \mathrm{i}\psi_l^+\gamma^0\gamma^\mu\left(\frac{1+\gamma_5}{2}\right)^2\partial_\mu\psi_l \\ &= \sum_l \mathrm{i}\bar{\psi}_{\nu_l}\gamma^\mu\frac{1-\gamma_5}{2}\partial_\mu\psi_{\nu_l} + \sum_l \mathrm{i}\bar{\psi}_l\gamma^\mu\frac{1-\gamma_5}{2}\partial_\mu\psi_l \\ &\quad + \sum_l \bar{\psi}_l\gamma^\mu\frac{1+\gamma_5}{2}\partial_\mu\psi_l \\ &= \sum_l \mathrm{i}\bar{\psi}_{\nu_l}\gamma^\mu\frac{1-\gamma_5}{2}\partial_\mu\psi_{\nu_l} + \sum_l \mathrm{i}\bar{\psi}_l\gamma^\mu\partial_\mu\psi_l \quad . \end{aligned} \tag{3}$$

(3) The Lepton–Gauge boson interaction was given in (4.100). By using (4.93), (4.94), and (4.106) it reads

$$\begin{aligned} L_{\text{int}}^{\text{LG}} &= \sum_l \frac{g}{2\sqrt{2}}\left(J_-^{(l)\mu}W_\mu^{(-)} + J_+^{(l)\mu}W_\mu^{(+)} + J_0^{(l)\mu}Z_\mu\right) - \sum_l eJ_{E\mu}^{(l)\mu}A_\mu \\ &= \sum_l \frac{g}{2\sqrt{2}}\left(\bar{\psi}_l\gamma^\mu(1-\gamma_5)\psi_{\nu_l}W_\mu^{(-)} + \bar{\psi}_{\nu_l}\gamma^\mu(1-\gamma_5)\psi_l W_\mu^{(+)}\right) \\ &\quad + \sum_l \frac{g}{2\sqrt{2}}\frac{1}{\sqrt{2}\cos\theta}\left(\bar{\psi}_{\nu_l}\gamma^\mu(1-\gamma_5)\psi_{\nu_l} - \bar{\psi}_l\gamma^\mu(g_V' - \gamma_5)\psi_l\right)Z_\mu \\ &\quad - \sum_l e\bar{\psi}_l\gamma^\mu\psi_l A_\mu \quad . \end{aligned} \tag{4}$$

(4) The Higgs boson self-interaction in the unitary gauge (4.117) reads

$$\begin{aligned} L^{\text{Higgs}} &= -U(\Phi) = \mu^2|\Phi|^2 - h|\Phi|^4 = \frac{\mu^2}{2}(\lambda+\chi)^2 - \frac{h}{4}(\lambda+\chi)^4 \\ &= \frac{1}{2}h\lambda^2(\lambda^2 + 2\lambda\chi + \chi^2) - \frac{1}{4}h(\lambda^4 + 4\lambda^3\chi + 6\lambda^2\chi^2 + 4\lambda\chi^3 + \chi^4) \\ &= \frac{1}{2}h\lambda^4 - h\lambda^2\chi^2 - h\lambda\chi^3 - \frac{1}{4}h\chi^4 \end{aligned} \tag{5}$$

where the condition $\mu^2 = h\lambda^2$ was used. The term linear in the field χ has dropped out since $\chi = 0$ corresponds to the minimum of the potential.

(5) The Higgs–Lepton coupling is given in (4.123):

$$L_{\text{int}}^{\text{HL}} = -\sum_l f_l(\lambda + \chi)\bar{\psi}_l \psi_l \quad , \tag{6}$$

where f_l is related to the lepton mass by $m_l = f_l \lambda$.

(6) The Higgs–Gauge boson interaction results from the minimal-coupling terms in (4.112):

$$L_{\text{int}}^{\text{HG}} = \left| \left(\mathrm{i}\partial_\mu + g\hat{\boldsymbol{T}} \cdot \boldsymbol{A}_\mu + \frac{g'}{2} B_\mu Y \right) \Phi \right|^2 - |\mathrm{i}\partial_\mu \Phi|^2 \quad . \tag{7}$$

By using (4.99), (4.113) and $\hat{T}_\pm = (\hat{T}_1 \pm \mathrm{i}\hat{T}_2)$ the gauge-field terms can be expressed as the following isospin matrix

$$\begin{aligned} g\hat{\boldsymbol{T}} \cdot \boldsymbol{A}_\mu + \frac{g'}{2} B_\mu \hat{Y} &= g\hat{T}^3 A^3 + \frac{g'}{2} B_\mu \hat{Y} + \frac{g}{\sqrt{2}} (\hat{T}_+ W_\mu^{(+)} + \hat{T}_- W_\mu^{(-)}) \\ &= \begin{pmatrix} eA\mu + g\dfrac{\cos 2\theta}{2\cos\theta} Z_\mu & \dfrac{g}{\sqrt{2}} W_\mu^{(+)} \\ \dfrac{g}{\sqrt{2}} W_\mu^{(-)} & -\dfrac{g}{2\cos\theta} Z_\mu \end{pmatrix} \quad . \end{aligned} \tag{8}$$

Writing out the square in (7), we find that the mixed terms involving $\partial_\mu \Phi$ cancel each other because of the factor i. The remaining term is

$$\begin{aligned} L_{\text{int}}^{\text{HG}} &= \frac{1}{2}(\lambda + \chi)^2 \left[\left(g\hat{\boldsymbol{T}} \cdot \boldsymbol{A}_\mu + \frac{g'}{2} B_\mu \hat{Y} \right) \begin{pmatrix} 0 \\ 1 \end{pmatrix} \right]^+ \left[\left(g\hat{\boldsymbol{T}} \cdot \boldsymbol{A}^\mu + \frac{g'}{2} B^\mu \hat{Y} \right) \begin{pmatrix} 0 \\ 1 \end{pmatrix} \right] \\ &= \frac{1}{2}(\lambda + \chi)^2 \begin{pmatrix} \dfrac{g}{\sqrt{2}} W_\mu^{(+)} \\ \dfrac{g}{2\cos\theta} Z_\mu \end{pmatrix}^\dagger \begin{pmatrix} \dfrac{g}{\sqrt{2}} W_\mu^{(+)} \\ -\dfrac{g}{2\cos\theta} Z_\mu \end{pmatrix} \\ &= \frac{g^2}{8}(\lambda + \chi)^2 \left(\frac{1}{\cos^2\theta} Z_\mu Z^\mu + 2W_\mu^{(+)} W^{(+)\mu} \right) \end{aligned} \tag{9}$$

where $(W^{(+)})^\dagger = W^{(-)}$ was used.

EXERCISE

4.11 Masses of the Vector Bosons

Problem. Determine the eigenvalues of the mass matrix of the vector bosons in the Salam–Weinberg theory without fixing the vacuum expectation value of the Higgs field.

Solution. The kinetic energy of the Higgs field, including the minimally coupled vector fields, is given by (4.112),

$$\left| \left(\mathrm{i}\partial_\mu + g\hat{\boldsymbol{T}} \cdot \boldsymbol{A}_\mu + \frac{1}{2} g' Y B_\mu \right) \Phi \right|^2 \quad , \tag{1}$$

Exercise 4.11

with $Y = +1$ being the weak hypercharge of the Higgs field. Constant Higgs fields imply that (1) takes the form

$$\begin{aligned}\Phi^\dagger \left| g\hat{\boldsymbol{T}} \cdot \boldsymbol{A}_\mu + \frac{1}{2} g' Y B_\mu \right|^2 \Phi \\ = A^i_\mu A^\mu_k g^2 (\Phi^\dagger \hat{T}_i \hat{T}_k \Phi) + g g' A^i_\mu B^\mu Y (\Phi^\dagger \hat{T}_i \Phi) + \frac{1}{4} g'^2 Y^2 B_\mu B^\mu |\Phi|^2 \quad . \qquad (2)\end{aligned}$$

The potential term of the Higgs field (4.111)

$$V(\phi) = -\mu^2 |\Phi|^2 + h|\Phi|^4 \qquad (3)$$

is minimized by

$$\left|\langle 0|\hat{\Phi}|0\rangle\right|^2 = \frac{1}{2}\lambda^2 \quad , \quad \lambda^2 = \frac{\mu^2}{h} \quad . \qquad (4)$$

We now rewrite the first term in (2) with the help of the commutation relations of the isospin matrices.

$$\hat{T}_i \hat{T}_k = \frac{1}{4}\delta_{ik} + \frac{\mathrm{i}}{2}\varepsilon_{ikl}\hat{T}_l \quad , \qquad (5)$$

obtaining

$$\Phi^\dagger \hat{T}_i \hat{T}_k \Phi = \frac{1}{4}\delta_{ik}|\Phi|^2 + \frac{\mathrm{i}}{2}\varepsilon_{ikl}(\Phi^\dagger \hat{T}_l \Phi) \quad . \qquad (6)$$

Since the factor $A^i_\mu A^\mu_k$ is symmetric with respect to the indices i, k, the second term in (6) does not contribute to (2), which yields the mass matrix

$$\left(\frac{1}{4} g^2 \boldsymbol{A}_\mu \cdot \boldsymbol{A}^\mu + \frac{1}{4} g'^2 Y^2 B_\mu B^\mu\right) |\Phi|^2 + g g' \boldsymbol{A}_\mu B^\mu Y \cdot (\Phi^\dagger \hat{\boldsymbol{T}} \Phi) \quad . \qquad (7)$$

Owing to (4) the vacuum expectation values of the Higgs terms in (7) are given by

$$\left|\langle 0|\hat{\Phi}|0\rangle\right|^2 = \frac{1}{2}\lambda^2 \quad , \quad \langle 0|\hat{\Phi}^\dagger \hat{\boldsymbol{T}} \hat{\Phi}|0\rangle = \frac{1}{2}\lambda^2 \boldsymbol{T}_{\mathrm{vac}} \quad , \qquad (8)$$

where we have defined an isospin vector of the Higgs vacuum by

$$\boldsymbol{T}_{\mathrm{vac}} = \frac{\langle 0|\hat{\Phi}^\dagger \hat{\boldsymbol{T}} \hat{\Phi}|0\rangle}{|\langle 0|\hat{\Phi}|0\rangle|^2} \quad , \quad |\boldsymbol{T}_{\mathrm{vac}}|^2 = \frac{1}{4} \quad . \qquad (9)$$

The norm of $\boldsymbol{T}_{\mathrm{vac}}$ will be explicitly calculated in Exercise 4.12.

Now we split up the isovector field $\boldsymbol{A}_\mu$ into an isospin component $A^\parallel_\mu$ parallel to $\boldsymbol{T}_{\mathrm{vac}}$ and a component $\boldsymbol{A}^\perp_\mu$ perpendicular to $\boldsymbol{T}_{\mathrm{vac}}$,

$$\boldsymbol{A}_\mu = 2A^\parallel_\mu \boldsymbol{T}_{\mathrm{vac}} + \boldsymbol{A}^\perp_\mu \quad , \qquad (10)$$

with the relations

$$A^\parallel_\mu = 2\boldsymbol{A}_\mu \cdot \boldsymbol{T}_{\mathrm{vac}} \quad , \quad \boldsymbol{A}^\perp_\mu \cdot \boldsymbol{T}_{\mathrm{vac}} = 0 \quad . \qquad (11)$$

Exercise 4.1

Clearly $2\boldsymbol{T}_{\text{vac}}$ is a unit vector in isospin space. This result is already contained in the second part of (9). Obviously

$$\boldsymbol{A}_\mu \cdot \boldsymbol{A}^\mu = A^{\|}_\mu A^\mu_{\|} + \boldsymbol{A}^{\perp}_\mu \cdot \boldsymbol{A}^\mu_{\perp} \quad , \tag{12}$$

is valid. Thus we can separate components with respect to $\boldsymbol{T}_{\text{vac}}$ in the mass term (7):

$$\begin{aligned}
&\frac{1}{8}\lambda^2[g^2 \boldsymbol{A}^{\perp}_\mu \cdot \boldsymbol{A}^\mu_{\perp} + g^2 A^{\|}_\mu A^\mu_{\|} + 2gg' Y A^{\|}_\mu B^\mu + g'^2 B_\mu B^\mu Y^2] \\
&= \frac{1}{8}\lambda^2 g^2 \left[\boldsymbol{A}^{\perp}_\mu \cdot \boldsymbol{A}^\mu_{\perp} + \left(A^{\|}_\mu + \frac{g'}{g} Y B_\mu \right)^2 \right] \\
&= \frac{1}{8}(\lambda g)^2 \left[\boldsymbol{A}^{\perp}_\mu \cdot \boldsymbol{A}^\mu_{\perp} + Z_\mu Z^\mu \left(1 + \frac{g'^2}{g^2} \right) \right] \quad .
\end{aligned} \tag{13}$$

By choosing $2\boldsymbol{T}_{\text{vac}}$ as a unit vector in the 3-direction in isospin space we get

$$\boldsymbol{A}^{\perp}_\mu = \{A^1_\mu, A^2_\mu\} \quad . \tag{14}$$

With

$$W^{(\pm)}_\mu = \frac{A^1_\mu \mp \mathrm{i} A^2_\mu}{\sqrt{2}} \tag{15}$$

it follows that

$$\boldsymbol{A}^{\perp}_\mu \cdot \boldsymbol{A}^{\perp\mu} = A^1_\mu A^{1\mu} + A^2_\mu A^{2\mu} = 2 W^{(+)}_\mu W^{(-)\mu} \quad . \tag{16}$$

Substituting the last expression in (13) we obtain

$$\frac{1}{8}(\lambda g)^2 \left[2 W^{(+)}_\mu W^{(-)\mu} + Z_\mu Z^\mu \left(1 + \frac{g'^2}{g^2} \right) \right] , \tag{17}$$

which corresponds to the last term of (4.129). In (13) we have inserted the linear combination

$$Z_\mu = \frac{g A^{\|}_\mu + g' Y B_\mu}{\sqrt{g^2 + g'^2}} \quad , \tag{18}$$

where the denominator is needed for correct normalization. Again as in (4.101), we use

$$e = g \sin\theta = g' \cos\theta \quad , \tag{19}$$

yielding

$$1 + \frac{g'^2}{g^2} = 1 + \tan^2\theta = \frac{1}{\cos^2\theta} \quad . \tag{20}$$

Hence the masses of the $\boldsymbol{W}_\mu$ and Z_μ fields are

$$M_{\mathrm{W}} = \frac{1}{2}\lambda g \quad , \quad M_{\mathrm{Z}} = \frac{M_{\mathrm{W}}}{\cos\theta} \quad , \tag{21}$$

whereas the field orthogonal to Z_μ,

$$A_\mu = \frac{Yg' A_\mu^{\|} - g B_\mu}{\sqrt{g^2 + g'^2}} \quad , \tag{22}$$

remains massless. Note that we did not fix the direction of the $\hat{\Phi}$ vacuum expectation value in isospin space. The distinction between the fields $\boldsymbol{W}^\mu = (W^{(+)\mu}, W^{(-)\mu})$ and Z^μ is fixed relative to the Higgs vacuum by (10) and (11)!

EXERCISE

4.12 The Norm of $\boldsymbol{T}_{\mathrm{vac}}$

Problem. Show that the vector

$$\boldsymbol{T}_{\mathrm{vac}} = (\Phi^\dagger \hat{\boldsymbol{T}} \Phi)(\Phi^\dagger \Phi)^{-1}$$

has length 1/2.

Solution. By representing Φ in the unitary gauge (4.117) we obviously have

$$\Phi^\dagger \hat{T}_3 \Phi = (0, \tilde{v}) \begin{pmatrix} 1/2 & 0 \\ 0 & -1/2 \end{pmatrix} \begin{pmatrix} 0 \\ \tilde{v} \end{pmatrix} = -\frac{1}{2}\tilde{v}^2 \tag{1}$$

with $\tilde{v} = \frac{1}{\sqrt{2}}(\lambda + \chi)$ and $\Phi^\dagger \hat{T}_1 \Phi = \Phi^\dagger \hat{T}_2 \Phi = 0$. Furthermore, $\Phi^\dagger \Phi = \tilde{v}^2$, yielding

$$\boldsymbol{T}_{\mathrm{vac}} = -\frac{1}{2}\boldsymbol{e}_3 \quad ; \tag{2}$$

$\boldsymbol{e}_3$ being the unit vector in the 3-direction in isospin space. In order to calculate $\boldsymbol{T}_{\mathrm{vac}}$ in the general case we introduce the 2×2 matrix $\Phi\Phi^\dagger$ whose isospin indices α and β are given by

$$(\Phi\Phi^\dagger)_{\alpha\beta} = \Phi_\alpha \Phi^*_\beta \quad . \tag{3}$$

For every matrix $\hat{F}$ we have

$$\Phi^\dagger \hat{F} \Phi = \Phi^*_\alpha F_{\alpha\beta} \Phi_\beta = F_{\alpha\beta} \Phi_\beta \Phi^*_\alpha = \mathrm{Tr}\{\hat{F}\Phi\Phi^\dagger\} \quad . \tag{4}$$

In particular the special choice of $\hat{F}$ as the unit matrix yields

$$|\Phi|^2 = \Phi^\dagger \Phi = \mathrm{Tr}\{\Phi\Phi^\dagger\} \quad . \tag{5}$$

Moreover,

$$(\Phi\Phi^\dagger)^2 = \Phi\Phi^\dagger\Phi\Phi^\dagger = \Phi(\Phi^\dagger\Phi)\Phi^\dagger = |\Phi|^2 \Phi\Phi^\dagger \quad . \tag{6}$$

$\Phi\Phi^\dagger$ is a Hermitian matrix because $(\Phi\Phi^\dagger)^\dagger = \Phi\Phi^\dagger$. Now every Hermitian 2×2 matrix can be represented as a linear combination of the unit matrix and the Pauli matrices, that is, the matrices $\hat{T}_i$,

$$\Phi\Phi^\dagger = a + b_i \hat{T}_i \tag{7}$$

Exercise 4.1

with a and b_i real valued. Since the $\hat{T}_i$ are traceless we get

$$|\Phi|^2 = \mathrm{Tr}\{\Phi\Phi^\dagger\} = 2a \quad , \tag{8}$$

because the trace of the 2×2 unit matrix is just 2. In addition,

$$(\Phi\Phi^\dagger)^2 = (a + b_i\hat{T}_i)^2 = a^2 + 2ab_i\hat{T}_i + (b_i\hat{T}_i)^2 \quad . \tag{9}$$

Using Exercise 4.11 (5), we get

$$\begin{aligned}(b_i\hat{T}_i)^2 = \hat{T}_i\hat{T}_k b_i b_k &= \left(\frac{1}{4}\delta_{ik} + \frac{\mathrm{i}}{2}\varepsilon_{ikl}T_l\right) b_i b_k \\ &= \frac{1}{4}\sum_i b_i^2 \ ,\end{aligned} \tag{10}$$

since $b_i b_k$ is symmetric, and therefore the term proportional to ε_{ikl} vanishes. Thus it remains that

$$(\Phi\Phi^\dagger)^2 = a^2 + \frac{1}{4}\sum_i b_i^2 + 2ab_i\hat{T}_i = |\Phi|^2(a + b_i\hat{T}_i) \tag{11}$$

as a result of (6). On comparing the coefficients we again find that $|\Phi|^2 = 2a$ in accordance with (8), and furthermore that

$$a^2 + \frac{1}{4}\sum_i b_i^2 = a|\Phi|^2 \ . \tag{12}$$

Combining (8) and (12) leads to

$$\frac{1}{4}\sum_i b_i^2 = \frac{1}{4}|\Phi|^4 \quad . \tag{13}$$

By choosing $\hat{F} = \hat{T}_i$ in (4) we obtain

$$\begin{aligned}\Phi^\dagger\hat{T}_i\Phi &= \mathrm{Tr}\{\hat{T}_i\Phi^\dagger\Phi\} = \mathrm{Tr}\{\hat{T}_i(a + b_j\hat{T}_j)\} \\ &= b_j\,\mathrm{Tr}\{\hat{T}_i\hat{T}_j\} = b_j\frac{1}{2}\delta_{ij} = \frac{1}{2}b_i \ ,\end{aligned} \tag{14}$$

because $\mathrm{Tr}\{\hat{T}_i\} = 0$ and $\mathrm{Tr}\{\hat{T}_i\hat{T}_j\} = (1/2)\delta_{ij}$, according to Exercise 4.11, (5). We can therefore derive

$$(\Phi^\dagger\hat{\boldsymbol{T}}\Phi)^2 = \sum_i(\Phi^\dagger\hat{T}_i\Phi)^2 = \frac{1}{4}\sum_i b_i^2 = \frac{1}{4}|\Phi|^4 \quad , \tag{15}$$

yielding

$$(\boldsymbol{T}_{\mathrm{vac}})^2 = \left(\frac{\Phi^\dagger\hat{\boldsymbol{T}}\Phi}{\Phi^\dagger\Phi}\right)^2 = \frac{1}{4} \quad \Rightarrow \quad |\boldsymbol{T}_{\mathrm{vac}}| = \frac{1}{2} \ . \tag{16}$$

Thus we have proved the norm of $\boldsymbol{T}_{\mathrm{vac}}$ to be 1/2, independent of the gauge.

4.6 Hidden SU(2) × U(1) Gauge Invariance

In this section we shall consider the Brout–Higgs mechanism of symmetry breaking in the standard model (GSW model) of weak interactions from a different point of view. In Sect. 4.1 we discussed the basic idea of the Brout–Higgs mechanism for generating masses in a gauge-invariant way in the context of an Abelian gauge theory. In the previous section we studied spontaneous symmetry breaking within the GSW model in some detail.

Let us repeat the main aspects of spontaneous symmetry breaking of a U(1) gauge theory. We started with a massless vector field A_μ with two degrees of freedom and then introduced a complex scalar Higgs field ϕ with two additional degrees of freedom. One of the two degrees of freedom of the Higgs field was "eaten" by the massless A_μ field, in order to transform the latter to a massive field A'_μ. The second component, χ, remained as a massive scalar field (see (4.35)).

In the case of the standard model we initially have a massless isotriplet gauge field

$$\boldsymbol{A}_\mu = \{A^a_\mu\} \quad , \tag{4.145}$$

coupled to the neutral and charged weak transition currents (see (4.96), or Example 4.8, (15)). The gauge field $\boldsymbol{A}_\mu$ was derived by demanding gauge invariance with respect to SU(2) gauge transformations. It must be massive (induced by appropriate spontaneous symmetry breaking) in order to ensure the short range of weak interactions. In addition, there has to exist a U(1) symmetry and a related vector field B_μ which stays massless, describing the photon in the framework of a unified theory.

In order to satisfy all these requirements the best approach is to generalize the scalar Higgs field introduced in Sect. 4.1 to a complex scalar isodoublet field Φ. This was the idea in the thinking of Weinberg and Salam. We denote the Higgs doublet by

$$\Phi = \frac{1}{\sqrt{2}}\begin{pmatrix}\phi_1 + \mathrm{i}\phi_2 \\ \phi_3 + \mathrm{i}\phi_4\end{pmatrix} \quad , \quad T_3 = \pm\frac{1}{2} \quad , \quad T = \frac{1}{2} \quad , \tag{4.146}$$

with the four real functions $\phi_1, \ldots, \phi_4$. The weak isospin $\boldsymbol{T}$ is already fixed here, but the weak hypercharge remains unknown. We shall fix it later.

The mechanism that generates the masses of the three initially massless gauge bosons through a non-vanishing vacuum expectation value of the Φ field is completely analogous to the case of U(1) symmetry. One wants to obtain a vacuum-induced current $\langle \boldsymbol{j}_\mu \rangle_0$ that has a part proportional to $\boldsymbol{A}_\mu$, in order to transform the wave equation of the massless vector field,

$$\Box \boldsymbol{A}^\mu - \partial^\mu \partial_\nu \boldsymbol{A}^\nu = \langle \boldsymbol{j}^\mu \rangle_0 \quad , \tag{4.147}$$

together with the relation

$$\langle \boldsymbol{j}^\mu \rangle_0 = -M^2 \boldsymbol{A}^\mu \quad , \tag{4.148}$$

which expresses the vacuum screening, into an equation of a massive vector field,

$$(\Box + M^2)\boldsymbol{A}^\mu - \partial^\mu \partial_\nu \boldsymbol{A}^\nu = 0 \quad . \tag{4.149}$$

The bold notation of $\boldsymbol{A}_\mu$ denotes the isovector character of that gauge field. It was the special choice of phase of the scalar field Φ, that is, $\Phi_0 = \lambda/\sqrt{2}\exp(\mathrm{i}\Theta(x))\binom{0}{1}$, which seemed to violate the gauge symmetry. Actually, the gauge invariance is naturally not broken, it is just hidden. This can be seen from the fact that the masses of the gauge fields are directly connected to the equilibrium value (expectation value of the ground state) of the field Φ. A relation of this kind (for example $M^2 = g^2\lambda^2$, g being the coupling constant) does not otherwise exist!

Without loss of generality we write our actual Higgs field (4.146) in the form

$$\Phi = \exp(\mathrm{i}\boldsymbol{\alpha}(x)\cdot\hat{\boldsymbol{T}})\begin{pmatrix}0\\ H(x)\end{pmatrix} \tag{4.150}$$

with $\boldsymbol{\alpha}(x) = \{\alpha_1(x), \alpha_2(x), \alpha_3(x)\}$ being three real phase fields and $H(x)$ an additional real field. These four fields are equivalent to the fields $\phi_1(x), \ldots, \phi_4(x)$ in (4.146). The x dependence of the "angle of rotation" $\boldsymbol{\alpha}(x)$ signifies that there is an independent rotation along the three isospin directions at each space-time point x. Demanding invariance of the theory with respect to these transformations establishes gauge invariance, in our case SU(2) gauge invariance. $\hat{\boldsymbol{T}} = \{\hat{T}_1, \hat{T}_2, \hat{T}_3\}$ are the generators of this symmetry group.

In order to implement the concept described in (4.147)–(4.149) we need to know the current $\boldsymbol{j}^\mu(\boldsymbol{A}_\nu, \Phi)$. It has to be an isovector, and each of three isospin components must be a four-vector. In the case of a single scalar field ϕ, $j^\mu(\phi)$ is given by[17]

$$j^\mu = -\mathrm{i}q\left[\phi^*(\partial^\mu\phi) - (\partial^\mu\phi^*)\phi\right] = q\left[\phi^*(-\mathrm{i}\partial^\mu\phi) + (-\mathrm{i}\partial^\mu\phi)^*\phi\right] \quad , \tag{4.151}$$

q denoting the charge of the scalar field. When coupled to the electromagnetic field A_μ, the current j^μ can be defined gauge invariantly by replacing ∂_μ with the U(1)-gauge-invariant derivatives

$$\partial_\mu \to D_\mu = \partial_\mu - \mathrm{i}qA_\mu = \mathrm{i}(-\mathrm{i}\partial_\mu - qA_\mu) \quad . \tag{4.152}$$

Considering the weak isodoublet Φ in (4.146), we see that the current must be an isovector; which is to say, it must be composed of three currents, one for each isospin component. It therefore must have the form

$$\begin{aligned}
\boldsymbol{j}^\mu = \{j^{a\mu}\} &= -\mathrm{i}g\{\Phi^\dagger\hat{\boldsymbol{T}}(\partial^\mu\Phi) - (\partial^\mu\Phi^\dagger)\hat{\boldsymbol{T}}\Phi\} \\
&= g\{\Phi^\dagger\hat{\boldsymbol{T}}(-\mathrm{i}\partial^\mu\Phi) + (-\mathrm{i}\partial^\mu\Phi)^\dagger\hat{\boldsymbol{T}}\Phi\} \\
&= -\mathrm{i}g\{\Phi^\dagger\hat{T}^a(\partial^\mu\Phi) - (\partial^\mu\Phi)^\dagger\hat{T}^a\Phi\} \\
&= g\{\Phi^\dagger\hat{T}^a(-\mathrm{i}\partial^\mu\Phi) + (-\mathrm{i}\partial^\mu\Phi)^\dagger\hat{T}^a\Phi\} \\
&= \{j^{a\mu}\} \ , \quad a = 1,2,3 \quad , \quad \mu = 0,1,2,3 \quad .
\end{aligned} \tag{4.153}$$

In the standard model of weak interactions with an SU(2) × U(1) gauge symmetry, there have to exist as many as four currents, namely three weak isospin currents $j^{a\mu}$ $(a = 1, 2, 3)$ and a current of hypercharge $j^{Y\mu}$. As stated by Noether's theorem, for every group generator of a symmetry group (here $\hat{T}_1, \hat{T}_2, \hat{T}_3, \hat{Y}$) a corresponding

[17] See W. Greiner: *Relativistic Quantum Mechanics – Wave Equations*, 3rd ed. (Springer, Berlin, Heidelberg, 2000).

current exists:

$$
\begin{aligned}
j^{Y\mu} &= -\mathrm{i}\frac{g'}{2}\left[\Phi^\dagger \hat{Y}(\partial^\mu \Phi) - (\partial^\mu \Phi^\dagger)\hat{Y}\Phi\right] \\
&= \frac{g'}{2}\left[\Phi^\dagger \hat{Y}(-\mathrm{i}\partial^\mu \Phi) + (-\mathrm{i}\partial^\mu \Phi^\dagger)\hat{Y}\Phi\right] \\
&= -\mathrm{i}\frac{g'}{2}Y\left[\Phi^\dagger(\partial^\mu \Phi) - (\partial^\mu \Phi^\dagger)\Phi\right] \;.
\end{aligned}
\tag{4.154}
$$

Again, by analogy with the steps from (4.151) to (4.153), the gauge invariance of the current with respect to SU(2) × U(1) can be ensured if we replace ∂_μ in (4.153) and (4.154) by the gauge-covariant derivative D^μ,

$$
\begin{aligned}
\partial^\mu \to D^\mu &= \partial^\mu - \mathrm{i}g\hat{\boldsymbol{T}}\cdot \boldsymbol{A}^\mu - \mathrm{i}\frac{g'}{2}\hat{Y}B^\mu \\
&= \mathrm{i}\left[-\mathrm{i}\partial^\mu - g\hat{\boldsymbol{T}}\cdot \boldsymbol{A}^\mu - g\frac{1}{2}\hat{Y}B^\mu\right] \\
&= \partial^\mu - \mathrm{i}g\hat{T}^b\cdot A^{b\mu} - \mathrm{i}\frac{g'}{2}\hat{Y}B^\mu \;.
\end{aligned}
\tag{4.155}
$$

Performing the replacement in (4.153) yields

$$
\begin{aligned}
j^{a\mu}(\Phi, \boldsymbol{A}_\mu, B_\mu) &= g\{\Phi^\dagger \hat{T}^a(-\mathrm{i}D^\mu \Phi) + (-\mathrm{i}D^\mu\Phi)^\dagger \hat{T}^a\Phi\} \\
&= -\mathrm{i}g\{\Phi^\dagger \hat{T}^a(D^\mu\Phi) - (D_\mu\Phi)^\dagger \hat{T}^a\Phi\} \\
&= -\mathrm{i}g\left\{\Phi^\dagger \hat{T}^a\left(\partial^\mu - \mathrm{i}g\sum_b \hat{T}^b A^{\mu b} - \mathrm{i}\frac{g'}{2}\hat{Y}B^\mu\right)\Phi\right. \\
&\qquad \left. - \Phi^\dagger\left(\overleftarrow{\partial}^\mu + \mathrm{i}g\sum_b \hat{T}^b A^{\mu b} + \mathrm{i}\frac{g'}{2}\hat{Y}B^\mu\right)\hat{T}^a\Phi\right\} \\
&= -\mathrm{i}g\left[\Phi^\dagger \hat{T}^a(\partial^\mu\Phi) - (\partial^\mu\Phi)^\dagger \hat{T}^a\Phi\right] \\
&\qquad + \mathrm{i}^2 g^2\left[\Phi^\dagger\sum_b(\hat{T}^a\hat{T}^b + \hat{T}^b\hat{T}^a)A^{\mu b}\Phi\right] + \mathrm{i}^2 gg'\Phi^\dagger\hat{T}^a\hat{Y}B^\mu\Phi \\
&= -\mathrm{i}g\left[\Phi^\dagger \hat{T}^a(\partial^\mu\Phi) - (\partial^\mu\Phi)^\dagger \hat{T}^a\Phi\right] \\
&\qquad - \frac{g^2}{2}[\Phi^\dagger\Phi A^{a\mu}] - gg'Y\Phi^\dagger\hat{T}^a\Phi B^\mu
\end{aligned}
\tag{4.156}
$$

where we have applied the well-known relation $\hat{T}^a\hat{T}^b + \hat{T}^b\hat{T}^a = \frac{1}{2}\delta^{ab}$ and the fact that $\hat{T}^a$ and $\hat{Y}$ are hermitian matrices. The operator $\overleftarrow{\partial}^\mu$ indicates that the differentiation is on the left-sided quantity, i.e. $\phi^\dagger\overleftarrow{\partial}^\mu \equiv (\partial^\mu\phi^\dagger)$.

For the current of the hypercharge we analogously replace the normal derivative by the gauge covariant one in (4.154) and obtain

$$
\begin{aligned}
j^{Y\mu}(\Phi, \boldsymbol{A}_\mu, B_\mu) &= \frac{g'}{2}\left[\Phi^\dagger\hat{Y}(-\mathrm{i}D^\mu\phi) + (-\mathrm{i}D^\mu\Phi)^\dagger\hat{Y}\Phi\right] \\
&= -\mathrm{i}\frac{g'}{2}Y\left\{\Phi^\dagger\left(\partial^\mu - \mathrm{i}g\sum_a \hat{T}^a A^{a\mu} - \mathrm{i}\frac{g'}{2}\hat{Y}B^\mu\right)\Phi\right.
\end{aligned}
$$

$$-\Phi^\dagger\left(\overleftarrow{\partial}^\mu+\mathrm{i}g\sum_a\hat{T}^aA^{a\mu}+\mathrm{i}\frac{g'}{2}\hat{Y}B^\mu\right)\Phi\Bigg\}$$

$$=-\mathrm{i}\frac{g'}{2}Y\left[\Phi^\dagger(\partial^\mu\Phi)-(\partial^\mu\Phi)^\dagger\Phi\right]$$

$$-gg'Y\Phi^\dagger\hat{\boldsymbol{T}}\Phi\cdot\boldsymbol{A}^\mu-\frac{g'^2}{2}Y^2\Phi^\dagger\Phi B^\mu \quad . \tag{4.157}$$

Here we have used $[\hat{Y},\hat{T}^a]_-=0$. In order to obtain the screening currents of the vacuum, we replace Φ by its vacuum expectation value in the usual manner. Without loss of generality we can choose

$$\Phi_0=\langle\hat{\Phi}\rangle_0=\begin{pmatrix}0\\ \lambda/\sqrt{2}\end{pmatrix} \tag{4.158}$$

by analogy with (4.150), which was originally proposed by Weinberg. Here we have removed the phase factor $\exp(\mathrm{i}\boldsymbol{\alpha}(x)\cdot\hat{\boldsymbol{T}})$ through SU(2) gauging and the phase factor $\exp(\mathrm{i}\hat{Y}\alpha(x))$,

$$\begin{pmatrix}0\\ H(x)\end{pmatrix}=\exp(\mathrm{i}\hat{Y}\alpha(x))\begin{pmatrix}0\\ \lambda/\sqrt{2}\end{pmatrix} \quad , \tag{4.159}$$

through U(1) gauging. On inserting (4.158) into (4.156) and (4.157) we get

$$=\langle j^{a\mu}(\Phi,\boldsymbol{A}_\mu,B_\mu)\rangle_0$$

$$=-\frac{g^2}{2}\frac{\lambda^2}{2}A^{a\mu}+gg'Y\frac{\lambda^2}{2}\frac{1}{2}\delta^{a3}B^\mu \tag{4.160}$$

and

$$j_0^{Y\mu}=\langle j^{Y\mu}(\Phi,\boldsymbol{A}^\mu,B^\mu)\rangle_0$$

$$=gg'Y\frac{\lambda^2}{2}\frac{1}{2}\delta^{a3}A^{a\mu}-\frac{g'^2}{2}Y^2\frac{\lambda^2}{2}B^\mu \quad . \tag{4.161}$$

The terms with δ^{a3} originate from the relation $\Phi^\dagger T^a\Phi\sim\delta^{a3}$; for $a=1,2$ the expectation value vanishes. Hence we obtain the wave equation (4.147) of the $A^{a\mu}$ fields in the form

$$(\Box A^{a\mu}+M_A^2A^{a\mu})-\partial^\mu\partial_\nu A^{a\nu}=\frac{1}{4}gg'Y\lambda^2\delta^{a3}B^\mu \quad , \tag{4.162}$$

where

$$M_A=\frac{1}{2}g\lambda \quad . \tag{4.163}$$

The field B^μ obeys an equation analogous to (4.147):

$$(\Box B^\mu+M_B^2B^\mu)-\partial^\mu\partial_\nu B^\nu=\frac{1}{4}gg'Y\lambda^2\,A^{3\mu} \quad , \tag{4.164}$$

$$M_B=\frac{1}{2}g'\lambda Y \quad . \tag{4.165}$$

Thus the $A^{a\mu}$ and B^{μ} fields have obtained masses. However, two facts are still unsatisfactory. First, we do not know which value we should choose for Y, and second, on the right-hand side of (4.162) and (4.164) there are inhomogeneities. The latter do not exist for the fields $A^{1\mu}$ and $A^{2\mu}$, but only for $A^{3\mu}$ and B^{μ}. Here the following consideration is helpful. We do not want electromagnetic screening currents to show up in the vacuum, since the photon should stay massless. On the other hand, a third massive boson should arise (mainly) from the field $A^{3\mu}$, which mediates the neutral weak current interaction. We note again the equations of the fields $A^{3\mu}$ and B^{μ} explicitly:

$$\begin{aligned}\Box A^{3\mu} - \partial^{\mu}\partial_{\nu}A^{3\nu} &= -M_A^2 A^{3\mu} + M_A M_B B^{\mu} \quad ,\\ \Box B^{\mu} - \partial^{\mu}\partial_{\nu}B^{\nu} &= -M_B^2 B^{\mu} + M_A M_B A^{3\mu} \quad .\end{aligned} \tag{4.166}$$

Obviously, in the vacuum state the fields $A^{3\mu}$ and B^{μ} are coupled for our choice of Φ_0 (4.158), since on the right-hand side of the equation for $A^{3\mu}$ there is a term proportional to B_{μ}, and vice versa. Usually we would transfer these linear terms to the left-hand side of the equation and interpret them as mass terms. However, it is just these mass terms that are coupled! In physical terms, neither the field $A^{3\mu}$ nor B^{μ} has definite mass. *Therefore they cannot be regarded as the proper physical fields.*

We first have to diagonalize the *mass matrix*. From (4.160) and (4.161) we see that

$$g' j_0^{3\mu} + g j_0^{Y\mu} = 0 \quad , \tag{4.167}$$

if we set $Y = 1$ for the Higgs doublet. Considering the Gell-Mann–Nishijima relation,

$$Q = T_3 + \frac{Y}{2} \quad , \tag{4.168}$$

we see that the lower component of Φ must then have electric charge zero. The upper component has to be charged positively. Hence the general Higgs field has the structure

$$\Phi = \begin{pmatrix} \phi^{(+)} \\ \phi^{(0)} \end{pmatrix} \quad , \tag{4.169}$$

as we demanded outright in (4.109). Here that structure follows from requiring that (4.167) be satisfied. Thus the combination

$$A^{\mu} \propto g' A^{3\mu} + g B^{\mu} \tag{4.170}$$

decouples the field equations (4.166). Moreover, in the general case ($Y \neq 1$) we have just to multiply the first of the equations by M_B and the second by M_A. The sum of both yields

$$(\Box \delta^{\mu}{}_{\nu} - \partial^{\mu}\partial_{\nu})(M_B A^{3\nu} + M_A B^{\nu}) = 0 \quad , \tag{4.171}$$

and we see that this combination of fields stays massless; it can be identified with the photon field A^{μ}.

After normalization we write

$$A^{\mu} = \sin\theta \; A^{3\mu} \; + \; \cos\theta \; B^{\mu} \tag{4.172}$$

with

$$\tan\theta = \frac{M_B}{M_A} = \frac{g'Y}{g} \quad ,$$

$$\sin\theta = \frac{g'Y}{\sqrt{(g^2+g'^2Y^2)}} \quad , \tag{4.173}$$

$$\cos\theta = \frac{g}{\sqrt{(g^2+g'^2Y^2)}} \quad .$$

The field A^μ fulfills the Maxwell equation,

$$(\Box\delta^\mu{}_\nu - \partial^\mu\partial_\nu)A^\nu = 0 \quad . \tag{4.174}$$

The corresponding orthogonal combination

$$Z^\mu = \cos\theta\ A^{3\mu} - \sin\theta\ B^\mu \tag{4.175}$$

obeys the equation

$$(\Box\delta^\mu{}_\nu - \partial^\mu\partial_\nu + M_Z^2\delta^\mu{}_\nu)Z^\nu = 0 \quad , \tag{4.176}$$

with

$$\begin{aligned} M_Z^2 = M_A^2 + M_B^2 &= \frac{1}{4}\lambda^2[(g'Y)^2+g^2] \\ &= M_A^2\left[1+\left(\frac{M_B}{M_A}\right)^2\right] = \frac{M_A^2}{\cos^2\theta} \quad . \end{aligned} \tag{4.177}$$

Thus the Z boson turns out to be massive. On inverting (4.172) and (4.175) we get

$$\begin{aligned} A^{3\mu} &= \sin\theta A^\mu + \cos\theta Z^\mu \quad , \\ B^\mu &= \cos\theta A^\mu - \sin\theta Z^\mu \quad . \end{aligned} \tag{4.178}$$

Inserting this into (4.155), we find that the SU(2) × U(1) covariant derivative results in the form

$$\begin{aligned} D^\mu &= \partial^\mu - \mathrm{i}g(\hat{T}^1A^{1\mu} + \hat{T}^2A^{2\mu}) - \mathrm{i}g\hat{T}^3A^{3\mu} - \frac{1}{2}\mathrm{i}g'\,\hat{Y}B^\mu \\ &= \partial^\mu - \mathrm{i}g(\hat{T}^1A^{1\mu} + \hat{T}^2A^{2\mu}) - \mathrm{i}g\hat{T}^3(\sin\theta\ A^\mu + \cos\theta\ Z^\mu) \\ &\quad - \frac{1}{2}\mathrm{i}g'\hat{Y}(\cos\theta\ A^\mu - \sin\theta\ Z^\mu) \\ &= \partial^\mu - \mathrm{i}g(\hat{T}^1A^{1\mu} + \hat{T}^2A^{2\mu}) - \mathrm{i}gA^\mu\left(\sin\theta\ \hat{T}^3 + \frac{g'Y}{2g}\cos\theta\right) \\ &\quad - \mathrm{i}gZ^\mu\left(\cos\theta\ \hat{T}^3 - \frac{g'Y}{2g}\sin\theta\right) \\ &= \partial^\mu - \mathrm{i}g(\hat{T}^1A^{1\mu} + \hat{T}^2A^{2\mu}) - \mathrm{i}g\ \sin\theta\ A^\mu\left(\hat{T}^3 + \frac{1}{2}\right) \\ &\quad - \frac{\mathrm{i}g}{\cos\theta}\ Z^\mu\left(\cos^2\theta\ \hat{T}^3 - \frac{1}{2}\sin^2\theta\right) \quad . \end{aligned} \tag{4.179}$$

The hypercharge is completely eliminated! Nonetheless, the choice $Y = 1$ for the Higgs field is essential. Only in the case $Y = 1$ can the term proportional to A^μ in (4.179) be written in the form

$$-ig\ \sin\theta\left(\hat{T}^3 + \frac{1}{2}\hat{Y}\right)A^\mu \quad , \quad (Y = 1) \quad . \tag{4.180}$$

Since A^μ is supposed to describe the photon field and $T_3 + \frac{1}{2}Y = Q$ is the charge in units of e, we derive the condition

$$g\ \sin\theta = e \quad . \tag{4.181}$$

From here we are directly led to the relations derived earlier:

$$\begin{aligned}
M_A &= M_\mathrm{W} = \frac{1}{2}g\lambda \quad , \\
\frac{G}{\sqrt{2}} &= \frac{g^2}{8{M_\mathrm{W}}^2} \quad , \\
\frac{G}{\sqrt{2}} &= \frac{1}{2\lambda^2} \quad \text{with} \quad \lambda = 250\ \text{GeV} \quad , \\
\sin^2\theta &\approx 0.23 \pm 0.001 \quad , \\
M_\mathrm{W} &= \frac{e}{2^{5/4}\sqrt{G}\sin\theta} \approx \frac{37.3}{\sin\theta}\ \text{GeV} \approx 80\ \text{GeV} \quad , \\
M_\mathrm{Z} &= \frac{M_\mathrm{W}}{\cos\theta} = 90\ \text{GeV} \quad .
\end{aligned}$$

Finally, we wish to emphasize three points once more.

1. The Gell-Mann–Nishijima relation (4.168) of weak interactions is an additional condition which is of great importance for the GSW theory but is not included in the basic principle of SU(2) × U(1) gauge invariance.
2. At first glance it seems to be curious that we obtain just one massive and one massless field from the initially massless fields $A^{3\mu}$ and B^μ through the Higgs mechanism. This can be understood by considering the special choice

$$\Phi_0 \equiv \langle\Phi\rangle_0 = \begin{pmatrix} 0 \\ \lambda/\sqrt{2} \end{pmatrix} \quad . \tag{4.182}$$

If we choose the vacuum Higgs field, Φ_0, to break the symmetry in some way, we always get a massive gauge boson. This is just the same as in Sect. 4.1 regarding U(1) symmetry. However, if the symmetry group is larger than U(1), as in our case SU(2) × U(1), it may happen that for a special choice of Φ_0 a part of the (total) symmetry group stays unbroken. We may alternatively say that Φ_0 can be invariant with respect to some of the symmetry transformations, that is, generators or combinations of generators. The corresponding gauge fields (gauge bosons) stay massless. In our case, because

$$\hat{Y}\Phi_0 = \Phi_0 \quad , \quad \hat{T}_3\Phi_0 = -\frac{1}{2}\,\Phi_0 \tag{4.183}$$

we have

$$\left(\hat{T}_3 + \frac{1}{2}\hat{Y}\right)\Phi_0 = 0 \quad , \tag{4.184}$$

and therefore

$$\Phi_0 \to \Phi_0' = \exp\left[\mathrm{i}\alpha(x)\left(\hat{T}_3 + \frac{1}{2}\hat{Y}\right)\right]\Phi_0 = \Phi_0 \quad . \tag{4.185}$$

This implies that Φ_0 is not changed by a transformation belonging to this special subgroup of gauge transformations. Hence the related gauge field, a linear combination of $A^{3\mu}$ (belonging to $\hat{T}_3$) and B^{μ} (belonging to $\hat{Y}$), acquires no mass. In (4.184) the generator $\hat{T}_3 + \frac{1}{2}\hat{Y} = \hat{Q}$ is obviously a linear combination of the SU(2) generator $\hat{T}_3$ and the U(1) generator $\hat{Y}$. According to Gell-Mann and Nishijima, $\hat{Q}$ represents the charge operator. Therefore the transformation (4.185) is just the U(1) phase invariance of the neutral electromagnetic field. The photon has no mass!

3. Lastly we mention that our discussion took place in the framework of the *unitary gauge*. In a general gauge the propagator of the massive vector boson has the form

$$\frac{-g^{\mu\nu} + \frac{(1-\zeta)k^{\mu}k^{\nu}}{k^2-\zeta M^2}}{k^2 - M^2} \quad . \tag{4.186}$$

The gauge symmetry, which was initially present, is now hidden. It is implied by the fact that $M = \frac{1}{2}g\lambda$ is connected with the vacuum expectation value of Φ and is therefore changed if we regauge the field Φ. The propagator (4.186) has a pole at $k^2 = M^2$ and a second pole at $k^2 = \zeta^2 M^2$. The former is physical, the latter is not. The second pole corresponds to an unphysical Higgs particle. Therefore we have to take care to suppress these gauge-dependent, unphysical contributions. In our discussions applying the unitary gauge ($\zeta \to \infty$), this problem does not arise.

4.7 Biographical Notes

DEBYE, Peter, physical chemist, *24.1.1884 in Maastricht (Netherlands), †2.11.1966 in Ithaca (New York), received his Ph.D. in physics at the University of Munich under Sommerfeld in 1908. In 1911 Debye was successor of Einstein as professor in Zürich, later became director of the Kaiser Wilhelm Institute for Physics in Berlin. In 1939 he refused to become a German citizen and returned first to the Netherlands, then emigrated to the United States. Since 1940 professor of chemistry at Cornell University in Ithaca. Debye developed the theory of dipolar materials and of electric screening in strong electrolytes. He also discovered X-ray diffraction from powdered substances, and pioneered the study of polymers. He was awarded the Nobel Prize for chemistry in 1936.

FADDEEV, Ludvig Dmitrievich, *23.3.1934 in Leningrad (USSR), Russian mathematician and theoretical physicist. His main contributions are on the quantum mechanical three-body problem (Faddeev equations), quantization of Yang–Mills equations (Faddeev–Popov method), the inverse problem in scattering theory, the theory of solitons and quantum groups. Director of the St. Petersburg branch of the Steklov Mathematical Institute of the Russian Academy of Sciences. He founded the International Institute of Theoretical Physics near St. Petersburg.

GOLDSTONE, Jeffrey, theoretical physicist, *3.9.1933 in Manchester, studied at Cambridge, since 1977 professor at the Massachusetts Institute of Technology, did fundamental work in particle physics and quantum field theory.

MILLS, Robert Laurence, theoretical physicist, *15.4.1927 in Englewood (New Jersey), †27.10.1997 in East Charleston (Vermont, USA), received Ph.D. at Columbia University in 1955. Since 1956 professor at Ohio State University, his work is in quantum field theory, many-body physics and the theory of alloys.

YUKAWA, Hideki, theoretical physicist, *23.1.1907, †8.9.1981 in Kyoto (Japan). He earned his Ph.D. at Osaka University in 1938; since 1939 he was professor and later director of the Research Institute for Fundamental Physics at the University of Kyoto. In 1935 Yukawa developed the meson theory of nuclear forces and predicted the existence of the pion, which was discovered in 1947. He also predicted the decay of an atomic nucleus by capture of an inner-shell electron. He received the Nobel Prize for physics in 1949.

5 Some Properties of the Salam–Weinberg Theory of Leptons

One major success of the standard model is the accurate prediction of the masses of the intermediate bosons, starting from the operator structure of the neutral current. This made a specific experimental search for these particles possible, to which end it is important to know the possible creation and decay mechanisms. We see immediately that in an electron–positron storage ring it is considerably easier to produce the neutral Z^0 bosons than the charged $W^{\pm}$ bosons. In the case of Z^0, simple pair annihilation suffices (Fig. 5.1a), whereas charged bosons can only be produced in higher-order processes (for example, Figs. 5.1b and 5.2). In particular, decays involving a neutrino are experimentally characteristic, because a large fraction of the energy present in the scattering is transferred to the neutrino and is therefore not seen in the detectors.

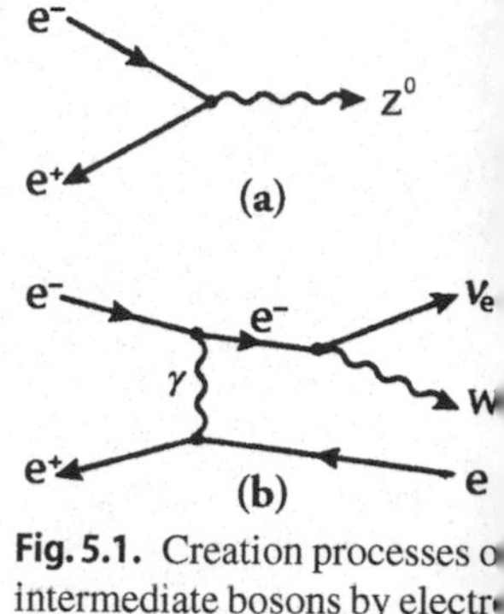

Fig. 5.1. Creation processes o intermediate bosons by electr positron scattering. (**a**) Z^0 ca be produced by pair annihila tion, (**b**) charged intermediat bosons are produced in e^+e^- scattering

5.1 Decay of the Charged Boson W⁻

In the following we first discuss the decay of the negatively charged boson W^-, induced by the first term in $L^{(3L)}_{\mathrm{SW}}$, (4.139). The lowest-order scattering matrix element is:

$$S(W^- \to \ell\bar{\nu}_\ell) = -\mathrm{i}\int \mathrm{d}^4x \frac{(-g)}{2\sqrt{2}}\bar{\psi}_\ell(x)\gamma^\mu(1-\gamma_5)\psi_{\nu_\ell}(x)W_\mu^{(-)}(x) \quad . \tag{5.1}$$

With the notation of Fig. 5.2c the wave functions of the incoming and outgoing particles are

$$\begin{aligned} W_\mu^{(-)}(x) &= (2Vk_0)^{-1/2}\varepsilon_\mu(k,\lambda)\mathrm{e}^{-\mathrm{i}k\cdot x} \quad , \\ \psi_\ell(x) &= (2Vq_0)^{-1/2}u_\ell(q,s)\mathrm{e}^{-\mathrm{i}q\cdot x} \quad , \\ \psi_{\nu_\ell}(x) &= (2Vq_0')^{-1/2}v_\nu(q',s')\mathrm{e}^{+\mathrm{i}q'\cdot x} \quad . \end{aligned} \tag{5.2}$$

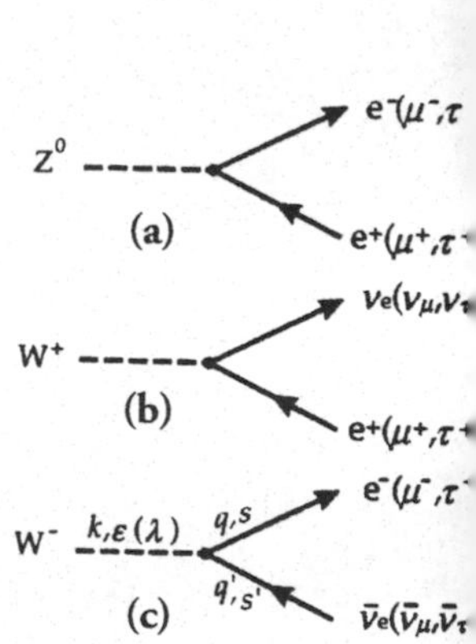

Fig. 5.2. Leptonic decay mech anisms of intermediate bosons

$\varepsilon_\mu(k,\lambda)$ is the polarization vector of the W^- boson. All we need to know about ε_μ is that summation over all three directions of polarization λ yields (see Appendix A.3)

$$\sum_{\lambda=1}^{3}\varepsilon_\mu(k,\lambda)\varepsilon_\nu(k,\lambda) = -g_{\mu\nu} + \frac{k_\mu k_\nu}{M_W{}^2} \quad . \tag{5.3}$$

Now we follow the scheme developed in the treatment of muon decay, Sect. 2.2, and perform the space-time integration

$$\int \mathrm{d}^4x\ \exp(\mathrm{i}q\cdot x + \mathrm{i}q'\cdot x - \mathrm{i}k\cdot x) = (2\pi)^4\delta^4(q+q'-k) \quad . \tag{5.4}$$

W. Greiner, B. Müller, *Gauge Theory of Weak Interactions*,
DOI 10.1007/978-3-540-87843-8_5, © Springer-Verlag Berlin Heidelberg 2009

Squaring the S matrix element by using (2.16), summing or integrating over final states, averaging over the polarization of the incoming boson, and finally dividing by the time T yields the decay rate

$$W = \frac{1}{T} V \int \frac{d^3q}{(2\pi)^3} V \int \frac{d^3q'}{(2\pi)^3} \frac{1}{3} \sum_{\lambda} \sum_{s,s'} |S(W^- \to \ell\bar{\nu}_\ell)|^2$$
$$= \frac{g^2}{8} \frac{1}{(2\pi)^2} \int \frac{d^3q}{2q_0} \int \frac{d^3q'}{2q'_0} \frac{1}{2k_0} \frac{1}{3} \sum_{\lambda,s,s'} |M_{\lambda ss'}|^2 \delta^4(q+q'-k) \quad , \tag{5.5}$$

with

$$M_{\lambda ss'} = \varepsilon_\mu(k,\lambda)\bar{u}_\ell(q,s)\gamma^\mu(1-\gamma_5)v_\nu(q',s') \quad . \tag{5.6}$$

Using the rules for calculating traces of Dirac spinors and γ matrices (see Appendix A.1) we get the following result:

$$\sum_{\lambda,s,s'} |M_{\lambda ss'}|^2 = \sum_{\lambda} \varepsilon_\mu(k,\lambda)\varepsilon_\nu(k,\lambda) \Big\{ \sum_{s,s'} \bar{u}_\ell(q,s)\gamma^\mu(1-\gamma_5)$$
$$\times v_\nu(q',s')\bar{v}_\nu(q',s')\gamma^\nu(1-\gamma_5)u_\ell(q,s) \Big\}$$
$$= \left(-g_{\mu\nu} + \frac{k_\mu k_\nu}{M_W{}^2}\right) \mathrm{Tr}\{(\not{q} + m_\ell)\gamma^\mu(1-\gamma_5)\not{q}'\gamma^\nu(1-\gamma_5)\} \quad . \tag{5.7}$$

The Dirac trace becomes

$$\mathrm{Tr}\{(\not{q} + m_\ell)\gamma^\mu(1-\gamma_5)\not{q}'\gamma^\nu(1-\gamma_5)\}$$
$$= 2\,\mathrm{Tr}\{(\not{q} + m_\ell)\gamma^\mu\not{q}'\gamma^\nu(1-\gamma_5)\}$$
$$= 2q_\alpha q'_\beta \,\mathrm{Tr}\{\gamma^\alpha\gamma^\mu\gamma^\beta\gamma^\nu(1-\gamma_5)\}$$
$$= 8q_\alpha q'_\beta (g^{\alpha\mu}g^{\beta\nu} + g^{\alpha\nu}g^{\beta\mu} - g^{\alpha\beta}g^{\mu\nu} + \mathrm{i}\varepsilon^{\alpha\mu\beta\nu})$$
$$= 8\left[q^\mu q'^\nu + q^\nu q'^\mu - (q\cdot q')g^{\mu\nu} + \mathrm{i}\varepsilon^{\alpha\mu\beta\nu}q_\alpha q'_\beta\right] \quad . \tag{5.8}$$

Here we have used (5.3) and (2.29). One may wonder why we are allowed to sum over the neutrino spin s' in (5.7), although we learned that only left-handed neutrinos exist. The reason is simply that the V–A coupling $\gamma^\mu(1-\gamma_5)$ vanishes when applied to right-handed neutrinos:

$$(1-\gamma_5)(1+\gamma_5)v_\nu = 0 \quad . \tag{5.9}$$

Inserting the result of the Dirac trace into (5.7) we see that the last term does not contribute, because (5.3) is symmetric with respect to the indices μ and ν, whereas $\varepsilon^{\alpha\mu\beta\nu}$ is totally antisymmetric. From the energy–momentum relation $k^2 = M_W^2$ we find that

$$\sum_{\lambda ss'} |M_{\lambda ss'}|^2 = \frac{8}{M_W^2}\left[2(q\cdot k)(q'\cdot k) + M_W{}^2(q\cdot q')\right] \quad . \tag{5.10}$$

Because of the delta function in (5.5) the second term can be expressed in terms of the rest masses:

$$M_W^2 = k^2 = (q+q')^2 = q^2 + q'^2 + 2(q\cdot q') = m_\ell^2 + 2(q\cdot q') \quad . \tag{5.11}$$

If we now insert all terms into (5.5) we obtain the decay rate

$$W = \frac{g^2}{48\pi^2}\frac{1}{k_0 M_W^2}\int\frac{d^3q}{q_0}\int\frac{d^3q'}{q_0'}\delta^4(q+q'-k) \times\left[(q\cdot k)(q'\cdot k) + \frac{1}{4}M_W^2(M_W^2 - m_\ell^2)\right] \quad . \tag{5.12}$$

For the calculation of the momentum-space integral we refer to results obtained earlier. Equation (17) in Exercise 2.6 states that

$$\begin{aligned} k^\alpha k^\beta \int\frac{d^3q}{q_0}\int\frac{d^3q'}{q_0'} q_\alpha q'_\beta \delta^4(q+q'-k) &= \frac{\pi}{6}\left(1-\frac{m_\ell^2}{k^2}\right)^2\left[k^4\left(1-\frac{m_\ell^2}{k^2}\right)+2\left(1+2\frac{m_\ell^2}{k^2}\right)k^4\right]\Theta(k^2-m_\ell^2) \\ &= \frac{\pi}{2}M_W^4\left(1-\frac{m_\ell^2}{M_W^2}\right)^2\left(1+\frac{m_\ell^2}{M_W^2}\right)\Theta(M_W-m_\ell) \quad . \end{aligned} \tag{5.13}$$

The second part can be done in the following way. We integrate in the rest frame of the W$^-$ boson, where $k_\alpha = (M_W, \boldsymbol{O})$, yielding

$$\begin{aligned} \int\frac{d^3q}{q_0}\int\frac{d^3q'}{q_0'}\delta(\boldsymbol{q}+\boldsymbol{q}')\delta(q_0+q_0'-M_W) &= \int\frac{d^3q}{q_0|\boldsymbol{q}|}\delta(q_0+|\boldsymbol{q}|-M_W) \\ &= 4\pi\int_0^\infty \frac{|\boldsymbol{q}|\,d|\boldsymbol{q}|}{\sqrt{\boldsymbol{q}^2+m_\ell^2}}\delta\left(\sqrt{\boldsymbol{q}^2+m_\ell^2}+|\boldsymbol{q}|-M_W\right) \\ &= 4\pi\int_{m_\ell}^\infty \frac{dx}{2x^2}(x^2-m_\ell^2)\delta(x-M_W) \\ &= 2\pi\left(1-\frac{m_l^2}{M_W^2}\right)\Theta(M_W-m_\ell) \quad , \end{aligned} \tag{5.14}$$

where we have substituted $x = |\boldsymbol{q}| + \sqrt{\boldsymbol{q}^2+m_l^2}$ (see Exercise 2.6).

By inserting (5.13), (5.14) into (5.12) we obtain the final result of the decay rate of the W$^-$ in its rest frame:

$$W = \frac{1}{48\pi}g^2 M_W\left(1-\frac{m_\ell^2}{M_W^2}\right)^2\left(1+\frac{m_\ell^2}{2M_W^2}\right)\Theta(M_W-m_\ell) \quad . \tag{5.15}$$

If we use the connection between g and the Fermi coupling constant G (4.131) and further exploit the fact that all known leptons are much lighter than the intermediate boson, the result is

$$W(\mathrm{W}^- \to \ell\,\bar{\nu}_\ell) \approx \frac{G}{6\pi\sqrt{2}} M_\mathrm{W}^3 \approx 225\ \mathrm{MeV} \approx 3.5 \times 10^{23}\ \mathrm{s}^{-1} \quad , \tag{5.16}$$

where we have taken the value of G as that given in (2.63). Taking into account the three leptonic decay channels ($\ell = \mathrm{e}, \mu, \tau$) we can estimate the lifetime of the charged intermediate bosons:

$$\tau_\mathrm{W} \le \frac{1}{3W} \approx \frac{2\pi\sqrt{2}}{GM_\mathrm{W}^3} \approx 10^{-24}\ \mathrm{s} \quad . \tag{5.17}$$

This is a remarkably long time for an elementary particle with a mass of more than 80 GeV, but it is far too short to be measured directly. A direct observation of W bosons, for example, as a track in a bubble chamber, seems to be impossible.

However, the decay of W bosons can be easily detected experimentally since the charged lepton and the neutrino are emitted in opposite directions, with momenta of the order of 40 GeV/c. The neutrino escapes all detectors, so that a highly energetic lepton should be observed whose corresponding recoil momentum is missing. When the W boson has been created through some mechanism, it yields a clear experimental sign of its decay. However, we shall see later in Chap. 6 that the W boson also couples to quarks. It can therefore also decay into hadrons, which make up about two thirds of all final states. The probability for the decay into any one of the leptonic decay channels amounts to only about 10 percent.

EXERCISE

5.1 Decay of the Z^0 Boson

Problem. Calculate the decay processes $Z^0 \to \ell^+\ell^-$ and $Z^0 \to \nu\bar{\nu}$.

Solution. (a) First we consider the decay of the Z^0 boson into charged leptons $\ell^+\ell^-$. Their rest mass can be neglected, since it is much smaller than the mass of the Z^0. According to (4.139) the scattering matrix element is given by

$$S(Z^0 \to \ell^+\ell^-) = \mathrm{i} \int \mathrm{d}^4x \frac{(-g)}{4\cos\theta} \bar{\psi}_\ell(x)\gamma^\mu(g'_\mathrm{V} - \gamma_5)\psi_\ell(x) Z_\mu(x) \tag{1}$$

with

$$g'_\mathrm{V} = 1 - 4\sin^2\theta \ll 1 \quad . \tag{2}$$

If we adopt the same notation for momenta and spins as in Fig. 5.1 and (5.2), the decay rate is again given by the result of (5.5), (5.6):

$$W = \frac{g^2}{16\cos^2\theta} \frac{1}{(2\pi)^2} \int \frac{\mathrm{d}^3q}{2q_0} \int \frac{\mathrm{d}^3q'}{2q'_0} \frac{1}{6k_0} \sum_{\lambda s s'} |M_{\lambda s s'}|^2 \delta^4(q + q' - k) \tag{3}$$

with

$$M_{\lambda s s'} = \varepsilon_\mu(k,\lambda)\bar{u}(q,s)\gamma^\mu(g'_\mathrm{V} - \gamma_5)v(q',s') \quad . \tag{4}$$

Exercise 5.

If we neglect the lepton mass, the averaged matrix element yields the Dirac trace (cf. (5.7)):

$$
\begin{aligned}
&\mathrm{Tr}\{\not{q}\gamma^\mu(g'_V-\gamma_5)\not{q}\gamma^\nu(g'_V-\gamma_5)\}\\
&=\mathrm{Tr}\{\not{q}\gamma^\mu\not{q}'\gamma^\nu(g'^2_V+1-2g'_V\gamma_5)\}\\
&=4(g'^2_V+1)[q^\mu q'^\nu+q^\nu q'^\mu-(q\cdot q')g^{\mu\nu}]+\mathrm{i}8g'_V\varepsilon^{\alpha\mu\beta\nu}q^\alpha q'^\beta \quad . \qquad (5)
\end{aligned}
$$

The last term again does not contribute, owing to antisymmetry, yielding

$$
\begin{aligned}
\sum_{\lambda ss'}|M_{\lambda ss'}|^2&=(-g_{\mu\nu}+k_\mu k_\nu/M_Z^2)\,\mathrm{Tr}\{\not{q}\gamma^\mu(g'_V-\gamma_5)\not{q}'\gamma^\nu(g'_V-\gamma_5)\}\\
&=\frac{4}{M_Z^2}(g'_V+1)[2(q\cdot k)(q'\cdot k)+M_Z^2(q\cdot q')]\\
&\simeq\frac{8}{M_Z^2}(g'^2_V+1)\left[(q\cdot k)(q'\cdot k)+\frac{1}{4}M_Z^4\right] \quad , \qquad (6)
\end{aligned}
$$

where we have used the analogue of (5.11),

$$
M_Z^2=2m_\ell^2+2(q\cdot q')\approx 2(q\cdot q') \quad . \qquad (7)
$$

If we substitute the result (6) into (3) and perform the integrals in momentum space, as in (5.13), (5.14), we obtain the decay rate

$$
\begin{aligned}
W(Z^0\to\ell^+\ell^-)&\approx\frac{g^2(g'^2_V+1)}{192\pi\cos^2\theta}M_Z\\
&=\frac{G(g'^2_V+1)}{\sqrt{2}\cdot 24\pi\cos^3\theta}M_W^3\approx 83\ \mathrm{MeV} \quad . \qquad (8)
\end{aligned}
$$

(b) According to (4.139) the decay process $Z^0\to\nu\bar{\nu}$ can be described by (1) with $g'_V=1$. Therefore we can copy the resulting decay rate from (8):

$$
\begin{aligned}
W(Z^0\to\nu\bar{\nu})&\approx\frac{g^2(1+1)}{192\pi\cos^2\theta}M_Z\\
&=\frac{G}{\sqrt{2}\,12\pi\cos^3\theta}M_W^3=167\ \mathrm{MeV} \quad . \qquad (9)
\end{aligned}
$$

Like (8) this result can also be written in the form

$$
\frac{g^2(g'^2_V+1)}{192\pi\cos^2\theta}M_Z \quad ,
$$

because according to V–A coupling one has $g'_V=1$. For the three leptonic generations e, μ, τ together we get

$$
W(Z^0\to\text{leptons})=751.5\ \mathrm{MeV} \quad . \qquad (10)
$$

In particular, every massless neutrino species contributes about 100 MeV to the decay rate of the Z^0 boson. Hence an experimental determination of its width allows for a determination of the number of neutrinos in a model-independent way! (Recent experimental progress is discussed in Example 5.3.)

5.2 The Process $e^+e^- \to Z^0 \to \mu^+\mu^-$

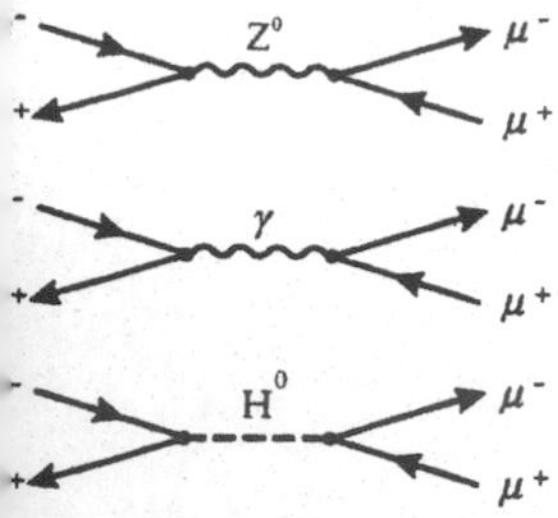

g. 5.3. The three processes ontributing to the scattering $^+e^- \to \mu^+\mu^-$. The interediate particle can be repsented by the vector boson ^{0}, the photon $\gamma(A_\mu)$, and e Higgs particle H^0

Since the short lifetime of the intermediate bosons prohibits their direct observation, it is convenient to consider the creation and decay process of the Z^0 boson as a single process (see Figs. 5.1 and 5.2). Here we shall consider the special process $e^+e^- \to Z^0 \to \mu^+\mu^-$. If we study the Salam–Weinberg Lagrangian (4.138)–(4.142) carefully, we recognize that there are two other processes that yield the same final state $\mu^+\mu^-$. They are, first, pair annihilation into a virtual photon ($e^+e^- \to \gamma \to \mu^+\mu^-$) and, second, formation of an intermediate Higgs particle ($e^+e^- \to H^0 \to \mu^+\mu^-$). According to the Feynman rules for vector bosons, derived in Sect. 4.3, the total scattering amplitude consists of three parts, which we write directly in momentum space:

$$S(e^+e^- \to \mu^+\mu^-) = \frac{(2\pi)^4\delta^4(p'+k'-p-k)}{4V^2(p_0k_0k'_0p'_0)^{1/2}}(A_\gamma + A_Z + A_H) \quad . \tag{5.18}$$

The three amplitudes are

$$A_\gamma = -i\bar{u}_\mu(p',s')i(-e)\gamma^\alpha v_\mu(k',t')\frac{g_{\alpha\beta} - ((k+p)_\alpha(k+p)_\beta/(k+p)^2)}{(k+p)^2}$$
$$\times\, \bar{v}_e(k,t)i(-e)\gamma^\beta u_e(p,s) \quad , \tag{5.19a}$$

$$A_Z = -i\bar{u}_\mu(p',s')\frac{i(-)g\gamma^\alpha}{4\cos\theta}(g'_V - \gamma_5)v_\mu(k',t')\frac{g_{\alpha\beta} - ((k+p)_\alpha(k+p)_\beta/M_Z^2)}{(k+p)^2 - M_Z^2}$$
$$\times\, \bar{v}_e(k,t)\frac{i(-)g\gamma^\beta}{4\cos\theta}(g'_V - \gamma_5)u_e(p,s) \quad , \tag{5.19b}$$

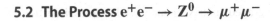

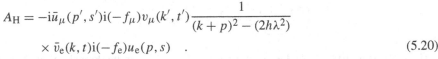

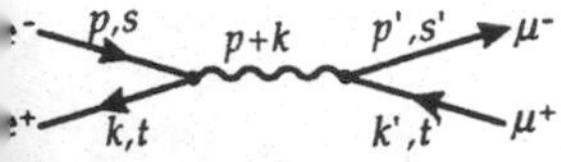

ig. 5.4. Notation of momenta in the process $e^+e^- \to \mu^+\mu^-$

$$A_H = -i\bar{u}_\mu(p',s')i(-f_\mu)v_\mu(k',t')\frac{1}{(k+p)^2 - (2h\lambda^2)}$$
$$\times\, \bar{v}_e(k,t)i(-f_e)u_e(p,s) \quad . \tag{5.20}$$

Here we have used the notation $g'_V = 1 - 4\sin^2\theta$ and the couplings from (4.139) and (4.142).

We first compare the relative magnitudes of the coupling constants in the three matrix elements:

$$A_\gamma \sim e^2 \sim g^2\sin^2\theta \quad , \tag{5.21a}$$

$$A_Z \sim \frac{g^2}{\cos^2\theta} \quad , \tag{5.21b}$$

$$A_H \sim f_\mu f_e = \frac{m_\mu m_e}{\lambda^2} = g^2\frac{m_\mu m_e}{4M_W^2} \approx 10^{-8}g^2 \quad , \tag{5.21c}$$

where we have made use of (4.143), (4.144). Obviously the contribution from the Higgs particle is totally negligible, whereas the matrix elements A_γ and A_Z are of the same order of magnitude – at least at scattering energies in the range of the mass of the intermediate boson. The reason for this is simple: we have used the Higgs field to generate masses of the intermediate bosons as well as of the leptons. Since the strength of the coupling between the Higgs particle and intermediate bosons is given by g (see (4.112)), the interaction between leptons and the Higgs field should be gm_i/M_W.[1]

[1] Owing to the large mass difference between leptons and intermediate bosons, the whole procedure seems to be very artificial. It is therefore appropriate to view the Higgs mechanism as a theoretical

Let us return to the evaluation of the scattering amplitude (5.18), where we can restrict ourselves to the contributions of the photon and Z^0 boson as discussed above. The electron–positron scattering experiments are usually performed with colliding beams, which implies that the laboratory frame is simultaneously the centre-of-mass frame of the e^+e^- pair. Hence our calculation can be most conveniently performed in the centre-of-mass frame. There we have $k_0 = p_0$, $\boldsymbol{k} = -\boldsymbol{p}$, and also $k_0' = p_0'$, $\boldsymbol{k}' = -\boldsymbol{p}'$. The square of the centre-of-mass energy is denoted by the variable

$$s = (k+p)^2 = 4p_0^2 = 4k_0^2 \quad . \tag{5.22}$$

According to the formula stated in Appendix A.3, the particle current in the centre-of-mass frame is given by

$$J = \frac{\sqrt{(k\cdot p)^2 - m_e^2\, m_e^2}}{k_0\, p_0\, V} = \frac{\sqrt{(p_0^2 + |\boldsymbol{p}|^2)^2 - m_e^4}}{p_0^2 V} \quad . \tag{5.23}$$

From the relation

$$\begin{aligned}(p_0^2 + |\boldsymbol{p}|^2)^2 - m_e^4 &= (2|\boldsymbol{p}|^2 + m_e^2)^2 - m_e^4 \\ &= 4|\boldsymbol{p}|^2(|\boldsymbol{p}|^2 + m_e^2) = 4|\boldsymbol{p}|^2 p_0^2 \quad ,\end{aligned} \tag{5.24}$$

J can be reduced to

$$J = \frac{2|\boldsymbol{p}|}{p_0 V} \quad . \tag{5.25}$$

JV is just the relative velocity of the colliding particles. The differential cross section in the centre-of-mass frame can be obtained by squaring the scattering matrix element (5.18) and dividing the result by VT to obtain the transition probability per volume, by J, and by the particle density $\rho = V^{-1}$. Finally we have to integrate over the final states for a fixed emission angle $d\Omega$. Let us denote this symbolically by

$$d\sigma \sim \frac{|S_{fi}|^2}{VT}\frac{1}{J}\frac{1}{\rho}V\int\frac{d^3p'}{(2\pi)^3}V\int\frac{d^3k'}{(2\pi)^3} \sim \frac{1}{V^4}\frac{1}{VT}\frac{1}{1/V}\frac{1}{1/V}(2\pi)^4\delta^4(0)V\cdot V$$

As should happen, all V and T factors drop out. Averaging over the polarizations of the incoming particles, we get

$$\begin{aligned}\frac{d\sigma}{d\Omega}(e^+e^- \to \mu^+\mu^-) &= \frac{(2\pi)^4}{32k_0|\boldsymbol{p}|}\frac{1}{4}\sum_{s,s',t,t'}\int_0^\infty\frac{|\boldsymbol{p}'|^2 d|\boldsymbol{p}'|}{(2\pi)^3 p_0'} \\ &\times\int\frac{d^3k'}{(2\pi)^3 k_0'}\delta^4(p'+k'-p-k)|A_\gamma + A_Z|^2 \quad .\end{aligned} \tag{5.26}$$

The spin-averaged square of the matrix elements A_γ and A_Z can be rewritten as traces of Dirac matrices in the usual manner. Using the formulas of Appendix A.2 we find

---

tool, which must eventually be replaced by a more fundamental theory for the generation of rest masses. Such an underlying theory may well yield an explanation for the large mass ratios. The development of a theory of this kind is one of the important tasks of particle physics.

that

$$\begin{aligned}
&\sum_{s,s',t,t'} |A_\gamma + A_\mathrm{Z}|^2 \\
&= \frac{\mathrm{e}^4}{s^2}\left[g_{\alpha\beta} - \frac{(k+p)_\alpha (k+p)_\beta}{s}\right]\left[g_{\bar{\alpha}\bar{\beta}} - \frac{(k+p)_{\bar{\alpha}}(k+p)_{\bar{\beta}}}{s}\right] \\
&\times \mathrm{Tr}\{(\not{p}' + m_\mu)\gamma^\alpha(\not{k}' - m_\mu)\gamma^{\bar{\alpha}}\}\,\mathrm{Tr}\{(\not{k} - m_\mathrm{e})\gamma^\beta(\not{p} + m_\mathrm{e})\gamma^{\bar{\beta}}\} \\
&+ \frac{\mathrm{e}^2 g^2}{16\cos^2\theta \times s(s - M_\mathrm{Z}^2)}\left[g_{\alpha\beta} - \frac{(k+p)_\alpha(k+p)_\beta}{s}\right] \\
&\times \left[g_{\bar{\alpha}\bar{\beta}} - \frac{(k+p)_{\bar{\alpha}}(k+p)_{\bar{\beta}}}{M_\mathrm{Z}^2}\right] \\
&\times \mathrm{Tr}\{(\not{p}' + m_\mu)\gamma^\alpha(\not{k}' - m_\mu)\gamma^{\bar{\alpha}}(g'_\mathrm{V} - \gamma_5)\} \\
&\times \mathrm{Tr}\{(\not{k} - m_\mathrm{e})\gamma^\beta(\not{p} + m_\mathrm{e})\gamma^{\bar{\beta}}(g'_\mathrm{V} - \gamma_5)\} \\
&+ \frac{\mathrm{e}^2 g^2}{16\cos^2\theta \times s(s - M_\mathrm{Z}^2)}\left[g_{\alpha\beta} - \frac{(k+p)_\alpha(k+p)_\beta}{M_\mathrm{Z}^2}\right] \\
&\times \left[g_{\bar{\alpha}\bar{\beta}} - \frac{(k+p)_{\bar{\alpha}}(k+p)_{\bar{\beta}}}{s}\right] \\
&\times \mathrm{Tr}\{(\not{p}' + m_\mu)\gamma^\alpha(g'_\mathrm{V} - \gamma_5)(\not{k}' - m_\mu)\gamma^{\bar{\alpha}}\} \\
&\times \mathrm{Tr}\{(\not{k} - m_\mathrm{e})\gamma^\beta(g'_\mathrm{V} - \gamma_5)(\not{p} + m_\mathrm{e})\gamma^{\bar{\beta}}\} \\
&+ \frac{g^4}{256\cos^4\theta \times (s - M_\mathrm{Z}^2)^2}\left[g_{\alpha\beta} - \frac{(k+p)_\alpha(k+p)_\beta}{M_\mathrm{Z}^2}\right] \\
&\times \left[g_{\bar{\alpha}\bar{\beta}} - \frac{(k+p)_{\bar{\alpha}}(k+p)_{\bar{\beta}}}{M_\mathrm{Z}^2}\right] \\
&\times \mathrm{Tr}\{(\not{p}' + m_\mu)\gamma^\alpha(g'_\mathrm{V} - \gamma_5)(\not{k}' - m_\mu)\gamma^{\bar{\alpha}}(g'_\mathrm{V} - \gamma_5)\} \\
&\times \mathrm{Tr}\{(\not{k} + m_\mathrm{e})\gamma^\beta(g'_\mathrm{V} - \gamma_5)(\not{p} + m_\mathrm{e})\gamma^{\bar{\beta}}(g'_\mathrm{V} - \gamma_5)\} \quad .
\end{aligned} \tag{5.27}$$

In order to simplify our calculation we take into account that the most interesting scattering energies are far above 10 GeV. Therefore we can safely neglect all terms including lepton masses. Consequently all terms in the Dirac traces proportional to m_e or m_μ are omitted. The simplification goes even further: we may omit all contributions in the numerators of the boson propagators containing the four-vector $(k+p)$. To wit, since $k^2 = p^2 = m_\mathrm{e}^2$, we get, for instance,

$$\begin{aligned}
(k+p)_\beta\,\mathrm{Tr}\{(\not{k} - m_\mathrm{e})\gamma^\beta(\not{p} + m_\mathrm{e})\gamma^{\bar{\beta}}\} &\approx \mathrm{Tr}\{\not{k}(\not{k} + \not{p})\not{p}\gamma^{\bar{\beta}}\} \\
&= \mathrm{Tr}\{k^2\not{p}\gamma^{\bar{\beta}}\} + \mathrm{Tr}\{\not{k}p^2\gamma^{\bar{\beta}}\} \\
&= m_\mathrm{e}^2\,\mathrm{Tr}\{\not{p}\gamma^{\bar{\beta}}\} + m_\mathrm{e}^2\,\mathrm{Tr}\{\not{k}\gamma^{\bar{\beta}}\} \\
&\approx 0 \quad ,
\end{aligned} \tag{5.28}$$

because $\not{k}^2 = k^2 = m_e^2$, $\not{p}^2 = p^2 = m_e^2$, and so on (see Appendix A.2). Since the δ function ensuring energy and momentum conservation in (5.26) implies that $(p+k) = (p'+k')$, the latter result is also valid for traces containing the four-momenta of the muons in the final state (p', k').

In the limit of high scattering energies, that is for $s \gg m_e^2, m_\mu^2$, only the contributions of the form $g_{\alpha\beta}g_{\bar{\alpha}\bar{\beta}}$ remain, yielding a considerably simplified form of (5.27),

$$\begin{aligned}\sum_{s,s',t,t'} |A_\gamma + A_Z|^2 \approx{}& \frac{e^4}{s^2}\mathrm{Tr}\{\not{p}'\gamma^\alpha \not{k}'\gamma^\beta\}\mathrm{Tr}\{\not{k}\gamma_\alpha \not{p}\gamma_\beta\} \\ &+ \frac{e^2 g^2}{2\cos^2\theta \times s \times (s - M_Z^2)}\mathrm{Tr}\{\not{p}'\gamma^\alpha \not{k}'\gamma^\beta(g_V' - \gamma_5)\} \\ &\times \mathrm{Tr}\{\not{k}\gamma_\alpha \not{p}\gamma_\beta(g_V' - \gamma_5)\} \\ &+ \frac{g^4}{256\cos^4\theta \times (s - M_Z^2)^2}\mathrm{Tr}\{\not{p}'\gamma^\alpha(g_V' - \gamma_5)\not{k}'\gamma^\beta(g_V' - \gamma_5)\} \\ &\times \mathrm{Tr}\{\not{k}\gamma_\alpha(g_V' - \gamma_5)\not{p}\gamma_\beta(g_V' - \gamma_5)\} \ . \end{aligned} \tag{5.29}$$

Here we have used the identity of both interference terms (in the limit $m_e, m_\mu = 0$!) resulting from the anticommutativity of γ_5 and γ_μ. The calculation of the remaining Dirac traces can be performed by complete analogy with the case of muon decay, discussed at length in Sect. 2.2. Here we perform the calculation of the interference term (the term proportional to e^2g^2) in detail; the evaluation of the other two terms is left to the reader as an exercise.

With the help of (2.29) we get (see also Appendix A.2)

$$\begin{aligned}&\mathrm{Tr}\{\not{p}'\gamma^\alpha \not{k}'\gamma^\beta(g_V' - \gamma_5)\} \\ &= p'_\mu k'_\nu \mathrm{Tr}\{\gamma^\mu\gamma^\alpha\gamma^\nu\gamma^\beta(g_V' - \gamma_5)\} \\ &= 4p'_\mu k'_\nu(g^{\mu\alpha}g^{\nu\beta} + g^{\mu\beta}g^{\nu\alpha} - g^{\mu\nu}g^{\alpha\beta})g_V' + 4\mathrm{i}\varepsilon^{\mu\alpha\nu\beta}p'_\mu k'_\nu \\ &= 4g_V'\left[p'^\alpha k'^\beta + p'^\beta k'^\alpha - (p'\cdot k')g^{\alpha\beta}\right] + 4\mathrm{i}\varepsilon^{\mu\alpha\nu\beta}p'_\mu k'_\nu \end{aligned} \tag{5.30}$$

and analogously

$$\begin{aligned}\mathrm{Tr}\{\not{k}\gamma_\alpha \not{p}\gamma_\beta(g_V' - \gamma_5)\} ={}& 4g_V'\left[k_\alpha p_\beta + k_\beta p_\alpha - (p\cdot k)g_{\alpha\beta}\right] \\ &+ 4\mathrm{i}\varepsilon_{\bar{\mu}\alpha\bar{\nu}\beta}k^{\bar{\mu}}p^{\bar{\nu}} \ . \end{aligned} \tag{5.31}$$

Since $\varepsilon^{\mu\alpha\nu\beta}$ is totally antisymmetric, the terms combined with expressions symmetric in α, β vanish. With the help of the relation (see Exercises 2.3 and 2.4)

$$\varepsilon^{\mu\alpha\nu\beta}\varepsilon_{\bar{\mu}\alpha\bar{\nu}\beta} = 2(g_{\bar{\nu}}^\mu g_{\bar{\mu}}^\nu - g_{\bar{\mu}}^\mu g_{\bar{\nu}}^\nu) \ , \tag{5.32}$$

the product of the two Dirac spinors yields

$$\begin{aligned}&32g_V'^2[(p'\cdot p)(k'\cdot k) + (p'\cdot k)(k'\cdot p)] + 32[(p'\cdot k)(k'\cdot p) - (p'\cdot p)(k'\cdot k)] \\ &= 32(g_V'^2 - 1)(p'\cdot p)(k'\cdot k) + 32(g_V'^2 + 1)(p'\cdot k)(k'\cdot p) \ . \end{aligned} \tag{5.33}$$

The main details of the calculation can be found in the solution of Exercise 2.3. Together with the two other contributions we get

$$\sum_{s,s',t,t'} |A_\gamma + A_Z|^2 = \left[32\frac{e^4}{s^2} + 4\frac{e^2 g^2 (g_V'^2+1)}{s(s-M_Z^2)\cos^2\theta} + \frac{g^4(g_V'^4+6g_V'^2+1)}{8(s-M_Z^2)^2\cos^4\theta}\right] \times (p'\cdot k)(k'\cdot p) + \left[32\frac{e^4}{s^2} + 4\frac{e^2 g^2 (g_V'^2-1)}{s(s-M_Z^2)\cos^2\theta} + \frac{g^4(g_V'^2-1)}{8(s-M_Z^2)^2\cos^4\theta}\right] \times (p'\cdot p)(k'\cdot k). \tag{5.34}$$

Here, all terms of the order m_e^2/s, m_μ^2/s have been neglected. With the help of the relations $e = g\sin\theta$ and $\sin 2\theta = 2\sin\theta\cos\theta$ we can simplify the expression

$$\sum_{s,s',t,t'} |A_\gamma + A_Z|^2 \approx 32\frac{e^4}{s^2}\left[1+\frac{1}{2}(g_V'^2+1)R(s) + \frac{1}{16}(g_V'^4+6g_V'^2+1)R(s)^2\right] \times (p'\cdot k)(k'\cdot p) + 32\frac{e^4}{s^2}\left[1+\frac{1}{2}(g_V'^2-1)R(s) + \frac{1}{16}(g_V'^4-1)R(s)^2\right] \times (p'\cdot p)(k'\cdot k) \tag{5.35}$$

with the resonance factor

$$R(s) = \frac{s}{(s-M_Z^2)\sin^2 2\theta} \quad . \tag{5.36}$$

Lastly, we must integrate over the final states in momentum space. By neglecting the particle rest masses compared to the centre-of-mass energy $\sqrt{s}$, that is by approximating

$$k^2 = p^2 = m_e^2 \approx 0 \quad , \quad k'^2 = p'^2 = m_\mu^2 \approx 0 \quad , \tag{5.37}$$

we have

$$p_0 \approx |\boldsymbol{p}| \quad , \quad k_0 \approx |\boldsymbol{k}| \quad , \quad p_0' \approx |\boldsymbol{p}'| \quad , \quad k_0' \approx |\boldsymbol{k}'| \quad . \tag{5.38}$$

Furthermore, in the centre-of-mass frame

$$\boldsymbol{p} = -\boldsymbol{k} \quad , \quad \boldsymbol{p}' = -\boldsymbol{k}' \tag{5.39}$$

yields $p_0 \approx k_0$, $p_0' \approx k_0'$. Hence energy conservation implies (see (5.22)):

$$p_0 \approx k_0 \approx p_0' \approx k_0' \approx \frac{1}{2}\sqrt{s} \tag{5.40}$$

and

$$\begin{aligned} \boldsymbol{p}'\cdot\boldsymbol{p} &\approx \frac{s}{4}\cos\vartheta \quad , \quad \boldsymbol{k}'\cdot\boldsymbol{k} \approx \frac{s}{4}\cos\vartheta \quad , \\ \boldsymbol{p}'\cdot\boldsymbol{k} &\approx -\frac{s}{4}\cos\vartheta \quad , \quad \boldsymbol{k}'\cdot\boldsymbol{p} \approx -\frac{s}{4}\cos\vartheta \quad , \end{aligned} \tag{5.41}$$

where ϑ denotes the scattering angle in the centre-of-mass frame (Fig. 5.5). Thus we get

$$(p' \cdot p) = p_0' \cdot p_0 - \boldsymbol{p}' \cdot \boldsymbol{p} \approx \frac{s}{4}(1 - \cos\vartheta) \quad ,$$
$$(k' \cdot k) = k_0' \cdot k_0 - \boldsymbol{k}' \cdot \boldsymbol{k} \approx \frac{s}{4}(1 - \cos\vartheta) \quad ,$$
$$(p' \cdot k) = p_0' \cdot k_0 - \boldsymbol{p}' \cdot \boldsymbol{k} \approx \frac{s}{4}(1 + \cos\vartheta) \quad ,$$
$$(k' \cdot p) = k_0' \cdot p_0 - \boldsymbol{k}' \cdot \boldsymbol{p} \approx \frac{s}{4}(1 + \cos\vartheta) \quad . \tag{5.42}$$

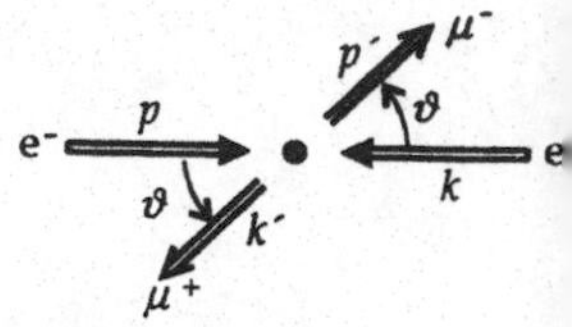

Fig. 5.5. Schematic illustratic of scattering in the center-o mass frame, which here coin cides with the laboratory fram

Applying these approximations in (5.26) and setting $\boldsymbol{p} + \boldsymbol{k} = 0$ we find that the remaining integral yields

$$\int_0^\infty \frac{|\boldsymbol{p}'|^2 \mathrm{d}|\boldsymbol{p}'|}{(2\pi)^3 p_0'} \int \frac{\mathrm{d}^3 k'}{(2\pi)^3 k_0'} \delta^3(\boldsymbol{p}' + \boldsymbol{k}')\delta(p_0' + k_0' - \sqrt{s})$$
$$\approx \int_0^\infty \frac{p_0' \mathrm{d}p_0'}{(2\pi)^6 p_0'} \delta(2p_0' - \sqrt{s}) = \frac{1}{2(2\pi)^6} \quad . \tag{5.43}$$

Altogether the differential cross section reads ($\alpha = e^2/4\pi$)

$$\frac{\mathrm{d}\bar{\sigma}}{\mathrm{d}\Omega}(e^+e^- \to \mu^+\mu^-) \approx \frac{\alpha^2}{4s}\left\{(1+\cos^2\vartheta)\left[1 + \frac{1}{2}g_V'^2 R(s) + \frac{1}{16}(g_V'^2 + 1)^2 R(s)^2\right] + \cos\vartheta\left[R(s) + \frac{1}{2}g_V'^2 R(s)^2\right]\right\} \tag{5.44}$$

with $R(s)$ given by (5.36). The total cross section is

$$\bar{\sigma}(e^+e^- \to \mu^+\mu^-) = 2\pi \int_{-1}^{+1} \mathrm{d}(\cos\vartheta)\frac{\mathrm{d}\bar{\sigma}}{\mathrm{d}\Omega}$$
$$\approx \frac{4\pi\alpha^2}{3s}\left[1 + \frac{1}{2}g_V'^2 R(s) + \frac{1}{16}(g_V'^2 + 1)^2 R(s)^2\right] \quad . \tag{5.45}$$

Before 1989 the maximum available energy was considerably less than $s \ll M_Z$. So long as this is true, it is a good approximation to set

$$R(s) \approx -\frac{s}{M_Z^2 \sin^2 2\theta} \quad , \quad R(s)^2 \approx 0 \quad . \tag{5.46}$$

The differential cross section (5.44) then takes the approximate form

$$\frac{\mathrm{d}\bar{\sigma}}{\mathrm{d}\Omega} \approx \frac{\alpha^2}{4s}\left[(1+\cos^2\vartheta)\left(1 - \frac{g_V'^2 s}{2M_Z^2 \sin^2 2\theta}\right) - \frac{s}{M_Z^2 \sin^2 2\theta}\cos\vartheta\right] \quad , \tag{5.47}$$

with

$$g_V' = 1 - 4\sin^2\theta \ll 1 \tag{5.48}$$

for the experimentally derived value of the Weinberg angle (4.108). In (5.47) both terms containing a factor $1/M_Z^2$ originate from weak currents, the main correction to the electromagnetic scattering cross section being a term proportional to $\cos\vartheta$. In the range of forward scattering, that is, for $0 < \vartheta < \pi/2$, the scattering cross section is suppressed, whereas in the case of backward scattering, for $\pi/2 < \vartheta < \pi$, it is enhanced. Obviously, the neutral weak current causes an asymmetry of the angular distribution around $\vartheta = 90°$.

This asymmetry was first clearly observed in experiments at the accelerator PETRA at the DESY laboratory in Hamburg, Germany (Fig. 5.6). The full line represents the prediction of Glashow–Salam–Weinberg theory, whereas the dashed line shows the prediction of pure quantum electrodynamics, the so-called *Bhabha scattering*.[2] Although there are also small asymmetric contributions in the framework of quantum electrodynamics when higher-order Feynman graphs are considered, these effects are of the order of 1% and therefore much smaller than the effect of the neutral weak current (at $\sqrt{s} = 34$ GeV). On the other hand, experimental values of the vector and

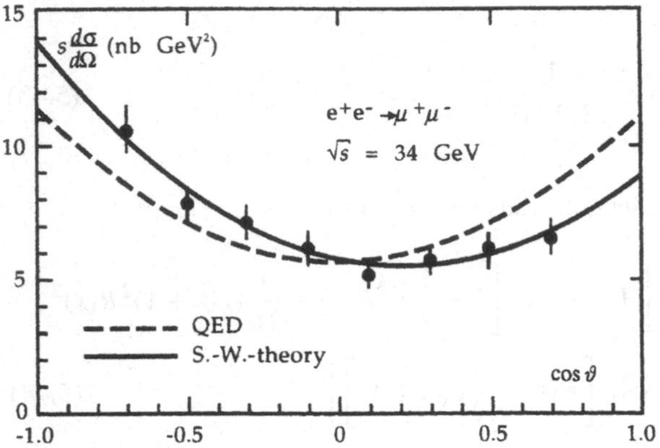

Fig. 5.6. Angular distribution of $e^+e^- \to \mu^+\mu^-$ scattering for a centre-of-mass energy $\sqrt{s} = 34$ GeV. The asymmetry corresponds to a Weinberg angle $\sin^2\theta \approx 0.25$ (*solid line*)

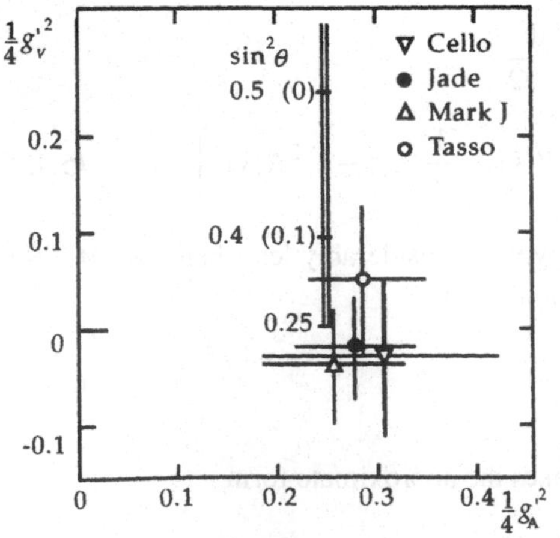

Fig. 5.7. Experimental values of vector and axial-vector coupling constants[3]

[2] A good survey of Bhabha scattering can be found in A. Scherdin, J. Reinhardt, W. Greiner, and B. Müller: Rep. Prog. Phys. **54**, 1 (1991) and in G. Salvini and A. Silverman: Phys. Rep. **171**, 231 (1988).

[3] M. Althoff et al. (TASSO collaboration), Z. Phys. C**22**, 13 (1984).

axial-vector coupling constants of the neutral weak current can be extracted from the measured angular asymmetry. As seen in Fig. 5.7, the measured values are in good agreement with the predictions of the Glashow–Salam–Weinberg theory

$$g'_V = 1 - 4\sin^2\theta \quad , \quad g'_A = 1$$

for $\sin^2\theta \approx 0.25$. The best value of the Weinberg angle determined in this way,

$$\sin^2\theta = 0.25 \pm 0.07 \quad , \tag{5.49}$$

is in good agreement with the value obtained from other experiments. Since for this value of θ we have $g'_V \approx 0$, the formula for the total cross section of muon pair production (5.45) attains the simple form

$$\bar{\sigma}(e^+e^- \to \mu^+\mu^-) \approx \frac{4\pi\alpha^2}{3s}\left[1 + \frac{1}{16\sin^4 2\theta}\frac{s^2}{(s - M_Z^2)^2}\right] \quad . \tag{5.50}$$

In the vicinity of $s = M_Z^2$ we expect a strong increase in the cross section. This is the typical sign of a resonance; in other words, in the process $e^+e^- \to Z^0 \to \mu^+\mu^-$ the intermediate Z boson acts like a resonance. In reality, the scattering cross section of course does not diverge at $s = M_Z^2$, because the Z boson itself decays and therefore has an intrinsic decay width Γ_Z. The correct expression of $R(s)$, replacing (5.36), is

$$R(s) = \frac{s}{\sin^2 2\theta[s - |M_Z - \frac{i}{2}\Gamma_Z|^2]} \quad . \tag{5.51}$$

In the scattering cross section a factor $|R(s)|^2$ occurs. After some calculation we obtain

$$\bar{\sigma} \approx \frac{4\pi\alpha^2}{3s}\left[1 + \frac{1}{16\sin^4 2\theta}\frac{s^2}{\left(s - M_Z^2 + \frac{\Gamma_Z^2}{4}\right)^2 + M_Z^2\Gamma_Z^2}\right] \quad . \tag{5.52}$$

At the point of resonance, $s = M_Z^2 - \frac{\Gamma_Z^2}{4}$, we have the cross section

$$\bar{\sigma}_{max} \approx \frac{4\pi\alpha^2}{3\Gamma_Z^2} \quad , \tag{5.53}$$

which depends on the decay width of the Z boson. This is illustrated in Fig. 5.8, where the ratio R between $\bar{\sigma}$ from (5.52) and the pure QED result $\bar{\sigma}_{QED} = 4\pi\alpha^2/3s$ is shown. R peaks at the value

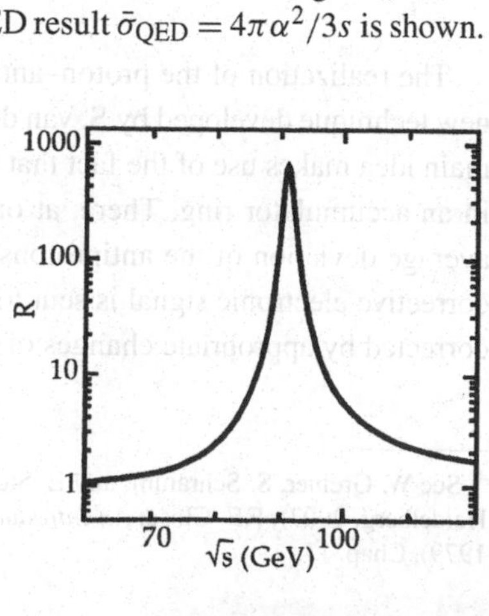

Fig. 5.8. Theoretical resonance curve of the Z boson in the case of electron–positron scattering

$$\frac{\bar{\sigma}_{\max}}{\sigma_{\mathrm{QED}}} \approx \frac{M_Z^2}{\Gamma_Z^2} \gg 1 \quad . \tag{5.54}$$

From a precise measurement of the resonance curve the mass and the width of the neutral boson can be determined with great accuracy. The experiments performed in 1989/90 at the CERN Large Electron–Positron Collider (LEP) and the Stanford Linear Collider (SLC) have confirmed the predictions of the GSW theory with unprecedented precision and constitute a new triumph of the "standard model" (see Example 5.3).

EXAMPLE

5.2 The Discovery of the Intermediate Vector Bosons

In 1977 Carlo Rubbia proposed converting the super proton synchrotron (SPS) at the European Centre for Nuclear Research (CERN), near Geneva, into a proton–antiproton storage ring. The basic principles of this type of storage ring have already been discussed in connection with the discovery of the τ lepton (Example 2.13). At CERN the total energy of the colliding particles was 540 GeV. As was known from other experiments,[4] half of the momentum of the protons is carried by its neutral constituents (gluons), the remaining half being distributed to the three constituent quarks (uud). Effectively, one would assume a total energy of the order of $\frac{1}{6} \times 540\,\mathrm{GeV} = 90\,\mathrm{GeV}$ per quark–antiquark collision. Therefore it should be possible to create a Z boson in a quark–antiquark collision, since the W or Z mass is predicted to be around 80–90 GeV by the Glashow–Salam–Weinberg theory, as we discussed before. The creation of a W boson should show up by its characteristic decay into a high-energy electron and neutrino, as shown in Fig. 5.9.

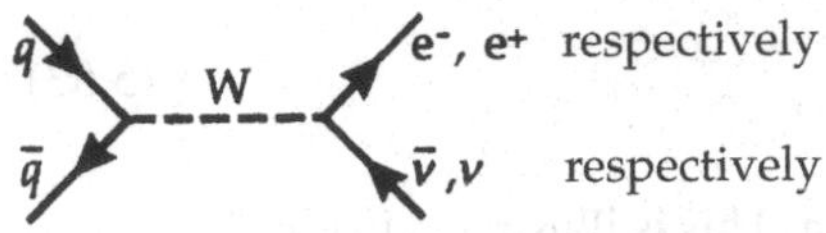

Fig. 5.9. Creation and decay of a W boson in a quark–antiquark collision

The realization of the proton–antiproton storage ring became possible thanks to a new technique developed by S. van der Meer, which is known as *stochastic cooling*. Its main idea makes use of the fact that the antiprotons produced in a target are collected in an accumulator ring. There, at one point (point "A" in Fig. 5.10) of the ring, the average deviation of the antiprotons from the desired trajectory is measured. Then a corrective electronic signal is sent to the other side of the ring where the beam can be corrected by appropriate changes of the field (position "B").

[4] See W. Greiner, S. Schramm, and E. Stein: *Quantum Chromodynamics*, 3rd ed. (Springer, Berlin, Heidelberg, 2007); F.E. Close: *An Introduction to Quarks and Partons* (Academic Press, New York, 1979), Chap. 11.

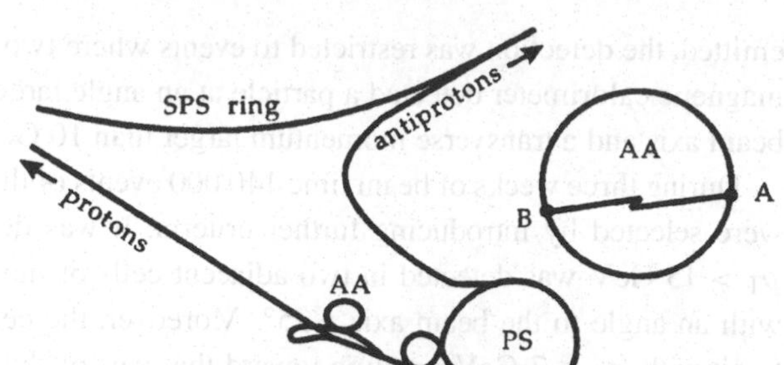

Fig. 5.10. Schematic sketc of the experimental setup prc ducing, collecting, and accel erating antiprotons and prc tons at the SPS collider c CERN

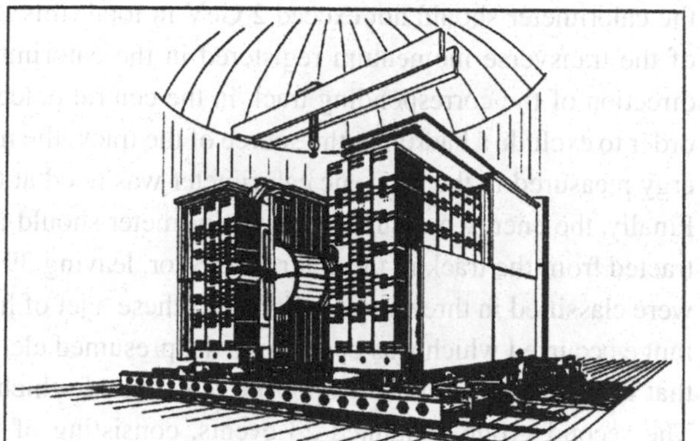

Fig. 5.11. The detector UA used for the search for the W and Z bosons

When the antiproton beam has a sufficient density, it is injected into the storage ring. The particle beams cross at two points of the storage ring; around these points two detectors are installed. The setup of the larger one, UA1, with dimensions 5 m × 5 m × 10 m (see Fig. 5.11), will be discussed in the following. Immediately surrounding the collision point is a central detector, consisting of a system of drift chambers. Around this detector, which produces a high-resolution image of all the tracks of charged particles, a strong magnetic field points in the direction of the particle beams. The momenta of the particles emerging from the reaction can now be determined from the curvature of their tracks, as observed in the central detector. The central detector is surrounded by an electromagnetic *calorimeter*, alternate layers of heavy material, where the particles lose energy, and layers of scintillator, which measure the energy loss. The electromagnetic calorimeter, in turn, is surrounded by a hadronic calorimeter. The outer part consists of a system of drift chambers detecting muons.

It is important that the calorimeter covers nearly the full (4π) solid angle, to allow detection of all produced electrons and hadrons and accurate determination of their energy. This is essential, because the neutrino created in the decay of the W boson cannot be detected but only determined indirectly by measuring a missing part in the balance of momenta of all other particles. In the search for the missing momenta, only the momentum components orthogonal to the beam direction were considered. Since no calorimeters could be installed within an angular range of 0.2° around the beam axis, particles which escape into that direction would distort the balance of momentum in the beam direction. Since the only events of interest were those where electrons with high transverse momentum p_T (that is, perpendicular to the beam direction) were

Example 5.2

emitted, the detection was restricted to events where two adjacent cells of the electromagnetic calorimeter detected a particle at an angle larger than 5° with respect to the beam axis and a transverse momentum larger than 10 GeV.

During three weeks of beam time 140 000 events of this kind were registered. They were selected by introducing further criteria. It was demanded that a particle with $p_T > 15$ GeV was detected in two adjacent cells of the electromagnetic calorimeter with an angle to the beam axis >25°. Moreover, the central detector should show a track with $p_T > 7$ GeV pointing toward that part of the calorimeter where the particle with $p_T > 15$ GeV was found. When these constraints were applied, 1106 events remained. The transverse momentum of all other tracks pointing to the same cells of the calorimeter should not exceed 2 GeV in total (this left 276 events). The direction of the transverse momentum registered in the calorimeter should coincide with the direction of the corresponding track in the central detector (167 events were left). In order to exclude a hadron as the source of the track, the maximum allowed value of energy measured in the hadronic calorimeter was fixed at 600 MeV (leaving 72 events). Finally, the energy measured in the calorimeter should agree with the momentum extracted from the track in the central detector, leaving 39 events. The remaining events were classified in three groups. For 11 of these a jet of hadrons within a small angular range occurred which was opposite to the presumed electron track. Here it is probable that in reality there are two opposite jets, one of which resembles an electron track. The second group contained 23 events, consisting of two jets where the measured electron was part of one jet or events where the electron clearly originates from the Dalitz decay $\pi^0 \to \gamma + e^+e^-$. It is important to note that for all 34 events of these two groups the balance of momenta was fulfilled within experimental accuracy, that is, the sum of the momenta of all measured particles was zero, as expected from momentum conservation.

The five remaining events exhibited no hadronic jets, and their balance of momentum was not reconciled. The missing momentum not registered by the detector (denoted as *missing energy* because the calorimeter measures the energy of the particles) was found to be opposite in direction to the electron track with great accuracy. The sole known explanation for such events is that the missing energy was carried away by one or more neutrinos that remain undetected, as it would be in the decay $W \to e\nu$. By fitting the angle and energy of the electron and the transverse energy of the neutrino it was possible to obtain a value for the mass of the W boson.

$$M_W = 81 \pm 5 \text{ GeV} \quad , \tag{1}$$

which is in good agreement with the prediction of the GSW theory. Extensive measurements at CERN, as well as at the Tevatron collider at the Fermi National Accelerator Laboratory (FNAL) near Chicago, have improved the accuracy of the mass determination considerably. A recent spectrum of the transverse component of the energy emitted in W decays is shown in the following figure. The high-energy peak corresponds to W bosons emitted perpendicular to the beam; its location determines M_W. At present (1993) the best value is:

$$M_W = 80.10 \pm 0.42 \text{ GeV} \quad . \tag{2}$$

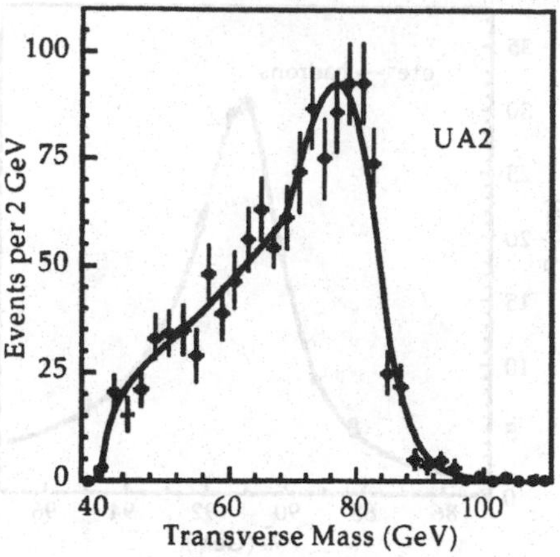

Fig. 5.12. Mass spectrum for the W particle, from the UA2 experiment at CERN's proton–antiproton collider. The solid line represents a fit to the data with $M_W = 80.49$ GeV

The Z boson was discovered with the same experimental technique, but now searching for events with two highly energetic electrons (e^+e^-) being emitted with opposite momenta. Although these events are much rarer in proton–antiproton collisions, a clear Z^0 signal was detected, yielding the mass

$$M_Z = 91.16 \text{ GeV} \quad . \tag{3}$$

The new precision measurements of the properties of the Z^0 boson are discussed in Example 5.3.

EXAMPLE

5.3 Precision Measurement of the Z^0 Boson

In the summer of 1989 two new electron–positron colliders started operation: the Stanford Linear Collider (SLC) and the Large Electron–Positron Collider (LEP) at CERN. Available beam energies of up to about 50 GeV allowed for detailed experimental investigation of the Z^0 resonance region in electron–positron scattering. Here we shall not discuss the technical details of the large detector systems employed in these studies, because they are based on similar principles as those used in the original discovery of the intermediate bosons at CERN (see Example 5.2). In total there are four detector collaborations at LEP (ALEPH, DELPHI, L3, and OPAL) and two at SLC. These groups have studied various aspects of the Z^0 resonance in the centre-of-mass energy region around $\sqrt{s} \approx 91$ GeV, in particular, the decays $Z^0 \to \mu^+\mu^-$, e^+e^- and $Z^0 \to$ hadrons (see Figs. 5.13, 5.14). The most significant early result was a precise determination of the mass and the total width of the Z boson:[5]

$$\begin{aligned} M_Z &= 91.16 \pm 0.03 \text{ GeV} \quad , \\ \Gamma_Z &= 2.492 \pm 0.025 \text{ GeV} \quad . \end{aligned} \tag{1}$$

[5] L3 collaboration, B. Adera et al.: Phys. Lett. **B231**, 509 (1989); ALEPH collaboration, D. Decamp et al.: Phys. Lett. **B231**, 519 (1989); OPAL collaboration, M.Z. Akrawy et al.: Phys. Lett. **B231**, 530 (1989); DELPHI collaboration, P. Aarino et al.: Phys. Lett. **B231**, 539 (1989).

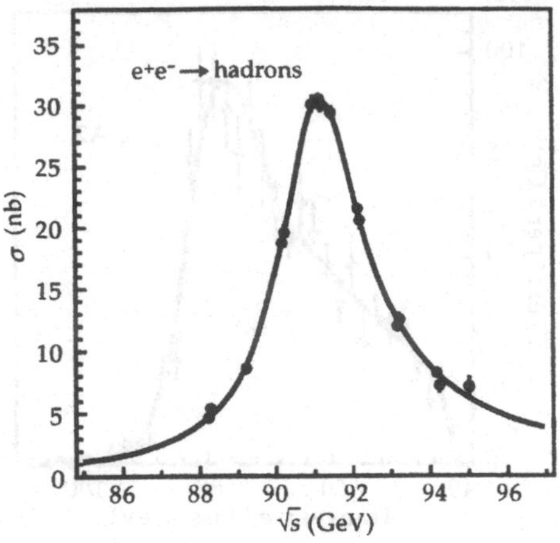

Fig. 5.13. The measured cross sections for $e^+e^- \to$ hadrons as a function of centre-of-mass energy $\sqrt{s}$. The *solid line* is a fit to a formula analogous to (5.52)

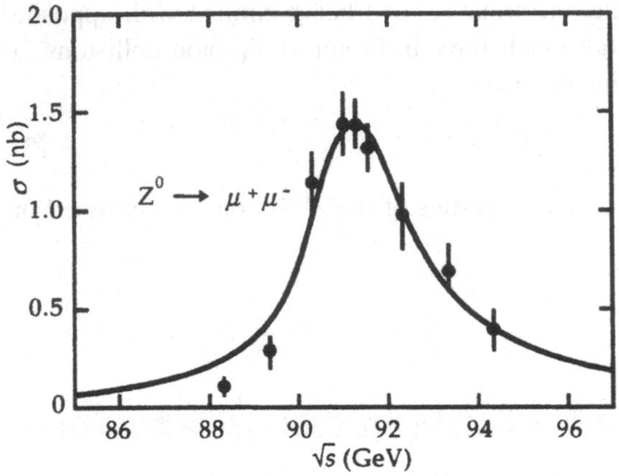

Fig. 5.14. Z_0 resonance in the reaction $e^+e^- \to \mu^+\mu^-$

The measured shape of the Z^0 resonance in the reaction $e^+e^- \to \mu^+\mu^-$ is shown in Fig. 5.14. M_Z and Γ_Z can be obtained from fits to the shape of the measured energy distribution in these reactions. The absolute yield in the various decay channels of the Z^0 boson also allows for a separate determination of the partial decay widths:[6]

$$\text{hadronic decays: } \Gamma_Z^{\text{had}} = 1748 \pm 35\ \text{MeV} \quad , \tag{2}$$

$$\text{charged lepton decays: } \Gamma_Z^{\ell\ell} = 83 \pm 2\ \text{MeV} \quad . \tag{3}$$

The difference between the total width and the combined measured widths must then be attributed to decays of the Z^0 into neutrinos, $Z^0 \to \nu\bar{\nu}$:

$$\Gamma_Z^{(\nu)} = \Gamma_Z - \Gamma_Z^{\text{had}} - 3\Gamma_Z^{\ell\ell} = 494 \pm 32\ \text{MeV} \quad . \tag{4}$$

In Exercise 5.1, (a), we derived the decay width for the decay $Z^0 \to \nu\bar{\nu}$:

$$\Gamma_Z^{\nu\bar{\nu}} = 166\ \text{MeV} \quad . \tag{5}$$

[6] B. Adeva et al. [L3 Collaboration]: Phys. Lett. **B249** (1990) 341–352. A more recent determination of the Z^0 boson resonance parameters is given by The ALEPH Collaboration et al.: Phys. Rept. **427** (2006) 257; e-Print: hep-ex/0509008.

Comparison with (4) leads to a *measurement of the number of neutrinos*:

$$N_\nu = \Gamma_Z^{(\nu)}/\Gamma_Z^{\nu\nu} = 2.96 \pm 0.11 \quad .$$

This is a very important result, because it states that *there are at most three species of low-mass neutrinos in nature* (ν_e, ν_μ, ν_τ). This result is in full agreement with simulations of the evolution of the early universe, which yield the limit $N_\nu \leq 4$. It is assumed that the still unknown neutrinos have masses smaller than half the mass of the Z^0. Neutrinos with larger masses have no effect on Γ_Z because the energy of the $\nu\bar{\nu}$ pair into which the Z^0 might decay has to be smaller than M_Z. A complete table of experimental results from LEP for the Z^0 parameters is shown below.

Table 5.1. Comparison of the experimental results (L3 detector) with standard model predictions

	Experiment	Prediction
M_Z (GeV)	$91.161 \pm 0.13 \pm 0.3$	–
Γ_Z (GeV)	2.492 ± 0.025	2.492
$\Gamma_{\ell\ell}$ (GeV)	0.0832 ± 0.0015	0.0838
Γ_{had} (GeV)	1.748 ± 0.035	1.740
$\Gamma_{\nu\bar{\nu}}$ (GeV)	0.494 ± 0.032	0.501
$\Gamma_{had}/\Gamma_{\ell\ell}$	21.02 ± 0.62	20.77

The agreement between the data and the predictions of the GSW theory (standard model) was so good that it became possible to set limits on the masses of the then two still undetected particles, the top quark (m_t) and the Higgs boson (M_H). Experimental limits for the masses of these two particles, from direct searches at CERN and FNAL, are

$$m_t \geq 89 \text{ GeV} \quad , \tag{6a}$$

$$M_H \geq 41.6 \text{ GeV} \quad . \tag{6b}$$

The reason the data are sensitive to the masses of these particles is that they contribute to the predicted mass values of the W and Z bosons through vacuum polarization effects, such as are indicated by the following Feynman diagrams:

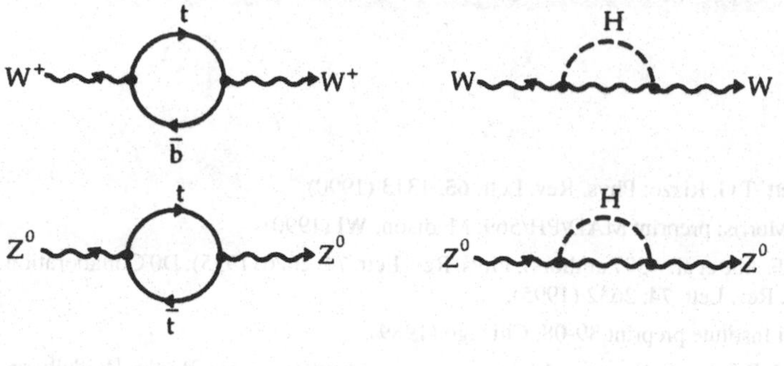

The magnitude of the contributions from these diagrams depends on the masses of the virtual particles in the loops. Because m_t enters in a different way into the W

Example 5.3

loop (once) and the Z loop (twice), the value of M_W/M_Z is especially sensitive to it. A detailed analysis[7] showed that the experimental value

$$M_W/M_Z = 0.8801 \pm 0.0037 \tag{7}$$

leads to an upper bound

$$m_t \leq 200\ \text{GeV} \tag{8}$$

with 95% confidence. The most likely value is around $m_t \approx 150$ GeV. Remarkably, an estimate of m_t can also be obtained[8] from the precise measurement of the muon lifetime τ_μ (2.61). The theoretical prediction for τ_μ is again sensitive to m_t through radiative corrections, giving a value $m_t = 189 \pm 47$ GeV, Meanwhile, in 1995 the top quark was unambiguously identified experimentally by two collaborations at the Tevatron (Fermi Laboratory near Chicago) and the following masses m_t were reported:[9]

$$\begin{aligned} m_t &= 176.0 \pm 6.5\ \text{GeV (CDF)} \quad , \\ m_t &= 172.1 \pm 7.1\ \text{GeV (D0)} \end{aligned} \tag{9}$$

For the Higgs boson the data pose less severe restrictions, because its coupling to the W and Z bosons is universal. Nonetheless, fits to the data with different values of M_H point towards a Higgs mass below about 350 GeV. The predicted large mass of the top quark gives rise to the fascinating speculation[10] that the Higgs field may actually be an effective description of a *top-quark condensate*, in much the same way as the London wavefunction of a superconductor effectively describes the distribution of Cooper pairs. The speculation is based on the observation that the coupling between the top quark and the Higgs field is predicted to be (see (4.143))

$$f_t = \frac{m_t}{\lambda} \approx \frac{189\ \text{GeV}}{246\ \text{GeV}} = 0.77 \quad ,$$

that is, the coupling strength is of the order of 1. For such couplings a top-quark–antiquark state would be supercritically bound,[11] and the top-quark vacuum would contain an infinite number of such $(t\bar{t})$ bound states. This "condensate" of $(t\bar{t})$ pairs acts very much like a Higgs field, giving mass to the W and Z bosons.[12]

---

[7] V. Barger, J.L. Hewett, T.G. Rizzo: Phys. Rev. Lett. **65**, 1313 (1990).

[8] F. Halzen and D.A. Morris: preprint MAD/PH/569, Madison, WI (1990).

[9] CDF Collaboration (F. Abe et al. – 397 authors), Phys. Rev. Lett. **74**, 2626 (1995); D0 Collaboration (S. Abachi et al.) Phys. Rev. Lett. **74**, 2632 (1995).

[10] Y. Nambu: E. Fermi Institute preprint 89-08, Chicago (1989).

[11] See W. Greiner and J. Reinhardt: *Quantum Electrodynamics*, 4th ed. (Springer, Berlin, Heidelberg, 2009), Chap. 7, for a discussion of supercritical binding.

[12] W.A. Bardeen, C.T. Hill, M. Lindner: Phys. Rev. **D41**, 1647 (1990); X.Y. Pham: Phys. Lett. **B241**, 111 (1990).

5.3 High-Energy Behavior of the GSW Theory

Apart from the experimentally proven existence of neutral weak currents the high-energy behavior gave a reason to reject the V–A Fermi interaction as the fundamental theory of weak interactions. In the previous section we studied some effects of the intermediate neutral boson in electron–positron annihilation. Now we discuss the behavior of the Glashow–Salam–Weinberg theory at very high energies. In particular we shall investigate whether the quadratic increase of the scattering cross section with respect to the centre-of-mass energy $E = \sqrt{s}$ is really damped as we had hoped. Furthermore, we have to discuss whether divergences occurring in calculations of higher-order processes are softened in a way that they can be compensated by renormalization of masses and coupling constants. We start with the high-energy behavior. Here 'high energy' implies that the scattering energy in the centre-of-mass frame $\sqrt{s}$ is considerably larger than the masses of the intermediate bosons, that is

$$\sqrt{s} \gg 100\ \text{GeV} \quad .$$

A first hint at the behavior of scattering cross sections at these energies is given by (5.45), which describes the process $e^+e^- \to \mu^+\mu^-$. According to (5.36), in the high-energy limit we have

$$\lim_{s\to\infty} R(s) = \frac{1}{\sin^2 2\theta} \ , \tag{5.55}$$

and therefore

$$\bar{\sigma}(e^+e^- \to \mu^+\mu^-) \underset{s\to\infty}{\longrightarrow} \frac{4\pi\alpha^2}{3s}\left(1 + \frac{g_V'^2}{2\sin^2 2\theta} + \frac{(g_V'^2+1)^2}{16\sin^4 2\theta}\right) \quad . \tag{5.56}$$

Thus we see that the cross section decreases asymptotically like $1/s$. The obvious explanation is given by the asymptotic behavior of $R(s)$, which approaches a constant instead of increasing linearly with s, as is the case for $s \ll M_Z^2$ (cf. (5.36)).

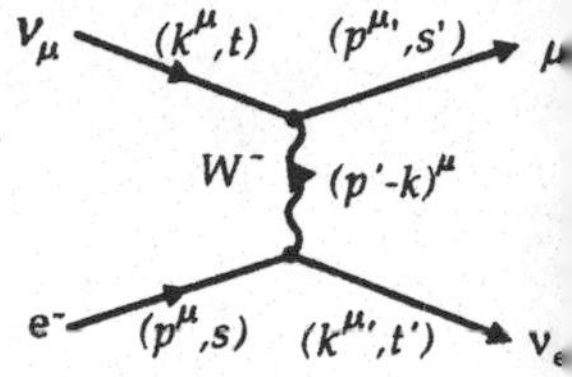

Fig. 5.15. Muon–neutrin[o] electron scattering through exchange of the intermediate bo[son] W⁻

In Sect. 3.4 we based our argumentation mainly on electron–neutrino scattering mediated by charged currents. Now we study the same process once more in the framework of gauge theory, especially considering the behavior with respect to s. We concentrate our discussion on the process $\nu_\mu e^- \to \mu^- \nu_e$, because in a purely electromagnetic process the neutral current also contributes (cf. the remark at the end of Exercise 3.3). The corresponding Feynman diagram is shown in Fig. 5.15, whereas two diagrams are possible in the process $\nu_e e^- \to e^- \nu_e$ (Fig. 5.16), whose contributions must be added coherently. The scattering $\nu_\mu \to \mu^- \nu_e$ through W-boson exchange is described by the first two terms of the Lagrangian (4.139). The term

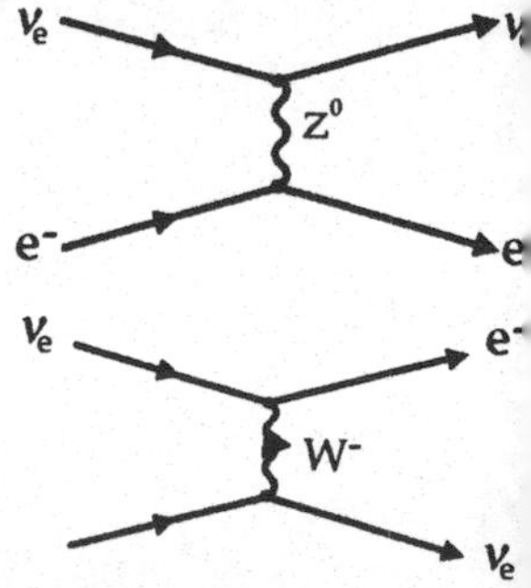

Fig. 5.16. Two diagrams describing lowest-order $\nu_e e^-$ scattering

$$-\frac{g}{2\sqrt{2}} \bar{\psi}_\mu(x)\gamma_\mu^\alpha(1-\gamma_5)\psi_{\nu_\mu}(x) W_\alpha^{(-)}(x) \tag{5.57}$$

describes the conversion of a muon neutrino into a negative muon absorbing a W^- boson, whereas

$$\frac{-g}{2\sqrt{2}} \bar{\psi}_{\nu_e}(x)\gamma^\alpha(1-\gamma_5)\psi_e(x) W_\alpha^{(+)}(x) \tag{5.58}$$

implies the transition of an electron into an electron neutrino, emitting a W^- boson. Applying the usual Feynman rules we find that the scattering matrix element corresponding to the diagram in Fig. 5.16 reads

$$S(\nu_\mu e^- \to \mu^- \nu_e) = \frac{(2\pi)^4 \delta^4(p' + k' - p - k)}{4V^2 (p_0 k_0 p'_0 k'_0)^{1/2}} A \tag{5.59}$$

with the invariant amplitude

$$\begin{aligned} A = &-\mathrm{i}\frac{g^2}{8} \bar{u}_\mu(p',s') \mathrm{i}\gamma^\alpha (1-\gamma_5) u_{\nu_\mu}(k,t) \left(\frac{g_{\alpha\beta} - (p'-k)_\alpha (p'-k)_\beta / M_W^2}{(p'-k)^2 - M_W^2} \right) \\ &\times \bar{u}_\nu(k',t') \mathrm{i}\gamma^\beta (1-\gamma_5) u_e(p,s) \quad . \end{aligned} \tag{5.60}$$

For low energies the propagator of the W boson can be approximated by $-g_{\alpha\beta}/M_W^2$, and the matrix element reduces to that in Exercise 3.3, (3), (4), since, owing to (4.144), we have

$$\frac{g^2}{8M_W^2} = \frac{G}{\sqrt{2}} \quad .$$

However, here we are interested in the high-energy behavior. We therefore simplify the calculation by neglecting all terms which include powers of the leptonic masses m_e and m_μ. The squared matrix element (5.60) is averaged over the initial spins (note that the neutrino has only a single spin direction!) and summed over the final spins, yielding

$$\begin{aligned} \frac{1}{2} \sum_{s,s',t,t'} |A|^2 = &\frac{g^4}{128} \left(g_{\alpha\beta} - \frac{(p'-k)_\alpha (p'-k)_\beta}{M_W^2} \right) \left(g_{\bar{\alpha}\bar{\beta}} - \frac{(p'-k)_{\bar{\alpha}} (p'-k)_{\bar{\beta}}}{M_W^2} \right) \\ &\times \frac{1}{[(p'-k)^2 - M_W^2]^2} \mathrm{Tr}\{(\not{p}' + m_\mu)\gamma^\alpha (1-\gamma_5) \not{k} \gamma^{\bar{\alpha}} (1-\gamma_5)\} \\ &\times \mathrm{Tr}\{\not{k}' \gamma^\beta (1-\gamma_5)(\not{p} + m_e) \gamma^{\bar{\beta}} (1-\gamma_5)\} \quad . \end{aligned} \tag{5.61}$$

According to what we said above we neglect the mass terms in the Dirac traces. Within this approximation the contribution proportional to $(p'-k)$ in the numerators of the boson propagators also vanishes. This is analogous to our consideration in (5.28) ($\not{p}'^2 = p'^2 = m_\mu^2 \approx 0$, $k^2 = k'^2 = 0$) and one obtains

$$\begin{aligned} &(p'-k)_\alpha \mathrm{Tr}\{(\not{p}' \gamma^\alpha (1-\gamma_5) \not{k} \gamma^{\bar{\alpha}} (1-\gamma_5)\} \\ &= \mathrm{Tr}\{p'^2 (1-\gamma_5) \not{k} \gamma^\alpha (1-\gamma_5) - \not{p}' (1-\gamma_5) k^2 \gamma^{\bar{\alpha}} (1-\gamma_5)\} \\ &= m_\mu^2 \mathrm{Tr}\{(1-\gamma_5) \not{k} \gamma^{\bar{\alpha}} (1-\gamma_5)\} \approx 0 \quad , \end{aligned} \tag{5.62}$$

and so on. Thereby (5.61) is simplified considerably to

$$\begin{aligned} \frac{1}{2} \sum_{s,s',t,t'} |A|^2 &\approx \frac{g^4}{128[(p'-k)^2 - M_W^2]^2} \mathrm{Tr}\{\not{p}' \gamma^\alpha (1-\gamma_5) \not{k} \gamma^\beta (1-\gamma_5)\} \\ &\quad \times \mathrm{Tr}\{\not{k}' \gamma_\alpha (1-\gamma_5) \not{p} \gamma_\beta (1-\gamma_5)\} \\ &= 2g^4 \frac{(k \cdot p)(p' \cdot k')}{[(p'-k)^2 - M_W^2]^2} \quad , \end{aligned} \tag{5.63}$$

where we have adopted the result of the trace calculations from Exercise 3.3, (8). Furthermore, the high-energy limit implies that

$$s = (k+p)^2 = k^2 + p^2 + 2(k\cdot p) \approx 2(k\cdot p) \quad ,$$
$$s' = (k'+p')^2 = m_\mu^2 + 2(k'\cdot p') \approx 2(k'\cdot p') \quad , \tag{5.64}$$

and

$$(p'-k)^2 = m_\mu^2 - 2(p'\cdot k) \approx -2(p'\cdot k) \quad . \tag{5.65}$$

As in Exercise 3.2, making use of four-momentum conservation, we get

$$\begin{aligned}(p'\cdot k) &= (p+k-k')\cdot k = (p\cdot k) - (k'\cdot k)\\ &\approx \frac{s}{2} - k_0 k_0'(1-\cos\vartheta) \quad ,\end{aligned} \tag{5.66}$$

where ϑ is the scattering angle of the neutrino. Neglecting rest masses in the centre-of-mass frame, each particle energy being $k_0 \approx p_0 \approx k_0' \approx p_0' \approx \frac{1}{2}\sqrt{s}$, and furthermore with

$$(p'\cdot k) \approx \frac{1}{4}s(1+\cos\vartheta) \quad , \tag{5.67}$$

we obtain as the result

$$\frac{1}{2}\sum_{s,s',t,t'} |A|^2 \approx \frac{2g^4 s^2}{[s(1+\cos\vartheta)+2M_W^2]^2} \quad . \tag{5.68}$$

Finally we integrate over the phase space of the outgoing particles. The considerations leading to the averaged differential cross section are analogous to our previous discussion. In the centre-of-mass frame ($\boldsymbol{p}+\boldsymbol{k}=\boldsymbol{0}$, $p_0+k_0=\sqrt{s}$), using Exercises 3.2, (5) and 3.1, (3)–(5), as well as $[(2\pi)^4\delta^4(p'+k'-p-k)]^2 = VT(2\pi)^4\delta^4(p'+k'-p-k)$, we get

$$\begin{aligned}\frac{d\bar\sigma}{d\Omega}(\nu_\mu e^- \to \mu^-\nu_e) &= \frac{V^2}{(2\pi)^6}\int d^3p' \int_0^\infty |\boldsymbol{k}'|^2 d|\boldsymbol{k}'| \frac{k_0 p_0 V}{(k\cdot p)T}\frac{1}{2}\sum_{s,s',t,t'}|S|^2\\ &= \frac{1}{64\pi^2}\int\frac{d^3p'}{p_0'}\int_0^\infty \frac{|\boldsymbol{k}'|^2 d|\boldsymbol{k}'|}{k_0'(k\cdot p)}\delta^4(p'+k'-p-k)\frac{1}{2}\sum_{s,s',t,t'}|A|^2\\ &= \frac{g^4}{16\pi^2}\frac{s}{[s(1+\cos\vartheta)+2M_W^2]^2}\int\frac{d^3p'}{p_0'}\int_0^\infty k_0' dk_0' \delta^3(\boldsymbol{p}'+\boldsymbol{k}')\\ &\quad\times\delta(p_0'+k_0'-\sqrt{s}) \quad ,\end{aligned} \tag{5.69}$$

where we have used (5.59), (5.64), (5.68) and the relation $VT = (2\pi)^4\delta^4(0)$. Remember also that $\int d^3k'/(2\pi)^3 = \int |\boldsymbol{k}'|^2 d|\boldsymbol{k}'| d\Omega/(2\pi)^3$. Furthermore we have taken $k_0' = |\boldsymbol{k}'|$ owing to the vanishing mass of the neutrino. The integration over p' is removed by the spatial δ^3 function. What we are left with is

$$\int_0^\infty \frac{k_0' \mathrm{d}k_0'}{\sqrt{k_0'^2 + m_\mu^2}} \delta\left(\sqrt{k_0'^2 + m_\mu^2} + k_0' - \sqrt{s}\right) = \frac{s - m_\mu^2}{2s} \approx \frac{1}{2} \quad , \tag{5.70}$$

by analogy with Exercise 3.3, (10) and by using

$$\int A(x)\delta(f(x))\mathrm{d}x = \frac{A(x_0)}{\left.\frac{\mathrm{d}f}{\mathrm{d}x}\right|_{x=x_0}} \quad ,$$

x_0 being the argument for which $f(x_0) = 0$.

For scattering energies well above the muon mass we thus have

$$\frac{\mathrm{d}\bar{\sigma}}{\mathrm{d}\Omega}(\nu_\mu \mathrm{e}^- \to \mu^- \nu_\mathrm{e}) \approx \frac{g^4}{32\pi^2} s\left[s(1 + \cos\vartheta) + 2M_\mathrm{W}^2\right]^{-2} \quad . \tag{5.71}$$

We have to compare this result with that of Fermi theory, Exercise 3.3, (11),

$$\frac{\mathrm{d}\bar{\sigma}}{\mathrm{d}\Omega}(\nu_\mu \mathrm{e}^- \to \mu^- \nu_\mathrm{e}) \approx \frac{G^2}{4\pi^2} s \quad . \tag{5.72}$$

At energies where we can neglect s compared to M_W^2, (5.71) together with relation $G/\sqrt{2} = g^2/8M_\mathrm{W}^2$, (4.144), can be reduced to the following:

$$\frac{\mathrm{d}\bar{\sigma}}{\mathrm{d}\Omega} \to \frac{g^4}{128\pi^2 M_\mathrm{W}^4} s = \frac{G^2}{4\pi^2} s \quad (s \ll M_\mathrm{W}^2) \quad .$$

In the case of asymptotically high energies, however, the additional s dependence results in a considerably smaller increase in the scattering cross section. This can be seen most clearly by considering the total cross section,

$$\begin{aligned} \bar{\sigma} &= 2\pi \int_{-1}^{+1} \mathrm{d}(\cos\vartheta) \frac{\mathrm{d}\bar{\sigma}}{\mathrm{d}\Omega} \approx \frac{g^4}{16\pi} \int_{-s}^{s} \mathrm{d}x [x + s + 2M_\mathrm{W}^2]^{-2} \\ &= \frac{g^4}{16\pi} \left[\frac{-1}{x + s + 2M_\mathrm{W}^2}\right]_{x=-s}^{x=s} = \frac{g^4 s}{(32\pi M_\mathrm{W}^2)(s + M_\mathrm{W}^2)} = \frac{G^2 s M_\mathrm{W}^2}{\pi(s + M_\mathrm{W}^2)} \quad . \end{aligned} \tag{5.73}$$

For $s \ll M_\mathrm{W}^2$ this accords with (3.33), the result of Fermi theory; at energies where $s \gg M_\mathrm{W}^2$ the cross section approaches a constant value,

$$\frac{1}{\pi}(GM_\mathrm{W})^2 \approx 3 \times 10^{-35}\ \mathrm{cm}^2 = 30\ \mathrm{pb} \quad .$$

That is, even at extremely high energies the cross section remains relatively small compared to the total cross section of proton–antiproton scattering of around 60 mb at $\sqrt{s} = 540$ GeV. Thus we conclude that the weak interaction described by the Glashow–Salam–Weinberg theory, remains 'weak' even at asymptotically high energies!

As we know from Sect. 3.4 the comparison of total cross sections is not the decisive criterion, because, apart from the interaction strength, its range also enters. The cross section of proton–antiproton scattering is larger than that of neutrino scattering, partly because the range of strong interactions is of the order of 1 fm, whereas in the Salam–Weinberg theory the range of the weak interaction is given by the Compton wavelength of the intermediate bosons,

$$M_Z^{-1}, M_W^{-1} \approx 0.0025 \text{ fm} \quad . \tag{5.74}$$

The real criterion for whether the theory is consistent or not requires us to check whether a partial wave exceeds the unitarity limit, as discussed in Sect. 3.4. The calculation performed in Exercise 5.4 yields the partial cross section of an s wave,

$$\sigma_0 = \frac{G^2 M_W^4}{\pi s}\left[\ln\left(1+\frac{s}{M_W^2}\right)\right]^2 \quad , \tag{5.75}$$

whereas according to (3.38) the unitarity limit yields the value

$$\sigma_0^{\text{lim}} = \frac{4\pi}{s} \quad . \tag{5.76}$$

Obviously, even in the framework of the Salam–Weinberg theory, this limit may be exceeded (within the Born approximation), but only at ultra-high energies,

$$\sqrt{s} \approx M_W \exp\left(\frac{\pi}{G M_W^2}\right) \approx M_W \exp\left(\frac{\sqrt{2}\sin^2\theta}{\alpha}\right) \approx 7\times 10^8 \text{ GeV} \quad . \tag{5.77}$$

Here the relations (4.181) were utilized, i.e. $g\sin\theta = e$, $M_W = \frac{1}{2}g\lambda$, $G/\sqrt{2} = g^2/8M_W^2 = 1/2\lambda^2$. This corresponds to a laboratory energy of the incoming neutrino of

$$E_\nu^{\text{lab}} = \frac{s}{2m_e} = 5\times 10^{20} \text{ GeV} \quad . \tag{5.78}$$

This relation follows from $(k+p)^2 \equiv s = k^2 + p^2 + 2k\cdot p = p^2 + 2k\cdot p \approx 2k\cdot p = 2(k_0 p_0 - \boldsymbol{k}\cdot\boldsymbol{p})$. In the laboratory the electron is at rest ($\boldsymbol{p} = 0$) and therefore $p_0 = m_e$. The energy (5.78) is far above energies attainable in the near future. However, in principle, centre-of-mass energies of the order of 10^9 GeV may become accessible one day. They certainly occurred in collisions of elementary particles in the plasma shortly after the big bang of the universe. In this regime the violation of the unitary limit within the Born approximation requires further study of scattering processes, as discussed in Sect. 3.4. The Feynman graphs of some of the involved processes are shown in Fig. 5.17. If one applies the theory of renormalization (see Chap. 9), it can be shown that the main effect of the higher-order Feynman graphs can be described by replacing the coupling constant g^2 by an effective 'running' coupling constant

$$g_{\text{eff}}^2(s) \approx \frac{g^2}{C\ln(s/M_W^2)} \quad (s\to\infty) \quad , \tag{5.79}$$

where C is a numerical constant depending on the number of fundamental particles which interact weakly (leptons, quarks, and so on). From (5.75), expressing the Fermi coupling constant G by the gauge coupling g, we get

$$\sigma_0 \approx \frac{g_{\text{eff}}^4}{32\pi s}\left(\ln\frac{s}{M_W^2}\right)^2 \approx \frac{g^4}{32\pi c^2 s} = \frac{\pi\alpha^2}{2\sin^4\theta c^2 s} \ll \frac{4\pi}{s} \tag{5.80}$$

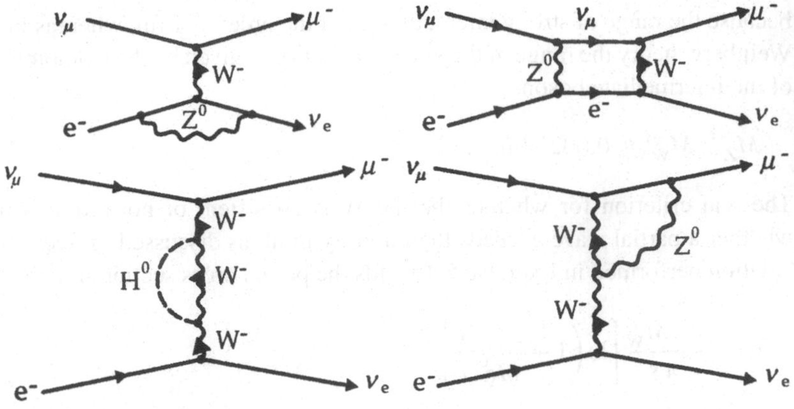

ig. 5.17. Typical higher-order processes of $\nu_\mu e^- \to \mu^- \nu_e$ scattering

at very high energies. Thus the incorporation of higher-order processes removes the difficulties of the Salam–Weinberg theory concerning the unitarity limit.

EXERCISE

5.4 The s-Wave Contribution to Lepton–Neutrino Scattering

Problem. Show that (5.75) describes the s-wave ($l = 0$) contribution to the total scattering cross section of the process $\nu_\mu e^- \to \mu^- \nu_e$.

Solution. For scattering without spin flip (the neutrino is always left-handed!) the differential cross section can be written in the form

$$\frac{d\sigma}{d\Omega} = \frac{1}{s}\left|\sum_{l=0}^{\infty}\left(\frac{4\pi}{2l+1}\right)^{1/2} T_l Y_{l0}(\vartheta)\right|^2 , \tag{1}$$

according to (3.34), (3.37). Comparing (1) with the averaged cross section (5.71) yields

$$\frac{g^2 s e^{i\delta}}{4\sqrt{2}\pi[s(1+\cos\vartheta)+2M_W^2]} = \sum_{l=0}^{\infty}\left(\frac{4\pi}{2l+1}\right)^{1/2} T_l Y_{l0} \; ; \tag{2}$$

$e^{i\delta}$ being an unknown phase factor. We obtain the s-wave part ($l = 0$) by multiplying both sides with $Y_{00} = 1/\sqrt{4\pi}$ and integrating over the full solid angle, exploiting the orthogonality of the spherical harmonics:

$$\begin{aligned}
\sqrt{4\pi}\, T_0 &= \int d\Omega\, Y_{00} \sum_{l=0}^{\infty}\left(\frac{4\pi}{2l+1}\right)^{1/2} T_l Y_{l0}(\vartheta) \\
&= \frac{g^2 s \;\; e^{i\delta}}{4\sqrt{2}\pi} 2\pi \int_{-1}^{+1} \frac{1}{\sqrt{4\pi}} \frac{d(\cos\vartheta)}{s(1+\cos\vartheta)+2M_W^2} \\
&= \frac{g^2}{4\sqrt{2}\pi} e^{i\delta} \int_{-s}^{s} \frac{dx}{x+s+2M_W^2} \\
&= \frac{g^2}{4\sqrt{2}\pi} e^{i\delta} \ln\frac{s+M_W^2}{M_W^2} \; .
\end{aligned} \tag{3}$$

It follows that

$$|T_0|^2 = \frac{g^4}{128\pi^2}\left[\ln\left(1+\frac{s}{M_W^2}\right)\right]^2$$
$$= \frac{G^2 M_W^4}{4\pi^2}\left[\ln\left(1+\frac{s}{M_W^2}\right)\right]^2 \quad , \tag{4}$$

and this together with relation (3.38) provides the desired result.

$$\sigma_0 = \frac{4\pi}{s}|T_0|^2 \quad . \tag{5}$$

5.4 Biographical Notes

BHABHA, Homi Jehangir, physicist, *30.10.1909 in Bombay (India), †24.1.1966. He received his Ph.D. in Cambridge in 1932, then became professor in Bangalore and, since 1945, director of the Tata Institute in Bombay. Bhabha made many contributions to the physics of cosmic rays, where he together with Heitler developed the theory of particle cascades.

GLASHOW, Sheldon, theoretical physicist, *5.12.1932 in New York. He received the Ph.D. at Harvard University in 1958, where he is also professor since 1966. Glashow shared in the 1979 Nobel Prize in physics with Salam and Weinberg for his prediction of weak neutral currents and his contribution to the development of a unified theory of weak and electromagnetic interactions. He also predicted, with Iliopoulos and Maiani, the existence of the charmed quark.

6 Semi-Leptonic Interactions of Hadrons

6.1 The World of Hadrons

All strongly interacting particles are called *hadrons*. One distinguishes *baryons* (baryon number $B = \pm 1$), which are fermions and carry spin $\frac{1}{2}, \frac{3}{2}, \dots$, and *mesons* (baryon number $B = 0$), which always have integer spin. The lightest hadrons, with equal spin (and equal parity), can be arranged in simple multiplets, where two further quantum numbers serve as order criteria: the *isospin* T, and its third component T_3, and the *strangeness* S, or alternatively the so-called strong *hypercharge* $Y = B + S$. These quantum numbers are characterized by the fact that they are exactly conserved under strong interactions. Conservation of strangeness is broken by weak interactions, which leads to decays of, for example, the Λ particle. The most important multiplets[1] are depicted in Figs. 6.1–6.4.

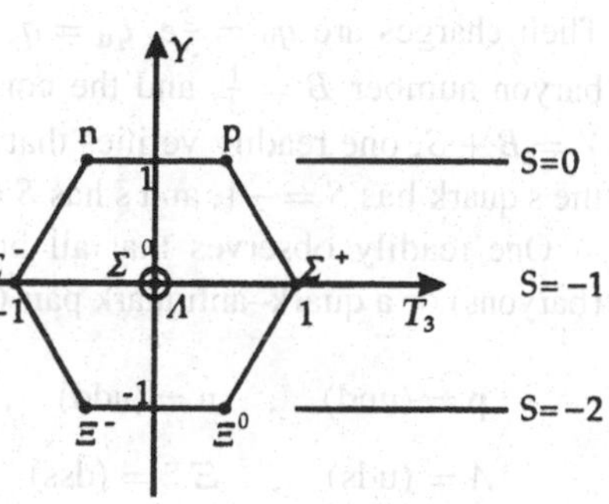

Fig. 6.1. Octet of spin-
baryons. All particles carr
baryon number $B = 1$

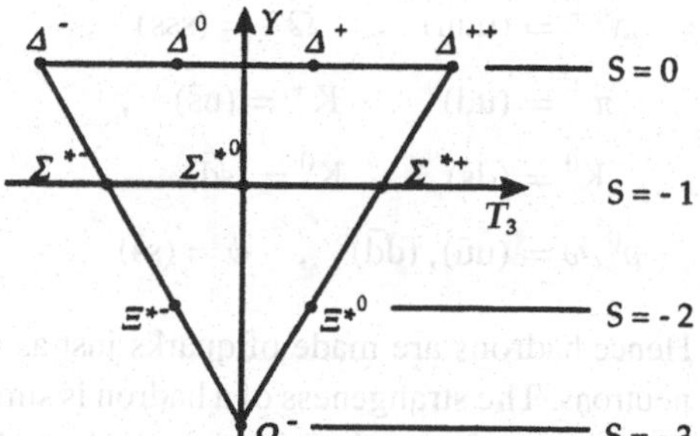

Fig. 6.2. Decuplet of spin-
baryons. All particles carr
baryon number $B = 1$

[1] See W. Greiner and B. Müller: *Quantum Mechanics – Symmetries*, 2nd ed. (Springer, Berlin, Heidelberg, 1994).

W. Greiner, B. Müller, *Gauge Theory of Weak Interactions*,
DOI 10.1007/978-3-540-87843-8_6, © Springer-Verlag Berlin Heidelberg 2009

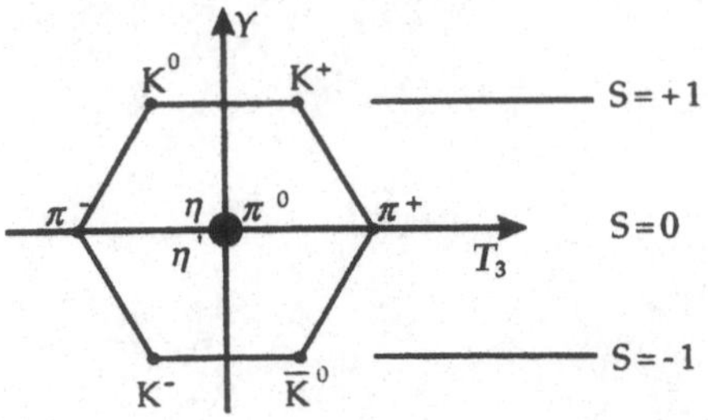

Fig. 6.3. Nonet of spin-0 mesons (pseudoscalar mesons). All mesons carry baryon number $B = 0$

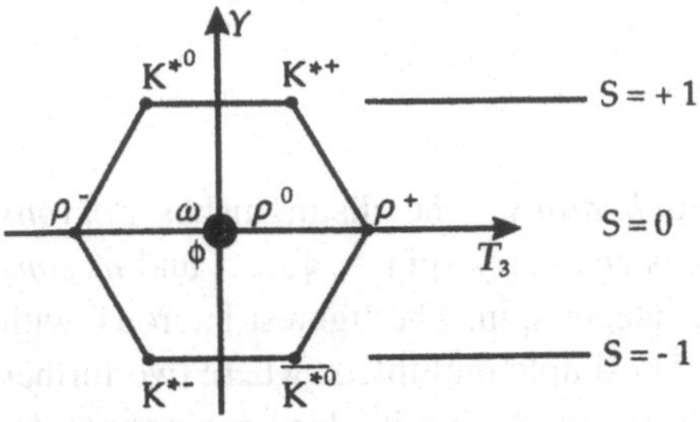

Fig. 6.4. Nonet of spin-1 mesons (vector mesons)

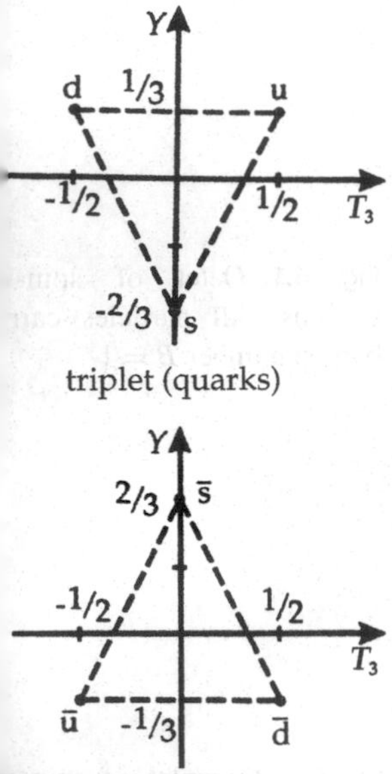

Fig. 6.5. The fundamental representations of SU(3)

For each baryonic multiplet there exists a corresponding multiplet of antiparticles, which is obtained from the former by reflection at the origin. These multiplets are just the eigenvalues belonging to the simple representations of the group SU(3). *Gell-Mann* and *Zweig* therefore postulated[2] that the particles belonging to the two fundamental representations (triplet, antitriplet) should also exist; these were called *quarks*. Their charges are $q_u = \frac{2}{3}e$, $q_d = q_s = -\frac{1}{3}e$ (Fig. 6.5). The u, d, and s quarks have baryon number $B = \frac{1}{3}$, and the corresponding antiquarks have $B = -\frac{1}{3}$. By using $Y = B + S$, one readily verifies that the u, d, ū, and d̄ quarks carry zero strangeness, the s quark has $S = -1$, and s̄ has $S = +1$.

One readily observes that all hadrons can be made up of either three quarks (baryons) or a quark–antiquark pair (mesons). Some examples are:

$$
\begin{aligned}
&\mathrm{p} = (\mathrm{uud}) \quad , \quad \mathrm{n} = (\mathrm{udd}) \quad , \\
&\Lambda = (\mathrm{uds}) \quad , \quad \Xi^- = (\mathrm{dss}) \quad , \quad \Xi^0 = (\mathrm{uss}) \quad , \\
&\Sigma^+ = (\mathrm{uus}) \quad , \quad \Sigma^- = (\mathrm{dds}) \quad , \quad \Sigma^0 = (\mathrm{uds}) \quad , \\
&\Delta^{++} = (\mathrm{uuu}) \quad , \quad \Omega^- = (\mathrm{sss}) \quad , \\
&\pi^+ = (\mathrm{u\bar{d}}) \quad , \quad \mathrm{K}^+ = (\mathrm{u\bar{s}}) \quad , \\
&\mathrm{K}^0 = (\mathrm{d\bar{s}}) \quad , \quad \bar{\mathrm{K}}^0 = (\mathrm{s\bar{d}}) \quad , \\
&\rho^0, \omega = (\mathrm{u\bar{u}}), (\mathrm{d\bar{d}}) \quad , \quad \phi = (\mathrm{s\bar{s}}) \quad .
\end{aligned}
$$

Hence hadrons are made of quarks just as an atomic nucleus is made of protons and neutrons. The strangeness of a hadron is simply given by the number of s̄ quarks minus the number of s quarks within the hadron. Although quarks have never been observed

[2] M. Gell-Mann: Phys. Lett. **8**, 214 (1964); G. Zweig: CERN Report No. 8182/TH 401. (This latter work could not be published in a scientific journal, which shows that sometimes strong resistance has to be overcome before a new idea gains common acceptance.)

as free particles, their existence inside hadrons must be considered firmly established. We list a few arguments.[3]

(1) The hadronic mass spectrum can be explained with the help of just a few parameters, if one regards quarks as (nearly) freely moving particles which are enclosed in a small space volume of about 1 fm radius (*"the bag model"*, see Fig. 6.6).
(2) High-energy deep-inelastic electron–nucleon scattering can only be interpreted by assuming that the electrons scatter off point-like constituents, the so-called *partons*, inside the hadrons (Fig. 6.7).

The radius of these partons must be smaller than 10^{-16} cm. Their charges can be determined from the measured cross sections and are in good agreement with the quark model. For the proton we have, for example,

$$\sum_p q_i^2 = e^2(2q_\mathrm{u}^2 + q_\mathrm{d}^2) = e^2\left(2\times\frac{4}{9}+\frac{1}{9}\right) = e^2 \quad ,$$

and for the neutron

$$\sum_n q_i^2 = e^2(q_\mathrm{u}^2 + 2q_\mathrm{d}^2) = e^2\left(\frac{4}{9}+2\times\frac{1}{9}\right) = \frac{2}{3}e^2 \quad .$$

(3) The quark model received special support from the discovery of relatively long-lived, very heavy mesons in electron–positron scattering. The excitation spectrum of these mesons and their lifetime can be explained if one regards them as hydrogen-like or positronium-like states of quark–antiquark pairs of new quark species (*"flavors"*). These mesons are the *charmonium*, made of quarks with "charm" (c) and the *upsilon* particles ($\Upsilon, \Upsilon', \Upsilon'', \ldots$), which are made of "bottom" quarks (b) (Fig. 6.8).

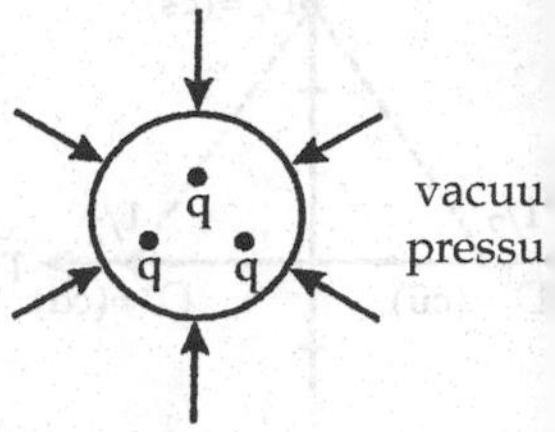

Fig. 6.6. The bag model of ha ron structure

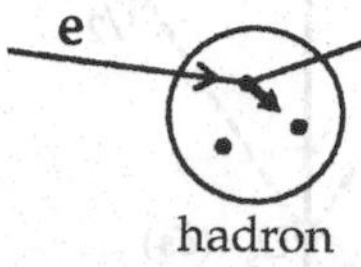

Fig. 6.7. Deep-inelastic elec tron–nucleon scattering

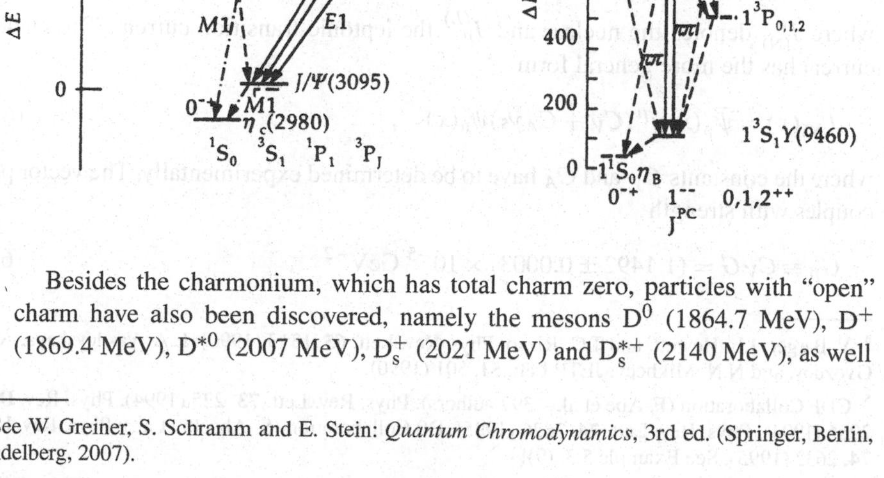

Fig. 6.8. The mass spectrur of Ψ (charmonium) and Υ (bottomonium) mesons

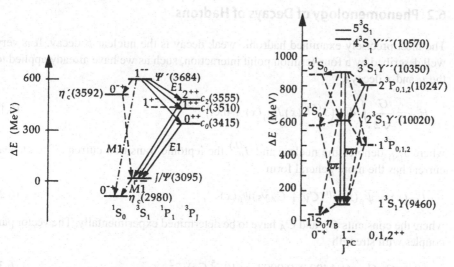

Besides the charmonium, which has total charm zero, particles with "open" charm have also been discovered, namely the mesons D^0 (1864.7 MeV), D^+ (1869.4 MeV), D^{*0} (2007 MeV), D_s^+ (2021 MeV) and $\mathrm{D}_\mathrm{s}^{*+}$ (2140 MeV), as well

[3] See W. Greiner, S. Schramm and E. Stein: *Quantum Chromodynamics*, 3rd ed. (Springer, Berlin, Heidelberg, 2007).

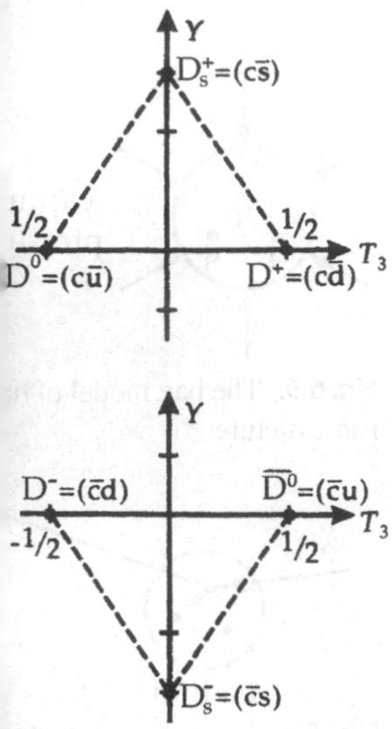

ig. 6.9. Mesons with open harm

as their antiparticles D^-, $\bar{D}^0$, D^{*-}, D_s^- and D_s^{*-}, so that the charm hypothesis is very well confirmed today (Fig. 6.9).

From the cross sections, which are proportional to q_i^2, we deduce that $q_c = (2/3)e$ and $q_b = -(1/3)e$. A table of quarks and their properties according to our present knowledge is given in Table 6.1. Up to masses of about 40 000 MeV no additional quark flavors have been discovered, but the existence of at least one more heavy quark ("top" or "truth", symbolized by "t") has been postulated, with electric charge $q_t = (2/3)e$. For the top quark one expects the following properties:[4] $T_3 = S = C = B = 0$, electric charge $q_t = +(2/3)e$, mass $m_t \simeq 135$–150 GeV. Meanwhile the first experimental evidence for the top quark has been reported.[5]

Table 6.1. Properties of experimentally observed quarks (1995). Note that T_3 and Y refer to the *strong* isospin and hypercharge

Quark flavor	u	d	s	c	b	t
Isospin T_3	1/2	−1/2	0	0	0	0
Strangeness S	0	0	−1	0	0	0
Hypercharge Y	1/3	1/3	−2/3	1/3	1/3	1/3
Charm C	0	0	0	1	0	0
Baryon number B	1/3	1/3	1/3	1/3	1/3	1/3
Beauty b	0	0	0	0	1	0
Charge (e)	+2/3	−1/3	−1/3	+2/3	−1/3	+2/3
Mass (MeV)	4	7.5	150	1200	4700	174000

6.2 Phenomenology of Decays of Hadrons

The most precisely examined hadronic weak decay is the nuclear β decay. It is very well described by a four-fermion point interaction, such as we have already applied to the μ and τ decays,

$$H_{\text{int}} = \frac{G}{\sqrt{2}} \int \mathrm{d}^3x \left(J_\mu^{(L)\dagger}(x) J_{(\mathrm{N})}^\mu(x) + \text{h.c.} \right) \quad , \tag{6.1}$$

where $J_{(\mathrm{N})}^\mu$ denotes the nuclear and $J_\mu^{(L)}$ the leptonic transition current. The nuclear current has the more general form

$$J_{(\mathrm{N})}^\mu(x) = \overline{\psi}_{\mathrm{p}}(x)\gamma^\mu(C_V + C_A\gamma_5)\psi_{\mathrm{n}}(x) \quad , \tag{6.2}$$

where the constants C_V and C_A have to be determined experimentally. The vector part couples with strength

$$G_\beta \equiv C_V G = (1.1492 \pm 0.0003) \times 10^{-5}\ \text{GeV}^{-2} \quad , \tag{6.3}$$

[4] V. Barger, J.L. Hewett, and T.G. Rizzo: Phys. Rev. Lett. **65**, 1313 (1990); L.A. Vasilevskaya, A.A. Gvozdov, and N.N. Mikheev: JETP Lett. **51**, 501 (1990).

[5] CDF Collaboration (F. Abe et al. – 397 authors): Phys. Rev. Lett. **73**, 225 (1994), Phys. Rev. **D50**, 2966 (1994), Phys. Rev. Lett. **74**, 2626 (1995). D0 Collaboration (S. Abachi et al.): Phys. Rev. Lett. **74**, 2632 (1995). See Example 5.3, (9)!

compared to

$$G = (1.16632 \pm 0.00002) \times 10^{-5}\ \mathrm{GeV}^{-2} \quad . \tag{6.4}$$

It therefore holds that

$$C_V = \frac{G_\beta}{G} = 0.9858 \pm 0.0011 \quad , \tag{6.5}$$

which is clearly distinct from 1, considering the size of the error. The axial-vector strength is determined by the experimental value

$$\frac{C_A}{C_V} = -1.255 \pm 0.006 \quad . \tag{6.6}$$

It is not really surprising that we have no exact V–A coupling for nucleons. Owing to strong interactions between nucleons, there exist corrections to the p–n–$W^\pm$ coupling, induced by simultaneous pion exchange. A few of these processes are depicted in the following Feynman diagrams (see Fig. 6.10).

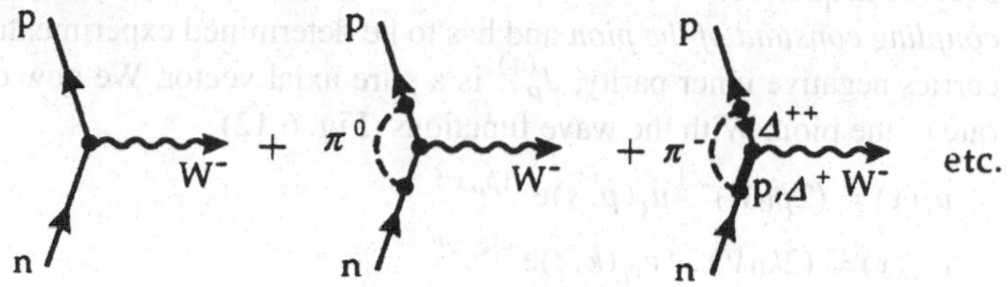

Fig. 6.10. Corrections to th p–n–W coupling

Similar corrections exist for the leptonic weak interactions in form of radiation effects, such as

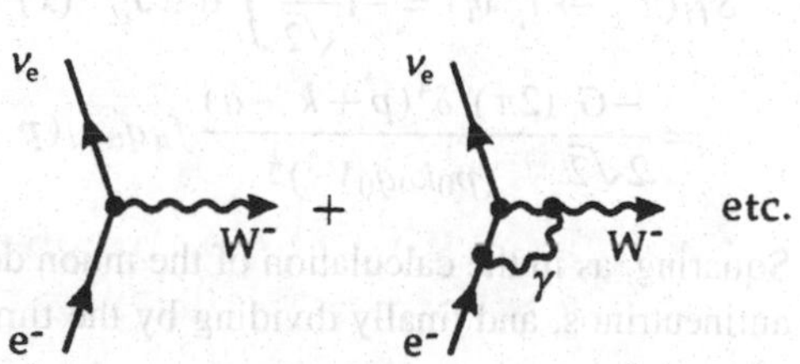

Fig. 6.11. Corrections to th e–ν–W coupling

However, as discussed in the context of muon decay, these are of the order $\alpha \simeq 10^{-2}$. In contrast to this, the pion–nucleon coupling constant is $g^2_{\pi N}/4\pi \simeq 14$. The axial-vector constant is therefore considerably modified. On the other hand, the coupling constant of the vector part obeys special selection principles, since the weak vector current has the same form as the electromagnetic current. Therefore, because of the strict validity of the charge conservation law there should appear no renormalization of the coupling constant due to strong interactions (*the hypothesis of conserved vector current*, "CVC"). This contradicts the very small, but not insignificant, deviation from the value $C_V = 1$. How can this be understood? To this end we first have to consider the phenomenology of weak hadronic decays, where we distinguish strangeness-conserving ($\Delta S = 0$) and strangeness-violating ($\Delta S \neq 0$) decays.

Strangeness-Conserving Decays ($\Delta S = 0$). In these cases the baryonic isospin T_3 always changes by one unit: $|\Delta T_3| = 1$. Examples are

$$\mathrm{n} \to \mathrm{p}e^-\bar{\nu}_e \quad , \quad \Sigma^+ \to \Lambda e^+\nu_e \quad , \quad \pi^+ \to \mu^+\nu_\mu, e^+\nu_e \quad .$$

The total isospin $T_3 = T_3^{(B)} + T_3^{(L)}$ for these decays, that is, the sum of the baryonic isospin $T_3^{(B)}$ and weak isospin $T_3^{(L)}$ of the leptons, is conserved. All these decays are proportional to the same coupling constant, G_β, as the nuclear β decay. The vector character of weak interactions is readily checked by investigating the pion decay into two leptons. The leptonic current

$$J_\mu^{(i)}(x) = \bar{u}_i(x)\gamma_\mu(1-\gamma_5)u_{\nu_i}(x) \qquad (i = \mathrm{e}, \mu) \tag{6.7}$$

must interact with the pionic current. Since the pion carries no spin there exists only a single way of constructing a 4-vector current for the decay $\pi^- \to \mathrm{e}^-\bar{\nu}_\mathrm{e}$ or $\mu^-\nu_\mu$,

$$J_\mu^{(\pi)}(x) = \mathrm{i}\sqrt{2}\, f_\pi q_\mu (2q_0V)^{-\frac{1}{2}}\mathrm{e}^{\mathrm{i}q_\mu x^\mu} \quad . \tag{6.8}$$

Here q_μ is the 4-momentum of the pion. The sign of the argument of the exponential function is chosen in such a way that the product of all plane waves in the transition matrix element yields the conservation of 4-momentum. This corresponds to assigning the character of an antiparticle to the negative pion, while the positive pion has the character of a particle. A better argument can be given at a later stage, when we become acquainted with the weak transition current of quarks. f_π is called the *axial coupling constant of the pion* and has to be determined experimentally. Since the pion carries negative inner parity, $J_\mu^{(\pi)}$ is a pure axial vector. We now calculate the decay rate of the pion. With the wave functions (Fig. 6.12)

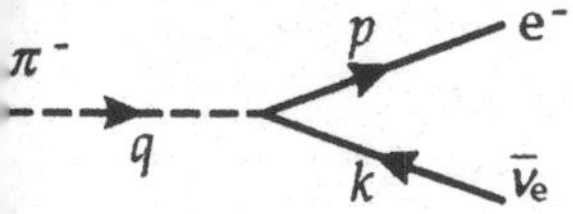

Fig. 6.12. Momentum notation in pion decay

$$u_i(x) = (2p_0V)^{-\frac{1}{2}}u_i(\boldsymbol{p},s)\mathrm{e}^{-\mathrm{i}p_\mu x^\mu} \quad , \tag{6.9a}$$

$$u_{\nu_i}(x) = (2k_0V)^{-\frac{1}{2}}v_{\nu_i}(\boldsymbol{k},t)\mathrm{e}^{+\mathrm{i}k_\mu x^\mu} \quad , \tag{6.9b}$$

we obtain ($i = \mathrm{e}, \mu$):

$$\begin{aligned} S_{fi}(\pi^- \to l_i^-\bar{\nu}_i) &= -\mathrm{i}\frac{G}{\sqrt{2}}\int \mathrm{d}^4x\, J_\alpha^{(\pi)}(x)^\dagger J_{(i)}^\alpha(x) \\ &= \frac{-G}{2\sqrt{2}}\frac{(2\pi)^4\delta^4(p+k-q)}{(p_0k_0q_0V^3)^{\frac{1}{2}}} f_\pi q_\alpha \bar{u}_i(\boldsymbol{p},s)\gamma^\alpha(1-\gamma_5)v_{\nu_i}(\boldsymbol{k},t) \quad . \end{aligned} \tag{6.10}$$

Squaring, as in the calculation of the muon decay, summing over spins of leptons and antineutrinos, and finally dividing by the time T, we obtain for the decay rate of the pion

$$\begin{aligned} &W(\pi^- \to l_i^-\bar{\nu}_i) \\ &= G^2(2\pi)^4 f_\pi^2 \int \frac{\mathrm{d}^3p}{(2\pi)^3p_0}\int\frac{\mathrm{d}^3k}{(2\pi)^3k_0}\frac{m_i^2}{q_0}\delta^4(p+k-q)(p\cdot k) \quad . \end{aligned} \tag{6.11}$$

Intermediate steps are performed in Exercise 6.1. Because of the four-dimensional δ function it holds that

$$m_\pi^2 = q^2 = (p+k)^2 = m_i^2 + 2(p\cdot k) \quad . \tag{6.12}$$

In the pion rest frame we furthermore have $q_0 = m_\pi$ and $\boldsymbol{q} = 0$. Hence we obtain because $|\boldsymbol{p}| = |\boldsymbol{k}| = k_0$,

$$\begin{aligned} W &= \frac{G^2}{8\pi^2} f_\pi^2 \frac{m_i^2}{m_\pi}(m_\pi^2 - m_i^2)\int\frac{\mathrm{d}^3p}{p_0}\int\frac{\mathrm{d}^3k}{k_0}\delta^3(\boldsymbol{p}+\boldsymbol{k})\delta(p_0+k_0-m_\pi) \\ &= \frac{G^2}{8\pi^2} f_\pi^2 \frac{m_i^2}{m_\pi}(m_\pi^2 - m_i^2)4\pi\int_0^\infty \frac{k_0\mathrm{d}k_0}{\sqrt{k_0^2+m_i^2}}\delta\left(k_0+\sqrt{k_0^2+m_i^2}-m_\pi\right) \end{aligned}$$

$$= \frac{G^2}{2\pi} f_\pi^2 \frac{m_i^2}{m_\pi}(m_\pi^2 - m_i^2)\frac{m_\pi^2 - m_i^2}{2m_\pi^2}$$

$$= \frac{G^2}{4\pi} f_\pi^2 m_\pi m_i^2 \left(1 - \frac{m_i^2}{m_\pi^2}\right)^2 \quad . \tag{6.13}$$

Here we have made use of (5.70) in the penultimate step. We observe that the decay probability depends crucially on the mass m_i of the charged lepton. Owing to the value of the charged pion mass (139.57 MeV), decay into an electron or muon can occur. The ratio between the two decay channels is

$$R_{e/\mu} = \frac{W(\pi^- \to e^- \bar{\nu}_e)}{W(\pi^- \to \mu^- \bar{\nu}_\mu)} = \frac{m_e^2}{m_\mu^2} \frac{(m_\pi^2 - m_e^2)^2}{(m_\pi^2 - m_\mu^2)^2} = 1.232 \times 10^{-4} \quad . \tag{6.14}$$

Taking radiative corrections into account, one obtains[6] the slightly modified value $R = 1.284 \times 10^{-4}$. R is called *the branching ratio* for the decay modes. The experimental value is

$$R_{\mathrm{exp}} = (1.218 \pm 0.014) \times 10^{-4} \quad , \tag{6.15}$$

in excellent agreement with the theoretical prediction. From the measured lifetime of the pion,

$$\tau_{\pi^\pm} = (2.6030 \pm 0.0023) \times 10^{-8}\ \mathrm{s} \quad , \tag{6.16}$$

one obtains the value of the *pion decay constant*,

$$f_\pi = (0.6504 \pm 0.0002) m_\pi = 90.8\ \mathrm{MeV} \quad . \tag{6.17}$$

The very small value of the branching ratio R is a consequence of the vector character, or axial-vector character, of weak interactions. Normally one would expect that the decay into the electronic leptons is slightly favored, since more kinetic energy is released in this case, and therefore the allowed phase-space volume is larger for the particles in the final state. For a scalar or pseudoscalar interaction one indeed finds a branching ratio of the order of 1. Hence the pion decay into two leptons is the best experimental evidence for the vector or axial-vector character of weak interactions. The argument is presented in more detail in Exercise 6.2.

EXERCISE

6.1 The $\pi^\pm$ Decay Rate

Problem. Derive (6.11) for the decay rate of the pion.

Solution. Substituting $(2\pi)^4\delta^4(0) = VT$ and starting from the scattering matrix elements (6.10), one obtains for the transition probability per unit time

6 S.M. Berman: Phys. Rev. Lett. **1**, 468 (1958); T. Kinoshita: Phys. Rev. Lett. **2**, 477 (1959).

xercise 6.1

$$W = \frac{V}{T} \int \frac{d^3k}{(2\pi)^3} V \int \frac{d^3p}{(2\pi)^3} \sum_{s,t} |S|^2$$

$$= \frac{G^2}{8} V^3 \int \frac{d^3p}{(2\pi)^3} \int \frac{d^3k}{(2\pi)^3} \frac{(2\pi)^4 \delta^4(p+k-q)}{p_0 k_0 q_0 V^3} f_\pi^2 \sum_{s,t} |M_{st}|^2 \quad (1)$$

with the matrix element

$$M_{st} = q_\alpha \bar{u}_i(\boldsymbol{p}, s) \gamma^\alpha (1-\gamma_5) v_\nu(\boldsymbol{k}, t) \quad . \quad (2)$$

Applying the Dirac trace rules of Appendix A.2 we obtain in established manner, using $(1-\gamma_5)^2 = 2(1-\gamma_5)$

$$\sum_{s,t} |M_{st}|^2 = \sum_{s,t} q_\alpha \bar{v}_\nu(\boldsymbol{k}, t) \gamma^\alpha (1-\gamma_5) u_i(\boldsymbol{p}, s) q_\beta \bar{u}_i(\boldsymbol{p}, s) \gamma^\beta (1-\gamma_5) v_\nu(\boldsymbol{k}, t)$$

$$= \mathrm{Tr}\{q_\alpha \gamma^\alpha (1-\gamma_5)(\not{p} + m_i) q_\beta \gamma^\beta (1-\gamma_5) \not{k}\}$$

$$= \mathrm{Tr}\{\not{q}(1-\gamma_5)(\not{p} + m_i) \not{q}(1-\gamma_5) \not{k}\}$$

$$= 2\,\mathrm{Tr}\{\not{q}(\not{p} + m_i) \not{q}(1-\gamma_5) \not{k}\}$$

$$= 2\,\mathrm{Tr}\{\not{q}\not{p}\not{q}\not{k}(1+\gamma_5)\} \quad . \quad (3)$$

This we could evaluate directly, as we have done for μ decay. In this case, however, there exists a more elegant way. Exploiting the momentum conservation guaranteed by the delta function, we can replace q by $(p+k)$. Using the relations

$$\not{p}^2 = p^2 = m_i^2 \quad , \quad \not{k}^2 = k^2 = 0 \quad , \quad (4)$$

we immediately obtain

$$\not{q}\not{p}\not{q}\not{k} = (\not{p} + \not{k})\not{p}(\not{p} + \not{k})\not{k}$$

$$= (\not{p}^2 + \not{k}\not{p})(\not{p}\not{k} + \not{k}^2)$$

$$= (m_i^2 + \not{k}\not{p})\not{p}\not{k} = m_i^2 \not{p}\not{k} \quad . \quad (5)$$

Thus, the Dirac trace can be obtained,

$$\sum_{s,t} |M_{st}|^2 = 2m_i^2\, \mathrm{Tr}\{\not{p}\not{k}(1+\gamma_5)\}$$

$$= 8m_i^2\, (p \cdot k) \quad , \quad (6)$$

because of (A.22), (A.25). Insertion of this result into (1) yields (6.11) directly.

EXERCISE

6.2 Concerning V–A Coupling in Pion Decay

Problem. (a) One often quotes the strong suppression of the decay $\pi^- \to e^- \bar{\nu}_e$ compared to $\pi^- \to \mu^- \bar{\nu}_\mu$ as proof of V–A coupling (see Exercise 6.3). Show that the suppression appears proportional to $(m_e/m_\mu)^2$ for *every* mixing of vector and axial-vector couplings.

(b) Show that the electronic decay channel would not be suppressed for scalar coupling.

Solution. (a) For arbitrary mixing of the coupling types V and A the pionic currents (6.8) would remain unchanged, while the leptonic currents assume the general form

$$J_\mu^{(i)}(x) = \bar{u}_i(x)\gamma_\mu(C_V + C_A\gamma_5)\,u_{\nu_i}(x) \quad . \tag{1}$$

The calculation of the transition matrix element in Exercise 6.1 would be modified in such a way that the leptonic matrix element now reads

$$M_{st} = q_\alpha \bar{u}_i(\boldsymbol{p}, s)\gamma^\alpha(C_V + C_A\gamma_5)v_{\nu_i}(\boldsymbol{k}, t) \quad . \tag{2}$$

Summing over final-state spins and using

$$(C_V + C_A\gamma_5)^2 = C_V^2 + C_A^2 + 2C_V C_A\gamma_5 \tag{3}$$

we would obtain the following result:

$$\begin{aligned}\sum_{s,t}|M_{st}|^2 &= \mathrm{Tr}\{\not{q}(C_V + C_A\gamma_5)(\not{p} + m_i)\not{q}(C_V + C_A\gamma_5)\not{k}\} \\ &= \mathrm{Tr}\{\not{q}\not{p}\not{q}\not{k}(C_V^2 + C_A^2 - 2C_V C_A\gamma_5)\} \quad .\end{aligned} \tag{4}$$

Equation (5) of Exercise 6.1 still holds:

$$\not{q}\not{p}\not{q}\not{k} = m_i^2\not{p}\not{k} \quad , \tag{5}$$

hence it results that

$$\sum_{s,t}|M_{st}|^2 = 4(C_V^2 + C_A^2)m_i^2(p\cdot k) \quad , \tag{6}$$

that is, the matrix element would in any case be proportional to the lepton mass squared. Therefore we would immediately obtain

$$R = \frac{m_e^2}{m_\mu^2}\frac{(m_\pi^2 - m_e^2)^2}{(m_\pi^2 - m_\mu^2)^2} \quad , \tag{7}$$

which is identical to (6.14), the result of the V–A theory. With the relation

$$q_\alpha = p_\alpha + k_\alpha \tag{8}$$

one can also see this directly by investigating the matrix element (2). In other words, the leptonic spinors are solutions of the free Dirac equation (A.19):

$$\bar{u}_i(\boldsymbol{p}, s)(\not{p} + m_i) = 0 \quad , \tag{9a}$$

$$\not{k}v_{\nu_i}(\boldsymbol{k}, t) = 0 \quad . \tag{9b}$$

Hence because of (8) it holds that

$$\begin{aligned}M_{st} &= \bar{u}_i(\boldsymbol{p}, s)(\not{p} + \not{k})(C_V + C_A\gamma_5)v_{\nu_i}(\boldsymbol{k}, t) \\ &= -m_i\bar{u}_i(\boldsymbol{p}, s)(C_V + C_A\gamma_5)v_{\nu_i}(\boldsymbol{k}, t)\end{aligned} \tag{10}$$

and thus necessarily $|M_{st}|^2 \sim m_i^2$.

(b) For scalar coupling (S or P) one would make the ansatz

$$J^{(\pi)}(x) = f'_\pi (q_0 V)^{-1/2} \mathrm{e}^{\mathrm{i}q_\mu x^\mu} \tag{11}$$

for the pionic current, while the leptonic current takes the form

$$J^{(i)}(x) = \bar{u}_i(x)(C_S + C_P\gamma_5)u_{\nu_i}(x) \quad . \tag{12}$$

Then the reduced matrix element reads

$$M_{st} = \bar{u}_i(p,s)(C_S + C_P\gamma_5)u_{\nu_i}(x) \quad . \tag{13}$$

Summation over the spins would yield

$$\begin{aligned}\sum_{s,t}|M_{st}|^2 &= \mathrm{Tr}\{(\not p + m_i)(C_S + C_P\gamma_5)\not k(C_S + C_P\gamma_5)\} \\ &= \mathrm{Tr}\left[\not p\not k(C_S^2 - C_P^2)\right] \\ &= 4(C_S^2 - C_P^2)(p\cdot k) \quad .\end{aligned} \tag{14}$$

Obviously, the factor m_i^2, which appears for the vectorial couplings in (6), is missing. Equation (14) differs from the result of the V–A theory (Exercise 6.1, (6)) by a factor

$$\frac{1}{2m_i^2}(C_S^2 - C_P^2) \quad ,$$

hence the branching ratio would be given by

$$R = \frac{m_\pi^2 - m_\mathrm{e}^2}{m_\pi^2 - m_\mu^2} \simeq 5.5 \quad . \tag{15}$$

Since the experimental value (6.15) agrees well with the result (7) of the vector coupling theory, we can conclude that any contribution from a coupling of scalar type (S or P) must be weaker by a factor of at least 10^{-6}!

EXERCISE

6.3 Suppression of the Electronic Decay Channel in Pion Decay

Problem. Discuss the balance of angular momentum for pion decay and show, by considering the helicities of the participating particles, that the electronic decay channel must be strongly suppressed.

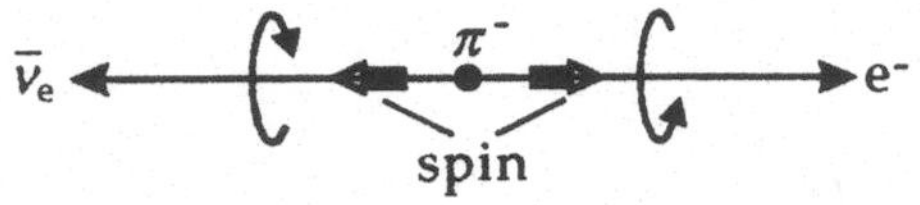

Fig. 6.13. Momenta and helicities for pion decay

Solution. The very small value of the branching ratio R (6.14) can be readily understood by way of the following considerations. The pion has zero spin; therefore the two leptons, which are emitted in opposite directions, must have the same helicity in the rest frame of the pion, so that their spins add to zero. Since the massless antineutrino can only have positive helicity, the electron must be emitted with positive helicity, as well. Weak interactions, however, couple only to the left-handed chiral component $(1-\gamma_5)u_e$, so that this decay of the pion would be completely forbidden in the limit $m_e = 0$. Since $m_e \neq 0$, $(1-\gamma_5)u_e$ contains a small part of positive helicity (see (1.21)), but the decay process is suppressed by a factor $(m_e/m_\pi)^2$. For muonic decay the same argument holds, but here the corresponding factor $(m_\mu/m_\pi)^2$ is much larger.

Remark. This argument provides no special support for the V–A theory, because the factor $(m_e/m_\mu)^2$ appears for any vectorial coupling, as was shown in Exercise 6.2!

Strangeness-Violating Decays ($\Delta S = \pm 1$). For weak decays, which change strangeness, one always observes a simultaneous change of the hadronic charge,

$$\Delta S = \Delta Q \quad . \tag{6.18}$$

Hence the Σ^- particle can, for example, decay into a neutron, since it carries strangeness (-1) and therefore $\Delta S = \Delta Q = -1$. On the other hand the Σ^+, which also has strangeness (-1), cannot decay into a neutron, since this decay would have $\Delta S = -1$, but $\Delta Q = +1$. Experimentally one finds that

$$\frac{W(\Sigma^+ \to \mathrm{n}e^+\nu_e)}{W(\Sigma^- \to \mathrm{n}e^-\bar{\nu}_e)} < 0.005 \quad . \tag{6.19}$$

The space-time structure of strangeness-changing weak interactions is readily studied by means of the decay of the charged kaons. By analogy with pion decay we select the two modes

$$\mathrm{K}^- \to e^-\bar{\nu}_e \quad , \quad \mathrm{K}^- \to \mu^-\bar{\nu}_\mu \quad .$$

Note that for these decays a hadron (in this case a meson) disappears completely, while this is not so for the Σ decay cited above, where a hadron in the initial state is converted into another hadron (in this case a neutron) in the final state. For all these decays, however, the baryon number B is conserved. Since the kaon has zero spin, its transition current is given, assuming vector coupling, by

$$J_\mu^{(\mathrm{K})}(x) = \mathrm{i}\sqrt{2} f_\mathrm{K} q_\mu (2q_0 V)^{-1/2} \mathrm{e}^{\mathrm{i}q_\lambda x^\lambda} \quad , \tag{6.20}$$

by analogy with (6.8). The calculation of the scattering matrix element ($i = e, \mu$)

$$S(\mathrm{K}^- \to e_i^- \bar{\nu}_i) = -\mathrm{i}\frac{G}{\sqrt{2}} \int \mathrm{d}^4x\, J_\mu^{(\mathrm{K})\dagger}(x) J_{(i)}^\mu(x) \tag{6.21}$$

proceeds exactly as for pion decay; one only has to replace m_π, f_π by m_K, f_K. The result is therefore

$$W(\mathrm{K}^- \to e_i^- \bar{\nu}_i) = \frac{G^2}{4\pi} f_\mathrm{K}^2 m_\mathrm{K} m_i^2 \left(1 - \frac{m_i^2}{m_\mathrm{K}^2}\right)^2 \quad , \tag{6.22}$$

and the branching ratio is (with $m_{\mathrm{K}} = 493.67$ MeV)

$$R_{\mathrm{e}/\mu} = \frac{W(\mathrm{K}^- \to \mathrm{e}^-\bar{\nu}_{\mathrm{e}})}{W(\mathrm{K}^- \to \mu^-\bar{\nu}_\mu)} = \frac{m_{\mathrm{e}}^2}{m_\mu^2}\frac{(m_{\mathrm{K}}^2 - m_{\mathrm{e}}^2)^2}{(m_{\mathrm{K}}^2 - m_\mu^2)^2} = 2.58 \times 10^{-5} \quad , \tag{6.23}$$

up to small radiative corrections. The experimental value

$$R_{\mathrm{exp}} = (2.425 \pm 0.012) \times 10^{-5} \tag{6.24}$$

confirms the vector coupling in an impressive way.

Finally we can determine the kaon decay constant from the measured lifetime,

$$\tau_{\mathrm{K}^\pm} = (1.237 \pm 0.003) \times 10^{-8}\ \mathrm{s} \quad . \tag{6.25}$$

In contrast to the charged pion, the kaon has competing hadronic decay modes (mainly $\mathrm{K}^\pm \to \pi^\pm + \pi^0$). To account for this the experimental lifetime has to be divided by the relative decay probability (branching ratio) into the muon channel

$$B_{\mathrm{K}\to\mu} = \frac{W(\mathrm{K}^- \to \mu^-\bar{\nu}_\mu)}{W(\mathrm{K}^- \to \text{anything})} = 0.6350 \pm 0.0016 \quad . \tag{6.26}$$

This leads to

$$\begin{aligned} f_{\mathrm{K}} &= \frac{m_{\mathrm{K}}^2}{m_\mu(m_{\mathrm{K}}^2 - m_\mu^2)G}\sqrt{\frac{4\pi B_{\mathrm{K}\to\mu}}{m_{\mathrm{K}}\tau_{\mathrm{K}}}} \\ &= (0.0505 \pm 0.0002)\, m_{\mathrm{K}} = 24.9\ \mathrm{MeV} \quad . \end{aligned} \tag{6.27}$$

Let us summarize the essential steps:

$$\tau = \frac{1}{W} = \frac{1}{\frac{G^2}{4\pi} f_{\mathrm{K}}^2 m_{\mathrm{K}} m_i^2 (1 - \frac{m_i^2}{m_{\mathrm{K}}^2})^2 + \cdots} = \frac{1}{W_{K\to\text{anything}}} \,,$$

$$B_{K\to\mu} = \frac{W_{K\to\mu}}{W_{K\to\text{anything}}} \quad ,$$

$$\frac{B_{K\to\mu}}{\tau} = \frac{W_{K\to\mu}}{W_{K\to\text{anything}}} \cdot W_{K\to\text{anything}} = \frac{G^2}{4\pi} f_{\mathrm{K}}^2 m_{\mathrm{K}} m_\mu^2 \left(1 - \frac{m_\mu^2}{m_{\mathrm{K}}^2}\right)^2 \quad ,$$

$$f_{\mathrm{K}}^2 = \frac{4\pi B_{K\to\mu}}{\tau} \frac{1}{m_{\mathrm{K}} m_\mu^2 (1 - \frac{m_\mu^2}{m_{\mathrm{K}}^2})^2} \frac{1}{G^2} \,,$$

$$\begin{aligned} f_{\mathrm{K}} &= \sqrt{\frac{4\pi B_{K\to\mu}}{\tau}} \frac{1}{G} \sqrt{\frac{m_{\mathrm{K}}^4}{m_{\mathrm{K}} m_\mu^2} \frac{1}{(m_{\mathrm{K}}^2 - m_\mu^2)^2}} \\ &= \sqrt{\frac{4\pi B_{K\to\mu}}{\mathrm{K}\tau}} \frac{1}{G} \frac{m_{\mathrm{K}}^2}{m_\mu^2 (m_{\mathrm{K}}^2 - m_\mu^2)} \quad . \end{aligned}$$

The relation to the pion decay constant (6.17) is

$$\frac{f_{\mathrm{K}}}{f_\pi} = 0.275 \quad . \tag{6.28}$$

Obviously the decay constant for strangeness-changing weak decays ($\Delta S = \pm 1$) is smaller by a factor of about 4 than that for decays with $\Delta S = 0$. The comparison with many other experimental data confirms this observation. An explanation is given by Cabibbo's theory, with which we will become acquainted in the next sections of this chapter.

6.3 Weak Interactions of Quarks

We have cast the interactions of leptons into a general scheme in the section concerning the GSW model in Chap. 4, which classified left-handed leptons (chirality -1) into three generations of weak isospin doublets, and right-handed leptons (chirality $+1$) into three generations of singlets:

$$\begin{pmatrix} \nu_e \\ e_L^- \end{pmatrix}, \quad \begin{pmatrix} \nu_\mu \\ \mu_L^- \end{pmatrix}, \quad \begin{pmatrix} \nu_\tau \\ \tau_L^- \end{pmatrix}, \quad e_R^-, \quad \mu_R^-, \quad \tau_R^- .$$

If we want to place quarks into a similar scheme, it is convenient also to arrange them in left-handed doublets and right-handed singlets. Here we must, however, note that there exists a small difference compared to the leptons: all quarks have non-vanishing rest mass. Consequently there are no quarks that have only left-handed components like neutrinos. If there were no right-handed component in a certain reference frame, this would be immediately different in another reference frame, because the helicity operator $\Lambda = \boldsymbol{J} \cdot \boldsymbol{p}/|\boldsymbol{p}|$ does not commute with the generators of Lorentz transformations. This can be readily understood: if, for example, the particle moves in the observer's rest frame with $|\boldsymbol{v}| < c$ in the direction of its spin, one only needs to consider a system that moves with velocity $\boldsymbol{v}_L = \alpha \boldsymbol{v}$, $\alpha > 1$, with respect to the observer's rest frame. In this frame the particle has the apparent velocity

$$\boldsymbol{v}' = -\frac{(\alpha - 1)\boldsymbol{v}}{1 - \alpha v^2} \equiv -\alpha' \boldsymbol{v} \ ,$$

that is, it moves opposite to the direction of its spin. Therefore the helicity changes its sign (see Fig. 6.14). Since we also want to describe electromagnetic interactions in the context of the Glashow–Salam–Weinberg theory extended to describe quarks, it is important to have the same charge structure in the quark doublets as in the lepton doublets. The charge of the lower components (for example, e_L^-) must be one unit (e) smaller than that of the upper component (for example, ν_e). Therefore quarks with charge $-(1/3)e$ should appear only in the lower components, those with charge $+(2/3)e$ only in the upper components.

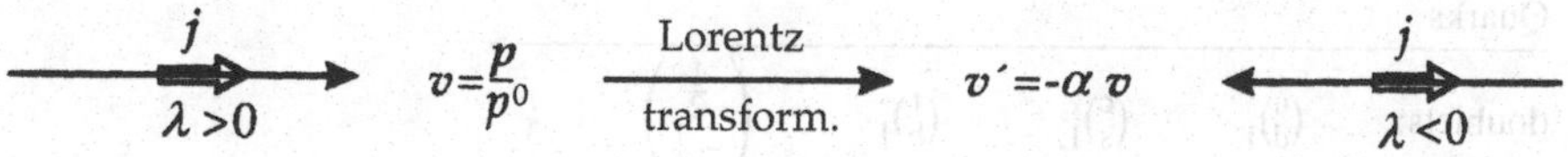

Fig. 6.14. Change of sign of the helicity for a Lorentzian transformation

One can see from Table 6.1 that the third quark with charge $-(1/3)e$, the b quark, still lacks a partner. The existence of a sixth quark was consequently postulated, the t quark (t = "top" or "truth"), with a mass $m_t \simeq 135$–150 GeV.[7] The charge of this quark

[7] CDF Collaboration (F. Abe et al. – 397 authors) Phys. Rev. Lett. **73**, 225 (1994), Phys. Rev. **D50**, 2966 (1994), Phys. Rev. Lett. **74**, 2626 (1995). D0 Collaboration (S. Abachi et al.) Phys. Rev. Lett. **74**, 2632 (1995). See Example 5.3, (9)!

must be $+(2/3)e$. Therefore we have to deal with three doublets and six singlets:

$$\begin{pmatrix} u_L \\ d_L \end{pmatrix} \begin{pmatrix} c_L \\ s_L \end{pmatrix} \begin{pmatrix} t_L \\ b_L \end{pmatrix} , \quad u_R \quad d_R \quad c_R \quad s_R \quad t_R \quad b_R \quad . \tag{6.29}$$

If we abbreviate doublets with L_i ($i = \mathrm{u, c, t}$), there are, according to the scheme known from the leptons (see (4.96)), the following couplings between particles belonging to the *same* doublet:

$$\begin{aligned} L_{\mathrm{int}}^{(i)} &= g\,\bar{L}_i \gamma^\mu \hat{\boldsymbol{T}} L_i \cdot \boldsymbol{A}_\mu + \frac{g'}{2}\left(\bar{L}_i \gamma^\mu \hat{Y}_i^{\mathrm{L}} L_i + \bar{R}_i \gamma^\mu \hat{Y}_i^{\mathrm{R}} R_i\right) B_\mu \\ &= g\bar{L}_i \gamma^\mu \hat{\boldsymbol{T}} L_i \cdot \boldsymbol{A}_\mu + \frac{g'}{2}\left(Y_i^{\mathrm{L}} \bar{L}_i \gamma^\mu L_i + Y_i^{\mathrm{R}} \bar{R}_i \gamma^\mu R_i\right) B_\mu \quad , \end{aligned} \tag{6.30}$$

where Y^{R}, Y^{L} are the weak hypercharges of left-handed and right-handed quarks, respectively. Here the hypercharge is again calculated according to the Gell-Mann–Nishijima relation

$$Y = 2(Q - T_3) \quad , \tag{6.31}$$

by which we obtain the following values. For the left-handed doublets:

$$Y^{\mathrm{L}} = 2\left(+\frac{2}{3} - \frac{1}{2}\right) = 2\left(-\frac{1}{3} + \frac{1}{2}\right) = \frac{1}{3} \quad , \tag{6.32a}$$

and for the right-handed singlets

$$Y^{\mathrm{R}}_{\mathrm{u,c,t}} = +2\,\frac{2}{3} = +\frac{4}{3} \quad , \quad Y^{\mathrm{R}}_{\mathrm{d,s,b}} = 2\left(-\frac{1}{3}\right) = -\frac{2}{3} \quad . \tag{6.32b}$$

Table 6.2 contains the *weak* isospin T_3 and the *weak* hypercharge Y of the particles in (6.29). We emphasize here *weak* isospin and *weak* hypercharge and distinguish it clearly from the *flavor* isospin and *flavor* hypercharge depicted in Figs. 6.1–6.5 and in Table 6.1. The form of (6.30) reduces for the leptons to that of (4.96) if the corresponding weak hypercharge for the lepton doublets are inserted.

Table 6.2. Values for the *weak* isospin and the *weak* hypercharge of the $SU(2)_L \times U(1)_Y$ theory. According to the Gell-Mann–Nishijima formula $Q = T_3 + \frac{1}{2}Y$ the physical charges of the particles can be calculated. Usually left- and right-handed particles (e.g. $\mathrm{e_L}$ and $\mathrm{e_R}$ or u_L and $u_R \ldots$) differ in their T_3 and Y quantum numbers.

Quarks					
doublets:	$\binom{u}{d}_L$	$\binom{c}{s}_L$	$\binom{t}{b}_L$	$\begin{pmatrix} \frac{1}{2} \\ -\frac{1}{2} \end{pmatrix}$	$\frac{1}{3}$
singlets:	u_R	c_R	t_R	0	$\frac{4}{3}$
	d'_R	s'_R	b'_R	0	$-\frac{2}{3}$
Leptons				T_3	Y
doublets:	$\binom{\nu_e}{e}_L$	$\binom{\nu_\mu}{\mu}_L$	$\binom{\nu_\tau}{\tau}_L$	$\begin{pmatrix} \frac{1}{2} \\ -\frac{1}{2} \end{pmatrix}$	-1
singlets:	e_R	μ_R	τ_R	0	-2

According to the couplings in (6.30) a d quark could only become a u quark, an s quark only a c quark, and so on. *For leptons, the analogous selection rules described the conservation of electron, muon, and tau number separately.* For quarks, however, this contradicts the observation in the decay of the negative Σ particle,

$$\Sigma^- \to \mathrm{n}\mathrm{e}^-\bar{\nu}_\mathrm{e} \; ,$$

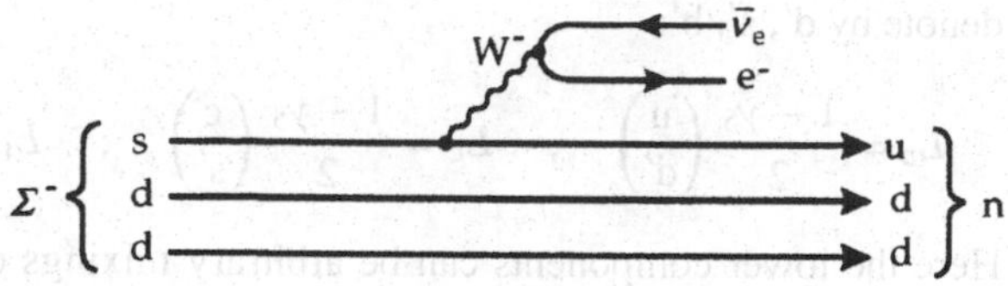

Fig. 6.15. Quark diagram of the decay $\Sigma^- \to \mathrm{n}\mathrm{e}^-\bar{\nu}_\mathrm{e}$

which at the quark level looks like

$$(\mathrm{sdd}) \to (\mathrm{udd}) + \mathrm{e}^- + \bar{\nu}_\mathrm{e} \quad . \tag{6.33a}$$

If one considers the quark content, an s quark obviously decays into a u quark. We found that strangeness-changing decay amplitudes are weaker by a factor of 4 or 5 than strangeness-conserving decays, for example of the proton into the neutron (cf. (6.28)). The same is true for the decay

$$\mathrm{K}^- \to \mu^- + \bar{\nu}_\mu \; ,$$

or according to the quark content

$$(\mathrm{s}\bar{\mathrm{u}}) \to \mu^- + \bar{\nu}_\mu \quad . \tag{6.33b}$$

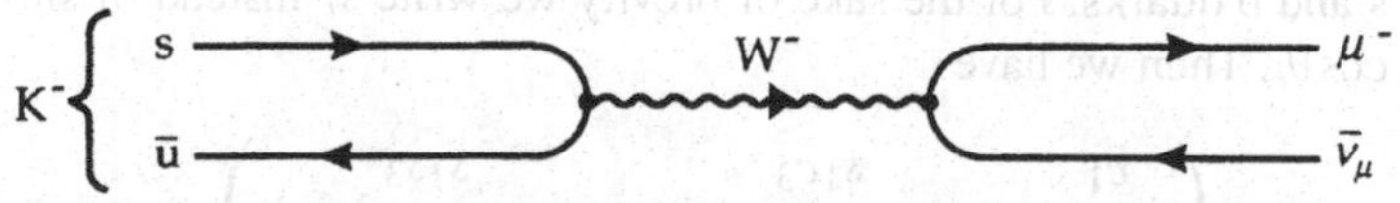

Fig. 6.16. Quark diagram of the decay $\mathrm{K}^- \to \mu^-\bar{\nu}_\mu$

Here an s quark decays into an u quark too, which then annihilates with the ū antiquark, so that no quarks at all appear in the outgoing channel of this reaction.

Let us summarize. The experiment shows that not only the d quark but also the s quark couples to the u quark via weak interactions; the s quark, however, couples 4–5 times more weakly. Hence we must either abandon the scheme originating from the leptons, that the coupling occurs only within the particular left-handed doublets, or we allow the left-handed u quark to couple to a *mixture* of left-handed d and s quarks. To obtain a uniform theory of weak interactions we reject the first possibility and make use of the second. At first sight this idea, to construct a doublet of the form

$$\begin{pmatrix} \mathrm{u_L} \\ \alpha\,\mathrm{d_L} + \beta\,\mathrm{s_L} \end{pmatrix} \; ,$$

appears to be somewhat far fetched. After some reflection, however, we see that this is not the case. It is by no means clear why weak interactions should distinguish particles by their masses, which also play no role for strong interactions. Since weak interactions cause the decay of d and s quarks, these cannot be eigenstates of the Hamiltonian of the weak interaction. In fact, the ansatz (6.29) shows that d and s quarks have the same quantum numbers Q, T_3, and Y. Therefore there is no deeper reason weak interactions should not affect a mixing of the states of d and s quarks. Of course, the question arises why the upper components of the doublets do not mix as well. This would have the same effect with respect to the decays discussed in (6.33) and could also explain the experimental observations.

In Exercise 6.4 we will convince ourselves that we have the freedom to commit ourselves to a mixing of either upper or lower components. We hence continue to denote doublets by their upper component, u, c, t, but the lower components we now denote by d′, s′, b′:

$$L_{\mathrm{u}} = \frac{1-\gamma_5}{2}\begin{pmatrix}\mathrm{u}\\ \mathrm{d}'\end{pmatrix} \quad , \quad L_{\mathrm{c}} = \frac{1-\gamma_5}{2}\begin{pmatrix}\mathrm{c}\\ \mathrm{s}'\end{pmatrix} \quad , \quad L_{\mathrm{t}} = \frac{1-\gamma_5}{2}\begin{pmatrix}\mathrm{t}\\ \mathrm{b}'\end{pmatrix} \quad . \tag{6.34}$$

Here the lower components can be arbitrary mixings of the mass eigenstates d, s, b known from the physics of strong interactions. Under the constraint of conservation of total probability, the connection between d′, s′, b′ and d, s, b must be given by a unitary 3×3 matrix $\hat{U}$,

$$\begin{pmatrix}\mathrm{d}'\\ \mathrm{s}'\\ \mathrm{b}'\end{pmatrix} = \hat{U}\begin{pmatrix}\mathrm{d}\\ \mathrm{s}\\ \mathrm{b}\end{pmatrix} \quad , \quad \hat{U}^{\dagger}\hat{U} = \mathbb{1} \quad . \tag{6.35}$$

One also speaks of the constraint of *universality of weak interactions*, since it ensures that the total strength of all charged hadronic currents remains unchanged.

A unitary 3×3 matrix has nine free real parameters; this number is reduced by five if one considers that the phase of each of the six quark wave functions can be chosen at will, but that the choice of the total phase has no effect on $\hat{U}$. According to Kobayashi and Maskawa[8] one uses as parameters the three Euler angles[9] $\theta_1, \theta_2, \theta_3$ for a (real) three-dimensional rotation and in addition a phase δ for the coupling in the space of s and b quarks. For the sake of brevity we write s_i instead of $\sin\theta_i$ and c_i instead of $\cos\theta_i$. Then we have

$$\begin{aligned}\hat{U} &= \begin{pmatrix} c_1 & s_1c_3 & s_1s_3 \\ -s_1c_2 & c_1c_2c_3 + s_2s_3\mathrm{e}^{\mathrm{i}\delta} & c_1c_2s_3 - s_2c_3\mathrm{e}^{\mathrm{i}\delta} \\ -s_1s_2 & c_1s_2c_3 - c_2s_3\mathrm{e}^{\mathrm{i}\delta} & c_1s_2s_3 + c_2c_3\mathrm{e}^{\mathrm{i}\delta}\end{pmatrix} \\ &= \begin{pmatrix}1 & 0 & 0\\ 0 & c_2 & -s_2\\ 0 & s_2 & c_2\end{pmatrix}\begin{pmatrix}c_1 & s_1 & 0\\ -s_1 & c_1 & 0\\ 0 & 0 & 1\end{pmatrix} \\ &\quad \times \begin{pmatrix}1 & 0 & 0\\ 0 & 1 & 0\\ 0 & 0 & \mathrm{e}^{\mathrm{i}\delta}\end{pmatrix}\begin{pmatrix}1 & 0 & 0\\ 0 & c_3 & s_3\\ 0 & -s_3 & c_3\end{pmatrix} \quad .\end{aligned} \tag{6.36}$$

The parameters are chosen in such a way that the choice $\theta_2 = \theta_3 = \delta = 0$ corresponds directly to the reduction to only two doublets (u, d') and (c, s'):

$$\hat{U}(\theta_2 = \theta_3 = \delta = 0) = \begin{pmatrix}\cos\theta_1 & \sin\theta_1 & 0\\ -\sin\theta_1 & \cos\theta_1 & 0\\ 0 & 0 & 1\end{pmatrix} \tag{6.37}$$

[8] M. Kobayashi and K. Maskawa: Prog. Theor. Phys. **49**, 652 (1973).

[9] For the definition of the Euler angles we refer to W. Greiner: *Theoretische Physik, Mechanik II*, 5th ed. (Harri Deutsch, Frankfurt, 1989) or H. Goldstein: *Classical Mechanics*, 2nd ed. (Addison-Wesley, Reading, MA, 1980).

Since the existence of the t quark has only recently been experimentally confirmed[10] we may limit the theory only to the first two doublets $\binom{u}{d'}$ and $\binom{c}{s'}$. The single angle θ_1, which appears in this case, is called the *Cabibbo*[11] *angle* θ_C. At first we want to restrict ourselves to this reduced theory. Historically, Cabibbo was first with the idea of mixing s and d quarks in weak interactions. Kobayashi and Maskawa simply generalized this idea about 10 years later.

EXERCISE

6.4 Mixing in Leptonic Families

Problem. Show that the introduction of a mixing matrix $\hat{U}$ for leptons does not yield any observable effects, provided that the neutrinos are massless.

Solution. For simplicity we restrict ourselves to two doublets $i = \mathrm{e}, \mu$. Hence we assume that the doublets involved in weak interactions are

$$L'_{\mathrm{e}} = \begin{pmatrix} \nu_{\mathrm{e}} \\ \mathrm{e}'^{-}_{L} \end{pmatrix} \quad , \quad L'_{\mu} = \begin{pmatrix} \nu_{\mu} \\ \mu'^{-}_{L} \end{pmatrix} \tag{1}$$

with the mixed lepton states

$$\begin{aligned} \begin{pmatrix} \mathrm{e}'_L \\ \mu'_L \end{pmatrix} &= \begin{pmatrix} \cos\theta & \sin\theta \\ -\sin\theta & \cos\theta \end{pmatrix} \begin{pmatrix} \mathrm{e}_L \\ \mu_L \end{pmatrix} \\ &= \hat{U} \begin{pmatrix} \mathrm{e}_L \\ \mu_L \end{pmatrix} \quad . \end{aligned} \tag{2}$$

According to (4.96) the contribution of the doublets to the total current is

$$\bar{L}'_{\mathrm{e}} \gamma^{\alpha} \boldsymbol{T} L'_{\mathrm{e}} + \bar{L}'_{\mu} \gamma^{\alpha} \boldsymbol{T} L'_{\mu} \quad , \tag{3a}$$

or

$$\bar{L}'_{\mathrm{e}} \gamma^{\alpha} L'_{\mathrm{e}} + \bar{L}'_{\mu} \gamma^{\alpha} L'_{\mu} \quad . \tag{3b}$$

In any case, the expression is of the form

$$\bar{L}'_{\mathrm{e}} \hat{O} L'_{\mathrm{e}} + \bar{L}'_{\mu} \hat{O} L'_{\mu} \quad , \tag{4}$$

where $\hat{O}$ is an operator that acts on spinor and isospin indices.

We now introduce two new doublets,

$$\begin{aligned} L''_{\mathrm{e}} &= L'_{\mathrm{e}} \cos\theta - L'_{\mu} \sin\theta \quad , \\ L''_{\mu} &= L'_{\mathrm{e}} \sin\theta + L'_{\mu} \cos\theta \quad . \end{aligned} \tag{5}$$

[10] Evidence for the top quark production in $\bar{p}p$ collisions at $\sqrt{s} = 1.8$ TeV has recently been reported: CDF collaboration (F. Abe et al. – 397 authors): Phys. Rev. Lett. **73**, 225 (1994), Phys. Rev. **D50**, 2966 (1994).

[11] N. Cabibbo: Phys. Rev. Lett. **10**, 531 (1963).

xercise 6.4

Note that here the doublets L'_{e}, L'_{μ}, mixed in the lower components, become altogether mixed. New doublets L''_{e}, L''_{μ} are combined from the doublets L'_{e}, L'_{μ}. For these the relation

$$\bar{L}''_{\mathrm{e}}\hat{O}L''_{\mathrm{e}} + \bar{L}''_{\mu}\hat{O}L''_{\mu} = \bar{L}'_{\mathrm{e}}\hat{O}L'_{\mathrm{e}} + \bar{L}'_{\mu}\hat{O}L'_{\mu} \quad , \tag{6}$$

now holds, since the mixed terms proportional to $\sin\theta\,\cos\theta$ just cancel each other for the orthogonal transformation (5). Hence it does not matter whether we arrange the left-handed particles in the doublets L'_i (1) or in the doublets L''_i (5)! With relation (2) the doublets L''_i have the following form:

$$\begin{aligned} L''_{\mathrm{e}} &= \begin{pmatrix} \nu_{\mathrm{e}}\cos\theta - \nu_{\mu}\sin\theta \\ \mathrm{e}'_L\cos\theta - \mu'_L\sin\theta \end{pmatrix} \\ &= \begin{pmatrix} \nu_{\mathrm{e}}\cos\theta - \nu_{\mu}\sin\theta \\ e_L\,(\cos^2\theta + \sin^2\theta) \end{pmatrix} = \begin{pmatrix} \nu'_{\mathrm{e}} \\ e_L \end{pmatrix} \quad , \\ L''_{\mu} &= \begin{pmatrix} \nu_{\mathrm{e}}\sin\theta + \nu_{\mu}\cos\theta \\ \mathrm{e}'_L\sin\theta + \mu'_L\sin\theta \end{pmatrix} = \begin{pmatrix} \nu'_{\mu} \\ \mu_L \end{pmatrix} \end{aligned} \tag{7}$$

with the mixed neutrino states

$$\begin{aligned} \begin{pmatrix} \nu'_{\mathrm{e}} \\ \nu'_{\mu} \end{pmatrix} &= \begin{pmatrix} \cos\theta & -\sin\theta \\ \sin\theta & \cos\theta \end{pmatrix} \begin{pmatrix} \nu_{\mathrm{e}} \\ \nu_{\mu} \end{pmatrix} \\ &= \hat{U}^{\dagger} \begin{pmatrix} \nu_{\mathrm{e}} \\ \nu_{\mu} \end{pmatrix} \quad . \end{aligned} \tag{8}$$

Hence we recognize that a mixing of the neutrinos with the matrix $\hat{U}^{\dagger}$ has exactly the same effect as a mixing of the charged leptonic states with the matrix $\hat{U}$. The advantage of the new doublets L''_i is that states of charged leptons with different masses appear unmixed. This is of some importance if we consider an "electron" with momentum $\boldsymbol{p}$. If we have mixing,

$$\begin{aligned} \psi^{(\boldsymbol{p})}_{\mathrm{e}'}(x) &= \psi^{(\boldsymbol{p})}_{\mathrm{e}}(x)\cos\theta + \psi^{(\boldsymbol{p})}_{\mu}(x)\sin\theta \\ &= \big[(2p_0V)^{-1/2}u_{\mathrm{e}}(\boldsymbol{p},s)\cos\theta\,\mathrm{e}^{-\mathrm{i}p_0t} \\ &\quad + (2p'_0V)^{-1/2}u_{\mu}(\boldsymbol{p},s)\sin\theta\,\mathrm{e}^{-\mathrm{i}p'_0t}\big]\mathrm{e}^{\mathrm{i}\boldsymbol{p}\cdot\boldsymbol{x}} \end{aligned} \tag{9}$$

with

$$\begin{aligned} p_0 &= (p^2 + m_{\mathrm{e}}^2)^{1/2} \quad , \\ p'_0 &= (p^2 + m_{\mu}^2)^{1/2} \quad . \end{aligned} \tag{10}$$

Owing to the difference between the two masses each contribution has a different time dependence, that is, the momentum eigenstate cannot be an eigenstate of the Hamiltonian! On the other hand, an energy eigenstate would not have a defined momentum. If the neutrinos are massless, then this is different for a mixing of the neutrino states:

$$\begin{aligned} \psi^{(\boldsymbol{p})}_{\nu'_{\mathrm{e}}}(x) &= \psi^{(\boldsymbol{p})}_{\nu_{\mathrm{e}}}(x)\cos\theta + \psi^{(\boldsymbol{p})}_{\nu_{\mu}}(x)\sin\theta \\ &= \big[u_{\nu_{\mathrm{e}}}(\boldsymbol{p},s)\cos\theta + u_{\nu_{\mu}}(\boldsymbol{p},s)\sin\theta\big](2|\boldsymbol{p}|V)^{-1/2}\mathrm{e}^{-\mathrm{i}|\boldsymbol{p}|t+\mathrm{i}\boldsymbol{p}\cdot\boldsymbol{x}} \quad . \end{aligned} \tag{11}$$

Both parts have the same time dependence; the state is stationary. Since the neutrinos participate only in weak interactions, there is no possibility of distinguishing experimentally between the doublets

$$L_e = \begin{pmatrix} \nu_e \\ e_L \end{pmatrix} \quad , \quad L''_e = \begin{pmatrix} \nu'_e \\ e_L \end{pmatrix} \quad . \tag{12}$$

The electron neutrino is simply, by definition, the neutrino state involved in the interaction with the left-handed electron. Similar arguments hold for the muon neutrino. Theories that are based upon the two choices L_i or L''_i are completely equivalent and experimentally indistinguishable if the rest masses of the neutrinos vanish exactly. (If the neutrinos carry mass, so-called "neutrino oscillations" arise; these will be discussed in Example 6.5.)

EXAMPLE

6.5 Neutrino Oscillations

In the preceding sections we began with the assumption of exactly conserved electron, muon, and tau quantum numbers. For example, because of the conservation of the muon quantum number, a muon neutrino must be created in muon decay, as well as an electron antineutrino, to conserve the electron quantum number. The separate conservation of the different lepton quantum numbers alone forbids the otherwise possible decay

$$\mu^{\pm} \rightarrow e^{\pm} + \gamma \quad ,$$

for there is no other quantum number by which the particles ν_e, ν_μ differ. Experimentally it is found that

$$\frac{W(\mu^- \rightarrow e^- \gamma)}{W(\mu^- \rightarrow e^- \bar{\nu}_e \nu_\mu)} < 10^{-10} \quad .$$

If, however, one assumes that the leptonic quantum numbers are only approximately conserved, there exists the fascinating possibility of neutrino oscillations, that is, that different kinds of neutrino can transform into each other, like the conversion of the strange meson K^0 into its antiparticle $\bar{K}^0$, which we shall discuss in Chap. 8.

Expressed somewhat differently, this possibility implies that the electron neutrino, which originates from nuclear electron capture,

$$e^- + p \rightarrow n + \nu_e \quad ,$$

does not represent an eigenstate of the full Hamiltonian. Neutrino oscillations would also present a possible explanation for the fact that only about a third of the electron neutrinos, which are expected to come from the sun according to common solar models, are actually observed (the solar neutrino problem). According to the hypothesis of neutrino oscillations these could on the way to the earth convert into muon or τ neutrinos, so that after the long travel time only about a third of the neutrinos would be present as electron neutrinos at the earth. This would nicely explain the observed suppression factor of 3 (see also Chap. 7, Example 7.10 for details of the solar neutrino problem).

xample 6.5

For convenience we leave the τ neutrino out of the following and assume that only the electron and muon neutrino mix with each other. We denote the wave function of the neutrinos simply by ν_e and ν_μ, respectively. One can construct eigenstates of the Hamiltonian by convenient linear combinations of ν_e and ν_μ, which we shall denote by ν_1 and ν_2, and for which we make the ansatz

$$\begin{pmatrix} \nu_1 \\ \nu_2 \end{pmatrix} = \begin{pmatrix} \cos\theta & -\sin\theta \\ \sin\theta & \cos\theta \end{pmatrix} \begin{pmatrix} \nu_e \\ \nu_\mu \end{pmatrix} . \tag{1}$$

The orthogonality of the mixing matrix guarantees that the normalization of the final state is conserved:

$$\int d^3\boldsymbol{r}\{|\nu_e(\boldsymbol{r})|^2 + |\nu_\mu(\boldsymbol{r})|^2\} = 1$$
$$\Rightarrow \int d^3\boldsymbol{r}\{|\nu_1(\boldsymbol{r})|^2 + |\nu_2(\boldsymbol{r})|^2\} = 1 . \tag{2}$$

The inversion of (1) is obviously

$$\begin{pmatrix} \nu_e \\ \nu_\mu \end{pmatrix} = \begin{pmatrix} \cos\theta & \sin\theta \\ -\sin\theta & \cos\theta \end{pmatrix} \begin{pmatrix} \nu_1 \\ \nu_2 \end{pmatrix} . \tag{3}$$

First we assume that an electron neutrino is created with definite momentum $\boldsymbol{p}$ at point $\boldsymbol{x}$ at time $t = 0$. In the energy representation it holds that, for the time evolution of the state,

$$\begin{aligned} \begin{pmatrix} \nu_1(\boldsymbol{x},t) \\ \nu_2(\boldsymbol{x},t) \end{pmatrix} &= \begin{pmatrix} \nu_1(0)e^{-iE_1t} \\ \nu_2(0)\,e^{-iE_2t} \end{pmatrix} e^{i\boldsymbol{p}\cdot\boldsymbol{x}} \\ &= \begin{pmatrix} e^{-iE_1t} & 0 \\ 0 & e^{-iE_2t} \end{pmatrix} \begin{pmatrix} \nu_1(0) \\ \nu_2(0) \end{pmatrix} e^{i\boldsymbol{p}\cdot\boldsymbol{x}} , \end{aligned} \tag{4}$$

where

$$E_1 = \sqrt{p^2 + m_1^2} , \quad E_2 = \sqrt{p^2 + m_2^2} \tag{5}$$

are the energies and m_1 and m_2 are respectively the masses of the eigenstates ν_1 and ν_2. With the help of (1) and (3) we obtain from (4)

$$\begin{aligned} &\begin{pmatrix} \nu_e(\boldsymbol{x},t) \\ \nu_\mu(\boldsymbol{x},t) \end{pmatrix} \\ &= \begin{pmatrix} \cos\theta & \sin\theta \\ -\sin\theta & \cos\theta \end{pmatrix} \begin{pmatrix} e^{-iE_1t} & 0 \\ 0 & e^{-iE_2t} \end{pmatrix} \begin{pmatrix} \cos\theta & -\sin\theta \\ \sin\theta & \cos\theta \end{pmatrix} \begin{pmatrix} \nu_e(0) \\ \nu_\mu(0) \end{pmatrix} e^{i\boldsymbol{p}\cdot\boldsymbol{x}} \\ &= e^{i\boldsymbol{p}\cdot\boldsymbol{x}} \begin{pmatrix} \cos\theta & \sin\theta \\ -\sin\theta & \cos\theta \end{pmatrix} \begin{pmatrix} \cos\theta e^{-iE_1t} & -\sin\theta e^{-iE_1t} \\ \sin\theta e^{-iE_2t} & \cos\theta e^{-iE_2t} \end{pmatrix} \begin{pmatrix} \nu_e(0) \\ \nu_\mu(0) \end{pmatrix} \\ &= \begin{pmatrix} \cos^2\theta e^{-iE_1t} + \sin^2\theta e^{-iE_2t} & \sin\theta\cos\theta(e^{-iE_2t} - e^{-iE_1t}) \\ \sin\theta\cos\theta(e^{-iE_2t} - e^{-iE_1t}) & \cos^2\theta e^{-iE_2t} + \sin^2\theta e^{-iE_1t} \end{pmatrix} \\ &\quad \times e^{i\boldsymbol{p}\cdot\boldsymbol{x}} \begin{pmatrix} \nu_e(0) \\ \nu_\mu(0) \end{pmatrix} . \end{aligned} \tag{6}$$

Example 6.

According to the assumption that purely neutrinos of the electronic type are emitted at the source, $\nu_e(0) = 1$ and $\nu_\mu(0) = 0$. Hence the probability of finding a muon neutrino at time t is

$$\begin{aligned}|\nu_\mu(\boldsymbol{x},t)|^2 &= \left|\sin\theta\cos\theta(\mathrm{e}^{-\mathrm{i}E_2t} - \mathrm{e}^{-\mathrm{i}E_1t})\right|^2 \\ &= \frac{1}{4}\sin^2(2\theta)|\mathrm{e}^{\mathrm{i}(E_2-E_1)t} - 1|^2 \\ &= \frac{1}{2}\sin^2(2\theta)\left[1-\cos((E_2-E_1)t)\right] \\ &= \sin^2(2\theta)\sin^2\frac{(E_2-E_1)t}{2} \quad . \end{aligned} \tag{7}$$

Under the assumption that the masses m_1 and m_2 are very small compared to p we find that

$$\begin{aligned} E_2 - E_1 &= \sqrt{m_2^2+p^2} - \sqrt{m_1^2+p^2} \\ &\simeq \frac{m_2^2-m_1^2}{2p} \quad . \end{aligned} \tag{8}$$

Furthermore, we can assume that the electron neutrino moves practically at the speed of light, and we therefore put $x = t$, where x is the distance of the neutrino source from the location where the neutrino is detected. Then we obtain

$$\begin{aligned} |\nu_\mu(\boldsymbol{x},t=x)|^2 &\simeq \sin^2(2\theta)\sin^2\frac{\Delta m^2 x}{4p} \\ &= \sin^2(2\theta)\sin^2\frac{\pi x}{l} \ , \end{aligned} \tag{9}$$

with

$$\Delta m^2 = m_2^2 - m_1^2 \ , \qquad l = \frac{4\pi p}{\Delta m^2} \ .$$

The quantity l is called the *oscillation length.* The same result is obtained if we start from a state of definite energy E and make the ansatz

$$\begin{pmatrix}\nu_1(\boldsymbol{x})\\ \nu_2(\boldsymbol{x})\end{pmatrix} = \begin{pmatrix}\mathrm{e}^{\mathrm{i}p_1x}\nu_1(0)\\ \mathrm{e}^{\mathrm{i}p_2x}\nu_2(0)\end{pmatrix} \tag{10}$$

for the spatial wave functions, with

$$p_1 = \sqrt{E^2-m_1^2} \ , \qquad p_2 = \sqrt{E^2-m_2^2} \ , \tag{11}$$

and again set $\nu_e = 1$ and $\nu_\mu = 0$ for $x = 0$. It results that

$$|\nu_\mu(x)|^2 = \sin^2(2\theta)\sin^2\frac{(p_2-p_1)x}{2} \quad . \tag{12}$$

As long as the energy is large compared to the neutrino masses, one obtains

$$p_2 - p_1 = \sqrt{E^2-m_2^2} - \sqrt{E^2-m_1^2} \simeq \frac{m_2^2-m_1^2}{2E} \quad . \tag{13}$$

xample 6.5

Since $E = p$ for massless neutrinos, this expression is identical to (8), and we conclude that the result is independent of the assumptions concerning the preparation of the quantum mechanical state.

In the past few years, several experiments have been searching for neutrino oscillations. One type of experiments was based on the fact that electron antineutrinos with energy of about 4 MeV are produced in nuclear reactors. For $\Delta m^2 \simeq 1\ (\mathrm{eV})^2$, (9) yields an oscillation length of about 10 m; one should then observe a decrease in the amount of electron antineutrinos a few meters from the reactor core. They are detected either through the inverse β-decay reaction

$$\bar{\nu}_{\mathrm{e}} + \mathrm{p} \to \mathrm{n} + \mathrm{e}^+ \quad ,$$

where the neutrons are measured, or, for example, through the branching ratio

$$\bar{\nu}_{\mathrm{e}} + \mathrm{d} \begin{array}{l} \nearrow \quad \mathrm{n} + \mathrm{n} + \mathrm{e}^+ \quad \text{(charged)} \\ \\ \searrow \quad \mathrm{n} + \mathrm{p} + \bar{\nu}_{\mathrm{e}} \quad \text{(neutral)} \end{array} \quad ,$$

since the cross section for neutral currents in practice does not depend on the kind of neutrino. No conclusive evidence for neutrino oscillations was found in reactor experiments.

Another possibility in looking for neutrino oscillations is the fact that muon neutrinos are generated in accelerators; for example, in the decay of K or π mesons. This method has some advantages compared to reactor experiments. One can more accurately determine the number of muon neutrinos created and, since the neutrinos have higher energy, they are more easily detected. A further advantage is the fact that the energy of the particle beam in the accelerator can be varied. For this kind of experiment the neutrino energy is of the order of a few GeV. For a beam energy of 2.5 GeV a mass splitting of $\Delta m^2 = 1\ (\mathrm{eV})^2$ corresponds to an oscillation length of about 1 km. The first such measurements have been reported[12] from the Liquid Scintillator Neutrino Detector (LSND) at the Los Alamos Meson Physics Facility (LAMPF). Muon neutrinos were generated in the decay of pions ($\pi^+ \to \mu^+ + \nu_\mu$), which are produced by stopping an intense 800-MeV proton beam in matter. A large tank filled with 167 tons of scintillation fluid 30 meters behind the neutrino source was used to search for the emergence of electron neutrinos from the oscillation $\nu_\mu \leftrightarrow \nu_{\mathrm{e}}$. In a second experiment the pions were stopped and a beam of muon antineutrinos was prepared, originating from the decay $\mu^+ \to \mathrm{e}^+ + \nu_{\mathrm{e}} + \bar{\nu}_\mu$. The detector was set to search for electron antineutrinos from $\bar{\nu}_\mu \leftrightarrow \bar{\nu}_{\mathrm{e}}$. Positive evidence for oscillations was found in both cases, with an oscillation probability $P_{\nu_\mu \to \nu_{\mathrm{e}}} = (3.1 \pm 1.5) \times 10^{-3}$. Figure 6.17a shows the range of $\sin^2 2\theta$ and Δm^2 values compatible with this result.

Even more compelling evidence for neutrino oscillations, and thus for finite neutrino masses, comes from an experiment detecting atmospheric neutrinos. High-energy cosmic rays (mostly protons) are stopped in the upper atmosphere and generate hadronic showers. The decay of pions ($\pi^+ \to \mu^+ + \nu_\mu \to \mathrm{e}^+ + \bar{\nu}_\mu + \nu_{\mathrm{e}} + \nu_\mu$ and similarly for π^-) leads to a mixture of neutrinos with electronic and muonic flavors bombarding the Earth from all directions. If oscillations occur, the composition of this mixture will change as a function of traveling distance according to (9).

[12] C. Athanassopoulos et al.: Phys. Rev. Lett. **77**, 3082 (1996) and Phys. Rev. Lett. **81**, 1774 (1998).

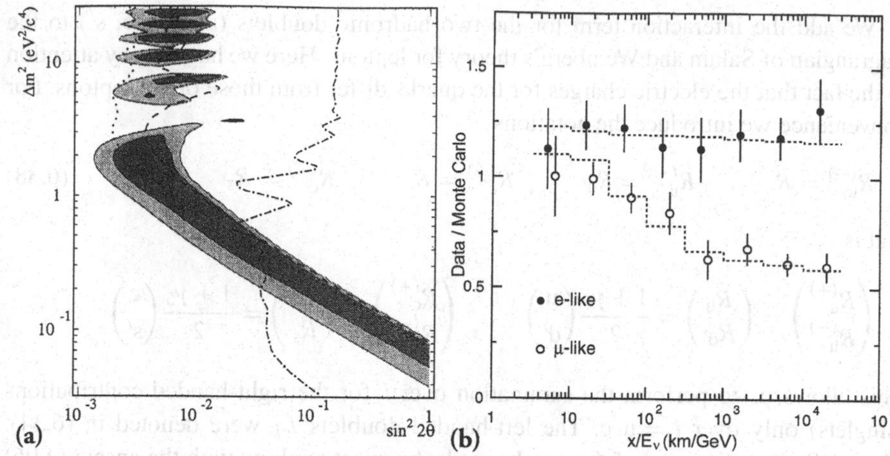

Fig. 6.17. (**a**) Allowed values for the mass splitting and the mixing angle for $\nu_\mu \leftrightarrow \nu_e$ neutrino oscillations according to the LSND experiment. The *light* and *dark shadings* correspond to 99% and 90% confidence levels, respectively. Values to the right of the *dashed–dotted line* are excluded from a reactor experiment. (**b**) Results from the Super-Kamiokande experiment for the ratio of measured to expected neutrino events as a function of the parameter x/E_ν. *Full circles*: ν_e and $\bar{\nu}_e$. *Open circles*: ν_μ and $\bar{\nu}_\mu$

The *Super-Kamiokande*[13] experiment in Japan detects the neutrinos in a huge detector filled with 50 000 tons of purified water, situated deep under ground in a mine. An array of ca 12 000 photomultiplier tubes detecting Cherenkov radiation provides information on the type and direction of neutrinos absorbed in the detector. Depending on the angle of incidence, this allows a large range of traveling distances between $x = 15$ km (atmospheric height) and $x = 13\,000$ km (Earth's diameter) to be probed.

The Super-Kamiokande group found that the ratio of muon–neutrino to electron–neutrino was lower than the expected value. As a function of the variable x/E_ν the number of muon neutrinos was found to drop by a factor of two while the number of electron neutrinos stayed constant; see Fig. 6.17b. This can be explained if flavor oscillations of the kind $\nu_\mu \leftrightarrow \nu_\tau$ occur. (Since tauonic processes are not detected in the experiment, ν_τ could also stand for some other type at neutrinolike particle.)

A detailed analysis of the experimental data indicates nearly complete mixing, $\sin^2 2\theta_{\mu\tau} \simeq 1$, and a small mass difference in the range 5×10^{-4} eV^2 $< \Delta m^2_{\mu\tau} <$ 6×10^{-3} eV^2 (90% confidence limit).

The present status of neutrino-oscillation results is rather complex, pointing to the coexistence of different flavor mixings. Further experiments are under way to clarify this situation.

6.4 Cabibbo's Theory of Flavor Mixing

Cabibbo recognized that the most apparent irregularities of the hadronic weak interactions could be explained by introducing the mixing angle $\theta_1 = \theta_C$ between the d and the s quark. We now want to examine Cabibbo's considerations.

13 Y. Fukuda et al., Phys. Rev. Lett. **81**, 1562 (1998).

We add the interaction term for the two hadronic doublets (u, d′), (c, s′) to the Lagrangian of Salam and Weinberg's theory for leptons. Here we have to pay attention to the fact that the electric charges for the quarks differ from those of the leptons. For convenience we introduce the notations

$$R_{\mathrm{u}}^{(+)} = R_{\mathrm{u}} \quad , \quad R_{\mathrm{u}}^{(-)} = R_{\mathrm{d}'} \quad , \quad R_{\mathrm{c}}^{(+)} = R_{\mathrm{c}} \quad , \quad R_{\mathrm{c}}^{(-)} = R_{\mathrm{s}'} \quad , \tag{6.38}$$

that is,

$$\begin{pmatrix} R_{\mathrm{u}}^{(+)} \\ R_{\mathrm{u}}^{(-)} \end{pmatrix} = \begin{pmatrix} R_{\mathrm{u}} \\ R_{\mathrm{d}'} \end{pmatrix} = \frac{1+\gamma_5}{2} \begin{pmatrix} \mathrm{u} \\ \mathrm{d}' \end{pmatrix} \quad , \quad \begin{pmatrix} R_{\mathrm{c}}^{(+)} \\ R_{\mathrm{c}}^{(-)} \end{pmatrix} = \begin{pmatrix} R_{\mathrm{c}} \\ R_{\mathrm{s}'} \end{pmatrix} = \frac{1+\gamma_5}{2} \begin{pmatrix} \mathrm{c} \\ \mathrm{s}' \end{pmatrix} \quad .$$

This allows us to perform the summation over i for the right-handed contributions (singlets) only over $i = \mathrm{u}, \mathrm{c}$. The left-handed doublets L_i were denoted in (6.34). Then the interaction part of the quarks reads, by exact analogy with the ansatz (4.96) or (15) of Example 4.8 for the leptons,

$$L_{\mathrm{int}}^{\mathrm{quark}} = \sum_{i=\mathrm{u,c}} \Bigg\{ g \bar{L}_i \gamma^\mu \boldsymbol{T} L_i \cdot \boldsymbol{A}_\mu + \frac{1}{2} g' \left(\frac{1}{3} \bar{L}_i \gamma^\mu L_i + \frac{4}{3} \bar{R}_i^{(+)} \gamma^\mu R_i^{(+)} - \frac{2}{3} \bar{R}_i^{(-)} \gamma^\mu R_i^{(-)} \right) B_\mu \Bigg\} \quad . \tag{6.39}$$

The factors $(\frac{1}{3}, \frac{4}{3}, -\frac{2}{3})$ in front of the couplings to the isosinglet field B_μ are the weak hypercharges – see Table 6.2 and also (15) of Example 4.8. They are determined by the electric charges of the quarks, which are multiples of $\frac{1}{3}$; see (6.31), (6.32).

Note here that the interaction for leptons denoted in (4.96) can also be written in the form

$$L_{\mathrm{int}}^{(\ell)} = g(\bar{L}_\ell \gamma^\alpha \boldsymbol{T} L_\ell) \boldsymbol{A}_\alpha + \frac{1}{2} g' \left[Y^{\mathrm{D}} (\bar{L}_\ell \gamma^\mu L_\ell) + Y^{\mathrm{S}} (\bar{R}_\ell \gamma^\mu R_\ell) \right] \quad , \tag{6.39a}$$

where $Y^{\mathrm{D}} = -1$, $Y^{\mathrm{S}} = -2$ are the hypercharges for the leptonic doublet and singlet respectively. This is the some form as used in (6.39)!

If we replace the fields A_μ^3 and B_μ by Z_μ and A_μ, according to (4.98), introduce $\hat{T}_\pm = \hat{T}_1 \pm \mathrm{i}\hat{T}_2$ and $W_\mu^{(\pm)} = \frac{1}{\sqrt{2}}(A_\mu^1 \mp \mathrm{i}A_\mu^2)$ (see (4.99)), then

$$\begin{aligned} L_{\mathrm{int}} = L_{\mathrm{int}}^{(\ell)} + \sum_{i=\mathrm{u,c}} \Bigg\{ & \frac{g}{\sqrt{2}} \bar{L}_i \gamma^\mu (\hat{T}_- W_\mu^{(-)} + \hat{T}_+ W_\mu^{(+)}) L_i \\ & + \Bigg[g \cos\theta \bar{L}_i \gamma^\mu \hat{T}_3 L_i - \frac{1}{3} g' \sin\theta \left(\frac{1}{2} \bar{L}_i \gamma^\mu L_i + 2 \bar{R}_i^{(+)} \gamma^\mu R_i^{(+)} \right. \\ & \left. - \bar{R}_i^{(-)} \gamma^\mu R_i^{(-)} \right) \Bigg] Z_\mu \\ & + \Bigg[\frac{1}{3} g' \cos\theta \left(\frac{1}{2} \bar{L}_i \gamma^\mu L_i + 2 \bar{R}_i^{(+)} \gamma^\mu R_i^{(+)} - \bar{R}_i^{(-)} \gamma^\mu R_i^{(-)} \right) \\ & + g \sin\theta \bar{L}_i \gamma^\mu \hat{T}_3 L_i \Bigg] A_\mu \Bigg\} \end{aligned} \tag{6.40}$$

with the Weinberg angle θ, by analogy to (4.100).

Charged Currents. From the interaction Hamiltonian one can immediately read off the weak currents of quarks as factors to the physical bosons $W^\pm$, Z_μ, A_μ. We rewrite these currents directly by means of the relations

$$\begin{aligned} d' &= \cos\theta_C d + \sin\theta_C s \ , \\ s' &= -\sin\theta_C d + \cos\theta_C s \ , \end{aligned} \tag{6.41}$$

which result from the reduced mixing matrix (6.37), in a form involving the "true" d and s quarks in the sense of mass eigenstates. We start with the charged weak currents, the coefficients of the field $W^{\mu(-)}$ and $W^{\mu(+)}$ respectively:

$$\begin{aligned} J_\mu^- &= 2\sum_{i=\mathrm{u,c}} \bar{L}_i \gamma_\mu \hat{T}_- L_i = \bar{\psi}_{\mathrm{d'}} \gamma_\mu (1-\gamma_5)\psi_\mathrm{u} + \bar{\psi}_{\mathrm{s'}} \gamma_\mu (1-\gamma_5)\psi_\mathrm{c} \\ &= \cos\theta_C \bar{\psi}_\mathrm{d} \gamma_\mu (1-\gamma_5)\psi_\mathrm{u} + \sin\theta_C \bar{\psi}_\mathrm{s} \gamma_\mu (1-\gamma_5)\psi_\mathrm{u} \\ &\quad - \sin\theta_C \bar{\psi}_\mathrm{d} \gamma_\mu (1-\gamma_5)\psi_\mathrm{c} + \cos\theta_C \bar{\psi}_\mathrm{s} \gamma_\mu (1-\gamma_5)\psi_\mathrm{c} \ , \end{aligned} \tag{6.42a}$$

$$J_\mu^+ = (J_\mu^-)^\dagger \ . \tag{6.42b}$$

The part responsible for the decay of the neutron is the term $\cos\theta_C[\bar{\psi}_\mathrm{d}\gamma_\mu(1-\gamma_5)\psi_\mathrm{u}]^\dagger$, since here one of the d quarks in the neutron (udd) decays into a u quark to form a proton (uud). This term appears in the current $J_\mu^+ = 2\sum_{\mathrm{u,c}} \bar{L}_i\gamma_\mu \hat{T}_+ L_i$ and is, according to (6.40), coupled with the (incoming) $\mathrm{W}^{(+)}$ boson. In fact, the field $W_\mu^{(+)} = \frac{1}{\sqrt{2}}(A_\mu^1 - \mathrm{i}A_\mu^2)$ corresponds either to an incoming $\mathrm{W}^{(+)}$ boson or to an outgoing $\mathrm{W}^{(-)}$ boson. In the four-fermion coupling approximation (owing to the low transition energy of 1.9 MeV, this approximation is excellent) the effective interaction for the β decay of the d quark is

$$\begin{aligned} H_\mathrm{int} &= \frac{G}{\sqrt{2}} \int \mathrm{d}^3x\, J_\mu^{(\mathrm{u,d})\dagger}(x) J_{(\mathrm{e})}^\mu(x) \\ &= \frac{G}{\sqrt{2}} \cos\theta_C \int \mathrm{d}^3x \left[\bar{\psi}_\mathrm{u}(x)\gamma_\mu(1-\gamma_5)\psi_\mathrm{d}(x)\right]^\dagger \\ &\quad \times \left[\bar{\psi}_\mathrm{e}(x)\gamma^\mu(1-\gamma_5)\psi_{\nu_\mathrm{e}}(x)\right] \ . \end{aligned} \tag{6.43}$$

A comparison with (6.1), (6.2) shows immediately the relationship

$$C_V = \cos\theta_C(1+\varepsilon_\mathrm{rad}) \ , \tag{6.44}$$

where $\varepsilon_\mathrm{rad} \simeq 0.012$ denotes the electromagnetic radiative corrections. Hence the Cabibbo angle θ_C, which has the experimental value

$$\begin{aligned} \cos\theta_C &= 0.9751 \pm 0.0003 \ , \\ \sin\theta_C &= 0.222 \pm 0.001 \ , \\ \theta_C &= 12.8° \pm 0.1° \ , \end{aligned} \tag{6.45}$$

describes the small decrease of the weak interaction constant in nuclear β decay.

The coupling strength for the decay s → u is proportional to $\sin\theta_C$, which "explains" (or "parametrizes", as we should say), owing to the relative smallness of the angle θ_C, the weakness of the strangeness-violating decays found in the section on the phenomenology of weak hadronic decays. The ratio between the coupling constants f_K of the K decay and f_π of the π decay is immediately given by (s–u transition to d–u transition)

$$\frac{f_K}{f_\pi} = \tan\theta_C(1+\varepsilon'_\mathrm{rad}) \tag{6.46}$$

with small radiative corrections. A detailed analysis gives $\sin\theta_C = 0.2655 \pm 0.0006$, while one finds from the decay of the strange baryons Λ, Σ, and so on the value $\sin\theta_C = 0.230 \pm 0.003$. When judging these numbers one has to take into consideration that they are not directly derived from decays of the quarks, but from decays of more complex particles. Therefore they only reflect the decay properties of quarks to such a degree that the particles consist simply of a quark–antiquark pair (mesons) or of three quarks (baryons), the so-called *valence quarks*. In reality the circumstances are more complex, since the quark vacuum is strongly polarized inside the particles, owing to the interaction between quarks. With a certain probability one therefore finds additional quark–antiquark pairs besides the valence quarks; one also speaks of the *sea* of virtual quark–antiquark pairs. For a proton, which has a valence-quark structure of (uud), the probability of finding it in each of the configurations

$$\mathrm{uud\bar{u}u} \quad , \quad \mathrm{uud\bar{d}d} \quad , \quad \mathrm{uud\bar{s}s}$$

is approximately 5%. To this extent the simple picture of the proton consisting of three quarks is not satisfied. The fact that this violation is relatively small explains the success of the quark model and of SU(3) symmetry when interpreting the low-energy properties of baryons, for example the mass formulas.

For mesons SU(3) symmetry is not nearly as successful. This concerns especially the pion, which fits only poorly into the mass formulas for mesons. One can understand this since, for mesons, which consist of a valence quark–antiquark pair, the admixture of virtual quark–antiquark pairs is essentially stronger. It is likely that configurations of the kind[14]

$$\mathrm{u\bar{d}u\bar{u}} \quad , \quad \mathrm{u\bar{d}d\bar{d}} \quad , \quad \mathrm{u\bar{d}u\bar{u}d\bar{d}}$$

make up a considerable fraction of the pion wave function (here for the π^+) and thus are jointly responsible for the lower mass of the pion. It is therefore not very surprising that by comparing the decay of the pion and the kaon one deduces a value of the Cabibbo angle experimentally that is 15% larger. Most probably this does not mean that Cabibbo theory provides a poor description of the weak interactions among quarks, but only that the quark structure of the pion is considerably more complex than that of other hadrons. This presumption is supported by the good agreement between the values for the Cabibbo angle obtained from the decay of the neutron and the strange baryons. Hence we assume in the following that the value (6.45) obtained from nuclear β decay represents the mixing angle between d and s quarks.

Neutral Weak Currents. Next we consider the coefficient of the field Z^μ, that is the neutral weak current of the quarks (see e.g. second line of (6.40)),

$$\begin{aligned}
J_\mu^{(0)} &= \sum_{i=\mathrm{u,c}} \Bigg[g\cos\theta \bar{L}_i \gamma_\mu \hat{T}_3 L_i \\
&\quad - \frac{1}{3} g' \sin\theta \left(\frac{1}{2} \bar{L}_i \gamma_\mu L_i + 2\bar{R}_i^{(+)} \gamma_\mu R_i^{(+)} - \bar{R}_i^{(-)} \gamma_\mu R_i^{(-)} \right) \Bigg] \\
&= \frac{g}{4\cos\theta} \Bigg[-\bar{\psi}_{\mathrm{d}'} \gamma_\mu \left(1 - \frac{4}{3}\sin^2\theta - \gamma_5 \right) \psi_{\mathrm{d}'} - \bar{\psi}_{\mathrm{s}'} \gamma_\mu \left(1 - \frac{4}{3}\sin^2\theta - \gamma_5 \right) \psi_{\mathrm{s}'}
\end{aligned}$$

[14] See W. Greiner and B. Müller: *Quantum Mechanics – Symmetries*, 2nd ed. (Springer, Berlin, Heidelberg, 1994).

$$+\bar{\psi}_{\mathrm{u}}\gamma_\mu\left(1-\frac{8}{3}\sin^2\theta-\gamma_5\right)\psi_{\mathrm{u}}+\bar{\psi}_{\mathrm{c}}\gamma_\mu\left(1-\frac{8}{3}\sin^2\theta-\gamma_5\right)\psi_{\mathrm{c}}\Bigg]$$

$$=\frac{g}{4\cos\theta}\left[-\bar{\psi}_{\mathrm{d}}\gamma_\mu\left(1-\frac{4}{3}\sin^2\theta-\gamma_5\right)\psi_{\mathrm{d}}-\bar{\psi}_{\mathrm{s}}\gamma_\mu\left(1-\frac{4}{3}\sin^2\theta-\gamma_5\right)\psi_{\mathrm{s}}\right.$$

$$\left.+\bar{\psi}_{\mathrm{u}}\gamma_\mu\left(1-\frac{8}{3}\sin^2\theta-\gamma_5\right)\psi_{\mathrm{u}}+\bar{\psi}_{\mathrm{c}}\gamma_\mu\left(1-\frac{8}{3}\sin^2\theta-\gamma_5\right)\psi_{\mathrm{c}}\right] \quad , \tag{6.47}$$

applying (4.101), i.e. $e = g\sin\theta = g'\cos\theta$. The details of the calculation are the subject of Exercise 6.8. The angle θ which appears here is the Weinberg angle already known to us. In addition there appear d′ and s′ quarks, which are Cabibbo-mixed (see (6.41)). The transformation to the d and s quarks appearing in strong interactions is readily performed using (6.41) and leads, as the last step shows, to formally identical expressions. All mixed terms of the form $\sin\theta_{\mathrm{C}}\cos\theta_{\mathrm{C}}\bar{\psi}_{\mathrm{d}}\gamma_\mu\psi_{\mathrm{s}}$, and so on cancel, as is shown in Exercise 6.8. It is worth emphasizing that undesirable terms in the neutral current of the doublet L_{u} cancel such terms in the current of the doublet L_{c}. Thus the neutral current contains only couplings which are diagonal in the quark flavors. In particular, *there is no part of the neutral current which changes strangeness.* This selection rule, $\Delta S = \Delta Q = 0$, was one of the important experimental demands on the theory (see Sect. 6.2 on the phenomenology of weak hadronic decays). Indeed, the requirement of vanishing strangeness-changing neutral current was the motivation for the (purely theoretical) introduction of the fourth quark c by Glashow, Iliopoulos, and Maiani.[15] Therefore one also speaks of the GIM mechanism.

This consideration can be easily extended to the two additional quarks t and b. For this we need only consider the quarks with charge $Q = -(1/3)$, which are mixed by the unitary matrix $\hat{U}$ (6.35). Their contribution to the neutral current is, in a generalization of (6.47),

$$J_\mu^{(0)}(\mathrm{d}',\mathrm{s}',\mathrm{b}') = -\sum_{i=\mathrm{d}',\mathrm{s}',\mathrm{b}'}\left[\frac{1}{4}g\cos\theta\bar{\psi}_i\gamma_\mu(1-\gamma_5)\psi_i\right]$$

$$+\left[\frac{1}{6}g'\sin\theta\left(\frac{1}{2}\bar{\psi}_i\gamma_\mu(1-\gamma_5)\psi_i-\bar{\psi}_i\gamma_\mu(1+\gamma_5)\psi_i\right)\right]$$

$$=-\frac{g}{4\cos\theta}\sum_{i'=\mathrm{d}',\mathrm{s}',\mathrm{b}'}\bar{\psi}_{i'}\gamma_\mu\left(1-\frac{4}{3}\sin^2\theta-\gamma_5\right)\psi_{i'} \quad , \tag{6.48}$$

where we have used relation (4.101), $g\sin\theta = g'\cos\theta$ for the coupling constants g and g'. Owing to the unitarity of $\hat{U}$ it holds for every operator $\hat{O}$ that

$$\sum_{i'=\mathrm{d}',\mathrm{s}',\mathrm{b}'}\bar{\psi}_{i'}\hat{O}\psi_{i'} = \sum_{i'=\mathrm{d}',\mathrm{s}',\mathrm{b}'}\left(\sum_{k=\mathrm{d,s,b}}\bar{\psi}_k U^*_{i'k}\right)\hat{O}\left(\sum_{l=\mathrm{d,s,b}}U_{i'l}\psi_l\right)$$

$$=\sum_{k=\mathrm{d,s,b}}\bar{\psi}_k\hat{O}\psi_k \quad , \tag{6.49}$$

[15] S.L. Glashow, J.C. Iliopoulos, and L. Maiani: Phys. Rev. **D2**, 1285 (1970). The name "charm" is to be understood as "magic", since the charmed quark eliminates the undesirable terms.

since

$$\sum_{i'} U^*_{i'k} U_{i'l} = \sum_{i'} U^\dagger_{ki'} U_{i'l} = (\hat{U}^\dagger \hat{U})_{kl} = \delta_{kl} \quad . \tag{6.50}$$

Hence we have shown in general that *no mixed-flavor terms appear in the neutral current*:

$$\begin{aligned} J^{(0)}_\mu(\mathrm{d}', \mathrm{s}', \mathrm{b}') &= -\frac{g}{4\cos\theta} \sum_{k=\mathrm{d,s,b}} \bar{\psi}_k \gamma_\mu \left(1 - \frac{4}{3}\sin^2\theta - \gamma_5\right)\psi_k \\ &= J^{(0)}_\mu(\mathrm{d}, \mathrm{s}, \mathrm{b}) \quad . \end{aligned} \tag{6.51}$$

Weak interactions do not lead to transitions of d, s, or b quarks among each other. The same holds also for the three quarks with charge $Q = +(2/3)$, that is, the quarks u, c, and t. Thus we can assert, as a general demand of the Salam–Weinberg theory extended to quarks that *changes of flavor quantum numbers always involve change of charge*. This is a generalization of the rule $\Delta S = \Delta Q = 0$ for neutral currents to $\Delta S = \Delta Q$ for all currents appearing in the theory of electroweak interactions.

The factors

$$-\left(1 - \frac{4}{3}\sin^2\theta - \gamma_5\right) \quad , \quad \left(1 - \frac{8}{3}\sin^2\theta - \gamma_5\right) \quad , \tag{6.52}$$

which appear in the neutral current, can be understood as follows: the sign is determined by the sign of the (weak) isospin T_3, that is, (+) for the upper quarks u, c, t and (−) for the lower components d, s, b. Formally we can therefore write the factors (6.52) in the form

$$2(T_3 - 2Q\sin^2\theta - T_3\gamma_5) = 2T_3(1 - \gamma_5) - 4Q\sin^2\theta \quad . \tag{6.53}$$

This expression also correctly describes the corresponding factors in the neutral current of the leptons (4.106):

$$\begin{aligned} 2(T_3 - 2Q\sin^2\theta - T_3\gamma_5)_\nu &= (1 - \gamma_5) \quad , \\ 2(T_3 - 2Q\sin^2\theta - T_3\gamma_5)_\mathrm{e} &= -(1 - 4\sin^2\theta - \gamma_5) \quad . \end{aligned} \tag{6.54}$$

Hence the contribution of the neutral weak current to the Lagrangian has the general form

$$\frac{g}{2\cos\theta}\left[\sum_k \bar{\psi}_k \gamma^\mu (T_3 - 2Q\sin^2\theta - T_3\gamma_5)_k \psi_k\right] Z_\mu \quad , \tag{6.55}$$

where the sum runs over all leptons and quarks.

EXERCISE

6.6 Proof of (6.55)

Problem. Starting with (6.30) and (6.31), show the general validity of (6.55).

Exercise 6.

Solution. The fermion coupling to the fields A_μ^3 and B_μ is, according to (6.30),

$$\frac{1}{2}g\bar{\psi}\gamma^\mu(1-\gamma_5)T_3\psi A_3^\mu + \frac{1}{4}g'\left[Y^L\bar{\psi}\gamma^\mu(1-\gamma_5)\psi + Y^R\bar{\psi}\gamma^\mu(1+\gamma_5)\psi\right]B_\mu \quad . \quad (1)$$

We obtain the current which couples to the neutral intermediate boson Z_μ using (4.98):

$$B_\mu = A_\mu\cos\theta - Z_\mu\sin\theta \quad , \quad (2)$$

$$A_\mu^3 = Z_\mu\cos\theta + A_\mu\sin\theta \quad . \quad (3)$$

Collecting all terms and putting $g' = g\tan\theta$, we get

$$\frac{1}{2}g\bar{\psi}\gamma^\mu\Big\{(1-\gamma_5)T_3\cos\theta - \frac{1}{2}\tan\theta\left[Y^L(1-\gamma_5)+Y^R(1+\gamma_5)\right]\sin\theta\Big\}\psi Z_\mu \quad , \quad (4)$$

$$Y^L = 2(Q-T_3) \quad , \quad Y^R = +2Q \quad , \quad (5)$$

and obtain the following for the factor in square brackets:

$$(1-\gamma_5)T_3\cos\theta + (1-\gamma_5)T_3\tan\theta\sin\theta - 2Q\tan\theta\sin\theta = \frac{1}{\cos\theta}\left[(1-\gamma_5)T_3 - 2Q\sin^2\theta\right] \quad . \quad (6)$$

Inserting this into (4), we obtain the desired result

$$\frac{g}{2\cos\theta}\bar{\psi}\gamma^\mu\left[(1-\gamma_5)T_3 - 2Q\sin^2\theta\right]\psi Z_\mu \quad . \quad (7)$$

EXAMPLE

6.7 Absence of Flavor-Changing Neutral Currents

If there were a contribution to the neutral current (6.47), which couples d and s quarks, that is,

$$J_\mu^{(0)\mathrm{fc}} = \frac{g}{4\cos\theta}\left[\bar{\psi}_\mathrm{d}\gamma_\mu(C_V^\mathrm{fc} + C_A^\mathrm{fc}\gamma_5)\psi_\mathrm{s} + h.c.\right] \quad , \quad (1)$$

the neutral K mesons could easily decay into a muon pair. This is illustrated in the following diagram:

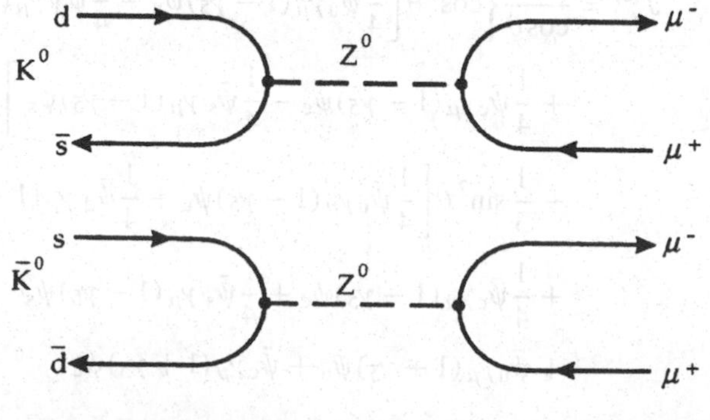

Fig. 6.18. Possible decay scheme of the neutral K mesons if a flavor-changing neutral current exists

xample 6.7

Experimentally, this decay is conveniently studied by investigating the long-lived K_L^0 state (see Chap. 7), for which a branching ratio

$$\frac{\Gamma(K_L^0 \mapsto \mu^+\mu^-)}{\Gamma(K_L^0)} = 9 \times 10^{-9} \tag{2}$$

was measured. From this we obtain a bound of

$$|C_V^{fc}|^2 + |C_A^{fc}|^2 < 10^{-8} \quad . \tag{3}$$

This limit is still further reduced by taking account of the fact that the experimentally measured value is well described by radiative corrections to processes which result from second-order effects of the charged weak current, for example,

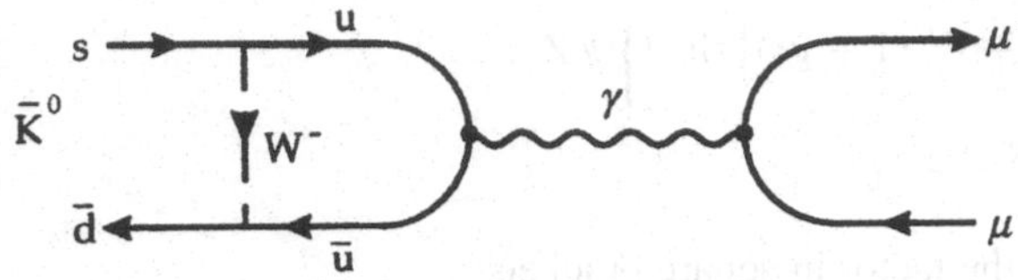

The value (2) is therefore a strong indication for the complete absence of fundamental flavor-changing weak currents.

EXERCISE

6.8 Vanishing of Mixed Currents Between d and s Quarks

Problem. Check the result (6.47) for the neutral currents and show that no mixed terms coupling s and d quarks arise through Cabibbo mixing.

Solution. With (4.101) we have

$$g' \sin\theta = \frac{g \sin^2\theta}{\cos\theta} \quad , \tag{1}$$

and hence, from (6.47),

$$\begin{aligned} J_\mu^{(0)} = \frac{g}{\cos\theta}\Big\{ & \cos^2\theta \Big[\frac{1}{4}\bar{\psi}_u\gamma_\mu(1-\gamma_5)\psi_u - \frac{1}{4}\bar{\psi}_{d'}\gamma_\mu(1-\gamma_5)\psi_{d'} \\ & + \frac{1}{4}\bar{\psi}_c\gamma_\mu(1-\gamma_5)\psi_c - \frac{1}{4}\bar{\psi}_{s'}\gamma_\mu(1-\gamma_5)\psi_{s'}\Big] \\ & - \frac{1}{3}\sin^2\theta\Big[\frac{1}{4}\bar{\psi}_u\gamma_\mu(1-\gamma_5)\psi_u + \frac{1}{4}\bar{\psi}_{d'}\gamma_\mu(1-\gamma_5)\psi_{d'} \\ & + \frac{1}{4}\bar{\psi}_c\gamma_\mu(1-\gamma_5)\psi_c + \frac{1}{4}\bar{\psi}_{s'}\gamma_\mu(1-\gamma_5)\psi_{s'} \\ & + \bar{\psi}_u\gamma_\mu(1+\gamma_5)\psi_u + \bar{\psi}_c\gamma_\mu(1+\gamma_5)\psi_u \end{aligned}$$

Exercise 6.

$$-\frac{1}{2}\bar{\psi}_{d'}\gamma_\mu(1+\gamma_5)\psi_{d'}-\frac{1}{2}\bar{\psi}_{s'}\gamma_\mu(1+\gamma_5)\psi_{s'}\Big]\Big\}$$

$$=\frac{g}{4\cos\theta}\Big[\bar{\psi}_u\gamma_\mu\left(1-\frac{3}{8}\sin^2\theta-\gamma_5\right)\psi_u+\bar{\psi}_c\gamma_\mu\left(1-\frac{3}{8}\sin^2\theta-\gamma_5\right)\psi_c$$

$$-\bar{\psi}_{d'}\gamma_\mu\left(1-\frac{4}{3}\sin^2\theta-\gamma_5\right)\psi_{d'}-\bar{\psi}_{s'}\gamma_\mu\left(1-\frac{4}{3}\sin^2\theta-\gamma_5\right)\psi_{s'}\Big] \quad . \tag{2}$$

Here we have made use of the intermediate steps

$$\cos^2\theta(1-\gamma_5)-\frac{1}{3}\sin^2\theta[(1-\gamma_5)+4(1+\gamma_5)]$$
$$=\cos^2\theta-\frac{5}{3}\sin^2\theta-\gamma_5(\cos^2\theta+\sin^2\theta)$$
$$=1-\frac{8}{3}\sin^2\theta-\gamma_5 \quad , \tag{3a}$$

$$-\cos^2\theta(1-\gamma_5)-\frac{1}{3}\sin^2\theta[(1-\gamma_5)-2(1+\gamma_5)]$$
$$=-\cos^2\theta+\frac{1}{3}\sin^2\theta+\gamma_5(\cos^2\theta+\sin^2\theta)$$
$$=-1+\frac{4}{3}\sin^2\theta+\gamma_5 \quad . \tag{3b}$$

The last two terms in (2) we rewrite in terms of the quark flavors d, s, that is, the mass eigenstates, with the help of (6.41). For an arbitrary operator $\hat{A}$ it holds that

$$\bar{\psi}_{d'}\hat{A}\psi_{d'}+\bar{\psi}_{s'}\hat{A}\psi_{s'}$$
$$=(\cos\theta_C\bar{\psi}_d+\sin\theta_C\bar{\psi}_s)\hat{A}(\cos\theta_C\psi_d+\sin\theta_C\psi_s)$$
$$+(\cos\theta_C\bar{\psi}_s-\sin\theta_C\bar{\psi}_d)\hat{A}(\cos\theta_C\bar{\psi}_s-\sin\theta_C\psi_d)$$
$$=(\cos^2\theta_C+\sin^2\theta_C)\bar{\psi}_d\hat{A}\psi_d+(\cos\theta_C\sin\theta_C-\sin\theta_C\cos\theta_C)\bar{\psi}_d\hat{A}\psi_s$$
$$+(\sin\theta_C\cos\theta_C-\cos\theta_C\sin\theta_C)\bar{\psi}_s\hat{A}\psi_d+(\sin^2\theta_C+\cos^2\theta_C)\bar{\psi}_s\hat{A}\psi_s$$
$$=\bar{\psi}_d\hat{A}\psi_d+\bar{\psi}_s\hat{A}\psi_s \quad . \tag{4}$$

This is simply a consequence of the orthogonality of the Cabibbo transformation. Altogether we have

$$J_\mu^{(0)}=\frac{g}{4\cos\theta}\sum_{i=u,d,s,c}\varepsilon_i\bar{\psi}_i\gamma_\mu(C_i-\gamma_5)\psi_i \tag{5}$$

with

$$\varepsilon_u=\varepsilon_c=1 \quad , \quad \varepsilon_d=\varepsilon_s=-1 \quad , \tag{6a}$$

$$C_u=C_d=1-\frac{8}{3}\sin^2\theta \quad ,$$
$$C_d=C_s=1-\frac{4}{3}\sin^2\theta \tag{6b}$$

which agrees exactly with (6.47).

Electromagnetic Current. Finally we consider the electromagnetic current, which is given by (see (6.40))

$$\begin{aligned}J_\mu^{(\mathrm{em})} &= \sum_{i=\mathrm{u,c}}\Big[g\sin\theta\,\bar{L}_i\gamma_\mu T_3 L_i + \frac{1}{3}g'\cos\theta \\ &\quad\times\left(\frac{1}{2}\bar{L}_i\gamma_\mu L_i + 2\bar{R}_i^{(+)}\gamma_\mu R_i^{(+)} - \bar{R}_i^{(-)}\gamma_\mu R_i^{(-)}\right)\Big] \\ &= \frac{2}{3}e(\bar{\psi}_\mathrm{u}\gamma_\mu\psi_u + \bar{\psi}_\mathrm{c}\gamma_\mu\psi_\mathrm{c}) - \frac{1}{3}e(\bar{\psi}_{\mathrm{d}'}\gamma_\mu\psi_{\mathrm{d}'} + \bar{\psi}_{\mathrm{s}'}\gamma_\mu\psi_{\mathrm{s}'}) \\ &= \frac{2}{3}e(\bar{\psi}_\mathrm{u}\gamma_\mu\psi_\mathrm{u} + \bar{\psi}_\mathrm{c}\gamma_\mu\psi_\mathrm{c}) - \frac{1}{3}e(\bar{\psi}_\mathrm{d}\gamma_\mu\psi_\mathrm{d} + \bar{\psi}_\mathrm{s}\gamma_\mu\psi_\mathrm{s}) \quad . \end{aligned} \tag{6.56}$$

As can be seen, it again contains no mixed terms in d and s quarks. Regarding the sign one should note that $e = -|e|$ is the charge of the electron. Of course, the factors coupling to the isospin-singlet field B_μ were chosen in such a way that the electromagnetic current comes out correctly. Since the coupling to the isotriplet field $\boldsymbol{A}_\mu$ and the singlet field B_μ can be chosen completely independently, this expresses no deeper physical understanding. After fixing the coupling of the charged weak currents (6.42) and the electromagnetic current (6.56), however, one has no options left for the neutral weak current $J_\mu^{(0)}$. The exact form (6.47) of $J_\mu^{(0)}$ must thus be regarded as a definite *prediction* of the extended Glashow–Salam–Weinberg theory.

Therefore we consider the neutral currents of u and d quarks in more detail. When splitting them into left-handed and right-handed parts one obtains for the u quarks

$$\begin{aligned}\frac{1}{g}\cos\theta\,J_\mu^{(0)}(\mathrm{u}) &= \bar{\psi}_\mathrm{u}\gamma_\mu\left(\frac{1}{4} - \frac{2}{3}\sin^2\theta - \frac{1}{4}\gamma_5\right)\psi_\mathrm{u} \\ &= \left(\frac{1}{2} - \frac{2}{3}\sin^2\theta\right)\bar{\mathrm{u}}_L\gamma_\mu\mathrm{u}_L - \frac{2}{3}\sin^2\theta\,\bar{\mathrm{u}}_R\gamma_\mu\mathrm{u}_R \quad ,\end{aligned} \tag{6.57}$$

and for the d quarks

$$\begin{aligned}\frac{1}{g}\cos\theta\,J_\mu^{(0)}(\mathrm{d}) &= -\bar{\psi}_\mathrm{d}\gamma_\mu\left(\frac{1}{4} - \frac{1}{3}\sin^2\theta - \frac{1}{4}\gamma_5\right)\psi_\mathrm{d} \\ &= \left(-\frac{1}{2} + \frac{1}{3}\sin^2\theta\right)\bar{\mathrm{d}}_L\gamma_\mu\mathrm{d}_L + \frac{1}{3}\sin^2\theta\,\bar{\mathrm{d}}_R\gamma_\mu\mathrm{d}_R \quad .\end{aligned} \tag{6.58}$$

With the value $\sin^2\theta \approx 0.25$ from (4.108) we obtain Table 6.3 for the coupling constants of the weak neutral current for u and d quarks. For comparison, values extracted from an analysis of neutrino–quark scattering are also given in Table 6.3 (see Example 6.9). We can assert good agreement in the range of experimental errors, especially regarding the isospin dependence. A precise check of the structure of the neutral

Table 6.3. The weak neutral-current couplings of the up and down quark

	Coupling constant	Theory	Experiment
u_L	$1/2 - 2/3\sin^2\theta$	0.33	0.35 ± 0.07
u_R	$-2/3\sin^2\theta$	−0.17	-0.19 ± 0.06
d_L	$-1/2 + 1/3\sin^2\theta$	−0.42	-0.40 ± 0.07
d_R	$1/3\sin^2\theta$	0.08	0.00 ± 0.11

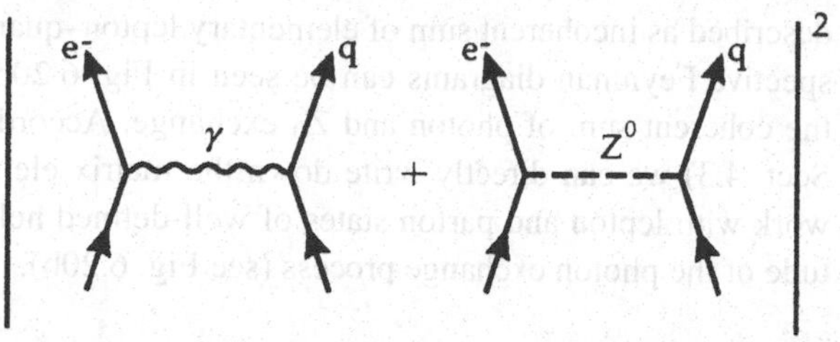

Fig. 6.19. Matrix element electron–quark scattering

quark current was performed at the linear electron accelerator in Stanford, California (SLAC). In this experiment the degree of parity violation in the scattering of polarized electrons at deuterons was investigated. Here the most important contribution stems from the interference term between the neutral weak current and the electromagnetic current (Fig. 6.19), as in the case of the angular asymmetry in muon pair creation (see Sect. 5.2). The experiment is discussed in detail in Example 6.9. From a comparison with the theoretically predicted degree of parity violation, there results a rather accurate value for the Weinberg angle,[16]

$$\sin^2\theta = 0.230 \pm 0.005 \quad , \tag{6.59}$$

which is in good agreement with the result obtained from purely leptonic scattering experiments. In conclusion, we remark that the isospin structure of the neutral current in the Glashow–Salam–Weinberg theory can be regarded as experimentally well confirmed.

EXAMPLE

6.9 Parity Violation in Inelastic Lepton–Nucleon Scattering

The substructure of the nucleon is investigated in deep inelastic lepton–nucleon scattering experiments. Hereby the energy–momentum transfer is large enough to break up the nucleon and the scattering can be described as elementary quark–nucleon scattering.

In order to investigate the degree of parity violation in this process, polarized electrons (muons) are scattered at unpolarized nuclear targets. In the experiment one measures the following asymmetry

$$\Delta = \frac{d\sigma_R - d\sigma_L}{d\sigma_R + d\sigma_L} \quad , \tag{1}$$

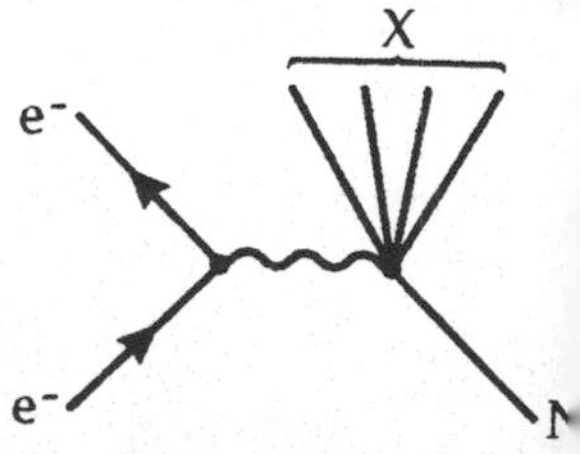

Fig. 6.20a. Deep inelastic electron–nucleon scattering

with $\sigma_{R/L}$ being the scattering cross section between a right/left-handed lepton and the target $e_{R/L} + N \to e_{R/L} + X$. The X are the fragments of the target which are not observed in the experiment and over which a sum has to be performed (see Fig. 6.20a).

A non-vanishing asymmetry can be interpreted as proof for a parity-violating process. In order to predict the asymmetry in the framework of the GSW model, we use the parton model of the nucleon. In this model the nucleon is composed of free non-interacting partons (quarks). Each lepton–nucleon scattering process can then be

[16] P. Langacker and A.K. Mann: Physics Today, **42**, 22 (1989).

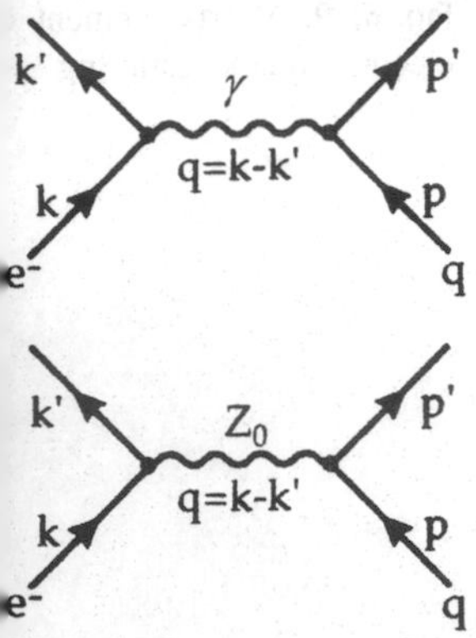

ig. 6.20b. Elementary Feynnan graphs of electron–quark cattering

described as incoherent sum of elementary lepton–quark scattering processes. The respective Feynman diagrams can be seen in Fig. 6.20b. Parity violation results from the coherent sum of photon and Z_0 exchange. According to the Feynman rules (see Sect. 4.3) we can directly write down the matrix elements. However, it is useful to work with lepton and parton states of well-defined helicity. We obtain for the amplitude of the photon exchange process (see Fig. 6.20b).

$$A_\gamma^{hH} = -i\,Q_e Q_q e^2 \bar{u}_e^h(k')\gamma_\alpha u_e^h(k)\frac{1}{q^2}\bar{u}_q^H(p')\gamma^\alpha u_q^H(p) \quad , \tag{2}$$

where Q_q is the electric charge numbers of the parton, Q_e that of the electron. h, H are the helicities of the electron and parton, respectively, which in the high-energy limit can be replaced by chiralities. For instance the spinor of a right-handed electron is then given by (see Sect. 1.2)

$$u_e^R(k) = \hat{P}' u_e(k) = \frac{1+\gamma_5}{2} u_e(k) \quad . \tag{3}$$

Note that any vector interaction preserves chirality, so that the initial and final values of h and H, respectively, are the same. In order to obtain the amplitude of the Z_0 exchange, we use the neutral weak current of (6.55):

$$\frac{g}{2\cos\theta}\left[\bar{\psi}\gamma_\mu(\underbrace{T_3 - 2Q\sin^2\theta}_{c_V} - \underbrace{T_3}_{c_A}\gamma_5)\psi\right] \tag{4}$$

and rewrite the following term

$$c_V - c_A\gamma_5 = \underbrace{(c_V - c_A)}_{c_R}\frac{1}{2}(1+\gamma_5) + \underbrace{(c_V + c_A)}_{c_L}\frac{1}{2}(1-\gamma_5) \quad . \tag{5}$$

Here we have introduced charges for the right- and left-handed leptons and partons:

$$c_R = -2Q\sin^2\theta \quad , \tag{6}$$

$$c_L = 2T_3 - 2Q\sin^2\theta \quad . \tag{7}$$

In analogy to (2) we obtain for the amplitude of the Z_0 exchange

$$A_Z^{hH} = -i\frac{g^2}{4\cos^2\theta} c_H^q c_h^e \bar{u}_e^h(k')\gamma^\alpha u_e^h(k)\frac{g_{\alpha\beta} - q_\alpha q_\beta/M_Z^2}{q^2 - M_Z^2}\bar{u}_q^H(p')\gamma^\beta u_q^H(p) \quad . \tag{8}$$

For energies which are small compared to the mass of the Z_0, we can approximate the equation by

$$\begin{aligned} A_Z^{hH} &= +i\frac{g^2}{4\cos^2\theta M_Z^2} c_H^q c_h^e \bar{u}_e^h(k')\gamma^\alpha u_e^h(k)\bar{u}_q^H(p')\gamma_\alpha u_q^H(p) \\ &= i\sqrt{2}G c_H^q c_h^e \bar{u}_e^h(k')\gamma^\alpha u_e^h(k)\bar{u}_q^H(p')\gamma_\alpha u_q^H(p) \quad . \end{aligned} \tag{9}$$

In order to obtain the cross section, both amplitudes have to be added coherently. Due to the choice of working with spinors of well-defined helicity, the interference term depends only on the strength of the vertices and the different propagators. The resulting cross section is

$$\mathrm{d}\sigma_{Hh} \sim \left| \frac{Q_{\mathrm{e}} Q_q e^2}{q^2} - \sqrt{2} G c_H^q c_h^{\mathrm{e}} \right|^2 \int \frac{\mathrm{d}^3 p'}{(2\pi)^3} \frac{\mathrm{d}^3 k'}{(2\pi)^3} (2\pi)^4 \delta^4(p + k - p' - k')$$
$$\times \left| \bar{u}_{\mathrm{e}}^h(k') \gamma^\alpha u_{\mathrm{e}}^h(k) \bar{u}_q^H(p') \gamma_\alpha u_q^H(p) \right|^2 \quad . \tag{10}$$

The contraction of the currents and the solution of the integrals are performed in the same fashion as in Sect. 5.2. All fermions are considered to be massless. To evaluate the squared matrix element in (10) we sum over all spin combinations, picking out the correct helicities with the help of chiral projection operators. This leads to

$$M_{Hh} = |\bar{u}_h(k') \gamma^\alpha u_h(k) \bar{u}_H(p') \gamma_\alpha u_H(p)|^2$$
$$= \sum_{s,s',S,S'} |\bar{u}(k' s') \gamma^\alpha P'_h u(k, s) \bar{u}(p' S') \gamma_\alpha P'_H u(p, S)|^2$$
$$= \frac{1}{2m_e 2m_q} \mathrm{Tr}(\not{k}' \gamma^\alpha P'_h \not{k} \gamma^\beta) \, \mathrm{Tr}(\not{p}' \gamma_\alpha P'_H \not{p} \gamma_\beta) \quad . \tag{11}$$

Evaluation of the traces (compare (5.30)–(5.32)) leads to the result (we drop constant factors)

$$M_{Hh} \sim k \cdot p k' \cdot p' (1 + hH) + k \cdot p' k' \cdot p (1 - hH) \quad . \tag{12}$$

Different results are obtained for the case of equal ($h = H = \pm 1$) or opposite ($h = -H = \pm 1$) helicities. In the center of momentum (c.m.) frame we have $k \cdot p = k' \cdot p' = 2E^2$ and $k \cdot p' = k' \cdot p = E^2(1 + \cos\vartheta)$, where ϑ is the scattering angle. Thus we obtain the following qualitative behavior for the cross sections of the different processes:

$$\frac{\mathrm{d}\sigma_{\mathrm{RR}}}{\mathrm{d}\cos\vartheta} \sim \left| \frac{Q_{\mathrm{e}} Q_q e^2}{q^2} - \sqrt{2} G c_{\mathrm{R}}^q c_{\mathrm{R}}^{\mathrm{e}} \right|^2 \quad , \tag{13}$$

$$\frac{\mathrm{d}\sigma_{\mathrm{LL}}}{\mathrm{d}\cos\vartheta} \sim \left| \frac{Q_{\mathrm{e}} Q_q e^2}{q^2} - \sqrt{2} G c_{\mathrm{L}}^q c_{\mathrm{L}}^{\mathrm{e}} \right|^2 \quad , \tag{14}$$

$$\frac{\mathrm{d}\sigma_{\mathrm{RL}}}{\mathrm{d}\cos\vartheta} \sim \left| \frac{Q_{\mathrm{e}} Q_q e^2}{q^2} - \sqrt{2} G c_{\mathrm{R}}^q c_{\mathrm{L}}^{\mathrm{e}} \right|^2 \frac{1}{4} (1 + \cos\vartheta)^2 \quad , \tag{15}$$

$$\frac{\mathrm{d}\sigma_{\mathrm{LR}}}{\mathrm{d}\cos\vartheta} \sim \left| \frac{Q_{\mathrm{e}} Q_q e^2}{q^2} - \sqrt{2} G c_{\mathrm{L}}^q c_{\mathrm{R}}^{\mathrm{e}} \right|^2 \frac{1}{4} (1 + \cos\vartheta)^2 \quad . \tag{16}$$

Usually the cross section is not given differentially towards the non-covariant variable ϑ. Instead, the covariant variable $y = (p_N \cdot q)/(p_N \cdot k)$ is introduced. p_N is the four-momentum of the nucleon. In the parton model it depends in a trivial fashion on the four-momentum p of the parton: $p = x p_N$. The x is the Bjorken x, which denotes the fraction of the total momentum carried by the scattered parton. Both, x and y can only have values between 0 and 1. In the rest frame of the target we have

$$y = \frac{p_N \cdot q}{p_N \cdot k} = \frac{p_N \cdot (k - k')}{p_N \cdot k} = \frac{E - E'}{E} \quad . \tag{17}$$

Example 6.9

In the c.m. system the scattering angle ϑ can be expressed via y:

$$1-y=\frac{p_N\cdot k'}{p_N\cdot k}=\frac{p\cdot k'}{p\cdot k}=\frac{p_0^2+\boldsymbol{k}\cdot\boldsymbol{k}'}{p_0^2+\boldsymbol{p}^2}=\frac{1}{2}(1+\cos\vartheta)\quad . \tag{18}$$

Here we have used in the c.m. system of massless particles $p=(p^0,\boldsymbol{p})$, $k=(k^0,\boldsymbol{k})=(p^0,-\boldsymbol{p})$ and $k'=(k^0,\boldsymbol{k}')$.

Now we want to transfer the results of the lepton–parton scattering to lepton–nucleon scattering. We can directly use the previous results if we are aware of the following simplifications: by neglecting the sea-quarks we do not take any scattering between leptons and anti-quarks into account.

To simplify matters the target is regarded as an isoscalar and therefore is supposed to contain the same number of *up* and *down* quarks. This is, e.g., the case for the deuteron. To evaluate the asymmetry Δ the cross sections (13)–(16) have to be summed over the hadron helicities H and the quark flavors $q=u,d$. The latter summation leads to, e.g.,

$$\begin{aligned}\partial\sigma_{RR}&=\left|\frac{Q_eQ_ue^2}{q^2}-\sqrt{2}Gc_R^uc_R^e\right|^2\left|\frac{Q_eQ_de^2}{q^2}-\sqrt{2}Gc_R^dc_R^e\right|^2\\&\simeq\frac{Q_e^2e^4}{q^4}(q_u^2+Q_d^2)-2\frac{\sqrt{2}Ge^2}{q^2}Q_ec_R^e(Q_uc_R^u+Q_dc_R^d)\quad .\end{aligned}\tag{19}$$

The squared Z_0 exchange term has been dropped since it is very small. The numerator in (1) results from the $\gamma-Z_0$ interference term, while the denominator is dominated by the electromagnetic cross section.

Insertion of (13) and the corresponding expressions for the other helicity combinations leads to the following asymmetry

$$\Delta=-q^2\frac{\sqrt{2}G}{e^2}\frac{6}{5}\left(c_1+c_2\frac{1-(1-y)^2}{1+(1-y)^2}\right)\quad , \tag{20}$$

where the charge numbers $Q_u=2/3$, $Q_d=-1/3$, $Q_e=-1$ have been inserted. The coefficients read

$$\begin{aligned}c_1&=c_A^e(2c_V^u-c_V^d)\quad ,\\c_2&=c_V^e(2c_A^u-c_A^d)\quad .\end{aligned}\tag{21}$$

According to the definition in (4)

$$\begin{aligned}c_V^e&=-\frac{1}{2}+2\sin^2\theta\quad ,\\c_A^e&=-\frac{1}{2}\quad ,\end{aligned}\tag{22}$$

$$\begin{aligned}c_V^u&=+\frac{1}{2}-\frac{4}{3}\sin^2\theta\quad ,\\c_A^u&=+\frac{1}{2}\quad ,\end{aligned}\tag{23}$$

$$\begin{aligned}c_V^d&=-\frac{1}{2}+\frac{2}{3}\sin^2\theta\quad ,\\c_A^d&=-\frac{1}{2}\quad .\end{aligned}\tag{24}$$

Example 6.

Using $\sqrt{2}G = e^2/(4\mu_z^2 \sin^2\theta\cos^2\theta)$, (14) can also be written as

$$\Delta = \left(\frac{1}{2\sin\theta\cos\theta}\right)^2 \frac{q^2}{\mu_z^2}\frac{9}{10}\left[\left(1-\frac{20}{9}\sin^2\theta\right)+(1-4\sin^2\theta)\frac{1-(1-y)^2}{1+(1-y)^2}\right]$$

The first experiment looking for such asymmetry was run at the Stanford Linear Accelerator (SLAC) in 1979, where a polarized electron beam was available with an energy of 19.4 GeV. The scattered electrons were measured in the energy range 10.2–16.3 GeV. Since the asymmetry Δ is of the order of 10^{-4}, the experiment set high demands on the accuracy, especially regarding the correction of systematic errors.[17] The result, $\Delta(y)$, is shown in Fig. 6.21. The experiments clearly prove the presence of a parity-violating effect in electron–nuclear scattering. Its sign and magnitude agree well with the prediction of the standard model using the Weinberg angle

$$\sin^2\theta = 0.2259 \pm 0.0046 \quad ,$$

which coincides with results from other experiments. By additional measurement of the asymmetry in the scattering on protons, the weak coupling constants of u and d quarks can be determined separately. They are given in Table 6.3.

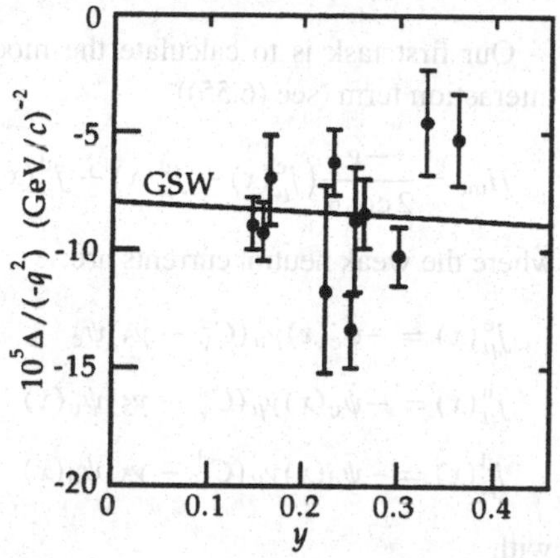

Fig. 6.21. Measured asym metry in the scatterin of polarized electrons a deuterons

EXAMPLE

6.10 Parity Violation in Atoms

Quantum electrodynamics, that is, the theory of electromagnetic interactions, is surely the most accurately checked theory in physics. The highest accuracy is reached in atomic-physics experiments. For example, one can think of the measurement of the Lamb shift in hydrogen atoms.[18] The Glashow–Salam–Weinberg theory predicts that the Z boson couples to the same particles as the photon. Besides virtual photons, virtual Z bosons are also exchanged between the atomic nucleus and the bound electrons.

[17] E.D. Commins and P.H. Bucksbaum: Ann. Rev. Nucl. Part. Sci. **30**, 1 (1980).
[18] See W. Greiner and J. Reinhardt: *Quantum Electrodynamics*, 4th ed. (Springer, Berlin, Heidelberg, 2009).

xample 6.10

This additional interaction leads to a modification of the wave function of bound electrons.

Although these effects are very small, it has been possible to detect them[19] because the interaction through neutral currents violates parity. The exchange of virtual Z^0 bosons modifies the Coulomb potential. The parity violation connected with this exchange causes the atomic levels to acquire a small admixture of the "wrong" parity (see Fig. 6.22).

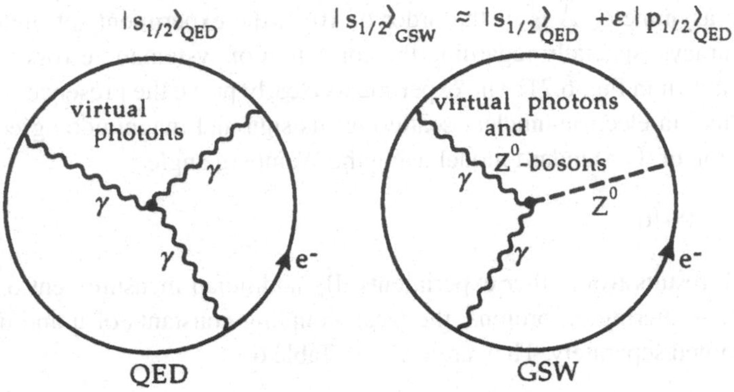

ig. 6.22. Electron–nucleus nteraction by means of a hoton and Z^0

Our first task is to calculate the modified potential. To do so we start with the interaction term (see (6.55))

$$H_{\text{int}} = \frac{-g}{2\cos\theta}\left(j^{\text{e}}_\mu(x) + j^{\text{u}}_\mu(x) + j^{\text{d}}_\mu(x)\right)Z^\mu \quad , \tag{1}$$

where the weak neutral currents are:

$$j^{\text{e}}_\mu(x) = -\bar{\psi}_{\text{e}}(x)\gamma_\mu(C^{\text{e}}_V - \gamma_5)\psi_{\text{e}} \quad , \tag{2a}$$

$$j^{\text{u}}_\mu(x) = -\bar{\psi}_{\text{u}}(x)\gamma_\mu(C^{\text{u}}_V - \gamma_5)\psi_{\text{u}}(x) \quad , \tag{2b}$$

$$j^{\text{d}}_\mu(x) = -\bar{\psi}_{\text{d}}(x)\gamma_\mu(C^{\text{d}}_V - \gamma_5)\psi_{\text{d}}(x) \quad , \tag{2c}$$

with

$$C^{\text{e}}_V = 1 - 4\sin^2\theta \quad , \tag{3a}$$

$$C^{\text{u}}_V = 1 - \frac{8}{3}\sin^2\theta \quad , \tag{3b}$$

$$C^{\text{d}}_V = 1 - \frac{4}{3}\sin^2\theta \quad . \tag{3c}$$

For simplicity, we consider only the valence quarks of the atomic nuclei. For virtual Z bosons in the low-energy limit (that is, $k^2 \ll M_Z^2$) this interaction is, according to (4.131) and (4.132), equivalent to the following Fermi-type current–current interaction:

$$H^{\text{eff}}_{\text{qe}} = \frac{G}{2\sqrt{2}}\, j^{\text{e}}_\mu(x)\left(j^{\text{u}\mu}(x) + j^{\text{d}\mu}(x)\right) \quad , \tag{4}$$

[19] M.A. Bouchiat, J. Guena, L. Hember, L. Pottier: Phys. Lett. **B117**, 358 (1982); P. Bucksbaum, E. Commins, L. Hember: Phys. Rev. Lett. **46**, 640 (1981); J.H. Mollister, G.R. Apperson, L.L. Lewis, T.P. Emmons, T.G. Vold, E.N. Fortson: Phys. Rev. Lett. **46**, 643 (1981).

Example 6.1

where $G = 1.166 \times 10^{-5}\,\mathrm{GeV}^{-2}$ is the Fermi coupling constant. In principle (4) has to be evaluated by using the wave functions for the quarks inside a nucleon, which can only be done in an approximate and model-dependent way.[20] As a simple prescription, which turns out to be sufficient for the present purpose, one replaces the quark by the nucleon currents, weighted by the number of valence quarks present and remembering that the u and d quarks have opposite electrical charge. Then the weak neutral currents of the proton and neutron are given, respectively, by

$$\begin{aligned} j^{\mathrm{p}}_{\mu}(x) &\simeq 2j^{\mathrm{u}}_{\mu}(x) + j^{\mathrm{d}}_{\mu}(x) \\ &= \bar{\psi}_{\mathrm{p}}(x)\gamma_{\mu}(2C^{\mathrm{u}}_{V} - C^{\mathrm{d}}_{V} - 2\gamma_5 + \gamma_5)\psi_{\mathrm{p}}(x) \\ &= \bar{\psi}_{\mathrm{p}}(x)\gamma_{\mu}(C^{\mathrm{p}}_{V} - \gamma_5)\psi_{\mathrm{p}}(x) \quad , \\ j^{\mathrm{n}}_{\mu}(x) &\simeq j^{\mathrm{u}}_{\mu}(x) + 2j^{\mathrm{d}}_{\mu}(x) \\ &= \bar{\psi}_{\mathrm{n}}(x)\gamma_{\mu}(C^{\mathrm{u}}_{V} - 2C^{\mathrm{d}}_{V} - \gamma_5 + 2\gamma_5)\psi_{\mathrm{n}}(x) \\ &= -\bar{\psi}_{\mathrm{n}}(x)\gamma_{\mu}(1 - \gamma_5)\psi_{\mathrm{n}}(x) \quad . \end{aligned} \tag{5}$$

The effective interaction between electron and nucleus is now given by

$$H^{\mathrm{eff}}_{\mathrm{Ne}} = \frac{G}{2\sqrt{2}} j^{\mathrm{e}}_{\mu}(x)\left(\sum_{i=1}^{Z} j^{\mathrm{p}\mu}_{i}(x) + \sum_{i=1}^{N} j^{\mathrm{n}\mu}_{i}(x)\right) \quad , \tag{6}$$

where Z and N are, respectively, the number of protons and neutrons contained in the nucleus. For the nuclear wave functions we can restrict ourselves to a non-relativistic approximation. Then we have (see Chap. 1, (1.3)–(1.5), and Exercise 1.3)

$$\begin{aligned} \bar{\psi}_{\mathrm{p}}\boldsymbol{\gamma}\gamma_5\psi_{\mathrm{p}} &\mapsto \Phi^{\dagger}_{\mathrm{p}}\boldsymbol{\sigma}\Phi_{\mathrm{p}} \quad , \\ \bar{\psi}_{\mathrm{p}}\gamma^0\gamma_5\psi_{\mathrm{p}} &\mapsto 0 \quad , \\ \bar{\psi}_{\mathrm{p}}\boldsymbol{\gamma}\psi_{\mathrm{p}} &\mapsto 0 \quad , \\ \bar{\psi}_{\mathrm{p}}\gamma^0\psi_{\mathrm{p}} &\mapsto \Phi^{\dagger}_{\mathrm{p}}\Phi_{\mathrm{p}} = \rho_{\mathrm{p}}(x) \quad , \end{aligned} \tag{7}$$

where ρ_{p} is the proton density. Analogous relations hold for the neutrons. As long as the spins of protons and neutrons are not aligned in the nucleus, the first spin-dependent term yields a much smaller contribution than the last term when we sum over all protons and neutrons. Thus we can write approximately

$$\begin{aligned} \frac{G}{2\sqrt{2}}\left(\sum_{i=1}^{N} j^{\mathrm{n}}_{\mu}(x) + \sum_{i=1}^{Z} j^{\mathrm{p}}_{\mu}(x)\right) &\simeq \delta_{\mu 0}\big(C^{\mathrm{p}}_{V} Z\rho_{\mathrm{p}}(x) - N\rho_{\mathrm{n}}(x)\big)\frac{G}{2\sqrt{2}} \\ &=: \delta_{\mu 0}\rho(x) \quad , \end{aligned} \tag{8}$$

where $\rho(x)$ is proportional (but not identical!) to the mass-density distribution in the nucleus because both $\rho_{\mathrm{p}}(x)$ and $\rho_{\mathrm{n}}(x)$ are approximately proportional to the mass-density. Then (6) reads

$$H^{\mathrm{eff}}_{\mathrm{Ne}} = \left[C^{\mathrm{e}}_{V}\psi^{\dagger}_{\mathrm{e}}(x)\psi_{\mathrm{e}}(x)\rho(x) - \psi^{\dagger}_{\mathrm{e}}(x)\gamma_5\psi_{\mathrm{e}}(x)\rho(x)\right] \quad . \tag{9}$$

[20] For a general discussion see, e.g., J.F. Donoghue, E. Golowich, B.R. Holstein: *Dynamics of the Standard Model* (Cambridge Univ. Press, Cambridge, 1997), Chap. 12.

xample 6.10

The Coulomb potential V_C is therefore modified as follows (note that $\rho(x)$ contains the Fermi constant!):

$$V_C \quad \mapsto \quad V_C + C_V^e \rho(x) - \gamma_5 \rho(x) \quad . \tag{10}$$

The term $C_V^p \rho(x)$ does not violate parity and hence is practically unobservable owing to its smallness. The term $\gamma_5 \rho(x)$, however, violates parity symmetry ($\psi_e^\dagger \gamma_5 \psi_e \mapsto -\psi_e^\dagger \gamma_5 \psi_e$ under a space reflection) and thus leads to observable effects. Moreover, one can assume that the nucleon density is radially symmetric, that is, $\rho(x)$ can be replaced by $\rho(r)$. Thus the effect of weak interactions in atomic physics can be effectively described by replacing the neutral Coulomb potential by

$$V_C(r) \quad \to \quad V_C(r) - \gamma_5 \rho(r) \quad . \tag{11}$$

The bound electrons are described by the Dirac equation, which with the potential (11) is

$$\begin{aligned} E\psi &= H\psi(r, \Omega) \\ &= \left(\boldsymbol{\alpha} \cdot \boldsymbol{p} + \beta m + V_C(r) - \rho(r)\gamma_5\right)\psi(r, \Omega) \quad , \end{aligned} \tag{12}$$

where Ω represents the angular coordinates. The total angular momentum

$$\boldsymbol{J} = \boldsymbol{r} \times \boldsymbol{p} + \frac{1}{2}\boldsymbol{\Sigma} \tag{13}$$

commutes with this Hamiltonian:

$$\begin{aligned} [\boldsymbol{J}, H]_- &= [\boldsymbol{J}, \rho(r)\gamma_5] \\ &= -\mathrm{i}\gamma_5 \boldsymbol{r} \times \left(\frac{\boldsymbol{r}}{r}\frac{\partial}{\partial r}\rho(r)\right) + \frac{1}{2}[\Sigma, \gamma_5]\rho(r) \\ &= 0 \quad . \end{aligned} \tag{14}$$

Together with J^2, $\kappa^2 = (J + 1/2)^2$ is also a good quantum number. Hence one can make the following general ansatz for the solution of (12):

$$\Psi(r, \Omega) = \frac{1}{r}\begin{pmatrix} \mathrm{i}g_\kappa(r)\chi_\kappa^\mu(\Omega) - g_{-\kappa}(r)\chi_{-\kappa}^\mu(\Omega) \\ f_{-\kappa}(r)\chi_{-\kappa}^\mu(\Omega) - \mathrm{i}f_\kappa(r)\chi_\kappa^\mu(\Omega) \end{pmatrix} \quad , \tag{15}$$

where $\chi_\kappa^\mu(\Omega)$ are the spinor spherical harmonics.[21] Then the radial functions are solutions of the following coupled equations:

$$\begin{aligned} \frac{\partial}{\partial r} g_\kappa &= -\frac{\kappa}{r} g_\kappa - (E - V_C(r) + m) f_\kappa - \rho(r) g_{-\kappa} \quad , \\ \frac{\partial}{\partial r} f_\kappa &= \frac{\kappa}{r} f_\kappa + (E - V_C(r) - m) g_\kappa + \rho(r) f_{-\kappa} \quad , \end{aligned} \tag{16}$$

$$\begin{aligned} \frac{\partial}{\partial r} g_{-\kappa} &= \frac{\kappa}{r} g_{-\kappa} + (E - V_C(r) + m) f_{-\kappa} + \rho(r) g_\kappa \quad , \\ \frac{\partial}{\partial r} f_{-\kappa} &= -\frac{\kappa}{r} f_{-\kappa} - (E - V_C(r) - m) g_{-\kappa} - \rho(r) f_\kappa \quad . \end{aligned} \tag{17}$$

[21] See W. Greiner: *Relativistic Quantum Mechanics – Wave Equations*, 3rd ed. (Springer, Berlin, Heidelberg, 2000).

Example 6.1

One observes that for $\rho(r)=0$ both pairs of equations (16) and (17) decouple. Then one obtains eigenfunctions not only for the quantum number κ^2 but also for κ. These have fixed parity $P=(-1)^\kappa\,\mathrm{sgn}(\kappa)$. If $\rho(r)\neq 0$, this is no longer true, that is, one no longer has eigenfunctions with defined parity. Therefore the parity selection rules, such as those for multipole radiation, are no longer exactly valid (since $\rho(r)$ is very small, they are of course still valid to a good approximation). Let us, for example, consider the transition $7\mathrm{p}_{1/2}\to 6\mathrm{p}_{1/2}$ in the element Tl (thallium). We choose this atom, because here the $6\mathrm{p}_{1/2}$ and the $6\mathrm{s}_{1/2}$ are nearly degenerate, so that one gets an appreciable $P=+1$ admixture to the $6\mathrm{p}_{1/2}$ $P=-1$ state. Instead of a pure M1 transition one now has a mixture of M1 and E1 matrix elements. This coherent superposition leads, for example, to a circular polarization of emitted photons.

Let us denote by $p_\pm$ the probability of emission of a left or right circular polarized photon, and the reduced transition matrix elements for electric and magnetic dipole radiation by $\mathcal{M}(E1,\mu)$ and $\mathcal{M}(M1,\mu)$, respectively, where $\mu=\mu_i-\mu_f=\pm 1$ is the photon helicity. Because of their different parity properties these matrix elements behave differently when the signs of the magnetic quantum numbers are inverted. One finds

$$\mathcal{M}(E1,\mu)=+\mathcal{M}(E1,-\mu)\;,$$
$$\mathcal{M}(M1,\mu)=-\mathcal{M}(M1,-\mu)\;.$$

By defining $\mathcal{M}(E1)=\mathcal{M}(E1,|\mu|)$ the circular degree of polarization can be written as[22]

$$\begin{aligned}
p_\mathrm{c} &= \frac{p_+-p_-}{p_++p_-}\\
&= \frac{|\mathcal{M}(E1)+\mathrm{i}\mathcal{M}(M1)|^2-|\mathcal{M}(E1)-\mathrm{i}\mathcal{M}(M1)|^2}{|\mathcal{M}(E1)+\mathrm{i}\mathcal{M}(M1)|^2+|\mathcal{M}(E1)-\mathrm{i}\mathcal{M}(M1)|^2}\\
&= \mathrm{i}\frac{\mathcal{M}(E1)^*\mathcal{M}(M1)-\mathcal{M}(E1)\mathcal{M}(M1)^*}{|\mathcal{M}(E1)|^2+|\mathcal{M}(M1)|^2}\\
&= \mathrm{Im}\frac{\mathcal{M}(E1)^*\mathcal{M}(M1)}{|\mathcal{M}(E1)|^2+|\mathcal{M}(M1)|^2}\;.
\end{aligned}$$

Obviously, $p_\mathrm{c}\neq 0$ is possible only if the same transition can occur by means of both $\mathcal{M}(E1)$ and $\mathcal{M}(M1)$ radiation. Normally this is forbidden by parity selection rules, but it becomes possible owing to the action of neutral weak currents, as we have seen.

The photon emission of individual atoms is too weak to be investigated directly, but one can measure the collective effect of many atoms. One sends, for example, linearly polarized light of a frequency corresponding to the $7\mathrm{p}_{1/2}$–$6\mathrm{p}_{1/2}$ transition through a test-tube filled with Tl vapor. Linearly polarized light can be decomposed into equally large parts of right and left circularly polarized light. Owing to the different refractive indices of the vapor, they are shifted with respect to each other in phase and amplitude such that the outgoing light is elliptically polarized and the polarization plane is rotated. This rotation of the polarization plane is measured. A schematic picture of the experimental set-up is shown in Fig. 6.23.

22 M.A. Bouchiat and C. Bouchiat: Journal de Physique **35**, 899 (1974).

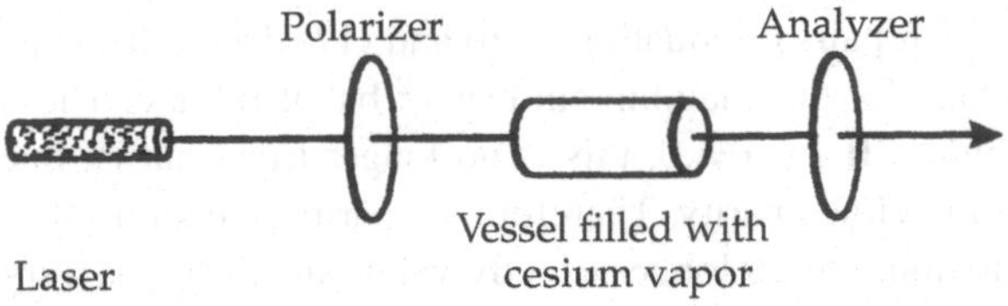

ig. 6.23. Experimental setup for measuring the rotation f the polarization plane

To conclude this chapter, we wish to state two points of view which show the special importance of these atomic-physics experiments:

1. Unlike in all other experiments, one probes here the low-energy behavior of the contribution of the neutral weak current. Hence one can check the energy dependence of the effective coupling constant of weak interactions, which is – through higher-order Feynman diagrams – determined by the interaction of intermediate vector bosons among each other (see also Sect. 6.2). These couplings are determined by the gauge invariance of the theory but are still not experimentally confirmed.
2. These experiments test the quark substructure of nuclei. This idea was developed as a result of deep-inelastic scattering experiments for large energies and is based on the assumption of nearly independent quarks. If quark–quark correlations became important in the low-energy experiments considered here, this might lead to a discrepancy between theory and experiment, from which one could draw conclusions concerning these correlations. Therefore experiments dealing with parity violation in atoms also probe the substructure of atomic nuclei.

6.5 Biographical Notes

GELL-MANN, Murray, physicist, *15.9.1929 in New York, professor at the California Institute of Technology in Pasadena. He worked on the theory of elementary particles, especially on form factors, symmetry groups and Regge poles. At the same time as Y. Ne'eman, G.-M. developed the eight-fold way model of baryons and mesons. For this work he received the Nobel Prize in 1969.

CABIBBO, Nicola, theoretical physicist, *30.4.1935 in Rome, since 1965 professor of Theoretical Physics at the University of Rome in Italy, president of the Institutio Nazionale di Fisica Nucleare. He mainly worked in theoretical elementary particle physics. In 1963 he was the first to formulate the universality of the weak interaction in terms of the mixing between strangeness-changing and strangeness-preserving processes (the Cabibbo mixing-angle).

7 Nuclear Beta Decay

7.1 The MIT Bag Model

Up to now we have developed the theory of the weak nuclear interaction at the level of the quarks, that is to say, of the constituents of the hadrons. In order to describe the weak interactions of the hadrons themselves we must in addition know how the hadrons are made up of quarks, that is, we must know the wave functions of the quarks within the hadrons. This problem has not yet been completely solved, but there are a number of models which reflect some properties of the hadrons quite well.[1] One of the best known of these models is the so-called MIT bag model,[2] which we shall now use to calculate the ratio g_A/g_V for the nucleon. In the case of the MIT bag model, one assumes that the quarks can move freely within a sphere of radius R (Fig. 7.1), subject to some boundary conditions at the surface of the bag. Since the nucleons represent the lowest baryon states, the quarks should have wave functions without angular or radial nodes. The most general solution of the Dirac equation,

$$\Phi = N\begin{pmatrix}\Phi' \\ \dfrac{\sigma\cdot\hat{p}}{E+m}\Phi'\end{pmatrix} \quad \text{with} \quad \Phi' = \begin{pmatrix}\Phi_1 \\ \Phi_2\end{pmatrix} , \tag{7.1}$$

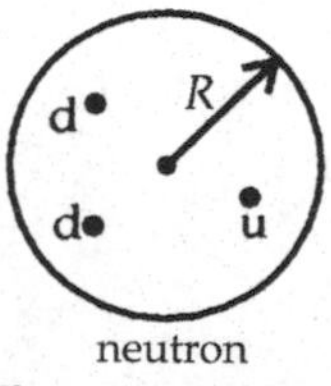

Fig. 7.1. Schematic view of the neutron in the MIT bag model

is given in this case by

$$\Phi_\pm(\boldsymbol{r}) = N\begin{pmatrix} j_0(kr)\chi_\pm \\ \dfrac{ik}{(E+m)} j_1(kr)\sigma_r\chi_\pm \end{pmatrix} , \tag{7.2}$$

where $\sigma_r = \boldsymbol{\sigma}\cdot\boldsymbol{r}/r$, $\sigma_r^2 = 1$, $E^2 = k^2 + m^2$ and $\chi_\pm$ are the unit two-component spinors (see Exercise 7.1). The spherical Bessel functions are explicitly

$$j_0(x) = \frac{\sin x}{x} \quad , \quad j_1(x) = \frac{\sin x}{x^2} - \frac{\cos x}{x} \quad . \tag{7.3}$$

To restrict the wave functions to the volume of the sphere (quark confinement), we need a suitable boundary condition. The latter should imply that the current of the

---

[1] R.F. Alvarez-Estrada, F. Fernandez, J.L. Sanchez-Gomez, V. Vento: *Models of Hadron Structure Based on Quantum Chromodynamics*, Lect. Notes Phys., Vol. 259 (Springer, Berlin, Heidelberg, New York, 1986).

[2] A. Chodos, R.C. Jaffe, K. Johnson, C.B. Thorn, V.F. Weisskopf: Phys. Rev. **D9**, 3471 (1974); A. Chodos, C.B. Thorne, V.F. Weisskopf: Phys. Rev. **D12**, 2733 (1975).

W. Greiner, B. Müller, *Gauge Theory of Weak Interactions*,
DOI 10.1007/978-3-540-87843-8_7, © Springer-Verlag Berlin Heidelberg 2009

particles perpendicular to the surface vanishes, that is, that the quarks cannot leave the hadronic volume:

$$\boldsymbol{n} \cdot \bar{\Phi} \boldsymbol{\gamma} \Phi|_s = 0 \quad , \quad \boldsymbol{n} = \boldsymbol{r}/r \quad . \tag{7.4}$$

This is ensured by the linear condition, the so-called *bag-equation*

$$\boldsymbol{n} \cdot \boldsymbol{\gamma} \Phi|_s = \mathrm{i}\Phi|_s \quad , \tag{7.5}$$

since together with the adjoint equation

$$\bar{\Phi} \boldsymbol{n} \cdot \boldsymbol{\gamma}|_s = -\mathrm{i}\bar{\Phi}|_s \quad , \tag{7.6}$$

multiplication by $\bar{\Phi}$ and Φ yields

$$+\mathrm{i}\bar{\Phi}\Phi|_s = \bar{\Phi}\boldsymbol{\gamma} \cdot \boldsymbol{n}\Phi|_s = -\mathrm{i}\bar{\Phi}\Phi|_s \quad . \tag{7.7}$$

The expression must hence vanish identically:

$$\bar{\Phi}\Phi|_s = \boldsymbol{n} \cdot \bar{\Phi}\boldsymbol{\gamma}\Phi|_s = 0 \quad . \tag{7.8}$$

In the standard representation of the γ matrices (see Appendix A.2) the boundary condition (7.4) for the solutions of the Dirac equation in a spherical volume of radius R given in (7.2) is:

$$N \begin{pmatrix} 0 & \sigma_r \\ -\sigma_r & 0 \end{pmatrix} \begin{pmatrix} j_0(kR)\chi_\pm \\ \dfrac{\mathrm{i}k}{(E+m)} j_1(kR)\sigma_r \chi_\pm \end{pmatrix} = \mathrm{i}N \begin{pmatrix} j_0(kR)\chi_\pm \\ \dfrac{\mathrm{i}k}{(E+m)} j_1(kR)\sigma_r \chi_\pm \end{pmatrix} \quad . \tag{7.9}$$

Independent of the spin direction this implies the condition

$$j_0(kR) = \frac{k}{(E+m)} j_1(kR) \quad . \tag{7.10}$$

In the bag model of the hadronic particles one assumes that u and d quarks have only a very small rest mass but a large kinetic energy owing to their confinement within a small volume. In fact the mass of a hadron is mainly due to the zero-point motion of the quarks. In the limit $m \ll 1/R$ condition (7.10) assumes the form

$$j_0(kR) = j_1(kR) \quad . \tag{7.11}$$

Insertion of the analytic expressions (7.3) for the Bessel functions yields

$$\tan(kR) = \frac{kR}{1-kR} \quad . \tag{7.12}$$

The lowest solution of this equation is

$$k = x_0/R \quad , \quad x_0 \approx 2.04 \quad . \tag{7.13}$$

The size of the hadronic volume is fixed by the requirement that the kinetic pressure of the confined quarks equals the volume pressure exerted from the exterior onto the surface. One assumes that the "true" vacuum (of quantum chromodynamics) is some kind of medium which refuses the penetration of quarks carrying color. Within the hadronic volume the quarks must expel the true vacuum, which requires energy. Because of the thermodynamic relation

$$E_{\mathrm{vac}} = -P_{\mathrm{vac}} V \quad , \tag{7.14}$$

where E_{vac} is the energy and V the volume, a pressure P_{vac} can be attributed to the vacuum. From the view of the hadron this pressure is negative, because if there were no quarks, the volume would shrink to zero. We set

$$P_{\mathrm{vac}} = -B \quad , \quad \text{i.e.} \quad E_{\mathrm{vac}} = BV \quad , \tag{7.15}$$

B is called the *bag constant.* The value of B is the actual parameter of the MIT bag model and must be fixed by comparison with the hadronic mass spectrum. The requirement that at the surface of the hadron there is a balance between internal quark pressure and external vacuum pressure is equivalent to the fact that an infinitesimal displacement of the surface requires no energy. If we restrict ourselves to spherical forms, this means that the equilibrium radius R must fulfill the condition

$$\frac{\mathrm{d}}{\mathrm{d}r} E_{\mathrm{tot}}(r) = \frac{\mathrm{d}}{\mathrm{d}r}\left(E_{\mathrm{vac}} + \sum_i E(Q_i) \right)\Bigg|_{r=R} = 0 \quad . \tag{7.16}$$

Here $E(Q_i)$ denotes the energy eigenvalue of the quark number i. According to (7.13) it holds that

$$E(Q_i) = \sqrt{k^2 + m_i^2} \approx k = \frac{x_0}{r} \tag{7.17}$$

for the u and d quarks in the ground state, so that, also using (7.14), (7.15), the condition

$$\frac{\mathrm{d}}{\mathrm{d}r}\left(\frac{4\pi}{3} B r^3 + N_Q \frac{x_0}{r} \right)\Bigg|_{r=R} = 0 \tag{7.18}$$

results, where N_Q is the number of quarks or antiquarks in the hadron. For baryons $N_Q = 3$ and for mesons $N_Q = 2$.

So far we have considered no interaction at all except that of the quark confinement. Comparison with the mass spectra of baryons and mesons reveals that one must introduce an additional contribution of the form

$$E_{\mathrm{int}} = -\frac{z_0}{r} \tag{7.19}$$

in the energy balance. The numerical constant z_0 has a value of about 1.5–2. The corrected equation (7.18) then yields an equilibrium radius R of the hadron of

$$R = \left(\frac{x_0 N_Q - z_0}{4\pi B} \right)^{1/4} \quad . \tag{7.20}$$

For the nucleon, that is, $N_Q = 3$, one obtains, together with the standard value of the bag constant

$$B^{1/4} = 145\ \mathrm{MeV} = (1.36\ \mathrm{fm})^{-1} \quad , \tag{7.21}$$

the nucleon radius

$$R_{\mathrm{N}} \approx 0.78 B^{-1/4} = 1.05\ \mathrm{fm} \quad . \tag{7.22}$$

Using this for the mean square radius, one thus obtains the value (see Exercise 7.2)

$$\langle r^2\rangle^{1/2} \approx 0.73 R_{\mathrm{N}} \approx 0.75\ \mathrm{fm} \quad , \tag{7.23}$$

which is in good agreement with the value obtained by electron scattering at the proton.

EXERCISE

7.1 The Ground State in the MIT Bag

Problem. Show that (7.2) is a solution of the Dirac equation (7.1).

Solution. We set

$$\Phi'(x) = \Phi'(r) = j_0(kr)\chi_\pm \quad . \tag{1}$$

Then we get for the lower part of the four-spinor

$$\begin{aligned}(\boldsymbol{\sigma}\cdot\hat{\boldsymbol{p}})\Phi'(r) &= -\mathrm{i}\boldsymbol{\sigma}\cdot\frac{\boldsymbol{r}}{r}\frac{\mathrm{d}}{\mathrm{d}r}j_0(kr)\chi_\pm \\ &= -\mathrm{i}\sigma_r k j_0'(kr)\chi_\pm = \mathrm{i}\sigma_r k j_1(kr)\chi_\pm \quad ,\end{aligned} \tag{2}$$

where we have set $\sigma_r = \boldsymbol{\sigma}\cdot\boldsymbol{r}/r$. Furthermore, we have used the coordinate representation of the momentum operator, $\hat{\boldsymbol{p}} = -\mathrm{i}\nabla$. After division by $(E+m)$ this yields the spinor of (7.2).

EXERCISE

7.2 The Mean Square Radius of a Nucleon

Problem. Calculate the mean square radius of a nucleon.

Solution. First we must calculate the normalization constant N in the wave function (7.2):

$$\begin{aligned}1 &= \int \mathrm{d}^3x\,\Phi^\dagger(x)\Phi(x) \\ &= 4\pi N^2 \int_0^R r^2\mathrm{d}r\left(j_0(kr)^2 + j_1(kr)^2\right) \quad ,\end{aligned} \tag{1}$$

where we have set $m=0$ and performed the integration over the solid angle. In addition we have used

$$\chi_\pm^\dagger\chi_\pm = \chi_\pm^\dagger\sigma_r^2\chi_\pm = 1 \quad , \tag{2}$$

because

$$\sigma_r^2 = 1 \quad . \tag{3}$$

If we insert the explicit expressions for the Bessel functions (7.3), then (1) together with $x_0 = kR$ yields

$$\begin{aligned}
1 &= 4\pi (N/k)^2 \int_0^R \mathrm{d}r \left(1 + \frac{\sin^2 kr}{(kr)^2} - 2\frac{\sin kr}{kr}\cos kr\right) \\
&= \frac{4\pi N^2}{k^3} \int_0^{kR} \mathrm{d}x \left[1 - \frac{\mathrm{d}}{\mathrm{d}x}\left(\frac{\sin^2 x}{x}\right)\right] \\
&= \frac{4\pi N^2}{k^3}\left(kR - \frac{\sin^2 kR}{kR}\right) \\
&= \frac{4\pi N^2}{k^3}\left(x_0 - \frac{\sin^2 x_0}{x_0}\right) \quad ,
\end{aligned} \tag{4}$$

which determines N. The expectation value of r^2 is then

$$\begin{aligned}
\langle r^2 \rangle &= \int \mathrm{d}^3 x\, x r^2 \Phi^\dagger(x)\Phi(x) \\
&= 4\pi \left(\frac{N}{k}\right)^2 \int_0^R r^2 \mathrm{d}r \left[1 - \frac{\mathrm{d}}{\mathrm{d}(kr)}\left(\frac{\sin^2 kr}{kr}\right)\right] \\
&= \frac{4\pi N^2}{k^5} \int_0^{kR} x^2 \mathrm{d}x \left[1 - \frac{\mathrm{d}}{\mathrm{d}x}\left(\frac{\sin^2 x}{x}\right)\right] \\
&= \frac{4\pi N^2}{k^5}\left[\left(\frac{x^3}{3} - x\sin^2 x\right)\bigg|_0^{kR} + 2\int_0^{kR} \mathrm{d}x \sin^2 x\right] \\
&= \frac{4\pi N^2}{k^5}\left[\frac{1}{3}(kR)^3 - kR\sin^2 kR + kR - \sin kR \cos kR\right] \\
&= \frac{4\pi N^2}{k^5}\left[\frac{1}{3}x_0^3 + x_0\cos^2 x_0 - \sin x_0 \cos x_0\right] \quad ,
\end{aligned} \tag{5}$$

where we have performed a partial integration. Using condition (7.11) this can be recast in the form

$$\langle r^2 \rangle = \frac{4\pi N^2}{k^5}\left(\frac{1}{3}x_0^3 - x_0 \cos x_0 \sin x_0\right) \quad . \tag{6}$$

If we divide by (4), the normalization constant drops out:

$$\langle r^2 \rangle = R^2 \frac{\frac{1}{3}x_0^2 - \sin x_0 \cos x_0}{x_0^2 - \sin^2 x_0} \quad . \tag{7}$$

Exercise 7.2

If we insert $x_0 \approx 2.04$ and take the square root, we obtain

$$\langle r^2 \rangle^{1/2} \approx 0.73R \quad . \tag{8}$$

This proves (7.23).

EXERCISE

7.3 Parameter Fit to the Hadronic Mass Spectrum

Problem. Determine the constants B and z_0 of the MIT bag model by trying to explain the masses of the non-strange baryons. Base this "fit" on the masses of the nucleon ($M_{\rm N} = 939$ MeV), Δ particle ($M_\Delta = 1232$ MeV), η meson ($M_\eta = 549$ MeV) and ω meson ($M_\omega = 783$ MeV).

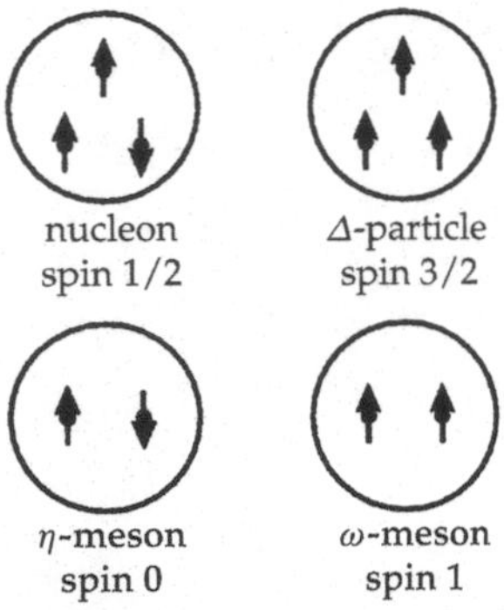

Fig. 7.2. Spin configuration of N, Δ, η and ω

Solution. Because of (7.18), (7.19) the mass of a hadronic particle, to lowest order, has the value

$$M = E_{\rm vac} + E_{\rm Q} + E_{\rm int} = \frac{4\pi}{3} B R^3 + \frac{x_0 N_Q - z_0}{R} \quad . \tag{1}$$

Insertion of the equilibrium radius (7.20) leads to the formula

$$M = \frac{4}{3}(x_0 N_Q - z_0)^{3/4} (4\pi B)^{1/4} \quad . \tag{2}$$

Now the example of the nucleon and delta reveals that the mass obviously also depends on the orientation of the quark spins: antiparallel spins lead to a lower mass, while parallel spins lead to a higher mass (one calls this the "chromomagnetic" interaction of the quark spins). The mass difference between the η and ω mesons confirms this principle.

The degenerate states we obtained in our simple version of the MIT bag model are thus split by a spin–spin interaction. The same is true for the degenerate states of the meson. The mean baryon mass is thus

$$M_{\rm B} = \frac{1}{2}(M_{\rm N} + M_\Delta) = 1085.5 \text{ MeV} \quad , \tag{3}$$

likewise we obtain for the meson mass

$$M_M = \frac{1}{2}(M_\eta + M_\omega) = 666\ \text{MeV} \quad . \tag{4}$$

The mass formula (2) therefore yields

$$\frac{M_M}{M_B} = \left(\frac{2x_0 - z_0}{3x_0 - z_0}\right)^{3/4} = 0.614 \quad . \tag{5}$$

If we insert $x_0 \approx 2.04$ and solve for z_0, we get the value

$$z_0 \approx 1.86 \quad . \tag{6}$$

Insertion of this constant into the mass formula (2) yields for the bag constant

$$B^{1/4} = \left[4\pi(3x_0 - z_0)^{-3/4}\right]\frac{3}{4}M_B \approx 145\ \text{MeV} \quad . \tag{7}$$

7.2 Beta Decay of the Neutron

Now we want to study the decay of the neutron into a proton at the level of quarks. Our most important aim will be to find an explanation for the ratio between the axial vector and vector coupling strength, C_A/C_V (6.6). Microscopically, in the decay of the neutron, one of the d quarks becomes a u quark under emission of an electron and an antineutrino:

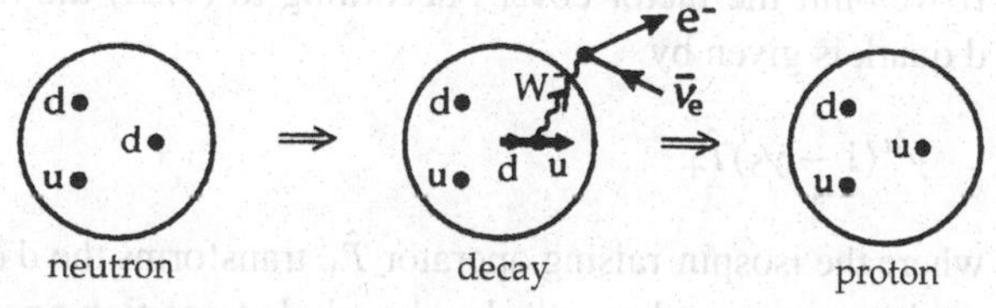

Fig. 7.3. Beta decay of the neutron

That part of the charged quark current (6.42), which transforms a d quark into a u quark is responsible for the decay

$$J_\mu^{(+)}(\text{d} \to \text{u}) = \cos\theta_C\left[\bar{u}\gamma_\mu(1-\gamma_5)d\right] \quad . \tag{7.24}$$

We must now consider the transition between two three-particle wave functions, which are constructed in the right way to describe particles of spin 1/2 and isospin 1/2. Spin and isospin are both described by the group SU(2); one can combine them in the group SU(4).[3] The complete spin–isospin wave function of a neutron with spin up is[3,4]

$$\begin{aligned}|\text{n}\uparrow\rangle = &-1/(6\sqrt{2})\left[|u(1)d(2)d(3)\rangle + |d(1)u(2)d(3)\rangle - 2|d(1)d(2)u(3)\rangle\right]\\ &\times\left[|\uparrow(1)\downarrow(2)\uparrow(3)\rangle + |\downarrow(1)\uparrow(2)\uparrow(3)\rangle - 2|\uparrow(1)\uparrow(2)\downarrow(3)\rangle\right]\end{aligned}$$

[3] See W. Greiner and B. Müller: *Quantum Mechanics – Symmetries*, 2nd ed. (Springer, Berlin, Heidelberg, 2001), Exercise 62.1.

[4] See also F.E. Close: *Introduction to Quarks and Partons* (Academic Press, London, 1979).

$$+\frac{1}{2\sqrt{2}}\left[|u(1)d(2)d(3)\rangle - |d(1)u(2)d(3)\rangle\right]$$
$$\times\left[|\uparrow(1)\downarrow(2)\uparrow(3)\rangle - |\downarrow(1)\uparrow(2)\uparrow(3)\rangle\right] \quad . \tag{7.25}$$

Here the numbers 1, 2 and 3 in parenthesis denote the three quarks. Since the numbers of the respective quarks can also be recognized by their order, we also write (7.25) in short as

$$|\mathrm{n}\uparrow\rangle = -\frac{1}{6\sqrt{2}}|udd + dud - 2ddu\rangle|\uparrow\downarrow\uparrow + \downarrow\uparrow\uparrow - 2\uparrow\uparrow\downarrow\rangle$$
$$+\frac{1}{2\sqrt{2}}|udd - dud\rangle|\uparrow\downarrow\uparrow - \downarrow\uparrow\uparrow\rangle \quad . \tag{7.26}$$

The spin part of the wave function of a proton with spin up is the same as this; in the isospin part one of the d quarks is replaced by a u quark. The wave function in the short form is

$$|\mathrm{p}\uparrow\rangle = \frac{1}{6\sqrt{2}}|udu + duu - 2uud\rangle|\uparrow\downarrow\uparrow + \downarrow\uparrow\uparrow - 2\uparrow\uparrow\downarrow\rangle$$
$$+\frac{1}{2\sqrt{2}}|udu - duu\rangle|\uparrow\downarrow\uparrow - \downarrow\uparrow\uparrow\rangle \quad . \tag{7.27}$$

The arrows in the spinor part of the wave function stand for the Pauli spinors contained in the Dirac spinors (7.2)

$$|\uparrow\rangle = \chi_+(\boldsymbol{r}) \quad , \quad |\downarrow\rangle = \chi_-(\boldsymbol{r}) \quad . \tag{7.28}$$

If we omit the factor $\cos\theta_{\mathrm{C}}$, according to (7.24) the transition operator for a single d quark is given by

$$\gamma^\mu(1-\gamma_5)\hat{T}_+ \quad , \tag{7.29}$$

where the isospin-raising operator $\hat{T}_+$ transforms the d quark into a u quark. Since the nucleon contains three quarks, the whole transition operator reads

$$\sum_{i=1}^{3}\gamma^\mu_{(i)}(1-\gamma_5^{(i)})\hat{T}_+^{(i)} \quad . \tag{7.30}$$

First we want to check which four-vector components we have to calculate. The index "+" or "−" of the spinor (7.2) is denoted as $\varepsilon = \pm 1$. For the sake of generality we first write the spinor with undetermined radial functions $f(r)$ and $g(r)$:

$$\Phi_\varepsilon = \begin{pmatrix} f(r)\chi_\varepsilon \\ \mathrm{i}g(r)\sigma_r\chi_\varepsilon \end{pmatrix} \tag{7.31}$$

For the quarks in the MIT bag there are in particular, according to (7.2), the functions

$$f(r) = Nj_0(kr) \quad , \quad g(r) = \frac{Nk}{(E+m)}j_1(kr) \quad . \tag{7.32}$$

Now we consider the components of the transition operator (7.29):

$$\text{(i)}\quad \bar{\Phi}_\varepsilon \gamma^0 \Phi_\varepsilon = \Phi_\varepsilon^\dagger \Phi_\varepsilon = f(r)^2 + g(r)^2 \quad ,$$

$$\text{(ii)}\quad \bar{\Phi}_\varepsilon \boldsymbol{\gamma} \Phi_\varepsilon = \bar{\Phi}_\varepsilon \gamma_0 \boldsymbol{\gamma} \Phi_\varepsilon = \Phi_\varepsilon^\dagger \boldsymbol{\alpha} \Phi_\varepsilon = \Phi_\varepsilon^\dagger \begin{pmatrix} 0 & \boldsymbol{\sigma} \\ \boldsymbol{\sigma} & 0 \end{pmatrix} \Phi_\varepsilon \quad . \tag{7.33}$$

If we do not consider spin-flip processes, that is, if the quark in the initial state has the same spin as in the final state, only the component σ_3 contributes. For this component we find, with $\sigma_3 \chi_\varepsilon = \varepsilon \chi_\varepsilon$, that

$$\begin{aligned} \Phi_\varepsilon^\dagger \begin{pmatrix} 0 & \sigma_3 \\ \sigma_3 & 0 \end{pmatrix} \Phi_\varepsilon &= -\mathrm{i} f(r) g(r) \chi_\varepsilon^\dagger \sigma_3 \sigma_r \chi_\varepsilon \\ &\quad - \mathrm{i} f(r) g(r) \chi_\varepsilon^\dagger \sigma_r \sigma_3 \chi_\varepsilon \\ &= -\mathrm{i} f(r) g(r) \varepsilon [\chi_\varepsilon^\dagger \sigma_r \chi_\varepsilon - \chi_\varepsilon^\dagger \sigma_r \chi_\varepsilon] = 0 \end{aligned} \tag{7.34}$$

The spatial part of the vector coupling thus does not contribute.

$$\begin{aligned} \text{(iii)}\quad \bar{\Phi}_\varepsilon \gamma^0 \gamma^5 \Phi_\varepsilon &= \Phi_\varepsilon^\dagger \gamma_5 \Phi_\varepsilon = \Phi_\varepsilon^\dagger \begin{pmatrix} 0 & I \\ I & 0 \end{pmatrix} \Phi_\varepsilon \\ &= \mathrm{i} f(r) g(r) \chi_\varepsilon^\dagger \sigma_r \chi_\varepsilon + (-\mathrm{i}) f(r) g(r) \chi_\varepsilon^\dagger \sigma_r \chi_\varepsilon \\ &= 0 \quad . \end{aligned} \tag{7.35}$$

The zero component of the axial vector thus does not contribute either.

$$\text{(iv)}\quad \bar{\Phi}_\varepsilon \boldsymbol{\gamma} \gamma_5 \Phi_\varepsilon = \Phi_\varepsilon^\dagger \boldsymbol{\alpha} \gamma_5 \Phi_\varepsilon = \Phi_\varepsilon^\dagger \begin{pmatrix} \boldsymbol{\sigma} & 0 \\ 0 & \boldsymbol{\sigma} \end{pmatrix} \Phi_\varepsilon \quad . \tag{7.36}$$

Again only σ_3 contributes:

$$\Phi_\varepsilon^\dagger \begin{pmatrix} \sigma_3 & 0 \\ 0 & \sigma_3 \end{pmatrix} \Phi_\varepsilon = f(r)^2 \chi_\varepsilon^\dagger \sigma_3 \chi_\varepsilon + g(r)^2 \chi_\varepsilon^\dagger \sigma_r \sigma_3 \sigma_r \chi_\varepsilon \quad . \tag{7.37}$$

Using the Pauli algebra one finds

$$\begin{aligned} \sigma_r \sigma_3 \sigma_r &= 1/r^2 (\boldsymbol{\sigma} \cdot \boldsymbol{r}) \sigma_3 (\boldsymbol{\sigma} \cdot \boldsymbol{r}) = 1/r^2 (\boldsymbol{\sigma} \cdot \boldsymbol{r})(\sigma_3 z - \sigma_1 x - \sigma_2 y) \sigma_3 \\ &= 1/r^2 \big[z^2 - x^2 - y^2 + (\sigma_1 \sigma_3 - \sigma_3 \sigma_1) x z + (\sigma_2 \sigma_3 - \sigma_3 \sigma_2) y z \big] \sigma_3 \\ &= 1/r^2 (z^2 - x^2 - y^2) \sigma_3 + 2z/r^2 (x \sigma_1 + y \sigma_2) \end{aligned} \tag{7.38}$$

because of the anti-commuting of the Pauli matrices,

$$(\sigma_1 \sigma_3 - \sigma_3 \sigma_1) \sigma_3 = \sigma_1 \sigma_3^2 + \sigma_3^2 \sigma_1 = 2\sigma_1 \quad , \quad \text{etc.} \tag{7.39}$$

Integration over the angles yields

$$\begin{aligned} \int \mathrm{d}\Omega (\sigma_r \sigma_3 \sigma_r) &= \int_0^{2\pi} \mathrm{d}\varphi \int_{-1}^{+1} \mathrm{d}(\cos\theta) \big[\sigma_3 (\cos^2\theta - \sin^2\theta) \\ &\quad + 2\cos\theta \sin\theta (\cos\varphi \sigma_1 + \sin\varphi \sigma_2) \big] \\ &= 2\pi \sigma_3 \int_{-1}^{+1} \mathrm{d}(\cos\theta)(2\cos^2\theta - 1) = -2\pi \cdot \frac{2}{3} \sigma_3 \quad . \end{aligned} \tag{7.40}$$

On the other hand, $\int d\Omega\sigma_3 = 4\pi\sigma_3$, and we thus obtain for the angle-integrated matrix element

$$\int d\Omega \Phi_\varepsilon^\dagger \begin{pmatrix} \sigma_3 & 0 \\ 0 & \sigma_3 \end{pmatrix} \Phi_\varepsilon = 4\pi \left(f(r)^2 - \frac{1}{3} g(r)^2 \right) \chi_\varepsilon^\dagger \sigma_3 \chi_\varepsilon$$
$$= 4\pi\varepsilon \left(f^2 - \frac{1}{3} g^2 \right) \quad . \tag{7.41}$$

For the spinors of the quarks confined to the nucleon, therefore, only the zero components of the vector part and the spatial (three-) component of the axial-vector part contribute to the transition. This is exactly the same structure as obtained in the non-relativistic limit (see (1.3) and Exercise 1.3). In the case of highly relativistic quarks, on the other hand, the two lower components of the four-spinor also give an essential contribution. Returning to the starting point of our consideration, we calculate the ratio of the transition amplitudes of the axial vector current for a single quark:

$$\delta = \frac{\int d^3x \bar{\psi}_\varepsilon \gamma^3 \gamma_5 \psi_\varepsilon}{\int d^3x \bar{\psi}_\varepsilon \gamma^0 \psi_\varepsilon} = \frac{4\pi\varepsilon \int_0^R r^2 dr (f^2 - \frac{1}{3} g^2)}{4\pi \int_0^R r^2 dr (f^2 + g^2)}$$
$$= \varepsilon \left[1 - \frac{4}{3} \frac{\int_0^R r^2 dr g^2}{\int_0^R r^2 dr (f^2 + g^2)} \right] \quad . \tag{7.42}$$

For non-relativistic quarks we would have $\delta = \pm 1$, but for the relativistic quarks of the MIT bag model we obtain a correction, which we now calculate. To this end we insert the expressions (7.32) for $f(r)$ and $g(r)$, setting $kr = x$. We then need the following elementary integrals:

$$I_0 = \int_0^R r^2 dr j_0(kr)^2 = k^{-3} \int_0^{kR} dx \sin^2 x = \frac{1}{2k^3}(kR - \sin kR \cos kR) \quad , \tag{7.43}$$

$$I_1 = \int_0^R r^2 dr j_1(kr)^2 = k^{-3} \int_0^{kR} dx \left(\frac{\sin x}{x} - \cos x \right)^2$$
$$= k^{-3} \int_0^{kR} dx \left[\cos^2 x - \frac{d}{dx} \left(\frac{\sin^2 x}{x} \right) \right]$$
$$= \frac{1}{2k^3} \left(kR + \sin kR \cos kR - 2 \frac{\sin^2 kR}{kR} \right) \quad . \tag{7.44}$$

Neglecting the rest mass, $k/(E+m) = 1$, and we obtain for the ratio (7.42)

$$|\delta| = 1 - \frac{4}{3}\frac{I_1}{I_0+I_1} = 1 - \frac{4}{3}\frac{kR+\sin kR\cos kR - 2\sin^2 kR/kR}{2kR - 2\sin^2 kR/kR}$$
$$= \frac{1}{3} - \frac{2}{3}\frac{\sin kR\cos kR - \sin^2 kR/kR}{kR - \sin^2 kR/kR}$$
$$= \frac{1}{3} - \frac{2}{3}\frac{\tan kR - \tan^2 kR/kR}{kR(1+\tan^2 kR) - \tan^2 kR/kR} \quad . \tag{7.45}$$

Here we have used the relation $(\cos^2 kR)^{-1} = 1 + \tan^2 kR$. The boundary condition (7.12) of the bag wave function is $\tan(kR) = kR/(1-kR)$. We can use this to eliminate $\tan(kR)$, yielding

$$|\delta| = \frac{1}{3} - \frac{2}{3}\frac{kR(1-kR) - kR}{kR(1-kR)^2 + (kR)^3 - kR} = \frac{kR}{3(kR-1)} = \frac{x_0}{3(x_0-1)} \quad . \tag{7.46}$$

For the value from (7.13), $x_0 = kR \approx 2.04$, we obtain

$$|\delta| \approx 0.655 \quad , \tag{7.47}$$

that is, for a single quark the strength of the axial-vector coupling is lowered by about $\frac{1}{3}$.

The strength of the vector coupling is, according to (7.33), given by

$$\left\langle p\uparrow \left| \sum_{i=1}^{3} \hat{T}_+^{(i)} \right| n\uparrow \right\rangle \quad , \tag{7.48}$$

while the strength of axial-vector coupling according to (7.41), (7.47) is given by the matrix element

$$-\delta \left\langle p\uparrow \left| \sum_{i=1}^{3} \sigma_3^{(i)} \hat{T}_+^{(i)} \right| n\uparrow \right\rangle \quad . \tag{7.49}$$

Here the bra vector simply denotes the adjoint wavefunction,

$$\langle p\uparrow| = |p\uparrow\rangle^\dagger \quad (\text{note } \overline{|p\uparrow\rangle} \equiv |p\uparrow\rangle^\dagger \gamma^0) \quad .$$

We first consider (7.48). The isospin-raising operator $\hat{T}_+^{(i)}$ first acts only on the ith quark. It transforms a d quark into a u quark and eliminates all parts of the wave function where the ith quark is already a u quark:

$$\hat{T}_+(i)|d(i)\rangle = |u(i)\rangle \quad , \quad \hat{T}_+(i)|u(i)\rangle = 0 \quad . \tag{7.50}$$

For the particular parts of the wave function (7.26) we thus have

$$\hat{T}_+(1)|u(1)d(2)d(3) + d(1)u(2)d(3) - 2d(1)d(2)u(3)\rangle$$
$$= |0 + u(1)u(2)d(3) - 2u(1)d(2)u(3)\rangle$$
$$= |uud - 2udu\rangle \quad , \tag{7.51a}$$

and further, in abbreviated notation,

$$\hat{T}_+(2)|udd + dud - 2ddu\rangle = |uud - 2duu\rangle \quad , \tag{7.51b}$$

$$\hat{T}_+(3)|udd + dud - 2ddu\rangle = |udu + duu\rangle \quad , \tag{7.51c}$$

$$\hat{T}_+(1)|udd - dud\rangle = -|uud\rangle \quad , \tag{7.52a}$$

$$\hat{T}_+(2)|udd - dud\rangle = |uud\rangle \quad , \tag{7.52b}$$

$$\hat{T}_+(3)|udd - dud\rangle = |udu - duu\rangle \quad . \tag{7.52c}$$

Combining all these contributions, we end up with the following result:

$$\begin{aligned}\sum_{i=1}^{3} \hat{T}_+(i)|n\uparrow\rangle &= -\frac{1}{6\sqrt{2}}|2uud - udu - duu\rangle|\uparrow\downarrow\uparrow + \downarrow\uparrow\uparrow + 2\uparrow\uparrow\downarrow\rangle \\ &\quad + \frac{1}{2\sqrt{2}}|udu - duu\rangle|\uparrow\downarrow\uparrow - \downarrow\uparrow\uparrow\rangle = |p\uparrow\rangle \quad ,\end{aligned} \tag{7.53}$$

as a comparison with the wave function (7.27) of the proton shows.

The matrix element (7.48) therefore has the value

$$\left\langle p\uparrow \left| \sum_{i=1}^{3} \hat{T}_+(i) \right| n\uparrow \right\rangle = 1 \quad . \tag{7.54}$$

In the case of the axial-vector matrix element we additionally need to determine the action of the Pauli matrices $\sigma_3(i)$ on the spin part of the wave function. The elementary formula for this reads

$$\sigma_3(i)|\uparrow(i)\rangle = |\uparrow(i)\rangle \quad , \quad \sigma_3(i)|\downarrow(i)\rangle = -|\downarrow(i)\rangle \quad . \tag{7.55}$$

Using this relation we obtain

$$\sigma_3(1)|\uparrow\downarrow\uparrow + \downarrow\uparrow\uparrow - 2\uparrow\uparrow\downarrow\rangle = |\uparrow\downarrow\uparrow - \downarrow\uparrow\uparrow - 2\uparrow\uparrow\downarrow\rangle \quad , \tag{7.56}$$

and so on. Combining this with (7.51), (7.52), we obtain

$$\begin{aligned}\sum_{i=1}^{3} \sigma_3^{(i)} \hat{T}_+^{(i)} |n\uparrow\rangle &= -\frac{1}{6\sqrt{2}}\big(|uud - 2udu\rangle|\uparrow\downarrow\uparrow - \downarrow\uparrow\uparrow - 2\uparrow\uparrow\downarrow\rangle \\ &\quad + |uud - 2duu\rangle|-\uparrow\downarrow\uparrow + \downarrow\uparrow\uparrow - 2\uparrow\uparrow\downarrow\rangle \\ &\quad + |udu + duu\rangle|\uparrow\downarrow\uparrow + \downarrow\uparrow\uparrow + 2\uparrow\uparrow\downarrow\rangle\big) \\ &\quad + \frac{1}{2\sqrt{2}}\big(-|uud\rangle|\uparrow\downarrow\uparrow + \downarrow\uparrow\uparrow\rangle + |uud\rangle| - \uparrow\downarrow\uparrow - \downarrow\uparrow\uparrow\rangle \\ &\quad + |udu - duu\rangle|\uparrow\downarrow\uparrow - \downarrow\uparrow\uparrow\rangle\big) \\ &= -\frac{1}{6\sqrt{2}}\big(-4|uud\rangle|\uparrow\uparrow\downarrow\rangle + |udu\rangle| - \uparrow\downarrow\uparrow + 3\downarrow\uparrow\uparrow + 6\uparrow\uparrow\downarrow\rangle \\ &\quad - |duu\rangle|3\uparrow\downarrow\uparrow - \downarrow\uparrow\uparrow + 6\uparrow\uparrow\downarrow\rangle\big) \\ &\quad - \frac{1}{2\sqrt{2}}\big(2|uud\rangle|\uparrow\downarrow\uparrow + \downarrow\uparrow\uparrow\rangle - |udu - duu\rangle|\uparrow\downarrow\uparrow - \downarrow\uparrow\uparrow\rangle\big)\end{aligned}$$

$$
\begin{aligned}
&= -\frac{1}{6\sqrt{2}}\big(2|uud\rangle|3\uparrow\downarrow\uparrow + 3\downarrow\uparrow\uparrow - 2\uparrow\uparrow\downarrow\rangle \\
&\quad + |udu\rangle|-\uparrow\downarrow\uparrow + 3\downarrow\uparrow\uparrow + 6\uparrow\uparrow\downarrow\rangle \\
&\quad + |duu\rangle|3\uparrow\downarrow\uparrow - \downarrow\uparrow\uparrow + 6\uparrow\uparrow\downarrow\rangle\big) \\
&\quad + \frac{1}{2\sqrt{2}}|udu - duu\rangle|\uparrow\downarrow\uparrow - \downarrow\uparrow\uparrow\rangle \quad .
\end{aligned}
\tag{7.57}
$$

Finally we must form the matrix element with $\langle p\uparrow|$. In doing this we must take care of the fact that every part in which a quark has a different spin or isospin on the left-hand and right-hand sides equals zero. As an example, we study the first term of (7.57). Here only that part of the wave function (7.27) which contains the three quarks in the isospin combination $|uud\rangle$ contributes because

$$
\begin{aligned}
\langle uud|uud\rangle &\equiv \langle u(1)u(2)d(3)|u(1)u(2)d(3)\rangle = 1 \quad , \\
\langle udu|uud\rangle &= \langle duu|uud\rangle = 0 \quad .
\end{aligned}
\tag{7.58}
$$

Thus we find that

$$
\begin{aligned}
&\langle p\uparrow||uud\rangle|3\uparrow\downarrow\uparrow + 3\downarrow\uparrow\uparrow - 2\uparrow\uparrow\downarrow\rangle \\
&= \frac{1}{6\sqrt{2}}(-2)\langle\uparrow\downarrow\uparrow + \downarrow\uparrow\downarrow - 2\uparrow\uparrow\downarrow|3\uparrow\downarrow\uparrow + 3\downarrow\uparrow\uparrow - 2\uparrow\uparrow\downarrow\rangle \\
&= -\frac{1}{3\sqrt{2}}(3+3+4) = -\frac{10}{3\sqrt{2}} \quad .
\end{aligned}
\tag{7.59}
$$

The evaluation of the other terms is done in the same way. Finally we obtain

$$
\begin{aligned}
\left\langle n\uparrow\left|\sum_{i=1}^{3}\sigma_3(i)\hat{T}_+(i)\right|n\uparrow\right\rangle &= \frac{1}{72}\big[4\langle uud|uud\rangle 10 \\
&\quad + \langle udu|udu\rangle(-10) + \langle duu|duu\rangle(-10)\big] \\
&\quad - \frac{1}{24}\big[\langle udu|udu\rangle(-4) - \langle duu|duu\rangle 4\big] \\
&\quad + \frac{1}{24}\big[\langle udu|udu\rangle 0 - \langle duu|duu\rangle 0\big] \\
&\quad + \frac{1}{8}\big[\langle udu|udu\rangle 2 + \langle duu|duu\rangle 2\big] \\
&= \frac{60}{72} + \frac{8}{24} + \frac{4}{8} = \frac{5}{3} \quad .
\end{aligned}
\tag{7.60}
$$

The ratio of axial-vector and vector coupling strengths is thus, according to (7.42),

$$
\frac{C_\mathrm{A}}{C_\mathrm{V}} = -\delta\frac{\langle p\uparrow|\sum_i\sigma_3(i)\hat{T}_+(i)|n\uparrow\rangle}{\langle p\uparrow|\sum_i\hat{T}_+(i)|n\uparrow\rangle} = -\frac{5}{3}\left(1 - \frac{4}{3}\frac{\int_0^R r^2\mathrm{d}r g^2}{\int_0^R r^2\mathrm{d}r(f^2+g^2)}\right) \quad . \tag{7.61}
$$

For the relativistic quarks of the MIT bag model this yields (in the limit $m \ll R^{-1}$) with (7.47) the numerical value

$$\frac{C_A}{C_V} = -\frac{5}{3} \times 0.655 = -1.09 \quad , \tag{7.62}$$

compared with the experimental value (6.6)

$$\left(\frac{C_A}{C_V}\right)_{\exp} = -1.255 \pm 0.006 \quad .$$

The correspondence is thus much better than it would have been for non-relativistic quarks. In other words, in the limit $m \gg E$ the lower component of the Dirac spinor is very small ($g(r) \ll f(r)$) and consequently we would have $\delta \approx 1$, or $(C_A/C_V)_{\rm nr} = -\frac{5}{3} = -1.67$.

It thus becomes apparent that the experimental value of the ratio C_A/C_V can be understood only in a relativistic quark model. The fact that the value (7.64) of the MIT bag model is about 15% too small indicates that in reality the motion of quarks in the nucleon is not as highly relativistic as is assumed in the MIT bag model. An improved model could be one where the boundary region is not sharp, as in the MIT bag model, but has a certain thickness, in which the quarks are slowed down and then reflected to the interior of the hadron. In the boundary region the motion of the quark would not be highly relativistic, so that the lower component $g(r)$ of the wave function would become smaller and the quantity δ correspondingly larger. Such models can be constructed by including the quarks in a deep potential pocket with a soft boundary[5] (see Fig. 7.4).

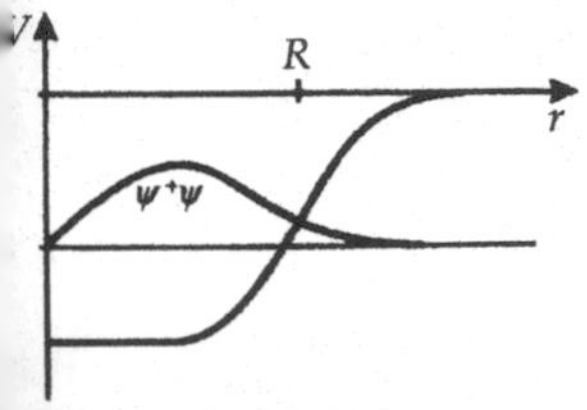

Fig. 7.4. Bag potential with soft boundary

7.3 Nuclear Beta Decay

In the last section we looked at how to describe the decay of a proton into a neutron at the level of the elementary constituents of these particles, that is, at the level of quarks. The effective hadronic transition current for the decay of the neutron, according to these discussions (see also (6.2)), is given by

$$J^\mu_{(N)}(x) = \bar{\psi}_p(x)(\gamma^\mu - C'_A\gamma^\mu\gamma_5)\psi_n(x) \quad , \tag{7.63}$$

where

$$C'_A = -C_A/C_V = +1.255 \quad . \tag{7.64}$$

Together with the current–current Hamiltonian (6.1), one obtains the following expression for the S-matrix element in nuclear beta decay:

$$\begin{aligned} S_{fi}(\mathrm{n} \to \mathrm{pe}^-\bar{\nu}) &= -\mathrm{i}\int \mathrm{d}t\, H_{\rm int} \\ &= -\mathrm{i}\frac{G}{\sqrt{2}}\cos\theta_C \int \mathrm{d}^4x \,(J^\mu_{(N)}(x) J^{(L)}_\mu(x)^\dagger + \mathrm{h.c.}) \end{aligned}$$

[5] L. Wilets: *Nontopological Solitons* (World Scientific, Singapore, 1989).

$$= -\mathrm{i}\frac{G}{\sqrt{2}}\cos\theta_\mathrm{C}\int \mathrm{d}^4x\left[\bar{\psi}_\mathrm{p}(x)\gamma^\mu(1-C_\mathrm{A}'\gamma_5)\psi_\mathrm{n}(x)\right]$$
$$\times\left[\bar{\psi}_\mathrm{e}\gamma_\mu(1-\gamma_5)\psi_\nu(x)\right] + \text{h.c.} \quad . \tag{7.65}$$

The Hermitian-conjugate expression describes the processes $\mathrm{p} \to \mathrm{ne}^+\nu$ and $\mathrm{pe}^- \to \mathrm{n}\nu$, which are called *inverse* beta decays. In (7.65) the wave functions $\psi_\mathrm{p}, \psi_\mathrm{n}, \psi_\mathrm{e}$ are, most generally, not the wave functions of free particles. Rather, the proton and neutron are in most cases bound in a nucleus, while the electron feels the charge of the atomic nucleus.

$$\psi_\mathrm{p}(x) = \psi_\mathrm{p}(\boldsymbol{x})\exp(-\mathrm{i}E_\mathrm{p}t) \tag{7.66a}$$

$$\psi_\mathrm{n}(x) = \psi_\mathrm{n}(\boldsymbol{x})\exp(-\mathrm{i}E_\mathrm{n}t) \tag{7.66b}$$

are thus localized wave functions in the atomic nucleus, to which no definite momentum can be attached. In the case of the electronic wave function

$$\psi_\mathrm{e}(x) = \psi_\mathrm{e}(\boldsymbol{x})\exp(-\mathrm{i}E_\mathrm{e}t) \quad , \tag{7.67}$$

the Coulomb distortion must be taken into account. As we already mentioned in Sect. 1.1, the nuclear transition energies

$$\Delta E_{fi} = E_\mathrm{n} - E_\mathrm{p} - E_\mathrm{e} - E_\nu \tag{7.68}$$

are usually of the order of several MeV, that is, larger than the electron mass. Both leptons therefore usually move relativistically, and their momenta are of the order of magnitude $p_\mathrm{e}, p_\nu \approx 1\,\mathrm{MeV}/c$. We can conclude from this that the recoil of the atomic nucleus can be neglected, because the transferred energy

$$E_\mathrm{recoil} \approx \frac{p_\mathrm{e}^2}{2M_\mathrm{N}A} \approx (2000A)^{-1}\,\mathrm{MeV} \tag{7.69}$$

is small compared to the transition energy. Here M_N is the nucleon mass and A the number of nucleons in the nucleus. Furthermore, the wave functions $\psi_\mathrm{e}(x), \psi_{\bar{\nu}}$ are practically constant over the nuclear volume, since R_{p_e} and $R_{p_{\bar{\nu}}} \ll 1$, R being the nuclear radius.

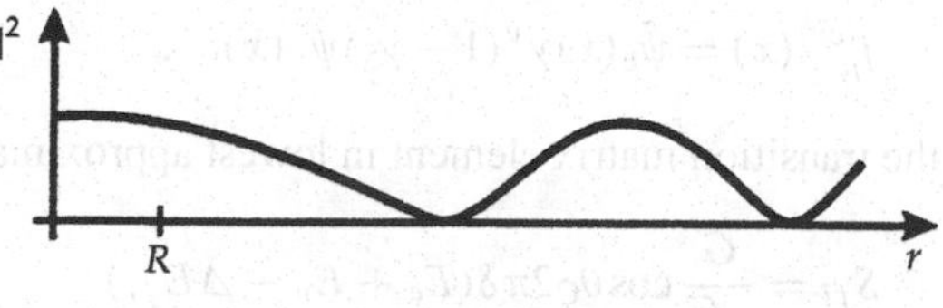

Fig. 7.5. Density distribution of the electron near the nucleus

Finally, the motion of the nucleons in the nucleus is to a good approximation non-relativistic, that is, $E_\mathrm{p} \approx E_\mathrm{n} \approx M_\mathrm{N}$. If we split the wave function of the nucleon into a large upper and a small lower component,

$$\psi(\boldsymbol{x}) = \begin{pmatrix} \chi(\boldsymbol{x}) \\ -\dfrac{\mathrm{i}\boldsymbol{\sigma}\cdot\nabla\chi(\boldsymbol{x})}{(2M_\mathrm{N})} \end{pmatrix} \approx \begin{pmatrix} \chi(\boldsymbol{x}) \\ 0 \end{pmatrix} \quad , \tag{7.70}$$

according to our considerations in Exercise 1.3, we can express those components of the nuclear transition term that survive in the non-relativistic limit by the upper component $\chi(\boldsymbol{x})$ alone:

$$\bar{\psi}_{\mathrm{p}}(\boldsymbol{x})\gamma^0\psi_{\mathrm{n}}(\boldsymbol{x}) \approx \chi_{\mathrm{p}}^\dagger(\boldsymbol{x})\chi_{\mathrm{n}}(\boldsymbol{x}) \quad , \tag{7.71}$$

$$\bar{\psi}_{\mathrm{p}}(\boldsymbol{x})\boldsymbol{\gamma}\gamma_5\psi_{\mathrm{n}}(\boldsymbol{x}) \approx \chi_{\mathrm{p}}^\dagger(\boldsymbol{x})\boldsymbol{\sigma}\chi_{\mathrm{n}}(\boldsymbol{x}) \quad . \tag{7.72}$$

This corresponds exactly to (1.4), (1.5), where we had already introduced the names *Fermi transitions* for decays due to the transition current (7.71), and *Gamow–Teller transitions* for decays according to (7.72).

In the case of the beta decay of a nucleus any neutron can transform into a proton. To take this into account, we write the total wave function of the nucleus as an antisymmetric product of the individual nucleon wave functions (Slater determinant):

$$\chi(\boldsymbol{x}_k) \equiv A[\chi_{p_1}(\boldsymbol{x}_1)\cdots\chi_{p_z}(\boldsymbol{x}_Z)\chi_{n_1}(\boldsymbol{x}_{z+1})\cdots\chi_{n_{A-Z}}(\boldsymbol{x}_A)] \quad , \tag{7.73}$$

where A denotes the antisymmetrization operator. Let $\chi_i(\boldsymbol{x}_1,\ldots,\boldsymbol{x}_A)$ be the wave function of the atomic nucleus in the initial state and $\chi_f(\boldsymbol{x}_1,\ldots,\boldsymbol{x}_A)$ that of the nucleus in the final state. Then the matrix elements of the many-body states corresponding to (7.71), (7.72) are

$$\langle f|\hat{T}_+|i\rangle = \sum_{k=1}^{A}\int \mathrm{d}^3x_1\cdots d^3x_A\chi_f^\dagger(\boldsymbol{x}_1\cdots\boldsymbol{x}_A)\hat{T}_+^{(k)}\chi_i(\boldsymbol{x}_1\cdots\boldsymbol{x}_A) \quad , \tag{7.74}$$

$$\langle f|\boldsymbol{\sigma}\hat{T}_+|i\rangle = \sum_{k=1}^{A}\int \mathrm{d}^3x_1\cdots d^3x_A\chi_f^\dagger(\boldsymbol{x}_1\cdots\boldsymbol{x}_A)\boldsymbol{\sigma}(\boldsymbol{k})\hat{T}_+^{(k)}\chi_i(\boldsymbol{x}_1\cdots\boldsymbol{x}_A) \quad . \tag{7.75}$$

Here $\hat{T}_+^{(k)}$ is the isospin-raising operator for the kth nucleon. If the latter is (initially) a neutron, $\hat{T}_+^{(k)}$ converts it into a proton; if it is already a proton, then $\hat{T}_+^{(k)}$ annihilates the whole wave function.

The time integral over the stationary wave functions in (7.65) yields

$$\int \mathrm{d}t\exp\big[\mathrm{i}(E_{\mathrm{p}}t - E_{\mathrm{e}}t + E_{\mathrm{n}}t + E_\nu t)\big] = 2\pi\delta(E_{\mathrm{e}} + E_\nu + E_{\mathrm{p}} - E_{\mathrm{n}}) \quad . \tag{7.76}$$

If we denote the leptonic transition current by $j_\mu^{(\mathrm{e}\nu)}(\boldsymbol{x})$,

$$j_\mu^{(\mathrm{e}\nu)}(\boldsymbol{x}) = \bar{\psi}_{\mathrm{e}}(\boldsymbol{x})\gamma^\nu(1-\gamma_5)\psi_\nu(\boldsymbol{x}) \quad , \tag{7.77}$$

the transition matrix element in lowest approximation takes the form

$$\begin{aligned} S_{fi} = {} & \frac{G}{\sqrt{2}}\cos\theta_{\mathrm{C}}2\pi\delta(E_{\mathrm{e}} + E_\nu - \Delta E_{fi}) \\ & \times\big[\langle f|T_+|i\rangle j^0_{(\mathrm{e}\nu)}(0) + C'_{\mathrm{A}}\langle f|\boldsymbol{\sigma}T_+|i\rangle\cdot\boldsymbol{j}_{(\mathrm{e}\nu)}(0)\big] \quad . \end{aligned} \tag{7.78}$$

Since the leptonic current, as mentioned above, varies only slightly in the region of the nucleus, we have replaced $j_\mu^{(\mathrm{e}\nu)}(\boldsymbol{x})$ in the spatial integral by $j_\mu^{(\mathrm{e}\nu)}(0)$ and put the constant value in front of the integral.

The transitions that can be described under the approximations discussed, are called *allowed transitions*. For practical purposes they constitute the most important class of beta decays. Disallowed or *forbidden transitions* are present if the leptons are emitted with higher angular momentum, such that $j_\mu^{(\mathrm{e}\nu)}(0) = 0$. In these cases the full spatial and momentum dependence of the transition operator must be taken into account. In such cases the decay probability per unit time is many orders of magnitude smaller than in the case of allowed decays.

As already mentioned, transitions due to the matrix element $\langle f|\hat{T}_+|i\rangle$ are called *Fermi* transitions. In their case the spins of the electron and neutrino are paired to the total spin $S=0$ (that is, the leptons are emitted in the singlet state $^1S_0$). Transitions due to the matrix element $\langle f|\sigma\hat{T}_+|i\rangle$ are called *Gamow–Teller* transitions; in their case the spins of the leptons are coupled to a total spin $S=1$ (triplet state $^3S_1$). For the nuclear decay, therefore, the following angular-momentum selection rules hold:

$$\text{Fermi transition:} \quad J_f = J_i \quad , \tag{7.79}$$

$$\text{Gamow–Teller transition:} \quad \begin{cases} J_f = J_i & (\text{except when } J_i = J_f = 0) \\ J_f = (J_i \pm 1) \end{cases} . \tag{7.80}$$

Since the lepton pair has no orbital angular momentum, the parity of the nucleus remains unchanged in any case.

According to (7.80), transitions between two nuclear 0^+ levels such as

$$^{14}\mathrm{O}(0^+) \to {}^{14*}\mathrm{N}(0^+) \quad ,$$

are pure Fermi transitions. Transitions with $|J_i - J_f| = 1$ are always pure Gamow–Teller transitions, such as

$$^{6}\mathrm{He}(0^+) \to {}^{6}\mathrm{Li}(1^+) \quad ,$$

or the decay of the nucleus ^{60}Co, which has been discussed extensively in Sect. 1.2. Transitions with $J_i = J_f \neq 0$ are, however, mostly mixed, among them also the fundamental decay process of the free neutron, $n \to p$.

7.4 Properties of Allowed Beta Decays

In the calculation of the leptonic transition current $j_\mu^{(e\nu)}(0)$ (7.77) it is important to take into account the "distortion" of the electron wave function due to the Coulomb field in the vicinity of the nucleus. The wave function of the antineutrino, on the other hand, can be considered a free wave, since the neutrino carries no charge. In order to determine the influence of the Coulomb potential on the electron wave, one solves the Dirac equation for continuum states in the potential of a point nucleus of charge Ze. The result is that the probability density near the nucleus is increased, namely by the factor[6]

$$\frac{|\psi_{e,Z}(r)|^2}{|\psi_{e,Z=0}(r)|^2} = 2(1+\gamma)(2pr)^{-2(1-\gamma)} \times \exp(\pi Z\alpha E/p)\frac{|\Gamma(\gamma - \mathrm{i}Z\alpha E/p)|^2}{\Gamma(2\gamma+1)^2} . \tag{7.81}$$

[6] See W. Greiner: *Relativistic Quantum Mechanics – Wave Equations*, 3rd ed. (Springer, Berlin, Heidelberg, 2000).

Here r denotes the distance from the centre of the atomic nucleus, and

$$p = (E^2 - m_e^2)^{1/2} \quad , \tag{7.82a}$$

$$\gamma = (1 - Z^2\alpha^2)^{1/2} \quad . \tag{7.82b}$$

Since $\gamma < 1$, the expression diverges in the limit $r \to 0$, which can be traced back to the fact that the Coulomb potential of a point-like charge distribution is singular. Considering the size of the nucleus, one finds that the electron density in practice stops increasing in the interior of the nucleus, so that the correct factor of increment is obtained by insertion of the nuclear radius R for r in (7.81), to a good approximation. As is common, we denote this factor by $F(E, Z)$:

$$F(E, Z) = \frac{|\Psi_{e,Z}(R)|^2}{|\Psi_{e,Z=0}(R)|^2} \quad . \tag{7.83}$$

The factor $(2pR)^{-2(1-\gamma)}$ is of importance only for heavy nuclei; in the case of light nuclei it can in practice be neglected. For instance, for ^{16}O, that is, $Z = 8$, $E = 1.1$ MeV, it has the value 1.012.

With the Coulomb correction factor, we obtain, starting with (7.78), the following expression for the nuclear decay probability per unit time:

$$\begin{aligned} dW &= \frac{1}{T}|S_{fi}|^2 \frac{d^3p}{(2\pi)^3}\frac{d^3k}{(2\pi)^3} \\ &= \frac{\delta(E_e + E_\nu - \Delta E_{fi})}{(2\pi)^5 2E_e 2E_\nu} F(E_e, Z)\,|A^0_{fi}|^2 d^3p\, d^3k \quad , \end{aligned} \tag{7.84}$$

with the matrix element

$$\begin{aligned} A^0_{fi} &= \frac{G}{\sqrt{2}} \cos\theta_C \big[\langle f|1|i\rangle \bar{u}_e(\boldsymbol{p}, s)\gamma^0(1-\gamma_5)v_\nu(\boldsymbol{k}, t) \\ &\quad + C'_A \langle f|\boldsymbol{\sigma}|i\rangle \cdot \bar{u}_e(\boldsymbol{p}, s)\boldsymbol{\gamma}(1-\gamma_5)v_\nu(\boldsymbol{k}, t)\big] \quad . \end{aligned} \tag{7.85}$$

Here $\boldsymbol{p}$ and s are the momentum and spin of the electron, $\boldsymbol{k}$ and t are the momentum and spin of the antineutrino, and $u_e(\boldsymbol{p}, s)$ and $v_\nu(\boldsymbol{k}, t)$ are the free spinors, respectively,

$$u_e(\boldsymbol{p}, s) = \begin{pmatrix} \chi_s \\ \dfrac{\boldsymbol{\sigma}\cdot\boldsymbol{p}}{E_e + m_e}\chi_s \end{pmatrix} \quad , \tag{7.86a}$$

$$v_\nu(\boldsymbol{k}, t) = \begin{pmatrix} \dfrac{\boldsymbol{\sigma}\cdot\boldsymbol{k}}{E_\nu}\chi_t \\ \chi_t \end{pmatrix} \quad . \tag{7.86b}$$

It is useful to distinguish between two classes of experiments. On the one hand, those which measure the polarization of the electrons in a probe of unpolarized atomic nuclei and, on the other, those in which the initial nuclei may be polarized but where the spins of the particles in the final state are not observed. The degree of longitudinal polarization of the electrons is independent of the structure of the nuclear transition; it is simply a result of the structure of the leptonic matrix element. Choosing as a basis the eigenstates of the helicity operator $\Lambda = \boldsymbol{\sigma} \cdot \boldsymbol{p}/p$, with $p = |\boldsymbol{p}|$, we obtain

$$\begin{aligned}
P_\pm &= u_e^\dagger(\lambda = \pm 1)\frac{1}{2}(1-\gamma_5)u_e(\lambda = \pm 1) \\
&= \frac{1}{2}\left(\chi^\dagger, \pm\frac{p}{E+m}\chi^\dagger\right)\begin{pmatrix} 1 & -1 \\ -1 & 1 \end{pmatrix}\begin{pmatrix} \chi \\ \pm\frac{p}{E+m}\chi \end{pmatrix} \\
&= \frac{1}{2}\left(1 \mp \frac{2p}{E+m} + \frac{p^2}{(E+m)^2}\right)\chi^\dagger\chi \\
&= \frac{1}{2}\left(1 \pm \frac{p}{E+m}\right)^2 = \frac{E \mp p}{E+m} \quad , \qquad (7.87)
\end{aligned}$$

where we have made use of the relativistic energy–momentum relation and of the normalization $\chi^\dagger\chi = 1$. The polarization is thus simply

$$P = \frac{P_+ - P_-}{P_+ + P_-} = -\frac{p}{E} = -\frac{v}{c} , \qquad (7.88)$$

independent of whether we deal with a Fermi transition, a Gamow–Teller transition, or a mixed transition. For β decays with positron emission, which are due to the elementary process $\mathrm{p} \to \mathrm{n e^+} \nu_e$, we have, correspondingly, $P = +v/c$, since the positron is emitted with right-handed chirality. (One simply replaces the projection operator in (7.87) by $(1+\gamma_5)/2$). As discussed in Sect. 1.1, the prediction (7.88) is confirmed by experimental data.[7]

In the second class of experiments we can average over the lepton spins, causing a reduction of the leptonic matrix elements to Dirac traces:

$$\begin{aligned}
\overline{|A^0_{fi}|^2}(G\cos\theta_C)^{-2} &= |\langle 1\rangle|^2 \mathrm{Tr}\{(\not p + m_e)\gamma^0(1-\gamma_5)\not k\gamma^0\} \\
&+ \sum_{i,j} |C'_A|^2 \langle\sigma_i\rangle^*\langle\sigma_j\rangle \mathrm{Tr}\{(\not p + m_e)\gamma^i(1-\gamma_5)\not k\gamma^j\} \\
&+ \sum_{i,j} C'_A\langle 1\rangle^*\langle\sigma_i\rangle \mathrm{Tr}\{(\not p + m_e)\gamma^i(1-\gamma_5)\not k\gamma^0\} \\
&+ \sum_{i,j} C'^*_A\langle 1\rangle\langle\sigma_i\rangle^* \mathrm{Tr}\{(\not p + m_e)\gamma^0(1-\gamma_5)\not k\gamma^i\} \quad . \qquad (7.89)
\end{aligned}$$

Here we have made use of the relation $(1-\gamma_5)^2 = 2(1-\gamma_5)$ and we have also abbreviated the transition matrix elements $\langle f|\hat{O}|i\rangle$ of the nucleus by $\langle\hat{O}\rangle$. If, as is usual, the polarization of the nucleus after the decay is not measured, the nuclear matrix elements must be summed over the final-state nuclear spins. The simplest case is that of a $0^+ \to 0^+$ transition, for which the Gamow–Teller matrix elements $\langle\sigma_i\rangle$ vanish. The Dirac trace of the first term reads

$$\begin{aligned}
\mathrm{Tr}\{(\not p + m_e)\gamma^0(1-\gamma_5)\not k\gamma^0\} &= \mathrm{Tr}\{\not p\gamma^0\not k\gamma^0\} = 4p_\mu k_\nu(2g^{\mu 0}g^{\nu 0} - g^{\mu\nu}g^{00}) \\
&= 8p^0k^0 - 4(p,k) = 4(p^0k^0 + \boldsymbol{p}\cdot\boldsymbol{k}) \quad , \qquad (7.90)
\end{aligned}$$

---

[7] D.M. Lazarus, J.S. Greenberg: Phys. Rev. **D2**, 45 (1970).

so that we obtain the result

$$d\overline{W}(0^+ \to 0^+) = (2\pi)^{-5}(G\cos\theta_C)^2\delta(E_e + E_\nu - \Delta E_{fi})F(Z, E_e) \times |\langle f|1|i\rangle|^2\left(1 + \frac{\boldsymbol{p}\cdot\boldsymbol{k}}{E_e E_\nu}\right)d^3p\, d^3k \quad , \tag{7.91}$$

where $p^0 = E_e$, $k^0 = E_\nu$, and the nuclear-spin summation is omitted. The direction-dependent factor can also be written in the form $(1 + \boldsymbol{\beta}_e \cdot \boldsymbol{n}_\nu)$, where

$$\boldsymbol{\beta}_e = \boldsymbol{p}/E_e \tag{7.92}$$

is the velocity of the electron and

$$\boldsymbol{n}_\nu = \boldsymbol{k}/E_\nu = \boldsymbol{k}/|\boldsymbol{k}| \tag{7.93}$$

is the unit vector in the direction of emission of the neutrino. If the nuclear spin of the final state is different from that of the initial state, we deal with a pure Gamow–Teller transition, and the Fermi matrix element $\langle f|1|i\rangle$ vanishes. Then only the second term in (7.85) contributes, the Dirac trace of which is, according to the standard rules (see Appendix A.2)

$$\begin{aligned}
&\mathrm{Tr}\{(\not{p} + m_e)\gamma^i(1-\gamma_5)\not{k}\gamma^j\} \\
&= \mathrm{Tr}\{\not{p}\gamma^i(1-\gamma_5)\not{k}\gamma^j\} \\
&= 4p_\mu k_\nu(g^{\mu i}g^{\nu j} - g^{\mu\nu}g^{ij} + g^{\mu j}g^{\nu i} - i\varepsilon^{\mu i\nu j}) \quad .
\end{aligned} \tag{7.94}$$

Because of the symmetry in the indices i and j the last term in (7.94) does not contribute, and after the summation over the nuclear spins in the final state one obtains

$$\begin{aligned}
&d\overline{W}(J_f = J_i \pm 1,\ J_f = J_i) \\
&= (2\pi)^{-5}(G\cos\theta_C)^2\delta(E_e + E_\nu - \Delta E_{fi})F(Z, E_e)|C'_A|^2 \\
&\quad \times |\langle f||\sigma||i\rangle|^2\left\{1 - \frac{1}{3}\boldsymbol{\beta}_e\cdot\boldsymbol{n}_\nu + \frac{2}{3}\kappa_i\hat{\boldsymbol{J}}\cdot(\boldsymbol{n}_\nu - \boldsymbol{\beta}_e)\right. \\
&\quad \left. + \Lambda_{fi}\left[\frac{1}{3}\boldsymbol{\beta}_e\cdot\boldsymbol{n}_\nu - (\hat{\boldsymbol{J}}\cdot\boldsymbol{\beta}_e)(\hat{\boldsymbol{J}}\cdot\boldsymbol{n}_\nu)\right]\right\}d^3p\, d^3k \quad .
\end{aligned} \tag{7.95}$$

Here $\hat{\boldsymbol{J}}$ is the unit vector in the direction of the polarization of the atomic nuclei before the decay, and $\langle f||\sigma||i\rangle$ is the so-called reduced nuclear matrix element, the value of which no longer depends on the spin projection. Here we have used the *Wigner–Eckart theorem*[8]

$$\langle J_f M_f|\sigma_\mu^{[1]}|J_i M_i\rangle = (J_i 1 J_f|M_i \mu M_f)\langle J_f||\sigma_\mu^{[1]}||J_i\rangle \quad .$$

The matrix element of the spin-one operator $\sigma_\mu^{[1]}$, given in spherical coordinates, can be expressed through a Clebsch–Gordan coefficient and a matrix element which is

[8] See W. Greiner and B. Müller: *Quantum Mechanics – Symmetries*, 2nd ed. (Springer, Berlin, Heidelberg, 1994).

independent of the spin projection μ. Λ_{fi} has a value dependent on the initial and final spin,

$$\Lambda_{fi} = \frac{3\langle M_i\rangle^2 - J_i(J_i+1)}{N_{fi}} \quad , \tag{7.96a}$$

with

$$N_{fi} = \begin{cases} J_i(2J_i-1) & (J_f = J_i - 1) \\ -J_i(J_i+1) & (J_f = J_i) \\ (J_i+1)(2J_i+3) & (J_f = J_i+1) \end{cases} \quad . \tag{7.96b}$$

Finally

$$\kappa_i = \begin{cases} 1 & (J_f = J_i - 1) \\ 1/(J_i+1) & (J_f = J_i) \\ -J_i/(J_i+1) & (J_f = J_i+1) \end{cases} \quad . \tag{7.96c}$$

The general case of a mixture of Fermi and Gamow–Teller transitions is quite complicated.[9] In the special case $J_i = J_f = 1/2$, which is important for the elementary process $\mathrm{n} \to \mathrm{pe}^-\bar{\nu}_\mathrm{e}$ and for the decay of the tritium nucleus, $^3\mathrm{H} \to {}^3\mathrm{He} + \mathrm{e}^- + \bar{\nu}_\mathrm{e}$, one obtains

$$\mathrm{d}\overline{W}\left(\frac{1}{2} \to \frac{1}{2}\right) = (2\pi)^{-5}(G\cos\theta_\mathrm{C})^2\delta(E_\mathrm{e}+E_\nu - \Delta E_{fi})F(Z,E_\mathrm{e})\mathrm{d}^3p\,\mathrm{d}^3k$$
$$\times \{a + b\boldsymbol{\beta}_\mathrm{e}\cdot\boldsymbol{n}_\nu + A\hat{\boldsymbol{J}}\cdot\boldsymbol{n}_\nu + B\hat{\boldsymbol{J}}\cdot\boldsymbol{\beta}_\mathrm{e}\} \tag{7.97}$$

with the coefficients

$$a = |\langle f\|1\|i\rangle|^2 + C_\mathrm{A}'^2|\langle f\|\sigma\|i\rangle|^2 \quad , \tag{7.98a}$$

$$b = |\langle f\|1\|i\rangle|^2 + \frac{1}{3}C_\mathrm{A}'^2|\langle f\|\sigma\|i\rangle|^2 \quad , \tag{7.98b}$$

$$A = -\frac{2}{3}C_\mathrm{A}'^2|\langle f\|\sigma\|i\rangle|^2 + \frac{2}{\sqrt{3}}C_\mathrm{A}'^2\langle f\|1\|i\rangle\langle f\|\sigma\|i\rangle \quad , \tag{7.98c}$$

$$B = \frac{2}{3}C_\mathrm{A}'^2|\langle f\|\sigma\|i\rangle|^2 + \frac{2}{\sqrt{3}}C_\mathrm{A}'^2\langle f\|1\|i\rangle\langle f\|\sigma\|i\rangle \quad . \tag{7.98d}$$

Here account has already been taken of the fact that C_A' is real; otherwise an additional term of the form

$$-2/\sqrt{3}I_m(C_\mathrm{A}')\langle 1\rangle\langle\sigma\rangle\hat{\boldsymbol{J}}\cdot(\boldsymbol{\beta}_\mathrm{e}\times\boldsymbol{n}_\nu) \tag{7.99}$$

would occur in the brackets of (7.97), which violates time-reversal invariance.

If we integrate over the emission directions of the leptons, the angular-dependent terms in (7.91), (7.95), (7.97) drop out and only the total strength factor (7.98a) remains:

$$a = |\langle f\|1\|i\rangle|^2 + C_\mathrm{A}'^2|\langle f\|\sigma\|i\rangle|^2 \quad . \tag{7.100}$$

[9] J.D. Jackson, S.B. Treiman, H.W. Wyld, Jr.: Nucl. Phys. **4**, 206 (1957).

If we further integrate over the unobserved energy of the neutrino, we obtain the expression for the energy spectrum of the electrons in the allowed β decay:

$$\begin{aligned}\frac{dW}{dE} &= \frac{(4\pi)^2}{(2\pi)^5}(G\cos\theta_C)^2 aF(Z,E)p^2\frac{dp}{dE}\int_0^\infty E_\nu^2 dE_\nu \delta(E+E_\nu-\Delta E_{fi}) \\ &= \frac{(G\cos\theta_C)^2}{2\pi^3} aF(Z,E)E(E^2-m_e^2)^{1/2}(\Delta E_{fi}-E)^2\theta(\Delta E_{fi}-E) \quad . \end{aligned} \tag{7.101}$$

Here we have used the fact that the neutrino is massless, hence $|\boldsymbol{k}| = E_\nu$, and we have dropped the index "e" for the electron energy. Usually the electron spectrum is represented in a *Kurie plot*, by division with the phase-space factor pE and taking the square root:

$$\begin{aligned}K(E) &= \left(\frac{dW/dE}{F(Z,E)pE}\right)^{1/2} \\ &= \left(\frac{a}{2\pi^3}\right)^{1/2} G\cos\theta_C(\Delta E_{fi}-E)\theta(\Delta E_{fi}-E) \quad . \end{aligned} \tag{7.102}$$

If one plots the experimental values for dW/dE in this manner, one expects a falling straight line, which ends exactly at the point corresponding to the total available energy (the energy E of the electron includes its rest mass!). However, this end point is only identical to the transition energy if the rest mass of the (anti)neutrino is zero.

Careful measurements of the electron spectrum in tritium decay,[10]

$$^3\text{H} \to {}^3\text{He} + e^- + \bar{\nu}_e \quad ,$$

led to the upper limit of the neutrino rest mass

$$m_{\bar{\nu}_e} < 20\ \text{eV} \quad .$$

For a more detailed account of these experiments see Example 7.4.

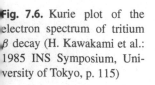

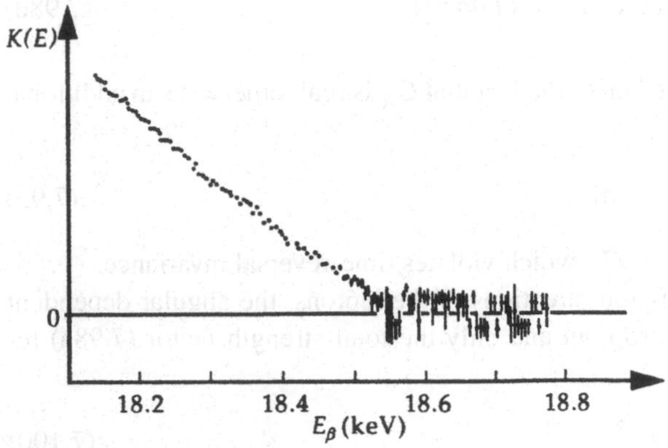

Fig. 7.6. Kurie plot of the electron spectrum of tritium β decay (H. Kawakami et al.: 1985 INS Symposium, University of Tokyo, p. 115)

[10] W. Kündig et al.: Nucl. Phys. **A478**, 425c (1988).

The total decay rate of the atomic nucleus is obtained if we integrate (7.101) over the energy of the electron. In this context there occurs the so-called *Fermi integral*

$$f_0(Z, \Delta E_{fi}) = \int_{m_e}^{\Delta E_{fi}} \mathrm{d}E\, F(Z,E)pE(\Delta E_{fi} - E)^2 \quad , \tag{7.103}$$

which cannot be expressed analytically in the case of $Z \neq 0$. For $Z = 0$ one obtains

$$f_0(0,\Delta) = \frac{1}{30} d^3(\Delta^2 + 4m_e^2) - \frac{1}{4} d\Delta^2 m_e^2 + \frac{1}{4} m_e^4 \Delta \ln\left(\frac{\Delta + d}{m_e}\right) \quad , \tag{7.104}$$

where we have used the abbreviation $\Delta = \Delta E_{fi}$ and $d = (\Delta^2 - m_e^2)^{1/2}$. The Fermi integral $f_0(Z, \Delta E_{fi})$ is tabulated over a wide range of Z and ΔE_{fi}.[11] With its aid we obtain for the decay rate

$$\frac{1}{\tau} = W = \int \mathrm{d}E \frac{\mathrm{d}W}{\mathrm{d}E} = \frac{a}{2\pi^3}(G \cos\theta_C)^2 f_0(Z, \Delta E_{fi}) \quad . \tag{7.105}$$

In real life the value of the Fermi integral must still be corrected by a few percent, in order to take into account effects such as finite nuclear size, electron screening, radiation corrections and contributions of "forbidden" transitions. Then an experimental value for $G \cos\theta_C$ can be deduced from the lifetime τ of the atomic nucleus, if one knows the transition strength a:

$$G \cos\theta_C = \left(\frac{2\pi^3}{a\tau f_0}\right)^{1/2} \quad . \tag{7.106}$$

While in general the nuclear matrix elements $\langle f|1|i\rangle$ and $\langle f|\sigma|i\rangle$ cannot be calculated precisely, some $0^+ \to 0^+$ decays represent an exception, for example,

$$^{10}\mathrm{C} \to {}^{10}\mathrm{B} \quad , \quad {}^{14}\mathrm{O} \to {}^{14}\mathrm{N} \quad , \quad {}^{34}\mathrm{Cl} \to {}^{34}\mathrm{S} \quad ,$$
$$^{42}\mathrm{Sc} \to {}^{42}\mathrm{Ca} \quad , \quad {}^{46}\mathrm{V} \to {}^{46}\mathrm{Ti} \quad , \quad {}^{54}\mathrm{Co} \to {}^{54}\mathrm{Fe} \quad .$$

In these cases one deals with pure Fermi transitions between two atomic nuclei which belong to an isospin triplet of nuclei with identical internal structure (for example, $^{10}\mathrm{Be} - {}^{10}\mathrm{B} - {}^{10}\mathrm{C}$ in the first transition above). The matrix element $\langle f|1|i\rangle$ then reduces to the isospin matrix element

$$\langle T = 1, T_3 = 0|\hat{T}_-|T = 1, T_3 = +1\rangle = \sqrt{2} \quad , \tag{7.107}$$

up to electromagnetic corrections to nuclear structure of less than one percent. Hence, in these cases $a = 2$ and $G \cos\theta_C$ can be precisely determined from the nuclear lifetime. Such measurements lead to the value (6.3)

$$G \cos\theta_C = G_\beta = (1.1492 \pm 0.0003) \times 10^{-5}\ \mathrm{GeV}^{-2} \quad . \tag{7.108}$$

[11] See, for example, National Bureau of Standards Applied Mathematic Series B, "Tables for Analysis of Beta Spectra", Chap. 2, 1952.

It is customary to give time the so-called *ft value* instead of the mean lifetime. By this one means the product of the Fermi integral and half-life period:

$$\text{“ft”} = f_0 t_{\frac{1}{2}} = f_0 \tau \ln 2 = \frac{2\pi^3 \ln 2}{(G\cos\theta_C)^2 a} \quad . \tag{7.109}$$

EXAMPLE

7.4 Determination of the Antineutrino Mass in Tritium Decay

If we do not set the mass m_ν of the antineutrino $\bar{\nu}_e$ equal to zero, the integral over the phase space of the neutrino in (7.101) is modified as follows:

$$\int_0^\infty E_\nu^2 \mathrm{d}E_\nu \delta(E + E_\nu - \Delta E_{fi}) \to \int_0^\infty k^2 \mathrm{d}k \delta(E + E_\nu(k) - \Delta E_{fi})$$

$$= \int_{m_\nu}^\infty (E_\nu^2 - m_\nu^2)^{1/2} E_\nu \mathrm{d}E_\nu \delta(E + E_\nu - \Delta E_{fi})^2$$

$$= (\Delta E_{fi} - E)\left[(\Delta E_{fi} - E)^2 - m_\nu^2\right]^{1/2} \Theta(\Delta E_{fi} - E - m_\nu) \quad . \tag{1}$$

The end point of the electron spectrum is thus shifted by the rest mass of the neutrino,

$$E < \Delta E_{fi} - m_\nu \quad , \tag{2}$$

and at the same time the form of the spectrum near the end point is modified:

$$K(E) = \left(\frac{a}{2\pi^3}\right)^{1/2} G\cos\theta_C (\Delta E_{fi} - E)\left[1 - \frac{m_\nu^2}{(\Delta E_{fi} - E)^2}\right]^{1/4} \quad . \tag{3}$$

The total transition energy available in tritium decay,

$$^3\mathrm{H} \to {}^3\mathrm{He} + \mathrm{e}^- + \bar{\nu}_e \quad ,$$

is

$$\Delta E_{fi} - m_e = 18.60 \text{ keV} \quad , \tag{4}$$

and a determination of the end point of the beta spectrum with an accuracy of 10^{-4} seems quite possible. This should enable the neutrino mass to be determined to within a few electron volts.

The problem in the experiment is essentially that the emitted electrons can lose some of their energy on their way from the nucleus to the detector by atomic and chemical effects. The electron which is emitted with high energy can ionize the atomic electron of the tritium and thus lose about 15–20 eV, or excite electrons of other atoms if the tritium is built into a chemical compound. (This is usually the case in many experiments, because in this way one can build compact solid targets, while free tritium is gaseous.) For some years a research group in Moscow (Russia) has been reporting

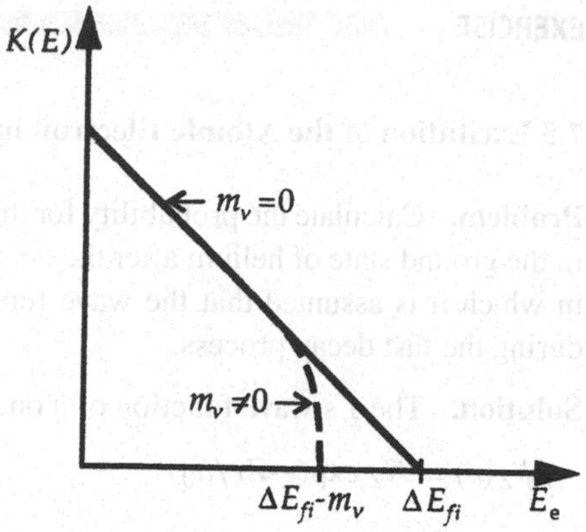

Fig. 7.7. A non-vanishing rest mass of the electron neutrino would modify the beta spectrum in the vicinity of the upper end point

values of the neutrino mass[12] in the range 20 eV $< m_\nu <$ 46 eV, but the validity of this result was doubted by many experts. Experiments at the Institute for Nuclear Studies in Tokyo,[13] Los Alamos National Laboratory,[14] and at the ETH in Zürich[15] have indicated that $m_\nu <$ 46 eV. More recent results of tritium beta-decay experiments of Mainz and Troitsk[16] have further reduced the upper limit for the mass of the electron neutrino to

$$m_{\nu_e} < 2.2\ \text{eV} . \tag{5}$$

The KATRIN experiment, being built at the Forschungszentrum Karlsruhe in Germany, is designed to measure the mass of the electron neutrino directly with a sensitivity of down to 0.2 eV.[17]

Information about neutrino masses can also be extracted from completely new sources. On 24 February 1987 a supernova of a star of mass $M \approx 20 M_\odot$ in the Large Magellanic Cloud (this is a small satellite galaxy of our Milky Way, about 160 000 light years away) was observed. Not only was an immense increase of luminosity observed ($\sim 10^6 L_\odot$), but also several neutrino events were registered in Kamioka (Japan) and in Ohio (Irvine–Michigan–Brookhaven detector). From the fact that the neutrinos were observed at all, and from an analysis of the different neutrino pulses (their time dispersion during the time of flight of 160 000 years) a value for the mass of the electron neutrino of $m_\nu \leq 10$ eV and a lifetime $\tau_\nu \geq 5 \times 10^{12} (m_\nu / E_\nu)$ s could be deduced.[18]

12 V.A. Lubimov, E.G. Novikov, V.Z. Nozik, E.F. Tretyakov, V.S. Kosik: Phys. Lett. **94B**, 266 (1980).

13 S. Kato et al.: Nucl. Phys. **A478**, 433c (1988).

14 R.G.H. Robertson, D.A. Knapp: Ann. Rev. Nucl. Part. Science **38**, 185 (1988).

15 W. Kündig et al.: Nucl. Phys. **A478**, 425c (1988).

16 Ch. Kraus et al. [Mainz Collaboration]: Eur. Phys. J. **C40**, 447 (2005); V.M. Lobashev et al. [Troitsk Collaboration]: Nucl. Phys. Proc. Suppl. **91**, 280 (2001).

17 KATRIN Collaboration, http://www-ik.fzk.de/tritium/index.html.

18 K. Sato and H. Suzuki: Phys. Rev. Lett. **58**, 2722–2725 (1987); K. Hirata et al.: Phys. Rev. Lett. **58**, 1490–1493 (1987).

EXERCISE

7.5 Excitation of the Atomic Electron in the β Decay of Tritium

Problem. Calculate the probability for the atomic electron of a tritium atom to remain in the ground state of helium after the decay. Use the so-called sudden approximation, in which it is assumed that the wave function of the atomic electron cannot change during the fast decay process.

Solution. The 1 s wave function of a one-electron atom with nuclear charge Z is

$$\psi_Z(r) = N_Z \exp(-Zr/a_0) \quad . \tag{1}$$

Here r is the distance from the nucleus and $a_0 = 0.529$ Å is the Bohr radius of the hydrogen atom. The normalization factor N_Z is determined by the requirement

$$\begin{aligned} 1 &= \int d^3r |\psi_Z(r)|^2 = 4\pi N_Z^2 \int_0^\infty r^2 dr \exp(-2Zr/a_0) \\ &= 4\pi N_Z^2 2\left(\frac{a_0}{2Z}\right)^3 = \pi N_Z^2 a_0^3/Z^3 \quad . \end{aligned} \tag{2}$$

If the wave function of the atomic electron after the beta decay is still given by $\psi_1(r)$, then the probability of finding it in the ground state $\psi_2(r)$ of the helium atom is given by

$$\begin{aligned} P &= \left| \int d^3r \psi_2^*(r)\psi_1(r) \right|^2 \\ &= \left| 4\pi N_2 N_1 \int_0^\infty r^2 dr \exp(-2r/a_0)\exp(-r/a_0) \right|^2 \\ &= \left[4\pi N_2 N_1 2(a_0/3)^3\right]^2 = (8/9)^3 \approx 0.702 \quad . \end{aligned} \tag{3}$$

In other words, we find the atomic electron in an excited state of helium or even ionized with a probability of 30%.

EXERCISE

7.6 Determination of C_A/C_V from the Lifetime of a Neutron

Problem. Determine the value of C'_A from the measured lifetime of the neutron

$$\tau_n = (898 \pm 16)\ \mathrm{s} \quad .$$

Solution. In the decay $n \to pe^-\bar{\nu}_e$ the "nucleus" in the initial and final state has spin 1/2 and isospin 1/2. The Fermi matrix element is thus

$$\begin{aligned} \langle f|1|i\rangle &= \left\langle T = \frac{1}{2}, T_3 = \frac{1}{2} \middle| \hat{T}_+ \middle| T = \frac{1}{2}, T_3 = -\frac{1}{2} \right\rangle \\ &= 1 \quad . \end{aligned} \tag{1}$$

The same is true for the isospin part of the Gamow–Teller matrix element. The spin part of the latter, on the other hand, is given by

$$
\begin{aligned}
|\langle f|\sigma|i\rangle|^2 &= \sum_{M_f} \left| \left\langle J=\frac{1}{2}, M_f \middle| \hat{\sigma} \middle| J=\frac{1}{2}, M_i \right\rangle \right|^2 \\
&= 4\left\langle J=\frac{1}{2}, M_i \middle| \left(\frac{1}{2}\hat{\sigma}\right)^2 \middle| J=\frac{1}{2}, M_i \right\rangle \\
&= 4 \times \frac{1}{2}\left(\frac{1}{2}+1\right) = 3 \quad .
\end{aligned} \tag{2}
$$

Together with (7.98a) we therefore have

$$a = |\langle f|1|i\rangle|^2 + C_A'^2 |\langle f|\sigma|i\rangle|^2 = 1 + 3C_A'^2 \quad . \tag{3}$$

Starting with (7.105) for the decay rate we obtain the relation

$$1 + 3C_A'^2 = \frac{2\pi^3}{\tau G_\beta^2 f_0} \quad , \tag{4}$$

where G_β is given by (7.108). Since the neutron carries no charge, we can use the result (7.104) for the Fermi integral, where we insert the following energy difference between the proton and neutron:

$$\Delta E_{fi} = m_n - m_p = 1.293\ \text{MeV} \quad . \tag{5}$$

This yields

$$f_0 \approx 0.057\ \text{MeV}^5 \quad . \tag{6}$$

We convert the units by insertion of $\hbar = 6.582 \times 10^{-22}$ MeV s and obtain

$$1 + 3C_A'^2 = \frac{2\pi^3 \times 6.582 \times 10^{-22}\ \text{MeV s}}{898\ \text{s} \times 1.15^2 \times 10^{-22}\ \text{MeV}^{-4} \times 0.057\ \text{MeV}^5} = 6.03 \quad . \tag{7}$$

This yields

$$|C_A'| \approx 1.29 \quad , \tag{8}$$

which, with radiative corrections, and so on, is still slightly modified to the presently accepted value[19]

$$|C_A'| = 1.25 \quad . \tag{9}$$

EXERCISE

7.7 An Astrophysical Limit to the Neutrino Mass

Problem. Use the fact that all neutrinos of the supernova 1987a appeared on earth with energies between 7.5 and 35 MeV in a time interval of 12.4 s, and the known

[19] A. Kropf, H. Paul: Z. Physik **267**, 129 (1974).

Exercise 7.7

distance of 50 kpc (1 pc = 3.26 light years = 3.08×10^{13} km), to deduce an upper limit for the mass of the electron neutrino.

Solution. The time of flight T needed to cover the distance L is

$$T(v) = L/v \quad , \tag{1}$$

where v is the velocity of the particle. Relativistically the latter is given by the ratio of momentum p and energy E:

$$v = \frac{pc^2}{E} = \sqrt{E^2 - m^2c^4}\,\frac{c}{E} = c\sqrt{1 - \frac{m^2c^4}{E^2}} \quad , \tag{2}$$

where m is the rest mass of the particle. In the case $E \gg mc^2$, which is surely true here, we have

$$v \approx c\left(1 - \frac{m^2c^4}{2E^2}\right) \quad , \tag{3}$$

or

$$T(v) \approx \frac{L}{c}\left(1 + \frac{m^2c^4}{2E^2}\right) \quad . \tag{4}$$

If two particles that have been simultaneously emitted and have energies E_1 and E_2, respectively, appear during a time interval ΔT, it holds that

$$c\Delta T \geq \frac{1}{2}Lm^2c^4\left|\frac{1}{E_1} - \frac{1}{E_2}\right| \tag{5}$$

and

$$mc^2 < \left(\frac{2c\Delta T}{L}\,\frac{E_1^2E_2^2}{|E_1^2 - E_2^2|}\right)^{1/2} \quad . \tag{6}$$

Insertion of the values given above yields a neutrino mass of

$$m_\nu c^2 < 17\ \text{eV} \quad . \tag{7}$$

The same method is, by the way, also applicable to deducing a limit for the mass of the photon.[20] In this case one uses, for example, the experimentally verified fact that visible light ($E_1 \sim 1$ eV) and radio radiation ($\nu_2 \sim 100$ MHz) from the Crab pulsar, which has a period of 33 ms and a distance from the earth of 2 kpc, pulse synchronously. With $E_2 = h\nu_2 = 4 \times 10^{-7}$ eV $\ll E_1 = 1$ eV this yields

$$\begin{aligned} m_\gamma c^2 &< \left(\frac{2c\Delta T}{L}\right)^{1/2} E_2 \\ &\approx 0.5 \times 10^{-6} \times 4 \times 10^{-7}\ \text{eV} = 2 \times 10^{-13}\ \text{eV} \quad . \end{aligned} \tag{8}$$

This implies a range of the Coulomb force of at least

$$\hbar/m_\gamma c > 1000\ \text{km} \quad . \tag{9}$$

[20] G. Feinberg: Science **166**, 879 (1969).

Experiments in the laboratory[21] yield a range of at least

$$\hbar/m_\gamma c > 3 \times 10^4\,\mathrm{km} \quad . \tag{10}$$

Exercise 7.

EXAMPLE

7.8 Double β Decay[20]

So far, our theoretical treatment of the weak interaction has been based entirely on first-order perturbation theory to the Fermi constant G, see (2.8). In fact, until recently, second-order effects in G were only observed in the neutral kaon system, which will be discussed in Chap. 8. In this system, perturbation theory is not applicable, owing to the exact degeneracy of the states $|\mathrm{K}^0\rangle$ and $|\bar{\mathrm{K}}^0\rangle$.

Any second-order effects in G are normally not observable in nuclear beta decay, because they are exceedingly small. There is, however, one exception to this rule, namely when the normal first-order beta decay is energetically forbidden, but a double beta decay is energetically allowed. This is illustrated in Fig. 7.8, where the lowest energy states of a neighboring triplet of isobaric nuclei are shown schematically. If the proton and neutron number, Z and N, of the parent nucleus are both even, the daughter nucleus in normal beta decay would be a double-odd nucleus. Such nuclei usually are not strongly bound and are highly unstable. The "granddaughter" nucleus $(Z+2, N-2)$ is again an even–even nucleus and therefore strongly bound. In many cases it is more strongly bound than the parent nucleus, for example, in ^{82}Se, ^{76}Ge, ^{100}Mo, ^{136}Xe, and ^{150}Nd.

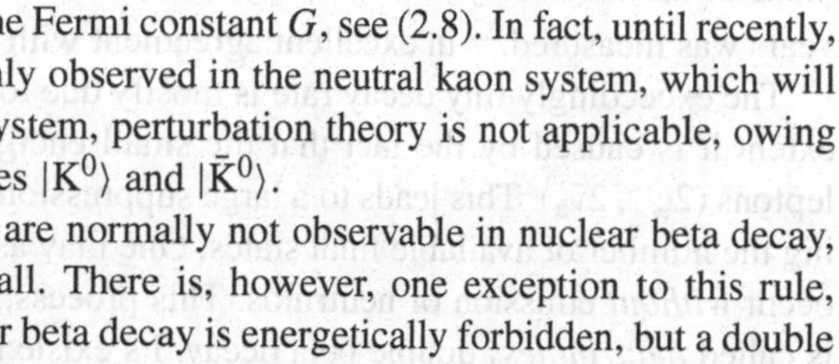

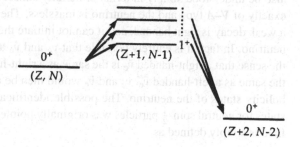

Fig. 7.8. Double beta decay with intermediate virtual states

Double beta decay can be understood as a normal effect of second-order perturbation theory, where the accessible 1^+ states of the daughter nucleus serve as intermediate states:

$$^{(Z,N)}0^+ \longrightarrow \underbrace{^{(Z+1,N-1)}1^+ + \mathrm{e}^- + \bar{\nu}_\mathrm{e}}_{\displaystyle\downarrow}$$
$$^{(Z+2,N-2)}0^+ + \mathrm{e}^- + \bar{\nu}_\mathrm{e} \quad .$$

[21] E.R. Williams, J.E. Faller, H.A. Hill: Phys. Rev. Lett. **26**, 721 (1971).

[20] A comprehensive discussion of this subject is provided by F. Boehm, P. Vogel: *Physics of Massive Neutrinos* (Cambridge University Press, Cambridge, 1987).

xample 7.8

Since each step corresponds to an allowed Gamow–Teller transition, the theory developed in Sects. 7.3, 7.4 applies. The calculation is only complicated by the necessity of summing over *all* 1^+ states of the daughter nucleus, most of which are not known experimentally. However, today quite accurate calculations are possible.[21]

Reflecting the smallness of G, the predicted half-lives for double beta decay are extremely long, for example,

$$\tau(^{82}\mathrm{Se} \to {}^{82}\mathrm{Kr}) = (1.1 \pm 0.1) \times 10^{20}\ \text{years} \quad .$$

Until recently, indirect evidence for the existence of double beta decay came only from geochemical analysis of isotopic element abundances, which showed an enhancement of the decay products. In 1987 the first precise observation of double beta decay was made in the laboratory for the decay $^{62}\mathrm{Se} \to {}^{82}\mathrm{Kr}$, and a half-life of $(1.1^{+0.8}_{-0.3})\times 10^{20}$ years was measured,[22] in excellent agreement with theory.

The exceedingly tiny decay rate is mostly due to the small value of G, but to some extent it is caused by the fact that the small energy release must be shared by four leptons ($2\mathrm{e}^-$, $2\bar{\nu}_\mathrm{e}$). This leads to a large suppression of the phase-space factor describing the number of available final states. One may ask whether the transition could not occur *without* emission of neutrinos. This process, which has not yet been observed, is called *neutrinoless* double beta decay. Its existence would imply that ν_e and $\bar{\nu}_\mathrm{e}$ are not distinct particles, and that the lepton number is not exactly conserved.

The difference between the neutrino ν_e and antineutrino $\bar{\nu}_\mathrm{e}$ was first demonstrated in neutrino scattering experiments: ν_e and $\bar{\nu}_\mathrm{e}$ trigger different reactions,

$$\bar{\nu}_\mathrm{e} + \mathrm{p} \longrightarrow \mathrm{e}^+ + \mathrm{n} \quad , \qquad \nu_\mathrm{e} + \mathrm{n} \longrightarrow \mathrm{e}^- + \mathrm{p} \quad ,$$

but an antineutrino does *not* react with a neutron. This was initially taken as proof that ν_e and $\bar{\nu}_\mathrm{e}$ are distinguished by a new quantum number (lepton number), but it can also be understood simply in terms of helicity conservation if the weak interaction is exactly of *V–A* type and the neutrino is massless. Then any antineutrino produced by a weak decay is right handed, and it cannot initiate the same reaction as a left-handed neutrino. In fact, it is entirely possible that ν_e and $\bar{\nu}_\mathrm{e}$ are one and the same particle, in the sense that a right-handed $\bar{\nu}_\mathrm{e}$ is the same as a right-handed ν_e and a left-handed ν_e is the same as a left-handed $\bar{\nu}_\mathrm{e}$. ν_e and $\bar{\nu}_\mathrm{e}$ would then be only fancy notations for the two helicity states of the neutrino. The possible identification of particle and antiparticle states for neutral spin-$\frac{1}{2}$ particles was originally pointed out by Majorana. The fermion field is simply defined as

$$\psi_M = \frac{1}{\sqrt{2}}(\psi + \psi^c) \quad , \tag{1}$$

where $\psi^c = \mathrm{i}\gamma_2\psi^*$ is the charge conjugated field. For massless neutrinos there is no distinction between the "normal" Dirac neutrino and a Majorana neutrino, because of exact helicity conservation. This would be different if the neutrinos had a small mass, because the chirality and helicity differ from each other by an amount of order $(m_\nu/E_\nu)^2$, as discussed in (1.19)–(1.22) and in a different context in Exercise 6.3. In

---

21 P. Vogel, M.R. Zirnbauer: Phys. Rev. Lett. **57**, 3148 (1986); O. Civitarese, A. Faessler, T. Tomoda: Phys. Lett. **B194**, 11 (1987).

22 S.R. Elliot, A.A. Hahn, M.K. Moe: Phys. Rev. **C36**, 2129 (1987).

this case a right-handed Majorana (anti)neutrino can behave like a left-handed Majorana neutrino with a small probability, and neutrinoless double beta decay becomes possible. Because phase space is only shared by two leptons in the reaction

$$(Z, N) \longrightarrow (Z+2, N-2) + 2e^- \quad , \tag{2}$$

the decay rate is much less suppressed. On the other hand, the helicity mixing factor $(m_\nu/E)^2$ reduces the expected rate for this decay mode drastically. The two effects balance each other for a Majorana neutrino mass of a few electron volts.

It is therefore possible to search for a small neutrino mass in double beta decay if the neutrino is a Majorana particle. The experimental signal would be rather striking: because there are no particles that escape undetected from reaction (2), the sum of the electron energies adds up to the full nuclear transition energy ΔE, whereas the sum energy spectrum yields a broad distribution peaking at $1/3\Delta E$ (see Fig. 7.9). Several experiments searching for neutrinoless double beta decay have been performed,[23] but the results have not been conclusive so far – see also the following Example 7.9. From the data an upper limit on the mass of a Majorana neutrino can be deduced:[24]

$$m_{\nu_e} \leq 0.35 \text{ eV} \quad . \tag{3}$$

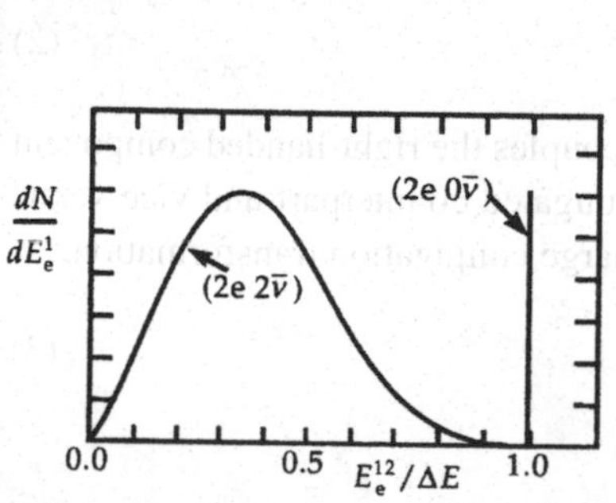

Fig. 7.9. Schematic sum energy spectrum of the two electrons emitted in double beta decay. The sum E_e^{12} of the electron energies is plotted in units of the nuclear transition energy ΔE. A broad spectrum is found for normal double beta decay ($2e2\bar{\nu}$); a narrow peak is predicted for the neutrinoless double beta decay ($2e0\bar{\nu}$)

This result is relevant in the context of the proposed unified gauge theories of strong and electroweak interactions (see Chap. 9), since many of these theories predict that the known neutrinos ν_e, ν_μ, ν_τ are in fact Majorana particles.[25] It is quite remarkable that low-energy nuclear physics provides one of the best tests for theories of this type, which are mostly concerned with the properties of elementary particles at energies of the order of 10^{14} GeV.

---

[23] See, for example, the reviews by F.T. Avignone, R.L. Brodzinski: Prog. Part. Nucl. Phys. **21**, 99 (1988); and A. Faessler: Prog. Part. Nucl. Phys. **21**, 183 (1988), and E. Fiorini: *Double beta decay: Experiments*, J. Phys. Conf. Ser. **39**, 243 (2006).

[24] H.V. Klapdor-Kleingrothaus et al. [Heidelberg–Moscow Collaboration]: Eur. Phys. J. **A12**, 147 (2001).

[25] J.W.F. Valle: Prog. Part. Nucl. Phys. **26**, 91 (1991).

EXAMPLE

7.9 The Majorana Neutrino

In the previous example we have learned that the neutrinoless double β decay is allowed only if helicity is non-conserved. This means that to allow this decay mode the neutrino has a non-vanishing mass and cannot be distinguished from an antineutrino, i.e. it has to be identical with its antiparticle. A Majorana neutrino has these properties. In this example we want to construct a mass term which allows the neutrinoless double β decay.

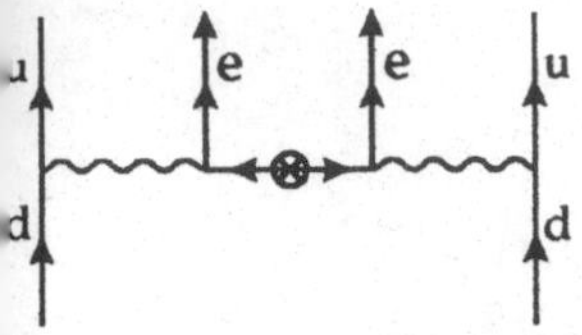

Fig. 7.10. Neutrinoless double β decay. The symbol $\otimes$ denotes a mass insertion of the Majorana type

In order to allow the process depicted in Fig. 7.10 the mass term must couple left-handed neutrinos to right-handed anti-neutrinos. If we consider a mass term of the Dirac type

$$\mathcal{L}_D = m_D(\bar{\psi}\psi) = m_d(\bar{\psi}_\mathrm{L}\psi_\mathrm{R} + \bar{\psi}_\mathrm{R}\psi_\mathrm{L}) \tag{1}$$

we realize that this term couples left- and right-handed components of the same field. The eigenstate of the mass is

$$\psi = \psi_\mathrm{L} + \psi_\mathrm{R} \quad . \tag{2}$$

However, we are interested in a mass term which couples the right-handed component of a field to the left-handed one of its charge-conjugated counterpart and vice versa. In this context we recall the conventions of the charge conjugation transformation.

$$\psi^c = C\gamma^0\psi^* = i\gamma^2\psi^* \quad ; \quad \bar{\psi}^c = \psi^T C \tag{3}$$

and we use the following abbreviation

$$\psi_\mathrm{L}^c := (\psi_\mathrm{L})^c = \frac{1}{2}(1+\gamma_5)\psi^c = (\psi^c)_\mathrm{R} \quad . \tag{4}$$

A mass term of the Majorana type couples the right- with the left-handed components of charge-conjugated fields

$$\begin{aligned} \mathcal{L}_{MA} &= m_A(\bar{\chi}\chi) = m_a(\bar{\psi}_\mathrm{L}^c\psi_\mathrm{L} + \bar{\psi}_\mathrm{L}^c\psi_\mathrm{L}) \quad , \\ \mathcal{L}_{MB} &= m_B(\bar{\omega}\omega) = m_b(\bar{\psi}_\mathrm{R}^c\psi_\mathrm{R} + \bar{\psi}_\mathrm{R}^c\psi_\mathrm{R}) \quad . \end{aligned} \tag{5}$$

The eigenstate of the mass is

$$\begin{aligned} \chi &= \psi_\mathrm{L} + \psi_\mathrm{L}^c \quad , \\ \omega &= \psi_\mathrm{R} + \psi_\mathrm{R}^c \quad . \end{aligned} \tag{6}$$

Obviously the fields are self-conjugated under charge conjugation:

$$\chi = \chi^c \quad ; \quad \omega = \omega^c \quad . \tag{7}$$

This means that a particle and its antiparticle are identical. If we consider a mass term which contains both Dirac and Majorana particles, we obtain a Lagrangian density

Example 7.

$$
\begin{aligned}
\mathcal{L}_{CM} &= m_D\bar{\psi}_{\rm L}\psi_{\rm R} + m_A\bar{\psi}^c_{\rm L}\psi_{\rm L} + m_B\bar{\psi}^c_{\rm R}\psi_{\rm R} + \text{h.c.} \\
&= \frac{1}{2}D(\bar{\chi}\omega + \bar{\omega}\chi) + A\bar{\chi}\chi + B\bar{\omega}\omega \\
&= (\bar{\chi},\bar{\omega})\begin{pmatrix} A & \frac{1}{2}D \\ \frac{1}{2}D & B \end{pmatrix}\begin{pmatrix} \chi \\ \omega \end{pmatrix}
\end{aligned} \tag{8}
$$

The matrix can be diagonalized and we obtain the eigenvalues for the mass

$$
M_{1,1} = \frac{1}{2}\left\{(A+B) \pm \sqrt{(A-B)^2 + D^2}\right\} \quad , \tag{9}
$$

which belongs to the Majorana eigenstates

$$
\begin{aligned}
\eta_1 &= \cos\theta\,\chi - \sin\theta\,\omega \quad , \\
\eta_2 &= \sin\theta\,\chi + \cos\theta\,\omega \quad ,
\end{aligned} \tag{10}
$$

with

$$
\tan(2\theta) = \frac{D}{A-B} \quad . \tag{11}
$$

The inversion of (9) leads to

$$
\begin{aligned}
D &= (M_1 - M_2)\sin(2\theta) \quad , \\
A &= M_1\cos^2\theta + M_2\sin^2\theta \quad , \\
B &= M_1\sin^2\theta + M_2\cos^2\theta \quad .
\end{aligned} \tag{12}
$$

The most general mass term (8) of a Dirac spinor therefore consists of two Majorana particles of different masses M_1 and M_2. The commonly known formalism for Dirac fields can be reconstructed for $A = B = 0$ and $\theta = \pi/4$.

Due to the coupling of different fields, the mass terms (5) violate the conservation of any additive number the field ψ carries, e.g. the electric charge. Therefore no elementary fermions, except the neutrinos, can have Majorana masses $A = B = 0$. In the case of the neutrino, the mass term (5) violates conservation of the lepton number ($\Delta L = 2$). If Majorana neutrinos exist, then the neutrinoless β decay and Kaon decays of the form $K^- \to \pi^+ \mathrm{e}^+ \mathrm{e}^-$ should be experimentally accessible.

There are speculations that the lepton number might not be conserved and that the corresponding symmetry is spontaneously broken. The breaking of the symmetry is treated in an analogous fashion to the Higgs mechanism. The spontaneous symmetry breaking leads to a massless Goldstone boson, which is called the majoron. Experiments searching for this majoron in the neutrinoless double β decay so far remain inconclusive.[26] The Heidelberg–Moscow experiment[27] established a limit for the neutrino–majoron coupling of $g_{\nu\chi} < 8.1 \times 10^{-5}$. The same collaboration gives a mass limit of the neutrino of $m_{\nu_e} < 0.35$ eV. Their claim to have observed actual neutrinoless double β decay events remains controversial.

[26] An overview of recent experiments can be found in E. Fiorini: J. Phys. Conf. Ser. **39**, 243 (2006).

[27] H.V. Klapdor-Kleingrothaus et al. [Heidelberg–Moscow Collaboration]: Eur. Phys. J. **A12**, 147 (2001).

EXAMPLE

7.10 The Solar Neutrino Problem

The energy radiated from the sun's surface is produced in the interior of the sun by a fusion of light nuclei to heavier, more strongly bound nuclei. The starting point of a whole chain of fusion processes are light hydrogen nuclei (protons), which in several steps, through deuterium nuclei, ^{2}H, fuse into helium ^{4}He. The complete chain is represented in Fig. 7.11, where the branching ratios, as calculated in the standard model of the structure of the sun, are also given.

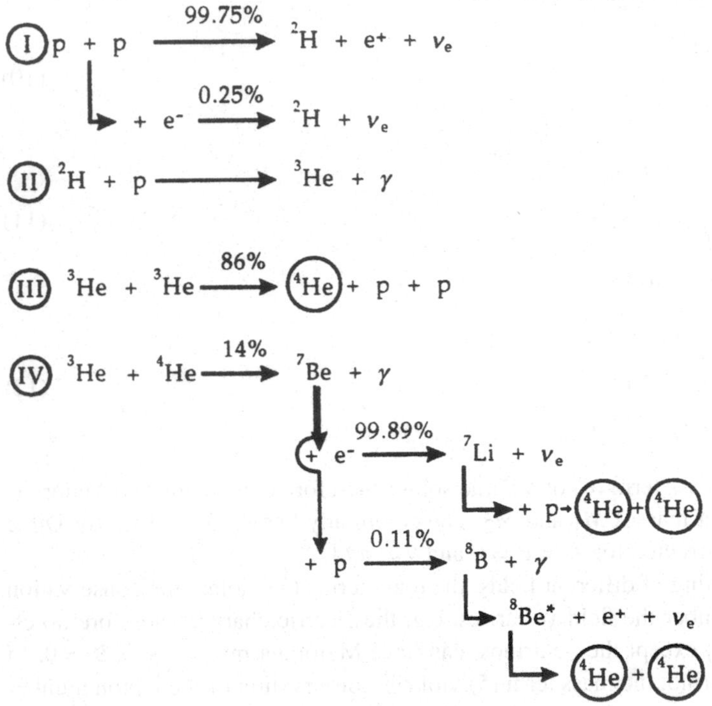

Fig. 7.11. Nuclear fusion chain in the sun

Since the cross sections for the single reactions strongly depend on energy, these ratios are very sensitive to the temperature in the interior of the sun. If one succeeded in measuring the relative frequency of the single reactions, then the astrophysicists would have a sensitive thermometer for the interior of a star. Unfortunately, the shell of the sun is so dense that all electrons and photons that are set free, and that could easily be measured, are absorbed. (This fact is, of course, very fortunate, since otherwise the earth would be bombarded by hard gamma radiation!) The only reaction products which reach the Earth's surface are the neutrinos produced in the nuclear reactions. Unfortunately, those of the most important reactions ($\mathrm{p}+\mathrm{p}\to {}^2\mathrm{H}+\mathrm{e}^+ +\nu_\mathrm{e}$, $^7\mathrm{Be}+\mathrm{e}^- \to {}^7\,\mathrm{Li}+\nu_\mathrm{e}$) have energies of less than 1 MeV. Even the neutrinos emitted in the reaction $\mathrm{p}+\mathrm{p}+\mathrm{e}^- \to {}^2\mathrm{H}+\nu_\mathrm{e}$ have a maximum energy of 1.44 MeV. The problem is simply that the reaction

$$^{37}\mathrm{Cl}+\nu_\mathrm{e}\to {}^{37}\mathrm{Ar}+\mathrm{e}^- \quad , \tag{1}$$

which is used for the detection of the neutrinos, has an *effective threshold* of 5.8 MeV and has a nearly vanishing response to all neutrinos of lower energy (below 0.814 MeV

Example 7.1

the reaction is completely forbidden). With this method one can detect only the neutrinos from the reaction ($^7Be + p \to {}^8B + \gamma$, $^8B + \gamma \to {}^8Be^* + e^+ + \nu_e$), the maximum energy of which is 14 MeV.

Since this reaction type only occurs with a frequency of 0.015%, the expected counting rate on earth is relatively low. According to the standard solar model,[28] a rate of

$$(5.8 \pm 0.7)\ \mathrm{SNU} \tag{2}$$

is expected, where the abbreviation SNU stands for *solar neutrino unit*:

$$1\ \mathrm{SNU} = 10^{-36}\ \mathrm{reactions}/(^{37}\mathrm{Cl\ atom} \cdot \mathrm{s}) \ . \tag{3}$$

The first experimental search for solar neutrinos has been carried out since 1965 by R. Davies and collaborators in the Homestake goldmine in South Dakota.[29] The total detector material of 615 tons of perchlorethylene (C_2Cl_4) is deep below the earth's surface, largely screened from cosmic radiation. On average, 0.47 ± 0.04 events per day are observed and 0.08 ± 0.03 events must be traced back to cosmic radiation. After subtracting this background, a reaction rate of

$$(2.0 \pm 0.3)\ \mathrm{SNU} \tag{4}$$

resulted, only about one third of the expected value!

This surprising result has for many years worried astrophysicists and nuclear and elementary-particle physicists alike, but a plausible explanation has been advanced by two Russian physicists, Mikheev and Smirnov.[30] At first it was thought that there was an error in the standard model of the structure of the sun, which yielded the predicted value (2). Steady improvements, however, have always confirmed the original value. On the nuclear-physics side, it was believed for a long time that the energy dependence of the reactions contributing to the fusion chain was not sufficiently well known. However, in this domain also the old data were confirmed by better experiments.

Finally, particle physicists were thought to have found a reasonable explanation for the phenomenon of neutrino oscillations (see Example 6.5). If one assumes that on their long journey from the sun to the earth, electron, muon, and tau neutrinos are completely mixed, then only one third of the originally emitted electron neutrinos should arrive as such on earth. One third of each would have been converted to muon and tau neutrinos, to which the detection reaction (1) does not respond. A mass difference $\sqrt{(\Delta m^2)} \approx 10^{-6}$ eV would already be enough for a sufficient mixture along the way. This is below the detection limit in laboratory experiments (10^{-1} eV). However, the angles of mixture between the different neutrino-mass eigenstates would have to be equal in magnitude (about 30°).[31]

---

28 J.N. Bahcall, W.F. Huebner, S.H. Lubow, P.D. Parker, R.G. Ulrich: Rev. Mod. Phys. **54**, 767 (1982).

29 R. Davies, Jr., D.S. Harmer, K.C. Hoffmann: Phys. Rev. Lett. **20**, 1205 (1968).

30 S.P. Mikheev, A.Yu. Smirnov: Sov. J. Nucl. Phys. **42**, 913 (1985).

31 An extended discussion of the attempts to explain this problem can be found in the article "The Solar Neutrino Puzzle" by W.C. Haxton: Comm. Nucl. Phys., as well as in H.V. Klapdor, B. Povh (eds.): *Neutrino Physics* (Springer, Berlin, Heidelberg, 1988).

Example 7.10

In order to find a solution of the problem, several experiments (SAGE, GALLEX, and GNO) have been prepared where ^{71}Ga instead of ^{37}Cl serves as a detector for the neutrinos. The advantage is that the reaction

$$^{71}\mathrm{Ga} + \nu_\mathrm{e} \to {}^{71}\mathrm{Ge} + \mathrm{e}^- \tag{5}$$

has a threshold far below 1 MeV and thus is also sensitive to the neutrinos of the reaction $\mathrm{p} + \mathrm{p} \to {}^2\mathrm{H} + \mathrm{e}^+ + \nu_\mathrm{e}, \mathrm{p} + \mathrm{p} + \mathrm{e}^- \to {}^2\mathrm{H} + \nu_\mathrm{e}$. The frequency with which these reactions occur, on the other hand, is nearly beyond any doubt, because they determine the energy balance of the sun. The observed deviation of the predicted event rate of 122 SNU has unambiguously proven that the original neutrinos had been lost on their way to earth.

As already mentioned, a plausible explanation for the experimental suppression factor has been given by S.P. Mikheev and A.Yu. Smirnov, which we will discuss shortly. In doing this we follow largely the presentation of **Hans A. Bethe.**[32] In this explanation one starts with the assumption that neutrinos are not massless and that the mass matrix is not diagonal with respect to electron and muon neutrino, but diagonal with respect to the linear combination ν_1 and ν_2, where the corresponding neutrino wave functions are connected as follows (see Example 6.5, (3)):

$$\begin{pmatrix} \psi_{\nu_\mathrm{e}} \\ \psi_{\nu_\mu} \end{pmatrix} = \begin{pmatrix} \cos\theta & \sin\theta \\ -\sin\theta & \cos\theta \end{pmatrix} \begin{pmatrix} \psi_{\nu_1} \\ \psi_{\nu_2} \end{pmatrix} \equiv \hat{U} \begin{pmatrix} \psi_{\nu_1} \\ \psi_{\nu_2} \end{pmatrix} \tag{6a}$$

and

$$(\psi^\dagger_{\nu_\mathrm{e}}, \psi^\dagger_{\nu_\mu}) = (\psi^\dagger_{\nu_1}, \psi^\dagger_{\nu_2})\hat{U}^\dagger \quad . \tag{6b}$$

If the mass matrix has eigenvalue m_1 for ν_1 and m_2 for ν_2, then one finds for the matrix of the squared masses in the representation of electron and muon neutrino that

$$\begin{aligned} &(\psi^\dagger_{\nu_1}, \psi^\dagger_{\nu_2}) \begin{pmatrix} m_1^2 & 0 \\ 0 & m_2^2 \end{pmatrix} \begin{pmatrix} \psi_{\nu_1} \\ \psi_{\nu_2} \end{pmatrix} \\ &= (\psi^\dagger_{\nu_\mathrm{e}}, \psi^\dagger_{\nu_\mu}) \hat{U} \begin{pmatrix} m_1^2 & 0 \\ 0 & m_2^2 \end{pmatrix} \hat{U}^\dagger \begin{pmatrix} \psi_{\nu_\mathrm{e}} \\ \psi_{\nu_\mu} \end{pmatrix} \quad , \end{aligned} \tag{7}$$

and furthermore that

$$\begin{aligned} &\frac{1}{2}(m_1^2 + m_2^2)\hat{U} \begin{pmatrix} 1 & 0 \\ 0 & 1 \end{pmatrix} \hat{U}^\dagger + \frac{1}{2}(m_1^2 - m_2^2)\hat{U} \begin{pmatrix} 1 & 0 \\ 0 & -1 \end{pmatrix} \hat{U}^\dagger \\ &= \frac{1}{2}(m_1^2 + m_2^2) \begin{pmatrix} 1 & 0 \\ 0 & 1 \end{pmatrix} + \frac{1}{2}(m_1^2 - m_2^2) \\ &\quad \times \begin{pmatrix} \cos\theta & \sin\theta \\ -\sin\theta & \cos\theta \end{pmatrix} \begin{pmatrix} 1 & 0 \\ 0 & -1 \end{pmatrix} \begin{pmatrix} \cos\theta & -\sin\theta \\ \sin\theta & \cos\theta \end{pmatrix} \\ &= \frac{1}{2}(m_1^2 + m_2^2) \begin{pmatrix} 1 & 0 \\ 0 & 1 \end{pmatrix} + \frac{1}{2}(m_1^2 - m_2^2) \\ &\quad \times \begin{pmatrix} \cos^2\theta - \sin^2\theta & -2\sin\theta\cos\theta \\ -2\sin\theta\cos\theta & -\cos^2\theta + \sin^2\theta \end{pmatrix} \\ &= \frac{1}{2}(m_1^2 + m_2^2) \begin{pmatrix} 1 & 0 \\ 0 & 1 \end{pmatrix} + \frac{1}{2}(m_2^2 - m_1^2) \begin{pmatrix} -\cos 2\theta & \sin 2\theta \\ \sin 2\theta & \cos 2\theta \end{pmatrix} \equiv \hat{M}^2 \quad . \end{aligned} \tag{8}$$

---

32 H.A. Bethe: Phys. Rev. Lett. **56**, 1305 (1986).

Example 7.1

Essential for the explanation of the small number of measured electron neutrinos presented here is the interaction between neutrinos and electrons of the solar matter. This is represented by the graph

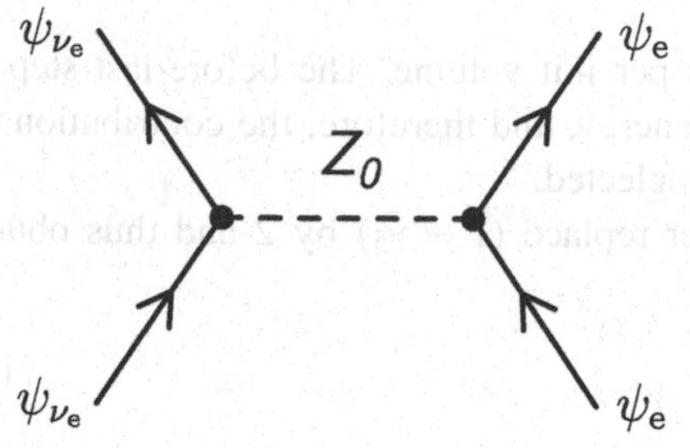

and described by the Hamiltonian density – see Exercise 4.10, (4) –

$$H_{\text{int}} = \left(\frac{-ig}{2\sqrt{2}}\right)^2 \left(\frac{1}{\sqrt{2}\cos\theta_{\text{W}}}\right)^2 \frac{1}{q^2 - M_{\text{Z}}^2}$$
$$\times 2\left(\bar{\psi}_{\nu_{\text{e}}}\gamma^{\mu}(1-\gamma_5)\psi_{\nu_{\text{e}}}\right)\cdot(-1)\cdot\left(\bar{\psi}_{\text{e}}\gamma_{\mu}(g'_{\text{V}}-\gamma_5)\psi_{\text{e}}\right). \tag{9}$$

The factor 2 stems from the fact that equation (4) of Exercise 4.10 yields twice the contribution represented by the above graph. Here, $g'_{\text{V}} = 1 - 4\sin^2\theta_{\text{W}} \approx 0$, and $1/(q^2 - M_{\text{Z}}^2)$ is the propagator of the Z_0, which in the low-energy limit approaches

$$\frac{1}{q^2 - M_{\text{Z}}^2} \xrightarrow{q\to 0} \frac{1}{-M_{\text{Z}}^2} = -\frac{\cos^2\theta_{\text{W}}}{M_{\text{W}}^2} \quad .$$

Hence, (9) becomes in the low-energy limit

$$H_{\text{int}} = \frac{g^2}{8M_{\text{W}}^2}\left(\bar{\psi}_{\nu_{\text{e}}}\gamma^{\mu}(1-\gamma_5)\psi_{\nu_{\text{e}}}\right)\left(\bar{\psi}_{\text{e}}\gamma_{\mu}\gamma_5\psi_{\text{e}}\right)$$
$$\approx \frac{G}{\sqrt{2}}\left(\bar{\psi}_{\nu_{\text{e}}}\gamma^{\mu}(1-\gamma_5)\psi_{\nu_{\text{e}}}\right)\left(\bar{\psi}_{\text{e}}\gamma_{\mu}\gamma_5\psi_{\text{e}}\right) \quad , \tag{10}$$

where we utilized (4.144). The interaction via the neutral current of (4.100),

$$H_{\text{neutral current}} \sim \frac{g}{2\sqrt{2}} J_0^{(e)\alpha} Z_{\alpha}$$

leads, by Z_0 exchange, to the current–current interaction

$$H_{\text{current-current}} \sim \left(\frac{g}{2\sqrt{2}}\right)^2 J_0^{(e)\alpha} J_{0\ \alpha}^{(e)}$$

which contains the contribution (9) of the interaction between electrons and electron neutrinos.

In the rest frame of the sun, the mean current of the electrons equals zero, that is,

$$\langle\bar{\psi}_{\text{e}}\gamma(1-\gamma_5)\psi_{\text{e}}\rangle = 0. \tag{11}$$

Therefore, in (9) only the term with $\mu = 0$ contributes. Now, in (11) we insert

$$\psi_{\text{e}} = \begin{pmatrix}\phi_{\text{e}}\\ \chi_{\text{e}}\end{pmatrix}$$

and replace

$$\bar{\psi}_e \gamma^0 (1-\gamma_5) \psi_e = (\phi_e^*, \chi_e^*) \begin{pmatrix} \mathbb{1} & -\mathbb{1} \\ -\mathbb{1} & \mathbb{1} \end{pmatrix} \begin{pmatrix} \phi_e \\ \chi_e \end{pmatrix} \approx \phi_e^* \phi_e = \rho_e \tag{12}$$

where ρ_e is the mean number of electrons per init volume. The before-last step is allowed because of the electrons (ϕ_e) low energy, and therefore, the contribution of the antiparticles (the positrons, χ_e) can be neglected.

For left-handed neutrinos we can further replace $(1-\gamma_5)$ by 2 and thus obtain finally

$$H_{\text{int}} = V \bar{\psi}_{\nu_e} \gamma_0 \psi_{\nu_e} \quad , \tag{13}$$

where

$$V = \sqrt{2} G \rho_e \quad . \tag{14}$$

The electrons present therefore act on the neutrinos like the time-like component of a four-potential. Hence we get the relation

$$k^2 + m^2 = (E-V)^2 \approx E^2 - 2EV \equiv E^2 - A \quad , \tag{15}$$

where we assume V to be very small, and where we can therefore neglect V^2. The electrons, for a neutrino ν_e of energy E, thus have the same effect as replacing m^2 by $m^2 + A$. We therefore write the matrix of the squared masses of ν_e and ν_μ as

$$\begin{aligned} \hat{M}^2 &\to \hat{M}^2 + \begin{pmatrix} A & 0 \\ 0 & 0 \end{pmatrix} \\ &= \frac{1}{2}(m_1^2 + m_2^2 + A) + \frac{1}{2} \begin{pmatrix} A - \Delta \cos 2\theta & \Delta \sin 2\theta \\ \Delta \sin 2\theta & -A + \Delta \cos 2\theta \end{pmatrix} \quad , \end{aligned} \tag{16}$$

where $\Delta = m_2^2 - m_1^2$. The eigenvalues of this matrix are given by

$$m_\nu^2 = \frac{1}{2}(m_1^2 + m_2^2 + A) \pm \frac{1}{2}\left[(\Delta \cos 2\theta - A)^2 + \Delta^2 \sin^2 2\theta\right]^{1/2} \quad . \tag{17}$$

The dependence of these eigenvalues on A is represented in the figure below. For $\theta = 0$, that is, if ν_e and ν_μ do not mix, these lie on the dashed lines, and at $A = \Delta$ there is a level crossing. This is different if there is an interaction between ν_e and ν_μ, that is, if $\theta \neq 0$. As one can see, in this latter case there is no level crossing; the levels, so to speak, repel each other. For

$$A_0 = 2\sqrt{2} G \rho_e E_0 = \Delta \cos 2\theta \tag{18}$$

the distance between the levels is a minimum. The idea now is the following. If in the interior of the sun a neutrino of energy $E > E_0$ is created, then in the figure below this lies on the upper curve, near the place marked by a cross. If this neutrino passes through the sun towards the sun's surface, the electron density, and thus the magnitude of A decreases. If this decrease occurs adiabatically, that is to say, if the electron density changes only slowly, the neutrino passes along the solid line, or, in other words, the wave function of the neutrino is always given by the linear combination that corresponds to the higher-energy eigenvalue. This has the following consequence.

Example 7.1

An electron neutrino of sufficiently high energy that is created in the interior of the sun appears at the surface as a neutrino eigenstate with the larger mass, m_2. If the mixing angle is small and, as expected, muon neutrinos have the larger mass, then the neutrino at the solar surface will essentially appear as a muon neutrino. Only those neutrinos which are created with $E < E_0$ can, with large probability, be detected as electron neutrinos on earth.

If this is the explanation for the missing solar neutrinos, one can set boundaries on the region in which E_0 can lie. E_0 cannot be much smaller than the threshold of the detector, because otherwise all observable neutrinos would be converted to muon neutrinos and one would not see any electron neutrinos at all. In fact one observes one third of the expected neutrinos, so that surely $E_0 \gtrsim 1$ MeV. On the other hand, the high-energy neutrinos resulting from the reaction $^8\mathrm{B} \to {}^8\mathrm{Be} + \mathrm{e}^+ + \nu_\mathrm{e}$ will be converted into muon neutrinos. This is only possible if E_0 is significantly lower than the maximum energy of 14 MeV, which is the maximum a neutrino of this reaction can have; thus $E_0 \lesssim 10$ MeV.

According to (18), one would have an explanation for the missing solar neutrinos if

$$E_0 = \frac{(m_2^2 - m_1^2)}{2\sqrt{2}G_F\rho_\mathrm{e}} \tag{19}$$

(for small θ we set $\cos 2\theta \approx 1$) were just of the described order of magnitude, which would of course be a special coincidence. If one estimates the electron density ρ_e in the interior of the sun in the region where the reaction with ^{8}B takes place according to the standard solar model to be about $4 \times 10^{25}/\mathrm{cm}^3$ (see the quoted references of Bahcall and Bethe), and if one assumes for E_0 a value of 6 MeV, one finds for the difference of the squared masses

$$m_2^2 - m_1^2 \approx 6 \times 10^{-5}\ \mathrm{eV}^2 \quad , \tag{20}$$

that is, for $m_1 \ll m_2$ we have

$$m_2 \approx 0.008\ \mathrm{eV} \tag{21}$$

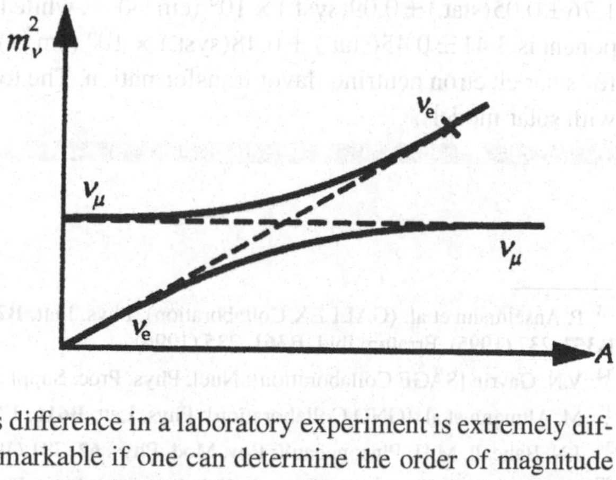

Fig. 7.12. Dependence of the masses of ν_e and ν_μ on the density of the surrounding matter

The proof of such a small mass difference in a laboratory experiment is extremely difficult. It would therefore be remarkable if one can determine the order of magnitude of the mass difference of the neutrinos by the missing solar neutrinos. An essential prediction of the presented model is that electron neutrinos of low energy are not

xample 7.10

converted into muon neutrinos and should therefore be observable on earth. If in the experiment with the gallium detector one observed an event rate of around the predicted 122 SNU, this would hint at the correctness of the model discussed above. If the event rate were significantly lower, one would have to find another explanation for the missing solar neutrinos.

The first results of the GALLEX experiment were published[33] in the summer of 1992. Averaged over an observation period of nearly one year, the experiment detected a neutrino flux of 83 ± (statistical errors) ± (systematic errors). This result, while marginally consistent with the presence of the full neutrino flux as predicted by the standard solar model, confirms the reduced fluxes of high-energy solar neutrinos observed in the previous experiments. The corresponding neutrino mass differences and mixing angles are restricted to a very narrow range:

$$\text{(a)}\ \sin^2 2\theta = 7\times 10^{-3} \quad , \quad \Delta m^2 \sim 6\times 10^{-6}\ \text{eV}^2 \quad ,$$

or

$$\text{(a)}\ \sin^2 2\theta = 0.6 \quad , \quad \Delta m^2 \sim 8\times 10^{-6}\ \text{eV}^2 \quad .$$

The initial 1992 GALLEX results have essentially been confirmed by subsequent data: The SAGE experiment, using 50 tons of gallium in metallic form, yielded a neutrino flux of[34] $66.9 \pm 3.9(\text{stat.}) \pm 3.6(\text{sys.})$ SNU, and the final, combined result of the GALLEX and GNO experiments give flux of[35]

$$\phi_{\nu_e}(\text{GALLEX/GNO}) = 69.3 \pm 5.5(\text{incl. sys.})\ \text{SNU}. \tag{22}$$

Calculations[36] and experimental data obtained by the Super-Kamiokande detector[37] confirm that the total flux of ^{8}B solar electron neutrinos amounts to only 36% of the flux predicted by standard theoretical solar models.

Convincing experimental proof that the reduced flux of solar electron neutrinos is caused by neutrino oscillations has been provided by the Sudbury Neutrino Observatory (SNO).[38] The SNO detector, a 1000 ton heavy water Cherenkov detector, is sensitive to both the ^{8}B electron neutrino flux and the total neutrino flux, averaged over all flavors. Data show that the electron neutrino component of the solar flux is $1.76 \pm 0.05(\text{stat.}) \pm 0.09(\text{syst.}) \times 10^6\ (\text{cm}^2\ \text{s})^{-1}$, while the non-electron neutrino component is $3.41 \pm 0.45(\text{stat.}) \pm 0.48(\text{syst.}) \times 10^6\ (\text{cm}^2\ \text{s})^{-1}$, providing strong evidence for solar electron neutrino flavor transformation. The total flux measured is consistent with solar models.

33 P. Anselmann et al. (GALLEX Collaboration), Phys. Lett. **B285**, 376 and 390 (1992); Phys. Lett **B357**, 237 (1995), Erratum ibid. **B361**, 235 (1996).

34 V.N. Gavrin [SAGE Collaboration]: Nucl. Phys. Proc. Suppl. **91**, 36 (2001).

35 M. Altmann et al. [GNO Collaboration]: Phys. Lett. **B616**, 174 (2005).

36 J.N. Bahcall, M.H. Pinsonneault: Rev. Mod. Phys. **67**, 781 (1995).

37 J. Hosaka et al. [Super-Kamiokande Collaboration]: Phys. Rev. D **73**, 112001 (2006).

38 Q.R. Ahmad et al. [SNO Collaboration]: Phys. Rev. Lett. **89**, 011301 (2002).

7.5 Biographical Notes

BETHE, Hans Albrecht, physicist, *2.7.1906 in Strasbourg, †6.3.2005 in Ithaca (New York, USA), professor at Cornell University in Ithaca. He was educated at the universities of Frankfurt and Munich, obtaining his Ph.D. in 1928 under Sommerfeld. He worked under Rutherford in Cambridge and Fermi at Rome, then taught physics at Munich and Tübingen until 1933, when he emigrated to England and later to the United States. Bethe's main contribution to science was working out the details of the nuclear mechanisms that power the stars, which he achieved in 1938, when Weizsäcker was independently reaching similar conclusions in Germany. He also contributed to the development of quantum electrodynamics, and was the first to estimate the Lamb shift in hydrogen. In 1967 he received the Nobel Prize for physics.

DAVIS, Raymond, Jr, *14.10.1914 in Washington, D.C. (USA), †31.5.2006 in Blue Point (New York, USA), was an American chemist and physicist. He graduated in chemistry from the University of Maryland in 1938, and obtained a Ph.D. from Yale University in physical chemistry in 1942. Upon his discharge from the army in 1946, Davis went to work at Monsanto Chemical Company's Mound Laboratory, doing applied radiochemistry of interest to the US Atomic Energy Commission. In 1948, he joined Brookhaven National Laboratory, where he remained until 1984. At Brookhaven, Davis developed a new form of neutrino detector, using a tank filled with fluid tetrachloroethylene. He correctly reasoned that neutrinos from the sun would react with the chlorine in the fluid to produce argon atoms. Running over more than 20 years, the detector successfully captured neutrinos and proved that fusion was the source of energy from the sun, but also evidenced the "solar neutrino puzzle." Davies was awarded the 2002 Nobel Prize for Physics for his contribution to the field of astrophysics, especially for his work in the detection of neutrinos.

MAJORANA, Ettore, *5.8.1906 in Catania (Italy), dropped out of sight in 1938, went to the classical secondary school of Catania until the final examination in 1923. Afterwards he studied engineering sciences in Rome until the last year of studies. 1928 transfer to the physics faculty and 1929 Ph.D. in theoretical physics under Fermi. Title of the thesis: "Quantum Theory of Radioactive Atomic Nuclei". In the subsequent years freelance collaborator at the Institute of Physics in Rome. In 1933 he went to Germany (Leipzig) for some years and worked with Heisenberg. This resulted in a publication on nuclear theory (Z. Phys. **82**, 137 (1933)). In 1937 he published "The Symmetric Theory of Electron and Positron" and four years after his disappearance the "Significance of Statistical Laws for Physics and Social Sciences" was published.

8 The Neutral Kaon System

8.1 The Particles K_S and K_L

There are four strange pseudoscalar mesons K^+, K^-, K^0 and $\bar{K}^0$, which are all eigenstates of the Hamiltonian of the strong interaction. Their quark content,[1]

$$\begin{aligned} K^+ &= (\bar{s}u) \quad , \quad K^0 = (\bar{s}d) \quad , \\ K^- &= (\bar{u}s) \quad , \quad \bar{K}^0 = (s\bar{d}) \quad , \end{aligned} \tag{8.1}$$

shows that K^+ and K^- form a conjugate particle–antiparticle pair, as does the pair K^0 and $\bar{K}^0$. We are used to thinking that the charge of an antiparticle is opposite to that of a particle, as is the case for the pair K^+, K^-. The pair $K^0, \bar{K}^0$, however, is neutral. Nevertheless, the K^0 differs from the $\bar{K}^0$ in the signs of other quantum numbers, namely isospin and strangeness:

$$\begin{aligned} K^0: \quad T &= \frac{1}{2} \quad , \quad T_3 = -\frac{1}{2} \quad , \quad S = +1 \quad , \\ \bar{K}^0: \quad T &= \frac{1}{2} \quad , \quad T_3 = +\frac{1}{2} \quad , \quad S = -1 \quad . \end{aligned} \tag{8.2}$$

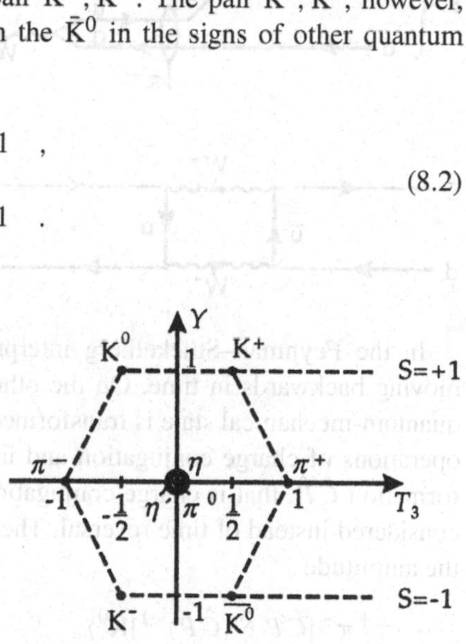

Fig. 8.1. The nonet of the spin-0 mesons

Hence K^0 and $\bar{K}^0$ are unambiguously different states. The situation is similar with the neutron n and its antiparticle $\bar{n}$. They differ in isospin and baryon number:

$$\begin{aligned} n(ddu): \quad T &= \frac{1}{2} \quad , \quad T_3 = -\frac{1}{2} \quad , \quad B = +1 \quad , \\ \bar{n}(\bar{d}\bar{d}\bar{u}): \quad T &= \frac{1}{2} \quad , \quad T_3 = +\frac{1}{2} \quad , \quad B = -1 \quad . \end{aligned} \tag{8.3}$$

[1] See W. Greiner and B. Müller: *Quantum Mechanics – Symmetries*, 2nd ed. (Springer, Berlin, Heidelberg, 1994).

W. Greiner, B. Müller, *Gauge Theory of Weak Interactions*,
DOI 10.1007/978-3-540-87843-8_8, © Springer-Verlag Berlin Heidelberg 2009

There is a subtle distinction between these two cases: according to the present state of experimental physics the baryon number is an absolutely conserved quantity of nature;[2] therefore a neutron can *never* transform into an antineutron. On the other hand, we know that isospin and strangeness conservation are violated by the weak interaction. Thus, despite the fact that K^0 and $\bar{K}^0$ are different particles at the level of the strong interaction, there is no reason why they should not be able to transform into each other under the influence of the weak interaction. We see that this is in fact possible, because K^0 as well as $\bar{K}^0$ may decay weakly into the two-pion state $(\pi^+\pi^-)$. Therefore it must be possible for the K^0 to transform into its antiparticle via the intermediate system $\pi^+\pi^-$ (see Fig. 8.2). It is easy to decompose the process $K^0 \leftrightarrow \bar{K}^0$ into reactions of the constituent quarks of the particles and to describe it in the framework of the Salam–Weinberg theory (Fig. 8.3), or more schematically by the graph shown in Fig. 8.4.

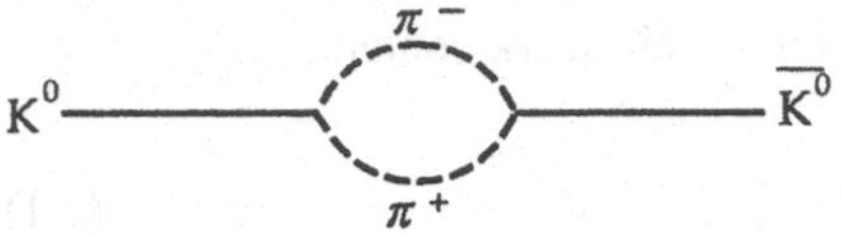

ig. 8.2. The transformation f a K^0 into a $\bar{K}^0$ via an in- ermediate $\pi^+\pi^-$ pair

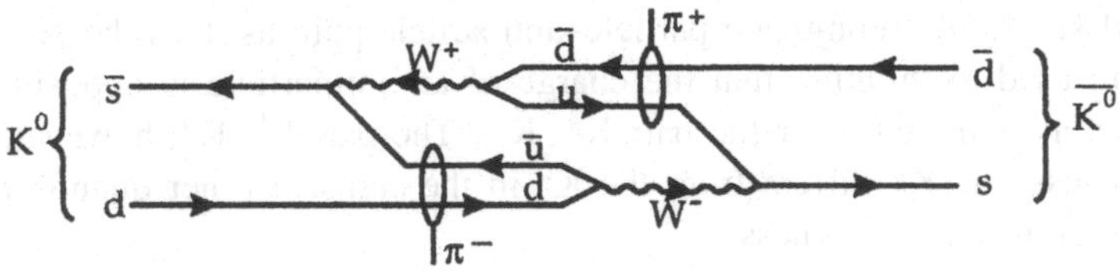

ig. 8.3. The transformation n microscopic view. The quark content of the meson is hown

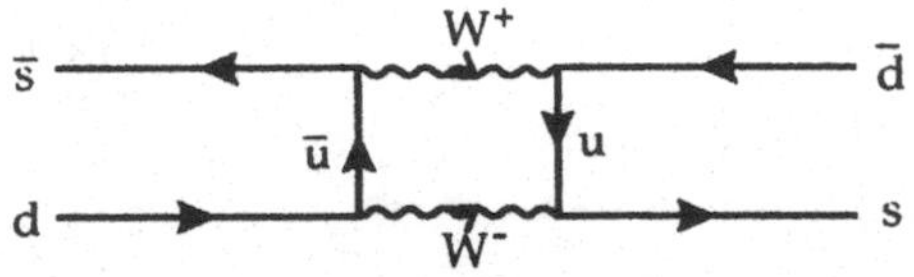

ig. 8.4. Schematic represen- ation of the K^0–$\bar{K}^0$ trans- ormation in the microscopic quark picture

In the Feynman–Stückelberg interpretation an antiparticle is simply a particle moving backwards in time. On the other hand, the CPT theorem[3] states that every quantum-mechanical state is transformed into itself under the combined action of the operations of charge conjugation and inversion of space and time. Hence the transformation $\hat{C}\hat{P}$, that is charge conjugation and simultaneous space inversion, may be considered instead of time reversal. The transition amplitude $\langle\pi^+\pi^-|S|K^0\rangle$ becomes the amplitude

$$-\langle\pi^+\pi^-|\hat{C}\hat{P}\,\hat{S}\,(\hat{C}\hat{P})^{-1}|\bar{K}^0\rangle$$

under the operation $\hat{C}\hat{P}$, because only π^+ and π^- exchange their roles on the left-hand side (the minus sign is connected with the internal parity (−) of the kaon (see Exercise 8.1)).

[2] For some time it has been considered theoretically that baryon-number conservation may be violated in some processes at a very low level. We shall discuss this in Chap. 9. Because of the known lower limit of the lifetime of the proton, $\tau_p > 10^{39}$ s, the effective coupling constant for such processes must be at least 24 orders of magnitude smaller than the Fermi constant of the weak interaction.

[3] G. Lüders: Danske Vid. Selskab Mat. Fys. Medd. **28**, 5 (1954); see also W. Greiner and J. Reinhardt: *Quantum Electrodynamics*, 2nd ed. (Springer, Berlin, Heidelberg, 1994).

If the physics of the kaon system is time-reversal invariant – we want to make this assumption here for the moment – then $\hat{C}\hat{P}\hat{S}(\hat{C}\hat{P})^{-1} = \hat{S}$ and

$$\langle \pi^+\pi^-|\hat{S}|\mathrm{K}^0\rangle = -\ \langle \pi^+\pi^-|\hat{S}|\bar{\mathrm{K}}^0\rangle \tag{8.4}$$

holds. To study this in greater detail we begin with

$$\hat{C}\hat{P}|K_0\rangle = -\hat{C}|K_0\rangle = -|\bar{K}_0\rangle \quad ,$$

because the inner parity of the K_0 is negative and $\hat{C}$ transforms a K_0 into a $\bar{K}_0$. Analogously it follows that

$$\hat{C}\hat{P}|\pi^+\pi^-\rangle = \hat{C}(-)^2|\pi^-\pi^+\rangle = +|\pi^+\pi^-\rangle \ .$$

Therefore if follows that

$$\langle \pi^+\pi^-|\hat{S}|K_0\rangle = -\langle \pi^+\pi^-|\hat{C}\hat{P}\hat{S}(\hat{C}\hat{P})^{-1}|\bar{K}_0\rangle \quad .$$

If we take $\hat{T}$ invariance into account,

$$\hat{C}\hat{P}\hat{S}(\hat{C}\hat{P})^{-1} = \hat{S} \quad ,$$

it follows that

$$\langle \pi^+\pi^-|\hat{S}|K_0\rangle = -\langle \pi^+\pi^-|\hat{S}|\bar{K}_0\rangle \quad .$$

Hence it also holds that

$$\begin{aligned}
\langle \mathrm{K}^0|\hat{S}|\bar{\mathrm{K}}^0\rangle &\approx \sum_{|\pi^+\pi^-\rangle} \langle \mathrm{K}^0|\hat{S}|\pi^+\pi^-\rangle\langle \pi^+\pi^-|\hat{S}|\bar{\mathrm{K}}^0\rangle \\
&= -\sum_{|\pi^+\pi^-\rangle} |\langle \pi^+\pi^-|\hat{S}|\mathrm{K}^0\rangle|^2 \\
&= \sum_{|\pi^+\pi^-\rangle} \langle \bar{\mathrm{K}}^0|\hat{S}|\pi^+\pi^-\rangle\langle \pi^+\pi^-|\hat{S}|\mathrm{K}^0\rangle \\
&\approx \langle \bar{\mathrm{K}}^0|\hat{S}|\mathrm{K}^0\rangle \quad ,
\end{aligned} \tag{8.5}$$

where the sum includes all momenta of both pions in the intermediate state. Of course, other intermediate states add minor contributions, but this does not affect the general argument, which is originally due to Gell-Mann and Pais.[4]

EXERCISE

8.1 CP Parity in Kaon Decay

Problem. Show that the two-pion system evolving from the kaon decay has positive CP parity and that

$$\langle \pi^+\pi^-|\hat{S}|\mathrm{K}^0\rangle = -\langle \pi^+\pi^-|\hat{C}\hat{P}\hat{S}(\hat{C}\hat{P})^{-1}|\bar{\mathrm{K}}^0\rangle \quad . \tag{1}$$

---

[4] M. Gell-Mann, A. Pais: Phys. Rev. **97**, 1387 (1955).

xercise 8.1

Solution. The two pions have opposite angular momenta in the centre-of-mass system of the kaon. Therefore the effect of the $\hat{C}\hat{P}$ operation can be derived particularly easily here. The most general ansatz for the two-pion system is

$$|\pi^+\pi^-\rangle = \sin\alpha|\pi^+(E, \boldsymbol{p})\,\pi^-(E, -\boldsymbol{p})\rangle + \cos\alpha|\pi^-(E, -\boldsymbol{p})\pi^+(E, \boldsymbol{p})\rangle \tag{2}$$

with an arbitrary mixing angle α. From (2) it follows that

$$\begin{aligned}\hat{P}|\pi^+\pi^-\rangle = P_\pi^2(&\sin\alpha|\pi^+(E, -\boldsymbol{p})\pi^-(E, \boldsymbol{p})\rangle \\ &+ \cos\alpha|\pi^-(E, \boldsymbol{p})\pi^+(E, -\boldsymbol{p})\rangle) \quad ,\end{aligned} \tag{3}$$

where $P_\pi = -1$ is the internal parity of the pions. We see that in general the two-pion system need not have good parity (this is only the case for $\alpha = \pi/4$ or $-\pi/4$). It is the weak interaction that violates parity. For CP parity

$$\begin{aligned}\hat{C}\hat{P}|\pi^+\pi^-\rangle &= (\sin\alpha\;|\pi^-(E, -\boldsymbol{p})\,\pi^+(E, \boldsymbol{p})\rangle + \cos\alpha\;|\pi^+(E, \boldsymbol{p})\,\pi^-(E, -\boldsymbol{p})\rangle) \\ &= +|\pi^+\pi^-\rangle\end{aligned} \tag{4}$$

holds, that is the CP parity of the two-pion state is always positive. On the other hand, we have

$$\hat{C}\hat{P}|\mathrm{K}^0\rangle = -\hat{C}|\mathrm{K}^0\rangle = -|\bar{\mathrm{K}}^0\rangle \quad , \tag{5}$$

because of the negative internal parity of the kaons. Relation (1) follows directly from (4) and (5).

We want to describe the time evolution of, for example, a K^0 meson. If $|\psi(0)\rangle$ is the initial state, then after a time t

$$|\psi(t)\rangle = \hat{U}(t,0)|\psi(0)\rangle \tag{8.6}$$

holds, with

$$\hat{U}(t,0) = \exp(-\mathrm{i}\hat{H}t) \quad ,$$

where $\hat{H}$ is the full Hamiltonian. We split $\hat{H}$ into a part $\hat{H}_0$ which contains the strong and electromagnetic interaction and has $|\mathrm{K}^0\rangle$ and $|\bar{\mathrm{K}}^0\rangle$ as eigenstates and a part $\hat{H}_\mathrm{W}$ (in the sense of the Fermi point interaction) which is responsible for the decay of these states and their transformation among each other:

$$\hat{H} = \hat{H}_0 + \hat{H}_\mathrm{W} \quad . \tag{8.7}$$

In the interaction picture

$$\hat{H}_\mathrm{W}(t) = \exp(-\mathrm{i}\hat{H}_0 t)\hat{H}_\mathrm{W}\exp(\mathrm{i}\hat{H}_0 t) \quad , \tag{8.8}$$

and

$$\hat{U}(t,0) = 1 - \mathrm{i}\int_0^t \hat{H}_\mathrm{W}(t')\mathrm{d}t' - \int_0^t \mathrm{d}t'\;\hat{H}_\mathrm{W}(t')\int_0^{t'}\mathrm{d}t''\;\hat{H}_\mathrm{W}(t'')\ldots \tag{8.9}$$

holds. The evolution of $|\psi(t)\rangle$ is described, inside the Hilbert space spanned by $|K^0\rangle$ and $|\bar{K}^0\rangle$, by an effective Hamiltonian which is a 2×2 matrix. The effective Hamiltonian in this subspace is determined by the virtual transitions to intermediate states $|n\rangle$ outside this subspace. It reads

$$\begin{aligned} \hat{H}_W^{\text{eff}} &= \hat{H}_W + \sum_n \frac{\hat{H}_W|n\rangle\langle n|\hat{H}_W}{m_K - E_n + i\varepsilon} + \cdots \\ &= \hat{H}_W + \sum_n \hat{H}_W|n\rangle\langle n|\hat{H}_W \left(P\frac{1}{m_K - E_n + i\varepsilon} - i\pi\delta(m_K - E_n)\right) + \cdots \end{aligned} \tag{8.10}$$

The resulting matrix in the K^0–$\bar{K}^0$ space is

$$\begin{pmatrix} \langle K^0|\hat{H}_W^{\text{eff}}|K^0\rangle & \langle K^0|\hat{H}_W^{\text{eff}}|\bar{K}^0\rangle \\ \langle \bar{K}^0|\hat{H}_W^{\text{eff}}|K^0\rangle & \langle \bar{K}^0|\hat{H}_W^{\text{eff}}|\bar{K}^0\rangle \end{pmatrix} \equiv \hat{M} - \frac{i}{2}\hat{\Gamma} \tag{8.11}$$

with ($m_K = m_{\bar{K}}$)

$$\hat{H}_0|K^0\rangle = m_K|K^0\rangle \quad , \quad \hat{H}_0|\bar{K}^0\rangle = m_K|\bar{K}^0\rangle \quad , \quad M = M^\dagger \quad , \quad \Gamma = \Gamma^\dagger \quad . \tag{8.12}$$

$\hat{M}$ and $\hat{\Gamma}$ in (8.11) are both 2×2 matrices. The sum in (8.10) runs over all intermediate states $|n\rangle$ states, with energies E_n distinct from $|K^0\rangle$ and $|\bar{K}^0\rangle$; the imaginary part is present because the kaons decay.

Let us repeat the basic idea of the calculation. We confine ourselves to two interacting states (particles) and take into account their interaction, which in second-order perturbation theory is given by the sum over excited intermediate states. The 2×2 matrix (8.11) is the effective Hamiltonian in this two-state (two-particle) subspace. Symbolically, we represent the state $|K^0\rangle$ by the unit vector $\binom{1}{0}$ and the state $|\bar{K}^0\rangle$ by the other unit vector $\binom{0}{1}$. The most general state in our restricted Hilbert space is then

$$|\psi\rangle = c|K^0\rangle + \bar{c}|\bar{K}^0\rangle = \begin{pmatrix} c \\ \bar{c} \end{pmatrix} \tag{8.13}$$

with independent coefficients c and $\bar{c}$; it evolves according to the Schrödinger equation

$$i\frac{\partial}{\partial t}|\psi(t)\rangle = \hat{H}_W^{\text{eff}}\,|\psi(t)\rangle \equiv \left(\hat{M} - \frac{i}{2}\hat{\Gamma}\right)|\psi(t)\rangle \quad . \tag{8.14}$$

Since particles and antiparticles behave in the same way (CPT invariance)

$$M_{11} = M_{22} \equiv M_0 \quad , \quad \Gamma_{11} = \Gamma_{22} \equiv \Gamma_0 \tag{8.15}$$

holds. If we assume time reversal invariance, then by consideration of (8.4), (8.5) the additional relations

$$M_{12} = M_{21} \equiv \tilde{M} \quad , \quad \Gamma_{12} = \Gamma_{21} \equiv \tilde{\Gamma} \tag{8.16}$$

hold, and because Γ_{12}, M_{12} are by definition real, so are $\tilde{M}$ and $\tilde{\Gamma}$. The same holds for M_0, Γ_0. The real diagonal part M_0 can be absorbed into the mass m_K; the imaginary part Γ_0 describes the decay of the kaons. The Schrödinger equation for the amplitudes is then

$$\mathrm{i}\frac{\mathrm{d}}{\mathrm{d}t}\begin{pmatrix} c \\ \bar{c} \end{pmatrix} = \begin{pmatrix} M_0 - \frac{\mathrm{i}}{2}\Gamma_0 & \tilde{M} - \frac{\mathrm{i}}{2}\tilde{\Gamma} \\ \tilde{M} - \frac{\mathrm{i}}{2}\tilde{\Gamma} & M_0 - \frac{\mathrm{i}}{2}\Gamma_0 \end{pmatrix}\begin{pmatrix} c \\ \bar{c} \end{pmatrix} \quad . \tag{8.17}$$

It is useful to introduce the linear combinations

$$c_L = \frac{1}{\sqrt{2}}(c + \bar{c}) \quad , \quad c_S = \frac{1}{\sqrt{2}}(c - \bar{c}) \tag{8.18}$$

corresponding to the representation

$$|\psi\rangle = c_L|\mathrm{K}_L^0\rangle + c_S|\mathrm{K}_S^0\rangle \tag{8.19}$$

with

$$\begin{aligned} |\mathrm{K}_L^0\rangle &= \frac{1}{\sqrt{2}}(|\mathrm{K}^0\rangle + |\bar{\mathrm{K}}^0\rangle) \quad , \\ |\mathrm{K}_S^0\rangle &= \frac{1}{\sqrt{2}}\,(|\mathrm{K}^0\rangle - |\bar{\mathrm{K}}^0\rangle) \quad . \end{aligned} \tag{8.20}$$

Then

$$\mathrm{i}\frac{\mathrm{d}}{\mathrm{d}t}\begin{pmatrix} c_L \\ c_S \end{pmatrix} = \begin{pmatrix} (M_0 + \tilde{M}) - \frac{\mathrm{i}}{2}(\Gamma_0 + \tilde{\Gamma}) & 0 \\ 0 & (M_0 - \tilde{M}) - \frac{\mathrm{i}}{2}(\Gamma_0 - \tilde{\Gamma}) \end{pmatrix}\begin{pmatrix} c_L \\ c_S \end{pmatrix} \quad . \tag{8.21}$$

From general considerations we can conclude that $\Gamma_0 > |\tilde{\Gamma}|$, hence both states decay. However, the state $|\mathrm{K}_L^0\rangle$ decays with the function $|c_L(t)|^2 = \exp(-(\Gamma_0 + \tilde{\Gamma})t)$; the state $|\mathrm{K}_S^0\rangle$ decays with $|c_S(t)|^2 = \exp(-(\Gamma_0 - \tilde{\Gamma})t)$.

In order to determine which of the states decays faster and which decays slower we have to consider the sign of $\tilde{\Gamma}$. According to the definition (8.11) and because of the sign in (8.4) we have

$$\begin{aligned} \tilde{\Gamma} = \Gamma_{12} &= -2\mathrm{Im}\,\langle \mathrm{K}^0|\hat{H}_\mathrm{W}^\mathrm{eff}|\bar{\mathrm{K}}_0\rangle \\ &\approx 2\pi\sum_n \delta(m_\mathrm{K} - E_n)\langle \mathrm{K}^0|\hat{H}_\mathrm{W}|n\rangle\langle n|\hat{H}_\mathrm{W}|\bar{\mathrm{K}}^0\rangle \\ &\approx -2\pi\sum_n \delta(m_\mathrm{K} - E_n)\langle \mathrm{K}^0|\hat{H}_\mathrm{W}|n\rangle\langle n|\hat{H}_\mathrm{W}|\mathrm{K}^0\rangle \\ &\approx -2\pi\sum_n \delta(m_\mathrm{K} - E_n)|\langle n|\hat{H}_\mathrm{W}|\mathrm{K}^0\rangle|^2 < 0 \quad , \end{aligned} \tag{8.22}$$

where we have inserted $\hat{H}_\mathrm{W}^\mathrm{eff}$ from (8.10). Therefore $\Gamma_0 - \tilde{\Gamma} \equiv \Gamma_S$ has the larger width and $\Gamma_0 + \tilde{\Gamma} \equiv \Gamma_L$ the narrower one, that is the state $|\mathrm{K}_S^0\rangle$ has a shorter life than the state $|\mathrm{K}_L^0\rangle$. On the other hand, we see directly from (8.10) that

$$\begin{aligned} \Gamma_0 = \Gamma_{11} &= -2\mathrm{Im}\,\langle \mathrm{K}^0|\hat{H}_\mathrm{W}^\mathrm{eff}|\mathrm{K}^0\rangle \\ &\approx 2\pi\sum_n \delta(m_\mathrm{K} - E_n)\langle \mathrm{K}^0|\hat{H}_\mathrm{W}|n\rangle\langle n|\hat{H}_\mathrm{W}|\mathrm{K}^0\rangle \\ &\approx 2\pi\sum_n \delta(m_\mathrm{K} - E_n)|\langle n|\hat{H}_\mathrm{W}|\mathrm{K}^0\rangle|^2 \approx -\tilde{\Gamma} > 0 \quad . \end{aligned} \tag{8.23}$$

This relation holds exactly only in second-order perturbation theory, that is when higher terms ($\hat{H}_W^3$ etc.) are neglected in (8.10). Hence we note that Γ_S is substantially larger than Γ_L:

$$\Gamma_S = \Gamma_0 - \tilde{\Gamma} \approx 2\Gamma_0 \gg \Gamma_0 + \tilde{\Gamma} = \Gamma_L \quad . \tag{8.24}$$

Thus we have shown that in reality (with the assumption of time-reversal invariance) not the states $|\mathrm{K}^0\rangle$ and $|\bar{\mathrm{K}}^0\rangle$ but the state vectors

$$|\mathrm{K}_L^0\rangle = \frac{1}{\sqrt{2}}(|\mathrm{K}^0\rangle + |\bar{\mathrm{K}}^0\rangle) \quad , \quad |\mathrm{K}_S^0\rangle = \frac{1}{\sqrt{2}}(|\mathrm{K}^0\rangle - |\bar{\mathrm{K}}^0\rangle) \tag{8.25}$$

are eigenstates of the Hamiltonian. One of the states, $|\mathrm{K}_S^0\rangle$, has a short lifetime,

$$\tau_{\mathrm{K}_S^0} = (\Gamma_0 - \tilde{\Gamma})^{-1} \quad , \tag{8.26}$$

whereas the other state lives much longer,

$$\tau_{\mathrm{K}_L^0} = (\Gamma_0 + \tilde{\Gamma})^{-1} \approx (2\Gamma_0)^{-1} \gg \tau_{\mathrm{K}_S^0} \quad . \tag{8.27}$$

However, in a reaction of elementary particles the eigenstates of the strong interaction are always created experimentally, for example

$$\pi^- + \mathrm{p} \rightarrow \mathrm{K}^0 + \Lambda \quad . \tag{8.28}$$

Then the initially formed state is

$$|\Psi(0)\rangle = |\mathrm{K}^0\rangle = \frac{1}{\sqrt{2}}(|\mathrm{K}_L^0\rangle + |\mathrm{K}_S^0\rangle) \quad , \tag{8.29}$$

that is the expansion coefficients are

$$c_S(0) = c_L(0) = \frac{1}{\sqrt{2}} \quad . \tag{8.30}$$

Hence, according to (8.21), the resulting time evolution is

$$|\psi(t)\rangle = \frac{1}{\sqrt{2}}\mathrm{e}^{-\mathrm{i}M_0 t}\left[\mathrm{e}^{\mathrm{i}\tilde{M}t - \frac{1}{2}\Gamma_S t}|\mathrm{K}_S^0\rangle + \mathrm{e}^{-\mathrm{i}\tilde{M}t - \frac{1}{2}\Gamma_L t}|\mathrm{K}_L^0\rangle\right] \quad , \tag{8.31}$$

which means that the short-lived component decays first, and a slowly decaying component $|\mathrm{K}_L^0\rangle$ remains for a longer period of time (Fig. 8.5). Experimentally one finds that

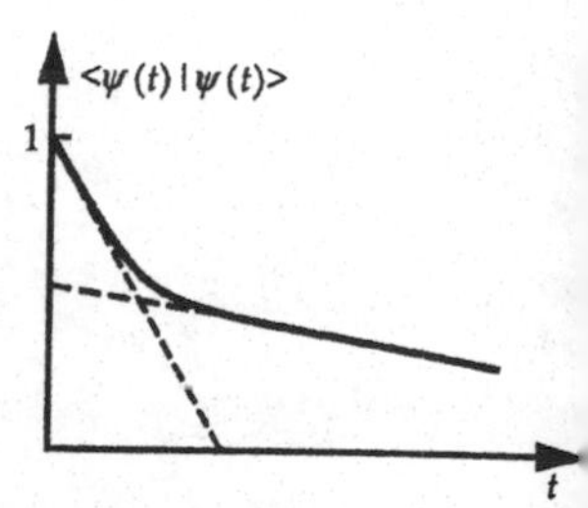

Fig. 8.5. Decay of an initially created K_0 as function of time. It seems to have two lifetimes

$$\tau_{\mathrm{K}_S^0} = 8.92 \times 10^{-11}\ \mathrm{s} \quad , \quad \tau_{\mathrm{K}_L^0} = 5.81 \times 10^{-8}\ \mathrm{s} \quad . \tag{8.32}$$

The large difference in the lifetimes, $\tau(\mathrm{K}_S^0)/\tau(\mathrm{K}_L^0) \approx 1.5 \times 10^{-3}$, is not surprising; we had already expected it owing to (8.24). Equation (8.23) in fact means that the decay width of the K_L^0 vanishes exactly in the framework of second-order perturbation theory. Therefore processes of higher order, such as final states with three pions, have to be responsible for the experimentally observed decay of the K_L^0. We shall discuss this in detail in the next section. Another important process is the so-called *regeneration*. If a beam of $|\mathrm{K}_L^0\rangle$ particles passes through matter, the kaons can scatter off the atomic nuclei. But the two components K^0 and $\bar{K}^0$ have entirely different properties, because the scattering is essentially effected by the strong interaction. Since there is no baryon with positive strangeness, the K_0 ($S=+1$) can scatter only elastically at low energies because of the conservation of strangeness, or it can transform into a K^+,

$$K^0 + \begin{pmatrix} n \\ p \end{pmatrix} \to K^0 + \begin{pmatrix} n \\ p \end{pmatrix} \quad , \quad K^0 + p \to K^+ + n \tag{8.33}$$

according to quark flow diagrams shown in Fig. 8.6.

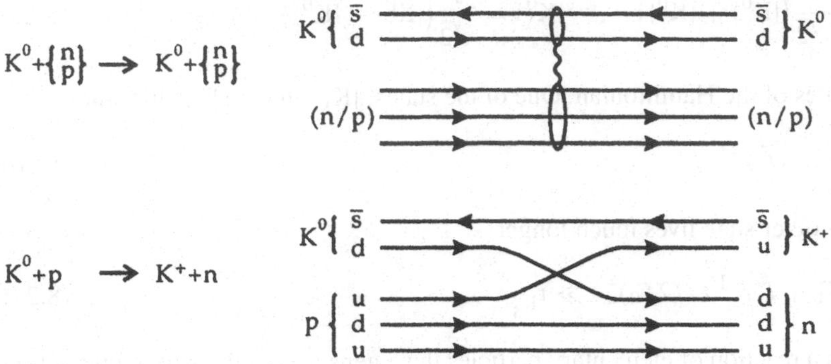

ig. 8.6. Quark flow diagrams or the two-kaon reactions $K^0 + \binom{n}{p} \to K^0 + \binom{n}{p}$ and $K^0 + p \to K^+ + n$

In contrast, the $\bar{K}^0$ meson ($S=-1$) can additionally scatter inelastically into a Λ particle (Fig. 8.7),

$$\bar{K}^0 + \begin{pmatrix} n \\ p \end{pmatrix} \to \bar{K}^0 + \begin{pmatrix} n \\ p \end{pmatrix} \quad , \quad \bar{K}^0 + n \to K^- + p \quad , \tag{8.34}$$

$$\bar{K}^0 + \begin{pmatrix} n \\ p \end{pmatrix} \to \Lambda + \begin{pmatrix} \pi^0 \\ \pi^+ \end{pmatrix} \quad . \tag{8.35}$$

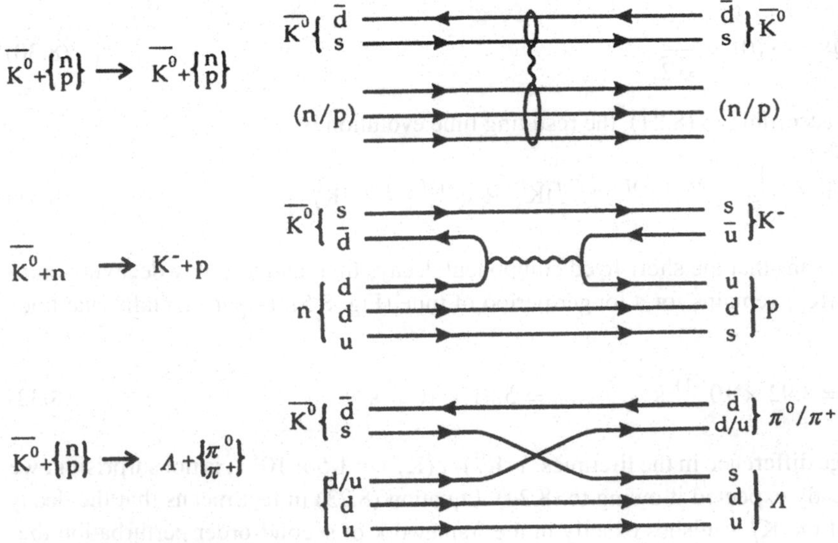

Fig. 8.7. Quark flow diagrams for different reactions of the $\bar{K}^0$ with nucleons

As a result, the forward scattering amplitude, which is related to the total scattering cross section by the optical theorem, is different for K^0 and $\bar{K}^0$ in nuclear matter:

$$f_{K^0}(0) \neq f_{\bar{K}^0}(0) \quad , \quad \mathrm{Im}\, f_{K^0}(0) < \mathrm{Im}\, f_{\bar{K}^0}(0) \quad . \tag{8.36}$$

Therefore, the two components propagate differently through matter, and after the passage the K_L^0 shows up as

$$|\psi\rangle = \frac{1}{2}\left(a|K^0\rangle + b|\bar{K}^0\rangle\right) = \frac{1}{2}\frac{a-b}{\sqrt{2}}|K_S^0\rangle + \frac{1}{2}\frac{a+b}{\sqrt{2}}|K_L^0\rangle \tag{8.37}$$

with $|a| > |b|$. This effect has to be carefully excluded if one wants to do experiments with a pure $|K_L^0\rangle$ beam.

The different behavior of K^0 and $\bar{K}^0$ in the scattering off matter can be used to measure the real part $\tilde{M}$ of the off-diagonal Hamiltonian matrix element, (8.16). The time evolution of a K^0 produced at $t = 0$ is given by $|\psi(t)\rangle$ in (8.31). Since only the $\bar{K}^0$ component can produce a Λ particle one measures in this way (see (8.31))

$$\begin{aligned}\bar{c}(t) &= \langle\bar{K}^0|\psi(t)\rangle = \frac{1}{\sqrt{2}}\langle K_S^0|\psi(t)\rangle - \frac{1}{\sqrt{2}}\langle K_L^0|\psi(t)\rangle \\ &= \frac{1}{2}e^{-iM_0t}\left(e^{i\tilde{M}t-\frac{1}{2}\Gamma_S t} - e^{-i\tilde{M}t-\frac{1}{2}\Gamma_L t}\right) \quad ,\end{aligned} \tag{8.38}$$

and hence, because of $\Gamma_S \gg \Gamma_L$,

$$\begin{aligned}|\bar{c}|^2 &= |\langle\bar{K}^0|\psi(t)\rangle|^2 = \frac{1}{4}\left|e^{i\tilde{M}t-\frac{1}{2}\Gamma_S t} - e^{-i\tilde{M}t}\right|^2 \\ &= \frac{1}{4}\left(1 + e^{-\Gamma_S t} - 2e^{-\frac{1}{2}\Gamma_S t}\cos 2\tilde{M}t\right) \quad .\end{aligned} \tag{8.39}$$

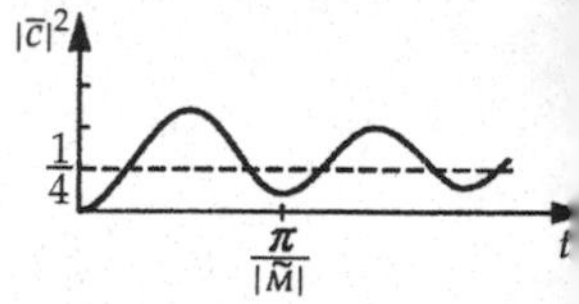

Fig. 8.8. Oscillations in Λ particle production owing to oscillations between K_S^0 and K_L^0

Experimentally one finds that

$$\Delta m = m_{K_L^0} - m_{K_S^0} = -2\tilde{M} = 0.535 \times 10^{-10}\ \mathrm{s}^{-1} = 3.5 \times 10^{-6}\ \mathrm{eV} \quad . \tag{8.40}$$

The probability $|\bar{c}|^2$ of finding a $\bar{K}^0$ in a K^0 beam oscillates (see Fig. 8.8) – a periodic shift from K^0 to $\bar{K}^0$ and vice versa occurs. This effect is called "regeneration".

8.2 CP Violation

During the general consideration which showed that the two states K_L^0 and K_S^0 have different lifetimes, we noticed the large difference between their lifetimes

$$\tau_{K_S^0}/\tau_{K_L^0} = 1.530 \times 10^{-3} \quad . \tag{8.41}$$

As we already discussed briefly in (8.31), there is a simple explanation for this: while the state $|K_S^0\rangle$ decays in two pions ($\pi^+\pi^-, \pi^0\pi^0$), the state $|K_L^0\rangle$ has to decay into at least three particles; the most important channels are

$$\begin{aligned}&\pi^0\pi^0\pi^0\ (21\%) \quad , \quad \pi^+\pi^-\pi^0\ (12\%) \quad , \\ &\pi^\pm\mu^\mp\nu_\mu\ (27\%) \quad , \quad \pi^\pm e^\mp\nu_e\ (39\%) \quad .\end{aligned} \tag{8.42}$$

The reason is once again a symmetry: the kaons belong to the pseudoscalar mesons, that is

$$\hat{P}|K^0\rangle = -|K^0\rangle \quad , \quad \hat{P}|\bar{K}^0\rangle = -|\bar{K}^0\rangle \tag{8.43}$$

holds under a space inversion. The operation of charge conjugation transforms each of the particles into its antiparticle:

$$\hat{C}|\mathrm{K}^0\rangle = |\bar{\mathrm{K}}^0\rangle \quad , \quad \hat{C}|\bar{\mathrm{K}}^0\rangle = |\mathrm{K}^0\rangle \quad . \tag{8.44}$$

Thus

$$\hat{C}\hat{P}|\mathrm{K}^0\rangle = -|\bar{\mathrm{K}}^0\rangle \quad , \quad \hat{C}\hat{P}|\bar{\mathrm{K}}^0\rangle = -|\mathrm{K}^0\rangle \tag{8.45}$$

holds. If the states $|\mathrm{K}_S^0\rangle$, $|\mathrm{K}_L^0\rangle$ are, respectively, the symmetric and antisymmetric superpositions of the initial particles, as discussed above, then

$$\begin{aligned}\hat{C}\hat{P}|\mathrm{K}_S^0\rangle &= \frac{1}{\sqrt{2}}(\hat{C}\hat{P}|\mathrm{K}^0\rangle - \hat{C}\hat{P}|\bar{\mathrm{K}}^0\rangle) = \frac{1}{\sqrt{2}}(-|\bar{\mathrm{K}}^0\rangle + |\mathrm{K}^0\rangle) = |\mathrm{K}_S^0\rangle \quad , \\ \hat{C}\hat{P}|\mathrm{K}_L^0\rangle &= \frac{1}{\sqrt{2}}(\hat{C}\hat{P}|\mathrm{K}^0\rangle + \hat{C}\hat{P}|\bar{\mathrm{K}}^0\rangle) = \frac{1}{\sqrt{2}}(-|\bar{\mathrm{K}}^0\rangle - |\mathrm{K}^0\rangle) = -|\mathrm{K}_L^0\rangle \quad .\end{aligned} \tag{8.46}$$

On the other hand, the two pion states $|\pi^+\pi^-\rangle$ and $|\pi^0\pi^0\rangle$ are left unchanged by the parity transformation as well as by charge conjugation, because $\hat{C}|\hat{\pi}^\pm\rangle = |\pi^\mp\rangle$ and $\hat{P}|\pi\rangle = -|\pi\rangle$:

$$\hat{C}\hat{P}|\pi^+\pi^-\rangle = |\pi^+\pi^-\rangle \quad , \quad \hat{C}\hat{P}|\pi^0\pi^0\rangle = |\pi^0\pi^0\rangle \quad . \tag{8.47}$$

Therefore, only the state $|\mathrm{K}_S^0\rangle$ can decay into two pions with the assumption of CP invariance. Because of the negative parity of the pions it holds that

$$\hat{C}\hat{P}|\pi^0\pi^0\pi^0\rangle = -|\pi^0\pi^0\pi^0\rangle \quad , \quad \text{etc.} \tag{8.48}$$

and hence the state $|\mathrm{K}_L^0\rangle$ can decay into three pions but the $|\mathrm{K}_S^0\rangle$ cannot. Of course, the two-pion decay is much faster, since one particle less has to be created and more energy is liberated.

EXERCISE

8.2 Transformation of Kaons Under Space Inversion and Charge Conjugation

Problem. Derive relations (8.43), (8.44) starting from the wave function of a quark–antiquark pair in a relative s state and with opposite spin (ψ is a 4×4 matrix!),

$$\psi(\boldsymbol{r}_1, \boldsymbol{r}_2) = \begin{pmatrix} \chi_1 & \dfrac{-\mathrm{i}\boldsymbol{\sigma}_2\cdot\boldsymbol{\nabla}_2}{E_2+m_2}\chi_2 \\ \dfrac{-\mathrm{i}\boldsymbol{\sigma}_1\cdot\boldsymbol{\nabla}_1}{E_1+m_1}\chi_1 & \chi_2 \end{pmatrix} f(|\boldsymbol{r}_1 - \boldsymbol{r}_2|) \quad . \tag{1}$$

The indices 1 and 2, respectively, denote the coordinates of the quark and antiquark. The two-spinors are

$$\chi_1 = \begin{pmatrix}1\\0\end{pmatrix} \quad , \quad \chi_2 = \begin{pmatrix}0\\1\end{pmatrix} \quad ,$$

since quark and antiquark have opposite spin in the kaon.

Solution. (a) The operator of space inversion is

$$\psi(\boldsymbol{r}_1,\boldsymbol{r}_2) \to \hat{P}\psi(\boldsymbol{r}_1,\boldsymbol{r}_2) = \beta_1\beta_2\psi(-\boldsymbol{r}_1,-\boldsymbol{r}_2) \quad , \tag{2}$$

where the Dirac matrices β_1, β_2 each act only on the corresponding Dirac spinors in (1). Hence we find for the various parts

$$f(|\boldsymbol{r}_1-\boldsymbol{r}_2|) \to f(|\boldsymbol{r}_1-\boldsymbol{r}_2|) \quad ,$$

$$\begin{pmatrix} \chi_1 \\ \dfrac{-\mathrm{i}\boldsymbol{\sigma}\cdot\nabla}{E+m}\chi_1 \end{pmatrix} \to \begin{pmatrix} 1 & 0 \\ 0 & -1 \end{pmatrix}\begin{pmatrix} \chi_1 \\ \dfrac{+\mathrm{i}\boldsymbol{\sigma}\cdot\nabla}{E+m}\chi_1 \end{pmatrix} = \begin{pmatrix} \chi_1 \\ \dfrac{-\mathrm{i}\boldsymbol{\sigma}\cdot\nabla}{E+m}\chi_1 \end{pmatrix} ,$$

$$\begin{pmatrix} \dfrac{-\mathrm{i}\boldsymbol{\sigma}\cdot\nabla}{E+m}\chi_2 \\ \chi_2 \end{pmatrix} \to \begin{pmatrix} 1 & 0 \\ 0 & -1 \end{pmatrix}\begin{pmatrix} \dfrac{+\mathrm{i}\boldsymbol{\sigma}\cdot\nabla}{E+m}\chi_2 \\ \chi_2 \end{pmatrix} = \begin{pmatrix} \dfrac{-\mathrm{i}\boldsymbol{\sigma}\cdot\nabla}{E+m}\chi_2 \\ \chi_2 \end{pmatrix} , \tag{3}$$

so that in total

$$\hat{P}\,\psi(\boldsymbol{r}_1,\boldsymbol{r}_2) = -\psi(\boldsymbol{r}_1,\boldsymbol{r}_2) \tag{4}$$

results.

(b) The charge conjugation operator is

$$\psi(\boldsymbol{r}_1,\boldsymbol{r}_2) \to \hat{C}\psi(\boldsymbol{r}_1,\boldsymbol{r}_2) = \mathrm{i}\gamma_2^{(1)}\,\mathrm{i}\gamma_2^{(2)}\,\psi^*(\boldsymbol{r}_1,\boldsymbol{r}_2) \quad . \tag{5}$$

Its action on the spinors is, using the relation $\sigma_2\boldsymbol{\sigma}^* = -\boldsymbol{\sigma}\sigma_2$,

$$\begin{aligned}
\begin{pmatrix} \chi_1 \\ \dfrac{-\mathrm{i}\boldsymbol{\sigma}\cdot\nabla_1}{E+m}\chi_1 \end{pmatrix} &\to \begin{pmatrix} 0 & \mathrm{i}\sigma_2 \\ -\mathrm{i}\sigma_2 & 0 \end{pmatrix}\begin{pmatrix} \chi_1 \\ \dfrac{\mathrm{i}\boldsymbol{\sigma}^*\cdot\nabla_1}{E+m}\chi_1 \end{pmatrix} \\
&= \mathrm{i}\begin{pmatrix} \dfrac{-\mathrm{i}\boldsymbol{\sigma}\cdot\nabla_1}{E+m}\sigma_2\chi_1 \\ -\sigma_2\chi_1 \end{pmatrix} \\
&= \begin{pmatrix} \dfrac{+\mathrm{i}\boldsymbol{\sigma}\cdot\nabla_1}{E+m}\chi_2 \\ \chi_2 \end{pmatrix} ,
\end{aligned} \tag{6a}$$

since $\sigma_2\chi_1 = \mathrm{i}\chi_2$, and similarly

$$\begin{aligned}
\begin{pmatrix} \dfrac{-\mathrm{i}\boldsymbol{\sigma}\cdot\nabla_2}{E+m}\chi_2 \\ \chi_2 \end{pmatrix} &\to \begin{pmatrix} 0 & \mathrm{i}\sigma_2 \\ -\mathrm{i}\sigma_2 & 0 \end{pmatrix}\begin{pmatrix} \dfrac{\mathrm{i}\boldsymbol{\sigma}^*\cdot\nabla_2}{E+m}\chi_2 \\ \chi_2 \end{pmatrix} \\
&= \mathrm{i}\begin{pmatrix} \sigma_2\chi_2 \\ \dfrac{\mathrm{i}\boldsymbol{\sigma}\cdot\nabla_2}{E+m}\sigma_2\chi_2 \end{pmatrix} \\
&= \begin{pmatrix} \chi_1 \\ \dfrac{\mathrm{i}\boldsymbol{\sigma}\cdot\nabla_2}{E+m}\chi_1 \end{pmatrix} ,
\end{aligned} \tag{6b}$$

with $\sigma_2\chi_2 = -\mathrm{i}\chi_1$. Eventually we have to consider that

$$\nabla_1 f(|\boldsymbol{r}_1-\boldsymbol{r}_2|) = -\nabla_2 f(|\boldsymbol{r}_1-\boldsymbol{r}_2|) \quad , \tag{7}$$

xercise 8.2

that is we can substitute ∇_1 by $-\nabla_2$ in (6a) and ∇_2 by ∇_1 in (6b). Hence (8.44) is proved, that is,

$$\hat{C}\,\psi(\boldsymbol{r}_1,\boldsymbol{r}_2) = \bar{\psi}(\boldsymbol{r}_1,\boldsymbol{r}_2) \quad , \tag{8}$$

where $\bar{\psi}$ denotes the wave function of the antimeson (the quark and antiquark are interchanged).

However, it has been found experimentally[5] that the $|K_L^0\rangle$ *can*, in fact, decay into two pions, albeit with very small probability:

$$\begin{aligned} B(K_L^0 \to \pi^+\pi^-) &= (0.203 \pm 0.005)\% \quad , \\ B(K_L^0 \to \pi^0\pi^0) &= (0.094 \pm 0.018)\% \quad . \end{aligned} \tag{8.49}$$

The partial widths have the following ratios compared to those of the state $|K_S^0\rangle$:

$$\frac{\Gamma(K_L^0 \to \pi^+\pi^-)}{\Gamma(K_S^0 \to \pi^+\pi^-)} = 5.1 \times 10^{-6} \quad , \quad \frac{\Gamma(K_L^0 \to \pi^0\pi^0)}{\Gamma(K_S^0 \to \pi^0\pi^0)} = 5.2 \times 10^{-6} \quad . \tag{8.50}$$

Hence, the two pion decay of the $|K_L^0\rangle$ is very strongly suppressed, but it does exist. Since the spin of the kaons is zero, the two pions can be emitted only in an s state ($l = 0$). Thus the parity of the states $|\pi^+\pi^-\rangle$, $|\pi^0\pi^0\rangle$ is definitely positive. The only possible conclusion is that the CP invariance (and thus, owing to CPT invariance, also the invariance under time reversal) is violated in the decay $|K^0\rangle \to |\pi\pi\rangle$. The violation has to be very weak, however, of relative magnitude 5×10^{-6}. Thus the statement of (8.5),

$$\langle K^0|\hat{S}|\bar{K}^0\rangle = \langle \bar{K}^0|\hat{S}|K^0\rangle \quad , \tag{8.51}$$

no longer holds exactly. In the same way it follows that Γ_{12}, Γ_{21} and M_{12}, M_{21} are distinct. M_{12}, M_{21}^* and $\Gamma_{12} = \Gamma_{21}^*$ still hold, however, owing to the Hermiticity of the matrices M and Γ. The two states $|K_S^0\rangle$, $|K_L^0\rangle$ result as eigenstates of the matrix $(M - \mathrm{i}\Gamma/2)$,

$$\begin{pmatrix} M_0 - \frac{\mathrm{i}}{2}\Gamma_0 & M_{12} - \frac{\mathrm{i}}{2}\Gamma_{12} \\ M_{12}^* - \frac{\mathrm{i}}{2}\Gamma_{12}^* & M_0 - \frac{\mathrm{i}}{2}\Gamma_0 \end{pmatrix} \begin{pmatrix} c \\ \bar{c} \end{pmatrix} = \lambda \begin{pmatrix} c \\ \bar{c} \end{pmatrix} \quad . \tag{8.52}$$

From the secular determinant it follows that

$$\begin{aligned} \left(M_0 - \frac{\mathrm{i}}{2}\Gamma_0 - \lambda\right)^2 &= \left(M_{12} - \frac{\mathrm{i}}{2}\Gamma_{12}\right)\left(M_{12}^* - \frac{\mathrm{i}}{2}\Gamma_{12}^*\right) \quad , \\ M_0 - \frac{\mathrm{i}}{2}\Gamma_0 - \lambda &= \pm\sqrt{\left(M_{12} - \frac{\mathrm{i}}{2}\Gamma_{12}\right)\left(M_{12}^* - \frac{\mathrm{i}}{2}\Gamma_{12}^*\right)} \quad . \end{aligned} \tag{8.53}$$

Furthermore,

$$\left(M_0 - \frac{\mathrm{i}}{2}\Gamma_0\right)c + \left(M_{12} - \frac{\mathrm{i}}{2}\Gamma_{12}\right)\bar{c} = \lambda c \tag{8.54}$$

[5] J.H. Christensen, J.W. Cronin, V.L. Fitch, R. Turlay: Phys. Rev. Lett. **13**, 138 (1964).

holds, and thus, because of (8.53), we find that

$$\frac{\bar{c}}{c} = \frac{\lambda - M_0 + \frac{i}{2}\Gamma_0}{M_{12} - \frac{i}{2}\Gamma_{12}} = \mp\sqrt{\frac{M_{12}^* - \frac{i}{2}\Gamma_{12}^*}{M_{12} - \frac{i}{2}\Gamma_{12}}} \quad , \tag{8.55}$$

with the normalization condition $|c|^2 + |\bar{c}|^2 = 1$. Hence the physical states are

$$\begin{aligned}
|K_S^0\rangle &= N\left(|K^0\rangle - \sqrt{\frac{M_{12}^* - \frac{i}{2}\Gamma_{12}^*}{M_{12} - \frac{i}{2}\Gamma_{12}}}|\bar{K}^0\rangle\right) \equiv N\big(|K^0\rangle - q|\bar{K}^0\rangle\big) \quad , \\
|K_L^0\rangle &= N\left(|K^0\rangle + \sqrt{\frac{M_{12}^* - \frac{i}{2}\Gamma_{12}^*}{M_{12} - \frac{i}{2}\Gamma_{12}}}|\bar{K}^0\rangle\right) \equiv N\big(|K^0\rangle + q|\bar{K}^0\rangle\big) \quad ,
\end{aligned} \tag{8.56}$$

with

$$\begin{aligned}
N^{-2} &= 1 + \sqrt{\frac{(M_{12}^* - \frac{i}{2}\Gamma_{12}^*)(M_{12} + \frac{i}{2}\Gamma_{12})}{(M_{12} - \frac{i}{2}\Gamma_{12})(M_{12}^* + \frac{i}{2}\Gamma_{12}^*)}} \\
&= 1 + \sqrt{\frac{|M_{12}|^2 + 1/4|\Gamma_{12}|^2 + \mathrm{Im}(\Gamma_{12}^* M_{12})}{|M_{12}|^2 + 1/4|\Gamma_{12}|^2 - \mathrm{Im}(\Gamma_{12}^* M_{12})}} \\
&\approx 2 + \frac{\mathrm{Im}(M_{12}\Gamma_{12}^*)}{|M_{12}|^2 + \frac{1}{4}|\Gamma_{12}|^2} \quad .
\end{aligned} \tag{8.57}$$

In the case of CP invariance, we recover $|K_{S/L}^0\rangle = 1/\sqrt{2}(|K^0\rangle \mp |\bar{K}^0\rangle)$. But if CP invariance is violated, M_{12} and Γ_{12} are complex, and thus

$$q \neq 1 \quad . \tag{8.58}$$

We now consider the components of $|K_{S/L}^0\rangle$ that may contribute to the two-pion decay. They are those components which do not change their sign under the transformation $\hat{C}\hat{P}$. To do this we apply the projection operator $(1/2)(\hat{C}\hat{P} + 1)$ and obtain using (8.45)

$$\begin{aligned}
\frac{1}{2}(1 + \hat{C}\hat{P})|K_S^0\rangle &= \frac{N}{2}(1 + \hat{C}\hat{P})\big(|K^0\rangle - q|\bar{K}^0\rangle\big) \\
&= \frac{N}{2}\big(|K^0\rangle - |\bar{K}^0\rangle - q|\bar{K}^0\rangle + q|K^0\rangle\big) \\
&= \frac{N}{2}(1 + q)\big(|K^0\rangle - |\bar{K}^0\rangle\big) \quad ,
\end{aligned} \tag{8.59}$$

$$\begin{aligned}
\frac{1}{2}(1 + \hat{C}\hat{P})|K_L^0\rangle &= \frac{N}{2}(1 + \hat{C}\hat{P})\big(|K^0\rangle + q|\bar{K}^0\rangle\big) \\
&= \frac{N}{2}\big(|K^0\rangle - |\bar{K}^0\rangle + q|\bar{K}^0\rangle - q|K^0\rangle\big) \\
&= N\frac{1-q}{2}\big(|K^0\rangle - |\bar{K}^0\rangle\big) \quad .
\end{aligned} \tag{8.60}$$

The ratio of the transition amplitudes of the processes $|K_S^0\rangle \to |\pi\pi\rangle$ and $|K_L^0\rangle \to |\pi\pi\rangle$ is therefore

$$\eta \equiv \frac{A(K_L^0 \to \pi\pi)}{A(K_S^0 \to \pi\pi)} = \frac{1-q}{1+q} \approx \frac{1-q}{2} \quad . \tag{8.61}$$

Experimentally the ratios of the amplitudes for $K_{L/S}^0 \to \pi^+\pi^-$ and $\pi^0\pi^0$, that is, η_{+-} and η_{00}, are measured independently. Since the kaons have isospin 1/2 and only isospin 0 or 2 can contribute from the two-pion system, $\eta_{+-} = \eta_{00}$ should hold in the absence of $\Delta I = 3/2$ transitions. We write $\eta_{+-} = |\eta_{+-}|\exp(\mathrm{i}\phi_{+-})$, $\eta_{00} = |\eta_{00}|\exp(\mathrm{i}\phi_{00})$. The phase can be determined from the interference of the decay processes as in (8.31). The experimental values are (see (8.49))

$$\begin{aligned} |\eta_{+-}| &= (2.274 \pm 0.022) \times 10^{-3} \quad , \quad \Phi_{+-} = (44.6 \pm 1.2)^\circ \quad , \\ |\eta_{00}| &= (2.33 \pm 0.08) \times 10^{-3} \quad , \quad \Phi_{00} = (54 \pm 5)^\circ \quad . \end{aligned} \tag{8.62}$$

In the range of measurement accuracy and for the case $\eta_{+-} = \eta_{00}$, a $\Delta I = 3/2$ component that may be present is small.

The question of the origin of CP violation remains. From the previous considerations this question means that we have to show why the matrix elements M_{12} and Γ_{12} are not real. We have already explained in Sect. 8.1 why the matrix elements for the transitions between $|K^0\rangle$ and $|\bar{K}^0\rangle$ do not vanish. However, there we used only the most simple quark diagram. In reality the physical d and s quarks are admixed to all three states d′, s′, b′ of the extended Salam–Weinberg theory via the mixing matrix $\hat{U}$ of (6.38). Thus instead of an intermediary u (or ū) quark, a c or t quark (or c̄, t̄) may equally well occur in the intermediate state (Fig. 8.9).

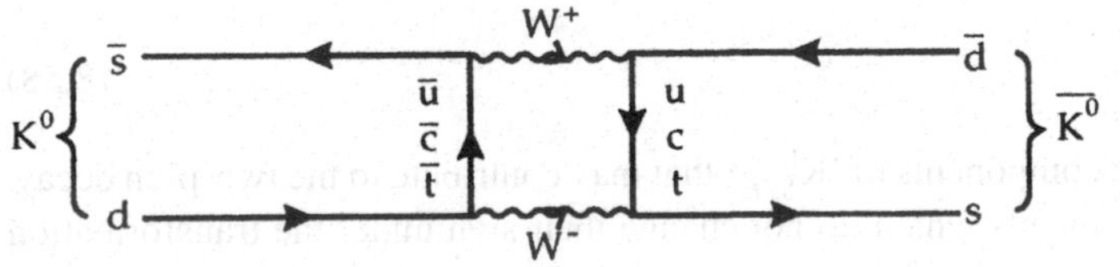

Fig. 8.9. Possible intermediary quarks in the intermediate state

The admixture of the quarks is given by

$$\begin{pmatrix} d \\ s \\ b \end{pmatrix} = \hat{U}^\dagger \begin{pmatrix} d' \\ s' \\ b' \end{pmatrix} \tag{8.63}$$

and that of the antiquarks by

$$\begin{pmatrix} \bar{d} \\ \bar{s} \\ \bar{b} \end{pmatrix} = \hat{U} \begin{pmatrix} \bar{d}' \\ \bar{s}' \\ \bar{b}' \end{pmatrix} \quad . \tag{8.64}$$

In the calculation we find that the three different quarks (u, c, t) in the intermediate state contribute to the matrix element M_{12} in the following way:

$$M_{12} = C(U^\dagger_{\mathrm{sd'}} U_{\mathrm{d'd}} m_u + U^\dagger_{\mathrm{ss'}} U_{\mathrm{s'd}} m_c + U^\dagger_{\mathrm{sb'}} U_{\mathrm{b'd}} m_t)^2 \quad , \tag{8.65}$$

where C contains the coupling constants and propagators of the W bosons and m_u, m_c, and m_t are the masses of the u, c, t quarks. We know that $m_\mathrm{u} \approx 5$ MeV, $m_\mathrm{c} \approx 1500$ MeV, and $m_\mathrm{t} > 89000$ MeV, that is,

$$m_\mathrm{u} \ll m_\mathrm{c}, m_\mathrm{t} \tag{8.66}$$

holds. Therefore we can neglect the first term between the parentheses. Further we consider that

$$U^\dagger_{\mathrm{sd}'} U_{\mathrm{d'd}} + U^\dagger_{\mathrm{ss}'} U_{\mathrm{s'd}} + U^\dagger_{\mathrm{sb}'} U_{\mathrm{b'd}} = 0 \tag{8.67}$$

holds, because of the unitarity of U. Thus it follows that

$$M_{12} = C\big[U^\dagger_{\mathrm{sd}'} U_{\mathrm{d'd}}(m_\mathrm{u} - m_\mathrm{c}) + U^\dagger_{\mathrm{sb}'} U_{\mathrm{b'd}}(m_\mathrm{t} - m_\mathrm{c})\big]^2 \quad . \tag{8.68}$$

Explicit insertion of the elements of the mixing matrix of (6.36) yields ($U^\dagger_{\mathrm{sd}'} = U^*_{\mathrm{d's}}$, etc.)

$$\begin{aligned} M_{12} &= C\big[s_1 c_3 c_1\,(m_\mathrm{u} - m_\mathrm{c}) + (c_1 s_2 c_3 - c_2 s_3 e^{-\mathrm{i}\delta})(-s_1 s_2)(m_\mathrm{t} - m_\mathrm{c})\big]^2 \\ &= C\big[s_1 c_1 c_3 (m_\mathrm{c} - m_\mathrm{u}) + (s_1 c_1 s_2 s_2 c_3 - s_1 s_2 c_2 s_3 e^{-\mathrm{i}\delta})(m_\mathrm{t} - m_\mathrm{c})\big]^2 \quad . \end{aligned} \tag{8.69}$$

Provided that the imaginary part is small, we can write

$$\mathrm{Im}\{M_{12}\} \approx -2 \frac{s_1 s_2 c_2 s_3 \sin\delta\,(m_\mathrm{t} - m_\mathrm{c})\, M_{12}}{s_1 c_1 c_3 (m_\mathrm{c} - m_\mathrm{u}) + (s_1 c_1 s_2 c_2 c_3 - s_1 s_2 c_2 s_3 \cos\delta)(m_\mathrm{t} - m_\mathrm{c})} \quad . \tag{8.70}$$

We see that the appearance of the complex phase factor $\exp(\mathrm{i}\delta)$ makes an imaginary part of the matrix elements possible. This factor, on the other hand, occurs only if there are at least three quark doublets, as first noticed by Kobayashi and Maskawa.[6] We want to estimate the order of magnitude of the CP violation. If we neglect the contribution of the width Γ_{12} (which may perhaps yield a factor of two), then according to (8.58), (8.61)

$$\begin{aligned} \eta &= \frac{1}{2}(1-q) = \frac{1}{2}\left(1 - \sqrt{\frac{M^*_{12}}{M_{12}}}\right) = \frac{1}{2}\left(1 - \sqrt{\frac{M_{12} - 2\mathrm{i}\,\mathrm{Im} M_{12}}{M_{12}}}\right) \\ &\approx \frac{1}{2}\left[1 - \left(1 - \mathrm{i}\frac{\mathrm{Im} M_{12}}{M_{12}}\right)\right] \approx \frac{\mathrm{i}}{2}\frac{\mathrm{Im} M_{12}}{M_{12}} \\ &\approx -\frac{s_1 s_2 c_2 s_3 \sin\delta\,(m_\mathrm{t} - m_\mathrm{c})}{s_1 c_1 c_3 (m_\mathrm{c} - m_\mathrm{u}) + s_1 s_2 (c_1 s_2 c_3 - c_2 s_3 \cos\delta)(m_\mathrm{t} - m_\mathrm{c})} \\ &\approx -\frac{s_2 s_3 \sin\delta\,(m_\mathrm{t}/m_\mathrm{c} - 1)}{c_1 c_3 / c_2 + s_2(-c_1 s_2 / c_2 c_3 + s_3 \cos\delta)(m_\mathrm{t}/m_\mathrm{c} - 1)} \quad , \end{aligned} \tag{8.71}$$

where we have set $m_\mu = 0$ in the last step. All angles $\theta_1, \theta_2, \theta_3, \delta$ are small ($\theta_1 = \theta_\mathrm{C} = 13°$); therefore for the estimate we can set $c_1 = c_2 = c_3 = \cos\delta = 1$. Furthermore, $m_\mathrm{t}/m_\mathrm{c} \gg 1$ and thus

$$\eta \approx \frac{\sin\theta_2 \sin\theta_3 \sin\delta m_t}{m_\mathrm{c} + m_\mathrm{t}\sin\theta_2} \quad . \tag{8.72}$$

[6] M. Kobayashi, K. Maskawa: Prog. Theor. Phys. **49**, 656 (1973).

The experimental value $\eta \approx 2 \times 10^{-3}$ can be explained in this way without difficulty. Unfortunately, the angles θ_3, δ are not yet known well experimentally. The bounds for the mixing angles between the quark flavours are[7]

$$\begin{aligned} \sin\theta_1 &= 0.227 \pm 0.011 \quad , \\ \sin\theta_2 &= 0.07 \pm 0.04 \quad , \\ \sin\theta_3 &\leq 0.06 \quad , \\ 90^\circ &\leq \delta \leq 180^\circ \quad . \end{aligned} \tag{8.73}$$

If $m_c \ll \sin\theta_2 m_t$, then

$$\eta \simeq \sin\theta_3 \sin\delta \simeq 2 \times 10^{-3} \tag{8.74}$$

should hold. This is compatible with the values of θ_3 and δ, but it does not yet represent a good test of the Kobayashi–Maskawa theory, owing to the large experimental uncertainty.

[7] A. Soni: Phys. Rev. Lett. **53**, 1407 (1984).

Unified Gauge Theories 9

9.1 Introduction: The Symmetry Group SU(5)

The Glashow–Salam–Weinberg theory combines the electromagnetic and weak interactions within the framework of the gauge group $\mathrm{SU}(2)_{\mathrm{L}} \times \mathrm{SU}(1)$. This theory treats leptons and quarks on the same footing. However, the fact that leptons carry integer charge while the charge of quarks is 1/3 is needed as a basic input. On the other hand, there are indications that this fact is due to some superior principle, that is a larger symmetry. For example, for leptons as well as for quarks, the sum of charge and baryon number, $Q + B$, is an integer:

$$\begin{aligned}
&\mathrm{e}: \quad Q = -1\,, B = 0 \quad \rightarrow \quad Q + B = -1\;, \\
&\nu: \quad Q = 0\,, B = 0 \quad \rightarrow \quad Q + B = 0\;, \\
&\mathrm{u}: \quad Q = +\tfrac{2}{3}\,, B = \tfrac{1}{3} \quad \rightarrow \quad Q + B = 1\;, \\
&\mathrm{d}: \quad Q = -\tfrac{1}{3}\,, B = \tfrac{1}{3} \quad \rightarrow \quad Q + B = 0\;.
\end{aligned}$$

The property that B and Q are multiples of 1/3 for quarks is connected to the principle that baryons consist of three quarks and have integer values of Q and B. Within the framework of quantum chromodynamics (QCD) this is understood in terms of the gauge group SU(3) of additional internal degrees of freedom for the quarks: the color. For example, in order to explain the Ω^- baryon it requires the assumption of three strange quarks all in the same 1s state. This is inconsistent with the Pauli principle, which is assumed as a fundamental property, unless the three s quarks differ in an additional quantum number. These quantum states are labeled by *color* and are distinguished as, for example red r, green g and blue b.

Similar relations to those above can also be stated in terms of the weak hypercharge. The respective quantum numbers are given in Table 6.2. The lepton doublets have $Y_L^{(\ell)} = -1$, while for quarks $Y_L^{(\mathrm{q})} = +\frac{1}{3}$. Owing to the color multiplicity the latter needs to be counted three times, leading to

$$Y_L^{(\ell)} + 3Y_L^{(\mathrm{q})} = 0\;.$$

Similar relations can be inferred for the right-handed singlets.

These properties lead to the hypothesis that the gauge group of the Glashow–Salam–Weinberg theory, $\mathrm{SU}(2) \times \mathrm{U}(1)$, and the SU(3) group of the strong interaction form part of a larger symmetry. The simplest group that incorporates the product $\mathrm{SU}(3) \times \mathrm{SU}(2) \times \mathrm{U}(1)$ as a subgroup is SU(5). As will be shown below, this group has irreducible representations of dimensions

$$5\,,\ 10\,,\ 15\,,\ 24\,,\ 35\,,\ 40\,,\ \ldots\;.$$

W. Greiner, B. Müller, *Gauge Theory of Weak Interactions*,
DOI 10.1007/978-3-540-87843-8_9, © Springer-Verlag Berlin Heidelberg 2009

Before discussing the details of this model of unification of the three interactions, which was proposed by **Georgi** and Glashow[1] in 1974, it is expedient to establish some basic mathematical properties of the group SU(5).[2]

SU(5) is a special case of the general groups SU(n) that are formed from unitary $n \times n$ matrices with determinant (+1). The corresponding groups without the constraint for the determinant are called U(n). An arbitrary unitary matrix can be represented in terms of an exponential of a Hermitian matrix $\hat{H}$:

$$\hat{U} = \exp(\mathrm{i}\hat{H}) \quad , \quad \hat{H}^\dagger = \hat{H} \quad . \tag{9.1}$$

$\hat{H}$ is called the *generating matrix* for $\hat{U}$. In case that the matrix $\hat{U}$ is not too different from the unit matrix $\mathbb{1}$, it holds

$$\hat{U} = \exp(\mathrm{i}\delta\hat{H}) \approx \mathbb{1} + \mathrm{i}\delta\hat{H} \quad . \tag{9.2}$$

The multiplication of two matrices U_1, U_2 corresponds to the sum of the infinitesimal Hermitian matrices,

$$\begin{aligned}\hat{U}_2\hat{U}_1 &= \mathrm{e}^{\mathrm{i}\delta\hat{H}_2}\mathrm{e}^{\mathrm{i}\delta\hat{H}_1} \approx (\mathbb{1} + \mathrm{i}\delta\hat{H}_2)(\mathbb{1} + \mathrm{i}\delta\hat{H}_1) \\ &\approx \mathbb{1} + \mathrm{i}\big(\delta\hat{H}_2 + \delta\hat{H}_1\big) \quad ,\end{aligned} \tag{9.3}$$

where quadratic terms have been neglected. A complete set of linearly independent, Hermitian matrices is termed a set of *generators* for the unitary matrices. Owing to the constraint $h_{ik} = h^*_{ki}$ for the elements of a Hermitian matrix $\hat{H}$, the group of the unitary $n \times n$ matrices contains n^2 generators. (Note that a general complex $n \times n$ matrix has $2n^2$ degrees of freedom.) The restriction $\det(\hat{U}) = 1$ that leads from the group U(n) to the group SU(n) results in traceless generating matrices $\hat{H}$. Since the diagonal elements of Hermitian matrices are real, $h_{kk} = h^*_{kk}$, only one degree of freedom is omitted. That is, the group SU(n) has $n^2 - 1$ generators.

The simplest representation of the generating matrices for the U(n) is formed by certain $n \times n$ matrices that contain a single nonvanishing matrix element with value 1, all other elements being zero. Such a matrix, where the matrix element with value 1 is given by the intersection of row α and column β as $\hat{C}_{\alpha\beta}$, is written

$$\hat{C}_{\alpha\beta} = \begin{pmatrix} & & \beta & & \\ & & | & & \\ \alpha & -- & 1 & -- & -- \\ & & | & & \\ & & | & & \end{pmatrix} \quad , \quad (C_{\alpha\beta})_{ik} = \delta_{\alpha i}\delta_{\beta k} \quad , \tag{9.4}$$

where the matrices

$$\hat{C}_{\alpha\beta} + \hat{C}_{\beta\alpha} \quad , \quad \frac{1}{\mathrm{i}}(\hat{C}_{\alpha\beta} - \hat{C}_{\beta\alpha}) \tag{9.5}$$

are Hermitian, since $\hat{C}^\dagger_{\alpha\beta} = \hat{C}_{\beta\alpha}$.

---

[1] H. Georgi, S.L. Glashow: Phys. Rev. Lett. **32**, 438 (1974).

[2] For an extended presentation of properties of unitary groups see W. Greiner and B. Müller: *Quantum Mechanics – Symmetries*, 2nd ed. (Springer, Berlin, Heidelberg, 1994).

Using the linear combinations of (9.5), we can alternatively employ the $\hat{C}_{\alpha\beta}$, $\alpha,\beta = 1,2,\dots,n$ as generators of the U(n). In order to infer the corresponding generators of the SU(n), we require the matrix $\hat{C}_{\alpha\alpha}$ to be traceless. This can be achieved by subtracting a multiple of the unit matrix:

$$\begin{aligned}\hat{C}'_{\alpha\alpha} &= \hat{C}_{\alpha\alpha} - \frac{1}{n}\mathbb{1} = \hat{C}_{\alpha\alpha} - \frac{1}{n}\sum_{\alpha=1}^{n}\hat{C}_{\alpha\alpha} \quad , \\ \hat{C}'_{\alpha\beta} &= \hat{C}_{\alpha\beta} \quad (\alpha \neq \beta) \quad .\end{aligned} \tag{9.6}$$

The n matrices $\hat{C}'_{\alpha\alpha}$ are linearly dependent, since

$$\sum_{\alpha=1}^{n}\hat{C}'_{\alpha\alpha} = \sum_{\alpha=1}^{n}\hat{C}_{\alpha\alpha} - \frac{1}{n}\sum_{\alpha=1}^{n}\mathbb{1} = 0 \quad . \tag{9.7}$$

To construct $n-1$ linearly independent, diagonal matrices, it is common to form $n-1$ linear combinations of the $\hat{C}'_{\alpha\alpha}$ as follows:

$$\hat{C}'_{11} - \hat{C}'_{22} \quad , \quad \hat{C}'_{11} + \hat{C}'_{22} \quad , \quad \hat{C}'_{33} - \hat{C}'_{44} \quad , \quad \hat{C}'_{33} + \hat{C}'_{44} \quad , \quad \text{etc.} \tag{9.8}$$

The commutation relations of the matrices $\hat{C}_{\alpha\beta}$ are

$$\begin{aligned}(\hat{C}_{\alpha\beta}\hat{C}_{\gamma\delta} - \hat{C}_{\gamma\delta}\hat{C}_{\alpha\beta})_{ik} &= \sum_l (\hat{C}_{\alpha\beta})_{il}(\hat{C}_{\gamma\delta})_{lk} - \sum_l (\hat{C}_{\gamma\delta})_{il}(\hat{C}_{\alpha\beta})_{lk} \\ &= \sum_l \delta_{\alpha i}\delta_{\beta l}\delta_{\gamma l}\delta_{\delta k} - \sum_l \delta_{\gamma i}\delta_{\delta l}\delta_{\alpha l}\delta_{\beta k} \\ &= \delta_{\alpha i}\delta_{\beta\gamma}\delta_{\delta k} - \delta_{\gamma i}\delta_{\delta\alpha}\delta_{\beta k} \\ &= \delta_{\beta\gamma}(\hat{C}_{\alpha\delta})_{ik} - \delta_{\delta\alpha}(\hat{C}_{\gamma\beta})_{ik} \quad ,\end{aligned} \tag{9.9}$$

or, in matrix notation,

$$[\hat{C}_{\alpha\beta}, \hat{C}_{\gamma\delta}] = \delta_{\beta\gamma}\hat{C}_{\alpha\delta} - \delta_{\alpha\delta}\hat{C}_{\gamma\beta} \quad . \tag{9.10}$$

Since the unit matrix commutes with any matrix, the commutation relations are the same for the matrices of the SU(n):

$$[\hat{C}'_{\alpha\beta}, \hat{C}'_{\gamma\delta}] = \delta_{\beta\gamma}\hat{C}'_{\alpha\delta} - \delta_{\alpha\delta}\hat{C}'_{\gamma\beta} \quad . \tag{9.11}$$

In the case of the group U(5) the generators are of the form

$$\hat{C}_{24} = \begin{pmatrix} 0 & 0 & 0 & 0 & 0 \\ 0 & 0 & 0 & 1 & 0 \\ 0 & 0 & 0 & 0 & 0 \\ 0 & 0 & 0 & 0 & 0 \\ 0 & 0 & 0 & 0 & 0 \end{pmatrix} \quad , \quad \text{etc.} \tag{9.12}$$

From these we construct Hermitian matrices according to the prescription (9.5):

$$\hat{C}_{24}+\hat{C}_{42}=\begin{pmatrix}0&0&0&0&0\\0&0&0&1&0\\0&0&0&0&0\\0&1&0&0&0\\0&0&0&0&0\end{pmatrix}\,,$$

$$\frac{1}{\mathrm{i}}(\hat{C}_{24}-\hat{C}_{42})=\begin{pmatrix}0&0&0&0&0\\0&0&0&-\mathrm{i}&0\\0&0&0&0&0\\0&\mathrm{i}&0&0&0\\0&0&0&0&0\end{pmatrix}\,,\quad \text{etc.} \tag{9.13}$$

The diagonal matrices have the following form:

$$\hat{C}_{11}=\begin{pmatrix}1&&&&\\&0&&&\\&&0&&\\&&&0&\\&&&&0\end{pmatrix}\,,\quad\dots\,,\quad \hat{C}_{55}=\begin{pmatrix}0&&&&\\&0&&&\\&&0&&\\&&&0&\\&&&&1\end{pmatrix}\,. \tag{9.14}$$

They are already Hermitian. Altogether there are $5^2=25$ generators.

In order to construct the generators of the group SU(5) we form new diagonal matrices that have zero trace as required:

$$\hat{C}'_{11}=\hat{C}_{11}-\frac{1}{5}\mathbb{1}=\begin{pmatrix}1&&&&\\&0&&&\\&&0&&\\&&&0&\\&&&&0\end{pmatrix}-\frac{1}{5}\begin{pmatrix}1&&&&\\&1&&&\\&&1&&\\&&&1&\\&&&&1\end{pmatrix}$$

$$=\begin{pmatrix}4/5&&&&\\&-1/5&&&\\&&-1/5&&\\&&&-1/5&\\&&&&-1/5\end{pmatrix}\,, \tag{9.15}$$

and so on. Choosing linear combinations as indicated in (9.8) we obtain four linearly independent matrices (which are normalized such that the traces of the squared matrices are equal to 2):

$$\hat{C}'_{11}-\hat{C}'_{22}=\frac{1}{5}\begin{pmatrix}4&&&&\\&-1&&&\\&&-1&&\\&&&-1&\\&&&&-1\end{pmatrix}-\frac{1}{5}\begin{pmatrix}-1&&&&\\&4&&&\\&&-1&&\\&&&-1&\\&&&&-1\end{pmatrix}$$

$$=\begin{pmatrix}1&&&&\\&-1&&&\\&&0&&\\&&&0&\\&&&&0\end{pmatrix}\,,$$

$$\sqrt{\frac{5}{3}}(\hat{C}'_{11}+\hat{C}'_{22})=\frac{1}{\sqrt{15}}\begin{pmatrix}3&&&&\\&3&&&\\&&-2&&\\&&&-2&\\&&&&-2\end{pmatrix}\,, \tag{9.16}$$

$$\hat{C}'_{33}-\hat{C}'_{44}=\begin{pmatrix}0&&&&\\&0&&&\\&&1&&\\&&&-1&\\&&&&0\end{pmatrix}$$

$$\sqrt{\frac{5}{3}}(\hat{C}'_{33}+\hat{C}'_{44})=\frac{1}{\sqrt{15}}\begin{pmatrix}-2&&&&\\&-2&&&\\&&3&&\\&&&3&\\&&&&-2\end{pmatrix}\,.$$

In addition there are $4\times 5=20$ non-diagonal generators, so that altogether we have 24 generators for the group SU(5).

The number of diagonal generators, that is, the number of mutually commuting generators, is called the *rank* of a group. Thus U(5) has rank 5, whereas SU(5) has only rank 4. The rank of a Lie group is equal to the number of its independent Casimir operators, that is the operators that commute with all generators of the group. The simplest and most important Casimir operator is a quadratic form of the $\hat{C}_{\alpha\beta}$, namely

$$\hat{C}_2=\sum_{\alpha,\beta}\hat{C}'_{\alpha\beta}\hat{C}'_{\beta\alpha}\quad. \tag{9.17}$$

That it is indeed a Casimir invariant is easily proved by using (9.11):

$$\begin{aligned}[\hat{C}_2,\hat{C}'_{\gamma\delta}]&=\sum_{\alpha,\beta}[\hat{C}'_{\alpha\beta}\hat{C}'_{\beta\alpha},\hat{C}'_{\gamma\delta}]=\sum_{\alpha,\beta}\hat{C}'_{\alpha\beta}[\hat{C}'_{\beta\alpha},\hat{C}'_{\gamma\delta}]+\sum_{\alpha,\beta}[\hat{C}'_{\alpha\beta},\hat{C}'_{\gamma\delta}]\hat{C}'_{\beta\alpha}\\&=\sum_{\alpha,\beta}\{\hat{C}'_{\alpha\beta}(\delta_{\alpha\gamma}\hat{C}'_{\beta\delta}-\delta_{\beta\delta}\hat{C}'_{\gamma\alpha})+(\delta_{\beta\gamma}\hat{C}'_{\alpha\delta}-\delta_{\alpha\delta}\hat{C}'_{\gamma\beta})\hat{C}'_{\beta\alpha}\}\\&=\sum_{\beta}\hat{C}'_{\gamma\beta}\hat{C}'_{\beta\delta}-\sum_{\alpha}\hat{C}'_{\alpha\delta}\hat{C}'_{\gamma\alpha}+\sum_{\alpha}\hat{C}'_{\alpha\delta}\hat{C}'_{\gamma\alpha}-\sum_{\beta}\hat{C}'_{\gamma\beta}\hat{C}'_{\beta\delta}=0\quad.\end{aligned}$$

To determine the irreducible representations of SU(5) it is convenient to use *Young diagrams*.[3] In terms of Young diagrams an irreducible representation of SU(5) is described by at most four rows of boxes, one upon the other, where the length of a row is required to be no greater than the length of the preceding row. With this convention the fundamental representation corresponds to a single box. An irreducible representation of SU(5) is labeled by four numbers, $h_1,\ldots,h_4$, that specify the number of boxes in the four rows of a Young diagram. Most commonly the combination

$$(h_1-h_2\,,\ h_2-h_3\,,\ h_3-h_4\,,\ h_4)$$

is employed.

[3] See W. Greiner and B. Müller: *Quantum Mechanics – Symmetries*, 2nd ed. (Springer, Berlin, Heidelberg, 1994).

Some important representations are:

$$(1,0,0,0),\quad (0,1,0,0),\quad (2,0,0,0),\quad (0,0,0,1),\quad (1,0,0,1),\quad (2,0,1,1).$$

The dimension of an irreducible representation of SU(5) is evaluated as follows: First draw the corresponding Young diagram twice, one above the other. Then fill the first row of the upper diagram with the numbers 5, 6, 7, ... from left to right; in the second row of the diagram start on the left-hand side with 4, 5, ..., in the third row with 3, and in the last one with 2. Next write in each box of the lower Young diagram the integer that results from summing the number of boxes further to the right and below plus 1. Finally multiply all numbers of the upper diagram and divide by the product of all numbers of the lower diagram.

Example: Evaluate the dimension of the irreducible representation of the SU(5) group, given by the following Young diagram:

Dimension 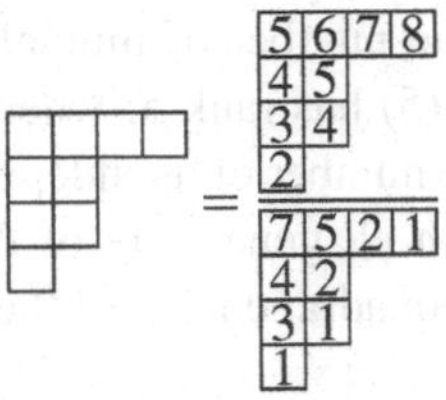

$$= \frac{5\times 6\times 7\times 8\times 4\times 5\times 3\times 4\times 2}{7\times 5\times 2\times 1\times 4\times 2\times 3\times 1\times 1}$$

$$= \frac{8\times 6\times 5\times 4}{2} = 480 \quad .$$

Other examples are

 $=5$, 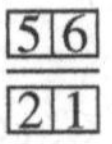 $= \frac{5\times 6}{2\times 1} = 15$,  $= \frac{5\times 4}{2\times 1} = 10$,

 $= \frac{5\times 4\times 3\times 2}{4\times 3\times 2\times 1} = 5$,  $= \frac{5\times 6\times 4\times 3\times 2}{5\times 1\times 3\times 2\times 1} = 24$.

The representations □ and 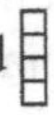 both have dimension 5. They are called *fundamental representations*. The dimension of the representation 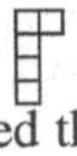 corresponds to the number of generators of the SU(5) group, that is 24; it is called the *regular* or *adjoint representation*.

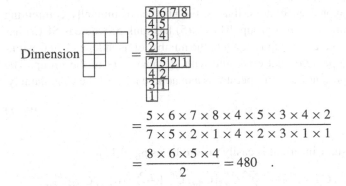

Representations of higher dimension are constructed by forming products with the *fundamental representation* □. It is understood that two boxes in one row denote the representation which is symmetric with respect to the corresponding factors; if the boxes belong to different rows the factors are antisymmetric. For example, the representation □□. stands for the *symmetric product* of two □ representations, while ⊟ denotes the *antisymmetric product*.

Any product of two representations, on the other hand, may be decomposed into irreducible representations by forming all possible combinations of the boxes of the two Young diagrams; all columns that contain five boxes or more are simply ignored.

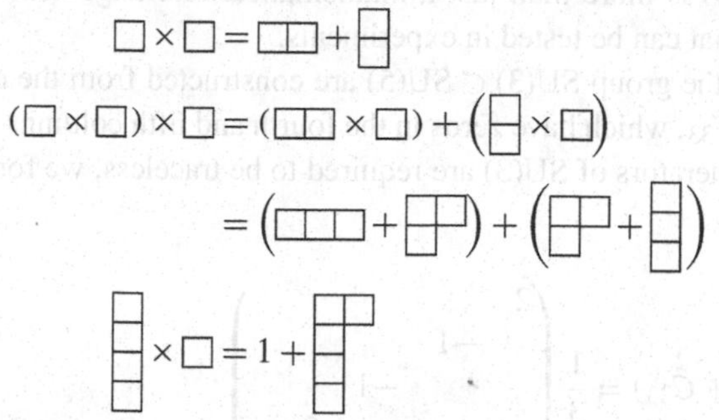

These reductions are often expressed in terms of the dimension of the representations,

$$\begin{aligned} [5] \times [5] &= [15] + [10] \quad , \\ [5] \times [5] \times [5] &= [35] + [40] + [40] + [\overline{10}] \quad , \\ [\bar{5}] \times [5] &= [1] + [24] \quad , \end{aligned} \tag{9.18}$$

In the following we will suppress the [ ]-brackets as long as no confusion arises. Only in special cases where the brackets are necessary for unambiguous notation will they be reintroduced.

9.2 Embedding SU(3)$_C$ × SU(2)$_L$ × U(1) into SU(5)

In order to realize that SU(5) is the smallest (semi-simple) unitary group incorporating all gauge groups of interactions, that is the color group SU(3)$_C$, the group of weak isospin of left-handed particles SU(2)$_L$, and the gauge group U(1) of the weak hypercharge, we simply need to enumerate the Casimir operators. U(1) and SU(2) have rank 1, and SU(3) has rank 2, so the unified gauge group at least must have rank 4. Indeed, SU(5) has rank 4 as discussed in the previous section.

To determine the explicit structure of SU(5) as the unified gauge group incorporating the above three groups of interactions, we consider some experimental facts. One is that the color group SU(3) is completely "*blind*" with respect to the weak interaction described by the Glashow–Salam–Weinberg group SU(2)$_L$×U(1): "red", "blue" and "green" quarks of the same flavor carry the same electric and weak charge. That implies that the groups SU(3) and SU(2)$_L$×U(1) need to commute with each other when unified within SU(5). This is only possible if the generators of the group SU(2)×U(1) behave as unit matrices or zero matrices with respect to the SU(3) generators, that is in the three-dimensional subspace of SU(3). On the other hand, the leptons are color singlets, that is to say, the generators of the SU(3) must have zero eigenvalues for these components.

The consequence of these considerations are built into the theory by reserving the first three rows and columns of the five-dimensional representation of the SU(5) generators for the color group SU(3), while the last two rows and columns are assigned to the weak group $SU(2)_L$. This arrangement completely specifies the structure of SU(5) and determines in which way U(1) is incorporated into SU(5). As we shall see, this enables us to calculate the Weinberg angle from SU(5) and exhibits that the postulate of a unified gauge group SU(5) is more than just a mathematical rearrangement: it contains physical predictions that can be tested in experiments.

The diagonal generators of the group SU(3) $\subset$ SU(5) are constructed from the diagonal matrices $\hat{C}_{11}$, $\hat{C}_{22}$ and $\hat{C}_{33}$, which have zeros in the fourth and fifth columns of the diagonal axis. Since the generators of SU(3) are required to be traceless, we form the following combinations:

$$\hat{C}''_{11} = \hat{C}_{11} - \frac{1}{3}(\hat{C}_{11} + \hat{C}_{22} + \hat{C}_{33}) = \frac{1}{3}\begin{pmatrix} 2 & & & & \\ & -1 & & & \\ & & -1 & & \\ & & & 0 & \\ & & & & 0 \end{pmatrix} ,$$

$$\hat{C}''_{22} = \hat{C}_{22} - \frac{1}{3}(\hat{C}_{11} + \hat{C}_{22} + \hat{C}_{33}) = \frac{1}{3}\begin{pmatrix} -1 & & & & \\ & 2 & & & \\ & & -1 & & \\ & & & 0 & \\ & & & & 0 \end{pmatrix} . \tag{9.19}$$

The combinations

$$\tilde{\lambda}_3 \equiv \hat{C}''_{11} - \hat{C}''_{22} = \begin{pmatrix} 1 & & & & \\ & -1 & & & \\ & & 0 & & \\ & & & 0 & \\ & & & & 0 \end{pmatrix} ,$$

$$\tilde{\lambda}_8 \equiv \sqrt{3}(\hat{C}''_{11} + \hat{C}''_{22}) = \frac{1}{\sqrt{3}}\begin{pmatrix} 1 & & & & \\ & 1 & & & \\ & & -2 & & \\ & & & 0 & \\ & & & & 0 \end{pmatrix} \tag{9.20}$$

then take the usual form of the SU(3) generators.[4] Similarly the diagonal generator of the isospin group SU(2) is constructed from the diagonal generators $\hat{C}_{44}$ and $\hat{C}_{55}$:

$$\tilde{\lambda}_{23} \equiv \tilde{\tau}_3 = \hat{C}''_{44} - \hat{C}''_{55} = \begin{pmatrix} 0 & & & & \\ & 0 & & & \\ & & 0 & & \\ & & & 1 & \\ & & & & -1 \end{pmatrix} . \tag{9.21}$$

[4] See W. Greiner and B. Müller: *Quantum Mechanics – Symmetries*, 2nd ed. (Springer, Berlin, Heidelberg, New York, 1994).

Finally the generator that corresponds to U(1) must be diagonal and of the form of a unit matrix with respect to SU(3) as well as SU(2); furthermore, it is required to be traceless. The only way to achieve this is the following combination (up to a factor):

$$\tilde{Y} \equiv \sqrt{\frac{5}{3}}(\hat{C}'_{44} + \hat{C}'_{55}) = \frac{1}{\sqrt{15}} \begin{pmatrix} -2 & & & & \\ & -2 & & & \\ & & -2 & & \\ & & & 3 & \\ & & & & 3 \end{pmatrix}, \tag{9.22}$$

where we have employed the matrices defined by (9.6). The complete set of SU(5) generators $\tilde{\lambda}_i$, $i = 1, 2, \ldots, 24$, will be determined in Exercise 9.1.

EXERCISE

9.1 The Generators of SU(3) × SU(2) × U(1)

Problem. Work out the explicit form for the SU(5) generators that determine the composition SU(5) ⊇ SU(3) × SU(2) × U(1).

Solution. With respect to our convention established in Sect. 9.2 we have assigned the first three rows and columns of the SU(5) generators to the subgroup SU(3) and the last two rows and columns to the subgroup SU(2). The subgroup U(1) has only one generator, namely $\tilde{Y}$ of (9.22). The diagonal operators of SU(3) and SU(2), that is $\tilde{\lambda}_3$, $\tilde{\lambda}_8$ and $\tilde{\tau}_3$, respectively, are given by (9.20), (9.21). The non-diagonal operators are constructed as prescribed in Sect. 9.1, (9.13).

We suppress rows and columns of zeros:

$$\tilde{\lambda}_1 = \hat{C}_{12} + \hat{C}_{21} = \begin{pmatrix} 0 & 1 & 0 & & \\ 1 & 0 & 0 & & \\ 0 & 0 & 0 & & \\ & & & & \\ & & & & \end{pmatrix},$$

$$\tilde{\lambda}_2 = \frac{1}{\mathrm{i}}(\hat{C}_{12} - \hat{C}_{21}) = \begin{pmatrix} 0 & -\mathrm{i} & 0 & & \\ \mathrm{i} & 0 & 0 & & \\ 0 & 0 & 0 & & \\ & & & & \\ & & & & \end{pmatrix},$$

$$\tilde{\lambda}_3 = \hat{C}''_{11} - \hat{C}''_{22} = \begin{pmatrix} 1 & & & & \\ & -1 & & & \\ & & 0 & & \\ & & & 0 & \\ & & & & 0 \end{pmatrix},$$

$$\tilde{\lambda}_4 = \hat{C}_{13} + \hat{C}_{31} = \begin{pmatrix} 0 & 0 & 1 & & \\ 0 & 0 & 0 & & \\ 1 & 0 & 0 & & \\ & & & & \\ & & & & \end{pmatrix},$$

xercise 9.1

$$\tilde{\lambda}_5 = \frac{1}{\mathrm{i}}(\hat{C}_{13} - \hat{C}_{31}) = \begin{pmatrix} 0 & 0 & -\mathrm{i} & & \\ 0 & 0 & 0 & & \\ \mathrm{i} & 0 & 0 & & \\ & & & & \\ & & & & \end{pmatrix} ,$$

$$\tilde{\lambda}_6 = \hat{C}_{23} + \hat{C}_{32} = \begin{pmatrix} 0 & 0 & 0 & & \\ 0 & 0 & 1 & & \\ 0 & 1 & 0 & & \\ & & & & \\ & & & & \end{pmatrix} ,$$

$$\tilde{\lambda}_7 = \frac{1}{\mathrm{i}}(\hat{C}_{23} - \hat{C}_{32}) = \begin{pmatrix} 0 & 0 & 0 & & \\ 0 & 0 & -\mathrm{i} & & \\ 0 & \mathrm{i} & 0 & & \\ & & & & \\ & & & & \end{pmatrix} ,$$

$$\tilde{\lambda}_8 = \sqrt{3}(\hat{C}''_{11} + \hat{C}''_{22}) = \frac{1}{\sqrt{3}} \begin{pmatrix} 1 & & & & \\ & 1 & & & \\ & & -2 & & \\ & & & 0 & \\ & & & & 0 \end{pmatrix} .$$

The other generators of SU(5) are

$$\tilde{\lambda}_9 = \hat{C}_{14} + \hat{C}_{41} = \begin{pmatrix} & & & 1 & 0 \\ & & & 0 & 0 \\ & & & 0 & 0 \\ 1 & 0 & 0 & & \\ 0 & 0 & 0 & & \end{pmatrix} ,$$

$$\tilde{\lambda}_{10} = \frac{1}{\mathrm{i}}(\hat{C}_{14} - \hat{C}_{41}) = \begin{pmatrix} & & & -\mathrm{i} & 0 \\ & & & 0 & 0 \\ & & & 0 & 0 \\ \mathrm{i} & 0 & 0 & & \\ 0 & 0 & 0 & & \end{pmatrix} ,$$

$$\tilde{\lambda}_{11} = \hat{C}_{24} + \hat{C}_{42} = \begin{pmatrix} & & & 0 & 0 \\ & & & 1 & 0 \\ & & & 0 & 0 \\ 0 & 1 & 0 & & \\ 0 & 0 & 0 & & \end{pmatrix} ,$$

$$\tilde{\lambda}_{12} = \frac{1}{\mathrm{i}}(\hat{C}_{24} - \hat{C}_{42}) = \begin{pmatrix} & & & 0 & 0 \\ & & & -\mathrm{i} & 0 \\ & & & 0 & 0 \\ 0 & \mathrm{i} & 0 & & \\ 0 & 0 & 0 & & \end{pmatrix} ,$$

Exercise 9.

$$\tilde{\lambda}_{13} = \hat{C}_{34} + \hat{C}_{43} = \begin{pmatrix} & & & 0 & 0 \\ & & & 0 & 0 \\ & & & 1 & 0 \\ 0 & 0 & 1 & & \\ 0 & 0 & 0 & & \end{pmatrix} ,$$

$$\tilde{\lambda}_{14} = \frac{1}{\mathrm{i}}(\hat{C}_{34} - \hat{C}_{43}) = \begin{pmatrix} & & & 0 & 0 \\ & & & 0 & 0 \\ & & & -\mathrm{i} & 0 \\ 0 & 0 & \mathrm{i} & & \\ 0 & 0 & 0 & & \end{pmatrix} ,$$

$$\tilde{\lambda}_{15} = \hat{C}_{15} + \hat{C}_{51} = \begin{pmatrix} & & & 0 & 1 \\ & & & 0 & 0 \\ & & & 0 & 0 \\ 0 & 0 & 0 & & \\ 1 & 0 & 0 & & \end{pmatrix} ,$$

$$\tilde{\lambda}_{16} = \frac{1}{\mathrm{i}}(\hat{C}_{15} - \hat{C}_{51}) = \begin{pmatrix} & & & 0 & -\mathrm{i} \\ & & & 0 & 0 \\ & & & 0 & 0 \\ 0 & 0 & 0 & & \\ \mathrm{i} & 0 & 0 & & \end{pmatrix} ,$$

$$\tilde{\lambda}_{17} = \hat{C}_{25} + \hat{C}_{52} = \begin{pmatrix} & & & 0 & 0 \\ & & & 0 & 1 \\ & & & 0 & 0 \\ 0 & 0 & 0 & & \\ 0 & 1 & 0 & & \end{pmatrix} ,$$

$$\tilde{\lambda}_{18} = \frac{1}{\mathrm{i}}(\hat{C}_{25} - \hat{C}_{52}) = \begin{pmatrix} & & & 0 & 0 \\ & & & 0 & -\mathrm{i} \\ & & & 0 & 0 \\ 0 & 0 & 0 & & \\ 0 & \mathrm{i} & 0 & & \end{pmatrix} ,$$

$$\tilde{\lambda}_{19} = \hat{C}_{35} + \hat{C}_{53} = \begin{pmatrix} & & & 0 & 0 \\ & & & 0 & 0 \\ & & & 0 & 1 \\ 0 & 0 & 0 & & \\ 0 & 0 & 1 & & \end{pmatrix} ,$$

$$\tilde{\lambda}_{20} = \frac{1}{\mathrm{i}}(\hat{C}_{35} - \hat{C}_{53}) = \begin{pmatrix} & & & 0 & 0 \\ & & & 0 & 0 \\ & & & 0 & -\mathrm{i} \\ 0 & 0 & 0 & & \\ 0 & 0 & \mathrm{i} & & \end{pmatrix} .$$

These generators describe transitions between multiplets that belong to different subgroups, that is between the subgroups SU(3) and SU(2) × U(1). Such transitions are mediated by the gauge bosons X and Y (see Sect. 9.3). The remaining SU(5) genera-

Exercise 9.1

tors are

$$\tilde{\lambda}_{21} \equiv \tilde{\tau}_1 = \hat{C}_{45} + \hat{C}_{54} = \begin{pmatrix} & & & & \\ & & & & \\ & & & & \\ & & & 0 & 1 \\ & & & 1 & 0 \end{pmatrix} ,$$

$$\tilde{\lambda}_{22} \equiv \tilde{\tau}_2 = \frac{1}{\mathrm{i}}(\hat{C}_{45} - \hat{C}_{54}) = \begin{pmatrix} & & & & \\ & & & & \\ & & & & \\ & & & 0 & -\mathrm{i} \\ & & & \mathrm{i} & 0 \end{pmatrix} ,$$

$$\tilde{\lambda}_{23} \equiv \tilde{\tau}_3 = \hat{C}''_{44} - \hat{C}''_{55} = \begin{pmatrix} 0 & & & & \\ & 0 & & & \\ & & 0 & & \\ & & & 1 & \\ & & & & -1 \end{pmatrix}$$

and finally

$$\tilde{\lambda}_{24} \equiv \tilde{Y} = \sqrt{\frac{5}{3}}(\hat{C}'_{44} + \hat{C}'_{55}) = \frac{1}{\sqrt{15}} \begin{pmatrix} -2 & & & & \\ & -2 & & & \\ & & -2 & & \\ & & & 3 & \\ & & & & 3 \end{pmatrix}$$

$$= \sqrt{\frac{3}{5}} \begin{pmatrix} -\frac{2}{3} & & & & \\ & -\frac{2}{3} & & & \\ & & -\frac{2}{3} & & \\ & & & 1 & \\ & & & & 1 \end{pmatrix} .$$

Altogether the generators $\tilde{\lambda}_1, \dots, \tilde{\lambda}_8$ form an SU(3) Lie algebra, while the generators $\tilde{\tau}_1$, $\tilde{\tau}_2$, $\tilde{\tau}_3$ determine a SU(2) algebra. The operators $\tilde{\lambda}_i$, $i = 1$–21, commute with all three operators $\tilde{\tau}_i$, implying that the subgroups SU(3) and SU(2) commute with each other. Of course, this is also true for the subgroup U(1) generated by $\tilde{Y}$, since $\tilde{Y}$ is diagonal with respect to the subgroups SU(3) and SU(2). The $\tilde{\lambda}_1, \dots, \tilde{\lambda}_8$ describe SU(3), $\tilde{\lambda}_{21}$, $\tilde{\lambda}_{22}$, $\tilde{\lambda}_{23}$ SU(2), and $\tilde{\lambda}_{24}$ U(1). Altogether these 12 generators represent the new group SU(3) × SU(2) × U(1). This group is a subgroup of the complete SU(5), which is formed by all 24 generators $\tilde{\lambda}_i$. The subgroup character of the SU(3) × SU(2) × U(1) is expressed by writing

$$\mathrm{SU}(5) \supseteq \mathrm{SU}(3) \times \mathrm{SU}(2) \times \mathrm{U}(1)$$

which means that the direct product group SU(3)×SU(2)×U(1) is *embedded* into SU(5). Formally speaking, the imbedding of a group A into a group B implies that the Lie algebra of the generators of A is a sub-algebra of the generators of B.

From the discussion of the Glashow–Salam–Weinberg model we recall that the hypercharge of the lepton doublet $\binom{\nu_e}{e^-}$ is -1; thus it is convenient to normalize the corresponding operator of the hypercharge $\tilde{Y} = \tilde{\lambda}_{24}$ appropriately:

$$\hat{Y} = \sqrt{\frac{5}{3}}\tilde{Y} = \begin{pmatrix} -\frac{2}{3} & & & & \\ & -\frac{2}{3} & & & \\ & & -\frac{2}{3} & & \\ & & & 1 & \\ & & & & 1 \end{pmatrix} \quad . \tag{9.23}$$

It is important to note that the left-handed lepton doublet $(\nu_e, e^-)_L$ belongs to the complex conjugate representation $\bar{5}$, in terms of which the operator for the hypercharge is given by $-\hat{Y}$. In order to see this, we consider the group operators rather than the generators:

$$\left[\exp(\mathrm{i}\alpha\hat{Y})\right]^* = \exp(-\mathrm{i}\alpha\hat{Y}) = \exp[\mathrm{i}\alpha(-\hat{Y})] \quad . \tag{9.24}$$

Only in this way, that is with respect to the $\bar{5}$ representation, is the hypercharge of the form

$$\begin{pmatrix} \frac{2}{3} & & & & \\ & \frac{2}{3} & & & \\ & & \frac{2}{3} & & \\ & & & -1 & \\ & & & & -1 \end{pmatrix}$$

with the appropriate values (-1) for the $(\nu_e, e^-)_L$ doublet. Taking into account the isospin operator (9.21), we find for the operator of the *electric charge*, in terms of the fundamental representation 5,

$$\hat{Q} = \hat{T}_3 + \frac{1}{2}\hat{Y} \equiv \frac{1}{2}\tilde{\tau}_3 + \sqrt{\frac{5}{12}}\tilde{Y} = \begin{pmatrix} -\frac{1}{3} & & & & \\ & -\frac{1}{3} & & & \\ & & -\frac{1}{3} & & \\ & & & 1 & \\ & & & & 0 \end{pmatrix} \quad . \tag{9.25}$$

According to (9.24) the charge operator for the representation $\bar{5}$ is given by $(-\hat{Q})$. The form of (9.25) exhibits the consistent unification of the elementary fermions within the quintuplet of SU(5). As will be shown in Example 9.2, we must assign the last two components to the lepton "antidoublet" $(\mathrm{e}^C, -\nu_\mathrm{e}^C)_R$, whereas the first three components, representing the color gauge group SU(3), correspond to particles of charge $-1/3$.

The latter can only be the color triplet (and isospin singlet) of the right-handed d quarks (see e.g. Table 6.2). Thus the five-dimensional representation is given by

$$[5] = (\psi_p)_R = \begin{pmatrix} d_r \\ d_b \\ d_g \\ e^+ \\ -\nu_\mathrm{e}^C \end{pmatrix}_R \quad , \quad p = 1, \ldots, 5 \quad . \tag{9.26}$$

The index $p = 1, 2, \ldots, 5$ characterizes the individual components of the SU(5) spinor

$$\begin{pmatrix} \psi_1 \\ \psi_2 \\ \psi_3 \\ \psi_4 \\ \psi_5 \end{pmatrix} \quad .$$

Note that the antiparticles corresponding to the SU(2) doublet $\binom{\nu_e}{e^-}_L$ are represented by the doublet $(e^+, -\bar{\nu}_e)_R$. The sign of the helicity is reversed for the antiparticles. This is essentially the result of Example 9.2, which is, except for the minus sign in the case of the anti neutrino, quite plausible.

EXAMPLE

9.2 Charge Conjugation

The fields of antiparticles with spin $\frac{1}{2}$ are obtained by charge conjugation

$$\psi^C = \hat{C}\bar{\psi}^T = \hat{C}\gamma^0\psi^* \quad , \quad \bar{\psi}^C = \overline{\psi^C} \quad , \tag{1}$$

where the operator of charge conjugation $\hat{C}$ is required to fulfill

$$\hat{C}\gamma^\mu\hat{C}^{-1} = -\gamma^{\mu T} \quad . \tag{2a}$$

This constraint is satisfied by choosing[5] for example,

$$\hat{C} = \mathrm{i}\gamma^2\gamma^0 \quad . \tag{2b}$$

In order to find the behavior of chirality eigenstates under charge conjugation, we consider

$$\begin{aligned}(\psi_L)^C &= \hat{C}\gamma^0(\psi_L)^* = \mathrm{i}\gamma^2\frac{1}{2}(1-\gamma_5)\psi^* \\ &= \frac{1}{2}(1+\gamma_5)\mathrm{i}\gamma^2\psi^* = \frac{1}{2}(1+\gamma_5)\hat{C}\gamma^0\psi^* \\ &= \frac{1}{2}(1+\gamma_5)\psi^C = (\psi^C)_R \end{aligned} \tag{3a}$$

and similarly

$$(\psi_R)^C = (\psi^C)_L \quad . \tag{3b}$$

Thus the charge-conjugated state of a right-handed electron is a left-handed positron, and so on.

Applying the operator of charge conjugation to isospin doublets, the situation becomes more complex. As we have already learned from the example of hyper-

[5] See Chap. 12 – W. Greiner: *Relativistic Quantum Mechanics – Wave Equations*, 3rd ed. (Springer, Berlin, Heidelberg, 2000).

charge (9.24), charge conjugation reverses the sign of the eigenvalues of all generators that generate symmetry transformations. This is due to the operation of complex conjugation which is part of charge conjugation. Hence, by naively applying charge conjugation to the particular components of a lepton doublet

$$L = \begin{pmatrix} \nu_e \\ e^- \end{pmatrix}_L \; , \tag{4a}$$

the resulting doublet

$$L' = \begin{pmatrix} \nu_e^C \\ e^+ \end{pmatrix}_R \tag{4b}$$

would have isospin $T_3 = -\frac{1}{2}$ and $T_3 = +\frac{1}{2}$ for the upper and lower components, respectively. Of course this cannot be right; however, the following simple prescription will yield the correct result. We only need to rotate L' around the 1 axis or 2 axis of the isospin space by 180°. It is common to choose the 2 axis to perform the rotation:

$$\begin{aligned} e^{i\pi \hat{T}_2} \hat{T}_3 e^{-i\pi \hat{T}_2} &= e^{i\frac{\pi}{2}\tau_2} \left(\frac{1}{2}\tau_3\right) e^{-i\frac{\pi}{2}\tau_2} \\ &= i\tau_2 \left(\frac{1}{2}\tau_3\right)(-i\tau_2) \\ &= \frac{1}{2}\tau_2\tau_3\tau_2 = -\frac{1}{2}\tau_3 = -\hat{T}_3 \end{aligned} \tag{5}$$

where we have used the relation

$$e^{i\frac{\pi}{2}\tau_2} = \cos\frac{\pi}{2} + i\tau_2 \sin\frac{\pi}{2} = i\tau_2 = \begin{pmatrix} 0 & +1 \\ -1 & 0 \end{pmatrix} . \tag{6}$$

Thus, the *correct charge-conjugated isospin doublet is given by the additional rotation* (6) *in isospin spaces, that is by*

$$L^e = i\tau_2 \begin{pmatrix} \nu_e^C \\ e^+ \end{pmatrix}_R = \begin{pmatrix} e^+ \\ -\nu_e^C \end{pmatrix}_R . \tag{7}$$

From our previous considerations, we might be tempted to combining the left-handed, charge-conjugated particles $(d_r, d_b, d_g)_L, e_L^+$ within a quintuplet. However, this is not possible, since the left-handed d quarks belong to an isospin doublet (see (6.29)),

$$\begin{pmatrix} u_r & u_b & u_g \\ d_r & d_b & d_g \end{pmatrix}_L \tag{9.27a}$$

whereas e_L^+, as the conjugated particle to e_R^-, is represented by an isospin singlet. The best we can do is to form a SU(5) antiquintuplet that contains the conjugated left-handed particles corresponding to (9.26),

$$[\bar{5}] = (\psi_p^C)_L = \begin{pmatrix} d_r^C \\ d_b^C \\ d_g^C \\ e^- \\ -\nu_e \end{pmatrix}_L \quad , \quad p = 1, \ldots, 5 \quad . \tag{9.28}$$

Note, however, that this is not an independent assignment. The superscript "C" denotes the antiparticles that correspond to the letters without a bar (for example, d^C indicates the d antiquark). The antiparticles of the d_R quarks are the left-handed d_L^C; both are SU(2) singlets. Frequently antiparticles are denoted by a bar over the letter, for example the d_L antiquark reads $\bar{d}_L$. This notation, however, may lead to misunderstanding with the adjoint spinor $\bar{d} = d^\dagger \gamma^0$. In order to prevent this we prefer to characterize the antiparticles by the superscript "C" and use the bar notation only where misunderstanding is remote.

We have up to now given SU(5) assignments to the right-handed d quarks, $(d_r)_R, (d_b)_R, (d_g)_R$, and to the right-handed positron $(e^+)_R$ and electron antineutrino $(\nu_e^C)_R$ – see (9.26) –, as well as to the left-handed d antiquarks, $(d_r^C)_L, (d_b^C)_L, (d_g^C)_L$, to the left-handed electron $(e^-)_L$ and the electron neutrino $(\nu_e)_L$ – see (9.28). The question remains open which SU(5) assignment has to be given to the left-handed fermions $\begin{pmatrix} u_r & u_b & u_g \\ d_r & d_b & d_g \end{pmatrix}_L$ – see (9.27a) – and to the singlets e_L^+ and $(u_r^C, u_b^C, u_g^C)_L$. To achieve this, i.e. to determine the classification of the elementary left-handed fermions, that is (9.27a), as well as the singlets

$$e_L^+, \quad (u_r^C, u_b^C, u_g^C)_L \quad , \tag{9.27b}$$

into further SU(5) multiplets, we need some additional tools from group theory. The problem is, how do we decompose the irreducible representations of SU(5) in terms of tensor products of representations of the groups SU(3) and SU(2)? We start from the trivial decomposition of the fundamental representation of SU(5),

$$5 \equiv \square_5 = (\square_3, 1_2) + (1_3, \square_2) \equiv (3, 1) + (1, 2) \quad , \tag{9.29}$$

where we have alternatively employed the notation of Young diagrams and the notation in terms of the dimensions of the representations. The indices "3" and "2" indicate whether the representation belongs to the SU(3) or SU(2), respectively. Using the distributive property, we obtain for the direct product of two fundamental representations

$$\begin{aligned} \square_5 \times \square_5 &= [(\square_3, 1_2) + (1_3, \square_2)] \times [(\square_3, 1_2) + (1_3, \square_2)] \\ &= (\square_3 \times \square_3, 1_2) + (\square_3, \square_2) + (\square_3, \square_2) \\ &\quad + (1_3, \square_2 \times \square_2) \\ &= (\square\!\square_3, 1_2) + \left(\begin{smallmatrix}\square\\\square\end{smallmatrix}_3, 1_2\right) + 2(\square_3, \square_2) \\ &\quad + (1_3, \square\!\square_2) + \left(1_3, \begin{smallmatrix}\square\\\square\end{smallmatrix}_2\right) \\ &\equiv (6, 1) + (\bar{3}, 1) + 2(3, 2) + (1, 3) + (1, 1) \quad . \end{aligned} \tag{9.30}$$

On the other hand (9.18) states that

$$5 \times 5 \equiv \Box_5 \times \Box_5 = \Box\Box_5 + \begin{smallmatrix}\Box\\ \Box\end{smallmatrix}_5 = 15 + 10 \quad .$$

Owing to the symmetry or antisymmetry, of the representations (boxes above one another correspond to antisymmetry, boxes next to one another to symmetry) we may immediately assign the product representations (6, 1), (1, 3) to the 15-dimensional representation and, similarly, $(\bar{3}, 1)$, (1, 1) to the 10-dimensional representation of SU(5). From the two representations (3, 2), we can form symmetric and antisymmetric combinations. We therefore obtain for the SU(3)×SU(2) decompositions

$$10 = (\bar{3}, 1) + (3, 2)_{\text{anti}} + (1, 1) \quad , \tag{9.31a}$$

$$15 = (6, 1) + (3, 2)_{\text{sym}} + (1, 3) \quad . \tag{9.31b}$$

EXERCISE

9.3 The Quintuplet of SU(5)

Problem. Consider the five basis states of the quintuplet of SU(5),

$$\psi_1 \equiv q_1 = \begin{pmatrix}1\\0\\0\\0\\0\end{pmatrix} \quad , \quad \psi_2 \equiv q_2 = \begin{pmatrix}0\\1\\0\\0\\0\end{pmatrix} \quad ,$$

$$\psi_3 \equiv q_3 = \begin{pmatrix}0\\0\\1\\0\\0\end{pmatrix} \quad , \quad \psi_4 \equiv \ell_1 = \begin{pmatrix}0\\0\\0\\1\\0\end{pmatrix} \quad ,$$

$$\psi_5 \equiv \ell_2 = \begin{pmatrix}0\\0\\0\\0\\1\end{pmatrix} \quad ,$$

(q = quark, ℓ = lepton), and work out the antisymmetric basis vectors of the 10-dimensional representation. Write down these basis vectors explicitly as 5×5 matrices.

Solution. The result of the general decomposition of the direct product [5] × [5], (9.30), shows that the 10-dimensional representation contains three basic components: a color antitriplet/isospin singlet, an antisymmetric color triplet/isospin doublet, and a complete singlet (see (9.31a)). The first and last can be expressed in terms of an

Exercise 9.3

antisymmetric tensor product of quark and lepton basis vectors, respectively:

$$\left.\begin{aligned}\psi_{12}&=\frac{1}{\sqrt{2}}(q_1q_2-q_2q_1)\\ \psi_{23}&=\frac{1}{\sqrt{2}}(q_2q_3-q_3q_2)\\ \psi_{31}&=\frac{1}{\sqrt{2}}(q_3q_1-q_1q_3)\end{aligned}\right\}\begin{array}{l}\text{SU(3) antitriplet}\\ \text{SU(2) singlet}\end{array}\,.$$

This corresponds to the $(\bar{3}, 1)$ on the right-hand side of (9.31a). Furthermore,

$$\psi_{45}=\frac{1}{\sqrt{2}}(\ell_1\ell_2-\ell_2\ell_1)\qquad\begin{array}{l}\text{SU(3) singlet}\\ \text{SU(2) singlet}\end{array}$$

is the singlet (1, 1) on the right-hand side of (9.31a). Similarly, the color triplet/isospin doublet can be described by a product of a quark and a lepton vector. There are exactly six possible ways to form such antisymmetric product states:

$$\left.\begin{aligned}&\left.\begin{aligned}\psi_{14}&=\frac{1}{\sqrt{2}}(q_1\ell_1-\ell_1q_1)\\ \psi_{24}&=\frac{1}{\sqrt{2}}(q_2\ell_1-\ell_1q_2)\\ \psi_{34}&=\frac{1}{\sqrt{2}}(q_3\ell_1-\ell_1q_3)\end{aligned}\right\}\begin{array}{l}\text{SU(3)}\\ \text{triplet}\end{array}\\ &\left.\begin{aligned}\psi_{15}&=\frac{1}{\sqrt{2}}(q_1\ell_2-\ell_2q_1)\\ \psi_{25}&=\frac{1}{\sqrt{2}}(q_2\ell_2-\ell_2q_2)\\ \psi_{35}&=\frac{1}{\sqrt{2}}(q_3\ell_2-\ell_2q_3)\end{aligned}\right\}\begin{array}{l}\text{SU(3)}\\ \text{triplet}\end{array}\end{aligned}\right\}\text{SU(2) doublet}\,.$$

These are exactly the states of the (3, 2) antisubmultiplet of the decomposition (9.31a). Altogether this makes a total of 10 basis vectors for the 10-dimensional representation. The tensor product of any two five-component vectors yields a 5×5 matrix. For example,

$$\psi_{12}=\frac{\psi_1\psi_2-\psi_2\psi_1}{\sqrt{2}}=\frac{1}{\sqrt{2}}\begin{pmatrix}0&1&0&0&0\\-1&0&0&0&0\\0&0&0&0&0\\0&0&0&0&0\\0&0&0&0&0\end{pmatrix}\,,$$

$$\vdots$$

$$\psi_{35}=\frac{\psi_3\psi_5-\psi_5\psi_3}{\sqrt{2}}=\frac{1}{\sqrt{2}}\begin{pmatrix}0&0&0&0&0\\0&0&0&0&0\\0&0&0&0&1\\0&0&0&0&0\\0&0&-1&0&0\end{pmatrix}\,.$$

From the considerations of Exercise 9.3 and (9.31a) it is now clear that the representation [10] contains one color triplet/isospin doublet, one colour antitriplet/isospin singlet, and finally a complete singlet. On the other hand, the decomposition of the adjoint fundamental representation,

$$[\bar{5}] = (\bar{3}, 1) + (1, \bar{2}) \quad , \tag{9.32}$$

represents the group-theoretical justification for the choice of the assignment (9.28), since the left-handed d antiquarks form an isospin singlet, whereas the color singlet particles $(e^-, \nu_e)_L$ belong to an isospin doublet. It is denoted in (9.32) as $(1, \bar{2})$, which is identical with $(1, 2)$ because for SU(2) the elementary doublet is identical with the elementary antidoublet. The particular components of the representation [10] (9.31a) then corresponds to the particles $(u^c_r, u^c_b, u^c_g)_L$ for the $(\bar{3}, 1)$, $\begin{pmatrix} u_r & u_b & u_g \\ d_r & d_b & d_g \end{pmatrix}_L$ for the $(3, 2)$, and to the singlet e^+_L for the $(1, 1)$.

The antisymmetric SU(5) representation [10] is conveniently expressed in terms of a 5×5 matrix that contains exactly 10 independent components (see Exercise 9.3). We shall denote these matrices by ψ_{ij}, $i, j = 1, \dots, 5$ $(\psi_{ij} = -\psi_{ji})$. The group operators of this representation are then simply expressed as tensor products of operators of the fundamental representation:

$$\begin{aligned} \psi_{kl} &= q_k q_l - q_l q_k \quad , \\ \psi'_{kl} &= \sum_{i,j} U^{[10]}_{kl,ij} \psi_{ij} = \sum_{i,j} (U^{[5]}_{ki} U^{[5]}_{lj}) \psi_{ij} \quad . \end{aligned} \tag{9.33}$$

Indeed, from $q'_k = \sum_i U^{[5]}_{ki} q_i$ it immediately follows that

$$\begin{aligned} \psi'_{kl} &= q'_k q'_l - q'_l q'_k = \sum_{i,j} (U^{[5]}_{ki} U^{[5]}_{lj} - U^{[5]}_{li} U^{[5]}_{kj}) q_i q_j \\ &= \sum_{i,j} U^{[5]}_{ki} U^{[5]}_{lj} (q_i q_j - q_j q_i) = \sum_{i,j} U^{[5]}_{ki} U^{[5]}_{lj} \psi_{ij} \quad . \end{aligned}$$

Now the multiplication of group operators corresponds to the addition of the generators. For a diagonal operator, such as the charge matrix $Q^{[5]}_{ki} = Q^{[5]}_i \delta_{ik}$, where the group rotation in the five-dimensional representation is $U^{[5]}_{kl} = \exp(\mathrm{i} Q^{[5]}_k) \delta_{kl}$, we therefore have in the 10-dimensional representation for the group operators $U^{[10]}_{kl,ij} = U^{[5]}_{ki} U^{[5]}_{lj} = \mathrm{e}^{\mathrm{i} Q^{[5]}_k} \delta_{ki} \mathrm{e}^{\mathrm{i} Q^{[5]}_l} \delta_{lj} = \mathrm{e}^{\mathrm{i} Q^{[10]}} \delta_{ki} \delta_{lj}$, and therefore the corresponding charge operator is obtained by addition:

$$\hat{Q}^{[10]} = Q^{[10]}_{kl,ij} = Q^{[5]}_k \delta_{ki} \delta_{lj} + Q^{[5]}_l \delta_{lj} \delta_{ki} = (Q^{[5]}_k + Q^{[5]}_l) \delta_{ki} \delta_{lj} \quad .$$

Consequently,

$$\begin{aligned} \hat{Q}^{[10]} \psi_{kl} &= \sum_{i,j} Q^{[10]}_{kl,ij} \psi_{ij} = \sum_{i,j} (Q^{[5]}_i + Q^{[5]}_j) \delta_{ki} \delta_{lj} \psi_{ij} \\ &= (Q^{[5]}_k + Q^{[5]}_l) \psi_{kl} \equiv Q_{kl} \psi_{kl} \quad . \end{aligned} \tag{9.34}$$

With the eigenvalues of the charge matrix (9.25) we obtain for the 5×5 matrix $(Q_k + Q_l)$

$$Q_{kl} = (Q_k + Q_l) = \begin{pmatrix} * & -\frac{2}{3} & -\frac{2}{3} & \frac{2}{3} & -\frac{1}{3} \\ -\frac{2}{3} & * & -\frac{2}{3} & \frac{2}{3} & -\frac{1}{3} \\ -\frac{2}{3} & -\frac{2}{3} & * & \frac{2}{3} & -\frac{1}{3} \\ \frac{2}{3} & \frac{2}{3} & \frac{2}{3} & * & 1 \\ -\frac{1}{3} & -\frac{1}{3} & -\frac{1}{3} & 1 & * \end{pmatrix} . \tag{9.35}$$

The fact that the charges Q_{kl} of the antisymmetric states $\psi_{kl} = q_k q_l - q_l q_k$ of (9.33) are given by the sum of the charges of the elementary quintuplet is physically highly plausible: the charge of a product of two states equals the sum of the charges of the individual states involved. Equation (9.33) states that the diagonal elements do not contribute since, owing to the antisymmetric character of the 10-dimensional representation, it holds that $\psi_{ii} = -\psi_{ii} = 0$. For this reason we have labeled them with a "$*$" in (9.35). The decomposition of the 10-dimensional representation in terms of its SU(2) and SU(3) content (9.31a) consequently leads to the arrangement of the remaining left-handed particles (9.27a), (9.27b) within the antisymmetric matrix ψ_{ij}, as follows (see Example 9.3):

$$10 = \left(\begin{array}{c|c} (\bar{3}, 1) & (3, 2) \\ \hline (3, 2) & (1, 1) \end{array} \right) . \tag{9.36}$$

From the color vector $(u_r^C, u_b^C, u_g^C)_L$ we form an antisymmetric 3×3 matrix by means of $\sum_k \varepsilon_{ijk} (u_k^C)_L$ and, similarly, from the singlet e_L^+ we construct an antisymmetric 2×2 matrix by $\varepsilon_{ij} e_L^+$. Hence, the matrix ψ_{ij} is of the following form:

$$\begin{aligned} \psi_{ij}^{[10]} &= \frac{1}{\sqrt{2}} \left(\begin{array}{ccc|cc} 0 & u_g^C & -u_b^C & -u_r & -d_r \\ -u_g^C & 0 & u_r^C & -u_b & -d_b \\ u_b^C & -u_r^C & 0 & -u_g & -d_g \\ \hline u_r & u_b & u_g & 0 & +e^+ \\ d_r & d_b & d_g & -e^+ & 0 \end{array} \right)_L \\ &= \frac{1}{\sqrt{2}} \begin{pmatrix} 0 & u_3^C & -u_2^C & -u_1 & -d_1 \\ -u_3^C & 0 & u_1^C & -u_2 & -d_2 \\ u_2^C & -u_1^C & 0 & -u_3 & -d_3 \\ u_1 & u_2 & u_3 & 0 & -e^C \\ d_1 & d_2 & d_3 & e^C & 0 \end{pmatrix}_L \end{aligned} \tag{9.37}$$

where a normalization factor $1/\sqrt{2}$ is introduced to take into account that every particle appears twice within the matrix ψ_{ij}. In the second version of ψ_{ij} in (9.37) we have relabeled the color indices r, b, g *by* $1, 2, 3$ which is more convenient for practical calculations.

In order to check the consistency of the above arrangement we apply the charge operator (9.35) onto the multiplet (9.37). Obviously this leads to $Q(\mathrm{u}) = -Q(\mathrm{u}^C) = +2/3$, $Q(\mathrm{d}) = -1/3$, and $Q(\mathrm{e}^+) = +1$. This is a remarkable result. By fixing a particular charge, for example the electron charge, within the gauge group, all other particle charges are completely determined, that is the charges of ν_e, u and d quarks, if we only arrange these particles within the multiplets 5 and 10 in correspondence with their color and isospin quantum numbers. In other words, *the quantization of charge is a direct consequence of the group* SU(5).[6]

In particular, the property of vanishing trace of the charge operator (9.25) (as a linear combination of the traceless generators) implies that the sum of charges of the elementary left-handed fermions within a particular multiplet is always zero:

$$[5]: \quad 3Q(\mathrm{d}) + Q(\mathrm{e}^+) + Q(\bar{\nu}_\mathrm{e}) = 0 \ , \tag{9.38a}$$

$$[10]: \quad \sum_{k,l} Q_{kl} = \sum_{k,l} (Q_k + Q_l) = 5\sum_k Q_k + 5\sum_l Q_l = 0 \ ,$$

$$\sum_{k,l} Q_{kl} = \sum_k Q_{kk} + \sum_{k\neq l} Q_{kl} = 0 + \sum_{k\neq l} Q_{kl} \ ,$$

$$\sum_{k\neq l} Q_{kl} = \sum_{k\neq l} (Q_k + Q_l) = 3Q(\mathrm{u}^C) + 3Q(\mathrm{u}) + 3Q(\mathrm{d}) + Q(\mathrm{e}^+)$$

$$= 3Q(\mathrm{d}) + Q(\mathrm{e}^+) = 0 \ , \tag{9.38b}$$

since $Q(\mathrm{u}^C) = -Q(\mathrm{u})$. In the second line above we used

$$\sum_k Q_{kk} = \sum_k (Q_k + Q_k) = 2\sum_k Q_k = 0.$$

Because of (9.38a) and (9.38b) we obtain

$$Q(\bar{\nu}_\mathrm{e}) = 0 \ , \quad Q(\mathrm{d}) = -\frac{1}{3}Q(\mathrm{e}^+) = \frac{1}{3}Q(\mathrm{e}^-) \ . \tag{9.39}$$

Furthermore, from the doublet character of the electroweak theory, it follows that the charge difference of the upper and lower components of the electroweak 2-spinor is $\Delta Q = q_0(Y/2 + T_3 = +1/2) - q_0(Y/2 + T_3 = -1/2) = q_0$, in other words it is equal for all 2-spinors, that is $1q_0$. Here q_0 denotes the reference charge which determines the scale of the charge. In particular, for the electroweak $\binom{\mathrm{u}}{\mathrm{d}}$ spinor and $\binom{\nu_\mathrm{e}}{\mathrm{e}^-}$ spinor one therefore finds $Q(\mathrm{u}) - Q(\mathrm{d}) = Q(\nu_\mathrm{e}) - Q(\mathrm{e}^-)$, or with (9.39)

$$Q(\mathrm{u}) = Q(\mathrm{d}) + [Q(\nu_\mathrm{e}) - Q(\mathrm{e}^-)] = -\frac{2}{3}Q(\mathrm{e}^-) \ , \tag{9.40}$$

which means that the charges of all particles can be expressed in terms of the electron charge. Another successful feature of SU(5) is that for every lepton doublet there exists exactly one quark doublet with three color states. Only for such a particular combination can the renormalization of the standard model be ensured. This statement is discussed in more detail in the field-theoretical digression (Supplement 9.11: Anomaly Freedom).

[6] The quantization of the electric charge can also be obtained from other semi-simple Lie groups that, instead of SU(5), could be chosen as the gauge groups of the unified interactions, for example SO(10), E_6, etc., cf. H. Saller: *Vereinheitlichte Feldtheorien der Elementarteilchen* (Springer, Berlin, Heidelberg, 1985).

EXAMPLE

9.4 SU(5) Classification of the Remaining Lepton and Quark Generations (Families)

In an analogous fashion to ν_e, e, u, d, also ν_μ, μ, c, s and ν_τ, τ, t, b have to be classified in the framework of SU(5). Embedding the first lepton and quark generation has already been thoroughly discussed. In this procedure we constructed the following multiplets consisting of 30 fermions, including the antiparticles. We identified those particles known as the left-handed SU(2) doublets and singlets,

$$\begin{pmatrix} \nu_e \\ e^- \end{pmatrix}_L \;, \quad \begin{pmatrix} u \\ d \end{pmatrix}_{c,L} \;, \quad e^+_L \;, \quad u^C_{\bar{c},L} \;, \quad d^C_{\bar{c},L} \;,$$

and right-handed SU(2) doublets and singlets,

$$\begin{pmatrix} e^+ \\ -\nu^C_e \end{pmatrix}_R \;, \quad \begin{pmatrix} d^C \\ -u^C \end{pmatrix}_{\bar{c},R} \;, \quad e^-_R \;, \quad u_{c,R} \;, \quad d_{c,R} \;,$$

with four SU(5) multiplets. The subscripts c and $\bar{c}$, denote the color degree of freedom of the quarks. We also proposed the following classification SU(5) quintuplet:

$$5 = \begin{pmatrix} d_1 & d_2 & d_3 & e^+ & -\nu^C_e \end{pmatrix}_R \;,$$

SU(5) antiquintuplet:

$$\bar{5} = \begin{pmatrix} d^C_1 & d^C_2 & d^C_3 & e^- & -\nu_e \end{pmatrix}_L \;,$$

SU(5) decuplet (antisymmetric representation):

$$10 = \frac{1}{\sqrt{2}} \begin{pmatrix} 0 & u^C_3 & -u^C_2 & -u_1 & -d_1 \\ -u^C_3 & 0 & u^C_1 & -u_2 & -d_2 \\ u^C_2 & -u^C_1 & 0 & -u_3 & -d_3 \\ u_1 & u_2 & u_3 & 0 & +e^+ \\ d_1 & d_2 & d_3 & -e^+ & 0 \end{pmatrix}_L \;,$$

SU(5) antidecuplet (antisymmetric representation):

$$\overline{10} = \frac{1}{\sqrt{2}} \begin{pmatrix} 0 & u_3 & -u_2 & -u^C_1 & -d^C_1 \\ -u_3 & 0 & u_1 & -u^C_2 & -d^C_2 \\ u_2 & -u_1 & 0 & -u^C_3 & -d^C_3 \\ u^C_1 & u^C_2 & u^C_3 & 0 & +e^- \\ d^C_1 & d^C_2 & d^C_3 & -e^- & 0 \end{pmatrix}_R \;.$$

One can see immediately that all the particles and antiparticles of the first fermion generation are involved.

Similarly, we can put the remaining two lepton and quark families into SU(5) schemes. The latter occurs in the following isospin doublets and singlets left-handed:

$$\begin{pmatrix} \nu_\mu \\ \mu^- \end{pmatrix} r_L \;, \quad \begin{pmatrix} c \\ s \end{pmatrix}_{c,L} \;, \quad \mu^+_L \;, \quad c^C_{\bar{c},L} \;, \quad s^C_{\bar{c},L} \;;$$

Example 9

right-handed:

$$\begin{pmatrix} \mu^+ \\ -\nu_\mu^C \end{pmatrix}_R \quad , \quad \begin{pmatrix} s^C \\ -c^C \end{pmatrix}_{\bar{c},R} \quad , \quad \mu_R^- \quad , \quad c_{c,R} \quad , \quad s_{c,R} \quad .$$

This yields the following SU(5) multiplets SU(5) quintuplet:

$$5 = \begin{pmatrix} s_1 & s_2 & s_3 & \mu^+ & -\nu_\mu^C \end{pmatrix}_R \quad ,$$

SU(5) antiquintuplet:

$$\overline{5} = \begin{pmatrix} s_1^c & s_2^c & s_3^c & \mu^- & \nu_\mu \end{pmatrix}_L \quad ,$$

SU(5) decuplet (antisymmetric representation):

$$10 = \frac{1}{\sqrt{2}} \begin{pmatrix} 0 & c_3^C & -c_2^C & -c_1 & -s_1 \\ -c_3^C & 0 & c_1^C & -c_2 & -s_2 \\ c_2^C & -c_1^C & 0 & -c_3 & -s_3 \\ c_1 & c_2 & c_3 & 0 & +\mu^+ \\ s_1 & s_2 & s_3 & -\mu^+ & 0 \end{pmatrix}_L \quad ,$$

SU(5) antidecuplet (antisymmetric representation):

$$\overline{10} = \frac{1}{\sqrt{2}} \begin{pmatrix} 0 & c_3 & -c_2 & -c_1^C & -s_1^C \\ -c_3 & 0 & c_1 & -c_2^C & -s_2^C \\ c_2 & -c_1 & 0 & -c_3^C & -s_3^C \\ c_1^C & c_2^C & c_3^C & 0 & +\mu^- \\ s_1^C & s_2^C & s_3^C & -\mu^- & 0 \end{pmatrix}_R \quad .$$

The third generation of leptons and quarks, which are represented by the isospin doublets and singlets, are

- left-handed:

$$\begin{pmatrix} \nu_\tau \\ \tau^- \end{pmatrix}_L \quad , \quad \begin{pmatrix} t \\ b \end{pmatrix}_{c,L} \quad , \quad \tau_L^+ \quad , \quad t_{\bar{c},L}^C \quad , \quad b_{\bar{c},L}^C \quad ;$$

- right-handed:

$$\begin{pmatrix} \tau^+ \\ -\nu_\tau^C \end{pmatrix}_R \quad , \quad \begin{pmatrix} b^C \\ -t^C \end{pmatrix}_{\bar{c},R} \quad , \quad \tau_R^- \quad ; \; t_{c,R} \quad , \quad b_{c,R} \quad .$$

They are now located within the SU(5) multiplets, namely the

- SU(5) quintuplet:

$$5 = \begin{pmatrix} b_1 & b_2 & b_3 & \tau^+ & -\nu_\tau^C \end{pmatrix}_R \quad ,$$

- SU(5) antiquintuplet:

$$\overline{5} = \begin{pmatrix} b_1^C & b_2^C & b_3^C & \tau^- & \nu_\tau \end{pmatrix}_L \quad ,$$

xample 9.4

- SU(5) decuplet (antisymmetric representation):

$$10=\frac{1}{\sqrt{2}}\begin{pmatrix} 0 & t_3^C & -t_2^C & -t_1 & -b_1 \\ -t_3^C & 0 & t_1^C & -t_2 & -b_2 \\ t_2^C & -t_1^C & 0 & -t_3 & -b_3 \\ t_1 & t_2 & t_3 & 0 & +\tau^+ \\ b_1 & b_2 & b_3 & -\tau^+ & 0 \end{pmatrix}_L ,$$

- SU(5) antidecuplet (antisymmetric representation):

$$\overline{10}=\frac{1}{\sqrt{2}}\begin{pmatrix} 0 & t_3 & -t_2 & -t_1^C & -b_1^C \\ -t_3 & 0 & t_1 & -t_2^C & -b_2^C \\ t_2 & -t_1 & 0 & -t_3^C & -b_3^C \\ t_1^C & t_2^C & t_3^C & 0 & +\tau^- \\ b_1^C & b_2^C & b_3^C & -\tau^- & 0 \end{pmatrix}_R .$$

This completes the classification of all known elementary fermions. There are no positions left in the four SU(5) multiplets of each generation. Furthermore, *particles which belong to different fermion families do not appear within the same* SU(5) *multiplet.* This classification implies that the numbers of leptons and quarks of each fermion generation are in proper agreement. Thus we expect that for each flavor of leptons there is exactly one flavor of quarks (with their color multiplicity). Above all, for the third generation this implies the existence of the t quark! It was experimentally discovered in 1995; see Sect. 6.1!

9.3 The SU(5) Gauge Theory

The gauge bosons belong to the adjoint (regular) representation of SU(5) with dimension $5^2-1=24$. According to (9.18) the adjoint representation is generated by the product $[5]\times[\bar{5}]=[24]+[1]$. Indeed we know from quantum mechanics[7] that mesons belong to the $3\times\bar{3}$ representation in the case of SU(3) and to the $4\times\bar{4}$ representation in the case of SU(4) flavor symmetry. The physical significance of this is that mesons are fermion–antifermion states, for example quark–antiquark states. Similarly the SU(5) gauge bosons can be considered equivalent to states of fermion–antifermion pairs, each belonging to the fundamental representation of SU(5). Hence it is logical to consider now the $5\times\bar{5}$ representation for the gauge bosons. With the knowledge of the $SU(3)\times SU(2)$ content of the fundamental representations that we have gained in the previous section (see (9.29)), it is now straightforward to decompose the adjoint representation in terms of the $SU(3)\times SU(2)$ portion:

$$\begin{aligned}\bar{5}\times 5 &= [\,(\bar{3},1)+(1,2)\,]\times[\,(3,1)+(1,2)\,] \\ &= \Big[\Big[\Big(\begin{smallmatrix}\Box\\\Box\end{smallmatrix}_3,1_2\Big)+(1_3,\Box_2)\Big]\times[(\Box_3,1_2)+(1_3,\Box_2)]\Big] \\ &= \Big(\begin{smallmatrix}\Box\\\Box\end{smallmatrix}_3\times\Box_3,1_2\Big)+\Big(\begin{smallmatrix}\Box\\\Box\end{smallmatrix}_3,\Box_2\Big)+(\Box_3,\Box_2)+(1_3,\Box_2\times\Box_2)\end{aligned}$$

---

[7] See W. Greiner and B. Müller: *Quantum Mechanics – Symmetries*, 2nd ed. (Springer, Berlin, Heidelberg, 1994).

$$
\begin{aligned}
&= \left(\ydiagram{2,1}_{\,3}, 1_2\right) + (1_3, 1_2) + \left(\ydiagram{1,1}_{\,3}, \square_2\right) + (\square_3, \square_2) \\
&\quad + (1_3, \ydiagram{2}_{\,2}) + (1_3, 1_2) \\
&= (8,1) + (\bar{3},2) + (3,2) + (1,3) + 2(1,1) \quad . \qquad (9.41)
\end{aligned}
$$

Therefore the decomposition of the adjoint representation is

$$[24]_5 = (8,1) + (\bar{3},2) + (3,2) + (1,3) + (1,1) \quad . \qquad (9.41a)$$

Since the gauge bosons of the color SU(3) group, the gluons, as well as the intermediate bosons and photons of SU(2) × U(1), are contained in (9.41) we can make the following assignments:

- There is a SU(3) octet of gluons G^a_μ $(a = 1, \ldots, 8) \to (8,1)$.
- There is an isovector of intermediate bosons W^i_μ $(i = 1,2,3) \to (1,3)$.
- There is an isoscalar field: the hypercharged boson $B_\mu \to (1,1)$.
- In addition, there are twelve more gauge bosons, belonging to the representations $(3,2)$ and $(\bar{3},2)$. These form an isospin doublet of bosons and their antiparticles, which are colored. It is common to use the following notation:

$$(3,2) = \begin{pmatrix} X_r & X_g & X_b \\ Y_r & Y_g & Y_b \end{pmatrix} \equiv \begin{pmatrix} X_1 & X_2 & X_3 \\ Y_1 & Y_2 & Y_3 \end{pmatrix} \quad . \qquad (9.42)$$

As we will prove in Exercise 9.5, the corresponding charges are $Q(\mathrm{X}) = +4/3$ and $Q(\mathrm{Y}) = +1/3$.

EXERCISE

9.5 The SU(5) Gauge Bosons

Problem. Evaluate the electric charges of the SU(5) gauge bosons.

Solution. The gauge bosons of the SU(5) belong to the 24-plet contained in the $[\bar{5}] \times [5]$ representation. Therefore the eigenvalues of the charge operator can be expressed as a sum of eigenvalues of the representations $[\bar{5}]$ and $[5]$, in the same way as for the 10-dimensional representation in (9.34):

$$Q^{[24]}_{kl} = Q^{[5]}_k + Q^{[\bar{5}]}_l = +Q^{[5]}_k - Q^{[5]}_l \quad . \qquad (1)$$

According to (9.25), the explicit form of the charge matrix is

$$Q^{[24]}_{kl} = \begin{pmatrix} 0 & 0 & 0 & -\frac{4}{3} & -\frac{1}{3} \\ 0 & 0 & 0 & -\frac{4}{3} & -\frac{1}{3} \\ 0 & 0 & 0 & -\frac{4}{3} & -\frac{1}{3} \\ +\frac{4}{3} & +\frac{4}{3} & +\frac{4}{3} & 0 & +1 \\ +\frac{1}{3} & +\frac{1}{3} & +\frac{1}{3} & -1 & 0 \end{pmatrix} \quad . \qquad (2)$$

In comparison with (9.44) it becomes obvious that the gluons as well as the bosons W_3 and B (respectively, Z and the photon) are electrically neutral particles, whereas the X and Y bosons carry charges $+\frac{4}{3}$ and $+\frac{1}{3}$, respectively.

xercise 9.5

Note that the "wave functions" of the $[5] \times [\bar{5}]$ representation are constructed as a direct product out of the wave functions of the [5] representation, that is

$$\begin{pmatrix} \psi_1 \\ \psi_2 \\ \psi_3 \\ \psi_4 \\ \psi_5 \end{pmatrix} ,$$

and of the $[\bar{5}]$ representation, that is $(\bar{\phi}_1, \bar{\phi}_2, \bar{\phi}_3, \bar{\phi}_4, \bar{\phi}_5)$. The order in which the product is taken is essential. The product of the wave functions of the $[\bar{5}]$ with those of the [5] yields a scalar state

$$[\bar{5}] \times [5] = (\bar{\phi}_1, \bar{\phi}_2, \bar{\phi}_3, \bar{\phi}_4, \bar{\phi}_5) \begin{pmatrix} \psi_1 \\ \psi_2 \\ \psi_3 \\ \psi_4 \\ \psi_5 \end{pmatrix} = \sum_i \bar{\phi}_i \psi_i \quad ,$$

while

$$[5] \times [\bar{5}] = \begin{pmatrix} \psi_1 \\ \psi_2 \\ \psi_3 \\ \psi_4 \\ \psi_5 \end{pmatrix} \times (\bar{\phi}_1, \bar{\phi}_2, \bar{\phi}_3, \bar{\phi}_4, \bar{\phi}_5)$$

$$= \begin{pmatrix} \psi_1\bar{\phi}_1 & \psi_1\bar{\phi}_2 & \psi_1\bar{\phi}_3 & \psi_1\bar{\phi}_4 & \psi_1\bar{\phi}_5 \\ \psi_2\bar{\phi}_1 & \psi_2\bar{\phi}_2 & \psi_2\bar{\phi}_3 & \psi_2\bar{\phi}_4 & \psi_2\bar{\phi}_5 \\ \psi_3\bar{\phi}_1 & \psi_3\bar{\phi}_2 & \psi_3\bar{\phi}_3 & \psi_3\bar{\phi}_4 & \psi_3\bar{\phi}_5 \\ \psi_4\bar{\phi}_1 & \psi_4\bar{\phi}_2 & \psi_4\bar{\phi}_3 & \psi_4\bar{\phi}_4 & \psi_4\bar{\phi}_5 \\ \psi_5\bar{\phi}_1 & \psi_5\bar{\phi}_2 & \psi_5\bar{\phi}_3 & \psi_5\bar{\phi}_4 & \psi_5\bar{\phi}_5 \end{pmatrix}$$

is obviously a tensor.

The charge operator for the $[5] \times [\bar{5}]$ states (pay attention to the operator sign ^ on the Q and to its absence!) is

$$\begin{aligned} \hat{Q}\psi_i\bar{\phi}_k &= (\hat{Q}^{[5]}\psi_i)\bar{\phi}_k + \psi_i(\bar{\phi}_k\hat{Q}^{[\bar{5}]}) = Q_i^{[5]}\psi_i\bar{\phi}_k + \psi_i\bar{\phi}_k Q_k^{[\bar{5}]} \\ &= (Q_i^{[5]} + Q_k^{[\bar{5}]})\psi_i\bar{\phi}_k = Q_{ik}\psi_i\bar{\phi}_k \quad . \end{aligned}$$

For the $[\bar{5}] \times [5]$ scalar, it follows similarly that

$$\begin{aligned} \hat{Q}\sum_i \bar{\phi}_i\psi_i &= \sum_i (\bar{\phi}_i\hat{Q}^{[\bar{5}]}\psi_i + \bar{\phi}_i\hat{Q}^{[5]}\psi_i) \\ &= \sum_i (Q_i^{[\bar{5}]} + Q_i^{[5]})\bar{\phi}_i\psi_i = 0 \quad , \end{aligned}$$

because the charges $Q_i^{[\bar{5}]} = -Q_i^{[5]}$. The scalar state carries no charge.

Now we proceed in full analogy to the gauge invariant formulation of $SU(2)_L \times U(1)$ of Example 4.8, but here for the case of SU(5). There the covariant derivative was

$$\begin{aligned}\hat{D}_\mu &= \partial_\mu - \mathrm{i}g A^i_\mu \hat{T}^i - \mathrm{i}\frac{g'}{2} B_\mu \hat{Y} \\ &= \partial_\mu - \mathrm{i}\frac{g}{2} A^i_\mu \hat{\tau}^i - \mathrm{i}\frac{g'}{2} B_\mu \hat{Y} \quad .\end{aligned} \tag{9.43}$$

The 24 gauge bosons A^i_μ $(i = 1, \dots, 24)$ are conveniently represented by a 5×5 matrix. Employing the 24 generators $\tilde{\lambda}_i$ of SU(5), we can write for the *gauge-boson operator*

$$\begin{aligned}\hat{A}_\mu &= \frac{1}{2}\sum_{a=1}^{24} A^a_\mu \tilde{\lambda}_a = \frac{1}{2}\left[\sum_{a=1}^{8} G^a_\mu \tilde{\lambda}_a + \sum_{a=9}^{20} A^a_\mu \tilde{\lambda}_a + \sum_{a=21}^{23} A^a_\mu \tilde{\lambda}_a + B_\mu \tilde{\lambda}_{24}\right] \\ &= \frac{1}{\sqrt{2}}\left(\begin{array}{ccc|cc} & & & X^C_{1\mu} & Y^C_{1\mu} \\ & \frac{1}{\sqrt{2}}\sum_a G^a_\mu \lambda_a & & X^C_{2\mu} & Y^C_{2\mu} \\ & & & X^C_{3\mu} & Y^C_{3\mu} \\ \hline X_{1\mu} & X_{2\mu} & X_{3\mu} & \frac{W^3_\mu}{\sqrt{2}} & W^+_\mu \\ Y_{1\mu} & Y_{2\mu} & Y_{3\mu} & W^-_\mu & -\frac{W^3_\mu}{\sqrt{2}}\end{array}\right) \\ &\quad + \frac{B_\mu}{2\sqrt{15}}\begin{pmatrix} -2 & & & & \\ & -2 & & & \\ & & -2 & & \\ & & & 3 & \\ & & & & 3\end{pmatrix} \quad ,\end{aligned} \tag{9.44}$$

where the λ_a are the familiar Gell-Mann matrices. Note that the singlet field B_μ is required to couple with the operator $\tilde{Y}$ (see (9.22)). The G^a_μ $(a = 1, \dots, 8)$ denote the *8 gluon fields* of SU(3), and $W^+_\mu, W^-_\mu, W^3_\mu, B_\mu$ are the W *and* B *bosons* of the SU(2)×U(1) standard model (see Chap. 4):

$$W^\pm_\mu = \frac{1}{\sqrt{2}}(A^{21}_\mu \mp \mathrm{i}A^{22}_\mu) \quad .$$

Note that in Chap. 4 the isovector fields were denoted as $A^{21}_\mu = A^1_\mu$, $A^{22}_\mu = A^2_\mu$, and $A^{23}_\mu = A^3_\mu$. They should not be confused with the first three fields A^a_μ $(a = 1, 2, 3)$, which are part of the gluon fields (see (9.44)). We shall therefore denote them here as

$$A^{21}_\mu \equiv W^1_\mu \quad , \quad A^{22}_\mu \equiv W^2_\mu \quad , \quad A^{23}_\mu \equiv W^3_\mu$$

or in the corresponding spherical representation W^+_μ, W^-_μ and $W^3_\mu = A^{23}_\mu$. Particular linear combinations of W^3_μ and B_μ (see (4.97)) describe the Z boson and the photon. These are

$$\begin{aligned}A_\mu &= \cos\theta_\mathrm{W} B_\mu + \sin\theta_\mathrm{W} A^3_\mu \quad , \\ Z_\mu &= -\sin\theta_\mathrm{W} B_\mu + \cos\theta_\mathrm{W} A^3_\mu\end{aligned}$$

where θ_W is the Wainberg angle. The X^i_μ and Y^i_μ as well as their charge-conjugated fields are new gauge bosons that occur in SU(5) and are not contained in the standard model.

Relation (9.44) reads in detail:

$$\hat{A}^\mu = \frac{1}{\sqrt{2}} \times \left(\begin{array}{ccc|cc} \frac{A^3_\mu}{\sqrt{2}} + \frac{A^8_\mu}{\sqrt{6}} - \frac{2B_\mu}{\sqrt{30}} & \frac{1}{\sqrt{2}}(A^1_\mu - \mathrm{i}A^2_\mu) & \frac{1}{\sqrt{2}}(A^4_\mu - \mathrm{i}A^5_\mu) & X^C_{1\mu} & Y^C_{1\mu} \\ \frac{1}{\sqrt{2}}(A^1_\mu + \mathrm{i}A^2_\mu) & -\frac{A^3_\mu}{\sqrt{2}} + \frac{A^8_\mu}{\sqrt{6}} - \frac{2B_\mu}{\sqrt{30}} & \frac{1}{\sqrt{2}}(A^6_\mu - \mathrm{i}A^7_\mu) & X^C_{2\mu} & Y^C_{2\mu} \\ \frac{1}{\sqrt{2}}(A^4_\mu + \mathrm{i}A^5_\mu) & \frac{1}{\sqrt{2}}(A^6_\mu + \mathrm{i}A^7_\mu) & -\sqrt{\frac{2}{3}}A^8_\mu - \frac{2B_\mu}{\sqrt{30}} & X^C_{3\mu} & Y^C_{3\mu} \\ \hline X_{1\mu} & X_{2\mu} & X_{3\mu} & \frac{W^3_\mu}{\sqrt{2}} + \frac{3B_\mu}{\sqrt{30}} & W^+_\mu \\ Y_{1\mu} & Y_{2\mu} & Y_{3\mu} & W^-_\mu & -\frac{W^3_\mu}{\sqrt{2}} + \frac{3B_\mu}{\sqrt{30}} \end{array}\right) \tag{9.44a}$$

where

$$\begin{aligned} W^3_\mu &= A^{23}_\mu \quad , \\ W^\pm_\mu &= \frac{1}{\sqrt{2}}(A^{21}_\mu \mp \mathrm{i}A^{22}_\mu) \quad , \\ X_{1\mu} &= \frac{1}{\sqrt{2}}(A^9_\mu + \mathrm{i}A^{10}_\mu) \quad , \quad X^C_{1\mu} = \frac{1}{\sqrt{2}}(A^9_\mu - \mathrm{i}A^{10}_\mu) \quad , \\ X_{2\mu} &= \frac{1}{\sqrt{2}}(A^{11}_\mu + \mathrm{i}A^{12}_\mu) \quad , \quad X^C_{2\mu} = \frac{1}{\sqrt{2}}(A^{11}_\mu - \mathrm{i}A^{12}_\mu) \quad , \\ X_{3\mu} &= \frac{1}{\sqrt{2}}(A^{13}_\mu + \mathrm{i}A^{14}_\mu) \quad , \quad X^C_{3\mu} = \frac{1}{\sqrt{2}}(A^{13}_\mu - \mathrm{i}A^{14}_\mu) \quad , \\ Y_{1\mu} &= \frac{1}{\sqrt{2}}(A^{15}_\mu + \mathrm{i}A^{16}_\mu) \quad , \quad Y^C_{1\mu} = \frac{1}{\sqrt{2}}(A^{15}_\mu - \mathrm{i}A^{16}_\mu) \quad , \\ Y_{2\mu} &= \frac{1}{\sqrt{2}}(A^{17}_\mu + \mathrm{i}A^{18}_\mu) \quad , \quad Y^C_{2\mu} = \frac{1}{\sqrt{2}}(A^{17}_\mu - \mathrm{i}A^{18}_\mu) \quad , \\ Y_{3\mu} &= \frac{1}{\sqrt{2}}(A^{19}_\mu + \mathrm{i}A^{20}_\mu) \quad , \quad Y^C_{3\mu} = \frac{1}{\sqrt{2}}(A^{19}_\mu - \mathrm{i}A^{20}_\mu) \quad . \end{aligned} \tag{9.44b}$$

With these ingredients we are now in a position to make a prediction for the Weinberg angle, within the framework of the unbroken SU(5). The essential point is that the gauge-covariant derivative of the SU(5),

$$\mathrm{i}\hat{D}_\mu = \mathrm{i}\partial_\mu + \frac{g_5}{2}\sum_i A^i_\mu \tilde{\lambda}_i = \mathrm{i}\partial_\mu + g_5 \hat{A}_\mu \quad , \tag{9.45}$$

contains only *a single* coupling constant g_5. This is easily understood by analogy with Example 4.8, (5).[8] From (9.44) we find that the coupling of the W bosons is given by

$$+\frac{g_5}{2}\left(\begin{array}{ccc|c} 0 & & & 0 \\ & 0 & & 0 \\ & & 0 & 0 \\ \hline 0 & 0 & 0 & \sum_i W^i_\mu \tau_i \end{array}\right) , \tag{9.46}$$

whereas the B boson couples to the hypercharge $\hat{Y}$ (9.23) according to

$$+\frac{g_5}{2}\sqrt{\frac{3}{5}}B_\mu \hat{Y} \quad . \tag{9.47}$$

To understand this relation let us consider

$$\begin{aligned}\sum_i W^i_\mu \tau_i &= W^1_\mu \tau_1 + W^2_\mu \tau_2 + W^3_\mu \tau_3 \\ &= W^1_\mu \begin{pmatrix} 0 & 1 \\ 1 & 0 \end{pmatrix} + W^2_\mu \begin{pmatrix} 0 & -\mathrm{i} \\ \mathrm{i} & 0 \end{pmatrix} + W^3_\mu \begin{pmatrix} 1 & 0 \\ 0 & -1 \end{pmatrix} \\ &= \begin{pmatrix} W^3_\mu & W^1_\mu - \mathrm{i}W^2_\mu \\ W^1_\mu + \mathrm{i}W^2_\mu & -W^3_\mu \end{pmatrix} \\ &= \begin{pmatrix} W^3_\mu & \sqrt{2}W^+_\mu \\ \sqrt{2}W^-_\mu & -W^3_\mu \end{pmatrix} \\ &= \sqrt{2}\begin{pmatrix} \frac{W^3_\mu}{\sqrt{2}} & W^+_\mu \\ W^-_\mu & -\frac{W^3_\mu}{\sqrt{2}} \end{pmatrix} \quad .\end{aligned}$$

By means of (9.45) and (9.44a) it therefore follows that, for the SU(2) submatrix,

$$(+g_5\hat{A}_\mu) = +\frac{g_5}{2}\left(\begin{array}{c|c} - & \\ \hline & \sum_i W^i_\mu \tau_i \end{array}\right) \tag{9.46a}$$

$$= +g_5\frac{1}{\sqrt{2}}\left(\begin{array}{c|cc} - & & \\ \hline & \frac{W^3_\mu}{\sqrt{2}} & W^+_\mu \\ & W^-_\mu & -\frac{W^3_\mu}{\sqrt{2}} \end{array}\right) , \tag{9.46b}$$

[8] See also W. Greiner, S. Schramm and E. Stein: *Quantum Chromodynamics*, 2nd ed. (Springer, Berlin, Heidelberg, New York, 2000).

which is the result stated in (9.44). Similarly the coupling of the B_μ field is obtained from (9.44a) and (9.45) – see also the last equation of Exercise 9.1 –

$$(+g_5\hat{A}_\mu)_{B_\mu\text{-coupling}} = \frac{g_5}{2}B_\mu\hat{\tilde{Y}} = +g_5\frac{B_\mu}{2\sqrt{15}}\begin{pmatrix} -2 & & & & \\ & -2 & & & \\ & & -2 & & \\ & & & 3 & \\ & & & & 3 \end{pmatrix}$$

$$= +g_5\frac{B_\mu}{2}\sqrt{\frac{3}{5}}\begin{pmatrix} -\frac{2}{3} & & & & \\ & -\frac{2}{3} & & & \\ & & -\frac{2}{3} & & \\ & & & 1 & \\ & & & & 1 \end{pmatrix}$$

$$= +\frac{g_5}{2}\sqrt{\frac{3}{5}}B_\mu\hat{Y} = \frac{g_5}{2}B_\mu\tilde{Y} \quad , \tag{9.47a}$$

where we have used (9.23).

The coupling of the fermion fields ψ is determined according to (9.45) via minimal coupling; namely

$$\bar{\psi}\gamma^\mu \mathrm{i}D_\mu\psi = \bar{\psi}\gamma^\mu(\mathrm{i}\partial_\mu + g_5\hat{A}_\mu)\psi = \bar{\psi}\gamma^\mu(\mathrm{i}\partial_\mu)\psi + g_5\bar{\psi}\gamma^\mu\hat{A}_\mu\psi \quad .$$

The last term represents the interaction. In correspondence with (9.46) and (9.47a) it contains the coupling to the W_μ and B_μ bosons, namely

$$L_{\text{int}} = +g_5\bar{\psi}\gamma^\mu\hat{A}_\mu\psi \to \left(\bar{\psi}\gamma^\mu\left[\underbrace{\frac{g_5}{2}\sum_i W^i_\mu\hat{\tau}_i}_{\text{W sector}} + \underbrace{\frac{g_5}{2}\sqrt{\frac{3}{5}}B_\mu\hat{Y}}_{\text{B sector}}\right]\psi\right)$$

$$= \left(\bar{\psi}\gamma^\mu\left[g_5\sum_i W^i_\mu\hat{T}_i + \frac{g_5}{2}\sqrt{\frac{3}{5}}B_\mu\hat{Y}\right]\psi\right) \quad .$$

From a comparison with (4.96), (4.113), or Example 4.8, (15) (note that in Chap. 4 the W^i_μ field is denoted by A^i_μ), the coupling constants are obtained. Those equations were obtained for the SU(2)×U(1) theory, with the corresponding gauge-covariant derivative $\hat{D}_\mu = \mathrm{i}\partial_\mu + g\hat{\boldsymbol{T}}\cdot\boldsymbol{A}_\mu + \frac{g'}{2}YB_\mu$ (see (4.112)).

Thus, we identify the coupling constant g and g' of the Glashow–Salam–Weinberg theory as

$$g = g_5 \quad , \quad g' = \sqrt{\frac{3}{5}}g_5 \quad . \tag{9.48}$$

Consequently the prediction of the Weinberg angle (4.103) within the SU(5) symmetric gauge theory is

$$\sin^2\theta_{\mathrm{W}} = \frac{g'^2}{g^2+g'^2} = \frac{3}{8} = 0.375 \quad . \tag{9.49}$$

Obviously this value deviates from the experimental result $\sin^2\theta_{\mathrm{W}}^{\exp} \approx 0.23$. In order to understand the considerable disagreement, note that we have obtained the Weinberg angle on the assumption of an unbroken SU(5) gauge theory. In reality this cannot be the case, since we already know that the SU(2)×U(1) symmetry of the GSW theory is spontaneously broken by the Higgs field.

To date, moreover, there are no experimental indications of the influence of the additional gauge fields X and Y. This fact leads to the hypothesis that these fields must have significantly larger masses than the bosons $W^{\pm}$ and Z^0, that is to say, the SU(5) symmetry is assumed to be realized only at very high energies. (As we shall see in the following sections, the corresponding breaking of the symmetry must happen at about 10^{15} GeV, in order not to conflict with the well-known life time of the proton.[9] In other words, below the energy threshold of about 10^{15} GeV SU(5) is a broken symmetry. Only for larger energies it is exact. However, in quantum field theory the coupling constant g depends on the considered energy (or more specifically on the momentum transfer), as we shall discuss below in Sect. 9.5. For energies lower then the mass of the X and Y bosons the energy dependence of the coupling is different for the particular subgroups SU(3), SU(2), and U(1), which consequently yields an energy-dependent Weinberg angle. Therefore it is quite possible, and indeed we shall confirm this in Sect. 9.5 – see (9.103) and Fig. 9.4 – that $\sin^2\theta_W \approx 0.23$ for the lower-energy region experimentally accessible to date, whereas for very large energies $\sin^2\theta_W$ can reach a value of $\frac{3}{8}$, since the symmetry breaking can then be neglected. From this requirement a prediction for the mass of the bosons X and Y can be obtained that must be compared with the bound from proton lifetime measurements.

In order to understand why the X and Y boson in principle can lead to the decay of the proton, let us consider, for example, the gauge field X_r $(= X_1)$. According to (9.44) it couples to the 5×5 matrix

$$\hat{M}(X_{r\mu}) = \begin{pmatrix} 0 & 0 & 0 & 0 & 0 \\ 0 & 0 & 0 & 0 & 0 \\ 0 & 0 & 0 & 0 & 0 \\ 1 & 0 & 0 & 0 & 0 \\ 0 & 0 & 0 & 0 & 0 \end{pmatrix} X_{r\mu} \; . \tag{9.50}$$

Applying this matrix to the elementary quintuplet of SU(5) – see (9.26), we obtain

$$g_5 \overline{\begin{pmatrix} d_r \\ d_b \\ d_g \\ e^+ \\ -\nu_e^c \end{pmatrix}} \gamma^\mu \hat{A}_\mu \begin{pmatrix} d_r \\ d_b \\ d_g \\ e^+ \\ -\nu_e^c \end{pmatrix} = \overline{\begin{pmatrix} d_r \\ d_b \\ d_g \\ e^+ \\ -\nu_e^c \end{pmatrix}}_R \gamma^\mu \hat{M}(X_{r\mu}) \begin{pmatrix} d_r \\ d_b \\ d_g \\ e^+ \\ -\nu_e^c \end{pmatrix}_R$$
$$= \begin{pmatrix} d_r \\ d_b \\ d_g \\ e^+ \\ -\nu_e^c \end{pmatrix}_R^\dagger \gamma_0 \gamma^\mu \hat{M}(X_{r\mu}) \begin{pmatrix} d_r \\ d_b \\ d_g \\ e^+ \\ -\nu_e^c \end{pmatrix}_R$$

---

[9] Recently obtained experimental data require an even higher limit, so that the SU(5) theory in its simple form described above is, in fact, ruled out. Nevertheless it is extremely interesting to see by the example of SU(5) how the idea of grand unification works.

$$= (\bar{d}_r, \bar{d}_b, \bar{d}_g, \bar{e}^+, -\bar{\nu}_e^c)_R \gamma^\mu \hat{M}(X_{r\mu}) \begin{pmatrix} d_r \\ d_b \\ d_g \\ e^+ \\ -\nu_e^c \end{pmatrix}_R$$

$$= \bar{e}_R^+ \gamma^\mu (X_{r\mu})_R (d_r)_R \quad . \tag{9.50a}$$

The d_r quark couples now directly to the positron, that is the d_r quark is transformed into a positron. To understand that the X_r field can transform a d_r quark into a positron, in diagrammatic representation, consider the reaction

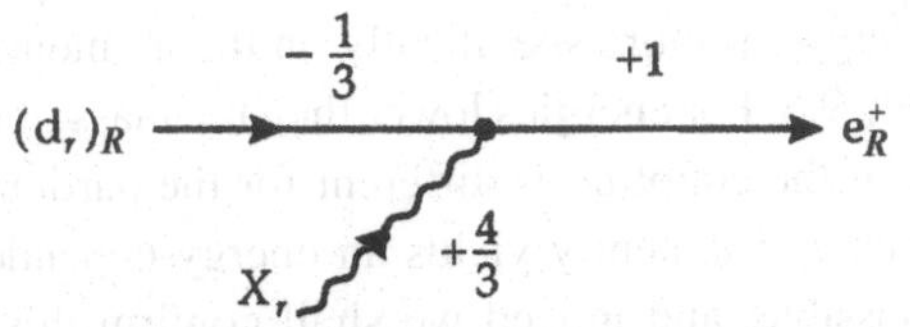

(9.50b)

The charges of the particles are indicated. They are conserved at the vertex. Many other similar processes are possible. Indeed, if we apply the $\hat{M}(X_r)$ matrix to the decuplet $\psi_{ij}^{[10]}$ of (9.37), we have (for convenience we drop here the space-time index μ at $X_{r\mu}$ and γ^μ from the coupling $g_5\gamma^\mu \hat{A}_\mu$ and insert them later again)

$$\hat{M}(X_r) \begin{pmatrix} 0 & u_g^C & -u_b^C & -u_r & -d_r \\ -u_g^C & 0 & u_r^C & -u_b & -d_b \\ u_b^C & -u_r^C & 0 & -u_g & -d_g \\ u_r & u_b & u_g & 0 & -e^+ \\ d_r & d_b & d_g & e^+ & 0 \end{pmatrix}_L$$

$$= X_r \begin{pmatrix} 0 & 0 & 0 & 0 & 0 \\ 0 & 0 & 0 & 0 & 0 \\ 0 & 0 & 0 & 0 & 0 \\ 0 & u_g^C & -u_b^C & -u_r & -d_r \\ 0 & 0 & 0 & 0 & 0 \end{pmatrix}_L \quad . \tag{9.51}$$

Next we evaluate the interaction as follows:

$$\mathrm{Tr}\{\bar{\psi}_L^{[10]} M \psi_L^{[10]}\}$$

$$= \mathrm{Tr} \begin{pmatrix} 0 & -\bar{u}_g^C & \bar{u}_b^C & \bar{u}_r & \bar{d}_r \\ \bar{u}_g^C & 0 & -\bar{u}_r^C & \bar{u}_b & \bar{d}_b \\ -\bar{u}_b^C & \bar{u}_r^C & 0 & \bar{u}_g & \bar{d}_g \\ -\bar{u}_r & -\bar{u}_b & -\bar{u}_g & 0 & \bar{e}^+ \\ -\bar{d}_r & -\bar{d}_b & -\bar{d}_g & -\bar{e}^+ & 0 \end{pmatrix}_L \begin{pmatrix} 0 & 0 & 0 & 0 & 0 \\ 0 & 0 & 0 & 0 & 0 \\ 0 & 0 & 0 & 0 & 0 \\ 0 & u_g^C & -u_b^C & -u_r & -d_r \\ 0 & 0 & 0 & 0 & 0 \end{pmatrix}_L X_r$$

$$= \left[0 + (\bar{u}_b)_L (u_g^C)_L - (\bar{u}_g)_L (u_b^C)_L + (\bar{e}^+)_L (d_r)_L\right] X_r \quad . \tag{9.52}$$

We shall justify this form of interaction in Example 9.6. Remember that we dropped the space-time index μ and the γ^μ. If we insert them, (9.52) becomes

$$\left[(\bar{u}_b)_L \gamma^\mu (u_g^C)_L - (\bar{u}_g)_L \gamma^\mu (u_b^C)_L + (e^+)\gamma^\mu (d_r)_L\right] X_{r\mu} \tag{9.52a}$$

The terms (9.52) correspond to the following diagrams:

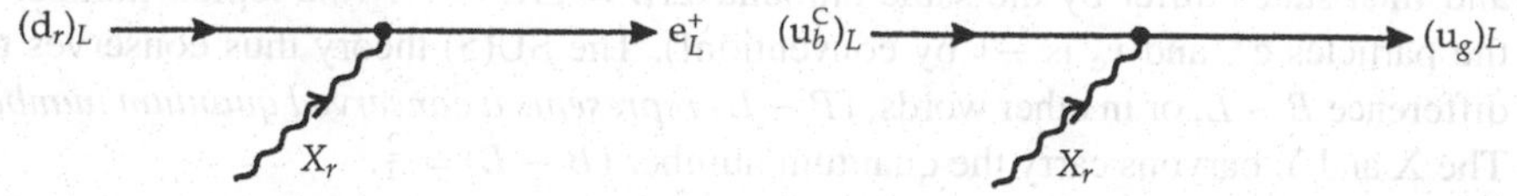

(9.51a)

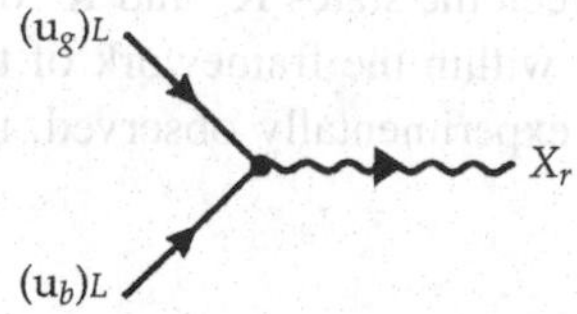

The first diagram, corresponding to the last term of (9.52), is identical to the process of (9.50b). As we shall see in Exercise 9.7, the $X_{i\mu}$ and $Y_{i\mu}$ bosons carry *anticolor* whereas the $X^c_{i\mu}$ and $Y^c_{i\mu}$ bosons carry color. With that in mind one easily verifies that at each vertex both the electric charges and the color charges are conserved. The fundamental reason why the X boson can transform a u quark into a u antiquark is that it couples to the SU(5) fermion decuplet, which contains quark fields as well as their charge-conjugate fields. These differ in their sign of the baryon number (+1/3 and −1/3, respectively), and therefore the decuplet representation is not an eigenstate of the baryon number operator. For the Y field we find by similar considerations the processes

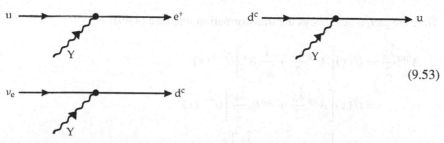

(9.53)

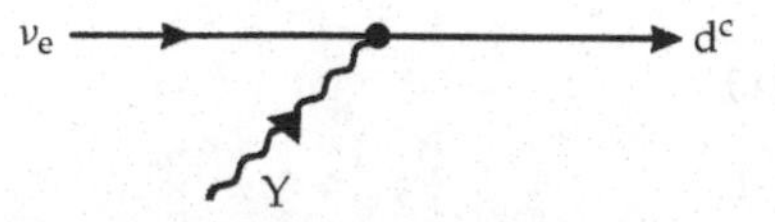

Since the X and Y bosons can transform quarks into leptons, they are sometimes also referred to as *leptoquarks*. By combining the particular interactions between quarks, leptons, and X and Y bosons (leptoquarks), it is straightforward to write down several processes that describe the decay of the proton (see Fig. 9.1). Let us consider the first graph of Fig. 9.1 more closely. The annihilation of two u quarks to form an X boson (charge +4/3) is contained in the first term of the interaction (9.52). One should notice here that the u^C field, a fermion field, annihilates u antiquarks and creates u quarks. The annihilation of the X leptoquark (charge +4/3) is similarly contained in the last term of (9.52). Note that fermion fields contain both particles and antiparticles. This will be explicitly described in Sect. 9.6 when we deal with proton decay. It is, in fact similar to the case of QED, where the interaction $(\bar{\psi}\gamma^\mu\psi)A_\mu$, ψ being the electron–positron field, contains all the following processes

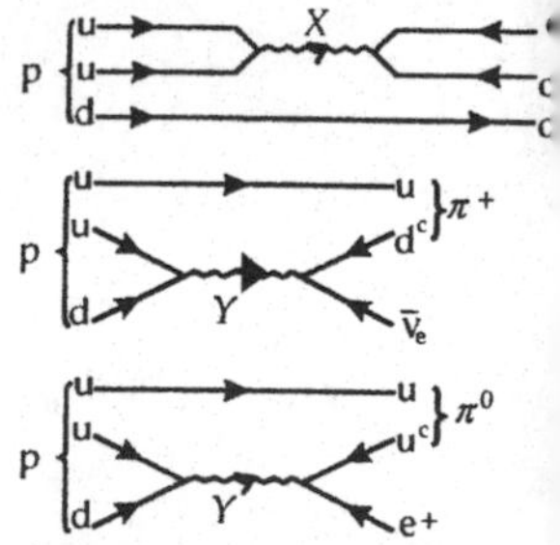

Fig. 9.1. Feynman diagrams c scribing proton decay

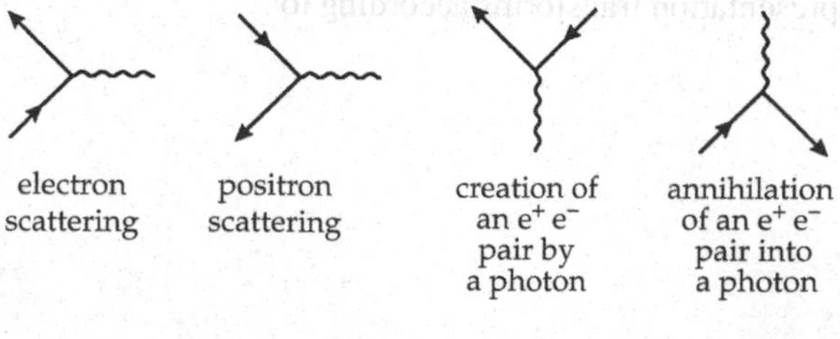

All these processes have in common that the baryon and lepton numbers of the initial and final states differ by the same amount: $\Delta B = \Delta L = -1$ (the lepton number of the particles e^+ and $\bar{\nu}_e$ is -1 by convention!). The SU(5) theory thus conserves the difference $B - L$, or in other words, *$(B - L)$ represents a conserved quantum number*. The X and Y baryons carry the quantum number $(B - L) = \frac{2}{3}$.

Processes such as the transformation of a neutron into an antineutron, so-called N–$\bar{\text{N}}$ oscillations (by analogy with the oscillations between the states K^0 and $\bar{K}^0$ due to the weak interaction), cannot therefore be described within the framework of the SU(5) theory. If the existence of such transitions was experimentally observed, the SU(5) theory must be rejected in principle.

EXAMPLE

9.6 Construction of the Lagrangian

We now have the ingredients necessary to formulate the Lagrangian of SU(5) gauge field theory. First, we consider transforming the [10] representation under a unitary transformation

$$\hat{U} = e^{+ig_5\theta_a(x)\tilde{\lambda}^a/2} \quad . \tag{1}$$

Here the gauge field obeys the transformation rule (see (4.49))

$$\begin{aligned} A'^{\mu a}\frac{\tilde{\lambda}_a}{2} &= \hat{U}(x)\left[A^{\mu b}\frac{\tilde{\lambda}_b}{2} + \frac{i}{g_5}\partial^\mu\right]\hat{U}^{-1}(x) \\ &= \hat{U}(x)\left[A^{\mu b}\frac{\tilde{\lambda}_b}{2} + \partial^\mu\theta_b\frac{\tilde{\lambda}_b}{2}\right]\hat{U}^{-1}(x) \\ &= \hat{U}(x)\left[(A^{\mu b} + \partial^\mu\theta_b)\frac{\tilde{\lambda}_b}{2}\right]\hat{U}^{-1}(x) \quad , \end{aligned} \tag{2}$$

which can also be written as

$$\hat{A}^{\mu\prime} = \hat{U}\hat{A}^\mu\hat{U}^{-1} + \frac{i}{g_5}\hat{U}\partial^\mu\hat{U}^{-1}(x) \quad .$$

Here $\hat{A}^\mu = A^{\mu a}\frac{\hat{\lambda}_a}{2}$ is the *vector field operator*. The transformation of the wave function in the [10] representation is determined by remembering that the [10] representation is based on the [5] representation,

$$\Psi^{[10]} = (\Psi^{[5]} \otimes \Psi^{[5]})_{\text{anti}} \quad . \tag{3}$$

The wave function in [5] representation transforms according to

$$\Psi^{[5]\prime} = \hat{U}(x)\Psi^{[5]} \quad , \tag{4}$$

or, in components,

$$\Psi^{[5]\prime}_i = U(x)_{ij}\Psi^{[5]}_j \quad . \tag{5}$$

Thus we get

$$
\begin{aligned}
\Psi^{[10]\prime}_{ik} &= (\Psi^{[5]\prime}_{i} \otimes \Psi^{[5]\prime}_{k})_{\text{anti}} \\
&= (\Psi^{[5]\prime}_{i}\Psi^{[5]\prime}_{k} - \Psi^{[5]\prime}_{k}\Psi^{[5]\prime}_{i}) \\
&= \left(U(x)_{ij}\Psi^{[5]}_{j}U(x)_{kl}\Psi^{[5]}_{l} - U(x)_{kl}\Psi^{[5]}_{l}U(x)_{ij}\Psi^{[5]}_{j}\right) \\
&= U(x)_{ij}\left(\Psi^{[5]}_{j} \otimes \Psi^{[5]}_{l}\right)_{\text{anti}} U(x)_{kl} \\
&= U(x)_{ij}\Psi^{[10]}_{jl}U^{\text{t}}_{lk}(x)
\end{aligned} \tag{6}
$$

for the wave function in [10] representation. This can be written more concisely as

$$
\Psi^{[10]\prime} = \hat{U}(x)\Psi^{[10]}\hat{U}^{\text{t}}(x) \quad , \tag{7}
$$

where the superscript "t" denotes matrix transposition. The covariant derivative is determined such that additional terms, because of the transformation of the gauge field (2), just cancel those terms that arise from application of the ordinary four-gradient to the primed wave function (7). To see this, we consider

$$
\begin{aligned}
\partial_\mu\Psi^{[10]\prime}(x) &= \partial_\mu(\hat{U}(x)\Psi^{[10]}(x)\hat{U}^{\text{t}}(x)) \\
&= (\partial_\mu\hat{U}(x))\Psi^{[10]}(x)\hat{U}^{\text{t}}(x) + \hat{U}(x)(\partial_\mu\Psi^{[10]})\hat{U}^{\text{t}}(x) \\
&\quad + \hat{U}(x)\Psi^{[10]}(x)\left(\partial_\mu\hat{U}^{\text{t}}(x)\right) \\
&= \hat{U}(x)\left(\partial_\mu\Psi^{[10]}(x)\right)\hat{U}^{\text{t}}(x) \\
&\quad + \hat{U}(x)\left\{\mathrm{i}g_5\partial_\mu\theta_a\frac{\tilde{\lambda}^a}{2}\Psi^{[10]} + \Psi^{[10]}\mathrm{i}g_5\partial_\mu\theta_a\frac{\tilde{\lambda}^{a\,\text{t}}}{2}\right\}\hat{U}^{\text{t}}(x) \quad .
\end{aligned} \tag{8}
$$

The expression in braces is compensated for by formulating the covariant derivative of the wave function in [10] representation as follows:

$$
D_\mu\Psi^{[10]} = \partial_\mu\Psi^{[10]} - \mathrm{i}g_5\left\{A^a_\mu\frac{\tilde{\lambda}_a}{2}\Psi^{[10]} + \Psi^{[10]}A^a_\mu\frac{\tilde{\lambda}^{\text{t}}_a}{2}\right\} \quad . \tag{9}
$$

The transformation properties of this covariant derivative under gauge transformation (1, 2) can easily be derived:

$$
\begin{aligned}
D'_\mu\Psi^{[10]\prime} &= \partial_\mu\Psi^{[10]\prime} - \mathrm{i}g_5\left\{A'^a_\mu\frac{\tilde{\lambda}_a}{2}\Psi^{[10]\prime} + \Psi^{[10]\prime}A'^a_\mu\frac{\tilde{\lambda}^{\text{t}}_a}{2}\right\} \\
&= \hat{U}(x)\left(\partial_\mu\Psi^{[10]}(x)\right)\hat{U}(x)^{\text{t}} \\
&\quad + \hat{U}(x)\left\{\underline{\mathrm{i}g_5\partial_\mu\theta_a\frac{\tilde{\lambda}^a}{2}\Psi^{[10]}} + \underline{\Psi^{[10]}\mathrm{i}g_5\partial_\mu\theta_a\frac{\tilde{\lambda}^{a\,\text{t}}}{2}}\right\}\hat{U}^{\text{t}}(x) \\
&\quad - \mathrm{i}g_5\left\{\hat{U}(x)\left[A^{\mu b}\frac{\tilde{\lambda}_b}{2} + \underline{\left(\partial^\mu\theta^b\frac{\tilde{\lambda}_b}{2}\right)}\right]\hat{U}^{-1}(x)\hat{U}(x)\Psi^{[10]}\hat{U}(x)^{\text{t}}\right. \\
&\quad \left. + \hat{U}(x)\Psi^{[10]}\hat{U}^{\text{t}}(x)\hat{U}^{-1\text{t}}(x)\left[A^{\mu b}\frac{\tilde{\lambda}^t_b}{2} + \underline{\left(\partial^\mu\theta^b\frac{\tilde{\lambda}^{\text{t}}_b}{2}\right)}\right]\hat{U}^{\text{t}}(x)\right\}
\end{aligned}
$$

Example 9.6

$$= \hat{U}(x)\left(\partial_\mu \Psi^{[10]} - \mathrm{i}g_5\left\{A_\mu^a \frac{\tilde{\lambda}_a}{2}\Psi^{[10]} + \Psi^{[10]}A_\mu^a\frac{\tilde{\lambda}_a^{\mathrm{t}}}{2}\right\}\right)\hat{U}^{\mathrm{t}}(x)$$
$$= \hat{U}(x)D_\mu\Psi^{[10]}\hat{U}^{\mathrm{t}}(x) \quad . \tag{10}$$

The _ _ _ and ... terms cancel. Thus, under gauge transformations, the covariant derivative $D_\mu\Psi^{[10]}$ transforms in the same way as $\Psi^{[10]}$ – compare (7) and (10). If the Lagrange function is defined as

$$\mathcal{L}_{\mathrm{kin}}^{[10]} = \mathrm{Tr}\{\bar{\Psi}^{[10]}\gamma^\mu D_\mu \Psi^{[10]}\} \quad , \tag{11}$$

and (10) is used, we get

$$\begin{aligned}\mathcal{L}_{\mathrm{kin}}^{[10]\prime} &= \mathrm{Tr}\{\bar{\Psi}^{[10]\prime}\, \not{D}'\Psi^{[10]\prime}\}\\ &= \mathrm{Tr}\left\{\left(\hat{U}(x)\Psi^{[10]}\hat{U}^t(x)\right)^\dagger\gamma_0\hat{U}(x)\,\not{D}\Psi^{[10]}\hat{U}(x)^t\right\}\\ &= \mathrm{Tr}\{\hat{U}^{-1t}(x)\Psi^{[10]\dagger}\gamma_0\hat{U}^{-1}(x)\hat{U}(x)\,\not{D}\Psi^{[10]}\hat{U}(x)^t\}\\ &= \mathrm{Tr}\{\bar{\Psi}^{[10]}\,\not{D}\Psi^{[10]}\} = \mathcal{L}_{\mathrm{kin}}^{[10]} \quad .\end{aligned} \tag{12}$$

Thus, the kinetic part of the SU(5) Lagrange function for coupling gauge fields to fermion wave functions in [10] representation is invariant under gauge transformation (1, 2). In our notation the Lagrange function for coupling the SU(5) quintuplet wave function to the gauge fields reads

$$\mathcal{L}_{\mathrm{kin}}^{[5]} = \bar{\Psi}^{[5]}\,\not{D}\Psi^{[5]} \quad , \tag{13}$$

and one easily checks its invariance under the gauge transformations (4) and (2). Indeed we know from (4) that

$$\Psi^{[5]\prime} = \hat{U}\Psi^{[5]}$$

and from (9.45)

$$\hat{D}_\mu = \partial_\mu - \mathrm{i}g_5\hat{A}_\mu \quad . \tag{14}$$

For the covariant derivation $\hat{D}_\mu\Psi^{[5]}$ it is requested that

$$(\hat{D}_\mu\Psi^{[5]})' = \hat{U}(\hat{D}_\mu\Psi^{[5]}) \quad , \tag{15}$$

where the prime indicates the gauge transformed quantity. This yields

$$\begin{aligned}\hat{D}'\Psi^{[5]\prime} &= (\partial_\mu - \mathrm{i}g_5\hat{A}'_\mu)\hat{U}\Psi^{[5]}\\ &= \hat{U}(\partial_\mu + \hat{U}^{-1}\partial_\mu\hat{U} - \mathrm{i}g_5\hat{U}^{-1}\hat{A}'_\mu\hat{U})\Psi^{[5]} \quad .\end{aligned} \tag{16}$$

Comparing this with (15) and (14) allows us to rewrite (16) in the form

$$\begin{aligned}D'_\mu\Psi^{[5]\prime} &= \hat{U}(D_\mu\Psi^{[5]})\\ &= \hat{U}\left((\partial_\mu - \mathrm{i}g_5\hat{A}_\mu)\Psi^{[5]}\right) \quad ,\end{aligned} \tag{16a}$$

and we conclude

$$\hat{A}'_\mu = \hat{U}\hat{A}_\mu\hat{U}^{-1} + \frac{\mathrm{i}}{g_5}\hat{U}\partial_\mu\hat{U}^{-1} \quad , \tag{17}$$

which is identical with (2), as it should be. Perhaps it is useful to follow the derivation of (17) in detail: Comparing (16) and (16a) yields

$$\hat{U}^{-1}\partial_\mu\hat{U} - \mathrm{i}g_5\hat{U}^{-1}\hat{A}'_\mu\hat{U} = -\mathrm{i}g_5\hat{A}_\mu \quad ,$$

and from that follows

$$-\mathrm{i}g_5\hat{U}^{-1}\hat{A}'_\mu\hat{U} = -\mathrm{i}g_5\hat{A}_\mu - \hat{U}^{-1}\partial_\mu\hat{U} \quad ,$$

$$\hat{U}^{-1}\hat{A}'_\mu\hat{U} = \hat{A}_\mu - \frac{\mathrm{i}}{g_5}\hat{U}^{-1}\partial_\mu\hat{U} \quad ,$$

$$\hat{A}'_\mu\hat{U} = \hat{U}\hat{A}_\mu - \frac{\mathrm{i}}{g_5}\partial_\mu\hat{U} \quad ,$$

$$\hat{A}'_\mu = \hat{U}\hat{A}_\mu\hat{U}^{-1} - \frac{\mathrm{i}}{g_5}(\partial_\mu\hat{U})\hat{U}^{-1} \quad .$$

Now, $\hat{U}\hat{U}^{-1} = \mathbb{1}$, and hence

$$(\partial_\mu\hat{U})\hat{U}^{-1} + \hat{U}(\partial_\mu\hat{U}^{-1}) = 0$$

or

$$(\partial_\mu\hat{U})\hat{U}^{-1} = -\hat{U}(\partial_\mu\hat{U}^{-1}) \quad ,$$

and thus,

$$\hat{A}'_\mu = \hat{U}\hat{A}_\mu\hat{U}^{-1} + \frac{\mathrm{i}}{g_5}\hat{U}(\partial_\mu\hat{U}^{-1}) \quad .$$

For the Lagrangian density $\mathcal{L}^{[5]\prime}_{\mathrm{int}}$ – see (13) – we now have

$$\begin{aligned}\mathcal{L}^{[5]\prime}_{\mathrm{int}} &= \overline{\Psi^{[5]\prime}}(\hat{D}\psi^{[5]})' \\ &= \overline{\hat{U}\Psi^{[5]}}\hat{U}\,\hat{\not{D}}\Psi^{[5]} \\ &= \bar{\Psi}^{[5]}\,\hat{\not{D}}\Psi^{[5]} \quad ,\end{aligned} \tag{18}$$

where $\hat{U}^{+} = \hat{U}^{-1}$ has been utilized. In total we then get

$$\mathcal{L}_{\mathrm{kin}} = \mathrm{Tr}\{\bar{\Psi}^{[10]}\,\not{D}\Psi^{[10]}\} + \bar{\Psi}^{[5]}\,\not{D}\Psi^{[5]} \quad . \tag{19}$$

Note that $\not{D}$ has a different form when it acts on $\Psi^{[10]}$ (see (9)) and on $\Psi^{[5]}$ (see (9.45)). In the following we will need the coupling terms of the interaction Lagrangian

$$\begin{aligned}\mathcal{L}_{\mathrm{int}} = &-\mathrm{i}g_5\mathrm{Tr}\left[\bar{\Psi}^{[10]}\,\not{A}_a\frac{\tilde{\lambda}^a}{2}\Psi^{[10]} + \bar{\Psi}^{[10]}\gamma_\mu\Psi^{[10]}\left(\frac{\tilde{\lambda}_a}{2}\right)^{\mathrm{t}}A^{\mu a}\right] \\ &- ig_5\bar{\Psi}^{[5]}\,\not{A}_a\frac{\tilde{\lambda}_a}{2}\Psi^{[5]} \quad .\end{aligned} \tag{20}$$

xample 9.6

Substituting the full expressions for the wave functions (see (9.28), (9.37), (9.44)) into $\mathcal{L}$, we get (only the part connected to the X and Y bosons is of interest here)

$$
\begin{aligned}
\mathcal{L}_{\text{int}} = -\frac{\mathrm{i}g_5}{\sqrt{8}}\mathrm{Tr}\Biggl\{ & \begin{pmatrix} 0 & \overline{u_3^C} & -\overline{u_2^C} & -\overline{u_1} & -\overline{d_1} \\ -\overline{u_3^C} & 0 & \overline{u_1^C} & -\overline{u_2} & -\overline{d_2} \\ \overline{u_2^C} & -\overline{u_1^C} & 0 & -\overline{u_3} & -\overline{d_3} \\ \overline{u_1} & \overline{u_2} & \overline{u_3} & 0 & -\overline{e^C} \\ \overline{d_1} & \overline{d_2} & \overline{d_3} & \overline{e^C} & 0 \end{pmatrix}_L^{\mathrm{T}} \\
& \times \Biggl[\begin{pmatrix} 0 & 0 & 0 & \not{X}_1^C & \not{Y}_1^C \\ 0 & 0 & 0 & \not{X}_2^C & \not{Y}_2^C \\ 0 & 0 & 0 & \not{X}_3^C & \not{Y}_3^C \\ \not{X}_1 & \not{X}_2 & \not{X}_3 & 0 & 0 \\ \not{Y}_1 & \not{Y}_2 & \not{Y}_3 & 0 & 0 \end{pmatrix} \begin{pmatrix} 0 & u_3^C & -u_2^C & -u_1 & -d_1 \\ -u_3^C & 0 & u_1^C & -u_2 & -d_2 \\ u_2^C & -u_1^C & 0 & -u_3 & -d_3 \\ u_1 & u_2 & u_3 & 0 & -e^C \\ d_1 & d_2 & d_3 & e^C & 0 \end{pmatrix}_L \\
& + \begin{pmatrix} 0 & u_3^C & -u_2^C & -u_1 & -d_1 \\ -u_3^C & 0 & u_1^C & -u_2 & -d_2 \\ u_2^C & -u_1^C & 0 & -u_3 & -d_3 \\ u_1 & u_2 & u_3 & 0 & -e^C \\ d_1 & d_2 & d_3 & e^C & 0 \end{pmatrix}_L \\
& \times \begin{pmatrix} 0 & 0 & 0 & \not{X}_1 & \not{Y}_1 \\ 0 & 0 & 0 & \not{X}_2 & \not{Y}_2 \\ 0 & 0 & 0 & \not{X}_3 & \not{Y}_3 \\ \not{X}_1^C & \not{X}_2^C & \not{X}_3^C & 0 & 0 \\ \not{Y}_1^C & \not{Y}_2^C & \not{Y}_3^C & 0 & 0 \end{pmatrix} \Biggr] \Biggr\} \\
& - \frac{\mathrm{i}g_5}{\sqrt{2}} \begin{pmatrix} \overline{d_1} & \overline{d_2} & \overline{d_3} & \overline{e^C} & -\overline{\nu_{\mathrm{e}}^C} \end{pmatrix}_R \\
& \times \begin{pmatrix} 0 & 0 & 0 & \not{X}_1^C & \not{Y}_1^C \\ 0 & 0 & 0 & \not{X}_2^C & \not{Y}_2^C \\ 0 & 0 & 0 & \not{X}_3^C & \not{Y}_3^C \\ \not{X}_1 & \not{X}_2 & \not{X}_3 & 0 & 0 \\ \not{Y}_1 & \not{Y}_2 & \not{Y}_3 & 0 & 0 \end{pmatrix} \begin{pmatrix} d_1 \\ d_2 \\ d_3 \\ e^C \\ -\nu_{\mathrm{e}}^C \end{pmatrix}_R \ .
\end{aligned} \tag{21}
$$

Carrying out matrix multiplications gives

$$
\begin{aligned}
\mathcal{L}_{\text{int}} = -\frac{\mathrm{i}g_5}{\sqrt{8}}\cdot 2\cdot \Bigl[& \overline{u_3^C}\not{X}_2^C u_1 - \overline{u_2^C}\not{X}_3^C u_1 - \overline{u_3^C}\not{X}_1^C u_2 \\
& + \overline{u_1^C}\not{X}_3^C u_2 + \overline{u_2^C}\not{X}_1^C u_3 - \overline{u_1^C}\not{X}_2^C u_3 \\
& - \overline{u_3}\not{X}_2 u_1^C + \overline{u_2}\not{X}_3 u_1^C + \overline{u_3}\not{X}_1 u_2^C \\
& - \overline{u_1}\not{X}_3 u_2^C - \overline{u_2}\not{X}_1 u_3^C + \overline{u_1}\not{X}_2 u_3^C \\
& + \overline{u_3^C}\not{Y}_2^C d_1 - \overline{u_2^C}\not{Y}_3^C d_1 - \overline{u_3^C}\not{Y}_1^C d_2
\end{aligned}
$$

Example 9.

$$
\begin{aligned}
&+\overline{u_1^C}\not{Y}_3^C d_2+\overline{u_2^C}\not{Y}_1^C d_3-\overline{u_1^C}\not{Y}_2^C d_3\\
&+\overline{d_1}\not{Y}_2 u_3^C-\overline{d_1}\not{Y}_3 u_2^C-\overline{d_2}\not{Y}_1 u_3^C\\
&+\overline{d_2}\not{Y}_3 u_1^C+\overline{d_3}\not{Y}_1 u_2^C-\overline{d_3}\not{Y}_2 u_1^C\\
&+\sum_j\left(\overline{u_j}\not{Y}_j^C e^C+\overline{e^C}\not{Y}_j u_j\right)\\
&-\sum_j\left(\overline{d_j}\not{X}_j^C e^C+\overline{e^C}\not{X}_j d_j\right)\Bigg]_L\\
&-\frac{ig_5}{\sqrt{2}}\sum_j[\bar{e}^C\not{X}_i d_i-\bar{\nu}_e^C\not{Y}_i d_i+\bar{d}_i\not{X}_i^C e^C-\bar{d}_i\not{Y}_i^C\nu_e^C]_R\quad. \qquad (22)
\end{aligned}
$$

To make this expression more readable, we have used the Feynman "daggers" in $\not{X}=\gamma_\mu X^\mu$ and $\not{Y}=\gamma_\mu Y^\mu$.

Using the relation $(\bar{u}_i^C\not{X}_k^C u_j)^*=\bar{u}_j\not{X}_k u_i^C$, which we verify (explicitly taking the γ matrices into account) through

$$
\begin{aligned}
(\bar{u}_i^C\not{X}_k^C u_j)^* &= (u_i^{C\dagger}\gamma_0\gamma_\mu u_j)^\dagger X_k^{C\mu *}\\
&=(\bar{u}_j\gamma_0\gamma_\mu^\dagger\gamma_0 u_i^C)X_k^\mu\\
&=(\bar{u}_j\gamma_\mu u_i^C)X_k^\mu\\
&=(\bar{u}_j\not{X}_k u_i^C)\quad,
\end{aligned}
$$

utilizing $\gamma_0\gamma_\mu^\dagger\gamma_0=\gamma_\mu$, $(u_i^{C\dagger})^\dagger=u_i^C$, and the antisymmetric tensor in three dimensions, ε_{ijk}, we get

$$
\begin{aligned}
\mathcal{L}_{\text{int}}=&-\frac{ig_5}{\sqrt{2}}\Big[-\varepsilon_{ijk}(\bar{u}_i\not{X}_k u_j^C)^*-\varepsilon_{ijk}(\bar{u}_i\not{X}_k u_j^C)\\
&+\varepsilon_{ijk}(\bar{d}_i\not{Y}_k u_j^C)^*+\varepsilon_{ijk}(\bar{d}_i\not{Y}_k u_j^C)\\
&-(\bar{e}^C\not{X}_j d_j)^*-(\bar{e}^C\not{X}_j d_j)+(\bar{e}^C\not{Y}_j u_j)^*+(\bar{e}^C\not{Y}_j u_j)\Big]_L\\
&-\frac{ig_5}{\sqrt{2}}\Big[(\bar{e}^C\not{X}_j d_j)^*+(\bar{e}^C\not{X}_j d_j)-(\bar{\nu}_e^C\not{Y}_j d_j)^*-(\bar{\nu}_e^C\not{Y}_j d_j)\Big]_R\quad. \qquad (23)
\end{aligned}
$$

To be very explicit, we note that the first two rows of (22) can be summarized as $-\varepsilon_{ijk}(\bar{u}_i\not{X}_k u_j^C)^*$; similarly, the third and fourth rows of (22) give $-\varepsilon_{ijk}(\bar{u}_i\not{X}_k u_j)$, the fifth and sixth rows give $+\varepsilon_{ijk}(\bar{d}_i\not{X}_k d_j^C)^*$, and the seventh and eighth rows $+\varepsilon_{ijk}(\bar{d}_i\not{X}_k d_j^C)$. If we stress the point that fermion wave functions originate from left-handed or right-handed multiplets by adding an index L or R, respectively, expression (23) then reads

$$
\begin{aligned}
\mathcal{L}_{\text{int}}=\frac{ig_5}{\sqrt{2}}\Big\{&+\varepsilon_{ijk}(\overline{u_{iL}}\not{X}_k u_{jL}^C)+(\overline{e_L^C}\not{X}_j d_{jL})-(\overline{e_R^C}\not{X}_j d_{jR})+\varepsilon_{ijk}(\overline{d_{iL}}\not{Y}_k u_{jL}^C)\\
&-(\overline{e_L^C}\not{Y}_j u_{jL})+(\overline{\nu_{eR}^C}\not{Y}_j d_{jR})+\text{h.c.}\Big\}\quad; \qquad (24)
\end{aligned}
$$

here h.c. signifies the Hermitian conjugate of the preceding expression.

PROBLEM

9.7 Color of the SU(5) Gauge Bosons

Determine the color of the X and Y bosons.

Solution. The 24 gauge bosons of the SU(5) group result from the direct product of the fundamental SU(5) quintuplet with the fundamental SU(5) antiquintuplet:

$$\begin{aligned} 5 &= (d_1, d_2, d_3, e^+, -\nu_e^C) \quad , \\ \bar{5} &= (d_1^C, d_2^C, d_3^C, e^-, -\nu_e) \quad . \end{aligned} \tag{1}$$

The fundamental quintuplet contains quarks that carry color whereas the antiquintuplet contains the respective antiquarks carrying anticolor. If we perform the direct product

$$\begin{aligned} [5] \times [\bar{5}] &= \begin{pmatrix} d_1 \\ d_2 \\ d_3 \\ e^+ \\ -\nu_e^C \end{pmatrix} \times \left(d_1^C, d_2^C, d_3^C, e^-, -\nu_e\right) \\ &= \begin{pmatrix} d_1 d_1^C & d_1 d_2^C & d_1 d_3^C & d_1 e^- & -d_1 \nu_e \\ d_2 d_1^C & d_2 d_2^C & d_2 d_3^C & d_2 e^- & -d_2 \nu_e \\ d_3 d_1^C & d_3 d_2^C & d_3 d_3^C & d_3 e^- & -d_3 \nu_e \\ e^+ d_1^C & e^+ d_2^C & e^+ d_3^C & e^+ e^- & -e^+ \nu_e \\ -\nu_e^C d_1^C & -\nu_e^C d_2^C & -\nu_e^C d_3^C & -\nu_e^C e^- & -\nu_e^C \nu_e \end{pmatrix} \quad , \end{aligned} \tag{2}$$

we formally obtain the matrix of the gauge bosons, which we have already seen in (9.44). Now we can directly extract the quantum numbers of the bosons from this representation. We see that, e.g., the upper 3×3 matrix must contain the QCD gauge bosons whereas the lower 2×2 matrix contains the gauge bosons of the SU(2). The remaining two 2×3 and 3×2 matrices contain the new X and Y bosons. Comparing this matrix with the defining matrix in (9.44) we find that the $X_{i\mu}$ and $Y_{i\mu}$ carry *anti*color, and the charge-conjugated $X_{i\mu}^C$ and $Y_{i\mu}^C$ carry color.

Also the X and Y bosons carry baryon number $B = -\frac{1}{3}$ and lepton number $L = -1$, thus $B - L = +\frac{2}{3}$. From the group-theoretical point of view this becomes apparent when reducing the product $[5] \times [\bar{5}]$ into its SU(3) × SU(2) subgroups.

$$[24]_5 = (8, 1) + (\bar{3}, 2) + (3, \bar{2}) + (1, 3) + (1, 1)$$

The multiplet $(\bar{3}, 2)$ contains the X, Y-bosons carrying *anti*color, and the multiplet $(3, \bar{2})$ its antiparticles, which carry color.

9.4 Spontaneous Breaking of the SU(5) Symmetry

We have already mentioned that SU(5) symmetry, if it exists at all, must be broken strongly, since quarks and leptons are entirely different particles under normal conditions and cannot be transformed among each other. The leptoquarks X and Y must have huge masses, which have to be much larger than the masses of the weak gauge bosons W^+, W^-, and Z^0, since the transformations of the neutrino and charged leptons or of quarks of different flavors mediated by those particles are easily detectable.

Symmetry breaking of the SU(5) gauge group must be done in two steps: first from the combined group SU(5) to the product group $SU(3)_C \times SU(2)_L \times U(1)_Y$, "$C$" denoting color, "$L$" denoting the weak isospin of the left-handed particles, and "Y" denoting (weak) hypercharge; and secondly the symmetry breaking of the Glashow–Salam–Weinberg theory must lead from the group $SU(2)_L \times U(1)_Y$ to the electromagnetic gauge group $U(1)_{EM}$. Altogether the scheme of symmetry breaking is

$$\mathrm{SU(5)} \overset{\mathrm{GUT}}{\longrightarrow} \mathrm{SU(3)}_C \times \underbrace{\mathrm{SU(2)}_L \times \mathrm{U(1)}_Y}_{\mathrm{GSW}\quad \mathrm{U(1)_{EM}}} \quad . \tag{9.54}$$

Only the gauge group of color $SU(3)_C$ and $U(1)_{EM}$ remain unbroken, since the gluons[10] and photons are massless. We will describe symmetry breaking as usual by scalar Higgs fields with a non-zero vacuum expectation value. The first symmetry breaking, denoted in (9.54) by "GUT" (grand unified theory) can be accomplished by a Higgs-field multiplet transforming under the 24-dimensional adjoint representation of SU(5). In the second, denoted by "SW", we shall use a Higgs multiplet $\hat{H}$ of the fundamental representation of SU(5). We start with the 24-fold Higgs field $\hat{\phi} = \sum_{i=1}^{24} \phi_i \tilde{\lambda}_i$. More precisely $\hat{\phi}$ is the Higgs-field operator. However, the short name Higgs field is commonly used. In order not to break the SU(5) gauge invariance of the Lagrangian explicitly but only "spontaneously" (by fixing a certain vacuum expectation value of the Higgs field), the Higgs potential can only be built from SU(5)-invariant functions that can be formed from powers of the Higgs field. All of these invariants can be written as $\mathrm{Tr}\{\hat{\phi}^n\}$, since the trace of a unitary matrix does not change under a unitary transformation. Since $\mathrm{Tr}\{\hat{A}\hat{B}\} = \mathrm{Tr}\{\hat{B}\hat{A}\}$

$$\begin{aligned} \mathrm{Tr}\{\hat{\phi}'^n\} &= \mathrm{Tr}\{(\hat{U}\hat{\phi}\hat{U}^\dagger)^n\} = \mathrm{Tr}\{\hat{U}\hat{\phi}\hat{U}^\dagger\hat{U}\hat{\phi}\hat{U}^\dagger \cdots \hat{U}\hat{\phi}\hat{U}^\dagger\} \\ &= \mathrm{Tr}\{\hat{U}\hat{\phi}^n\hat{U}^\dagger\} = \mathrm{Tr}\{\hat{U}^\dagger\hat{U}\hat{\phi}^n\} = \mathrm{Tr}\{\hat{\phi}^n\} \end{aligned} \tag{9.55}$$

holds for every unitary transformation $\hat{U}$. Under the assumption that $V(\hat{\phi})$ does not depend on the sign of $\hat{\phi}$ and is built only from powers up to order four (like the Higgs potential in the GSW theory), the most general form of the Higgs potential is

$$V(\hat{\phi}) = -\frac{1}{2}\mu^2 \mathrm{Tr}\{\hat{\phi}^2\} + \frac{a}{4}(\mathrm{Tr}\{\hat{\phi}^2\})^2 + \frac{b}{2}\mathrm{Tr}\{\hat{\phi}^4\} \quad . \tag{9.56}$$

Again we have chosen the coefficient of the quadratic term to be negative, so that the minimum is not at $\hat{\phi} = 0$ but the symmetry is spontaneously broken. Under the assumption that $a > -(7/15)b$, $b > 0$ we shall show in Exercise 9.8[11] that the minimum

[10] As no free gluons are detectable, this hypothesis is hard to prove experimentally. It is certain that the rest mass of the gluons is considerably smaller than 100 MeV. A mass much less than 10 MeV would probably not be detectable at present.

[11] A detailed discussion of possible symmetry breakings of SU(5) theory is given by M. Magg, Q. Shafi: Z. Phys. C **4**, 63 (1980).

of the potential $V(\hat{\phi})$ lies at the field configuration

$$\hat{\phi}_0 = \frac{3}{2} v \hat{Y} = \begin{pmatrix} -v & & & & \\ & -v & & & \\ & & -v & & \\ & & & \frac{3}{2}v & \\ & & & & \frac{3}{2}v \end{pmatrix} , \tag{9.57}$$

where $\hat{Y}$ is known from (9.23) and

$$v = \left(\frac{2\mu^2}{15a + 7b} \right)^{1/2} . \tag{9.58}$$

Certainly any other gauge-equivalent field $\hat{\phi} = \hat{U} \phi_0 \hat{U}^{-1}$ leads to the same value of the potential. We have chosen the diagonal representation for convenience. The value of the potential at the minimum is

$$V(\hat{\phi}_0) = -\frac{15}{8} \mu^2 v^2 < 0 . \tag{9.59}$$

Relations (9.57)–(9.59) will be calculated in Exercise 9.8.

EXERCISE

9.8 Minimum of the Higgs Potential

Problem. Show that

$$\hat{\phi}_0 = \frac{3}{2} \hat{Y} \left(\frac{2\mu^2}{15a + 7b} \right)^{1/2}$$

is a minimum (up to a gauge transformation) of the Higgs potential

$$V(\hat{\phi}) = -\frac{1}{2} \mu^2 \mathrm{Tr}\{\hat{\phi}^2\} + \frac{a}{4} \left(\mathrm{Tr}\{\hat{\phi}^2\} \right)^2 + \frac{b}{2} \mathrm{Tr}\{\hat{\phi}^4\} \tag{1}$$

if $b > -(15a/7)$, and show that $V(\phi_0) < 0$ holds.

Solution. As $V(\hat{\phi})$ is gauge invariant, we can choose a certain gauge for the Higgs field. We can use a diagonal 5×5 matrix for $\hat{\phi}$ because every unitary 5×5 matrix can be diagonalized by a unitary transformation. Formulated in group theoretical language, every element of the SU(5) Lie algebra can be rotated into the Cartan subalgebra of diagonal generators by an appropriate SU(5) rotation. The Cartan subalgebra of SU(5) is spanned by the operators $\tilde{\lambda}_3, \tilde{\lambda}_8, \tilde{\lambda}_{23} = \tilde{\tau}_3$ and $\tilde{\lambda}_{24} = \tilde{Y}$, which are explicitly denoted in Exercise 9.1. A general ansatz for $\hat{\phi}_0$ is

Exercise 9.

$$\hat{\phi} = \alpha\tilde{\lambda}_3 + \beta\sqrt{3}\hat{\lambda}_8 + \gamma\tilde{\lambda}_{23} + \frac{\sqrt{15}}{2}\delta\tilde{\lambda}_{24}$$

$$= \begin{pmatrix} \alpha+\beta-\delta &&&& \\ & -\alpha+\beta-\delta &&& \\ && -2\beta-\delta && \\ &&& \gamma+\frac{3}{2}\delta & \\ &&&& -\gamma+\frac{3}{2}\delta \end{pmatrix} . \quad (2)$$

By explicit evaluation we obtain

$$\mathrm{Tr}\{\hat{\phi}^2\} = (\alpha+\beta-\delta)^2 + (-\alpha+\beta-\delta)^2 + (2\beta+\delta)^2 + \left(\gamma+\frac{3}{2}\delta\right)^2 + \left(\gamma-\frac{3}{2}\delta\right)^2$$
$$= 2\alpha^2 + 6\beta^2 + 2\gamma^2 + \frac{15}{2}\delta^2 \quad (3)$$

and

$$\mathrm{Tr}\{\hat{\phi}^4\} = (\alpha+\beta-\delta)^4 + (-\alpha+\beta-\delta)^4 + (2\beta+\delta)^4 + \left(\gamma+\frac{3}{2}\delta\right)^4 + \left(\gamma-\frac{3}{2}\delta\right)^4$$
$$= 2\alpha^4 + 18\beta^4 + 2\gamma^4 + \frac{105}{8}\delta^4 + 12\alpha^2(\beta-\delta)^2 + 27\gamma^2\delta^2 + 24\beta^3\delta + 36\beta^2\delta^2 \quad . \quad (4)$$

The extrema of the potential $V(\hat{\phi})$ are fixed by the condition

$$\frac{\partial V}{\partial \alpha} = \frac{\partial V}{\partial \beta} = \frac{\partial V}{\partial \gamma} = \frac{\partial V}{\partial \delta} = 0 \quad . \quad (5)$$

Successively we get

$$0 = \frac{\partial V}{\partial \alpha} = \alpha\left[-2\mu^2 + 2a\mathrm{Tr}\{\hat{\phi}^2\} + 4b(\alpha^2 + 3(\beta-\delta)^2)\right] \quad (6)$$

and hence

$$\alpha_0 = 0 \quad , \quad (7)$$

and then

$$\left.\frac{\partial V}{\partial \beta}\right|_{\alpha=0} = \beta\left[-6\mu^2 + 6a\mathrm{Tr}\{\phi^2\} + 36b(\beta^2 + \beta\delta + \delta^2)\right] \quad (8)$$

with the solution

$$\beta_0 = 0 \quad , \quad (9)$$

and

$$\left.\frac{\partial V}{\partial \gamma}\right|_{\alpha=\beta=0} = \gamma\left[-2\mu^2 + 2a\mathrm{Tr}\{\hat{\phi}^2\} + b(4\gamma^2 + 27\delta^2)\right] \quad (10)$$

Exercise 9.8

with the solution

$$\gamma_0 = 0 \quad . \tag{11}$$

Finally we obtain the condition

$$\left.\frac{\partial V}{\partial \delta}\right|_{\alpha=\beta=\gamma=0} = \frac{15}{4}\delta(-2\mu^2 + 15a\delta^2 + 7b\delta^2) \quad . \tag{12}$$

A choice of $\delta = 0$ would lead to the trivial solution $\hat{\phi}_0 = 0$, so that we choose the second possibility,

$$\delta_0^2 = \frac{2\mu^2}{15a + 7b} \quad . \tag{13}$$

For

$$a > -\frac{7}{15}b \tag{14}$$

this is a solution of (5). With $\hat{Y} = \sqrt{\frac{5}{3}}\tilde{Y}$ (see (9.23)) we obtain under condition (14) the vacuum expectation value – see (9.58)

$$\phi_0 = \delta\frac{\sqrt{15}}{2}\tilde{Y} = \left(\frac{2\mu^2}{15a + 7b}\right)^{1/2}\frac{3}{2}\hat{Y} \equiv v\frac{3}{2}\hat{Y} \quad . \tag{15}$$

We calculate the second derivative in order to show that we have obtained a minimum of the Higgs potential:

$$\begin{aligned}\left.\frac{\partial^2 V}{\partial \delta^2}\right|_{\alpha=\beta=\gamma=0,\delta=\delta_0} &= \frac{15}{4}(-2\mu^2 + 3(15a + 7b)\delta_0^2)\\ &= \frac{15}{4}(-2\mu^2 + 6\mu^2)\\ &= 15\mu^2 > 0 \quad .\end{aligned} \tag{16}$$

So, assuming that (14) holds, we have obtained a minimum. If we insert (13) into the potential, we get

$$\begin{aligned}V(\hat{\phi}_0) &= -\frac{1}{2}\mu^2\frac{15}{2}\delta_0^2 + \frac{a}{4}\left(\frac{15}{2}\delta_0^2\right)^2 + \frac{b}{2}\frac{105}{8}\delta_0^4\\ &= -\frac{15}{4}\mu^2\delta_0^2 + \frac{15}{16}(15a + 7b)\delta_0^4\\ &= \left(-\frac{15}{4}\mu^2 + \frac{15}{16}2\mu^2\right)\delta_0^2\\ &= -\frac{15\mu^4}{4(15a + 7b)} < 0 \quad ,\end{aligned} \tag{17}$$

hence the minimum is smaller than $V(0) = 0$ if condition (14) is fulfilled.

The type of symmetry breaking by the Higgs field $\hat{\phi}_0$ is given by (9.57) directly. Since $\hat{\phi}_0$ is a unit matrix with respect to the first three rows and columns and last two rows and columns, all SU(5) generators $\tilde{\lambda}_1, \ldots, \tilde{\lambda}_8$ of the gluon fields, the generators of isospin $\tilde{\lambda}_{21} = \tilde{\tau}_1, \tilde{\lambda}_{22} = \tilde{\tau}_2, \tilde{\lambda}_{23} = \tilde{\tau}_3$, and of course the generator of hypercharge $\tilde{\lambda}_{24} = \tilde{Y}$ commute with $\hat{\phi}_0$. On the other hand, the generators of the leptoquark fields X_i and Y_i incorporate elements mixing the color and isospin degrees of freedom and hence do not commute with $\hat{\phi}_0$. Thus the gauge fields X_i and Y_i obtain a non-vanishing mass and the SU(5) symmetry is broken spontaneously to SU(3) × SU(2) × U(1) with massless gluons and W and B bosons.

EXERCISE

9.9 Kinetic Energy of the Higgs Field

Problem. Show that (9.60) is equivalent to the representation of the kinetic part of the Lagrange function

$$L_\phi^{(\mathrm{kin})} = (\hat{D}_\mu \phi_i)^\dagger (\hat{D}^\mu \phi_i) \quad , \tag{1}$$

where $\hat{D}_\mu = \partial_\mu - \mathrm{i}g_5 \hat{A}_\mu$ denotes the well-known covariant derivative. You may use the regular representation of the SU(5) $\tilde{\lambda}$ matrices.

Solution. The regular representation of the $\tilde{\lambda}$ matrices is defined by

$$\tilde{\lambda}_i \phi_j = C_{ijk} \phi_k \ , \tag{2}$$

where the coefficients C_{ijk} are the structure constants of the SU(5) algebra,

$$[\tilde{\lambda}_i, \tilde{\lambda}_j] = C_{ijk} \tilde{\lambda}_k \quad . \tag{3}$$

We use the Einstein summation convention. Furthermore, we have

$$\hat{A}_\mu = \frac{1}{2} A_{\mu a} \tilde{\lambda}_a \quad , \quad \hat{\phi} = \phi_a \tilde{\lambda}_a \quad . \tag{4}$$

Applying this to the Lagrange function (9.60) gives

$$\begin{aligned} L_\phi^{(\mathrm{kin})} &= \frac{1}{2} \mathrm{Tr}\{(\partial_\mu \hat{\phi} + \mathrm{i}g_5 [\hat{A}_\mu, \hat{\phi}])^\dagger (\partial^\mu \hat{\phi} + \mathrm{i}g_5 [\hat{A}^\mu, \hat{\phi}])\} \\ &= \frac{1}{2} \mathrm{Tr}\left\{ \left(\partial_\mu \tilde{\lambda}_c \phi_c + \frac{1}{2} \mathrm{i}g_5 A_{\mu a} \phi_b [\tilde{\lambda}_a, \tilde{\lambda}_b] \right)^\dagger \right. \\ &\quad \left. \times \left(\partial^\mu \tilde{\lambda}_i \phi_i + \frac{1}{2} \mathrm{i}g_5 A_j^\mu \phi_k [\tilde{\lambda}_j, \tilde{\lambda}_k] \right) \right\} \\ &= \frac{1}{2} \mathrm{Tr}\left\{ \left(\partial_\mu \tilde{\lambda}_c \phi_c + \frac{1}{2} \mathrm{i}g_5 A_{\mu a} \phi_b C_{abc} \tilde{\lambda}_c \right)^\dagger \right. \\ &\quad \left. \times \left(\partial^\mu \tilde{\lambda}_k \phi_k + \frac{1}{2} \mathrm{i}g_5 A_i^\mu \phi_j C_{ijk} \tilde{\lambda}_k \right) \right\} \quad . \end{aligned} \tag{5}$$

Exercise 9.9

We evaluate the trace by remembering that the $\tilde{\lambda}$ matrices are Hermitian

$$\mathrm{Tr}\{\tilde{\lambda}_c^\dagger\tilde{\lambda}_k\} = \mathrm{Tr}\{\tilde{\lambda}_c\tilde{\lambda}_k\} = 2\delta_{ck} \quad . \tag{6}$$

The factor $\frac{1}{2}$ cancels, and we are left with

$$L_\phi^{(\mathrm{kin})} = \left(\partial_\mu\phi_c + \frac{1}{2}\mathrm{i}g_5A_{\mu a}\phi_bC_{abc}\right)^\dagger\left(\partial^\mu\phi_c + \frac{1}{2}\mathrm{i}g_5A_i^\mu\phi_jC_{ijc}\right) \quad . \tag{7}$$

This expression can be rewritten using the definition of the regular representation. Taking into account the antisymmetry of the structure constants under exchange of two indices, we get

$$\phi_bC_{abc} = \phi_b(-C_{acb}) = -\tilde{\lambda}_a\phi_c \quad . \tag{8}$$

This yields

$$\begin{aligned} L_\phi^{(\mathrm{kin})} &= \left(\partial_\mu\phi_c - \frac{1}{2}\mathrm{i}g_5A_{\mu a}\tilde{\lambda}_a\phi_c\right)^\dagger\left(\partial^\mu\phi_c - \frac{1}{2}\mathrm{i}g_5A_i^\mu\tilde{\lambda}_i\phi_c\right) \\ &= \left[(\partial_\mu - \mathrm{i}g_5\hat{A}_\mu)\phi_c\right]^\dagger\left[(\partial^\mu - \mathrm{i}g_5\hat{A}^\mu)\phi_c\right] \quad . \end{aligned} \tag{9}$$

This last equation displays the relation already known to us.

Now we shall examine the mass terms of the leptoquarks in detail. Their origin is, as usual, the gauge-covariant derivation in the kinetic energy of the Higgs field,

$$L_\phi^{(\mathrm{kin})} = \frac{1}{2}\mathrm{Tr}\left\{(\partial_\mu\hat{\phi} + \mathrm{i}g_5[\hat{A}_\mu,\hat{\phi}])^\dagger(\partial^\mu\hat{\phi} + \mathrm{i}g_5[\hat{A}^\mu,\hat{\phi}])\right\} \quad . \tag{9.60}$$

As the Higgs field $\hat{\phi}$ and the gauge boson fields $\hat{A}_\mu$ are part of the same representation of SU(5), the matrix notation in (9.60) is natural. Thus the mass term of the gauge fields is

$$\begin{aligned} L_M^{(\phi)} &= \frac{1}{2}g_5^2\mathrm{Tr}\left\{[\hat{A}_\mu,\hat{\phi}_0]^\dagger[\hat{A}^\mu,\hat{\phi}_0]\right\} \\ &= \frac{9}{8}g_5^2v^2\mathrm{Tr}\left\{[\hat{A}_\mu,\hat{Y}]^\dagger[\hat{A}^\mu,\hat{Y}]\right\} \quad , \end{aligned} \tag{9.61}$$

where we have inserted (9.57). Denoting the eigenvalues of the matrix $\hat{Y}$ by y_i ($y_{ik} = y_i\delta_{ik}$, $y_1 = y_2 = y_3 = -2/3$, $y_4 = y_5 = 1$), the matrix element is

$$\begin{aligned} [\hat{A}^\mu,\hat{Y}]_{ik} &= A_{ik}^\mu y_k - y_iA_{ik}^\mu = (y_k - y_i)A_{ik}^\mu \quad , \\ [A^\mu,\hat{Y}]_{ik}^\dagger &= [\hat{A}^\mu,\hat{Y}]_{ki}^* = A_{ki}^{\mu*}y_i - y_kA_{ki}^{\mu*} = -(y_k - y_i)A_{ki}^{\mu*} = -(y_k - y_i)A_{ik}^\mu \end{aligned} \tag{9.62}$$

because $A_{ki}^{\mu*} = A_{ik}^\mu$ (hermiticity!) and thus

$$\mathrm{Tr}\left\{[\hat{A}_\mu,\hat{Y}]^\dagger[\hat{A}^\mu,\hat{Y}]\right\} = \sum_{i,k}[\hat{A}_\mu,\hat{Y}]_{ik}^*[\hat{A}^\mu,\hat{Y}]_{ik} = \sum_{i,k}|A_{ik}^\mu|^2(y_k - y_i)^2 \quad . \tag{9.63}$$

Only the pairs (i,k) where y_i and y_k are different contribute to the sum. These are $i = 1,2,3$ and $k = 4,5$, and their permutations. $|y_k - y_i| = 5/3$ holds in each case. Thus, we obtain the following mass term

$$L_M^{(\phi)} = \frac{25}{4} g_5^2 v^2 \left(|A_{14}^\mu|^2 + |A_{24}^\mu|^2 + |A_{34}^\mu|^2 + |A_{15}^\mu|^2 + |A_{25}^\mu|^2 + |A_{35}^\mu|^2\right) \quad . \tag{9.64}$$

Comparing this with the matrix of the gauge bosons $(\hat{A}_\mu)_{ik}$ (9.44) shows that L_M contains the leptoquark fields, so that the mass term has the form

$$L_M^{(\phi)} = M_X^2 \sum_{i=r,g,b} |X_i^M|^2 + M_Y^2 \sum_{i=r,g,b} |Y_i^M|^2 \tag{9.65}$$

with

$$M_X = M_Y = \frac{5}{2\sqrt{2}} g_5 v \quad . \tag{9.66}$$

It is important that $A_{14}^\mu = \frac{1}{\sqrt{2}} X^\mu$ (see (9.44)) and that there is no factor $\frac{1}{2}$ in (9.65), since X^μ and Y_i^μ are complex fields. Indeed, we recall that for complex fields the mass term in the Lagrange density is simply $M^2\phi^*\phi$, because the variation after ϕ or ϕ^* is performed independently, so that the wave equations finally contain the terms $M^2\phi$ (in the wave equation for ϕ) or $M^2\phi^*$ (in the wave equation for ϕ^*).

Now we shall examine the second part of the symmetry breaking and transfer the Higgs mechanism of the GSW theory to the SU(5) theory. We choose a Higgs field H, transforming under the fundamental 5-dimensional representation: $H = (h_1, h_2, h_3, h_4, h_5)$. If we simply choose a Higgs potential

$$V(H) = -\frac{1}{2} v^2 (H^\dagger H) + \frac{\lambda}{4} (H^\dagger H)^2 \quad , \tag{9.67}$$

as in the SU(2) × U(1) theory, the vacuum expectation value H_0 would have the value

$$h_0^0 = \sqrt{(H_0^\dagger H_0)} = \frac{v}{\sqrt{\lambda}} \tag{9.68}$$

in an arbitrary direction. All choices

$$H_0^{(1)} = h_0^0 \begin{pmatrix} 1 \\ 0 \\ 0 \\ 0 \\ 0 \end{pmatrix} \quad , \quad \ldots \quad , \quad H_0^{(5)} = h_0^0 \begin{pmatrix} 0 \\ 0 \\ 0 \\ 0 \\ 1 \end{pmatrix} \tag{9.69}$$

would be equivalent. But as SU(5) symmetry is already broken to SU(3) × SU(2) × U(1), the different possibilities would lead to entirely different physical consequences. While $H_0^{(4)}$ or $H_0^{(5)}$ would lead to the desired breaking of SU(2) symmetry (last two components), $H_0^{(1)}, \ldots, H_0^{(3)}$ would break the SU(3) symmetry in the first three components, which is undesirable, since the gauge group SU(3) of the strong interaction appears unbroken.

The solution consists in incorporating into the Higgs field H the fact that the Higgs field $\hat{\phi}$ has already broken SU(5) symmetry, and thus the color and isospin degrees of freedom already have different meanings. We can achieve this by adding a term to the potential (9.67) coupling both Higgs fields $\hat{\phi}$ and $\hat{H}$ in such a way that the matrix structure of the vacuum expectation value $\hat{\phi}_0$ is taken into account. Under the usual conditions (the interaction must be at most quartic in the fields, invariant under change of sign and SU(5) invariant) only

$$V_{\phi H} = \beta H^\dagger \hat{\phi}^2 H \tag{9.70}$$

is reasonable. The other possibility,

$$V'_{\phi H} = \alpha (H^\dagger H)\mathrm{Tr}\{\hat{\phi}^2\} \quad , \tag{9.71}$$

does not allow a coupling of the direction of the H field to the matrix structure of $\hat{\phi}$ and hence will not be considered further. As the difference of the scales of symmetry breaking of the fields $\hat{\phi}$ and H is expected to be large, we simply insert the vacuum expectation value $\hat{\phi}_0$ of (9.57) into (9.70) and neglect any effect of H on $\hat{\phi}_0$.

The total potential of the field H we want to minimize is

$$\begin{aligned} V_{\mathrm{eff}}(H) &= V(H) + V_{\phi H}(H, \hat{\phi}_0) \\ &= H^\dagger \left(-\frac{\nu^2}{2} + \frac{9}{4}\beta v^2 \hat{Y}^2 \right) H + \frac{\lambda}{4}(H^\dagger H)^2 \quad . \end{aligned} \tag{9.72}$$

The matrix in the square term is given by

$$\begin{aligned} &\frac{-\nu^2}{2} + \frac{9}{4}\beta v^2 \hat{Y}^2 \\ &= \frac{1}{2}\begin{pmatrix} -\nu^2 + 2\beta v^2 & & & & \\ & -\nu^2 + 2\beta v^2 & & & \\ & & -\nu^2 + 2\beta v^2 & & \\ & & & -\nu^2 + \frac{9}{2}\beta v^2 & \\ & & & & -\nu^2 + \frac{9}{2}\beta v^2 \end{pmatrix} , \end{aligned} \tag{9.73}$$

so that for $\beta < 0$ the last two components shift H away from $H = 0$ most strongly. This means that the additional potential $V_{\phi H}$ gives for every direction of H another effective value for the parameter ν in (9.68), which takes into account the coupling to the ϕ field. Comparing with (9.68), we can see immediately that the absolute value of the vacuum expectation value $H_0^{(i)}$ takes the values

$$h_0^{(i)} = \sqrt{(\nu^2 - 2\beta v^2)/\lambda} \quad \text{for} \quad i = 1, 2, 3 \quad , \tag{9.74a}$$

$$h_0^{(i)} = \sqrt{(\nu^2 - \frac{9}{2}\beta v^2)/\lambda} \quad \text{for} \quad i = 4, 5 \quad . \tag{9.74b}$$

The corresponding values of the potential are

$$V_{\mathrm{eff}}(H_0^{(i)}) = \begin{cases} -(\nu^2 - 2\beta v^2)^2/4\lambda \quad , & i = 1, 2, 3 \\ -(\nu^2 - \frac{9}{2}\beta v^2)^2/4\lambda \quad , & i = 4, 5 \end{cases} \quad .$$

Obviously for $\beta < 0$ the directions $i = 4, 5$ are preferred. Hence we choose $\beta < 0$ and use in the following

$$h_0 = \sqrt{\left(\nu^2 - \frac{9}{2}\beta v^2\right)/\lambda} \quad . \tag{9.76}$$

A choice of direction among the last two components has no physical importance; we set, following a general convention,

$$H_0 = H^{(5)} = h_0 \begin{pmatrix} 0 \\ 0 \\ 0 \\ 0 \\ 1 \end{pmatrix} . \tag{9.77}$$

The Higgs field H creates the mass terms of the bosons W^+, W^-, and Z^0 by coupling to the gauge fields in

$$L_H^{(\mathrm{kin})} = \frac{1}{2}(\partial_\mu H - \mathrm{i}g_5 \hat{A}_\mu H)^\dagger (\partial^\mu H - \mathrm{i}g_5 \hat{A}^\mu H) \quad . \tag{9.78}$$

Using (9.44a) we obtain

$$\begin{aligned} L_H(H) &= \frac{1}{2} g_5^2 (H_0^\dagger \hat{A}_\mu^\dagger \hat{A}^\mu H_0) \\ &= \frac{1}{4} g_5^2 h_0^2 \left\{ Y_{1\mu}^{C*}, Y_{2\mu}^{C*}, Y_{3\mu}^{C*}, W_\mu^{+*}, -\frac{1}{\sqrt{2}}\left(W_\mu^3 - \frac{3}{\sqrt{15}} B_\mu \right)^* \right\} \\ &\quad \times \begin{pmatrix} Y_1^{C\mu} \\ Y_2^{C\mu} \\ Y_3^{C\mu} \\ W^{+\mu} \\ -\frac{1}{\sqrt{2}}\left(W^{3\mu} - \frac{3}{\sqrt{15}} B_\mu \right) \end{pmatrix} \\ &= \frac{1}{4} g_5^2 h_0^2 \left\{ \sum_i Y_{i\mu} Y_i^{*\mu} + W_\mu^+ W^{\mu -} + \frac{1}{2}\left(W_3^\mu - \sqrt{\frac{3}{5}} B^\mu \right)^2 \right\} \end{aligned} \tag{9.79}$$

where we have used $Y_1^{\mu C} = Y_1^{\mu *}$. As $v \gg h_0$ the contribution to the masses of the Y bosons can be neglected, compared with the term in (9.65) (but breaking the degeneracy of M_X and M_Y!) and the only interesting term surviving is

$$\begin{aligned} L_M^{'(H)} &= \frac{1}{4} g_5^2 h_0^2 W^{+\mu} W_\mu^- + \frac{1}{5} g_5^2 h_0^2 \left(\sqrt{\frac{5}{8}} W_3^\mu - \sqrt{\frac{3}{8}} B^\mu \right)^2 \\ &\equiv M_W^2 W^{+\mu} W_\mu^- + \frac{1}{2} M_Z^2 Z^\mu Z_\mu \quad . \end{aligned} \tag{9.80}$$

The factors $\sqrt{\frac{5}{8}} = \cos\theta_W$ and $\sqrt{\frac{3}{8}} = \sin\theta_W$ in the second term are, according to (9.48), just the proper mixing coefficients of the W_3^μ and B^μ fields (see also (4.97) and (4.98)!). Hence mixing of the two neutral boson fields W_3^μ and B^μ occurs as desired, and the ratio of the masses of the neutral bosons is

$$\frac{M_W^2}{M_Z^2} = \frac{5}{8} = 1 - \sin^2\theta_W = \cos^2\theta_W \quad , \tag{9.81}$$

if we take (9.49) for the Weinberg angle into account. We conclude that the Higgs mechanism allows a breaking of SU(5) symmetry, reproducing the phenomenology of the $SU(2) \times U(1)$ gauge theory of electroweak interaction.

Finally we want to point to a conceptual difficulty. Since $V(\phi_0)$ is large (about 10^{15} GeV, as we shall see in Sect. 9.5), the parameter β in (9.76) has to be chosen very small, in order that h_0 has the value required by GSW theory (about 250 GeV). This "fine-tuning" of the Higgs-potential parameters is not appealing and leads to problems in quantum field theory, where the parameters of potentials have to be renormalized. This problem is related to the large difference between the scales of symmetry breaking. Several solutions have been suggested, but a conclusive answer to this problem, which is often called the *hierarchy problem*, is not yet known.

9.5 Determination of the Scale of SU(5) Symmetry Breaking

In Sect. 9.3 we saw that the coupling constants of g_C, g_L, g' of the $SU(3)_C \times SU(2)_L \times U(1)_Y$ gauge theory are determined by the coupling constant g_5 of the unified SU(5) gauge theory (see (9.48)):

$$g_C = g_L = g_5 \quad , \quad g' = \sqrt{\frac{3}{5}} g_5 \quad . \tag{9.82}$$

We describe the coupling to the gluons (that is, to the color fields) by the coupling constant g_C (the index C denotes color). The ratio g'/g_L determines the Weinberg angle (9.49).[12] But there is a problem, because the experimentally determined values

$$\begin{aligned} g_C &\sim 1.5 \quad , \\ g_L &= e/\sin\theta_W \sim 0.65 \quad , \\ g' &= e/\cos\theta_W \sim 0.34 \quad , \end{aligned} \tag{9.83}$$

do not coincide at all with the (9.82). However, as we shall soon see, this 'problem' turns out to be a means for the determination of the energy scale of SU(5) symmetry breaking. To understand this remark, we need to refer to a result of the quantum theory of gauge fields. In this theory[13] it can be shown that the value of the experimentally determined coupling constant depends on the energy – more precisely, on the square of the four-momentum transfer – at which the process is measured. In the framework of quantum field theory one has to take into account the effect of vacuum polarization during the exchange, in addition to the direct exchange of a gauge boson between two fermions. The vacuum polarization is described as the virtual creation of a particle–antiparticle pair, either of gauge bosons, or of fermions (quarks or leptons).

Since the probability for the creation of such a virtual pair depends on the transferred four-momentum q^2, the experimentally measured effective coupling constant is q^2 dependent too. Furthermore the computation of the "running coupling constant" leads to infinitely large terms. The cure of this behavior is commonly called

[12] Here we write for the sake of clarity g_L instead of g for the coupling constant of the isospin gauge group $SU(2)_L$.

[13] See W. Greiner, S. Schramm, and E. Stein: *Quantum Chromodynamics*, 3rd ed. (Springer, Berlin, Heidelberg, 2007).

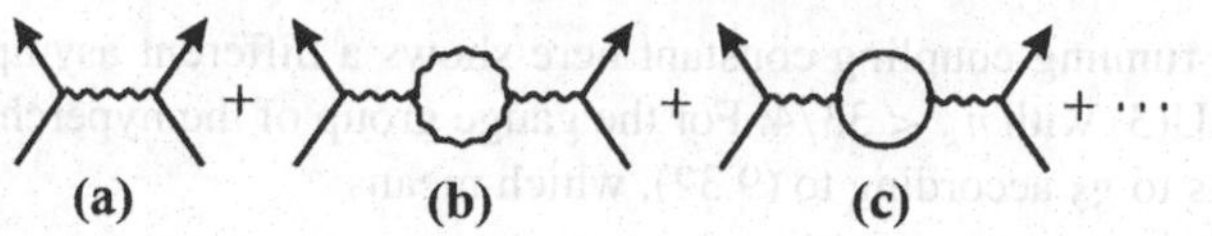

Fig. 9.2. Diagrams for direct exchange and vacuum polarization. **(a)** describes the direct exchange of a gauge boson, **(b)** contains an additional polarization ring due to the interaction of the gauge bosons among themselves and **(c)** contains an additional polarization ring due to the interaction between the gauge bosons and the fermions

the renormalization of the coupling constant. For energies above the threshold for the symmetry breaking, the SU(5) gauge theory guarantees that the coupling strengths of the SU(5) gauge fields change with q^2 in exactly the same manner for all fields. On the other hand, in the event that the energies are lower than the breaking scale, X and Y bosons can no longer be easily created as virtual quanta, since they have non-vanishing masses. Consequently the coupling constant of the still massless gluons and the coupling constant of the W and B bosons have a different q^2 dependence.

In general the functional dependence of the coupling constants is logarithmic in q^2 (as we shall see in Example 9.10):

$$\frac{1}{\alpha_i(q^2)} = \frac{1}{\alpha_i(M^2)} + b_i \ln \frac{q^2}{M^2} \quad , \tag{9.84}$$

where M is an arbitrary point of reference on the energy scale. Furthermore,

$$\alpha_i = \frac{g_i^2}{4\pi} \quad , \tag{9.85}$$

is a kind of fine-structure constant in analogy to $\frac{e^2}{4\pi}$ in QED and the b_i are numerical constants which specifically depend on the gauge group considered and on the number of coupled fermions. For example, if we assume n_g generations of lepton and quark doublets (where most likely $n_g = 3$, since this is the number of neutrino flavors contributing to the to Z^0 decay, see Example 5.3), we obtain for the non-abelian gauge group SU(n) a general expression for the constant b_i

$$b_i = \frac{11n - 4n_g}{12\pi} \quad , \quad n \geq 2$$

and therefore for the color gauge group SU(3)

$$b_C = \frac{33 - 4n_g}{12\pi} \tag{9.86a}$$

and for the isospin gauge group SU(2)

$$b_L = \frac{11 - 2n_g}{6\pi} \quad . \tag{9.86b}$$

In the case of the abelian gauge group U(1) the term proportional to n, which describes the self-interaction of the gluons, does not contribute and the coefficient of a U(1) group is generally

$$b = -\frac{1}{3\pi} n_g \quad . \tag{9.86c}$$

Note that the running coupling constant here shows a different asymptotic behavior than for the SU(3) with $n_g < 33/4$. For the gauge group of the hypercharge, we relate all coefficients to g_5 according to (9.32), which means

$$\alpha' \equiv \alpha_{g'} = \frac{g'^2}{4\pi} = \frac{3}{5}\frac{g_5^2}{4\pi} \quad ,$$

and therefore

$$\frac{1}{\alpha_{g'}(q^2)} = \frac{1}{\alpha_{g'}(M^2)} + b' \ln\left(\frac{q^2}{M^2}\right)$$

or

$$\frac{1}{\frac{g'^2}{4\pi}(q^2)} = \frac{1}{\frac{g'^2}{4\pi}(M^2)} + b' \ln\left(\frac{q^2}{M^2}\right) \quad ,$$

or, with (9.82),

$$\frac{1}{\frac{3}{5}\frac{g_5^2}{4\pi}(q^2)} = \frac{1}{\frac{3}{5}\frac{g_5^2}{4\pi}(M^2)} + b' \ln\left(\frac{q^2}{M^2}\right) \quad ,$$

or

$$\frac{1}{\frac{g_5^2}{4\pi}(q^2)} = \frac{1}{\frac{g_5^2}{4\pi}(M^2)} + \frac{3}{5} b' \ln\left(\frac{q^2}{M^2}\right) \quad .$$

This is valid only for $q^2 \lesssim M_W^2$, when the SU(2)×U(1) symmetry is valid. Now we have to keep attention: The coefficient $(3/5)b'$ should – for the U(1) group – be identical with $\bar{b} = -n_g/(3\pi)$ according to (9.86c). Hence, we get

$$\frac{3}{5} b' = \bar{b} \quad \text{or} \quad b' = \frac{5}{3}\bar{b} = \frac{5}{3}\left(-\frac{1}{3\pi} n_g\right) = -\frac{5}{9\pi} n_g \quad .$$

With $\alpha' \equiv \alpha_{g'}$, (9.86c) can be written as

$$\begin{aligned} \frac{1}{\alpha'(q^2)} &= \frac{1}{\alpha'(M^2)} + \frac{5}{3}\left(-\frac{1}{3\pi} n_g\right) \ln\left(\frac{q^2}{M^2}\right) \\ &= \frac{1}{\alpha'(M^2)} - \frac{5}{9\pi} n_g \ln\left(\frac{q^2}{M^2}\right) \quad . \end{aligned}$$

These running coupling constants are derived from quantum field theory.[14]

For the following considerations we shall choose the point of reference on the energy scale such that it determines the energy at which the SU(5) symmetry is spontaneously broken, that is, the mass of the X and Y bosons is $M \approx M_X \approx M_Y$. With this choice, at $q^2 = M^2$ the coupling constants g_C, g_L, g' satisfy the relations valid for unbroken SU(5), namely

$$\frac{g_C^2}{4\pi}(M^2) = \frac{g_L^2}{4\pi}(M^2) = \frac{g_5^2}{4\pi}(M^2) \quad \text{or}$$

---

[14] See e.g. W. Greiner, S. Schramm and E. Stein: *Quantum Chromodynamics*, 3rd ed. (Springer, Berlin, Heidelberg, 2007).

$$\alpha_C(M^2) = \alpha_L(M^2) = \frac{5}{3}\alpha'(M^2) \quad \text{where} \tag{9.87}$$
$$\alpha' \equiv \alpha_{g'} = \frac{g'^2}{4\pi} = \frac{3}{5}\frac{g_5^2}{4\pi} \; .$$

If the quantities α_C, α_L, and α' are known at a particular value of q^2 (for example, $q^2 = M_W^2 \approx (100\,\text{GeV})^2$), it is always possible to choose the energy scale M^2 of symmetry breaking such that one of the relations (9.87) is satisfied. The question whether the other relation is fulfilled simultaneously or not, therefore, is an important test for the concept of the unified gauge theory of all interactions.

In order to check the validity of (9.87), we proceed in two steps: first we determine M by calculating the coupling constant of the electromagnetic interaction, as is composed of α_L and α' (see (9.83)); and, second, we compare that with the known expression of the running coupling constant in QED. We have

$$\sin^2\theta_W + \cos^2\theta_W = \frac{e^2}{g_L^2} + \frac{e^2}{g'^2} = 1 \quad ,$$

and therefore

$$\frac{1}{\alpha} = \frac{4\pi}{e^2} = 4\pi\left(\frac{1}{g_L^2} + \frac{1}{g'^2}\right) = \frac{1}{\alpha_L} + \frac{1}{\alpha'} \tag{9.88}$$

By means of (9.84) and (9.86b) we can infer the q^2 dependence of α:

$$\begin{aligned}\frac{1}{\alpha(q^2)} &= \frac{1}{\alpha_L(q^2)} + \frac{1}{\alpha'(q^2)}\\ &= \frac{1}{\alpha_L(M^2)} + \frac{1}{\alpha'(M^2)} + (b_L + b')\ln\left(\frac{q^2}{M^2}\right)\\ &\equiv \frac{1}{\alpha(M^2)} + b\ln\frac{q^2}{M^2} \quad ,\end{aligned} \tag{9.89}$$

where

$$b = b_L + b' = \frac{33 - 16n_g}{18\pi} \quad ,$$

and

$$\frac{1}{\alpha(M^2)} = \frac{1}{\alpha_L(M^2)} + \frac{1}{\alpha'(M^2)} \; . \tag{9.90}$$

On the other hand, (9.87) requires that

$$\frac{1}{\alpha(M^2)} = \frac{1}{\alpha_L(M^2)} + \frac{1}{\alpha'(M^2)} = \frac{8}{3}\frac{1}{\alpha_C(M^2)} \quad . \tag{9.91}$$

Hence, the values of the constants α and α_C at any other value of q^2 are determined by the relation:

$$\begin{aligned}\frac{1}{\alpha(q^2)} - \frac{8}{3}\frac{1}{\alpha_C(q^2)} &= \left(b - \frac{8}{3}b_C\right)\ln\frac{q^2}{M^2}\\ &= -\frac{11}{2\pi}\ln\frac{q^2}{M^2} \quad .\end{aligned} \tag{9.92}$$

Note that the right-hand side of (9.92) is independent of the number of fermion generations n_g. This property is quite useful since we do not know for sure how many generations exist within the gap between presently accessible accelerator energies and the energy scale of the unified theory. We shall now use (9.92) for the determination of the point of reference on the energy scale. The value of α_C for $q^2 \approx (5\,\mathrm{GeV})^2$ is well known, namely

$$\alpha_C((5\,\mathrm{GeV})^2) \approx 0.175 \pm 0.01 \quad . \tag{9.93}$$

Using (9.84), we may extrapolate up to the energy scale of the electroweak theory, $q^2 \approx (100\ \mathrm{GeV})^2 \approx M_W{}^2$, and obtain, with $n_g = 3$,

$$\begin{aligned}
\frac{1}{\alpha_C((5\,\mathrm{GeV})^2)} &= \frac{1}{\alpha_C(M^2)} + b_C \ln \frac{(5\ \mathrm{GeV})^2}{M^2} \\
\frac{1}{\alpha_C(M_W^2)} &= \frac{1}{\alpha_C(M^2)} + b_C \ln \frac{M_W^2}{M^2} \\
&= \left(\frac{1}{\alpha_C((5\ \mathrm{GeV})^2)} - b_C \ln \frac{(5\ \mathrm{GeV})^2}{M^2} \right) + b_C \ln \frac{M_W^2}{M^2} \\
&= \frac{1}{0.175} + b_C \ln \left(\frac{M_W^2}{(5\,\mathrm{GeV})^2} \right)
\end{aligned}$$

and with $n_g = 3$:

$$\alpha_C(M_W^2) = \alpha_C((100\,\mathrm{GeV})^2) \approx 0.11 \quad . \tag{9.94}$$

For the electromagnetic coupling constant we find in QED:[15]

$$\frac{1}{\alpha(q^2)} = \frac{1}{\alpha(m_0)} - \frac{1}{3\pi} \sum_k e_k^2 \ln \frac{q^2}{m_0^2} \quad ,$$

and therefore, for $q^2 = M_W^2 \approx (100\ \mathrm{GeV})^2$,

$$\frac{1}{\alpha(M_W^2)} = \frac{1}{\alpha(0)} - \frac{1}{3\pi} \left(\sum_k e_k^2 \ln \frac{M_W^2}{m_0^2} \right) \quad , \tag{9.95}$$

where e_k denote the charges in units of the electron charge and m_0 is the mass of a light fermion fixing the renormalization point. Furthermore $1/\alpha(m_0) \approx 1/\alpha(0) = 137.036$ is the Sommerfeld constant. Taking into account the color degree of freedom for the quarks, which gives a factor of 3, we find that

$$\frac{1}{\alpha(M_W^2)} \approx 128 \quad . \tag{9.96}$$

The mass scale M is evaluated using (9.92):

$$\begin{aligned}
M &= M_W \exp \left[\frac{\pi}{11} \left(\frac{1}{\alpha(M_W^2)} - \frac{8/3}{\alpha_C(M_W^2)} \right) \right] \\
&\approx 7.4 \times 10^{14}\ \mathrm{GeV} \quad .
\end{aligned} \tag{9.97}$$

[15] W. Greiner and J. Reinhardt: *Quantum Electrodynamics*, 4th ed. (Springer, Berlin, Heidelberg, 2009).

Obviously the reason for this rather large value is the slow, logarithmic dependence of the coupling strengths on the momentum transfer q, which is itself connected to the renormalization procedure of the gauge theory. Therefore this exorbitant value of M is not a specific feature of the SU(5) model but is to be expected within the framework of any gauge theory of unified interaction. Having determined M we are now in position to check whether the relation between α_L and α' (9.87) is fulfilled at $q^2 = M_W^2$ and how well it agrees with experimental data.

Let us also investigate the Weinberg angle θ_W in dependence of q^2, i.e. $\theta_W(q^2)$. We are especially interested in the Weinberg angle at the mass of the W–Z bosons $\theta_W(M_W^2)$.

In terms of (9.83) the ratio of g' and g_L can be expressed as the tangent of the Weinberg angle θ_W. Supposing that the relation (9.82) also holds for the energy values of the unified scale M, we can therefore test whether the experimental value of the Weinberg angle at $q^2 \approx M_W^2$ is predicted correctly by the SU(5) theory. According to the second equation of (9.83) we have

$$\sin^2\theta_W(q^2) = \frac{e^2(q^2)}{g_L^2(q^2)} = \frac{\alpha(q^2)}{\alpha_L(q^2)} \qquad (9.98)$$

Using (9.88), the denominator can be expressed as

$$\frac{1}{\alpha_L(q^2)} = \frac{1}{\alpha(q^2)} - \frac{1}{\alpha'(q^2)} \quad ,$$

and therefore

$$\frac{1}{\alpha_L(q^2)} = \frac{5}{8}\frac{1}{\alpha_L(q^2)} + \frac{3}{8}\frac{1}{\alpha_L(q^2)} = \frac{5}{8\alpha_L(q^2)} + \frac{3}{8}\left(\frac{1}{\alpha(q^2)} - \frac{1}{\alpha'(q^2)}\right) \qquad (9.99)$$

Hence we obtain

$$\sin^2\theta_W(q^2) = \frac{3}{8} + \frac{\alpha(q^2)}{8}\left(\frac{5}{\alpha_L(q^2)} - \frac{3}{\alpha'(q^2)}\right) \quad . \qquad (9.100)$$

Because of the relation (9.87), the expression in parentheses vanishes for $q^2 = M^2$. To exhibit the explicit q^2 dependence we use (9.84), which yields

$$\begin{aligned} \sin^2\theta_W(q^2) &= \frac{3}{8} + \frac{\alpha(q^2)}{8}\left[\left(\frac{5}{\alpha_L(M^2)} - \frac{3}{\alpha'(M^2)}\right) + (5b_L - 3b')\ln\frac{q^2}{M^2}\right] \\ &= \frac{3}{8} + \frac{\alpha(q^2)}{8}(5b_L - 3b')\ln\frac{q^2}{M^2} \\ &= \frac{3}{8} + \frac{55}{48\pi}\alpha(q^2)\ln\frac{q^2}{M^2} \quad . \end{aligned} \qquad (9.101)$$

Here we have used (9.87), verifying that

$$5b_L - 3b' = \frac{5(11 - 2n_g)}{6\pi} + \frac{3 \times 5n_g}{9\pi} = \frac{55}{6\pi} \quad .$$

For $q^2 = M^2$ we recover the former relation (9.49), valid for the unbroken SU(5). However, with respect to the energy scale of the electroweak interaction, $q^2 \approx M_W^2$, we obtain with (9.96):

$$\sin^2\theta_W(M_W^2) \approx 0.206 \quad . \qquad (9.102)$$

This does not quite agree with the experimental value

$$\sin^2\theta_W^{\exp} = 0.2325 \pm 0.0008 \quad , \tag{9.103}$$

but is sufficiently close to represent a remarkable quantitative success of the SU(5) model of unified interactions.

Let us now consider the connection between the coupling constants from a different point of view, by extrapolating from presently known values (at $q^2 \approx (100\,\text{GeV})^2$) up to larger energies, via (9.84). As illustrated in Fig. 9.3, the three lines $\frac{5}{3}\alpha'(q^2)$, $\alpha_L(q^2)$, and $\alpha_C(q^2)$ meet at a single point, $q^2 \approx (10^{15}\,\text{GeV})^2$. Indeed, this result is an important condition, allowing us to describe all three interactions by a single, unified gauge group. For $q^2 > (10^{15}\,\text{GeV})^2$ the SU(5) symmetry is effectively unbroken, since the X and Y boson loops also give sizable contributions to the vacuum polarization, which causes the q^2 dependence of the effective coupling constants. Because of the underlying SU(5) symmetry, all coupling constants "run" in the same way when $q^2 \gg M_X^2, M_Y^2$; that is to say for $q^2 \gg M_X^2, M_Y^2$, there is, in effect, only a single running coupling constant $\alpha_5(q^2)$. Because of the large number of "additional" gauge bosons (X and Y), the q^2 dependence becomes much stronger:

$$\frac{1}{\alpha_5(q^2)} = \frac{1}{\alpha_5(M^2)} + b_5 \ln\frac{q^2}{M^2} \qquad (q^2 \gg M_X^2, M_Y^2) \quad ,$$

with – see (9.86) –

$$b_5 = \frac{55 - 4n_g}{12\pi} \quad .$$

Around $q^2 \approx M_X^2, M_Y^2$ there is a transition region, where the three different coupling constants α', α_L and α_C merge into a single one, α_5, as shown in Fig. 9.3. Figure 9.4 exhibits the dependence of the effective mixing angle θ_W of the GSW theory SU(2)×U(1) with respect to the energy, starting from the measured value $\sin^2\theta_W = 0.22$. Obviously the curve almost coincides with the predicted value $\frac{3}{8}$ at 10^{15} GeV. This agreement between SU(5) model predictions and experimental data is remarkable. Although the agreement is not precise, it is hard to imagine that these results are completely coincidental.

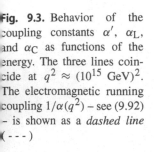

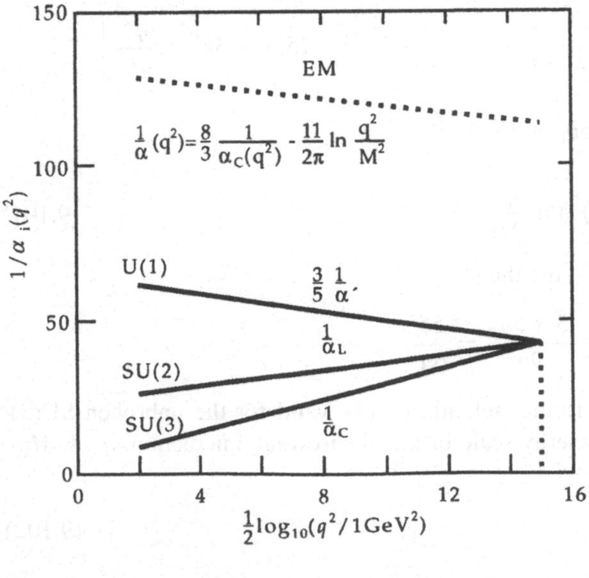

Fig. 9.3. Behavior of the coupling constants α', α_L, and α_C as functions of the energy. The three lines coincide at $q^2 \approx (10^{15}\ \text{GeV})^2$. The electromagnetic running coupling $1/\alpha(q^2)$ – see (9.92) – is shown as a *dashed line* (- - -)

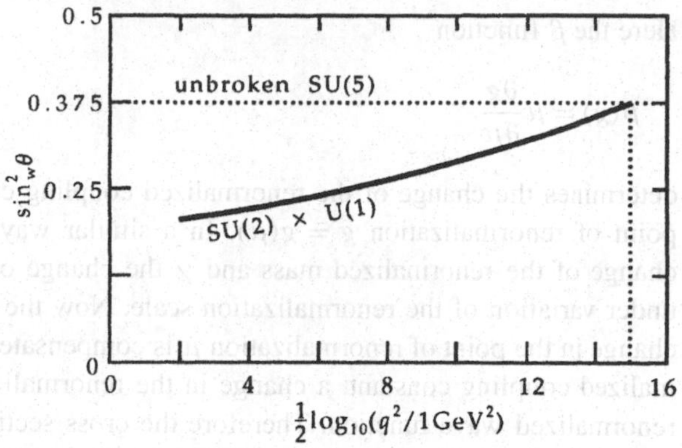

Fig. 9.4. Behavior of the mix-ing angle between the photo and Z^0 boson (Weinberg an-gle) as a function of the en-ergy

EXAMPLE

9.10 The Running Coupling Constant in Quantum Field Theory

When performing the transition from a free quantum field theory to a quantum field theory containing interaction we find that divergences appear in higher-order perturbation theory. These divergences are isolated in the framework of the renormalization programme and are subtracted from the physical parameters of the theory in such a way that these parameters retain a finite value. The divergences are regarded as self-energy and self-charge of the quanta due to their interaction with their own field of radiation. We have to keep in mind that self-energies and renormizable divergences also appear in classical electrodynamics. The process of renormalization in quantum field theory is unequivocal apart from *finite* renormalizations. For example, when performing a mass-renormalization we find that the quantity δm is unequivocally defined through a certain covariant, but diverging integral; however, any given *finite* values can be added to or subtracted from this integral. The physical mass is obtained from the bare (unrenormalized) mass and the renormalization contribution $m_{\text{phys}} = m_b + \delta m$. The same holds for the renormalized charge g: $g_{\text{phys}} = g_b + \delta g$. Apart from the mass and charge of the considered particle also its wave-function has to be renormalized. The renormalization is unequivocal if certain normalization constraints are imposed on the renormalized quantities. One demands that the renormalized physical quantities for a certain value of the external momenta, e.g. $p^2 = \mu^2$, take on a specific value. Ideally this value is known through an experiment. In QED, for example, the coupling constant at medium momentum transfers is normalized to the measured value of $\alpha(\text{QED}) = 1/137$. One might think that this method causes all renormalized quantities, supposedly even the physical observables, to be dependent on the renormalization point μ^2. This is not really the case, however. In reality the renormalizations of the charge, mass and wavefunction must balance in a way that physical observables, such as a cross section, no longer depend on the chosen point of renormalization. If $G(p^2, g, m, \mu^2)$ is a Green's function belonging to the theory, then its independence is described by the renormalization-group equation:

$$\left(\mu\frac{\partial}{\partial\mu} + \beta\frac{\partial}{\partial g} + \eta\frac{\partial}{\partial m} - \gamma_F\right) G(p^2, g, m, \mu^2) = 0 \quad . \tag{1}$$

xample 9.10

Here the β function

$$\beta(g) = \mu \frac{\partial g}{\partial \mu} \tag{2}$$

determines the change of the renormalized coupling constant under variation of the point of renormalization $g = g(\mu)$. In a similar way $\eta(g) = \mu \frac{\partial m}{\partial \mu}$ determines the change of the renormalized mass and γ the change of the particle's wave function under variation of the renormalization scale. Now the meaning of (1) is clear. Each change in the point of renormalization μ is compensated for by a change in the renormalized coupling constant, a change in the renormalized mass and a change in the renormalized wave function. Therefore the cross section, which is the physical observable actually being measured, is independent of μ^2. It follows from (1) that the coupling constant g fulfills a differential equation which, following (2) and using $\alpha = g^2/4\pi$, can be written as

$$\mu^2 \frac{\mathrm{d}\alpha(\mu^2)}{\mathrm{d}\mu^2} = \beta(\alpha) = -b_0\alpha^2 - b_1\alpha^3 + \cdots \quad . \tag{3}$$

ig. 9.5. Renormalization of he coupling constant $g_n \to g(\mu)$. The points symbolize he inclusion of radiative corrections in the propagators and he vertex

Here we have expanded the β function in terms of the coupling constant. b_0 is the coefficient of the β function in the one-loop approximation. We hereby determine the dependence of the coupling constant on the renormalization parameter by using perturbation theory. One obtains the coefficients b_i by calculating the graph (of Fig. 9.5). The radiative corrections are calculated in the one-loop approximation to generate β_0.

In the case of non-abelian SU(n) gauge theories one finds in this approximation

$$b_0 = \frac{11n - 4n_f}{12\pi} \quad , \quad n \geq 2 \quad , \tag{4}$$

and in the case of the abelian U(1) theory

$$b_0 = -\frac{n_f}{3\pi} \quad . \tag{5}$$

Here n_f is the number of active fermions.[16] The differential equation used to define the β function can be solved and we obtain for the running coupling constant

$$\alpha(Q^2) = \frac{\alpha(\mu^2)}{1 + \alpha(\mu^2) b_0 \ln(Q^2/\mu^2)} \quad . \tag{6}$$

This is exactly (9.84):

$$\frac{1}{\alpha_i(Q^2)} = \frac{1}{\alpha_i(\mu^2)} + b_0 \ln \frac{Q^2}{\mu^2} \quad . \tag{7}$$

The sign of b_0 obviously determines the asymptotic behavior of the coupling constant. For $b_0 < 0$ the coupling constant increases with increasing momentum transfer Q^2, whereas for $b_0 > 0$ it vanishes for $Q^2 \to \infty$. This behavior is known in QCD as *asymptotic freedom* and is a fundamental property of non-abelian gauge theories. The

16 For the explicit calculation of these coefficients we refer to W. Greiner, S. Schramm, and E. Stein: *Quantum Chromodynamics*, 3rd ed. (Springer, Berlin, Heidelberg, 2007); W. Greiner and J. Reinhardt: *Quantum Electrodynamics*, 4th ed. (Springer, Berlin, Heidelberg, 2009).

positive sign may occur if $11n > 4n_f$ in (4). The positive contribution proportional to n is caused by the self-coupling of the gauge bosons, which is only possible in non-abelian gauge theories. The different asymptotic behavior of the SU(3) and U(1) coupling constants is a necessary condition in order to unify them in a higher gauge group SU(5) in which the coupling constants should meet at the unification point at a higher Q^2 scale.

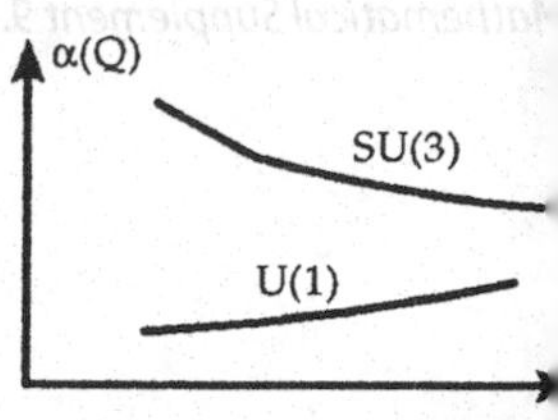

Fig. 9.6. Schematic Q deper dence of the renormalized co pling constants of QED an QCD

MATHEMATICAL SUPPLEMENT

9.11 Anomaly Freedom

Since about 1970 considerable interest in gauge theories has emerged, mainly because of the success of proving the renormalizability of such theories. However, in order to complete the proof of renormalizability an additional condition is required which cannot be fulfilled for arbitrary choices of particle multiplets. This fact leads to an important constraint for the standard model of the electroweak interaction.

The proof of renormalizability of the Glashow–Salam–Weinberg model only holds under the condition that the sum of the weak hypercharges of all left-handed fermions is zero, that is,

$$\sum_{\substack{\text{left-handed}\\ \text{quarks and leptons}}} Y_i = 0 \quad . \tag{1}$$

Have a look at Table 6.2! Since every lepton multiplet contributes a value of $Y = -2$ (there are two left-handed leptons, each of which carries $Y = -1$) and every quark multiplet a value of $+2/3$ (there are two left-handed quarks, each of which carries $Y = 1/3$), the sum only vanishes if

1. there are three colors, that is, every quark exists in three color versions, and
2. the number of quark flavors equals the number of lepton species.

Both conditions are an immediate consequence of the Georgi–Glashow model of the SU(5) gauge theory of unified interactions. Indeed, (1) is essential for the unification of the electroweak and color interactions, since only if there exists a principle of order exceeding the SU(3)×SU(2)×U(1) symmetry and relating color charges and weak charges can (1) be fulfilled. In the following we will briefly describe how the requirement for the validity of (1) emerges.[17] The technique for renormalizing gauge theories is the so-called dimensional regularization. This method makes extensive use of the rich possibilities of the theory of analytic functions to isolate the divergences that occur. The procedure is the following. First all relevant Feynman diagrams are evaluated for a general integer space-time dimension D. The expressions obtained, being functions of D, are then analytically continued to arbitrary, even complex values of D. For $D = 4$ these expressions diverge, and the residue for $D = 4$ is identified with the divergent contribution to be renormalized. The renormalizability of all n-point functions occurring is then iteratively proved by means of complete induction.

[17] W. Greiner, S. Schramm, and E. Stein: *Quantum Chromodynamics*, 3rd ed. (Springer, Berlin, Heidelberg, 2007); C. Itzykson, J.-B. Zuber: *Quantum Field Theory* (McGraw-Hill, New York, 1980).

Mathematical Supplement 9.11

An essential ingredient of the proof is the knowledge of certain relations between different n-point functions. These relations, the so-called Slavnov–Taylor identities (the generalizations of the Ward identities of quantum electrodynamics), follow from the gauge invariance of the theory. Since the proof for renormalizability is based on the validity of these relations, it is very important that the employed regularization scheme does not break the gauge invariance. The method of dimensional regularization satisfies this requirement. It has, however, the disadvantage that it is not applicable for certain Feynman diagrams. For example, the diagram in Fig. 9.7 in four dimensions is proportional to the anti-symmetric tensor $\varepsilon_{\mu\nu\lambda\alpha}$. Since this tensor carries exactly four indices, it cannot be continued to general dimensions.

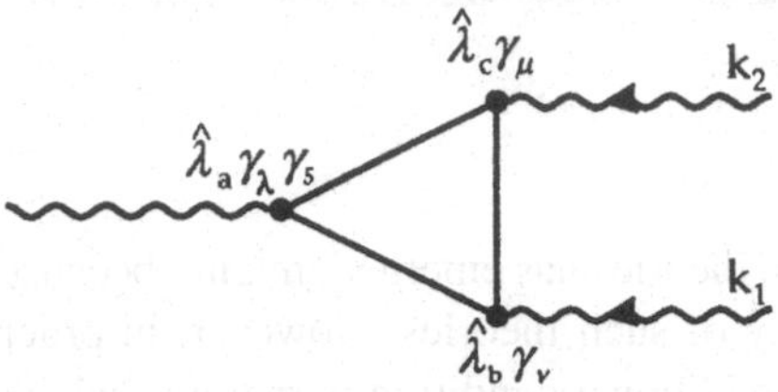

Fig. 9.7. For this diagram the dimensional regularization cannot be applied

This diagram is therefore excluded from the general proof and must be treated separately by means of a different regularization scheme, for example, the Pauli–Villars method. It has to be proved that this diagram and also all diagrams of higher order containing the ε tensor can be renormalized without destroying gauge invariance. If the scheme were not gauge invariant, the Slavnov–Taylor identities could not be established, and consequently there would be no basis for the proof for renormalizability in general. As it turns out, the gauge invariance is conserved only if

$$d_{abc} = \sum_{\text{fermions}} \mathrm{Tr}\{\hat{\lambda}^a\{\hat{\lambda}^b, \hat{\lambda}^c\}_+\} = 0 \quad . \tag{2}$$

This non-trivial additional constraint for the gauge group and the chosen representations is called *anomaly freedom*. We emphasize that gauge theories that do not satisfy the condition (2) could nevertheless be renormalizable. The statement is solely that by means of the known methods nothing can be said about the renormalizability, whereas for theories fulfilling (2) the renormalizability can be proved explicitly.

Let us now consider these statements and calculate the anomaly freedom. The great success of gauge theories in the past twenty years has been driven mainly by the discovery of a general proof that gauge theories can be renormalized. However there is an additional constraint that has to be fulfilled in order to apply this proof: the particular gauge theory must be *anomaly free*.

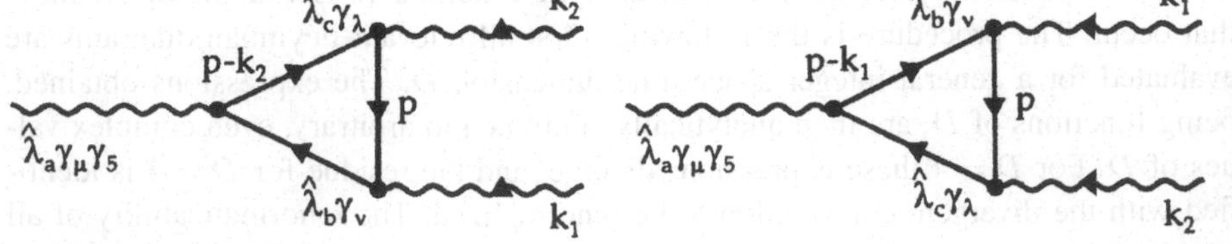

Fig. 9.8. The two diagrams that are important for the anomaly. The *left* one differs from the *right* one by the exchange of the two photons

The anomaly of gauge theories is a modification of the ordinary Ward identities and arises in the calculation of the triangle diagrams in Fig. 9.8. The major aspect is an odd number of axial-vector couplings where only the case of *one* γ_5 coupling

Mathematical Supplement 9.1

has to be considered since the diagram with three axial-vector couplings leads back to one γ_5 case by applying the commutation relations for γ_5. Triangle diagrams with an even number of axial-vector couplings imply an odd number of vector couplings and therefore vanish due to Furry's theorem. The $\hat{\lambda}^i$ are the generators of the gauge group under consideration.

The amplitude of the triangle diagram and of the respective exchange diagram may be denoted by following the Feynman rules:

$$T^{abc}_{\mu\nu\lambda} = -\mathrm{i}\int\frac{\mathrm{d}^4p}{(2\pi)^4}\left\{\mathrm{Tr}\left[\frac{\mathrm{i}}{\not{p}-\not{k}_2-m}\frac{\hat{\lambda}^c}{2}\gamma_\lambda\frac{\mathrm{i}}{\not{p}-m}\frac{\hat{\lambda}^b}{2}\gamma_\nu\frac{\mathrm{i}}{\not{p}+\not{k}_1-m}\frac{\hat{\lambda}^a}{2}\gamma_\mu\gamma_5\right] + \begin{pmatrix} k_1\leftrightarrow k_2\\ \nu\leftrightarrow\lambda\\ b\leftrightarrow c\end{pmatrix}\right\} . \quad (3)$$

The generators $\hat{\lambda}^i$ and the γ matrices act in different spaces and therefore the trace can be decomposed in the product of the trace over the $\hat{\lambda}$ and the trace over the γ:

$$T^{abc}_{\mu\nu\lambda} = \frac{1}{8}\mathrm{Tr}(\hat{\lambda}^a\hat{\lambda}^c\hat{\lambda}^b)\left(-\mathrm{i}\int\frac{\mathrm{d}^4p}{(2\pi)^4}\mathrm{Tr}\left[\frac{\mathrm{i}}{\not{p}-\not{k}_2-m}\gamma_\lambda\frac{\mathrm{i}}{\not{p}-m}\gamma_\nu\frac{\mathrm{i}}{\not{p}+\not{k}_1-m}\gamma_\mu\gamma_5\right]\right) + \frac{1}{8}\mathrm{Tr}(\hat{\lambda}^a\hat{\lambda}^b\hat{\lambda}^c)\begin{pmatrix} k_1\leftrightarrow k_2\\ \nu\leftrightarrow\lambda\end{pmatrix} . \quad (4)$$

A simple calculation shows that the trace over the γ is (in the limit $m\to 0$) symmetric under the exchange $(\nu\leftrightarrow\lambda, k_1\leftrightarrow k_2)$:

$$\begin{aligned}&\mathrm{Tr}[\gamma_\alpha\gamma_\lambda\gamma_\beta\gamma_\nu\gamma_\gamma\gamma_\mu\gamma_5](p-k_2)^\alpha p^\beta(p+k_1)^\gamma\\ &\overset{p\to -p}{\longrightarrow} -\mathrm{Tr}[\gamma_\alpha\gamma_\lambda\gamma_\beta\gamma_\nu\gamma_\gamma\gamma_\mu\gamma_5](p+k_2)^\alpha p^\beta(p-k_1)^\gamma\\ &= -\mathrm{Tr}[\gamma_5^t\gamma_\mu^t\gamma_\gamma^t\gamma_\nu^t\gamma_\beta^t\gamma_\lambda^t\gamma_\alpha^t](p+k_2)^\alpha p^\beta(p-k_1)^\gamma .\end{aligned}$$

We have used $\mathrm{Tr}(M)=\mathrm{Tr}(M^t)$. Note that the replacement $p\to -p$ is valid due to the integration over the total space. Inserting the unity operator $1=\hat{C}^{-1}\hat{C}$, with $\hat{C}$ being the charge-conjugation operator, and using $\hat{C}\gamma_\mu\hat{C}^{-1}=-\gamma_\mu^t$ and $\hat{C}\gamma_5\hat{C}^{-1}=+\gamma_5^t$ one obtains

$$-\mathrm{Tr}[\gamma_5\gamma_\mu\gamma_\gamma\gamma_\nu\gamma_\beta\gamma_\lambda\gamma_\alpha](p+k_2)^\alpha p^\beta(p-k_1)^\gamma , \quad (5)$$

which yields, after the renaming of the indices $\alpha\leftrightarrow\gamma$ and commutation of γ_5 with γ_μ – note that $\gamma_5\gamma^\mu=-\gamma^\mu\gamma_5$ (see (A.15′)) –

$$\mathrm{Tr}[\gamma_\alpha\gamma_\nu\gamma_\beta\gamma_\lambda\gamma_\gamma\gamma_\mu\gamma_5](p-k_1)^\alpha p^\beta(p+k_2)^\gamma \quad (6)$$

After the exchange $(\nu\leftrightarrow\lambda, k_1\leftrightarrow k_2)$ this simply reproduces our initial expression. Hence due to the symmetry we denote

$$T^{abc}_{\mu\nu\lambda} = \frac{1}{8}D^{abc}\left(-\mathrm{i}\int\frac{\mathrm{d}^4p}{(2\pi)^4}\mathrm{Tr}\left[\frac{\mathrm{i}}{\not{p}-\not{k}_2}\gamma_\lambda\frac{\mathrm{i}}{\not{p}}\gamma_\nu\frac{\mathrm{i}}{\not{p}-\not{k}_1}\gamma_\mu\gamma_5\right]\right) \quad (7)$$

with

$$D^{abc} = \sum_{\mathrm{f}}\{\mathrm{Tr}[\hat{\lambda}^a\hat{\lambda}^b\hat{\lambda}^c]+\mathrm{Tr}[\hat{\lambda}^a\hat{\lambda}^c\hat{\lambda}^b]\} = \sum_{\mathrm{f}}\mathrm{Tr}[\hat{\lambda}^a\{\hat{\lambda}^b,\hat{\lambda}^c\}_+] , \quad (8)$$

Mathematical Supplement 9.11

where the sums run over all fermions that contribute to the anomaly exactly in the same way.

The numerator of the integral expression is of order p^4 while the denominator is only of the order p^3. This leads to an *linearly divergent* integral expression that cannot be renormalized. Thus we cannot apply the proof of renormalizability to the gauge theory *unless the contribution of the anomaly vanishes*. Obviously this can be guaranteed only if the gauge group fulfills the requirement for anomaly freedom:

$$D^{abc} = \sum_{\mathrm{f}} \mathrm{Tr}[\hat{\lambda}^a \{\hat{\lambda}^b, \hat{\lambda}^c\}_+] = 0 \quad . \tag{9}$$

In this context we will stress once more that a gauge theory might be renormalizable, whether or not it is anomaly free. The general proof of renormalizability cannot be applied on its own.

EXERCISE

9.12 Is the Glashow–Salam–Weinberg Theory Anomaly Free?

Show that the Glashow–Salam–Weinberg model of electroweak interaction is anomaly free if there are three colors and the number of lepton families equals the number of quark families. Note that left-handed (lh) and right-handed (rh) fermions couple to the anomaly with opposite sign:

$$J^i_\mu(\mathrm{L}) \sim \bar{\psi}_\mathrm{L} \hat{\lambda}^i \gamma_\mu (1 - \gamma_5) \psi_\mathrm{L} \quad , \tag{1}$$

$$J^i_\mu(\mathrm{R}) \sim \bar{\psi}_\mathrm{R} \hat{\lambda}^i \gamma_\mu (1 + \gamma_5) \psi_\mathrm{R} \quad . \tag{2}$$

Solution. The GSW theory is based on the gauge group SU(2) × U(1) that is generated by the three Pauli matrices σ_i and the hypercharge Y:

$$\hat{\lambda}^i = \sigma_i \quad , \quad \text{for} \quad i = 1, 2, 3 \quad , \tag{3}$$

$$\hat{\lambda}^4 = \hat{Y} = 2(\hat{Q} - \hat{T}_3) \quad , \tag{4}$$

obeying the commutation relations

$$\{\sigma_i, \sigma_j\}_+ = 2\delta_{ij} \quad . \tag{5}$$

Equation (9) of Mathematical Supplement 9.11 ($i, j, k = 1, 2, 3$) reads

$$D^{ijk} = \sum_{\mathrm{f}} \mathrm{Tr}[\sigma_i \{\sigma_j, \sigma_k\}_+] = 2\delta_{jk} \underbrace{\mathrm{Tr}\,\sigma_i}_{=0} = 0 \quad . \tag{6}$$

In the case when two of the $\hat{\lambda}^i$ are the hypercharge $\hat{Y}$ we find that expression (8) in Mathematical Supplement 9.11 also vanishes due to $[\hat{Y}, \sigma_i] = 0$ and because hypercharge and Pauli matrices act in separate spaces:

$$D^{i44} = D^{4i4} = D^{44i} = \sum_{\mathrm{f}} \mathrm{Tr}\,[\sigma_i 2\hat{Y}^2] \sim \mathrm{Tr}\,\sigma_i = 0 \quad . \tag{7}$$

Exercise 9.1

For one $\hat{\lambda}$ being replaced by $\hat{Y}$ we get

$$D^{4jk} = D^{j4k} = D^{jk4} = \sum_{\mathrm{f}} \mathrm{Tr}(\hat{Y}\{\sigma_j, \sigma_k\}_+) = 2\delta_{jk} \sum_{\mathrm{f}} \mathrm{Tr}\hat{Y} = 2\delta_{jk} \sum_{\mathrm{f}} Y \tag{8}$$

where we used $[\sigma_i, \hat{Y}] = 0$ and cyclic permutation under the trace. The sum can be decomposed onto the sum over leptons and quarks where the left-handed and right-handed contributions enter with opposite signs due to (1) and (2)

$$\sum_{\mathrm{f}} Y = \sum_{\text{lh leptons}} Y + \sum_{\text{lh quarks}} Y - \left(\sum_{\text{rh leptons}} +Y \sum_{\text{rh quarks}} Y \right) . \tag{9}$$

For the left-handed lepton and quark states in one family[18] (ν, e_L and u_L, d_L, respectively) one obtains, using Table 9.1,

$$\sum_{\text{lh leptons}} Y = -1 + (-1) = -2 \quad , \tag{10}$$

$$\sum_{\text{lh quarks}} Y = 1/3 + 1/3 = 2/3 \quad , \tag{11}$$

while the right-handed contributions (e_R, and u_R, d_R) give

$$\sum_{\text{rh leptons}} Y = -2 \quad , \tag{12}$$

$$\sum_{\text{rh quarks}} Y = 4/3 + (-2/3) = 2/3 \quad . \tag{13}$$

Table 9.1. The quantum numbers of the first fermion family of the standard model

	T	T_3	$\hat{Y}$	Q
ν_e	$+\frac{1}{2}$	$+\frac{1}{2}$	-1	$+0$
e_L	$+\frac{1}{2}$	$-\frac{1}{2}$	-1	-1
e_R	$+0$	$+0$	-2	-1
u_L	$+\frac{1}{2}$	$+\frac{1}{2}$	$+\frac{1}{3}$	$+\frac{2}{3}$
$(d_\mathrm{C})_\mathrm{L}$	$+\frac{1}{2}$	$-\frac{1}{2}$	$+\frac{1}{3}$	$-\frac{1}{3}$
u_R	$+0$	$+0$	$+\frac{4}{3}$	$+\frac{2}{3}$
$(d_\mathrm{C})_\mathrm{R}$	$+0$	$+0$	$-\frac{2}{3}$	$-\frac{1}{3}$

Thus the contributions of leptons and quarks cancel if

- all quarks have three representations, i.e. there are three colors.
- the number of lepton families equals the number of quarks families.

[18] The extension to more than one family is trivial due to identical quantum numbers of the respective leptons and quarks in the families.

xercise 9.12

Finally we have to examine the case that all three $\hat{\lambda}$ are the hypercharge. Since this represents a singlet state the hypercharge operator simply reduces to a c number and we get

$$\mathrm{Tr}(\hat{Y}\hat{Y}\hat{Y}) \sim Y^3 \quad . \tag{14}$$

Due to left-handed and right-handed fermions having opposite sign we obtain

$$\sum_{\text{lh leptons}} Y^3 - \sum_{\text{rh leptons}} Y^3 = (-1)^3 + (-1)^3 - (-2)^3 = +6 \quad , \tag{15}$$

$$\sum_{\text{lh quarks}} Y^3 - \sum_{\text{rh quarks}} Y^3 = 3\big((1/3)^3 + (1/3)^3 - (4/3)^3 - (-2/3)^3\big) = -6 \quad , \tag{16}$$

where the factor of 3 is introduced to account for the three colors. We observe that the singlet contribution vanishes as well. Thus the standard model of electroweak interaction is anomaly free and therefore renormalizable.

9.6 Proton Decay

As discussed at the end of Sect. 9.3, the exchange of leptoquarks X or Y violates the conservation of the baryon number and therefore allows the proton to decay into a meson and a lepton. Figure 9.1 shows some of the decay channels

$$\mathrm{p} \to \begin{Bmatrix} \mathrm{d\bar{d}e^+} \\ \mathrm{u\bar{u}e^+} \end{Bmatrix} \to \begin{cases} \pi^0\mathrm{e}^+ \\ \rho^0\mathrm{e}^+ \\ \omega\mathrm{e}^+ \\ \eta\mathrm{e}^+ \end{cases} , \tag{9.104a}$$

$$\mathrm{p} \to \mathrm{u\bar{d}}\bar{\nu}_\mathrm{e} \to \begin{cases} \pi^+\bar{\nu}_\mathrm{e} \\ \rho^+\bar{\nu}_\mathrm{e} \end{cases} \quad . \tag{9.104b}$$

Note that we use for the antiparticles two notations, e.g. $\bar{d} \equiv d^\mathrm{C}$, $\bar{u} = u^\mathrm{C}$, $\bar{\nu}_\mathrm{e} = \nu_\mathrm{e}^\mathrm{C}$ etc., just as it seems convenient. The ρ and the π mesons have the same quark structure, as do the ω and the η mesons. Only their angular-momentum coupling differs, namely a total angular momentum of 1 for the ρ and the ω and 0 for the π and the η. See also Figs. 6.3 and 6.4. On the other hand, the fact that the elementary processes

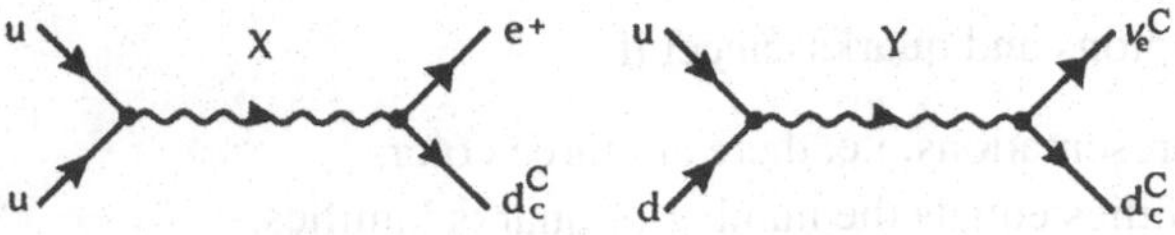

(9.105)

include the production of a "Cabibbo" d antiquark,

$$d_\mathrm{c}^\mathrm{C} = d^\mathrm{C}\cos\theta_\mathrm{c} + s^\mathrm{C}\sin\theta_\mathrm{c} \quad , \tag{9.106}$$

implies that decays into a lepton and K meson are also possible (for the quark content see the table opposite to Fig. 6.5):

$$\mathrm{p} \to \mathrm{d}\mathrm{s}^{\mathrm{C}}\mathrm{e}^+ \to \mathrm{K}^0\mathrm{e}^+ \quad , \tag{9.107a}$$

$$\mathrm{p} \to \mathrm{u}\mathrm{s}^{\mathrm{C}}\nu_{\mathrm{e}}^{\mathrm{C}} \to \mathrm{K}^+\nu_{\mathrm{e}}^{\mathrm{C}} \quad . \tag{9.107b}$$

Note that the superscript C indicates the antiparticle (charge conjugation) while the subscript C stands for "Cabibbo". Another possibility is the decay of the intermediate $\bar{\mathrm{X}}$ and $\bar{\mathrm{Y}}$ bosons into quarks and leptons of the second generation:

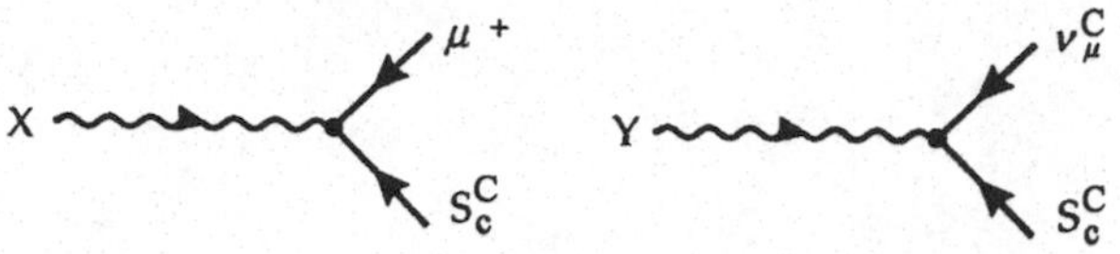

(9.108)

The decay channels (9.104), (9.107) therefore need to be supplemented by the following channels:

$$\left.\begin{array}{l} \mathrm{p} \to \mathrm{d}\mathrm{d}^{\mathrm{C}}\mu^+ \to \pi^0\mu^+ \\ \mathrm{p} \to \mathrm{u}\mathrm{d}^{\mathrm{C}}\bar{\nu}_\mu \to \pi^+\nu_\mu^{\mathrm{C}} \\ \mathrm{p} \to \mathrm{d}\mathrm{s}^{\mathrm{C}}\mu^+ \to \mathrm{K}^0\mu^+ \\ \mathrm{p} \to \mathrm{u}\mathrm{s}^{\mathrm{C}}\nu_\mu^{\mathrm{C}} \to \mathrm{K}^+\nu_\mu^{\mathrm{C}} \end{array}\right\} \quad , \tag{9.109}$$

where the last two decay processes are again "Cabibbo suppressed".

The calculation of the proton lifetime, as predicted by the SU(5) theory, is complicated by the fact that the quarks in the initial and final states are bound within the proton or a meson, respectively. Since the explicit form of the wave functions corresponding to these bound states is unknown, one has to rely on certain models. One possibility is to employ the MIT bag model (see Sect. 6.1).

In the following we shall instead sketch how to evaluate the decay of the proton within the non-relativistic quark model.[19] In this model the quarks are bound by an external potential. This picture is justified insofar as it allows us to describe weak decay processes quite well, for example,

$$\Lambda \to \mathrm{p} + \pi^- \quad ,$$

which involve the same difficulties as proton decay.[20] As an example we calculate the matrix element that corresponds to the annihilation of a left-handed positron and a left-handed d antiquark, that is,

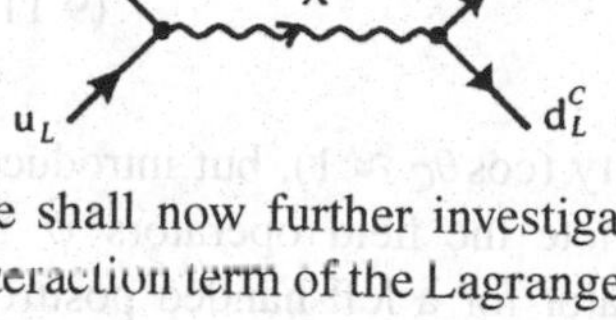

(9.110)

We shall now further investigate this diagram and its origin. First we consider the interaction term of the Lagrange density in (8) of Example 9.13 (left couples to right!):

$$(\bar{e}^{\mathrm{C}})_L\gamma^\mu(d)_L X_\mu - (\bar{u}^{\mathrm{C}})_L\gamma^\mu(u)_L X_\mu^* = \bar{e}^{\mathrm{C}}\gamma^\mu\frac{(1-\gamma_5)}{2}dX_\mu - \bar{u}^{\mathrm{C}}\gamma^\mu\frac{(1-\gamma_5)}{2}uX_\mu^* \quad .$$

---

[19] M.B. Gavela, A. Le Yauonc, L. Oliver, O. Pène, J.C. Raynal: Phys. Rev. **D23**, 1580 (1981).

[20] A. Le Yauonc, L. Oliver, O. Pène, J.C. Raynal: Phys. Lett. **B72**, 53 (1977).

Graphically we can describe this interaction as

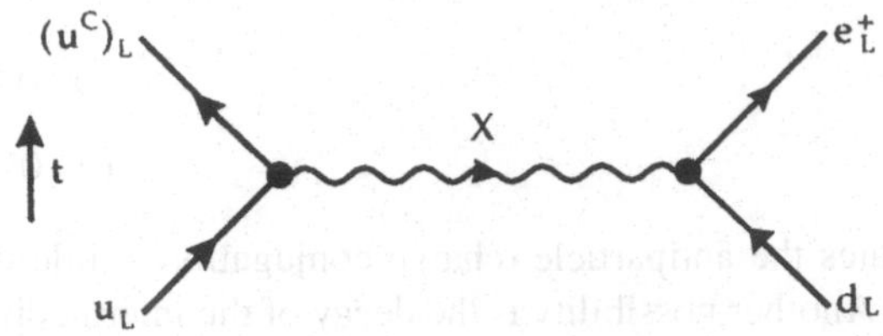

Helicity and charge are conserved at each vertex, due to

$$+\frac{2}{3} = \frac{4}{3} - \frac{2}{3} \quad ,$$
$$1 = -\frac{1}{3} + \frac{4}{3} \quad .$$

Now we consider the annihilation process

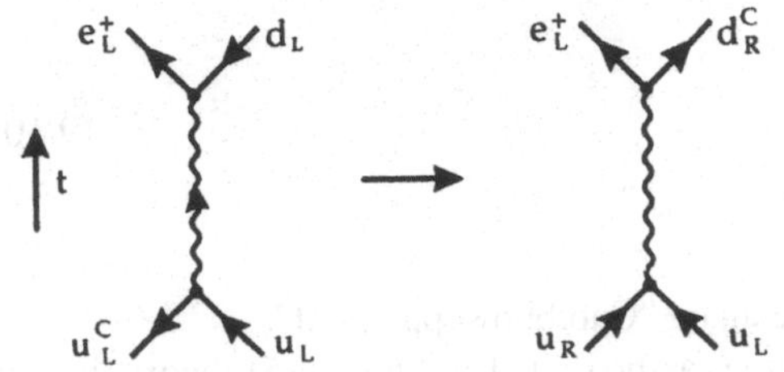

Again charge and helicity are conserved. Note that an outgoing left-handed antiparticle is equivalent to an incoming right-handed particle. Like the gluons, the $W^{\pm}$ and the Z^0 bosons are vector bosons. The same is true for the X and Y bosons: they are also vector bosons.

Recalling that the mass of the X bosons exceeds the W and Z boson masses by several orders of magnitude, one can describe the exchange of the X boson by an effective point-like interaction, analogous to the Fermi theory, by approximating the X-boson propagator as

$$\frac{g_{\mu\nu} - k_\mu k_\nu / M_X^2}{k^2 - M_X^2} \longrightarrow -\frac{g_{\mu\nu}}{M_X^2} \quad . \tag{9.111}$$

Hence, the effective Hamiltonian that describes the considered process is given by (see Exercise 9.13)

$$\hat{H}_I = \frac{g_5^2}{M_X^2} \varepsilon_{\alpha\beta\gamma} \hat{\bar{u}}^c_\gamma \gamma_\mu \frac{1-\gamma_5}{2} \hat{u}_\beta \hat{\bar{e}}^c_L \gamma^\mu \frac{1-\gamma_5}{2} \hat{d}_\alpha \quad , \tag{9.112}$$

where we have omitted the factor $\cos\theta_C$ for simplicity ($\cos\theta_C \approx 1$), but introduced the left-handedness of the interaction. $\hat{u}$ and $\hat{d}$ denote the field operators $\hat{\psi}$ for u and d quark, respectively, and $\hat{e}^c_L$ is the field operator for a left-handed positron. The indices α, β, γ label the color of the quarks, which are coupled by the $\varepsilon_{\alpha\beta\gamma}$ tensor to color zero. $\hat{\psi}^c$ is the charge conjugate of the field operator $\hat{\psi}$. The field operators are explicitly given by

$$\hat{\psi} = u\hat{b}_q + v\hat{d}_{\bar{q}}{}^\dagger \quad , \quad \hat{\bar{\psi}} = \bar{u}\hat{b}^\dagger_q + \bar{v}\hat{d}_{\bar{q}} \quad , \tag{9.113}$$

$$\hat{\psi}^c = u_c\hat{d}_{\bar{q}} + v_c\hat{b}_q{}^\dagger \quad , \quad \hat{\bar{\psi}}^c = \bar{u}_c\hat{d}_{\bar{q}}{}^\dagger + \bar{v}_c\hat{b}_q \quad ,$$

where $\hat{b}_q$ and $\hat{d}_{\bar{q}}$ are the annihilation operators for quarks and $\hat{b}_q{}^\dagger$ and $\hat{d}_{\bar{q}}{}^\dagger$ are the corresponding creation operators. Finally u and v denote the wave functions for the particle and antiparticle, while u_c and v_c are spinors for the antiparticle and particle, respectively. Between u_c and v, as well as between v_c and u, holds the well-known connection via charge conjugation, which will be written down explicitly in (9.117).

The matrix element corresponding to the process (9.110) then takes the following form (see Exercise 9.13):

$$\frac{g_5{}^2}{M_X^2}\varepsilon_{\alpha\beta\gamma}\left[\bar{v}_{\bar{u}\gamma}\gamma_\mu\left(\frac{1-\gamma_5}{2}\right)u_{u\beta}\right]\left[\bar{u}_{\bar{e}}\gamma^\mu\left(\frac{1-\gamma_5}{2}\right)v_{d\alpha}\right] \quad . \tag{9.114a}$$

The spinor $\bar{v}_{\bar{u}\gamma}$ physically describes an incoming u antiquark. The antiquark of a left-handed u quark is right-handed, that is, a right-handed u antiquark. We now perform a Fierz transformation, similar to that in Mathematical Supplement 2.12 and obtain

$$\frac{g_5^2}{M_X^2}\varepsilon_{\alpha\beta\gamma}\left[\bar{u}_{\bar{e}}\gamma_\mu\left(\frac{1-\gamma_5}{2}\right)u_{u\beta}\right]\cdot\left[\bar{v}_{\bar{u}\gamma}\gamma^\mu\left(\frac{1-\gamma_5}{2}\right)v_{d\alpha}\right] \quad . \tag{9.114b}$$

In the following we will label transposed spinors and matrices with a tilde (˜). Using this notation we have

$$\bar{v}_{\bar{u}\gamma}\gamma^\mu\left(\frac{1-\gamma_5}{2}\right)v_{d\alpha} = \tilde{v}_{d\alpha}\left(\frac{1-\gamma_5}{2}\right)\tilde{\gamma}^\mu\tilde{\bar{v}}_{\bar{u}\gamma} = \tilde{v}_{d\alpha}\tilde{\gamma}^\mu\left(\frac{1+\gamma_5}{2}\right)\tilde{\bar{v}}_{\bar{u}\gamma} \quad . \tag{9.115}$$

With respect to the standard representation of the γ matrices we have

$$\tilde{\gamma}^\mu = \begin{Bmatrix} +\gamma^\mu & \text{for} & \mu = 0,2 \\ -\gamma^\mu & \text{for} & \mu = 1,3 \end{Bmatrix} = \mathrm{i}\gamma^2\gamma^0\gamma^\mu\mathrm{i}\gamma^2\gamma^0 \quad . \tag{9.116}$$

Furthermore, we can express the spinor $u_{\bar{q}}(\boldsymbol{p},s)$ for an antiquark by the charge-conjugated spinor of a quark, because

$$\begin{aligned} \psi_c &= \hat{C}\bar{\psi}^{\mathrm{T}} = \mathrm{i}\gamma^2\gamma^0\gamma^{0\mathrm{T}}(\psi^\dagger)^{\mathrm{T}} \\ &= \mathrm{i}\gamma^2(\tilde{\psi}^*)^\sim = \mathrm{i}\gamma^2\psi^* \quad , \\ u_{\bar{q}}(\boldsymbol{p},s) &= \mathrm{i}\gamma^2 v_q^*(\boldsymbol{p},s) = \mathrm{i}\gamma^2\tilde{v}_q^\dagger(\boldsymbol{p},s) \quad . \end{aligned} \tag{9.117}$$

Since the negative energy spinor v_{q} corresponds to the antiparticle, we find, with $\gamma^{2\dagger} = -\gamma^2$, that

$$\begin{aligned} \tilde{v}_{d\alpha}\tilde{\gamma}^\mu\left(\frac{1+\gamma_5}{2}\right)\tilde{\bar{v}}_{\bar{u}\gamma} &= \tilde{v}_{d\alpha}\mathrm{i}\gamma^2\gamma^0\gamma^\mu\mathrm{i}\gamma^2\gamma^0\left(\frac{1+\gamma_5}{2}\right)\gamma^0\tilde{v}^\dagger_{\bar{u}\gamma} \\ &= \tilde{v}_{d\alpha}(\mathrm{i}\gamma^2)^\dagger\gamma^0\gamma^\mu\mathrm{i}\gamma^2\gamma^0\left(\frac{1+\gamma_5}{2}\right)\gamma^0\tilde{v}^\dagger_{\bar{u}\gamma} \\ &= \bar{u}^\dagger_{\bar{d}\alpha}\gamma^0\gamma^\mu\left(\frac{1+\gamma_5}{2}\right)\mathrm{i}\gamma^2\gamma^0\gamma^0\tilde{v}^\dagger_{\bar{u}\gamma} = \bar{u}_{\bar{d}\alpha}\gamma^\mu\left(\frac{1+\gamma_5}{2}\right)u_{u\gamma} \quad , \end{aligned} \tag{9.118}$$

so that the matrix element (9.114) can be written as

$$\frac{g_5^2}{M_X^2}\varepsilon_{\alpha\beta\gamma}\left[\bar{u}_{\bar{d}\alpha}\gamma^\mu\left(\frac{1+\gamma_5}{2}\right)u_{u\gamma}\right]\left[\bar{u}_{\bar{e}}\gamma_\mu\left(\frac{1-\gamma_5}{2}\right)u_{u\beta}\right] \quad . \tag{9.114c}$$

In the non-relativistic approximation the four-component spinors of the quarks reduce to

$$u_q = \begin{pmatrix} \chi_q \\ 0 \end{pmatrix} \quad , \tag{9.119}$$

where χ_q is a two-component Pauli spinor. The left-handed positron, on the other hand, can be approximated by its ultra-relativistic limit, that is, by neglecting its mass. Therefore we make the following ansatz for the positron wave function:

$$\begin{aligned} (u_{\bar{e}})_L &= \frac{(1-\gamma_5)}{2} \begin{pmatrix} \chi_{\bar{e}} \\ \dfrac{\boldsymbol{\sigma}\cdot\boldsymbol{p}}{p}\chi_{\bar{e}} \end{pmatrix} = \frac{1}{2}\left[\begin{pmatrix} 1 & 0 \\ 0 & 1 \end{pmatrix} - \begin{pmatrix} 0 & 1 \\ 1 & 0 \end{pmatrix}\right] \begin{pmatrix} \chi_{\bar{e}} \\ \dfrac{\boldsymbol{\sigma}\cdot\boldsymbol{p}}{p}\chi_{\bar{e}} \end{pmatrix} \\ &= \frac{1}{2} \begin{pmatrix} \left(1 - \dfrac{\boldsymbol{\sigma}\cdot\boldsymbol{p}}{p}\right)\chi_{\bar{e}} \\ -\left(1 - \dfrac{\boldsymbol{\sigma}\cdot\boldsymbol{p}}{p}\right)\chi_{\bar{e}} \end{pmatrix} \end{aligned}$$

or

$$(u_{\bar{e}})_L = \frac{1}{\sqrt{2}} \begin{pmatrix} \chi_{\bar{e},L} \\ -\chi_{\bar{e},L} \end{pmatrix} \quad , \quad \chi_{\bar{e},L} = \left(1 - \frac{\boldsymbol{\sigma}\cdot\boldsymbol{p}}{p}\right)\chi_{\bar{e}} \quad . \tag{9.120}$$

In the last step we normalized the spinor $u_{\bar{e}}$ to 1, assuming that the Pauli spinor $\chi_{\bar{e}L}$ is also normalized to 1.

EXERCISE

9.13 The Interaction Hamiltonian for Proton Decay

Problem. Motivate the structure of the interaction Hamilton operator (9.112) and its special form (9.114a) by considering (9.52).

Solution. From (9.52) we know the form from the interaction with one X_r leptoquark. It is written

$$\mathcal{L}'_{\text{int}} = \left[0 + (\overline{u}_b)_L (u_g^C)_L - (\overline{u}_g)_L (u_b^C)_L + (\overline{e}^+)_L (d_r)_L\right] X_r \quad . \tag{1}$$

We add to it the Hermitian-conjugated expression in order to get the complete interaction Hermitian as well. Considering that $(\overline{u}u^C)^\dagger = u^{C\dagger}\gamma^{0\dagger}(u^\dagger)^\dagger = u^{C\dagger}\gamma^{0\dagger}u = \overline{u}^C u$, we obtain

$$\begin{aligned} \mathcal{L}_{\text{int}} = &\left[+(\overline{u}_b)_L (u_g^C)_L - (\overline{u}_g)_L (u_b^C)_L + (\overline{e}^+)_L (d_r)_L\right] X_r \\ &+ \left[+(\overline{u}_g^C)_L (u_b)_L - (\overline{u}_b^C)_L (u_g)_L + (\overline{d}_r)_L (e^+)_L\right] X_r^* \quad . \end{aligned} \tag{2}$$

One can also get the second (Hermitian-conjugated to the first) component in (2) by calculating the contribution from one X_r^C leptoquark to the Lagrange density according to (9.50) and (9.51). Then one gets with (see (9.44)!)

$$M(X_{\mathrm{r}}^{\mathrm{C}}) = X_{\mathrm{r}}^{\mathrm{C}} \begin{pmatrix} 0 & 0 & 0 & 1 & 0 \\ 0 & 0 & 0 & 0 & 0 \\ 0 & 0 & 0 & 0 & 0 \\ 0 & 0 & 0 & 0 & 0 \\ 0 & 0 & 0 & 0 & 0 \end{pmatrix} , \tag{3}$$

as in (9.51),

$$\begin{aligned} M(X_{\mathrm{r}}^{\mathrm{C}})\psi^{[10]} &= M(X_{\mathrm{r}}^{\mathrm{C}}) \begin{pmatrix} 0 & u_{\mathrm{g}}^{\mathrm{C}} & -u_{\mathrm{b}}^{\mathrm{C}} & -u_{\mathrm{r}} & -d_{\mathrm{r}} \\ -u_{\mathrm{g}}^{\mathrm{C}} & 0 & u_{\mathrm{r}}^{\mathrm{C}} & -u_{\mathrm{b}} & -d_{\mathrm{b}} \\ u_{\mathrm{b}}^{\mathrm{C}} & -u_{\mathrm{r}}^{\mathrm{C}} & 0 & -u_{\mathrm{g}} & d_{\mathrm{g}} \\ u_{\mathrm{r}} & u_{\mathrm{b}} & u_{\mathrm{g}} & 0 & -e^{+} \\ d_{\mathrm{r}} & d_{\mathrm{b}} & d_{\mathrm{g}} & e^{+} & 0 \end{pmatrix}_{\mathrm{L}} \\ &= X_{\mathrm{r}}^{\mathrm{C}} \begin{pmatrix} u_{\mathrm{r}} & u_{\mathrm{b}} & u_{\mathrm{g}} & 0 & -e^{+} \\ 0 & 0 & 0 & 0 & 0 \\ 0 & 0 & 0 & 0 & 0 \\ 0 & 0 & 0 & 0 & 0 \\ 0 & 0 & 0 & 0 & 0 \end{pmatrix} , \end{aligned} \tag{4}$$

and therefore, according to (9.52),

$$\begin{aligned} &\mathrm{Tr}\{\overline{\psi}^{[10]} M(X_{\mathrm{r}}^{\mathrm{C}})\psi^{[10]}\} \\ &= X_{\mathrm{r}}^{\mathrm{C}} \mathrm{Tr} \begin{pmatrix} 0 & -\overline{u}_{\mathrm{g}}^{\mathrm{C}} & \overline{u}_{\mathrm{b}}^{\mathrm{C}} & \overline{u}_{\mathrm{r}} & \overline{d}_{\mathrm{r}} \\ \overline{u}_{\mathrm{g}}^{\mathrm{C}} & 0 & -\overline{u}_{\mathrm{r}}^{\mathrm{C}} & \overline{u}_{\mathrm{b}} & \overline{d}_{\mathrm{b}} \\ -\overline{u}_{\mathrm{b}}^{\mathrm{C}} & \overline{u}_{\mathrm{r}}^{\mathrm{C}} & 0 & \overline{u}_{\mathrm{g}} & \overline{d}_{\mathrm{g}} \\ -\overline{u}_{\mathrm{r}} & -\overline{u}_{\mathrm{b}} & -\overline{u}_{\mathrm{g}} & 0 & \overline{e}^{+} \\ -\overline{d}_{\mathrm{r}} & -\overline{d}_{\mathrm{b}} & -\overline{d}_{\mathrm{g}} & -\overline{e}^{+} & 0 \end{pmatrix}_{\mathrm{L}} \begin{pmatrix} u_{\mathrm{r}} & u_{\mathrm{b}} & u_{\mathrm{g}} & 0 & -e^{+} \\ 0 & 0 & 0 & 0 & 0 \\ 0 & 0 & 0 & 0 & 0 \\ 0 & 0 & 0 & 0 & 0 \\ 0 & 0 & 0 & 0 & 0 \end{pmatrix}_{\mathrm{L}} \\ &= X_{\mathrm{r}}^{\mathrm{C}} \left[(\overline{u}_{\mathrm{g}}^{\mathrm{C}})_{\mathrm{L}} (u_{\mathrm{b}})_{\mathrm{L}} - (\overline{u}_{\mathrm{b}}^{\mathrm{C}})_{\mathrm{L}} (u_{\mathrm{g}})_{\mathrm{L}} + (\overline{d}_{\mathrm{r}})_{\mathrm{L}} (e^{+})_{\mathrm{L}} \right] . \end{aligned} \tag{5}$$

The $X_{\mathrm{r}}^{\mathrm{C}}$ is given as a bosonic field from

$$X_{\mathrm{r}}^{\mathrm{C}} = X_{\mathrm{r}}^{*} . \tag{6}$$

In the Lagrange density $\mathcal{L}_{\mathrm{int}}$ from (2), which is completed as above, the terms

$$(\overline{e}^{+})_{\mathrm{L}} (d_{\mathrm{r}})_{\mathrm{L}} X_{\mathrm{r}} - (\overline{u}_{\mathrm{b}}^{\mathrm{C}})_{\mathrm{L}} (u_{\mathrm{g}})_{\mathrm{L}} X_{\mathrm{r}}^{*} \tag{7}$$

also appear. The second term, $-(\overline{u}_{\mathrm{b}}^{\mathrm{C}})_{\mathrm{L}}(u_{\mathrm{g}})_{\mathrm{L}}X_{\mathrm{r}}^{*}$, describes, for instance, the annihilation of two u quarks and the production of one X boson. The first term, $(\overline{e}^{+})_{\mathrm{L}}(d_{\mathrm{r}}^{+})_{\mathrm{L}}X_{\mathrm{r}}$, describes, for instance, the annihilation of the X_{r} boson and the production from one e^{+} and one d quark.

It is important that the X_{r} boson has the same color as the d_{r} quark, as follows when the X_{r} radiates into $(e^{+})_{\mathrm{L}}(d_{\mathrm{r}}^{+})_{\mathrm{L}}$. The X_{r} boson is a vector boson, and it is described by the four-vector $X_{\mathrm{r}\mu}$ ($\mu = 0, 1, 2, 3$). Thus the coupling in (7) had to be of vector type, and (7) is written in full as

$$(\overline{e}^{\mathrm{C}})_{\mathrm{L}} \gamma^{\mu} (d_{\mathrm{r}})_{\mathrm{L}} X_{\mathrm{r}\mu} - (\overline{u}_{\mathrm{b}}^{\mathrm{C}})_{\mathrm{L}} \gamma^{\mu} (u_{\mathrm{g}})_{\mathrm{L}} X_{\mathrm{r}\mu}^{*} . \tag{8}$$

Exercise 9.13

Because, for example,

$$(d_\mathrm{r})_\mathrm{L} = \frac{(1-\gamma_5)}{2} d_\mathrm{r}$$

and

$$\begin{aligned}(\overline{e}^\mathrm{C})_\mathrm{L} &= \left[\frac{(1-\gamma_5)}{2} e^\mathrm{C}\right]^\dagger \gamma^0 = e^{\mathrm{C}\dagger}\frac{(1-\gamma_5)^\dagger}{2}\gamma_0 \\ &= e^{\mathrm{C}\dagger}\gamma_0\frac{(1+\gamma_5)}{2} = \overline{e}^\mathrm{C}\frac{(1+\gamma_5)}{2} \quad , \end{aligned} \tag{9}$$

one can rewrite (8) in the following form:

$$\begin{aligned}&\left[\overline{e}^\mathrm{C}\frac{(1+\gamma_5)}{2}\gamma^\mu\frac{(1-\gamma_5)}{2}d_\mathrm{r}\right]X_{\mathrm{r}\mu} - \left[\overline{u}_\mathrm{b}^\mathrm{C}\frac{(1+\gamma_5)}{2}\gamma^\mu\frac{(1-\gamma_5)}{2}u_\mathrm{g}\right]X^*_{\mathrm{r}\mu} \\ &= \left[\overline{e}^\mathrm{C}\gamma^\mu\frac{(1-\gamma_5)}{2}d_\mathrm{r}\right]X_{\mathrm{r}\mu} - \left[\overline{u}_\mathrm{b}^\mathrm{C}\gamma^\mu\frac{(1-\gamma_5)}{2}u_\mathrm{g}\right]X^*_{\mathrm{r}\mu} \quad . \end{aligned} \tag{10}$$

For this derivation the well-known relation

$$\frac{(1+\gamma_5)}{2}\gamma^\mu\frac{(1-\gamma_5)}{2} = \gamma^\mu\frac{(1-\gamma_5)}{2}$$

was used.

For the reaction uu → X → e^+d, which is graphically represented as

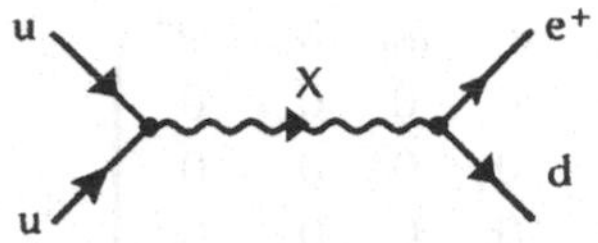

we get in lowest order the following Hamilton operator:

$$\begin{aligned}\hat{H}_\mathrm{int} &= -g_5^2\left[\overline{u}_b^\mathrm{C}\gamma^\mu\frac{(1-\gamma_5)}{2}u_g\right]\left[\frac{g_{\mu\nu}-\frac{k_\mu k_\nu}{M_\mathrm{X}^2}}{k^2-M_\mathrm{X}^2}\right]\left[\overline{e}^\mathrm{C}\gamma^\nu\frac{(1-\gamma_5)}{2}d_r\right] \\ &\Rightarrow -g_5^2\left[\overline{u}_b^\mathrm{C}\gamma^\mu\frac{(1-\gamma_5)}{2}u_g\right]\left[-\frac{g_{\mu\nu}}{M_\mathrm{X}^2}\right]\left[\overline{e}^\mathrm{C}\gamma^\nu\frac{(1-\gamma_5)}{2}d_r\right] \\ &= \frac{g_5^2}{M_\mathrm{X}^2}\left[\overline{u}_b^\mathrm{C}\gamma^\mu\frac{(1-\gamma_5)}{2}u_g\right][g_{\mu\nu}]\left[\overline{e}^\mathrm{C}\gamma^\nu\frac{(1-\gamma_5)}{2}d_r\right] \quad . \end{aligned} \tag{11}$$

Please note that the colors present are all combinations of the three-product $bgr \equiv$ white. Summation over all color permutations finally yields

$$\hat{H}_\mathrm{int} = \frac{g_5^2}{M_\mathrm{X}^2}\varepsilon_{\alpha\beta\gamma}\left[\hat{\overline{u}}_\gamma^\mathrm{C}\gamma_\mu\frac{(1-\gamma_5)}{2}\hat{u}_\beta\right]\left[\hat{\overline{e}}^\mathrm{C}\gamma^\mu\frac{(1-\gamma_5)}{2}\hat{d}_\alpha\right] \quad . \tag{12}$$

This is exactly (9.112). The operator sign above the spinors was introduced here to remind us that the spinors are field operators.

We want to understand now the form of (9.114a). The spinor fields u, d, e are second quantized according to (9.113) and comprise, in the usual way, particles and antiparticles. We can therefore write the field combinations which were introduced in (12) in the following form:

$$
\begin{aligned}
\hat{\bar{u}}^{C}_{\gamma} \cdots \hat{u}_{\beta} &= (\bar{u}_{\bar{u}\gamma}\hat{d}^{\dagger}_{\bar{u}\gamma} + \bar{v}_{u\gamma}\hat{b}_{u\gamma}) \cdots (u_{u\beta}\hat{b}_{u\beta} + v_{u\beta}\hat{d}^{\dagger}_{\bar{u}\beta}) \\
&= \bar{u}_{\bar{u}\gamma} \cdots u_{u\beta}\hat{d}^{\dagger}_{\bar{u}\gamma}\hat{b}_{u\beta} + \bar{u}_{\bar{u}\gamma} \cdots v_{q\beta}\hat{d}^{\dagger}_{\bar{u}\gamma}\hat{d}^{\dagger}_{\bar{u}\beta} \\
&\quad + \underset{\sim\sim\sim\sim\sim\sim\sim}{\bar{v}_{u\gamma} \cdots u_{u\beta}\hat{b}_{u\gamma}\hat{b}_{u\beta}} + \bar{v}_{\bar{u}\gamma} \cdots v_{u\beta}\hat{b}_{u\gamma}\hat{d}^{\dagger}_{\bar{u}\beta}
\end{aligned}
\tag{13a}
$$

and

$$
\begin{aligned}
\hat{\bar{e}}^{C} \cdots \hat{d}_{\alpha} &= [\bar{u}_{\bar{e}}\hat{d}^{\dagger}_{\bar{e}} + \bar{v}_{\bar{e}}\hat{b}_{\bar{e}}] \cdots [u_{d\alpha}\hat{b}_{d\alpha} + v_{d\alpha}\hat{d}^{\dagger}_{d\alpha}] \\
&= \bar{u}_{\bar{e}} \cdots u_{d\alpha}\hat{d}^{\dagger}_{\bar{e}}\hat{b}_{d\alpha} + \underset{\sim\sim\sim\sim\sim\sim\sim}{\bar{u}_{\bar{e}} \cdots v_{d\alpha}\hat{d}^{\dagger}_{\bar{e}}\hat{d}^{\dagger}_{d\alpha}} \\
&\quad + \bar{v}_{\bar{e}} \cdots u_{d\alpha}\hat{b}_{\bar{e}}\hat{b}_{d\alpha} + \bar{v}_{\bar{e}} \cdots v_{d\alpha}\hat{b}_{\bar{e}}\hat{d}^{\dagger}_{d\alpha} \quad .
\end{aligned}
\tag{13b}
$$

Here $u_{q\beta}$ describes an u spinor for an q quark of the color β, $v_{q\beta}$ a v spinor for an q quark of the color β, and $\bar{u}_{\bar{q}\gamma}$ an adjoint u spinor for an q antiquark of the color γ. The points $\cdots$ represent the vertex function $\gamma_{\mu}(1-\gamma_5)/2$. If one is interested in the annihilation process of two u quarks,

$$
\hat{b}^{\dagger}_{u\gamma}\hat{b}^{\dagger}_{u\beta}|0\rangle \quad , \tag{14a}
$$

and the production process of one e^{+} and one d antiquark,

$$
\langle 0|\hat{d}_{\bar{e}}\hat{d}_{d\alpha} \quad , \tag{14b}
$$

the process has the amplitude

$$
\langle 0|\hat{d}_{\bar{e}}\hat{d}_{d\alpha}\bar{H}_{\text{int}}\hat{b}^{\dagger}_{u\gamma}\hat{b}^{\dagger}_{u\beta}|0\rangle \quad . \tag{15}
$$

If one employs the interaction Hamilton operator $\hat{H}_{\text{int}}$ (12) and the field operators (13a), (13b), one obtains immediately that only the marked terms ($\sim\sim\sim$) of (13) contribute, with the result

$$
\begin{aligned}
&\langle 0|\hat{d}_{\bar{e}}\hat{d}_{d\alpha}\bar{H}_{\text{int}}\hat{b}^{\dagger}_{u\gamma}\hat{b}^{\dagger}_{u\beta}|0\rangle \\
&= \frac{g_5^2}{M_X^2}\varepsilon_{\alpha\beta\gamma}\left[\bar{v}_{\bar{u}\gamma}\gamma_{\mu}\frac{(1-\gamma_5)}{2}u_{u\beta}\right]\left[\bar{u}_{\bar{e}}\gamma^{\mu}\frac{(1-\gamma_5)}{2}v_{d\alpha}\right] \quad .
\end{aligned}
\tag{16}
$$

This is exactly (9.114a). It can also be seen as the translation of diagram (9.110) into formal language. One only has to keep in mind that an incoming u quark in the left vertex is equivalent to an outgoing anti-u^{C} quark and therefore a v spinor appears.

According to the standard representation we have

$$
\begin{aligned}
&\gamma^0 = \begin{pmatrix} 1 & 0 \\ 0 & -1 \end{pmatrix} \quad , \quad \gamma^5 = \begin{pmatrix} 0 & 1 \\ 1 & 0 \end{pmatrix} \quad , \quad \gamma^0\gamma^5 = \begin{pmatrix} 0 & 1 \\ -1 & 0 \end{pmatrix} \quad , \\
&\gamma^i = \begin{pmatrix} 0 & \sigma^i \\ -\sigma^i & 0 \end{pmatrix} \quad , \quad \gamma^i\gamma^5 = \begin{pmatrix} \sigma^i & 0 \\ 0 & -\sigma^i \end{pmatrix} \quad ,
\end{aligned}
\tag{9.121}
$$

so that with (9.119)

$$\bar{u}_{\bar{d}\alpha}\gamma^0\left(\frac{1+\gamma_5}{2}\right)u_{u\gamma} = u^\dagger_{\bar{d}\alpha}\left(\frac{1+\gamma^5}{2}\right)u_{u\gamma} = \frac{1}{2}(\chi^\dagger_{\bar{d}\alpha},0)\begin{pmatrix}1 & 1\\ 1 & 1\end{pmatrix}\begin{pmatrix}\chi_{u\gamma}\\ 0\end{pmatrix}$$
$$= \frac{1}{2}\chi^\dagger_{\bar{d}\alpha}\chi_{u\gamma} \,,$$
$$\bar{u}_{\bar{d}\alpha}\gamma^i\left(\frac{1+\gamma_5}{2}\right)u_{u\gamma} = u^\dagger_{\bar{d}\alpha}\gamma^0\gamma^i\left(\frac{1+\gamma^5}{2}\right)u_{u\gamma} = \frac{1}{2}u^\dagger_{\bar{d}\alpha}(\gamma^0\gamma^i+\gamma^0\gamma^i\gamma^5)u_{u\gamma} \tag{9.122}$$
$$= \frac{1}{2}(\chi^\dagger_{\bar{d}\alpha},0)\begin{pmatrix}\sigma_i & \sigma_i\\ \sigma_i & \sigma_i\end{pmatrix}\begin{pmatrix}\chi_{u\mu}\\ 0\end{pmatrix} = \frac{1}{2}\chi^\dagger_{\bar{d}\alpha}\sigma^i\chi_{u\gamma} \quad .$$

Furthermore, since the spinor $u_{\bar{e}}$ is purely left handed, it holds that

$$\bar{u}_{\bar{e}}\gamma_\mu\frac{1-\gamma_5}{2}u_{u\beta} = (\bar{u}_{\bar{e}})_L\gamma_\mu\left(\frac{1-\gamma_5}{2}\right)u_{u\beta} = (\bar{u}_{\bar{e}})_L\left(\frac{1+\gamma_5}{2}\right)\gamma_\mu u_{u\beta}$$
$$= (u^\dagger_{\bar{e}})_L\gamma_0\left(\frac{1+\gamma_5}{2}\right)\gamma_\mu u_{u\beta} = u^\dagger_{\bar{e}}\left(\frac{1-\gamma_5}{2}\right)\gamma_0\left(\frac{1+\gamma_5}{2}\right)\gamma_\mu u_{u\beta}$$
$$= u^\dagger_{\bar{e}}\left(\frac{1-\gamma_5}{2}\right)^2\gamma_0\gamma_\mu u_{u\beta} = u^\dagger_{\bar{e}}\left(\frac{1-\gamma_5}{2}\right)\gamma_0\gamma_\mu u_{u\beta}$$
$$= \bar{u}_{\bar{e}}\gamma_\mu u_{u\beta} \quad . \tag{9.123}$$

Hence we obtain with (9.119) and (9.120)

$$\bar{u}_{\bar{e}}\gamma_0\left(\frac{1-\gamma_5}{2}\right)u_{u\beta} = \frac{1}{2}\frac{1}{\sqrt{2}}(\chi^\dagger_{\bar{e},L},-\chi^\dagger_{\bar{e},L})\begin{pmatrix}1 & -1\\ -1 & 1\end{pmatrix}\begin{pmatrix}\chi_{u\beta}\\ 0\end{pmatrix}$$
$$= \frac{1}{\sqrt{2}}\chi^\dagger_{\bar{e},L}\chi_{u\beta} \,,$$
$$\bar{u}_{\bar{e}}\gamma_i\left(\frac{1-\gamma_5}{2}\right)u_{u\beta} = -\bar{u}_{\bar{e}}\gamma^i\left(\frac{1-\gamma_5}{2}\right)u_{u\beta} \tag{9.124}$$
$$= \frac{1}{2}\frac{1}{\sqrt{2}}(\chi^\dagger_{\bar{e},L},-\chi^\dagger_{\bar{e},L})\begin{pmatrix}0 & \sigma_i\\ \sigma_i & 0\end{pmatrix}\begin{pmatrix}\chi_{u\beta}\\ 0\end{pmatrix}$$
$$= \frac{1}{\sqrt{2}}\chi^\dagger_{\bar{e},L}\sigma^i\chi_{u\beta} \quad .$$

Altogether the matrix element (9.114) takes the form

$$\frac{g_5^2}{2\sqrt{2}M_X^2}\varepsilon_{\alpha\beta\gamma}\left\{(\chi^\dagger_{\bar{d}\alpha}\chi_{u\gamma})(\chi^\dagger_{\bar{e},L}\chi_{u\beta}) + (\chi^\dagger_{\bar{d}\alpha}\boldsymbol{\sigma}\chi_{u\gamma})(\chi^\dagger_{\bar{e},L}\boldsymbol{\sigma}\chi_{u\beta})\right\} \quad . \tag{9.114d}$$

We proceed by evaluating the color dependence of the matrix element (9.114d). The proton wave function consists of three quark wave functions coupled to a color singlet (see diagram of proton decay), that is the wave function has the form

$$\frac{1}{\sqrt{6}}\varepsilon_{\alpha\beta\gamma}\chi_{u_1\alpha}\chi_{u_2\beta}\chi_{d\gamma} \quad . \tag{9.125}$$

The normalization factor $1/\sqrt{6}$ originates from the fact that there are six different contributions to the wave function, depending on the distribution of the three colors for the quarks. The wave function of the meson (see diagram of proton decay) produced is also a color singlet and has the form

$$\frac{1}{\sqrt{3}}\delta^{\alpha\rho}\chi_{\bar{\mathrm{d}},\alpha}\chi_{\mathrm{u},\rho} = \frac{1}{\sqrt{3}}\chi_{\bar{\mathrm{d}}\alpha}\chi_{\mathrm{u}\alpha} \quad . \tag{9.126}$$

With this expression we can calculate the color factor of the proton-decay matrix element:

$$\frac{1}{\sqrt{6}}\epsilon_{\beta\gamma\rho}\frac{1}{\sqrt{3}}\delta_{\alpha\rho}\epsilon_{\alpha\beta\gamma} = \frac{1}{\sqrt{6}}\frac{1}{\sqrt{3}}\epsilon_{\alpha\beta\gamma}\epsilon_{\alpha\beta\gamma} = \frac{1}{\sqrt{18}}6 = \sqrt{2} \quad . \tag{9.127}$$

The first factor reflects the color coupling in the wave function of the proton (see figure), the second one reflects that in the meson wave function, the third reflects that of the interaction (9.114d). This interaction states that the colors β and γ of the two u quarks must couple to the color α of the created d^{C} quark. From now on we refrain from writing the color indices and multiply the matrix element with $\sqrt{2}$ instead.

The diagram for the proton decay now follows. The indices characterize the color.

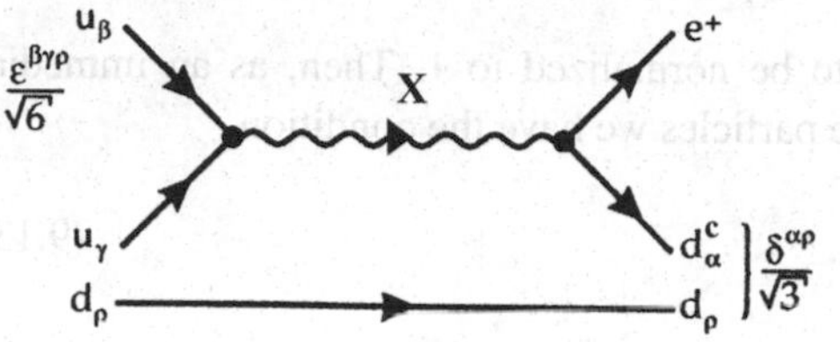

Instead of (9.114d), therefore, we obtain

$$\frac{g_5^2}{2M_{\mathrm{X}}^2}\{(\chi_{\bar{\mathrm{d}}}^\dagger\chi_{\mathrm{u}})(\chi_{\bar{\mathrm{e}},\mathrm{L}}^\dagger\chi_{\mathrm{u}}) + (\chi_{\bar{\mathrm{d}}}^\dagger\sigma\chi_{\mathrm{u}})(\chi_{\bar{\mathrm{e}},\mathrm{L}}^\dagger\sigma\chi_{\mathrm{u}})\} \quad , \tag{9.128}$$

where the particular spinors are normalized to 1.

A completely analogous consideration can be carried through for the diagram

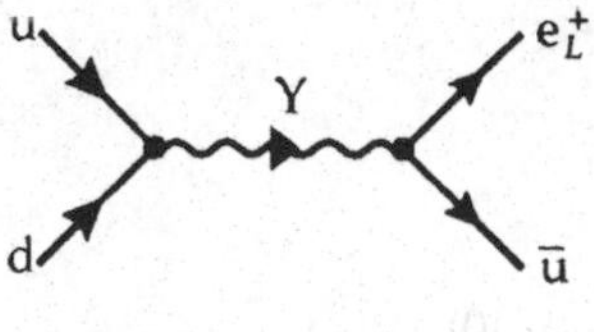

(9.129)

which yields, assuming $M_{\mathrm{X}} = M_{\mathrm{Y}}$,

$$-\frac{g_5^2}{2M_{\mathrm{X}}^2}\{(\chi_{\bar{\mathrm{u}}}^\dagger\chi_{\mathrm{d}})(\chi_{\bar{\mathrm{e}}\mathrm{L}}^\dagger\chi_{\mathrm{u}}) + (\chi_{\bar{\mathrm{u}}}^\dagger\sigma\chi_{\mathrm{d}})(\chi_{\bar{\mathrm{e}}\mathrm{L}}^\dagger\sigma\chi_{\mathrm{u}})\} \quad . \tag{9.130}$$

The relative minus sign of (9.130) as compared with (9.128) stems from the different signs of the coupling terms to X and Y bosons; see Example 9.6, (23). In order to evaluate the matrix elements (9.128) and (9.130) we need to specify the wave function of the proton. We shall assume all three quarks to be in the same s state, so that the spatial part of the wave function is separable. The fermion character of the quarks requires that the complete wave function is antisymmetric. Since we have already formed antisymmetric combinations with respect to the color indices, the spin and isospin contributions must together be symmetric. In order to construct the proton wave function, first imagine that the quarks would be three different, distinguishable particles, their states being characterized by spin and isospin quantum numbers. A physical state $|Z\rangle$ of the three quarks is given by

$$|Z\rangle = c_{ijk}|i, j, k\rangle = c_{ijk}\hat{a}^\dagger_{(1)i}\hat{a}^\dagger_{(2)j}\hat{a}^\dagger_{(3)k}|0\rangle \tag{9.131}$$

(where a sum over i, j, k is understood). The indices i, j, k represent the spin and isospin quantum numbers of the three quarks, $\hat{a}^\dagger_{(1)i}$ is the creation operator for the first quark being in the state $i, j, k, \ldots$, and $|0\rangle$ denotes the vacuum state, which is defined through

$$\hat{a}_{(1)i}|0\rangle = \hat{a}_{(2)i}|0\rangle = \hat{a}_{(3)i}|0\rangle = 0 \quad . \tag{9.132}$$

c_{ijk} is the spin and isospin part of the wave function. As mentioned before, we assume that the coefficients c_{ijk} are symmetric with respect to the permutation of any two indices,

$$c_{ijk} = c_{jik} = c_{ikj} = \cdots \quad . \tag{9.133}$$

We require the state vector (9.131) to be normalized to 1. Then, as an immediate consequence, for three distinguishable particles we have the condition

$$c^*_{ijk}c_{ijk} = 1 \quad . \tag{9.134}$$

However, for n distinguishable particles we also need to introduce a factor $1/\sqrt{n!}$. Let us illustrate the origin of this additional normalization factor for the case of two particles. As before we assume the coefficients c_{ij} to be symmetric with respect to a permutation of the indices i, j. By means of the relation

$$[\hat{a}_i, \hat{a}^\dagger_j] = \delta_{ij} \tag{9.135}$$

we find that

$$\begin{aligned}
\langle Z|Z\rangle &= c^*_{kl}c_{ij}\langle k, l|i, j\rangle = c^*_{kl}c_{ij}\langle 0|\hat{a}_k\hat{a}_l\hat{a}^\dagger_j\hat{a}^\dagger_i|0\rangle \\
&= c^*_{kl}c_{ij}\langle 0|\hat{a}_k\hat{a}^\dagger_j\hat{a}_l\hat{a}^\dagger_i + \hat{a}_k\hat{a}^\dagger_i\delta_{lj}|0\rangle \\
&= c^*_{kl}c_{ij}\langle 0|\hat{a}^\dagger_j\hat{a}_k\hat{a}_l\hat{a}^\dagger_i + \delta_{kj}\hat{a}_l\hat{a}^\dagger_i + \hat{a}^\dagger_i\hat{a}_k\delta_{lj} + \delta_{ik}\delta_{lj}|0\rangle \\
&= c^*_{kl}c_{ij}\langle 0|\delta_{kj}\delta_{li} + \delta_{ik}\delta_{lj}|0\rangle \\
&= c^*_{ji}c_{ij} + c^*_{ij}c_{ij} = 2\cdot c^*_{ij}c_{ij} \quad .
\end{aligned} \tag{9.136}$$

Similarly, in the case of n particles there are $n!$ contributions where the indices of the coefficients c^* are particular permutations of the indices of c. This must be compensated for by an additional normalization factor $1/\sqrt{n!}$.

However, this is not the whole story, since there are six different processes that may lead to the decay of the proton via (9.110), depending on which of the three quarks is the incoming left-handed u quark and which is the "spectator" d quark. Of course, in reality these processes are indistinguishable: the transition amplitude is the same for all six possibilities. We are therefore justified in considering only one of these processes and multiplying our result by a factor of 6.

Together with the normalization factor, which accounts for the fact that the quarks are indistinguishable particles, we therefore obtain an additional factor

$$\frac{1}{\sqrt{6}} \times 6 = \sqrt{6} \quad . \tag{9.137}$$

An analogous consideration can be carried through for (9.129). Finally we take into account the factor $\sqrt{2}$ from (9.127), which originates from the quark colors, so that the effective transition operator for the processes (9.110) and (9.129) is given by

$$\begin{aligned} &\sqrt{6}\frac{(1+\hat{\sigma}_1\cdot\hat{\sigma}_2)}{2}\frac{g_5^2}{M_X^2} \\ &\times [\hat{V}_1(\mathrm{u}\to\bar{\mathrm{e}}_\mathrm{L})\hat{V}_2(\mathrm{u}\to\bar{\mathrm{d}}) + \hat{V}_1(\mathrm{u}\to\bar{\mathrm{e}}_\mathrm{L})\hat{V}_2(\mathrm{d}\to-\bar{\mathrm{u}})] \quad , \end{aligned} \tag{9.138}$$

where we have used (9.128) and (9.130). σ_1 and σ_2 are the spin matrices for particle 1 and particle 2, respectively. The operator $\hat{V}_1(\mathrm{u}\to\bar{\mathrm{e}}_\mathrm{L})$ does not contribute if particle 1 is not a u quark, otherwise particle 1 is transformed into a left-handed positron.

In order to evaluate the matrix element of the operator (9.138) it is convenient to write the spin–isospin part of the proton wave function as

$$\frac{1}{\sqrt{2}}(\chi'\phi' + \chi''\phi'') \quad , \tag{9.139}$$

where ϕ' and ϕ'' denote the isospin wave functions corresponding to a total isospin $\frac{1}{2}$; specifically in ϕ' the two particles are coupled to isospin 0, whereas in ϕ'' they are coupled to isospin 1. The explicit forms of these wave functions are

$$\begin{aligned} \phi' &= \frac{1}{\sqrt{2}}(|udu\rangle - |duu\rangle) \quad , \\ \phi'' &= \frac{1}{\sqrt{6}}(2|uud\rangle - |udu\rangle - |duu\rangle) \quad . \end{aligned} \tag{9.140}$$

The spin wave functions χ' and χ'' are defined correspondingly, by simply replacing in (9.140) u by spin $\uparrow$ and d by spin $\downarrow$. Since the outgoing left-handed positron is in an eigenstate of helicity, that is, of the spin along the direction of the positron momentum, it is convenient to quantize the proton spin along this direction too. Thereby it is understood that a spin projection of $+1/2$ corresponds to a spin oriented opposite to the emission of the positron, that is, a left-handed positron then carries spin $+1/2$. That (9.139) is indeed the spin and isospin wave function of a proton is easily seen: the wave function corresponds to total spin 1/2 and total isospin 1/2 and furthermore is symmetric overall with respect to the three quarks. The latter property follows from

$$\begin{aligned} &\frac{1}{\sqrt{2}}(\chi'\phi' + \chi''\phi'') \\ &= \frac{1}{2\sqrt{2}}(|u\uparrow d\downarrow u\uparrow\rangle - |u\downarrow d\uparrow u\uparrow\rangle - |d\uparrow u\downarrow u\uparrow\rangle + |d\downarrow u\uparrow u\uparrow\rangle) \\ &\quad + \frac{1}{6\sqrt{2}}(4|u\uparrow u\uparrow d\downarrow\rangle - 2|u\uparrow u\downarrow d\uparrow\rangle - 2|u\downarrow u\uparrow d\uparrow\rangle \\ &\quad - 2|u\uparrow d\uparrow u\downarrow\rangle + |u\uparrow d\downarrow u\uparrow\rangle + |u\downarrow d\uparrow u\uparrow\rangle \\ &\quad - 2|d\downarrow u\uparrow u\uparrow\rangle + |d\uparrow u\downarrow u\uparrow\rangle + |d\downarrow u\uparrow u\uparrow\rangle) \end{aligned}$$

$$= \frac{1}{\sqrt{2}}\left(\frac{2}{3}|u\uparrow u\uparrow d\downarrow\rangle - \frac{1}{3}|u\uparrow u\downarrow d\uparrow\rangle - \frac{1}{3}|u\downarrow u\uparrow d\uparrow\rangle\right.$$
$$+\frac{2}{3}|u\uparrow d\downarrow u\uparrow\rangle - \frac{1}{3}|u\downarrow d\uparrow u\uparrow\rangle - \frac{1}{3}|u\uparrow d\uparrow u\downarrow\rangle$$
$$\left.+\frac{2}{3}|d\downarrow u\uparrow u\uparrow\rangle - \frac{1}{3}|d\uparrow u\downarrow u\uparrow\rangle - \frac{1}{3}|d\uparrow u\uparrow u\downarrow\rangle\right) \quad . \tag{9.141}$$

This result is identical to the wave function that we obtained in (7.27) when we discussed the beta decay of the neutron. The advantage of the particular representation of the proton wave function (9.139) is the property that χ' and χ'' are eigenfunctions of the operator $\boldsymbol{\sigma}_1 \cdot \boldsymbol{\sigma}_2$ that occurs in (9.138): if we couple two particles with spin $\frac{1}{2}$ to an angular momentum j we have

$$j(j+1) = \left(\frac{\boldsymbol{\sigma}_1}{2} + \frac{\boldsymbol{\sigma}_2}{2}\right)^2 = \frac{\boldsymbol{\sigma}_1^2}{4} + \frac{\boldsymbol{\sigma}_2^2}{4} + \frac{1}{2}\boldsymbol{\sigma}_1\cdot\boldsymbol{\sigma}_2 = \frac{3}{2} + \frac{1}{2}\boldsymbol{\sigma}_1\cdot\boldsymbol{\sigma}_2 \quad ,$$
$$\boldsymbol{\sigma}_1\cdot\boldsymbol{\sigma}_2 = 2j(j+1) - 3 \quad . \tag{9.142}$$

Now, in χ' the first two quarks are coupled to $j = 0$, whereas in χ'' they are coupled to $j = 1$. Therefore

$$\boldsymbol{\sigma}_1\cdot\boldsymbol{\sigma}_2\chi' = -3\chi' \quad , \quad \boldsymbol{\sigma}_1\cdot\boldsymbol{\sigma}_2\chi'' = \chi'' \quad . \tag{9.143}$$

If the final state contains a π^0, the remaining spectator quark and the produced antiquark, that is, particle 2 and particle 3 in (9.138), must be coupled to spin 0, while the positron carries the spin $\frac{1}{2}$ of the decaying proton. Thus, the spin part of the final-state wave function is

$$\chi_f = \frac{1}{\sqrt{2}}(|\uparrow\uparrow\downarrow\rangle - |\uparrow\downarrow\uparrow\rangle) \quad . \tag{9.144}$$

To proceed with the evaluation of the transition amplitude we need to consider

$$\chi_f{}^\dagger\chi' = \frac{1}{\sqrt{2}}(\langle\uparrow\uparrow\downarrow| - \langle\uparrow\downarrow\uparrow|)\frac{1}{\sqrt{2}}(|\uparrow\downarrow\uparrow\rangle - |\downarrow\uparrow\uparrow\rangle) = -\frac{1}{2} \quad ,$$
$$\chi_f{}^\dagger\chi'' = \frac{1}{\sqrt{2}}(\langle\uparrow\uparrow\downarrow| - \langle\uparrow\downarrow\uparrow|)\frac{1}{\sqrt{6}}(2|\uparrow\uparrow\downarrow\rangle - |\uparrow\downarrow\uparrow\rangle - |\downarrow\uparrow\uparrow\rangle)$$
$$= +\frac{\sqrt{3}}{2} \quad . \tag{9.145}$$

The flavor contribution of the π^0 wave function is $(\bar{u}u - \bar{d}d)/\sqrt{2}$, that is, altogether we have

$$\phi_f = \frac{1}{\sqrt{2}}\{|\bar{e}\,\bar{u}u\rangle - |\bar{e}\,\bar{d}d\rangle\} \quad . \tag{9.146}$$

With these ingredients we obtain

$$\phi_f{}^\dagger\hat{V}_1(\mathrm{u}\to\bar{\mathrm{e}})\big[\hat{V}_2(\mathrm{u}\to\bar{\mathrm{d}}) + \hat{V}_2(\mathrm{d}\to-\bar{\mathrm{u}})\big]\phi'$$
$$= \phi_f{}^\dagger\hat{V}_1(\mathrm{u}\to\bar{\mathrm{e}})[\hat{V}_2(\mathrm{u}\to\bar{\mathrm{d}}) + \hat{V}_2(\mathrm{d}\to-\bar{\mathrm{u}})]\frac{1}{\sqrt{2}}(|udu\rangle - |duu\rangle)$$
$$= \frac{1}{\sqrt{2}}(\langle\bar{e}\,\bar{u}u| - \langle\bar{e}\,\bar{d}d|)(-\frac{1}{\sqrt{2}}|\bar{e}\,\bar{u}u\rangle) = -\frac{1}{2} \tag{9.147}$$

and

$$\begin{aligned}
&\phi_f{}^\dagger \hat V_1(\mathrm{u}\to \bar{\mathrm{e}})\big[\hat V_2(\mathrm{u}\to\bar{\mathrm{d}})+\hat V_2(\mathrm{d}\to -\bar{\mathrm{u}})\big]\phi'' \\
&= \phi_f{}^\dagger \hat V_1(\mathrm{u}\to\bar{\mathrm{e}})\big[\hat V_2(\mathrm{u}\to\bar{\mathrm{d}})+\hat V_2(\mathrm{d}\to-\bar{\mathrm{u}})\big]\frac{1}{\sqrt6}(2|uud\rangle - |udu\rangle - |duu\rangle) \\
&= \frac{1}{\sqrt2}(\langle\bar e\,\bar u u| - \langle \bar e\,\bar d d|)\frac{1}{\sqrt6}(2|\bar e\,\bar d d\rangle + |\bar e\,\bar u u\rangle) \\
&= -\frac{1}{2\sqrt3}\,.
\end{aligned} \tag{9.148}$$

Finally the contributions of spin and isospin to the transition matrix element are obtained by using (9.138), (9.139), (9.143), (9.145), (9.147):

$$\begin{aligned}
&\sqrt6\frac{g_5{}^2}{M_X^2}\frac{1}{\sqrt2}\left[\frac{1-3}{2}\left(-\frac12\right)\left(-\frac12\right)+\frac{1+1}{2}\frac{\sqrt3}{2}\left(-\frac{1}{2\sqrt3}\right)\right] \\
&= -\frac{\sqrt3}{2}\frac{g_5{}^2}{M_X^2}\,.
\end{aligned} \tag{9.149}$$

The transition probability is proportional to the square of (9.149). We then have to average over the initial spin of the proton. Equation (9.148) has been derived for the particular case in which the proton spin is pointing in a direction opposite to that of the emission of the positron, so that the produced positron is left handed. This is impossible if the proton spin is parallel to the direction of emission of the positron. Therefore, averaging over the initial spin directions yields a factor of $\frac12$.

In order to get the total probability for the decay of the proton into a π^0 and a positron we also need to account for the process which involves the production of a right-handed positron. As it turns out, the corresponding amplitude is just half the amplitude for the production of a left-handed positron. This can be seen as follows: the latter process involves contributions of both diagrams (9.110) and (9.129), since $\bar e_L$, d_L, and u_L are all in the same SU(5) multiplet. As is obvious from the two terms in the brackets of (9.149), the corresponding two amplitudes are equal for these two processes. On the other hand, $\bar e_R$ and d_R are in the same multiplet, but not $\bar e_R$ and u_R (see for example Exercise 9.4 or Example 9.6). Therefore only the process

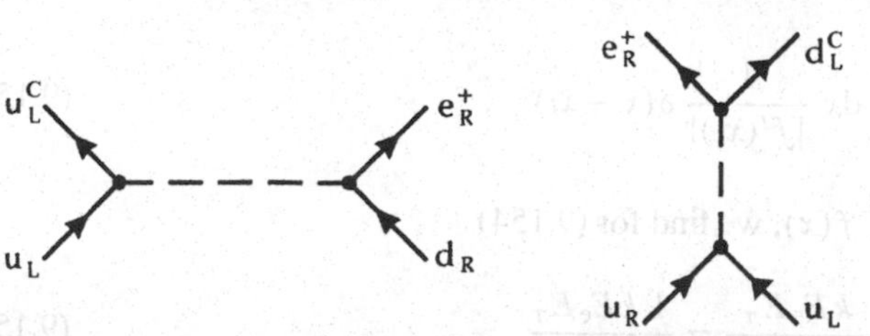

(9.150)

contributes to the production of a right-handed positron. Again, the corresponding amplitude equals the amplitudes for (9.110) and (9.149), that is, it is just half the value of the amplitude for the production of a left-handed positron. Consequently, the probability is 1/4 times the probability for the production of a left-handed positron. Hence, the total probability for the production of a positron is proportional to

$$\frac12\left(1+\frac14\right)\frac34\frac{g_5{}^4}{M_X^4} = \frac{15}{32}\frac{g_5{}^4}{M_X^4}\,, \tag{9.151}$$

where the factor $\frac{1}{2}$ originates from averaging over the initial spins.

To find the transition probability, we furthermore need to take into account the spatial part of the proton wave function. Since we assume a point-like interaction, the transition rate is proportional to the probability that both quarks are located at the same spatial point. Under this assumption the spatial part of the transition matrix elements is given by

$$\int d^3r_1 d^3r_2 \psi^*(\boldsymbol{r}_1)\psi^*(\boldsymbol{r}_2)\delta^3(\boldsymbol{r}_1-\boldsymbol{r}_2)\psi(\boldsymbol{r}_1)\psi(\boldsymbol{r}_2) \quad , \tag{9.152}$$

where $\psi(\boldsymbol{r})$ denotes the spatial wave function which, in accordance with our assumption, is the same for all three quarks; $\boldsymbol{r}_1$ and $\boldsymbol{r}_2$ are the coordinates of particles 1 and 2, respectively.

In the following we shall neglect the dependence of the matrix element on the wave function of the outgoing particles. This can be justified by the weak decay of the Λ particle, where the same approximation yields quite accurate results. However, we will not evaluate the integral (9.152) explicitly but rather assume a value of $2\times10^{-3}\,\mathrm{GeV}^3$, corresponding to $0.25\,\mathrm{fm}^{-3}$. This is reasonable insofar as, first, the calculation of the decay rate for the Λ and Ω particles within a similar model yields this value and, second, this particular value agrees quite well with results from bag-model calculations.

In order to obtain the final expression for the transition rate, the matrix element has to be multiplied by the δ function, which ensures energy and momentum conservation. Then we have to sum over the final states. These two steps lead to a factor

$$\int \frac{d^3k_e}{(2\pi)^3}\frac{d^3k_\pi}{(2\pi)^3}(2\pi)^4\delta^4(P-k_e-k_\pi) \quad . \tag{9.153}$$

In its own rest frame the proton four-momentum is given by $(M_p,0,0,0)$, and we obtain for (9.153)

$$\begin{aligned}&\frac{1}{(2\pi)^2}\int d^3k_e\, d^3k_\pi\, \delta(M_p-E_e-E_\pi)\delta^3(\boldsymbol{k}_e+\boldsymbol{k}_\pi)\\ &=\frac{4\pi}{(2\pi)^2}\int {k_\pi}^2 dk_\pi\, \delta\left(M_p-\sqrt{{m_e}^2+{k_\pi}^2}-\sqrt{{m_\pi}^2+{k_\pi}^2}\right) \quad .\end{aligned} \tag{9.154}$$

By means of the identity

$$\int dx\,\delta(f(x))=\sum_i\int dx\,\frac{1}{|f'(x_i)|}\,\delta(x-x_i) \quad , \tag{9.155}$$

with x_i being the zeros of $f(x)$, we find for (9.154)

$$\frac{k^2}{\pi}\frac{1}{k/E_e+k/E_\pi}=\frac{1}{\pi}\frac{kE_eE_\pi}{E_e+E_\pi}=\frac{1}{\pi}\frac{kE_eE_\pi}{M_p} \quad , \tag{9.156}$$

where k is determined by

$$(m_e^2+k^2)^{1/2}+(m_\pi^2+k^2)^{1/2}=M_p \quad . \tag{9.157}$$

Neglecting the electron mass m_e, we have

$$k=(M_p^2-m_\pi^2)/2M_p \quad , \tag{9.158}$$

which, substituted into (9.156), yields

$$\frac{kE_e E_\pi}{\pi M_p} = 0.122\,\frac{M_p^2}{\pi} \quad . \tag{9.159}$$

This completes the calculation for proton decay.

However, an accurate treatment of proton decay requires the calculation of radiative corrections from gluonic interactions, as illustrated in Fig. 9.9. These diagrams lead to a renormalization of the coupling constant, increasing the decay rate by a factor of

$$\begin{array}{ll} 12.5 & \text{for} \quad p \to \pi^0 + \bar{e}_L \quad , \\ 11 & \text{for} \quad p \to \pi^0 + \bar{e}_R \quad . \end{array} \tag{9.160}$$

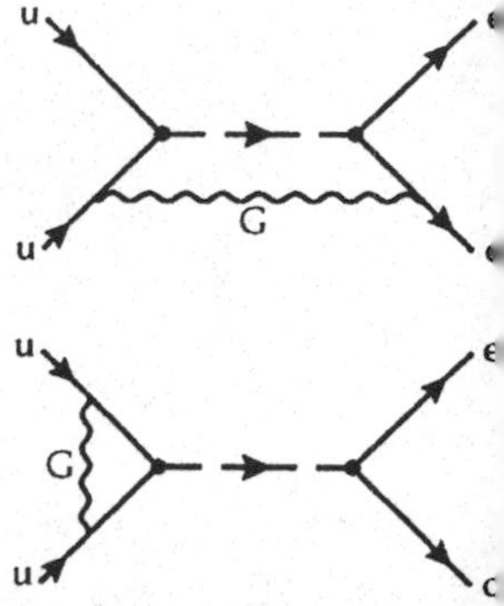

Fig. 9.9. Radiative corrections to the proton decay from gluonic interactions

As discussed before, the first process contributes four times more then the second, and the radiative corrections of Fig. 9.9 result in an overall factor of

$$\frac{1}{5}(4 \times 12.5 + 11) = 12.2 \quad . \tag{9.161}$$

The result for the decay rate is then $1.3 \times 10^{-4}\ M_X^4/\mathrm{GeV}^5$ for the decay of a proton into a neutral pion and a positron. However, this particular process contributes only 37% of the total decay rate. The branching ratios for the other possible decay channels are listed in Table 9.2. Of course, these values are model dependent, but they deviate only a few percent from MIT-bag-model predictions.

Table 9.2. Calculated branching ratios for the decay

Process	Branching ratio
$p \to e^+\pi^0$	37%
$p \to e^+\eta$	7%
$p \to e^+\rho_0$	2%
$p \to e^+\omega$	18%
$p \to \bar{\nu}_e\pi^+$	15%
$p \to \bar{\nu}_e\rho^+$	1%
$p \to \mu^+K^0$	19%

Altogether these different decay processes lead to a prediction for the lifetime of the proton of

$$\tau_p \sim 6 \times 10^{27} \left(\frac{M_X}{(10^{14}\ \mathrm{GeV})}\right)^4 \text{years} \quad . \tag{9.162}$$

Since the value for M_X as calculated within the two-loop approximation is $(6 \pm 3) \times 10^{14}$ GeV, the proton lifetime should be shorter than 4×10^{31} years.[21] In recent experiments[22] 3300 tons of water, or 2.2×10^{32} free protons, not counting those bound in oxygen nuclei, were observed for 417 days. In total 401 events were counted, of which

[21] J. Ellis, M. Gaillard, D. V. Nanopoulos, S. Rudaz: Nucl. Phys. **B176**, 61 (1980).

[22] G. Blewett et al.: Phys. Rev. Lett. **55**, 2114 (1985); W. Gajewski et al.: Phys. Rev. **D42**, 2974 (1990).

the largest part, if not all, originated from neutrinos of cosmic radiation. In order to reduce this large background contribution, only events with balanced momentum were taken into account, since proton decay occurs at rest. Furthermore, only those events have been considered which corresponded to a restricted region of total energy of the reaction products. For example, the process $p \to e^+\pi^0$ should occur at an energy of 940 MeV. Therefore the range of acceptance was restricted to 736–1100 MeV.

By this selective treatment of the observed events the number of relevant events was reduced to 11. These processes are consistent with the decay channels e^+ or $\mu^+ + K^0$, η^0, ρ^0 or ω^0, as well as $\nu + K^+$, ρ^+, or K^{*+}, and also $\mu^+\mu^+\mu^-$. However, these events might also have been produced by neutrinos. No event for the particular decay of the proton into $e^+\pi^0$ was observed. Assuming a lifetime of 4×10^{31} years for the proton and a branching ratio of 37% for the decay channel $e^+\pi^0$, and under the assumption that only the decay of free protons is observed, that is, that protons bound in oxygen nuclei do not decay at all, or at least that the products of such reactions are absorbed, then the probability for the decay of a proton within 417 days is about 90%. Together with the probability for the detection of the decay products of 80%, this implies a probability of 70% for the observation of a decay event $p \to e^+\pi^0$. However, this value is increased if a smaller value for M_X is assumed and the decay of protons bound in oxygen nuclei is taken into consideration. On the other hand, bag-model calculations predict a longer lifetime for the proton.

A detailed discussion of the uncertainties of lifetime calculations for the protons has been given by Ellis and co-workers. Improved experiments looking for the decay $p \to e^+ + \pi^0$ over a period of several years have yielded the much more stringent lower limit[23]

$$\tau_p \geq 6 \times 10^{32}\,\text{years} \times B(p \to e^+\pi^0) \tag{9.163}$$

for the proton lifetime. For the branching ratio given in Table 9.2 it is clear that the experimental limit is higher than the SU(5) prediction by more than a factor of 5. We can therefore say that the simplest version of a grand unified gauge theory, the SU(5) model of Georgi and Glashow, has been ruled out by experiment. Nonetheless, the ideas discussed at the beginning of this chapter appear quite compelling, and it may well be that a somewhat more sophisticated version of a unified gauge theory is realized in nature.[24]

9.7 Outlook: Extensions of the Standard Model

As we saw in the last section, the SU(5) model of "grand" unification of all gauge groups predicts a proton decay rate that seems incompatible with experimental limits.

23 W. Gajewski, in *Last Workshop on Grand Unification*, ed. by P. Frampton (World Scientific, Singapore, 1989), p. 18.

24 One compelling argument for a unified gauge theory, namely that it allows for proton decay, which is necessary to explain the matter–antimatter asymmetry in the universe, has become less compelling, since it was understood that there exists a complicated mechanism for the violation of baryon number conservation in the electroweak gauge theory (see, for example, P. Arnold, L. McLerran: Phys. Rev. **D36**, 581 (1987), **D37**, 1020 (1988); V. Kuzmin, V. Rubakov, M. Shaposhnikov: Phys. Lett. **155B**, 36 (1985)).

Since the predicted proton lifetime (9.162) depends on the fourth power of the X-boson mass, this can be understood as an indication that the SU(5) model predicts too small a value for the unification scale M_{GUT}, and hence for M_{X}. Moreover, the new high-precision data on the parameters of the electroweak gauge theory and quantum chromodynamics obtained in the experiments at LEP[25] (see Chap. 5) have revealed that the three coupling constants

$$\begin{aligned} \alpha_1(q^2) &= \frac{5}{3}\alpha'(q^2) \;, \\ \alpha_2(q^2) &= \alpha_{\text{L}}(q^2) \;, \\ \alpha_3(q^2) &= \alpha_{\text{C}}(q^2) \;, \end{aligned} \tag{9.164}$$

do not quite meet at a single point (see Fig. 9.10). Inspecting Fig. 9.10 one may speculate that α_2 and α_3 fall too rapidly with q^2, or that α_1 increases too rapidly, or both. This finding points in the same direction as the conclusion drawn from the absence of observed proton decay, because a softer q^2 dependence of the coupling constants implies a higher symmetry-breaking scale M_{GUT}.

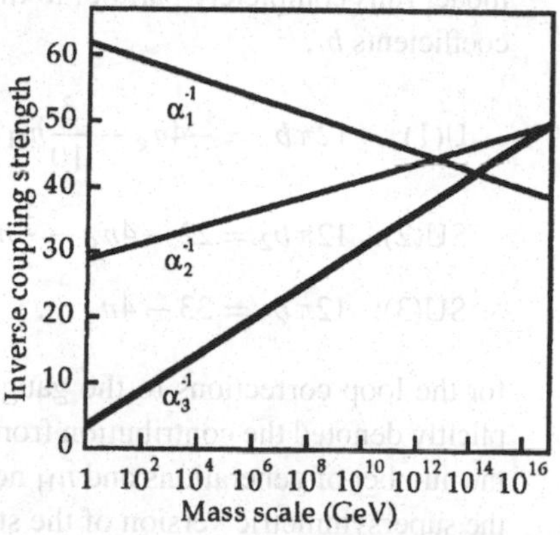

Fig. 9.10. The running coupling constants α_i fail to meet in a single point when they are extrapolated to high energies by computations incorporating the particle content of the minimal SU(5) model

Is this a fundamental flaw of the SU(5) model? Maybe not, because it can be remedied in a very elegant, yet simple, manner. The basic idea is that the contributions of fermions and spin-zero bosons to the coefficients b_i describing the strength of the q^2 dependence in (9.84) are negative and hence soften the "running" of the inverse coupling constants. Thus the discrepancy may be cured if one can find a method to include more fermions and scalar bosons into the SU(5) model in a natural way.[26] The simplest method would be to postulate the existence of more than $n_g = 3$ generations, but this possibly is excluded by the LEP data, which show that there are only three neutrino flavors (see Example 5.3).

Another, perhaps even more elegant, way is provided by the concept of *supersymmetry*, where one postulates that each species of fundamental particle is supplemented by a yet unknown species obeying the opposite particle statistics. In other words, for every boson species of the standard model (gauge and Higgs bosons), there should exist a corresponding species of fermion, and all fundamental fermions (quarks and

[25] DELPI collab. P. Abren et al.: Phys. Lett. **B247**, 167 (1990), Phys. Lett. **B252**, 149 (1990).

[26] There exist of course a wide variety of GUT theories, based on similar concepts as the SU(5) model; see e.g. R. Slansky: Group Theory for Unified Model Building, Phys. Rep. **79**, 1 (1981).

leptons) should have bosonic partners. A complete table of the components of the minimal supersymmetric extension of the standard model of electroweak and strong interactions is given below, see Table 9.3. Except for their spin, all supersymmetric partners carry the same quantum numbers as their related particles from the standard model. Although we consider here the supersymmetric version of the SU(5) gauge theory primarily because it contains the desired additional spin-$\frac{1}{2}$ and spin-0 particles, we mention that the supersymmetric gauge theory has several virtues in its own right.

1. The divergences of loop diagrams are generally less severe in the supersymmetric version of the theory, owing to cancellations between contributions from virtual bosons and fermions. This leads, among other things, to possible solutions of the hierarchy and fine-tuning problems associated with the different symmetry-breaking scales.
2. Supersymmetry gives rise to an elegant mathematical formalism, which allows one to include quantized gravity in a straightforward manner. Supersymmetric unified gauge theories arise naturally as a "low-energy" limit of theories of quantum gravity, such as supersymmetric string theories.[27] The analysis of the q^2 dependence of the coupling constants $\alpha_1, \alpha_2, \alpha_3$ in the supersymmetric version of the standard model runs completely parallel to that performed in Sect. 9.5. There we found the coefficients b_i,

$$\begin{aligned}
\text{U(1):}\quad & 12\pi b_1 = -4n_g - \frac{3}{10}n_\text{H} \quad ,\\
\text{SU(2):}\quad & 12\pi b_2 = 22 - 4n_g - \frac{1}{2}n_\text{H} \quad ,\\
\text{SU(3):}\quad & 12\pi b_3 = 33 - 4n_g \quad ,
\end{aligned} \tag{9.165}$$

for the loop corrections to the gauge boson propagators, where we now have explicitly denoted the contribution from virtual Higgs bosons. Here n_g again signifies the number of generations and n_H now stands for the number of Higgs doublets. In the supersymmetric version of the standard model, these are replaced by[28]

$$\begin{aligned}
12\pi\overline{b}_1 &= -6n_g - \frac{9}{10}n_\text{H} \quad ,\\
12\pi\overline{b}_2 &= 18 - 6n_g - \frac{3}{2}n_\text{H} \quad ,\\
12\pi\overline{b}_3 &= 27 - 6n_g \quad .
\end{aligned} \tag{9.166}$$

In the standard model, $n_g = 3$ and $n_\text{H} = 1$; the latter is replaced by $n_\text{H} = 2$ in the supersymmetric version. Let us compare the two sets of coefficients:

$$\begin{aligned}
12\pi b_1 &= -12.3 \quad , \quad & 12\pi\overline{b}_1 &= -19.8 \quad ,\\
12\pi b_2 &= +9.5 \quad , \quad & 12\pi\overline{b}_2 &= -3 \quad ,\\
12\pi b_3 &= +21 \quad , \quad & 12\pi\overline{b}_3 &= +9 \quad .
\end{aligned} \tag{9.167}$$

[27] See, for example, M. B. Green, A. S. Schwartz, E. Witten: *Superstring Theory* (Cambridge University Press, Cambridge, 1987).

[28] M.B. Einhorn, D.R.T. Jones: Nucl. Phys. **B196**, 475 (1982).

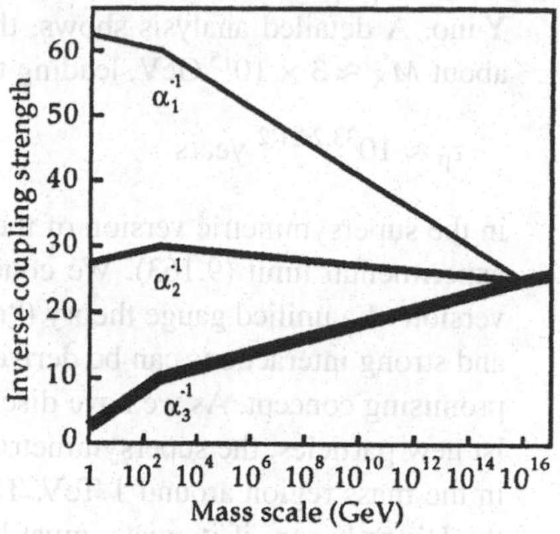

Fig. 9.11. The minimal supersymmetric SU(5) model causes the couplings α_i to meet in a single point

As we see, the coefficients $\bar{b}_i$ of the supersymmetric extension of the standard model are smaller than those of the standard model itself, and hence the coupling constants $\alpha_i(q^2)$ run more slowly, as desired. The surprising effect of these changes is that all three functions $\alpha_i(q^2)$ now really meet at a single point, as shown in Fig. 9.11, when the supersymmetric partners of the known particles are assigned masses of the order of 1 TeV (1000 GeV). The consequence is that the transition from the parameters (9.165) to the parameters (9.166) is being fitted at $M \sim 1000$ GeV. This transition causes the bending of the inverse coupling strength visible in Fig. 9.11. This result[29] is based on a precise analysis of the coupling constants α_i of the standard model at the mass scale of the Z boson:

$$\alpha_1(M_Z) = 0.016887 \pm 0.000040 \quad ,$$
$$\alpha_2(M_Z) = 0.03322 \pm 0.00025 \quad , \qquad (9.168)$$
$$\alpha_3(M_Z) = 0.108 \pm 0.005 \quad .$$

The lines intersect almost exactly at $q^2 = (10^{16}\ \text{GeV})^2$, that is, slightly more than an order of magnitude later than in the case of the standard model.

Table 9.3. Particle content of the minimal supersymmetric extension of the standard model

Particle (spin)	Supersym. partner (spin)
photon (1)	photino ($\frac{1}{2}$)
W, Z (1)	W-ino, Z-ino ($\frac{1}{2}$)
gluon (1)	gluino ($\frac{1}{2}$)
lepton ($\frac{1}{2}$)	slepton (0)
quark ($\frac{1}{2}$)	squark (0)
Higgs (0)	Higgsino ($\frac{1}{2}$)

This suggests the construction of a unified supersymmetric SU(5) gauge theory which contains, in addition to the particles listed in Table 9.3, very massive X and Y gauge bosons and their supersymmetric, fermionic partners, the X-ino and

[29] U. Amaldi, W. de Boer, H. Fürstenau: Phys. Lett. **260B**, 447 (1991); P. Longacker, U. Luo, Phys. Rev. **D44**, 817 (1991).

Y-ino. A detailed analysis shows, then, that the X boson should have a mass of about $M_X \approx 3 \times 10^{15}$ GeV, leading to the prediction of a proton lifetime of

$$\tau_p \approx 10^{33.2 \pm 1.2} \text{ years} \tag{9.169}$$

in the supersymmetric version of the SU(5) gauge theory. This is well beyond the experimental limit (9.163). We conclude that, despite the failure of the simplest version of a unified gauge theory ("minimal" SU(5)), the idea that all electroweak and strong interactions can be derived from a unified gauge theory remains a very promising concept. As we have discussed, there are indications that there may exist new particles, the supersymmetric partners of the known fermions and bosons, in the mass region around 1 TeV. There is also evidence from precision data that the Higgs boson, if it exists, must have a mass well below 1 TeV. It is therefore very likely that experiments at the next generation of particle accelerators (the CERN Large Hadron Collider and the U.S. Superconducting Super-Collider, usually known as LHC and SSC) will shed light on the prospects of unifying all forces of nature, except gravity, into a single gauge group. The search for the Higgs particle is in this context the most important task for the near future.[30,31]

9.8 Biographical Notes

GEORGI, Howard, theoretical physicist, *6.1.1947 in San Bernardino (California), received Ph.D. at Yale University. Since 1980 professor at Harvard University, working on unified models of elementary particle physics.

YOUNG, Alfred, mathematician, *16.4.1873 in Widnes (Lancashire, England), †15.12.1940 in Birdbrook (Essex, England). Young earned his living as a country clergyman in Birdbrook, near Cambridge, which left him with little time for mathematics. His ideas to use diagrams to classify the representations of the permutation group were developed in a series of articles 1901–02 and 1928–34, which aimed at developing invariant theory [cf. G. de B. Robinson: *Representation Theory of the Symmetric Group* (University of Toronto Press 1961)].

---

30 G.G. Ross: *Grand Unified Theories, Frontiers in Physics* (The Benjamin/Cummings Publ. Corp., 1984).

31 H.J.W. Müller-Kirsten, A. Wiedemann: *Supersymmetry* (World Scientific, Singapore, 1987).

Appendix

A.1 Conventions and "Natural" Units

The theory of weak interactions deals with the phenomena of conversion of one type of elementary particle into another. Such processes are naturally quantum jumps, i.e. the theory of weak interaction makes sense only as a quantum theory. In addition the theory must necessarily be formulated in a relativistically invariant way, because electrons with high energy and, above all, massless neutrinos, which move with the speed of light, participate in most processes. Hence powers of Planck's constant $\hbar$ and of the speed of light c are ubiquitous, rendering the equations considerably more complicated than they really are. This can be avoided if one uses the so-called natural units:

$$\hbar = c = 1 \quad . \tag{A.1}$$

This means that length and time are measured in the same units (such as Fermi = fm), as are mass, energy, frequency, etc. At any time one can go over to the standard (atomic) units by introducing factors of

$$\begin{aligned} c &= 3 \times 10^{23}\ \text{fm/s} \\ \hbar c &= 197.32\ \text{MeV} \times \text{fm} \end{aligned} \tag{A.2}$$

in the equations.

Example. The Compton wavelength $\hbar/mc$ of the electron in natural units simply reads $1/m$. The transformation goes like this:

$$m^{-1} \to \hbar c (mc^2)^{-1} = 197\ \text{MeV} \times \text{fm}\ (0.511\ \text{MeV})^{-1} = 386\ \text{fm} \quad , \tag{A.3}$$

where we have inserted the mass of the electron. Length and reciprocal mass thus have the same units in the measuring system $\hbar = c = 1$; 1 fm corresponds to $(197\ \text{MeV})^{-1} \approx 5\ \text{GeV}^{-1}$, and 1 s corresponds to 3×10^{23} fm or $1.52 \times 10^{21}\ \text{MeV}^{-1}$.

Example. The lifetime of a muon is calculated to be

$$\tau_\mu = 192\pi^3\ (G^2\, m_\mu^5)^{-1} \tag{A.4}$$

with Fermi's constant $G = 1.166 \times 10^{-11}\,\text{MeV}^{-2}$. With $m_\mu = 105.66$ MeV, τ_μ becomes

$$\begin{aligned} \tau_\mu &= 192\pi^3\ (1.166 \times 10^{-11})^{-2} \times (105.66)^{-5}\ \text{MeV}^{-1} \\ &= 3.3 \times 10^{15}\ \text{MeV}^{-1} = 2.2 \times 10^{-6}\ \text{s} \quad . \end{aligned} \tag{A.5}$$

W. Greiner, B. Müller, *Gauge Theory of Weak Interactions*,
DOI 10.1007/978-3-540-87843-8, © Springer-Verlag Berlin Heidelberg 2009

The invariant infinitesimal distance of two events in Minkowski space is

$$\begin{aligned} ds^2 &= dt^2 - dx^2 - dy^2 - dz^2 \\ &= (dx^0)^2 - \sum_{i=1}^{3} (dx^i)^2 \quad , \end{aligned} \tag{A.6}$$

(where we have put $c = 1$!). Besides the *contravariant* position vector $x^\mu = (t, x, y, z)$, the upper index of which runs from 0 to 3, we introduce the *covariant* position vector $x_\mu = (t, -x, -y, -z)$ with a lower index. Both forms are connected by the *metric* tensor of Minkowski space,

$$g_{\mu\nu} = \begin{pmatrix} 1 & 0 & 0 & 0 \\ 0 & -1 & 0 & 0 \\ 0 & 0 & -1 & 0 \\ 0 & 0 & 0 & -1 \end{pmatrix} \quad , \tag{A.7}$$

$$x_\mu = \sum_{\nu=0}^{3} g_{\mu\nu} x^\nu \equiv g_{\mu\nu} x^\nu \quad . \tag{A.8}$$

If the last expression above the summation over the index ν, which occurs twice, is implicitly assumed (*Einstein's summation convention*), we can write

$$ds^2 = g_{\mu\nu}\, dx^\mu dx^\nu = dx^\mu dx_\mu \quad . \tag{A.9}$$

This abbreviated style, which formally exhibits invariance under Lorentz transformations (the sum is over all four indices, one pair covariant and the second contravariant) will be often used.

A.2 The Dirac Equation

The Dirac equation for a free particle with rest mass m and spin $\frac{1}{2}$ is

$$\left(i\gamma^\mu \frac{\partial}{\partial x^\mu} - m \right) \psi(x) = 0 \quad , \tag{A.10}$$

with the Dirac matrices obeying the commutation relations

$$\gamma^\mu \gamma^\nu + \gamma^\nu \gamma^\mu = 2 g^{\mu\nu} \quad . \tag{A.11}$$

The three space-like Dirac matrices are also written in the form

$$\gamma^i = \gamma^0 \alpha_i \quad (\text{or } \boldsymbol{\gamma} = \gamma^0 \boldsymbol{\alpha}) \quad . \tag{A.12}$$

As a standard representation of the Dirac matrices we use the 4×4 matrices

$$\gamma^0 = \begin{pmatrix} 1 & 0 \\ 0 & -1 \end{pmatrix} \quad , \quad \gamma^i = \begin{pmatrix} 0 & \sigma_i \\ -\sigma_i & 0 \end{pmatrix} \quad , \quad \alpha^i = \begin{pmatrix} 0 & \sigma_i \\ \sigma_i & 0 \end{pmatrix} \quad , \tag{A.13}$$

where each element is a 2×2 matrix. The Pauli matrices σ_i are represented in the form

$$\sigma_1 = \begin{pmatrix} 0 & 1 \\ 1 & 0 \end{pmatrix} \quad , \quad \sigma_2 = \begin{pmatrix} 0 & -\mathrm{i} \\ \mathrm{i} & 0 \end{pmatrix} \quad , \quad \sigma_3 = \begin{pmatrix} 1 & 0 \\ 0 & -1 \end{pmatrix} \quad . \tag{A.14}$$

In addition, the following combinations occur frequently:

$$\gamma_5 = \mathrm{i}\gamma^0\gamma^1\gamma^2\gamma^3 = \begin{pmatrix} 0 & 1 \\ 1 & 0 \end{pmatrix} \quad , \tag{A.15}$$

$$\sigma^{\mu\nu} = \frac{\mathrm{i}}{2}[\gamma^\mu, \gamma^\nu] \quad , \tag{A.16}$$

with

$$\sigma^{ij} = \Sigma_k = \begin{pmatrix} \sigma_k & 0 \\ 0 & \sigma_k \end{pmatrix} \quad , \quad (ijk) = (123) \quad ,$$

$$\sigma^{0k} = \mathrm{i}\alpha_k \quad .$$

One verifies easily

$$\gamma^\mu\gamma_5 = -\gamma_5\gamma^\mu \quad . \tag{A.15$'$}$$

We denote scalar products of four-vectors with γ matrices by a slash, for example,

$$\gamma^\mu p_\mu = \gamma^0 p^0 - \boldsymbol{\gamma} \cdot \boldsymbol{p} = \not{p} \quad , \quad \text{etc.} \tag{A.17}$$

If we separate a plane-wave factor for particles

$$\psi^{(+)}_{p,s}(x) = (2p^0V)^{-1/2} u(p,s) \exp(-\mathrm{i}p_\mu x^\mu) \tag{A.18a}$$

and antiparticles

$$\psi^{(-)}_{p,s}(x) = (2p^0V)^{-1/2} v(p,s) \exp(\mathrm{i}p_\mu x^\mu) \quad , \tag{A.18b}$$

the Dirac equation (A.10) yields a matrix equation for the space-independent spinors:

$$(\not{p} - m)\, u(p,s) = 0 \quad , \tag{A.19a}$$

$$(\not{p} + m)\, v(p,s) = 0 \quad . \tag{A.19b}$$

Here s denotes the spin polarizations of the Dirac particle, which can assume the values $\pm 1/2$. Explicitly we have

$$u(p,s) = \sqrt{p^0 + m} \begin{pmatrix} \chi_s \\ \dfrac{\boldsymbol{\sigma} \cdot \boldsymbol{p}}{p^0 + m}\chi_s \end{pmatrix} \quad , \tag{A.20a}$$

$$v(p,s) = \sqrt{p^0 + m} \begin{pmatrix} \dfrac{\boldsymbol{\sigma} \cdot \boldsymbol{p}}{p^0 + m}\chi_s \\ \chi_s \end{pmatrix} \quad , \tag{A.20b}$$

with the two-spinors

$$\chi_{+1/2} = \begin{pmatrix} 1 \\ 0 \end{pmatrix} \quad , \quad \chi_{-1/2} = \begin{pmatrix} 0 \\ 1 \end{pmatrix} \quad . \tag{A.21}$$

The normalization of the wave functions (A.18) is chosen so that the volume V contains just one particle:

$$\int_V d^3x\, \psi^{(\varepsilon')\dagger}_{p',s'}(x)\, \psi^{(\varepsilon)}_{p,s}(x) = \begin{cases} 1 & \text{for } \boldsymbol{p}' = \boldsymbol{p},\ s' = s,\ \varepsilon' = \varepsilon \\ 0 & \text{otherwise} \end{cases} \quad , \tag{A.22a}$$

$$u^\dagger(p,s)u(p,s) = v^\dagger(p,s)v(p,s) = 2p^0 \quad . \tag{A.22b}$$

The adjoint spinor is defined as

$$\bar{\psi}(x) = \psi^\dagger(x)\gamma^0 \quad . \tag{A.23}$$

For the products of two Dirac spinors one has the following useful formulas, where for clarity we write down the spinor indices explicitly:

$$u_\alpha(p,s)\bar{u}_\beta(p,s) = \frac{1}{2}[(\not{p} + m)(1 + \gamma_5 \not{s})]_{\alpha\beta} \quad , \tag{A.24a}$$

$$v_\alpha(p,s)\bar{v}_\beta(p,s) = \frac{1}{2}[(\not{p} - m)(1 + \gamma_5 \not{s})]_{\alpha\beta} \quad . \tag{A.24b}$$

Here the four-vector s^μ is obtained from the rest frame spin $\hat{s}$ by a Lorentz transformation

$$s^\mu = \left(\frac{\boldsymbol{p} \cdot \hat{\boldsymbol{s}}}{m}, \hat{\boldsymbol{s}} + \frac{\boldsymbol{p}(\boldsymbol{p} \cdot \hat{\boldsymbol{s}})}{m(p^0 + m)} \right) \quad , \tag{A.25}$$

where the spin vector $\hat{\boldsymbol{s}}$ is given by

$$\hat{\boldsymbol{s}} = \pm \boldsymbol{e}_z \quad \text{for} \quad s = \pm 1 \quad . \tag{A.26}$$

Summing over the spin directions (A.24) yields

$$\sum_s u_\alpha(p,s)\bar{u}_\beta(p,s) = (\not{p} + m)_{\alpha\beta} \quad , \tag{A.27a}$$

$$\sum_s v_\alpha(p,s)\bar{v}_\beta(p,s) = (\not{p} - m)_{\alpha\beta} \quad . \tag{A.27b}$$

Computations with the Dirac matrices are simplified by a number of rules and identities concerning certain traces. One useful identity is

$$\not{a}\not{b} = (a \cdot b) - \mathrm{i}\sigma^{\mu\nu} a_\mu b_\nu \quad , \quad \text{with } (a \cdot b) = a_\mu b^\mu \quad . \tag{A.28}$$

This is the four-dimensional generalization of the relation

$$(\boldsymbol{\sigma} \cdot \boldsymbol{a})(\boldsymbol{\sigma} \cdot \boldsymbol{b}) = \boldsymbol{a} \cdot \boldsymbol{b} + \mathrm{i}\boldsymbol{\sigma} \cdot (\boldsymbol{a} \times \boldsymbol{b}) \quad . \tag{A.29}$$

The trace of a product of an odd number of Dirac matrices vanishes; in particular,

$$\mathrm{Tr}\{\gamma^\mu\} = 0 \quad . \tag{A.30}$$

Furthermore, one has the relations

$$\mathrm{Tr}(\gamma^\alpha \gamma^\beta) = 4g^{\alpha\beta} \quad , \tag{A.31}$$

$$\mathrm{Tr}(\gamma^\alpha \gamma^\beta \gamma^\mu \gamma^\nu) = 4(g^{\alpha\beta} g^{\mu\nu} - g^{\alpha\mu} g^{\beta\nu} + g^{\alpha\nu} g^{\beta\mu}) \quad . \tag{A.32}$$

For traces containing the matrix γ_5, one obtains the following rules:

$$\mathrm{Tr}(\gamma_5) = 0 \quad , \tag{A.33}$$

$$\mathrm{Tr}(\gamma_5\gamma^\alpha\gamma^\beta) = 0 \quad , \tag{A.34}$$

$$\mathrm{Tr}(\gamma_5\gamma^\alpha\gamma^\beta\gamma^\mu\gamma^\nu) = -4\mathrm{i}\varepsilon^{\alpha\beta\mu\nu} \quad , \quad \text{with} \tag{A.35}$$

$$\varepsilon^{0123} = 1 \quad . \tag{A.36}$$

A.3 Feynman Rules

The following rules are valid for reactions with two particles in the initial state and n particles in the final state (Fig. A.1). The cross section per unit of the phase-space volume of the particles in the final state is given by

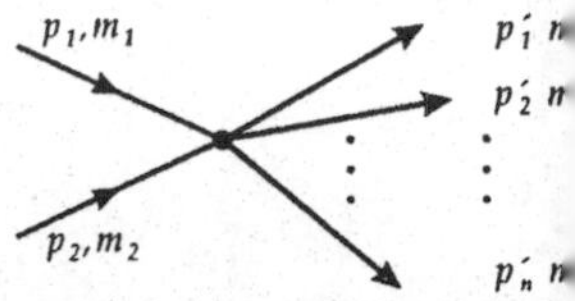

Fig. A.1. Reactions with two particles in the initial state and n particles in the final state

$$\mathrm{d}\sigma = (J_{12}\rho_2 VT)^{-1}|S_{fi}|^2\, V\frac{\mathrm{d}^3 p_1'}{(2\pi)^3}\cdots V\frac{\mathrm{d}^3 p_n'}{(2\pi)^3} \quad , \tag{A.37}$$

where for our normalization of the wave functions the incoming current corresponding to particle 1 takes the form

$$J_{12} = \frac{\sqrt{(p_1\cdot p_2)^2 - m_1^2 m_2^2}}{p_1^0 p_2^0 V} \quad , \tag{A.38}$$

while the density of particle 2 is given by

$$\rho_2 = 1/V \quad . \tag{A.39}$$

In reference frames where the momenta of the incoming particles are parallel (for example, in the centre-of-mass frame), the current J_{12} can be expressed in terms of the velocities of the particles 1 and 2:

$$J_{12} = \frac{|\boldsymbol{v}_1 - \boldsymbol{v}_2|}{V} \quad . \tag{A.40}$$

The matrix element S_{fi} of the scattering operator in general includes a factor

$$F_1 = (2\pi)^4\delta^4(p_1' + \cdots + p_n' - p_1 - p_2) \quad , \tag{A.41}$$

which results from the space-time integration over the plane waves, and, furthermore, a factor

$$F_2 = (2p_1^0 V \times 2p_2^0 V \times 2p_1'^0 V \times \cdots \times 2p_n'^0 V)^{-1/2} \tag{A.42}$$

from the normalizations of the wave functions. Separating these factors, we can introduce a reduced matrix element $\mathcal{M}_{fi}$ by defining

$$S_{fi} = F_1 F_2 \mathcal{M}_{fi} \quad . \tag{A.43}$$

Substituting $(2\pi)^4\delta^4(0)$ for VT, we then have for the scattering cross section

$$\mathrm{d}\sigma = \frac{|\mathcal{M}_{fi}|^2(2\pi)^4\delta^4(p'_1+\cdots+p'_n-p_1-p_2)}{4[(p_1\cdot p_2)^2-m_1^2m_2^2]^{1/2}}\frac{\mathrm{d}^3p'_1}{(2\pi)^3 2p'^0_1}\cdots\frac{\mathrm{d}^3p'_n}{(2\pi)^3 2p'^0_n}\,. \tag{A.44}$$

The cross section is invariant under Lorentz transformations.

The Feynman rules concern the calculation of the reduced scattering matrix element $\mathcal{M}_{fi}$. In detail, they depend on the properties of the particles involved, especially on their masses and spins. A *Feynman graph* describing a scattering process consists of three parts: (1) the *external lines* representing the wave functions of the incoming and outgoing particles, (2) the *internal lines* described by the propagators, and (3) the *vertices* representing the interactions between the particles.

In the following we quote some important cases of these rules:

(1) External Lines:

spin 0: 1.

spin 1/2: $u(p,s)$ or $v(p,s)$.

spin 1: $\varepsilon_\mu(p,\lambda)$ with $\varepsilon^2=-1$; for $m=0$ one further has $(\varepsilon\cdot p)=0$.

(2) Internal Lines:

spin 0: $\mathrm{i}\Delta(p)=\mathrm{i}(p^2-m^2+\mathrm{i}\varepsilon)^{-1}$.

spin 1/2: $\mathrm{i}S(p)=\mathrm{i}(\not p-m+\mathrm{i}\varepsilon)^{-1}=\mathrm{i}(\not p+m)\Delta(p)$.

spin 1:

$$\mathrm{i}D_{\mu\nu}(p)=\begin{cases}-\mathrm{i}(g_{\mu\nu}-\frac{p_\mu p_\nu}{m^2})\,(p^2+\mathrm{i}\varepsilon)^{-1} & \text{for } m=0\\ -\mathrm{i}(g_{\mu\nu}-\frac{p_\mu p_\nu}{p^2})\,\Delta(p) & \text{for } m\neq 0\end{cases}\,.$$

(3) Vertices:

There are many different kinds of vertex. We give only a few examples, namely those between a Dirac particle and bosons with spins 0 or 1, where g denotes the corresponding coupling constant.

$$\text{spin 0:}\ \begin{cases}-\mathrm{i}g & \text{for scalar bosons}\\ g\gamma_5 & \text{for scalar bosons}\end{cases}\,.$$

$$\text{spin 1:}\ \begin{cases}-\mathrm{i}g\gamma^\mu & \text{for vector bosons}\\ -\mathrm{i}g\gamma^\mu\gamma_5 & \text{for axial vector bosons}\end{cases}\,.$$

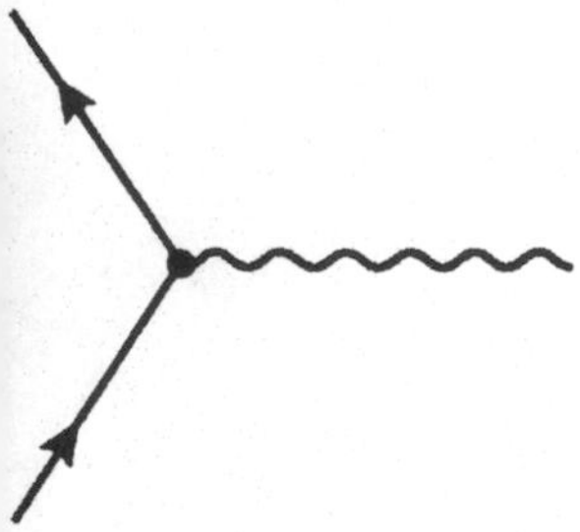

Fig. A.2.

The rules for the Yang–Mills theory are discussed in Sect. 4.3.

At each vertex the sum of all four-momenta is conserved. One integrates over the four-momenta of the internal lines that are not determined by these conservation laws: $\int \mathrm{d}^4p/(2\pi)^4$. For each closed fermion ring one obtains an additional factor -1.

For the external lines one often sums over the possible polarizations. For spin-1/2 particles one uses formula (A.27), while for particles with spin 1 the following relations hold:

$$m = 0: \qquad \sum_{\lambda=\pm 1} \varepsilon_\mu(p,\lambda)\varepsilon_\nu(p,\lambda) = -g_{\mu\nu} + \frac{p_\mu p_\nu}{p^2} \quad , \tag{A.45a}$$

$$m \neq 0: \qquad \sum_{\lambda=0,\pm 1} \varepsilon_\mu(p,\lambda)\varepsilon_\nu(p,\lambda) = -g_{\mu\nu} + \frac{p_\mu p_\nu}{m^2} \quad . \tag{A.45b}$$

If there are several particles of the same kind in the final state, one has to take into account the proper symmetrization or antisymmetrization of the wave function. With ν identical particles in the final state one has to divide the cross section by the statistical factor $(\nu!)$.

More details concerning the derivation of these rules and the problems connected with the renormalization of divergent graphs can be found in the literature.[1]

A.4 Symmetry Transformations

One classifies the elementary particles according to their quantum numbers, which are often related to certain symmetry principles. In addition to the quantum numbers arising from internal symmetries such as isospin, strangeness, lepton number, and so on, the quantum numbers associated with the symmetries related to space-time play an important role, in particular, *rest mass, spin*, and *parity*.

All transformations of the space-time four-vector

$$x^\mu \longrightarrow x'^\mu = a^\mu{}_\nu x^\nu + b^\mu \tag{A.46}$$

that leave the Minkowskian line element

$$\mathrm{d}s^2 = (\mathrm{d}x^0)^2 - \mathrm{d}\boldsymbol{x}^2 \tag{A.47}$$

invariant are called *inhomogeneous Lorentz transformations* or *Poincaré transformations*.[2] With each real transformation one associates a 4×4 matrix A and a four-vector b. The invariance of ds^2 is guaranteed if

$$\det(A) = \pm 1 \quad . \tag{A.48}$$

The Poincaré transformations can be divided into four disconnected classes which are characterized by the sign of $\det(A)$ and of the element $a^0{}_0$:

Component	$P_+^\uparrow$	$P_-^\uparrow$	$P_-^\downarrow$	$P_+^\downarrow$
det(A)	$+1$	1	-1	$+1$
a_0^0	$\geq +1$	$\geq +1$	≤ -1	≤ -1

[1] See W. Greiner and J. Reinhardt: *Quantum Electrodynamics*, 2nd ed. (Springer, Berlin, Heidelberg, 1994); J.D. Bjorken, S.D. Drell: *Relativistic Quantum Mechanics* (McGraw-Hill, New York, 1964).

[2] W. Greiner, J. Rafelski: *Theoretische Physik 3A, Spezielle Relativitätstheorie* (Harri Deutsch, Thun, Frankfurt am Main, 1989).

The transformations with $\det(A) = +1$ are called *proper Lorentz transformations*, whereas transformations with $a^0{}_0 \geq 0$ are called *orthochronous*. The particular components of the Poincaré group can be obtained from $P_+^\uparrow$ by discrete symmetry operations:

$P_-^\uparrow$ by space reflection: $\boldsymbol{x} \to -\boldsymbol{x}$,

$P_-^\downarrow$ by time reflection: $x^0 \to -x^0$,

$P_+^\downarrow$ by space-time reflection: $(\boldsymbol{x}, x^0) \to (-\boldsymbol{x}, -x^0)$.

The component $P_+^\uparrow$ is a group by itself, that is, the product of two elements from $P_+^\uparrow$ is again an element of $P_+^\uparrow$. $P_+^\uparrow$ is called the *restricted* Poincaré group.

Within the Poincaré group the transformations with $b = 0$ form a subgroup, the (homogeneous) *Lorentz group* L. Like the Poincaré group it can be divided into the four components $L_+^\uparrow$, $L_-^\uparrow$, $L_+^\downarrow$, and $L_-^\downarrow$, which are defined in the same way as above. The transformations with $A = \mathbb{1}$ also form a subgroup, which is called the group of translations T. Thus the Poincaré group is the direct product of L and T: $P = L \times T$.

As far as we know today, all laws of physics are covariant under the transformations of the restricted Poincaré group $P_+^\uparrow$. To each such transformation characterized by A and b there must exist a unitary transformation $\hat{U}(b, A)$ which describes the change of the physical quantum state under the action of a Poincaré transformation. This unitary transformation can be written as

$$\hat{U}(b, A) = \hat{U}(b, \mathbb{1})\,\hat{U}(0, A) \equiv \hat{U}(b)\,\hat{U}(A) \quad . \tag{A.49}$$

Using the ten Hermitian operators $\hat{P}_\mu$, $\hat{M}_{\mu\nu}$, we can write the transformations $\hat{U}(b)$ and $\hat{U}(A)$ in the form[3]

$$\hat{U}(b) = \exp(\mathrm{i}b^\mu \hat{P}_\mu) \quad , \tag{A.50}$$

$$\hat{U}(A) = \exp\left(\frac{\mathrm{i}}{2}\alpha^{\mu\nu}\hat{M}_{\mu\nu}\right) \quad , \tag{A.51}$$

where

$$\alpha^{\mu\nu} = [\ln(A)]^{\mu\nu} = -\alpha^{\mu\nu} \quad , \tag{A.52}$$

$$\hat{M}_{\mu\nu} = -\hat{M}_{\mu\nu} \quad . \tag{A.53}$$

From the group properties of the inhomogeneous Lorentz transformations the following commutation relations for the generators $\hat{P}_\mu$ and $\hat{M}_{\mu\nu}$ result:

$$[\hat{P}_\mu, \hat{P}_\nu] = 0 \quad , \tag{A.54a}$$

$$[\hat{M}_{\mu\nu}, \hat{P}_\alpha] = \mathrm{i}(g_{\nu\alpha}\hat{P}_\mu - g_{\mu\alpha}\hat{P}_\nu) \quad , \tag{A.54b}$$

$$[\hat{M}_{\mu\nu}, \hat{M}_{\alpha\beta}] = \mathrm{i}(g_{\mu\beta}\hat{M}_{\nu\alpha} + g_{\nu\alpha}\hat{M}_{\mu\beta} - g_{\nu\beta}\hat{M}_{\mu\alpha} - g_{\mu\alpha}\hat{M}_{\nu\beta}) \quad . \tag{A.54c}$$

[3] Here we consider active transformations, where the bodies are rotated but not the axes of the coordinate systems. Otherwise we would get an additional minus sign in (A.50), (A.51).

The operator $\hat{P}_\mu \hat{P}^\mu = \hat{P}^2$ commutes with all generators; thus it is a *Casimir* operator of the Poincaré group. Each irreducible representation of the Poincaré group is therefore characterized by a fixed eigenvalue m^2 of the operator $\hat{P}^2$. Since the operators $\hat{P}_\mu$ are identified as the operators of the four-momentum, m is the rest mass of the state. The spin of a state is characterized by the Pauli–Lubanski operator

$$\hat{W}^\alpha = \frac{1}{2}\varepsilon^{\mu\nu\alpha\beta}\hat{M}_{\mu\nu}\hat{P}_\beta \quad . \tag{A.55}$$

The operator $\hat{W}^\alpha \hat{W}_\alpha = \hat{W}^2$ also commutes with all $\hat{P}_\mu$ and $\hat{M}_{\mu\nu}$, so that it is a second Casimir operator. Its eigenstates for massive particles have the form $m^2 s(s+1)$, with $s = 0, \frac{1}{2}, 1, \frac{3}{2}, \ldots$, where s describes the spin of the state. Thus each massive quantum-mechanical state ($m \neq 0$) is described by the quantum numbers $[m, s]$.

Instead of the operators $\hat{M}_{\mu\nu}$ one often uses

$$\hat{J}_i = \frac{1}{2}\varepsilon^{ikl}\hat{M}_{kl} \quad , \tag{A.56}$$

with $(i, k, l) = (1, 2, 3)$ and

$$\hat{K}_i = \hat{M}_{i0} \quad . \tag{A.57}$$

$\hat{J}_i$ are the angular-momentum operators, whereas the operators $\hat{K}_i$ transform to uniformly moving frames ("boosts"). The new operators satisfy the following commutation relations, which can be derived from (A.54):

$$[\hat{J}_i, \hat{P}_k] = \mathrm{i}\varepsilon_{ikl}\hat{P}_l \quad , \quad [\hat{J}_i, \hat{P}_0] = 0 \quad , \tag{A.58a}$$

$$[\hat{K}_i, \hat{P}_k] = \mathrm{i}g_{ik}\hat{P}_0 \quad , \quad [\hat{K}_i, \hat{P}_0] = -\mathrm{i}\hat{P}_i \quad , \tag{A.58b}$$

$$[\hat{J}_i, \hat{K}_k] = \mathrm{i}\varepsilon_{ikl}\hat{K}_l \quad , \quad [\hat{K}_i, \hat{K}_k] = -\mathrm{i}\varepsilon_{ikl}\hat{J}_l \quad , \quad [\hat{J}_i, \hat{J}_k] = \mathrm{i}\varepsilon_{ikl}\hat{J}_l \quad . \tag{A.58c}$$

Expressed in terms of $\hat{J}_i$ and $\hat{K}_i$, the spin operator (A.55) takes the form

$$\hat{W}^\alpha = (\hat{\boldsymbol{J}} \cdot \hat{\boldsymbol{P}}, \hat{\boldsymbol{J}}\hat{P}^0 + \hat{\boldsymbol{K}} \times \hat{\boldsymbol{P}}) \quad . \tag{A.59}$$

The operator

$$\hat{\Lambda} = \hat{W}^0/|\hat{\boldsymbol{P}}| = \hat{\boldsymbol{J}} \cdot \hat{\boldsymbol{P}}/|\hat{\boldsymbol{P}}| \tag{A.60}$$

is the helicity operator. For particles with $\hat{P}^2 = m^2 = 0$, Λ commutes with all generators of the inhomogeneous Lorentz group. In this case the only possible eigenstates of $\hat{\Lambda}$ are $\lambda = \pm s$, so that the irreducible representations in the case $m = 0$ are characterized by this quantum number.

The operator for a rotation about an angle θ around the axis given by the unit vector $\boldsymbol{n}$ is given by

$$\hat{U}(\theta, \boldsymbol{n}) = \exp(-\mathrm{i}\theta\boldsymbol{n} \cdot \hat{\boldsymbol{J}}) \quad . \tag{A.61}$$

If the rotation is expressed in terms of the three Euler angles θ_1, θ_2, and θ_3, then the rotation operator reads

$$\hat{U}(\theta_1, \theta_2, \theta_3) = \exp(-\mathrm{i}\theta_1\hat{J}_z)\exp(-\mathrm{i}\theta_2\hat{J}_y)\exp(-\mathrm{i}\theta_3\hat{J}_z) \quad . \tag{A.62}$$

A Lorentz transformation which boosts a system from rest to a velocity $\boldsymbol{v}$ is described by the operator

$$\hat{U}(\xi, \boldsymbol{n}) = \exp(\mathrm{i}\xi \boldsymbol{n} \cdot \hat{\boldsymbol{K}}) \quad , \tag{A.63}$$

where

$$\boldsymbol{n} = \frac{\boldsymbol{v}}{|\boldsymbol{v}|} \quad , \quad \xi = \operatorname{arctanh} |\boldsymbol{v}| \quad . \tag{A.64}$$

The variable ξ is called the *rapidity*; in contrast to the velocity it is additive.

The simplest irreducible representations of the Poincaré group belong to spins $s = 0$ and $s = 1/2$. The generators for the transformations of the wave functions in this case are

$$\begin{aligned} s = 0: \quad & \hat{P}_\mu = \mathrm{i}\partial_\mu \quad , \\ & \hat{M}_{\mu\nu} = \mathrm{i}(x_\mu \partial_\nu - x_\nu \partial_\mu) \quad , \\ & \hat{\boldsymbol{J}} = -\mathrm{i}\boldsymbol{x} \times \nabla \quad , \\ & \hat{\boldsymbol{K}} = -\mathrm{i}\left(\boldsymbol{x}\frac{\partial}{\partial t} + t\nabla\right) \end{aligned} \tag{A.65}$$

$$\begin{aligned} s = 1/2: \quad & \hat{P}_\mu = \mathrm{i}\partial_\mu \quad , \\ & \hat{M}_{\mu\nu} = \mathrm{i}(x_\mu \partial_\nu - x_\nu \partial_\mu) + \frac{1}{2}\sigma_{\mu\nu} \quad , \\ & \hat{\boldsymbol{J}} = -\mathrm{i}\boldsymbol{x} \times \nabla + \frac{1}{2}\boldsymbol{\sigma} \quad , \\ & \hat{\boldsymbol{K}} = -\mathrm{i}\left(\boldsymbol{x}\frac{\partial}{\partial t} + t\nabla\right) + \frac{\mathrm{i}}{2}\boldsymbol{\alpha} \quad . \end{aligned} \tag{A.66}$$

Subject Index

W. Greiner, B. Müller, *Gauge Theory of Weak Interactions*,
DOI 10.1007/978-3-540-87843-8, © Springer-Verlag Berlin Heidelberg 2009